Single Entity Electrochemistry

York, United Kingdom

31 August – 2 September 2016

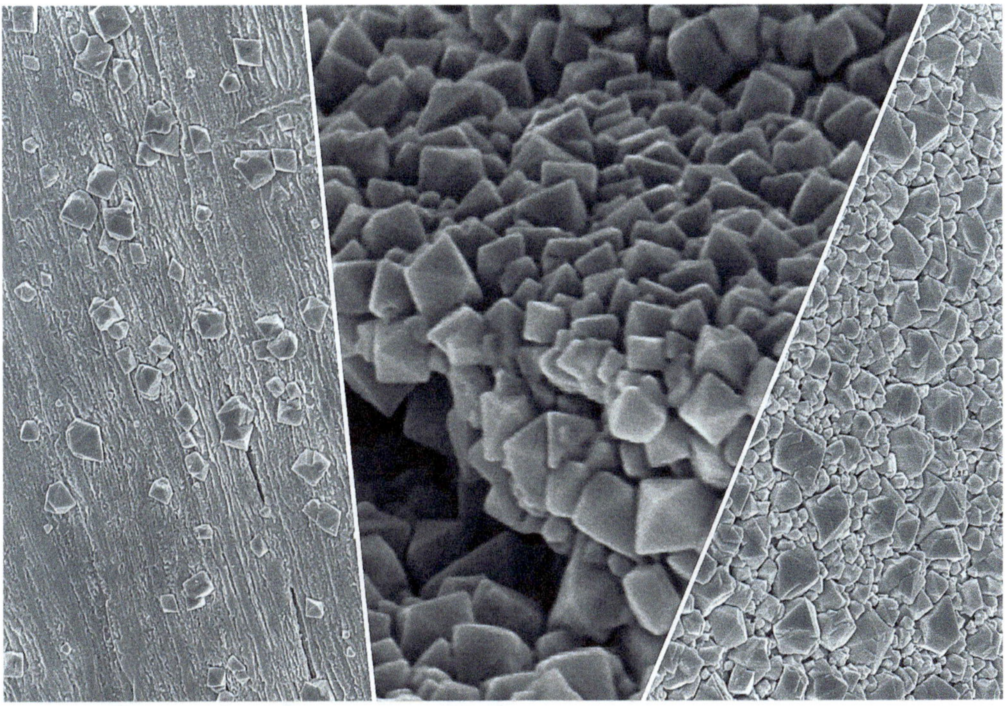

FARADAY DISCUSSIONS

Volume 193, 2016

ROYAL SOCIETY OF CHEMISTRY

The Faraday Division of the Royal Society of Chemistry, previously the Faraday Society, founded in 1903 to promote the study of sciences lying between Chemistry, Physics and Biology.

Faraday Discussions (print ISSN 1359-6640, electronic ISSN 1364-5498) is published 10 times a year by the Royal Society of Chemistry, Thomas Graham House, Science Park, Milton Road, Cambridge, UK CB4 0WF.

Volume 193 ISBN 13: 978-1-78262-481-3

2016 annual subscription price: print+electronic £929 US $1635; electronic only £883, US $1553. Customers in Canada will be subject to a surcharge to cover GST. Customers in the EU subscribing to the electronic version only will be charged VAT. All orders, with cheques made payable to the Royal Society of Chemistry, should be sent to RSC Order Department, Royal Society of Chemistry, Thomas Graham House, Science Park, Milton Road, Cambridge, CB4 0WF, UK.

Tel +44 (0) 1223 432398; E-mail **orders@rsc.org**

If you take an institutional subscription to any RSC journal you are entitled to free, site-wide web access to that journal. You can arrange access via Internet Protocol (IP) address at **www.rsc.org/ip**. Customers should make payments by cheque in sterling payable on a UK clearing bank or in US dollars payable on a US clearing bank.

Printed in the UK

MIX
Paper from responsible sources
FSC www.fsc.org FSC® C013604

Faraday Discussions are unique international discussion meetings that focus on rapidly developing areas of chemistry and its interfaces with other scientific disciplines.

Single Entity Electrochemistry

Faraday Discussions

www.rsc.org/faraday_d

A General Discussion on Single Entity Electrochemistry was held in York, UK on the 31st of August and 1st and 2nd of September 2016.

RSC Publishing is a not-for-profit publisher and a division of the Royal Society of Chemistry. Any surplus made is used to support charitable activities aimed at advancing the chemical sciences. Full details are available from www.rsc.org

CONTENTS

ISSN 1359-6640; ISBN: 978-1-78262-481-3

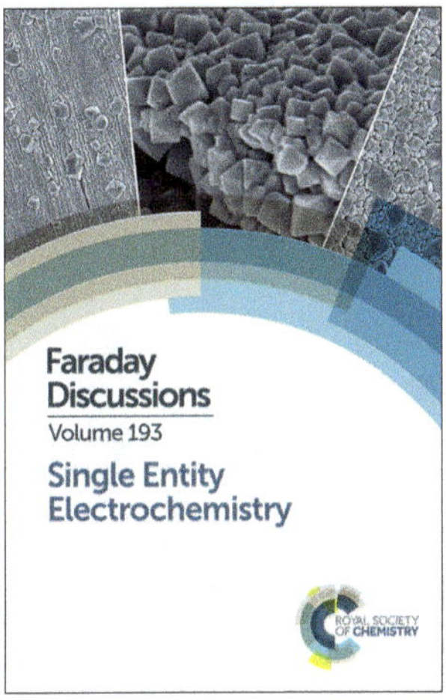

Cover
See Hersbach *et al.*, *Faraday Discuss.*, 2016, **193**, 207–222.

Anisotropic etching of rhodium and gold as the onset of nanoparticle formation by cathodic corrosion.

Image reproduced by permission of Thom Hersbach from *Faraday Discuss.*, 2016, **193**, 207–222.

INTRODUCTORY LECTURE

PAPERS AND DISCUSSIONS

SPONSORS

INTERNATIONAL SOCIETY OF ELECTROCHEMISTRY

EXHIBITORS

ec-lab

Metrohm Autolab U.K.

CONCLUDING REMARKS

ADDITIONAL INFORMATION

Faraday Discussions

PAPER

Emerging tools for studying single entity electrochemistry

Yixian Wang,†[a] Xiaonan Shan‡[a] and Nongjian Tao*[ab]

Received 28th August 2016, Accepted 20th September 2016
DOI: 10.1039/c6fd00180g

Electrochemistry studies charge transfer and related processes at various microscopic structures (atomic steps, islands, pits and kinks on electrodes), and mesoscopic materials (nanoparticles, nanowires, viruses, vesicles and cells) made by nature and humans, involving ions and molecules. The traditional approach measures averaged electrochemical quantities of a large ensemble of these individual entities, including the microstructures, mesoscopic materials, ions and molecules. There is a need to develop tools to study single entities because a real system is usually heterogeneous, *e.g.*, containing nanoparticles with different sizes and shapes. Even in the case of "homogeneous" molecules, they bind to different microscopic structures of an electrode, assume different conformations and fluctuate over time, leading to heterogeneous reactions. Here we highlight some emerging tools for studying single entity electrochemistry, discuss their strengths and weaknesses, and provide personal views on the need for tools with new capabilities for further advancing single entity electrochemistry.

1. Introduction

The dropping mercury electrode has played a unique role in developing the fundamentals of modern electrochemistry because of its uncontaminated and atomically flat surface. However, for most practical electrochemical systems, even for a single crystal electrode, atomic steps, islands and pits, and other microscopic and mesoscopic structures on the crystal surface must be considered in order to fully understand heterogeneous electrochemical reactions on the electrode. Electrochemical systems also include mesoscopic materials, including

[a]*Center for Biosensors and Bioelectronics, Biodesign Institute and School of Electrical, Computer and Energy Engineering, Arizona State University, Tempe, Arizona 85287, USA. E-mail: njtao@asu.edu*

[b]*State Key Laboratory of Analytical Chemistry for Life Science, School of Chemistry and Chemical Engineering, Nanjing University, Nanjing 210093, China*

† Current address: Department of Chemistry and Biochemistry, California State University, Los Angeles, Los Angeles, CA 90032, USA.

‡ Current address: Department of Electrical & Computer Engineering, University of Houston, Houston, TX 77024, USA.

nanoparticles, nanowires, viruses and vesicles, which are of great interest because of their existence in nature, and unique properties for various applications, including catalysis, energy storage and conversion and biosensors.[1–12] The traditional electrochemical methods measure an average quantity of the entire electrode or a large number of the mesoscopic entities. While this ensemble approach is valuable, averaging over many microstructures on an electrode, or mesoscopic entities in electrolytes, often washes out the unique properties of each microstructure or entity.

Central to chemistry are the reactions of individual molecules, and developing a capability to measure, analyze and control the electrochemical reactions of single molecules thus represents a frontier in electrochemistry. Unlike the mesoscopic materials, the individual molecules are "identical" or appear to be "homogeneous", but heterogeneity still arises due to, *e.g.*, different binding sites on an electrode surface, or different conformations of the molecules (especially for macromolecules, such as proteins). Furthermore, when measuring a single or a small number of molecules, stochastic behaviors become increasingly pronounced or even dominant. The traditional ensemble approach also washes out the dynamics of single molecules.

Single entities discussed or mentioned in this paper include microstructures on an electrode, mesoscopic materials, such as nanoparticles, nanowires, nanotubes, nanobubbles (see ref. 13 and DOI: 10.1039/c6fd00099a), viruses and vesicles, and also ions and molecules. Advances in single entity electrochemistry over the past several decades have been rapid, and are largely driven by the development of new theoretical and experimental tools. In this review, we focus on experimental tools only. We divide these tools into two categories: local probe-based methods (*e.g.*, nanoelectrodes and microelectrodes, nanopores, and various scanning probe-based imaging techniques), and optical methods (*e.g.*, fluorescence, plasmonic, and Raman scattering techniques). The tools in the former category provide high spatial resolution, but often at the expense of temporal resolution. In contrast, the tools in the latter offer high temporal resolution, but with compromised spatial resolution.

The field of single entity electrochemistry is diverse and dynamic; providing a comprehensive review of the entire field with the limited space and scope here is thus impossible. Instead, we summarize the most common tools used to study single entity electrochemistry, highlight some of the recent advances, and provide our personal views on the limitations of the existing tools and outlook into the future. We apologize for any inadvertent omission of many exciting works in literature. Fortunately, several excellent reviews on single entity electrochemistry and related topics have been published in recent years, which cover various aspects of the topics that are omitted here.[14–20]

We organize the remainder of the paper in the following order. First, we discuss tools based on local probes, including fixed probes (micro- and nanoelectrodes, nanopipettes, and nanopores) and scanning probes based on detecting various quantities (tunneling current, ionic conduction, electrochemical current, force *etc.*). We then highlight recent advances in optical imaging approaches (including electron microscopy). Finally, we conclude by mentioning the need for tools with new capabilities to further advance the field of single entity electrochemistry.

2. Micro- and nano-electrodes

Micro- and nano-electrodes provide local electrochemical information, which is especially important for studying biological or other heterogeneous systems.[14,21] Additionally, the mass transport of molecules and ions at micro- and nano-scales is fast, making it possible to measure fast electron transfer processes. Another important feature is the reduction of the background current associated with double layer charging and other sources, because the background current scales with the electrode area. This feature is extremely critical for single entity measurement because the electrochemical current from a single entity is typically small, and could be easily overwhelmed by a large background current. For these reasons, micro- and nano-electrodes have become a basic tool in the study of single entity electrochemistry.[14,22] There are two basic approaches to use micro- and nano-electrodes: detection of single entities (nanoparticles and molecules) immobilized on an electrode, and detection of transient collision of the entities in electrolyte with the electrode.

Detection of single entities immobilized on micro- or nano-electrodes

By attaching a single nanoparticle onto a micro- or nano-electrode, one can apply various electrochemical methods, such as cyclic voltammetry (CV) to study its electrochemical behaviors.[23–29] Chen *et al.*[23,24] measured oxygen reduction and hydrogen oxidation on a submicron-size Pt particle deposited on a carbon microelectrode. The small size of the particle led to efficient mass transport, allowing them to investigate fast electron transfer reaction kinetics on the particle. Zhang *et al.*[25] were able to study oxygen reduction on single Au nanoparticles with diameters as small as 14–24 nm, and observed oxygen reduction with catalytic activities.

One of the key tasks of this approach is to immobilize a single nanoparticle onto a micro- or nano-electrode. Mirkin *et al.*[28] attached Au nanoparticles onto carbon nanoelectrodes using several methods: direct adsorption of the nanoparticles onto the carbon surface, electrostatic attachment of the nanoparticles onto a positively charged polyphenylene film (aryl diazonium),[30] and formation of covalent C–Au bond *via* electrochemical reduction of diazonium group[31] (Fig. 1a and b). They found that different immobilization methods resulted in different catalytic activities, which underscores the importance of the nanoparticle–nanoelectrode interface in the electrochemical reactions of the nanoparticles. Bard *et al.*[32] grew single Pt nanoparticles with diameters of 1–40 nm onto the TiO_2 passivated microelectrode *via* electrodeposition (Fig. 1c), and obtained voltammograms of single Pt nanoparticles (Fig. 1d).

A widely used method to immobilize nanoparticles to an electrode is *via* an organic layer with different functional groups self-assembled on the electrode.[33] Experiments have shown that the nanoparticle–organic layer–electrode structure prepared with this method leads to fast electron transfer reactions of redox species in electrolyte, and the electron transfer rates appear to be independent of the thickness of the organic layer.[34–36] These findings seem puzzling if the organic layer is viewed as an insulating layer. Chazalviel and Allongue[37] modeled the electron transfer of the structure in two sequential steps: electron transfer between the redox species in the electrolyte and the nanoparticle, and between

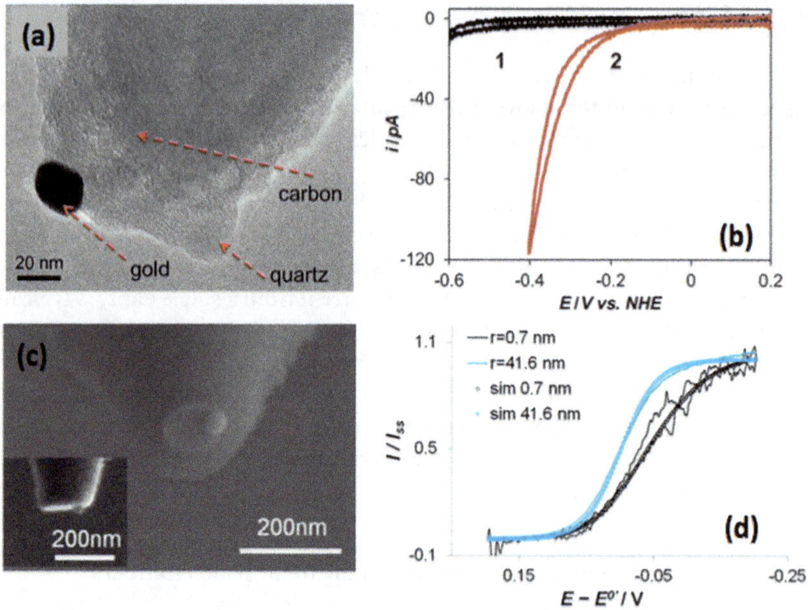

Fig. 1 Single nanoparticle immobilized on an electrode. (a) Transmission electron microscopy image of a carbon nanoelectrode with a 20 nm Au nanoparticle attached to its tip. (b) Voltammograms of the hydrogen reduction in 0.1 M $HClO_4$ of a blank carbon nanoelectrode (1) and a carbon nanoelectrode with a 3 nm-AuNP attached to it (2). Potential scanning rate: 100 mV s^{-1}. Adapted with permission from ref. 28. Copyright 2015, WILEY-VCH Verlag GmbH & Co. KGaA, Weinheim. (c) Scanning electron microscopy images of a 28.5 nm radius Pt nanoparticle on a TiO_2 passivated microelectrode. The inset shows a focused ion beam image. (d) Measured and simulated voltammograms of 0.7 and 41.6 nm Pt nanoparticle on TiO_2 passivated microelectrodes. Adapted with permission from ref. 32. Copyright 2016, American Chemical Society.

the nanoparticle and the electrode *via* the organic layer-tunneling barrier. This model explains the observed fast and thickness-independent electron transfer rates if it is assumed that the former step is much slower than the latter step. This assumption is justified because the electrical resistance of the nanoparticle–organic layer–electrode scales with the density of states of the nanoparticle at the Fermi level, which is large (proportional to the number of atoms in the nanoparticle). Gooding *et al.*[38] carried out a systematic experimental study of the system, which confirmed the model and helped establish the basics of electron transfer in the nanoparticle–organic layer–electrode structure. While the two step model explains the observations, detailed electron transfer kinetics requires consideration of the alignment among the Fermi levels of the electrode and nanoparticle, and chemical potential of the redox species.[39,40]

The above examples show that micro- and nano-electrodes enable direct measurement of the electrochemical reactions of single nanoparticles immobilized to the electrodes. In principle, one could also study the electrochemical reactions of a single redox molecule with this approach. By performing fast and repeated cyclic voltammetry, Watkins *et al.*[41] demonstrated detection of ~7000 molecules immobilized on a Pt nanoelectrode. However, directly measuring the electron transfer current in a single molecule without catalytic or enzymatic amplification remains a technical challenge.

Detecting transient electrochemical reactions on nano- and micro-electrodes

Instead of immobilizing a single nanoparticle or molecule on an electrode, one can also detect the transient electrochemical current arising from the collision of a single entity with the small electrode held at a potential. This amperometric detection is used for studying single cell secretion, including vesicular exocytosis of neurotransmitters and hormones,[42–45] and oxidative stress of reactive oxygen and nitrogen species[46] by placing a microelectrode close to a cell. When a vesicle containing electroactive species (*e.g.*, dopamine) reaches the microelectrode, a current spike occurs due to the reaction of the electroactive species, allowing real time analysis of cell secretion. Quantitative analysis of the transient current leads to deep insights into the mechanism of vesicle rupture and release of neurotransmitters on the electrode surface (see DOI: 10.1039/c6fd00102e).

Replacing a single microelectrode with an array of individually addressed microelectrodes offers high throughput and spatial information[47] for monitoring exocytosis from different locations of a cell *via* amperometry[48,49] or fast scan CV.[49] Decreasing the size of the microelectrode represents another way to improve the capability of this microelectrode approach for studying vesicle release in cells.[21] For example, Huang *et al.*[50] detected neurotransmitter release from individual synapses by inserting a nanoelectrode (diameter of 50 nm) into a single synapse (Fig. 2).

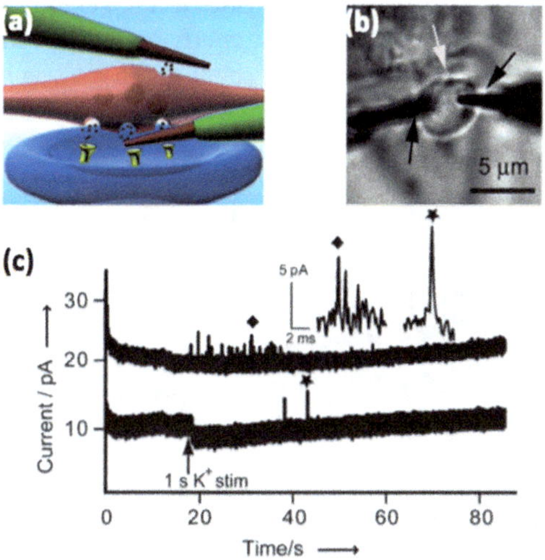

Fig. 2 Simultaneous monitoring of vesicular exocytosis under "semi-artificial synapse" and "inside synapse" configurations with two carbon fiber microelectrodes. (a and b) Schematic diagram (a) and optical micrograph (b) showing two microelectrodes (black arrows, the left one inside the synapse and the right one on the top of the synapse) for simultaneous detection of vesicular exocytosis from basal and apical pole of a single varicosity (white arrow). (c) Amperometric spikes, the upper and bottom traces represent current spikes recorded on the electrodes inside the synapse and at the apical pole. Adapted with permission from ref. 50. Copyright 2014, WILEY-VCH Verlag GmbH & Co. KGaA, Weinheim.

In addition to vesicles, transient electrochemical reactions of single nanoparticles or other entities during the collision with a microelectrode can be detected and analyzed. Bard *et al.*[51] showed that the collision of Pt nanoparticles on a microelectrode led to transient current changes arising from catalytic reaction of the nanoparticles. His and other groups have applied the strategy to study other single entities,[52–63] including alloy nanoparticles (see DOI: 10.1039/c6fd00112b), microdroplets,[61,64] viruses,[65,66] DNA,[62] and enzymes.[67] If the single entity has catalytic activity, an increase in the current is associated with the collision. Otherwise, the collision results in a decrease in the current as it blocks the background electrochemical current on the electrode. Furthermore, depending on whether the entity sticks to the microelectrode or not, the current change can be either stepwise or spike-like. When interpreting the transient current responses, aggregation of nanoparticles in the electrolyte must be considered, as discussed by Koper *et al.*[68] and Stevenson *et al.* (DOI: 10.1039/c6fd00121a).

In the transient detection method described above, the background current is reduced because of the small surface area of the electrodes. An alternative approach to reduce the background current is to use a scanning electrochemical cell microscope (SECCM),[69] in which the electrochemical cell volume is determined by the small meniscus at the end of a micropipette in contact with a substrate electrode (Fig. 3a). This approach offers exciting new capabilities. Recent examples include the studies of electrocatalytic reactions of N_2H_4 at Au nanoparticles[52] and H_2O_2 at ruthenium oxide (RuO_x) nanoparticles (Fig. 3b).[70]

Measuring transient current responses during the collision of individual entities with an ultramicroelectrode allows the study of the entities, one at a time. However, the amperometric detection provides limited chemical insight into the nature of the reaction.[71] In contrast, voltammetry measures current *vs.* potential, which offers better chemical specificity. Zhang *et al.*[71,72] scanned the electrode potential rapidly (50–500 V s^{-1}) while monitoring the catalytic reactions of single

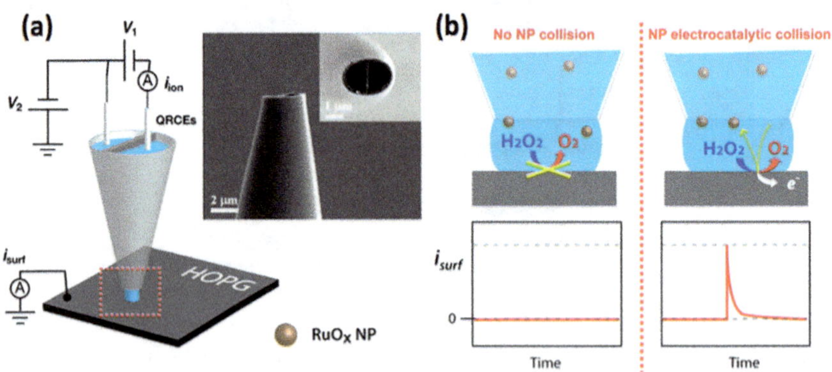

Fig. 3 SECCM for detecting single RuO_x nanoparticle collisions. (a) Schematic of the experimental setup with a pipette for meniscus contact and nanoparticle delivery to a working electrode (HOPG) substrate. Inset shows SEM images of the pipette. (b) Mechanism of detection: (left) no oxidation of H_2O_2 at the HOPG electrode surface, *i.e.*, no surface current (i_{surf}); (right) the electrocatalytic oxidation of H_2O_2 at the nanoparticle when it collides with the surface. Adapted with permission from ref. 70. Copyright 2015, American Chemical Society.

nanoparticles during the collision with an electrode (Fig. 4). Compton *et al.*[73] measured CVs of catalytic reactions of individual Pd-coated carbon nanotubes during the collision with a carbon microwire electrode. This approach allowed these authors to obtain time-resolved CVs, from which both amperometric (current *vs.* time at a given potential), and voltammetric (current *vs.* potential) information can be extracted. However, time-resolved voltammetry requires a fast potential scan rate to accommodate the short residence time of a nanoparticle on the electrode.

Detection of redox recycling of single entities in a confined space

Detection of single molecules with the above approaches has been a challenge. An exception reported recently is the detection of single enzymes because of the enzymatic amplification of electrochemical current (see DOI: 10.1039/c6fd00061d). Another method to achieve signal amplification is to trap redox molecules between two closely positioned electrodes so that the molecules can be repeatedly oxidized and reduced.[14,19,20] This redox-recycling method was proposed by Bard *et al.*[74] using a scanning electrochemical microscope (SECM), in which a ~10 nm-gap was created between a Pt–Ir electrode and a substrate electrode. Recently, Mirkin *et al.*[75] reported a zeptoliter cell by immersing a slightly recessed Pt nanoelectrode containing redox species into a Hg pool. These works demonstrated the detection of a single or a few redox molecules.

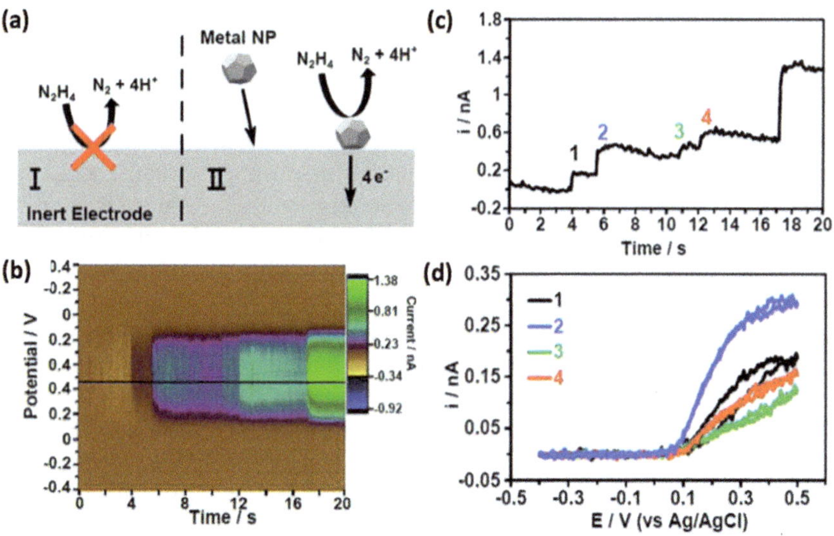

Fig. 4 Fast-scan cyclic voltammetry during single nanoparticle collision. (a) Schematic illustration showing that direct hydrazine oxidation is kinetically unfavorable (I) and the nanoparticle provides amplification of the hydrazine oxidation (II). (b) Time-resolved cyclic voltammograms during the collision of 12 nm Au nanoparticles on a carbon ultramicroelectrode. (c) Current *vs.* time trace taken from (b) at 0.465 V in the forward scan (along the black line). (d) Snapshots of single-particle cyclic voltammograms at the events marked in (c). Electrolyte: 10 mM hydrazine, 50 mM phosphate buffer (pH 7.4), and ~12.5 pM AuNPs (12 nm). Potential scan rate: 25 V s^{-1}. Adapted with permission from ref. 72. Copyright 2016, American Chemical Society.

Unwin *et al.*[76] developed a SECCM approach to achieve redox recycling. They positioned a four-channel micropipette on top of a substrate electrode, and formed a small meniscus at the end of the micropipette that defined the electrochemical cell. By filling two of four channels with carbon, they obtained two closely placed electrodes, and analyzed the anticorrelation of the currents from the two electrodes, which removed common background noise, and allowed them to achieve single molecule detection.

The works described above create a small nanogap between two electrodes for redox cycling with SECM and SECCM. Lemay *et al.*[77] developed a lithographic approach to fabricate a nanofluidic channel with top and bottom electrodes separated with a gap of a few tens of nm. This setup allowed them to detect single redox molecules from anticorrelated current,[78] and carry out detailed statistical analysis of a single molecule trapped in the nanogap (see DOI: 10.1039/c6fd00075d).

3. Nanopores, nanochannels and nanopipettes

The nanopore method measures ionic current through a nanoscale pore,[79] which is conceptually similar to the Coulter counter[80] and single ion channel recording technique[81,82] developed in the 50s and 70s, respectively. Various nanopores have been fabricated, such as solid-state nanopores,[83,84] artificial ion channels,[85–87] carbon nanotubes,[88,89] etched graphene nanopores[90] and glass nanopipettes.[91–99] Nanopores are also sometimes divided into nanopores, nanochannels and nanopipettes based on the geometries,[100] but the basic sensing mechanism is the same: when a molecule or nanoparticle passes through a nanopore, the ionic conduction through the nanopore is blocked, leading to a current (or resistance) spike. The frequency of the spike occurrence provides concentration information for the detected species, and the amplitude allows analysis of the size and identity of the species. Nanopore technology has many applications, including DNA sequencing,[101–103] biosensing,[83,89,104,105] ion-selective sensing,[106,107] single molecule analysis,[108–113] and single nanoparticle counting and size characterization.[114] Recent reviews provide detailed summaries of the fundamentals and applications of the technology,[100,111,115,116] and here we highlight only a few examples that are closely related to electrochemistry.

While the classical Coulter counter measures the size of micron-scale particles, most recent nanopore studies have focused on the analysis of single molecules. White *et al.*[117,118] have applied nanopores to study nanoparticles. They were able to move a nanoparticle in and out of a nanopore (orifice of a micropipette), and perform multiple measurements on the same nanoparticle. Furthermore, they showed that the size of nanoparticles could be determined with sub-angstrom resolution by detecting the current drop across the orifice.[114]

As the diameter of a nanopore decreases to a few nanometers, approaching the thickness of the electrical double layer, field driven migration[119–123] and field-induced accumulation or depletion[124–128] of charged entities in the nanopores become important. By controlling the surface charge of nanopore electrodes, the accumulation of oppositely charged ions leads to an enhancement in the ionic current through the nanopore.[129–131] Bohn *et al.* (see ref. 132 and DOI: 10.1039/c6fd00062b) fabricated nanopore arrays with embedded ring and disk electrodes, and demonstrated 100-fold current enhancement for $Ru(NH_3)_6{}^{2/3+}$ in water (without supporting electrolyte) in a nanopore electrode array with 100 nm inter-electrode spacing.[132] More recently, they observed 280-fold current enhancement, and attributed the large enhancement to both ion accumulation and ion migration effects (Fig. 5).[133]

For high throughput applications, such as sequencing DNA and analysis of molecules, fast detection is highly desirable. Albrecht *et al.* (DOI: 10.1039/c6fd00109b) demonstrated μs time resolution current detection with nanopipettes *via* optimized electronics. To fully describe the transport of single entities through the nanopore, microscopic details of molecules and ions must be considered. For example, Schmickler *et al.*[134–136] (DOI: 10.1039/c6fd00076b) developed models based on density functional theory to describe electrolytes in a carbon nanotube by including the interactions of ions with the nanotube, and effects on the electronic properties the nanotube. Eikerling *et al.* (DOI: 10.1039/c6fd00094k) modeled proton distribution and transport in water-filled Pt nanopores.

4. Scanning probe microscopes

The micro-/nano-electrodes and nanopores measure the electrochemical responses from single molecules and nanoparticles, but they offer limited spatial resolution and lack imaging capability. The scanning tunneling microscope (STM) and atomic force microscope (AFM)[137–144] can image electrochemical processes on well-defined electrode surfaces with atomic resolution. The scanning probe techniques can also create nanoscale structures on the electrode *via* electrodeposition and etching,[145,146] and image electrochemical reactions at the individually created nanostructures. Despite the high spatial resolution for structural studies, traditional STM and AFM do not directly measure electrochemical current related to the fundamentals of electron transfer reactions.

SECM was developed to induce and image local electrochemical reactions with high spatial resolution,[147–149] which has been applied to study the electrochemistry of single nanoparticles[150] as well as other mesoscopic structures.[151–153] One of

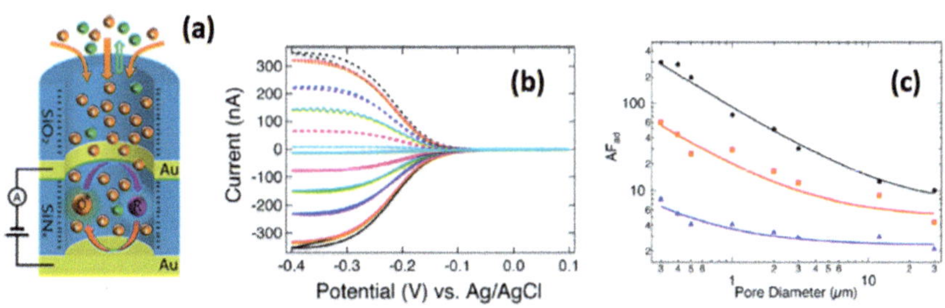

Fig. 5 (a) Schematics of a nanopore for studying ion transport. (b) Cyclic voltammograms of 10 μM $Ru(NH_3)_6^{3+}$ in the presence of 0 mM (black), 0.3 mM (red), 1 mM (blue), 3 mM (green), 10 mM (purple), and 100 mM (cyan) KCl. Current at the bottom electrode (solid) and top electrode (dashed). Triangular waveform is applied to the bottom electrode at a scan rate of 100 mV s^{-1} while the top electrode is held at 0.1 V *vs.* Ag/AgCl. (c) Current amplification (AF_{ad}) obtained for $Ru(NH_3)_6^{3+}$ solutions at 1 μM (black circle), 10 μM (red square), and 100 μM (blue triangle) concentration as a function of nanopore diameter. Adapted with permission from ref. 133. Copyright 2016, American Chemical Society.

the most critical tasks in SECM is the fabrication of the nanoelectrode probes. To achieve high spatial resolution, Mirkin *et al.*[154] fabricated probes with radii as small as a few nm and imaged electrochemical reactions at single nanoparticles. However, pushing SECM spatial resolution by reducing the probe size results in reduced current, which ultimately affects the imaging speed because of the conflicting requirements for a current amplifier with both high bandwidth and high gain. High spatial resolution also requires a small probe–sample separation, which demands a robust feedback system in order to avoid the tip crashing into the substrate. Despite the challenges, Mirkin *et al.*[155] were able to resolve the sharp edges of cubic Pd nanoparticles, demonstrating the high spatial resolution with nanoelectrode probes.

Unwin *et al.*[69] developed SECCM to image local electrochemical reactions. As mentioned before, SECCM has a small electrochemical cell volume defined by the meniscus between a micropipette probe and a substrate (Fig. 3). This strategy reduces the background current and helps resolve electrochemical current from the sample (*e.g.*, nanoparticles).[69] Moreover, SECCM exposes a sample to electrolyte only when the scanning probe is moved to the area of the sample, which prevents the sample from reacting with the electrolyte before it is imaged. SECCM has been successfully applied to image various nanoscale materials.[156–159] One recent example is the imaging of local redox activity and charging/discharging of $LiFePO_4$ nanoparticles (100–200 nm) on a Li-ion battery cathode (Fig. 6).[160]

Combining different scanning probe techniques into one system offers new capabilities. One such effort is to combine SECM with scanning ion conductance microscopy (SICM).[161,162] SECM is sensitive to the electrochemical reactions on a surface, but also to the topography of the surface. Thus, the local chemical reactivity and topography are convoluted and difficult to separate. In contrast, SICM images local ion conductance and maps surface topography.[162] Combining

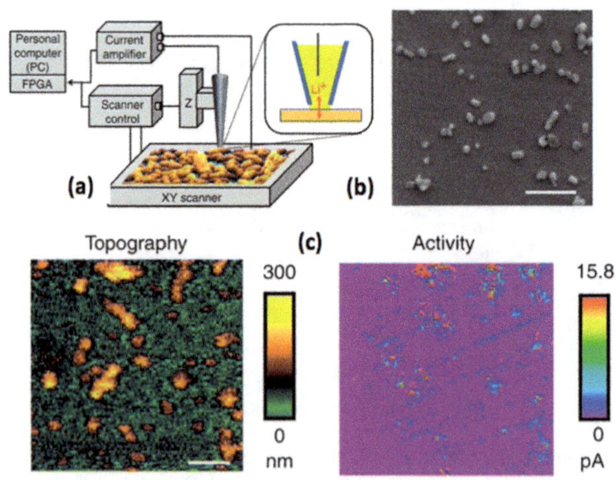

Fig. 6 Characterizing the activity and topography of individual $LiFePO_4$ nanoparticles with SECCM. (a) Schematic of SECCM. (b) SEM image of $LiFePO_4$ nanoparticles on a Pt collector electrode (scale bar, 1 mm). (c) Simultaneous SECCM topography (left) and current (right) images. Scan range is 5 × 5 mm (scale bar, 1 mm). The substrate potential was 0.65 V *vs.* Ag/AgCl QRCE. Adapted with permission from ref. 160. Copyright 2014, Nature Publishing Group.

SECM and SICM can thus help to isolate the topographical signal from the electrochemical signal. O'Connell *et al.*[163] used a double-barrel capillary as the scanning probe, where one barrel filled with carbon served as the SECM probe and the second barrel filled with electrolyte as the SICM probe (Fig. 7a).[163] Using the combined SECM and SICM, they studied the electrocatalytic reduction of oxygen at single Au nanoparticles, in which the substrate was biased to a potential to reduce oxygen and the SECM probe was biased to a higher potential to oxidize H_2O_2 generated on the substrate (Fig. 7a).[163,164] The SICM images resolve the individual Au nanoparticles (Fig. 7b), and the SECM images acquired at various substrate potentials reveal the formation of the intermediate H_2O_2 at the nanoparticles (Fig. 7c). The study showed that the Au nanoparticles with similar sizes could have different electrocatalytic activities. This variability was attributed to the different morphologies and crystal structures of the individual Au nanoparticles.

The SICM mentioned above provides high-resolution topography but typically lacks chemical information.[162] An effort to overcome the limitation is to combine SICM and nanopore method.[165] Recently, Shi *et al.* (DOI: 10.1039/c6fd00133e) used a dual-barrel ion channel SICM probe, consisting of one open barrel for ion conduction measurement, and a barrel with a membrane patch embedded with ion channels (nanopores). The open barrel provides a feedback mechanism for topography imaging and the membrane-patch barrel allows imaging of local chemical reactions.

Demaille *et al.*[166–171] combined SECM with AFM for simultaneous imaging of local topography and electrochemical activities using an AFM microcantilever-electrode probe, where the probe measures local electrochemical current arising from redox molecules, and also the force for AFM surface topography imaging. They successfully imaged the topography and electrochemical reactions of

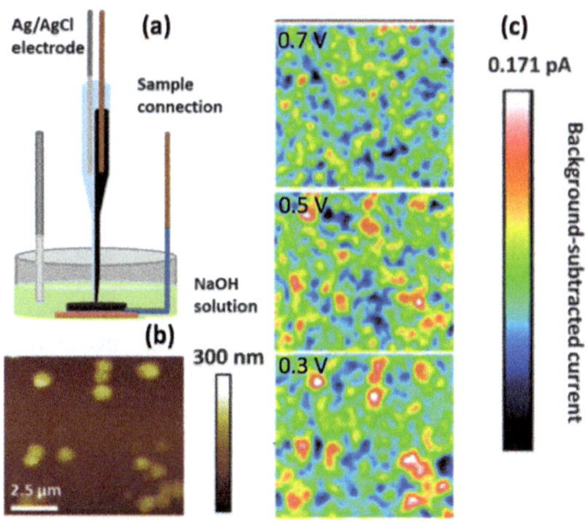

Fig. 7 Electrochemical imaging of hydrogen peroxide generation at individual gold nanoparticles with SICM/SECM. (a) Schematic of SICM/SECM for combined topographical–electrochemical imaging. (b) SICM and (c) SECM imaging of Au nanoparticles at various substrate potentials (*vs.* RHE). The SECM tip is biased at 1 V *vs.* RHE to oxidize hydrogen peroxide. Adapted with permission from ref. 164. Copyright 2015, Royal Society of Chemistry.

ferrocene-PEGylated Au nanoparticles with diameters of ~20 nm.[170] The high spatial resolution afforded by the combined AFM and SECM allows the resolution of the PEG corona and the Au core of the individual nanoparticles. Further improvement in the spatial resolution led to the image of single proteins.[171] Combining SECM with AFM also allows the study of mechanical properties of nanomaterials (see DOI: 10.1039/c6fd00128a).

5. Electrochemical STM break junction

Electrochemical STM provides sub-molecular resolution images of redox molecules on an electrode,[172–176] and the measured tunneling current carries direct electron transfer rate information.[177–179] However, quantitative analysis of electron transfer rate and conductance of the molecule from the tunneling current of STM is non-trivial. This is because the molecule–tip contact in STM is a tunneling gap filled with electrolyte, which is not well defined, and depends on the distance between the tip and the molecule, and also on the lateral position of the tip over the molecule. Because electron transport through the molecule is extremely sensitive to both the substrate–molecule and molecule–tip contacts, the uncertainty in the molecule–tip contact in STM leads to an extremely large uncertainty (orders of magnitude) in the measured conductance. One approach to overcome this limitation is to place a Au nanoparticle on the top of the molecule.[180] This approach greatly improves the conductance measurement, but the use of the nanoparticle introduces new complications.[181]

STM break junction was developed to address the need for defined contacts for single molecule conductance measurement and electron transfer rate quantification.[182–184] The method studies molecules terminated with two linker groups, *e.g.*, dithiols and amines[185] that can form well-defined contacts with both the substrate and tip electrodes. The STM tip is first brought into contact with molecules adsorbed on the substrate *via* one of the linkers, and then pulled away. During the brief moment of contact, the molecule may bind to the STM tip *via* its second linker group, which is reflected as a plateau in the conductance *vs.* time trace recorded during the pulling away stage. The process is repeated thousands of times, and statistical analysis of the plateau conductance leads to a conductance histogram, from which the conductance of the molecule is analyzed. This approach provides not only the average conductance of single molecules but also the variability arising from atomic scale differences in the molecule–electrode contacts, and the conformation of the molecule. More recently, this approach has been expanded to current (I)–voltage (V) characteristics measurement, providing the I–V histogram, transition voltage and tunneling barrier information of the molecular junctions.[186]

Haiss *et al.*[187,188] developed another approach to bridge a molecule between an STM tip and a substrate electrode. First, molecules terminated with two linkers are deposited onto the substrate surface, and then the STM tip is brought close to, but not into contact with, the molecules. When thermal fluctuations bring the second linker of a molecule to the tip and form a chemical bond with the tip, a sudden jump occurs in the tunneling current, which is detected and analyzed to determine the conductance of the molecule.

In both approaches, one can study electron transport through a redox molecule by controlling the potential of the electrodes with respect to a reference electrode.[189–193] This allows one to tune the energy levels of molecular redox states relative to the Fermi levels of the electrodes, and control the redox states of the molecule.[172,194,195] For example, Darwish *et al.*[196] studied charge transport in a single anthraquinone-based molecule using the STM break-junction technique (Fig. 8). Once a molecule is trapped between the STM tip (Au) and the Au substrate, the potential is scanned in order to alternate between the oxidized anthraquinone (AQ) and the reduced hydroanthraquinone (H_2AQ) forms. They showed that the conductance increases by an order of magnitude when AQ was switched to H_2AQ.

The electrochemical STM break junction method provides electrochemical gating to single molecule conductance, which is related to molecular electronics. Additionally, the measured current provides rich information about charge transport in single molecules. Experiments have revealed different transport mechanisms, including tunneling and hopping as well as an intermediate tunneling-hopping mechanism (see ref. 197 and 198 and DOI: 10.1039/c6fd00080k).

The charge transport data measured by the STM break junction method have been modeled in terms of rate equations, which describe charge transport through the molecule from one electrode (*e.g.*, tip) to another (*e.g.*, substrate), as individual charge transfer steps, and the overall current or conductance is determined by the balance of the individual steps.[199] Analysis of the experimental data using the theoretical models provides deep insights into electron transfer processes at the single molecule level.

6. Plasmonics based electrochemical current imaging

All the methods discussed above use a probe, which measures one entity at a time, thus have limited throughput and temporal resolution. Optical methods provide high throughput and temporal resolution. One example is plasmonics based electrochemical microscopy (P-ECM).[200–202] P-ECM relies on detecting surface plasmon resonance (SPR) on an electrode (*e.g.*, thin film), which is sensitive to both double-layer charging[201] and electrochemical reactions on the electrode.[203]

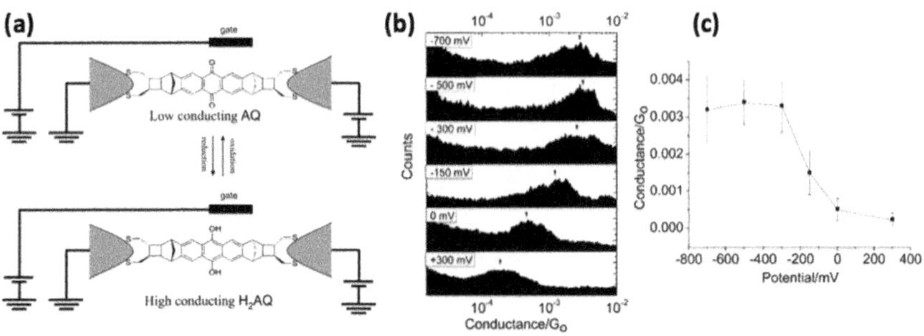

Fig. 8 (a) Schematic illustration of the electrochemical gated switching of 5AQ5 to $5H_2AQ5$. 5AQ5 is a completely rigid molecule possessing five sigma bonds on each side and an AQ moiety in the center. (b) Conductance histograms of 5AQ5 at different gating potentials. (c) Evolution of the conductance of 5AQ5 with changing potential. Adapted with permission from ref. 196. Copyright 2012, American Chemical Society.

The former is due to the surface plasmon frequency depending on the electrode surface charge density which changes with the charging and discharging processes. In contrast, the latter arises from the sensitive dependence of SPR on the refractive index of the medium near the electrode, which changes with the concentrations of the reactants and products when an electrochemical reaction takes place on the electrode. So P-ECM images contain quantitative information on both the double layer charging and faradaic currents,[204] and can perform amperometry (chronoamperometry) and voltammetry (cyclic and square-wave) in an analogous way to the traditional electrical methods.[205] Additionally, P-ECM provides the size information of the nanoparticles deposited on the electrode from the scattering of the surface plasmons by the nanoparticles.

P-ECM has been used to image and study various biological processes, including cell apoptosis and electroporation,[202] Ca^{2+} release from cells[206] and molecular binding to protein microarrays.[207] It has also found applications in imaging electrochemical reactions in nanoscale materials.[204,208,209] For example, Shan *et al.*[204] reported the P-ECM imaging of single Pt nanoparticles and monitored the hydrogen evolution reaction on the nanoparticles. A recent example is the imaging of electrochemical oxidation of single Ag nanoparticles with P-ECM (Fig. 9).[209] As the individual nanoparticles land on the electrode, they oxidize and dissolve, which is imaged in real time, allowing study of the variability in the oxidation rate for different nanoparticles, and analysis of particle size–reactivity relationships. The temporal resolution of P-ECM is fast, but its spatial resolution is limited by optical diffraction, which is a fraction of a μm.

7. Dark field microscopy

Dark field microscopy images light scattering from localized surface plasmon resonance (LSPR) in metal nanoparticles,[210–212] which is in contrast to P-ECM,

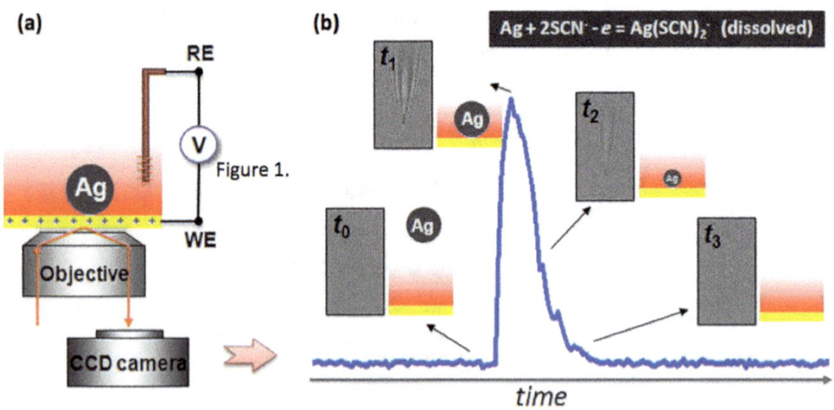

Fig. 9 Plasmonic imaging of electrochemical oxidation of single nanoparticles. (a) Schematic illustration of P-ECM imaging of single nanoparticle electrochemistry, gold chip as working electrode (WE), and Ag/AgCl as reference electrode (RE). (b) Transient collision and electrochemical oxidation of a single Ag nanoparticle correlated with the transient plasmonic image intensity (blue curve). Adapted with permission from ref. 209. Copyright 2014, American Chemical Society.

which measures SPR on a thin film electrode. Light scattering and absorption measured by dark field microscopy are sensitive to the size and shape of the metal nanoparticles, as well as on the local dielectric environment (refractive index) of each nanoparticle, thus providing sensitive detection of electrochemical processes at single nanoparticles. To date processes studied by dark field microscopy include charging/discharging of single gold nanorods (See ref. 213, DOI: 10.1039/c6fd00069j and DOI: 10.1039/c6fd00098c), surface oxidation of Au and Ag nanoparticles,[214] electrodeposition[215,216] and electrocatalytic reactions on Au nanoparticles,[217,218] and formation and stripping of single Au amalgam nanoalloy particles.[219]

Coupling dark field microscopy with a spectrometer offers spectroscopic information. Jing *et al.*[220] monitored the electrocatalytic oxidation of H_2O_2 on single gold nanorods by measuring shifts in the optical spectra. They observed different spectral shifts on different nanorods, and attributed the observation to heterogeneity in the structure and catalytic activity of the nanorods. Landes *et al.*[221] obtained CVs of Au nanoparticles from the time derivative of the scattered light intensity (Fig. 10). The CVs measured by the dark field microscopy resemble those obtained with the traditional electrical measurement. They further applied the technique to study the electro-adsorption and desorption of different anions, including sulfate, acetate, and perchlorate.

Both dark field microscopy and P-ECM rely on plasmonic resonance for the signal enhancement and surface sensitivity that are needed for studying surface electrochemistry. However, an important difference between them is that the former is based on LSPR at metallic nano-materials, while the latter excites SPR on a thin film electrode and images electrochemical reactions either on the planar electrode itself, or at the individual nanoscale entities on the electrode. Dark field microscopy

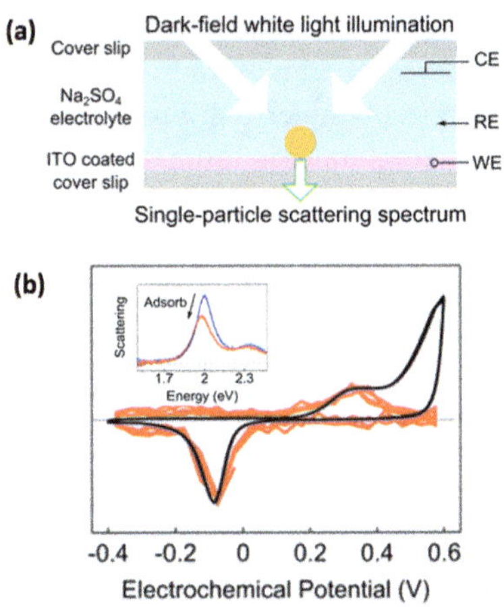

Fig. 10 Single-particle plasmon voltammetry (spPV) for detecting anion adsorption. (a) Schematic of a transparent spectroelectrochemical cell for single-nanoparticle spectroscopy with counter, reference, and working electrodes (CE, RE, and WE). (b) Comparison between conventional CV (black) and spPV (red) for perchlorate adsorption (three cycles shown). Inset shows the scattering spectrum of a 50 nm-diameter Au nanoparticle dimer on Au thin film substrate at cathodic (blue) and anodic (red) potentials. Adapted with permission from ref. 221. Copyright 2016, American Chemical Society.

is thus particularly suitable for imaging electrochemical processes at metallic nanoparticles, and P-ECM can image both metallic and non-metallic entities, including biological samples. Another difference is in the light intensity measured by the two techniques. Dark field optical microscopy measures scattered light (after blocking direct transmitted or reflected light), and P-ECM measures reflected light. Since the scattered light is usually weaker than the reflected light in optical microscopy, the exposure time in dark field microscopy, especially when coupled to a spectrometer, must be sufficiently long, which limits its temporal resolution for studying fast electrochemical reactions. In contrast, the high light intensity in P-ECM makes it possible to image fast (*e.g.*, μs) electrochemical processes.

8. Single-molecule fluorescence detection and imaging

Both P-ECM and dark-field microscopy are label-free optical imaging methods, which provide fast and high-throughput imaging of electrochemical reactions without labels. While sensitive, studying single molecule electrochemistry with the label-free imaging techniques is still difficult. This is especially the case for imaging small molecules as the label-free detection signal scales with the mass of the molecules. Fluorescence spectroscopy has been developed to image single molecules, and applied to electrochemistry.[222–224] Bard and Barbara[225,226] developed a single-molecule spectroelectrochemistry technique to study the oxidation of conjugated polymers deposited on an indium tin oxide (ITO) electrode. The technique measures fluorescence quenching in single molecules associated with oxidation. Statistical analysis of the potentials at which fluorescence quenching takes places for the individual molecules provides the distribution of half wave potential.[225,227] This approach has been extended to the study of small molecules.[228,229]

Single-molecule fluorescence has also been applied to study electrocatalytic reactions at single nanoparticles and nanorods.[16,230–235] The basic principle is to image fluorescent products (*e.g.* amplex red: resazurin) electrochemically generated at the nanoparticles or nanorods from non-fluorescent molecules (*e.g.* resorufin) (Fig. 11a and b). Super-resolution tracking of the positions[230] of the individual molecules allows the study of local photoelectrocatalytic reactions. For example, Zhang *et al.*[236] mapped the spatial distribution of catalytic activity on Sb-doped TiO_2 nanorods and triangular Au nanoplates. Chen *et al.*[230] recently reported on the mapping of both the electron- and hole-driven photoelectrocatalytic activities along single TiO_2 nanorods (Fig. 11c).

Although single-molecule fluorescence provides super-resolution, fluorescent labels are required. Another limitation is photostability, especially in aqueous solutions in the presence of oxygen. Finally, fluorescence emission is often weak so that sufficient exposure time is needed to ensure a good signal to noise ratio, which places a constraint in the temporal resolution, and thus the imaging speed.

9. Surface/tip enhanced Raman spectroscopy

Raman spectroscopy measures inelastic light scattering from vibrational and rational modes in molecules. Compared to elastic light scattering (*e.g.*, Rayleigh

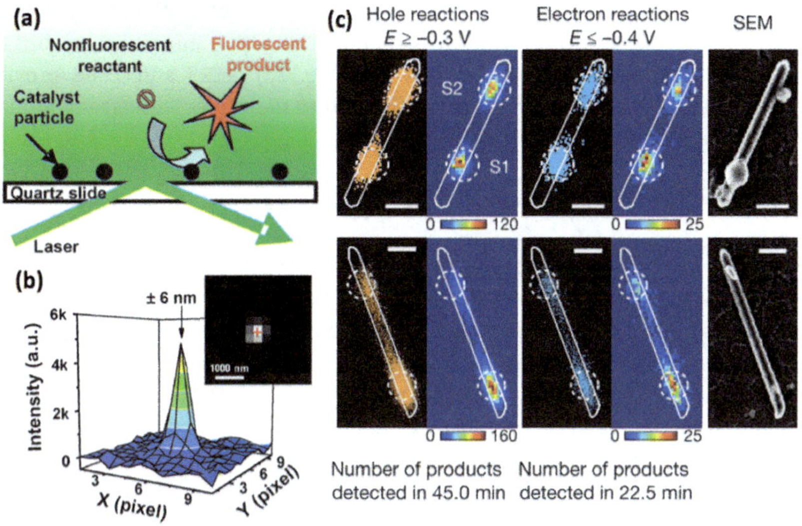

Fig. 11 (a) Schematic for imaging catalysis with single-molecule fluorescence imaging. (b) A 3D fluorescence plot of a single resorufin product molecule (each pixel ~267 nm). Adapted with permission from ref. 16. Copyright 2013, Royal Society of Chemistry. (c) Super-resolution mapping of hole and electron reactions on single TiO_2 nanorods. Adapted with permission from ref. 230. Copyright 2016, Nature Publishing Group.

scattering), Raman scattering from molecules on an electrode surface is usually too weak to measure. However, by roughening a metal (*e.g.*, Ag and Au) electrode surface, the Raman signal is greatly enhanced, leading to surface enhanced Raman spectroscopy (SERS) for measuring vibrational modes in molecules adsorbed on the surface.[237–241] SERS arises from two contributions, the large local electromagnetic field caused by LSPR in the nanoscale structures of the roughened electrode surface, and direct electron transfer between the molecules and the electrode surface.[242,243]

SERS is a powerful tool for studying single molecules on an electrode.[240,244] Cortés *et al.*[245] reported a SERS study of single molecule electrochemistry in Nile blue adsorbed on Ag nanoparticles, where the redox reaction of Nile blue was observed from the change in the vibration mode of the molecule at 590 cm^{-1}. Willets *et al.*[246] observed site-specific electrochemical reduction and oxidation potentials of Nile blue on a Ag nanoparticle using a SERS superlocalization technique. Recently, Wang *et al.*[247] investigated single electron transfer at a hemin–Ag nanoparticle interface by monitoring Raman shifts. Using dark-field-assisted SERS, Long *et al.*[248] studied local catalytic reactions on a nanowire.

A large enhancement in SERS requires a special electrode surface (*e.g.*, roughened Au or Ag). To overcome this limitation, Tian *et al.*[249] developed a shell-isolated nanoparticle approach, where the shell is a chemically inert layer, and the core is a metal (Au or Ag) to provide plasmonic enhancement. Another way to enhance Raman scattering is to excite plasmons at the STM tip–substrate gap, leading to tip enhanced Raman spectroscopy (TERS) with sub-nanometer spatial resolution.[250,251] Ren *et al.*[252] recently measured potential-dependent electrochemical TERS of (4′-(pyridin-4-yl)biphenyl-4-yl)methanethiol adsorbed on

Au(111) with a STM/Raman microscope, and van Duyne *et al.*[253] reported a TERS study of redox reactions of a few (less than 10) Nile blue molecules on an ITO electrode with an AFM/Raman microscope (Fig. 12). SERS and TERS provide single-molecule sensitivity, but they are limited to molecules with strong Raman scattering.

10. *In situ* transmission electron microscopy

In situ transmission electron microscopy (TEM)[254–262] has become possible recently for imaging electrochemical reactions in nanostructured materials using a sealed liquid cell (Fig. 13a).[263] Holtz *et al.*[264] applied it to image lithium insertion and deinsertion dynamics during a complete charging/discharging cycle, and to study degradation of $LiFePO_4$ in real time (Fig. 13b). TEM provides high spatial resolution for electrochemical processes in nanostructures, but the effects of the high-energy electron beam on the sample during a TEM experiment require further study.

11. Outlook

Integrating complementary capabilities of different tools

The tools discussed in this paper provide remarkable capabilities to detect and image electrochemical processes at the single entity level. Integrating different

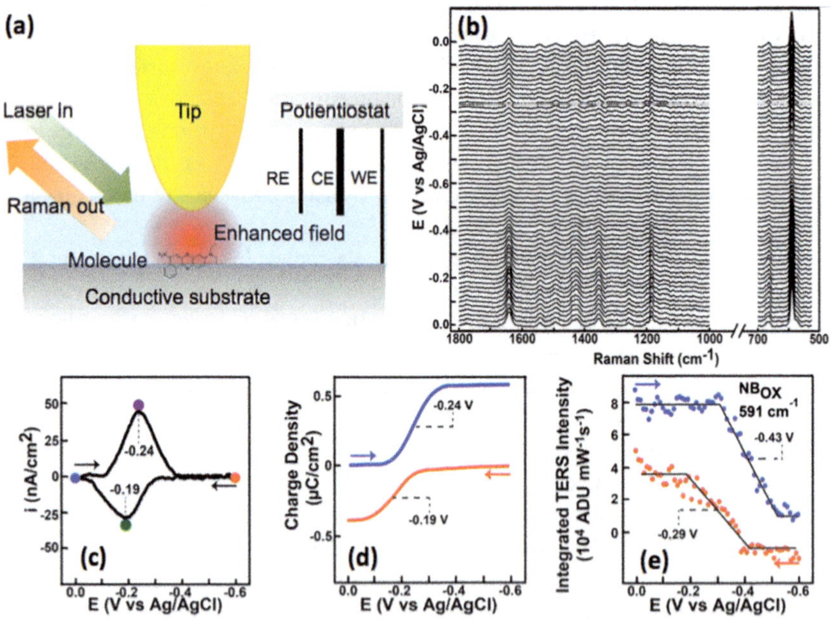

Fig. 12 TERS. (a) Schematic of an electrochemical TERS setup. (b) TER spectra of reversible reduction and oxidation of Nile blue acquired at different potentials during CV. (c) Background subtracted CV of Nile blue on ITO acquired concurrently with the TER spectra in (b). (d) Charge acquired by integrating the CV in (c). (e) Potential dependence of the integrated TERS intensity of the 591 cm^{-1} band of NB_{OX}. Adapted with permission from ref. 253. Copyright 2015, American Chemical Society.

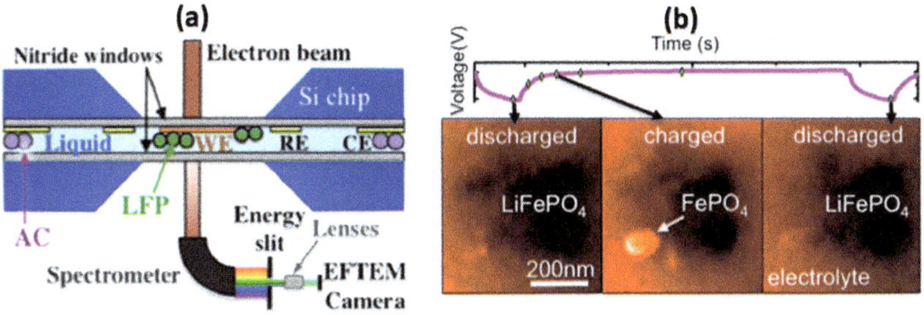

Fig. 13 *In situ* electrochemical TEM. (a) Schematic of an electrochemical cell sealed with a silicon nitride membrane, where the working electrode (WE) is made of carbon with $LiFePO_4$ (LFP) nanoparticles deposited on top, and the Pt counter electrode (CE) is coated with activated carbon (AC). (b) Snapshots of a $LiFePO_4/FePO_4$ cluster during one charging/discharging cycle, and the voltage profile is shown above the TEM snapshots. Adapted with permission from ref. 264. Copyright 2014, American Chemical Society.

tools into one system offers opportunities to create new capabilities. For example, simultaneous optical imaging and electrochemical current measurement have been carried out to track the individual collision events of nanoparticles with an ultramicroelectrode.[265,266] Plasmonic, bright field, dark field and fluorescence imaging capabilities can been combined in a single optical system to provide complementary information on molecular binding in single cells.[267] Other examples include TERS and the combination of plasmonic imaging with a micropipette probe.[268] Different scanning probe microscopes have been combined to expand the capabilities of the individual microscopes (*e.g.*, SICM/SECM, and AFM/SECM, mentioned earlier). Further development in this direction will lead to new research opportunities. For example, to determine the effects of crystal facets and microstructures on the electrochemical activities of single nanoparticles, it is desirable to integrate electrochemically specific detection tools (*e.g.*, nanoelectrodes, dark field and P-ECM) with high-resolution structural imaging techniques (*e.g. in situ* TEM and STM).

Pushing the speed limit

The essence of single entity measurement is to examine the electrochemistry of each entity without ensemble averaging. However, electrochemical processes of single entities involve thermal fluctuations at various time scales. Insufficient temporal resolution results in time averaging, which washes out detailed dynamics of the fluctuations. A particularly important example is thermal fluctuations in the electron transfer between a molecule and an electrode.[269,270] The fluctuations bring a molecule close to an electrode, change the conformation of the molecule, and re-orient solvent molecules surrounding the molecule, which help move the energy levels of the molecule into alignment with the Fermi energy level of the electrode, and create transient states with large electronic coupling between the donor and acceptor states for electron transfer.[271] Thermal fluctuations are not limited to molecules. Other single entities, such as nanoparticles, also experience thermal fluctuations, leading to stochastic phenomena, such as the nanoparticle–electrode collision events (see ref. 272, DOI: 10.1039/c6fd00075d and DOI: 10.1039/c6fd00100a). The underlying physical processes in thermal fluctuations can be complex and vary from system to system, and the associated time scales are often extremely broad and fast. For example, the conformational fluctuations of a protein range from ms to ns, and the vibrations and rotations of a small molecule can be even faster.[273,274] Fast electrochemical detection is thus needed to resolve the fast dynamic processes. This capability will also allow the study of transient (unstable) or intermediate reaction products in an electrochemical reaction.[275]

Scanning probe microscopy is typically slow because of the mechanical scanning of the probe. However, advances in technology have led to the development of video-speed STM, which can resolve atomic scale processes at ~100 ms (see ref. 276 and DOI: 10.1039/c6fd00086j). Unwin *et al.*[277] have pushed the speed of SECCM to 4 seconds per scan, orders of magnitude faster than the previous typical imaging speeds. Optical imaging techniques are usually faster (at the expense of spatial resolution). For example, plasmonic imaging speed can reach μs-time scale using fast CMOS imagers (*e.g.*, 10^5 to 10^6 frames per second commercially available).

Fast electrochemical measurements are often limited by voltage drop due to a finite electrolyte resistance, and time constant of an electrochemical cell, which both decrease with the size of the electrode.[275] By reducing the electrode size, sub-μs electrochemical measurements have been demonstrated.[278,279] Further improvement of temporal resolution with this strategy eventually faces a new obstacle: the finite stray capacitance of the system, including the electrodes, wiring and electronic circuit. Fast electrochemical measurements require not only fast response of the electrochemical cell, but also sufficient signal to noise ratio. Because electrochemical current scales with the electrode size, improving the time response by reducing the size results in a smaller current. Detecting the small current at high speed requires both high gain and broad bandwidth electronics, which are two conflicting requirements for a traditional electronic circuit. One effort to overcome this bandwidth-gain limit is to use an AC-coupled circuit, which detects fast signals with a high frequency amplifier. Using this strategy, Guo *et al.*[280] measured the fast breakdown of a Au–Au bond in a quantum point contact formed between two Au electrodes at the sub-ns time scale. The AC approach achieves fast detection speed, but loses information at frequencies outside of the bandwidth of the AC-coupled circuit.

It is challenging to directly measure fast microscopic processes electrically. Optical methods (*e.g.*, pump–probe) offer an alternative route towards ultrafast detection of single entity electrochemistry. Feldberg and Smalley[281] developed an optical triggering method, which measured the transient response of the open circuit potential by heating the electrode with a fast laser pulse. They achieved fast electrochemical detection without miniaturizing the electrode, but the temporal resolution was limited by the signal to noise ratio. Further development of the pump–probe optical approach may lead to even faster detection capabilities for studying the reaction dynamics of single entities.

Connecting the individual dots

Despite the technical challenges, it is safe to predict that new tools will continue to be developed in the future for studying single entity electrochemistry. However, a practical electrochemical system, consisting of a large number of known and unknown individual entities (*e.g.*, molecules, ions, and nanoparticles), is often too complex for us to obtain sufficient information for each entity. Even if we know the behaviors of every single entity, it does not automatically mean that we understand and can predict the electrochemical reactions of the whole system. Collective behaviors of a system cannot always be deduced from the behaviors of the individual entities in the system. For example, a complex and interesting reaction pattern in an electrochemical system may arise from collective behaviors of many entities, involving nonlinear and non-equilibrium processes (see ref. 282 and DOI: 10.1039/c6fd00115g). A multi-scale approach will be needed to understand electrochemical phenomena involving different length and time scales.

Acknowledgements

Works from the authors' laboratories were supported by Gordon and Betty Moore Foundation and National Natural Science Foundation of China (NSFC, Grant No. 21327008 and 21327902).

References

1 A. T. Bell, *Science*, 2003, **299**, 1688–1691.
2 A. Roucoux, J. Schulz and H. Patin, *Chem. Rev.*, 2002, **102**, 3757–3778.
3 R. Sardar, A. M. Funston, P. Mulvaney and R. W. Murray, *Langmuir*, 2009, **25**, 13840–13851.
4 I. I. Slowing, B. G. Trewyn, S. Giri and V. S. Y. Lin, *Adv. Funct. Mater.*, 2007, **17**, 1225–1236.
5 K. Saha, S. S. Agasti, C. Kim, X. Li and V. M. Rotello, *Chem. Rev.*, 2012, **112**, 2739–2779.
6 P. K. Jain, X. Huang, I. H. El-Sayed and M. A. El-Sayed, *Acc. Chem. Res.*, 2008, **41**, 1578–1586.
7 T. Sannomiya and J. Vörös, *Trends Biotechnol.*, 2011, **29**, 343–351.
8 J. N. Anker, W. P. Hall, O. Lyandres, N. C. Shah, J. Zhao and R. P. Van Duyne, *Nat. Mater.*, 2008, **7**, 442–453.
9 R. A. Sperling, P. R. Gil, F. Zhang, M. Zanella and W. J. Parak, *Chem. Soc. Rev.*, 2008, **37**, 1896–1908.
10 Q. A. Pankhurst, J. Connolly, S. K. Jones and J. Dobson, *J. Phys. D: Appl. Phys.*, 2003, **36**, R167–R181.
11 W. I. Katz and J. Wang, *Electroanalysis*, 2004, **16**, 19–44.
12 M.-C. Daniel and D. Astruc, *Chem. Rev.*, 2004, **104**, 293–346.
13 L. Luo and H. S. White, *Langmuir*, 2013, **29**, 11169–11175.
14 S. M. Oja, Y. Fan, C. M. Armstrong, P. Defnet and B. Zhang, *Anal. Chem.*, 2016, **88**, 414–430.
15 J. B. Sambur and P. Chen, *Annu. Rev. Phys. Chem.*, 2014, **65**, 395–422.
16 P. Chen, X. Zhou, N. M. Andoy, K.-S. Han, E. Choudhary, N. Zou, G. Chen and H. Shen, *Chem. Soc. Rev.*, 2014, **43**, 1107–1117.
17 S. E. F. Kleijn, S. C. S. Lai, M. T. M. Koper and P. R. Unwin, *Angew. Chem., Int. Ed.*, 2014, **53**, 3558–3586.
18 A. P. O'Mullane, *Nanoscale*, 2014, **6**, 4012–4026.
19 K. Mathwig, T. J. Aartsma, G. W. Canters and S. G. Lemay, *Annu. Rev. Anal. Chem.*, 2014, **7**, 383–404.

20 S. G. Lemay, S. Kang, K. Mathwig and P. S. Singh, *Acc. Chem. Res.*, 2013, **46**, 369–377.
21 J. Clausmeyer and W. Schuhmann, *TrAC, Trends Anal. Chem.*, 2016, **79**, 46–59.
22 J. N. A. Matthews, *Phys. Today*, 2011, **64**, 20.
23 S. Chen and A. Kucernak, *J. Phys. Chem. B*, 2004, **108**, 3262–3276.
24 S. Chen and A. Kucernak, *J. Phys. Chem. B*, 2003, **107**, 8392–8402.
25 Y. Li, J. T. Cox and B. Zhang, *J. Am. Chem. Soc.*, 2010, **132**, 3047–3054.
26 B. Zhang, L. Fan, H. Zhong, Y. Liu and S. Chen, *J. Am. Chem. Soc.*, 2013, **135**, 10073–10080.
27 J. Kim, B. K. Kim, S. K. Cho and A. J. Bard, *J. Am. Chem. Soc.*, 2014, **136**, 8173–8176.
28 Y. Yu, Y. Gao, K. Hu, P. Y. Blanchard, J. M. Noël, T. Nareshkumar, K. L. Phani, G. Friedman, Y. Gogotsi and M. V. Mirkin, *ChemElectroChem*, 2015, **2**, 58–63.
29 J. Clausmeyer, J. Masa, E. Ventosa, O. Dennis and W. Schuhmann, *Chem. Commun.*, 2016, **52**, 2408–2411.
30 H. Gehan, L. Fillaud, N. Felidj, J. Aubard, P. Lang, M. M. Chehimi and C. Mangeney, *Langmuir*, 2010, **26**, 3975–3980.
31 G. Liu, E. Luais and J. J. Gooding, *Langmuir*, 2011, **27**, 4176–4183.
32 J. Kim and A. J. Bard, *J. Am. Chem. Soc.*, 2016, **138**, 975–979.
33 E. Katz, I. Willner and J. Wang, *Electroanalysis*, 2004, **16**, 19–44.
34 J. B. Shein, L. M. H. Lai, P. K. Eggers, M. N. Paddon-Row and J. J. Gooding, *Langmuir*, 2009, **25**, 11121–11128.
35 D. Bethella, M. Brust, D. J. Schiffrin and C. Kiely, *J. Electroanal. Chem.*, 1996, **409**, 137–143.
36 C. R. Bradbury, J. Zhao and D. J. Fermín, *J. Phys. Chem. C*, 2008, **112**, 10153–10160.
37 J. N. Chazalviel and P. Allongue, *J. Am. Chem. Soc.*, 2011, **133**, 762–764.
38 A. Barfidokht, S. Ciampi, E. Luais, N. Darwish and J. J. Gooding, *Anal. Chem.*, 2013, **85**, 1073–1080.
39 C. M. Hill, J. Kim and A. J. Bard, *J. Am. Chem. Soc.*, 2015, **137**, 11321–11326.
40 M. D. Scanlon, P. Peljo, M. A. Mendez, E. Smirnov and H. H. Girault, *Chem. Sci.*, 2015, **6**, 2705–2720.
41 J. J. Watkins, J. Chen, H. S. White, H. D. Abruña, E. Maisonhaute and C. Amatore, *Anal. Chem.*, 2003, **75**, 3962–3971.
42 R. Trouillon, M. K. Passarelli, J. Wang, M. E. Kurczy and A. G. Ewing, *Anal. Chem.*, 2012, **85**, 522–542.
43 E. V. Mosharov and D. Sulzer, *Nat. Methods*, 2005, **2**, 651–659.
44 D. M. Cannon, N. Winograd and A. G. Ewing, *Annu. Rev. Biophys. Biomol. Struct.*, 2000, **29**, 239–263.
45 W. Wang, S.-H. Zhang, L.-M. Li, Z.-L. Wang, J.-K. Cheng and W.-H. Huang, *Anal. Bioanal. Chem.*, 2009, **394**, 17–32.
46 C. Amatore, S. Arbault, M. Guille and F. Lemaître, *Chem. Rev.*, 2008, **108**, 2585–2621.
47 J. Wang, R. Trouillon, Y. Lin, M. I. Svensson and A. G. Ewing, *Anal. Chem.*, 2013, **85**, 5600–5608.
48 B. Zhang, K. L. Adams, S. J. Luber, D. J. Eves, M. L. Heien and A. G. Ewing, *Anal. Chem.*, 2008, **80**, 1394–1400.
49 B. Zhang, M. L. A. V Heien, M. F. Santillo, L. Mellander and A. G. Ewing, *Anal. Chem.*, 2011, **83**, 571–577.

50 Y.-T. Li, S.-H. Zhang, L. Wang, R.-R. Xiao, W. Liu, X.-W. Zhang, Z. Zhou, C. Amatore and W.-H. Huang, *Angew. Chem., Int. Ed.*, 2014, **53**, 12456–12460.
51 X. Xiao and A. J. Bard, *J. Am. Chem. Soc.*, 2007, **129**, 9610–9612.
52 S. E. F. Kleijn, S. C. S. Lai, T. S. Miller, A. I. Yanson, M. T. M. Koper and P. R. Unwin, *J. Am. Chem. Soc.*, 2012, **134**, 18558–18561.
53 Y. Xiao, F. R. F. Fan, J. Zhou and A. J. Bard, *J. Am. Chem. Soc.*, 2008, **130**, 16669–16677.
54 S. E. F. Kleijn, B. Serrano-Bou, A. I. Yanson and M. T. M. Koper, *Langmuir*, 2013, **29**, 2054–2064.
55 R. Dasari, D. A. Robinson and K. J. Stevenson, *J. Am. Chem. Soc.*, 2013, **135**, 570–573.
56 H. Zhou, F. R. F. Fan and A. J. Bard, *J. Phys. Chem. Lett.*, 2010, **1**, 2671–2674.
57 S. J. Kwon, F. R. F. Fan and A. J. Bard, *J. Am. Chem. Soc.*, 2010, **132**, 13165–13167.
58 Y.-G. Zhou, N. V. Rees and R. G. Compton, *Angew. Chem., Int. Ed.*, 2011, **50**, 4219–4221.
59 Y.-G. Zhou, N. V Rees and R. G. Compton, *Chem. Commun.*, 2012, **48**, 2510–2512.
60 J. Dunevall, H. Fathali, N. Najafinobar, J. Lovric, J. Wigström, A. S. Cans and A. G. Ewing, *J. Am. Chem. Soc.*, 2015, **137**, 4344–4346.
61 B. K. Kim, J. Kim and A. J. Bard, *J. Am. Chem. Soc.*, 2015, **137**, 2343–2349.
62 J. E. Dick, C. Renault and A. J. Bard, *J. Am. Chem. Soc.*, 2015, **137**, 8376–8379.
63 J. Ellison, C. Batchelor-McAuley, K. Tschulik and R. G. Compton, *Sens. Actuators, B*, 2014, **200**, 47–52.
64 B. K. Kim, A. Boika, J. Kim, J. E. Dick and A. J. Bard, *J. Am. Chem. Soc.*, 2014, **136**, 4849–4852.
65 J. E. Dick, A. T. Hilterbrand, A. Boika, J. W. Upton and A. J. Bard, *Proc. Natl. Acad. Sci. U. S. A.*, 2015, **112**, 5303–5308.
66 L. Sepunaru, B. J. Plowman, S. V. Sokolov, N. P. Young and R. G. Compton, *Chem. Sci.*, 2016, **7**, 3892–3899.
67 A. N. Sekretaryova, M. Y. Vagin, A. P. F. Turner and M. Eriksson, *J. Am. Chem. Soc.*, 2016, DOI: 10.1021/jacs.5b13149.
68 S. E. F. Kleijn, B. Serrano-Bou, A. I. Yanson and M. T. M. Koper, *Langmuir*, 2013, **29**, 2054–2064.
69 N. Ebejer, M. Schnippering, A. W. Colburn, M. A. Edwards and P. R. Unwin, *Anal. Chem.*, 2010, **82**, 9141–9145.
70 M. Kang, D. Perry, Y. R. Kim, A. W. Colburn, R. A. Lazenby and P. R. Unwin, *J. Am. Chem. Soc.*, 2015, **137**, 10902–10905.
71 Z. Guo, S. J. Percival and B. Zhang, *J. Am. Chem. Soc.*, 2014, **136**, 8879–8882.
72 S. J. Percival and B. Zhang, *J. Phys. Chem. C*, 2016, DOI: 10.1021/acs.jpcc.5b11330.
73 X. Li, C. Batchelor-mcauley, S. A. I. Whitby, K. Tschulik, L. Shao and R. G. Compton, *Angew. Chem., Int. Ed. Engl.*, 2016, **55**, 4296–4299.
74 F.-R. F. Fan and A. J. Bard, *Science*, 1995, **267**, 871–874.
75 P. Sun and M. V. Mirkin, *J. Am. Chem. Soc.*, 2008, **130**, 8241–8250.
76 J. C. Byers, B. P. Nadappuram, D. Perry, K. Mckelvey, A. W. Colburn and P. R. Unwin, *Anal. Chem.*, 2015, **87**, 10450–10456.
77 M. A. G. Zevenbergen, P. S. Singh, E. D. Goluch, B. L. Wolfrum and S. G. Lemay, *Nano Lett.*, 2011, **11**, 2881–2886.

78 S. Kang, A. F. Nieuwenhuis, K. Mathwig, D. Mampallil and S. G. Lemay, *ACS Nano*, 2013, **7**, 10931–10937.
79 H. Bayley and C. R. Martin, *Chem. Rev.*, 2000, **100**, 2575–2594.
80 W. H. Coulter, Patents 2656508, 1953.
81 S. B. Hladky and D. A. Haydon, *Nature*, 1970, **225**, 451–453.
82 E. Neher and B. Sakmann, *Nature*, 1976, **260**, 799–802.
83 L. T. Sexton, L. P. Horne and C. R. Martin, *Mol. BioSyst.*, 2007, **3**, 667–685.
84 J. Griffiths, *Anal. Chem.*, 2008, **80**, 23–27.
85 J. J. Kasianowicz, E. Brandin, D. Branton and D. W. Deamer, *Proc. Natl. Acad. Sci. U. S. A.*, 1996, **93**, 13770–13773.
86 H. Bayley and P. S. Cremer, *Nature*, 2001, **413**, 226–230.
87 S. Majd, E. C. Yusko, Y. N. Billeh, M. X. Macrae, J. Yang and M. Mayer, *Curr. Opin. Biotechnol.*, 2010, **21**, 439–476.
88 T. Ito, L. Sun, R. R. Henriquez and R. M. Crooks, *Acc. Chem. Res.*, 2004, **37**, 937–945.
89 Y. Choi, L. A. Baker, H. Hillebrenner and C. R. Martin, *Phys. Chem. Chem. Phys.*, 2006, **8**, 4976–4988.
90 S. J. Heerema and C. Dekker, *Nat. Nanotechnol.*, 2016, **11**, 127–136.
91 S. Umehara, M. Karhanek, R. W. Davis and N. Pourmand, *Proc. Natl. Acad. Sci. U. S. A.*, 2009, **106**, 4611–4616.
92 Y. Wang, K. Kececi, M. Mirkin and V. Mani, *Chem. Sci.*, 2013, **4**, 655–663.
93 H. Cai, Y. Wang, Y. Yu, M. V. Mirkin, S. Bhakta, G. W. Bishop, A. A. Joshi and J. F. Rusling, *Anal. Chem.*, 2015, **87**, 6403–6410.
94 Y. Wang, H. Cai and M. V. Mirkin, *ChemElectroChem*, 2015, **2**, 343–347.
95 K. Hu, Y. Wang, H. Cai, M. V. Mirkin, Y. Gao, G. Friedman and Y. Gogotsi, *Anal. Chem.*, 2014, **86**, 8897–8901.
96 P. Terejánszky, I. Makra, P. Fürjes and R. E. Gyurcsányi, *Anal. Chem.*, 2014, **86**, 4688–4697.
97 W. Shi, N. Sa, R. Thakar and L. A. Baker, *Analyst*, 2015, **140**, 4835–4842.
98 A. P. Ivanov, P. Actis, P. Jönsson, D. Klenerman, Y. Korchev and J. B. Edel, *ACS Nano*, 2015, **9**, 3587–3594.
99 M. Karhanek, J. T. Kemp, N. Pourmand, R. W. Davis and C. D. Webb, *Nano Lett.*, 2005, **5**, 403–407.
100 D. G. Haywood, A. Saha-Shah, L. A. Baker and S. C. Jacobson, *Anal. Chem.*, 2015, **87**, 172–187.
101 G. F. Schneider and C. Dekker, *Nat. Biotechnol.*, 2012, **30**, 326–328.
102 Y. Feng, Y. Zhang, C. Ying, D. Wang and C. Du, *Genomics, Proteomics Bioinf.*, 2015, **13**, 4–16.
103 D. Branton, D. W. Deamer, A. Marziali, H. Bayley, S. A. Benner, T. Butler, M. Di Ventra, S. Garaj, A. Hibbs, X. Huang, S. B. Jovanovich, P. S. Krstic, S. Lindsay, X. S. Ling, C. H. Mastrangelo, A. Meller, J. S. Oliver, Y. V Pershin, J. M. Ramsey, R. Riehn, G. V Soni, V. Tabard-Cossa, M. Wanunu, M. Wiggin and J. A. Schloss, *Nat. Biotechnol.*, 2008, **26**, 1146–1153.
104 C. R. Martin and Z. S. Siwy, *Science*, 2007, **317**, 331–332.
105 S. K. Kannam, S. C. Kim, P. R. Rogers, N. Gunn, J. Wagner, S. Harrer and M. T. Downton, *Nanotechnology*, 2014, **25**, 155502.
106 J. H. Shim, J. Kim, G. S. Cha, H. Nam, R. J. White, H. S. White and R. B. Brown, *Anal. Chem.*, 2007, **79**, 3568–3574.
107 C. Song and B. Corry, *J. Phys. Chem. B*, 2009, **113**, 7642–7649.

108 R. Fan, R. Karnik, M. Yue, D. Li, A. Majumdar and P. Yang, *Nano Lett.*, 2005, **5**, 1633–1637.
109 M. Gershow and J. A. Golovchenko, *Nat. Nanotechnol.*, 2007, **2**, 775–779.
110 S. W. Kowalczyk, L. Kapinos, T. R. Blosser, T. Magalhaes, P. van Nies, L. Y. H. and C. Dekker, *Nat. Nanotechnol.*, 2011, **6**, 433–438.
111 S. Howorka and Z. Siwy, *Chem. Soc. Rev.*, 2009, **38**, 2360–2384.
112 E. A. Heins, Z. S. Siwy, L. A. Baker and C. R. Martin, *Nano Lett.*, 2005, **5**, 1824–1829.
113 A. E. P. Schibel, T. Edwards, R. Kawano, W. Lan and H. S. White, *Anal. Chem.*, 2010, **82**, 7259–7266.
114 M. A. Edwards, S. R. German, J. E. Dick, A. J. Bard and H. S. White, *ACS Nano*, 2015, **9**, 12274–12282.
115 I. Makra and R. E. Gyurcsányi, *Electrochem. Commun.*, 2014, **43**, 55–59.
116 W. Timp, A. M. Nice, E. M. Nelson, V. Kurz, K. McKelvey and G. Timp, *IEEE Access*, 2014, **2**, 1396–1408.
117 S. R. German, L. Luo, H. S. White and T. L. Mega, *J. Phys. Chem. C*, 2013, **117**, 703–711.
118 S. R. German, T. S. Hurd, H. S. White and T. L. Mega, *ACS Nano*, 2015, **9**, 7186–7194.
119 J. D. Norton, H. S. White and S. W. Feldberg, *J. Phys. Chem.*, 1990, **94**, 6772–6780.
120 J. B. Cooper and A. M. Bond, *J. Electroanal. Chem. Interfacial Electrochem.*, 1991, **315**, 143–160.
121 M. Ciszkowska and Z. Stojek, *Anal. Chem.*, 2000, **72**, 754A–760A.
122 J. Xiong, Q. Chen, M. A. Edwards and H. S. White, *ACS Nano*, 2015, **9**, 8520–8529.
123 E. J. F. Dickinson and R. G. Compton, *J. Electroanal. Chem.*, 2011, **661**, 198–212.
124 J. B. Cooper, A. M. Bond and K. B. Oldham, *J. Electroanal. Chem.*, 1992, **331**, 877–895.
125 K. B. Oldham, *J. Electroanal. Chem.*, 1992, **337**, 91–126.
126 Y. Zhang, B. Zhang and H. S. White, *J. Phys. Chem. B*, 2006, **110**, 1768–1774.
127 J. E. Bartelt, M. R. Deakin, C. Amatore and R. M. Wightman, *Anal. Chem.*, 1988, **60**, 2167–2169.
128 A. Bund and C. Kubeil, *Faraday Discuss.*, 2013, **164**, 339–348.
129 M. R. Newton, A. K. Bohaty, H. S. White and I. Zharov, *J. Am. Chem. Soc.*, 2005, **127**, 7268–7269.
130 G. Wang, A. K. Bohaty, I. Zharov and H. S. White, *J. Am. Chem. Soc.*, 2006, **128**, 13553–13558.
131 C.-Y. Li, Y.-W. Tian, W.-T. Shao, C.-G. Yuan, K. Wang and X.-H. Xia, *Electrochem. Commun.*, 2014, **42**, 1–5.
132 C. Ma, N. M. Contento and P. W. Bohn, *J. Am. Chem. Soc.*, 2014, **136**, 7225–7228.
133 C. Ma, W. Xu, W. R. A. Wichert and P. W. Bohn, *ACS Nano*, 2016, **10**, 3658–3664.
134 A. Goduljan, F. Juarez, L. Mohammadzadeh, P. Quaino, E. Santos and W. Schmickler, *Electrochem. Commun.*, 2014, **45**, 48–51.
135 L. Mohammadzadeh, A. Goduljan, F. Juarez, P. Quaino, E. Santos and W. Schmickler, *Electrochim. Acta*, 2015, **162**, 11–16.

136 L. Mohammadzadeh, A. Goduljan, F. Juarez, P. Quaino, E. Santos and W. Schmickler, *ChemPhysChem*, 2016, **17**, 78–85.

137 R. Sonnenfeld and P. K. Hansma, *Science*, 1986, **232**, 211–213.

138 H. Y. Liu, F. R. F. Fan, C. W. Lin and A. J. Bard, *J. Am. Chem. Soc.*, 1986, **108**, 3838–3839.

139 K. Itaya and E. Tomita, *Surf. Sci.*, 1988, **201**, L507–L512.

140 R. Christoph, H. Siegenthaler, H. Rohrer and H. Wiese, *Electrochim. Acta*, 1989, **34**, 1011–1022.

141 M. P. Green, K. J. Hanson, D. A. Scherson, X. Xing, M. Richter, P. N. Ross, R. Carr and I. Lindau, *J. Phys. Chem.*, 1989, **93**, 2181–2184.

142 F. F. Fan and A. J. Bard, *J. Electrochem. Soc.*, 1989, **136**, 3216–3222.

143 O. M. Magnussen, J. Hotlos, R. J. Nichols, D. M. Kolb and R. J. Behm, *Phys. Rev. Lett.*, 1990, **64**, 2929–2932.

144 S. Manne, P. K. Hansma, J. Massie, V. B. Elings and A. A. Gewirth, *Science*, 1991, **251**, 183–186.

145 R. M. Nyffenegger and R. M. Penner, *Chem. Rev.*, 1997, **97**, 1195–1230.

146 N. J. Tao, C. Z. Li and H. X. He, *J. Electroanal. Chem.*, 2000, **492**, 81–93.

147 P. Sun, F. O. Laforge and M. V. Mirkin, *Phys. Chem. Chem. Phys.*, 2007, **9**, 802–823.

148 M. V Mirkin, W. Nogala, J. Velmurugan and Y. Wang, *Phys. Chem. Chem. Phys.*, 2011, **13**, 21196–21212.

149 F. O. Laforge, J. Velmurugan, Y. Wang and M. V. Mirkin, *Anal. Chem.*, 2009, **81**, 3143–3150.

150 R. Tel-Vered and A. J. Bard, *J. Phys. Chem. B*, 2006, **110**, 25279–25287.

151 H. Xiong, D. A. Gross, J. Guo and S. Amemiya, *Anal. Chem.*, 2006, **78**, 1946–1957.

152 J. Kim, H. Xiong, M. Hofmann, J. Kong and S. Amemiya, *Anal. Chem.*, 2010, **82**, 1605–1607.

153 E. Kim, J. Kim and S. Amemiya, *Anal. Chem.*, 2009, **81**, 4788–4791.

154 T. Sun, Y. Yu, B. J. Zacher and M. V. Mirkin, *Angew. Chem., Int. Ed.*, 2014, **53**, 14120–14123.

155 P. Blanchard, T. Sun, Y. Yu, Z. Wei, H. Matsui and M. V. Mirkin, *Langmuir*, 2016, **32**, 2500–2508.

156 S. C. S. Lai, P. V. Dudin, J. V. MacPherson and P. R. Unwin, *J. Am. Chem. Soc.*, 2011, **133**, 10744–10747.

157 A. G. Güell, K. E. Meadows, P. V Dudin, N. Ebejer, J. C. Byers, J. V. Macpherson and P. R. Unwin, *Faraday Discuss.*, 2014, **172**, 439–455.

158 A. G. Güell, K. E. Meadows, P. V. Dudin, N. Ebejer, J. V. Macpherson and P. R. Unwin, *Nano Lett.*, 2014, **14**, 220–224.

159 J. C. Byers, A. G. Güell and P. R. Unwin, *J. Am. Chem. Soc.*, 2014, **136**, 11252–11255.

160 Y. Takahashi, A. Kumatani, H. Munakata, H. Inomata, K. Ito, K. Ino, H. Shiku, P. R. Unwin, Y. E. Korchev, K. Kanamura and T. Matsue, *Nat. Commun.*, 2014, **5**, 5450.

161 P. K. Hansma, B. Drake, O. Marti, S. A. Gould and C. B. Prater, *Science*, 1989, **243**, 641–643.

162 C.-C. Chen, Y. Zhou and L. A. Baker, *Annu. Rev. Anal. Chem.*, 2012, **5**, 207–228.

163 M. A. O'Connell and A. J. Wain, *Anal. Chem.*, 2014, **86**, 12100–12107.

164 M. A. O'Connell, J. R. Lewis and A. J. Wain, *Chem. Commun.*, 2015, **51**, 10314–10317.
165 J. Gorelik, Y. Gu, H. A. Spohr, A. I. Shevchuk, M. J. Lab, S. E. Harding, C. R. W. Edwards, M. Whitaker, G. W. J. Moss, D. C. H. Benton, D. Sánchez, A. Darszon, I. Vodyanoy, D. Klenerman and Y. E. Korchev, *Biophys. J.*, 2016, **83**, 3296–3303.
166 J. Abbou, A. Anne and C. Demaille, *J. Am. Chem. Soc.*, 2004, **126**, 10095–10108.
167 J. Abbou, A. Anne and C. Demaille, *J. Phys. Chem. B*, 2006, **110**, 22664–22675.
168 A. Anne, E. Cambril, A. Chovin and C. Demaille, *Anal. Chem.*, 2010, **82**, 6353–6362.
169 A. Anne, A. Chovin, C. Demaille and M. Lafouresse, *Anal. Chem.*, 2011, **83**, 7924–7932.
170 K. Hunag, A. Anne, M. A. Bahri and C. Demaille, *ACS Nano*, 2013, **7**, 4151–4163.
171 L. Nault, C. Taofifenua, A. Anne, A. Chovin, C. Demaille, J. Besong-Ndika, D. Cardinale, N. Carette, T. Michon and J. Walter, *ACS Nano*, 2015, **9**, 4911–4914.
172 N. J. Tao, *Phys. Rev. Lett.*, 1996, **76**, 4066–4069.
173 W. Han, E. N. Durantini, T. A. Moore, A. L. Moore, D. Gust, P. Rez, G. Leatherman, G. R. Seely, N. Tao and Lindsay, *J. Phys. Chem. B*, 1997, **101**, 10719–10725.
174 J. Zhang, Q. Chi, S. Dong and E. Wang, *Bioelectrochem. Bioenerg.*, 1996, **39**, 267–274.
175 E. P. Friis, J. E. T. Andersen, Y. I. Kharkats, A. M. Kuznetsov, R. J. Nichols, J.-D. Zhang and J. Ulstrup, *Proc. Natl. Acad. Sci. U. S. A.*, 1999, **96**, 1379–1384.
176 D. I. Gittins, D. Bethell, D. J. Schiffrin and R. J. Nichols, *Nature*, 2000, **408**, 67–69.
177 A. M. Kuznetsov, P. Sommer-Larsen and J. Ulstrup, *Surf. Sci.*, 1992, **275**, 52–64.
178 W. Schmickler, *Surf. Sci.*, 1993, **295**, 43–56.
179 J. Zhang, A. M. Kuznetsov, I. G. Medvedev, Q. Chi, T. Albrecht, P. S. Jensen and J. Ulstrup, *Chem. Rev.*, 2008, **108**, 2737–2791.
180 X. D. Cui, A. Primak, X. Zarate, J. Tomfohr, O. F. Sankey, A. L. Moore, T. A. Moore, D. Gust, G. Harris and S. M. Lindsay, *Science*, 2001, **294**, 571–574.
181 T. Morita and S. Lindsay, *J. Am. Chem. Soc.*, 2007, **129**, 7262–7263.
182 B. Xu and N. J. Tao, *Science*, 2003, **301**, 1221–1223.
183 J. Hihath and N. Tao, *Semicond. Sci. Technol.*, 2014, **29**, 54007.
184 R. J. Nichols and S. J. Higgins, *Annu. Rev. Anal. Chem.*, 2015, **8**, 389–417.
185 L. Venkataraman, J. E. Klare, I. W. Tam, C. Nuckolls, M. S. Hybertsen and M. L. Steigerwald, *Nano Lett.*, 2006, **6**, 458–462.
186 S. Guo, J. Hihath, I. Díez-Pérez and N. Tao, *J. Am. Chem. Soc.*, 2011, **133**, 19189–19197.
187 W. Haiss, H. van Zalinge, S. J. Higgins, D. Bethell, H. Höbenreich, D. J. Schiffrin and R. J. Nichols, *J. Am. Chem. Soc.*, 2003, **125**, 15294–15295.
188 W. Haiss, R. J. Nichols, H. van Zalinge, S. J. Higgins, D. Bethell and D. J. Schiffrin, *Phys. Chem. Chem. Phys.*, 2004, **6**, 4330–4337.
189 X. Xiao, X. Yang, L. Zang and Tao, *J. Am. Chem. Soc.*, 2005, **127**, 2386–2387.
190 F. Chen, J. He, C. Nuckolls, T. Roberts, J. E. Klare and S. Lindsay, *Nano Lett.*, 2005, **5**, 503–506.

191 I. V Pobelov, Z. Li and T. Wandlowski, *J. Am. Chem. Soc.*, 2008, **130**, 16045–16054.
192 Z. Li and E. Borguet, *J. Am. Chem. Soc.*, 2012, **134**, 63–66.
193 H. M. Osorio, S. Catarelli, P. Cea, J. B. G. Gluyas, F. Hartl, S. J. Higgins, E. Leary, P. J. Low, S. Martín, R. J. Nichols, J. Tory, J. Ulstrup, A. Vezzoli, D. C. Milan and Q. Zeng, *J. Am. Chem. Soc.*, 2015, **137**, 14319–14328.
194 X. Li, B. Xu, X. Xiao, X. Yang, L. Zang and N. Tao, *Faraday Discuss.*, 2006, **131**, 111–120.
195 W. Haiss, T. Albrecht, H. van Zalinge, S. J. Higgins, D. Bethell, H. Höbenreich, D. J. Schiffrin, R. J. Nichols, A. M. Kuznetsov, J. Zhang, Q. Chi and J. Ulstrup, *J. Phys. Chem. B*, 2007, **111**, 6703–6712.
196 N. Darwish, I. Díez-Pérez, S. Guo, N. Tao, J. J. Gooding and M. N. Paddon-Row, *J. Phys. Chem. C*, 2012, **116**, 21093–21097.
197 L. Xiang, J. L. Palma, C. Bruot, V. Mujica, M. A. Ratner and N. Tao, *Nat. Chem.*, 2015, **7**, 221–226.
198 C. Liu, L. Xiang, Y. Zhang, P. Zhang, D. N. Beratan, Y. Li and N. Tao, *Nat. Chem.*, 2016, DOI: 10.1038/nchem.2545.
199 A. Migliore and A. Nitzan, *Electrochim. Acta*, 2015, **160**, 363–375.
200 X. Shan, U. Patel, S. Wang, R. Iglesias and N. Tao, *Science*, 2010, **327**, 1363–1366.
201 K. J. Foley, X. Shan and N. J. Tao, *Anal. Chem.*, 2008, **80**, 5146–5151.
202 W. Wang, K. Foley, X. Shan, S. Wang, S. Eaton, V. J. Nagaraj, P. Wiktor, U. Patel and N. Tao, *Nat. Chem.*, 2011, **3**, 249–255.
203 S. Wang, X. Huang, X. Shan, K. J. Foley and N. Tao, 2010, **82**, 935–941.
204 X. Shan, I. Díez-Pérez, L. Wang, P. Wiktor, Y. Gu, L. Zhang, W. Wang, J. Lu, S. Wang, Q. Gong, J. Li and N. Tao, *Nat. Nanotechnol.*, 2012, **7**, 668–672.
205 X. Shan, S. Wang, W. Wang and N. Tao, *Anal. Chem.*, 2011, **83**, 7394–7399.
206 J. Lu and J. Li, *Angew. Chem., Int. Ed.*, 2015, **54**, 13576–13580.
207 J. Lu, W. Wang, S. Wang, X. Shan, J. Li and N. Tao, *Anal. Chem.*, 2012, **84**, 327–333.
208 X. Shan, S. Chen, H. Wang, Z. Chen, Y. Guan, Y. Wang, S. Wang, H.-Y. Chen and N. Tao, *Adv. Mater.*, 2015, **27**, 6213–6219.
209 Y. Fang, W. Wang, X. Wo, Y. Luo, S. Yin, Y. Wang, X. Shan and N. Tao, *J. Am. Chem. Soc.*, 2014, **136**, 12584–12587.
210 M. Hu, C. Novo, A. Funston, H. Wang, H. Staleva, S. Zou, P. Mulvaney, Y. Xia and G. V. Hartland, *J. Mater. Chem.*, 2008, **18**, 1949–1960.
211 J. Rodríguez-Fernández, C. Novo, V. Myroshnychenko, A. M. Funston, A. Sánchez-Iglesias, I. Pastoriza-Santos, J. Pérez-Juste, F. J. García de Abajo, L. M. Liz-Marzán and P. Mulvaney, *J. Phys. Chem. C*, 2009, **113**, 18623–18631.
212 C. L. Nehl, N. K. Grady, G. P. Goodrich, F. Tam, N. J. Halas and J. H. Hafner, *Nano Lett.*, 2004, **4**, 2355–2359.
213 C. Novo, A. M. Funston, A. K. Gooding and P. Mulvaney, *J. Am. Chem. Soc.*, 2009, **131**, 14664–14666.
214 A. Kuzma, M. Weis, S. Flickyngerova, J. Jakabovic, A. Satka, E. Dobrocka, J. Chlpik, J. Cirak, M. Donoval, P. Telek, F. Uherek and D. Donoval, *J. Appl. Phys.*, 2012, **112**, 103531.
215 C. M. Hill and S. Pan, *J. Am. Chem. Soc.*, 2013, **135**, 17250–17253.
216 M. Chirea, S. S. E. Collins, X. Wei and P. Mulvaney, *J. Phys. Chem. Lett.*, 2014, **5**, 4331–4335.

217 M. Eo, J. Baek, H. D. Song, S. Lee and J. Yi, *Chem. Commun.*, 2013, **49**, 5204–5206.
218 D. Seo, G. Park and H. Song, *J. Am. Chem. Soc.*, 2012, **134**, 1221–1227.
219 J.-G. Wang, J. S. Fossey, M. Li, T. Xie and Y.-T. Long, *ACS Appl. Mater. Interfaces*, 2016, **8**, 8305–8314.
220 C. Jing, F. J. Rawson, H. Zhou, X. Shi, W. H. Li, D. W. Li and Y.-T. Long, *Anal. Chem.*, 2014, **86**, 5513–5518.
221 C. P. Byers, B. S. Hoener, W.-S. Chang, S. Link and C. F. Landes, *Nano Lett.*, 2016, **16**, 2314–2321.
222 C. M. Hill, D. A. Clayton and S. Pan, *Phys. Chem. Chem. Phys.*, 2013, **15**, 20797–20807.
223 K. Mathwig, T. J. Aartsma, G. W. Canters and S. G. Lemay, *Annu. Rev. Anal. Chem.*, 2014, **7**, 383–404.
224 J. C. Bolinger, M. C. Traub, J. Brazard, T. Adachi, P. F. Barbara and D. A. Vanden Bout, *Acc. Chem. Res.*, 2012, **45**, 1992–2001.
225 R. E. Palacios, F.-R. F. Fan, A. J. Bard and P. F. Barbara, *J. Am. Chem. Soc.*, 2006, **128**, 9028–9029.
226 R. E. Palacios, F.-R. F. Fan, J. K. Grey, J. Suk, A. J. Bard and P. F. Barbara, *Nat. Mater.*, 2007, **6**, 680–685.
227 H. Yang, G. Luo, P. Karnchanaphanurach, T.-M. Louie, I. Rech, S. Cova, L. Xun and X. S. Xie, *Science*, 2003, **302**, 262–266.
228 C. Lei, D. Hu and E. J. Ackerman, *Chem. Commun.*, 2008, 5490–5492.
229 C. Lei, D. Hu and E. Ackerman, *Nano Lett.*, 2009, **9**, 655–658.
230 J. B. Sambur, T.-Y. Chen, E. Choudhary, G. Chen, E. J. Nissen, E. M. Thomas, N. Zou and P. Chen, *Nature*, 2016, **530**, 77–80.
231 X. Zhou, E. Choudhary, N. M. Andoy, N. Zou and P. Chen, *ACS Catal.*, 2013, **3**, 1448–1453.
232 X. Zhou, N. M. Andoy, G. Liu, E. Choudhary, K.-S. Han, H. Shen and P. Chen, *Nat. Nanotechnol.*, 2012, **7**, 237–241.
233 K. Naito, T. Tachikawa, M. Fujitsuka and T. Majima, *J. Am. Chem. Soc.*, 2009, 934–936.
234 T. Tachikawa, S. Yamashita and T. Majima, *J. Am. Chem. Soc.*, 2011, **133**, 7197–7204.
235 W. Xu, P. K. Jain, B. J. Beberwyck and A. P. Alivisatos, *J. Am. Chem. Soc.*, 2012, **134**, 3946–3949.
236 Y. Zhang, J. M. Lucas, P. Song, B. Beberwyck, Q. Fu, W. Xu and A. P. Alivisatos, *Proc. Natl. Acad. Sci. U. S. A.*, 2015, **112**, 8959–8964.
237 M. Fleischmann, P. J. Hendra and A. J. McQuillan, *Chem. Phys. Lett.*, 1974, **26**, 163–166.
238 D. L. Jeanmaire and R. P. Van Duyne, *J. Electroanal. Chem. Interfacial Electrochem.*, 1977, **84**, 1–20.
239 M. G. Albrecht and J. A. Creighton, *J. Am. Chem. Soc.*, 1977, **99**, 5215–5217.
240 J. Kneipp, H. Kneipp and K. Kneipp, *Chem. Soc. Rev.*, 2008, **37**, 1052–1060.
241 S. Schlücker, *Angew. Chem., Int. Ed.*, 2014, **53**, 4756–4795.
242 S. Nie and S. R. Emory, *Science*, 1997, **275**, 1102–1106.
243 K. Kneipp, Y. Wang, H. Kneipp, L. T. Perelman, I. Itzkan, R. R. Dasari and M. S. Feld, *Phys. Rev. Lett.*, 1997, **78**, 1667–1670.
244 Z.-Q. Tian and B. Ren, *Annu. Rev. Phys. Chem.*, 2004, **55**, 197–229.

245 E. Cortés, P. G. Etchegoin, E. C. Le Ru, A. Fainstein, M. E. Vela and R. C. Salvarezza, *J. Am. Chem. Soc.*, 2010, **132**, 10834–10837.
246 A. J. Wilson and K. A. Willets, *Nano Lett.*, 2014, **14**, 939–945.
247 Y. Wang, P. C. Sevinc, Y. He and H. P. Lu, *J. Am. Chem. Soc.*, 2011, **133**, 6989–6996.
248 X. Shi, H. Li, Y.-L. Ying, C. Liu, L. Zhang and Y.-T. Long, *Chem. Commun.*, 2015, **52**, 1044–1047.
249 J. F. Li, Y. F. Huang, Y. Ding, Z. L. Yang, S. B. Li, X. S. Zhou, F. R. Fan, W. Zhang, Z. Y. Zhou, W. Yin, B. Ren, Z. L. Wang and Z. Q. Tian, *Nature*, 2010, **464**, 392–395.
250 M. D. Sonntag, E. A. Pozzi, N. Jiang, M. C. Hersam and R. P. Van Duyne, *J. Phys. Chem. Lett.*, 2014, **5**, 3125–3130.
251 N. Jiang, D. Kurouski, E. A. Pozzi, N. Chiang, M. C. Hersam and R. P. Van Duyne, *Chem. Phys. Lett.*, 2016, **659**, 16–24.
252 Z.-C. Zeng, S.-C. Huang, D.-Y. Wu, L.-Y. Meng, M.-H. Li, T.-X. Huang, J.-H. Zhong, X. Wang, Z.-L. Yang and B. Ren, *J. Am. Chem. Soc.*, 2015, **137**, 11928–11931.
253 D. Kurouski, M. Mattei and R. P. van Duyne, *Nano Lett.*, 2015, **15**, 7956–7962.
254 M. J. Williamson, R. M. Tromp, P. M. Vereecken, R. Hull and F. M. Ross, *Nat. Mater.*, 2003, **2**, 532–536.
255 H. Zheng, R. K. Smith, Y. Jun, C. Kisielowski, U. Dahmen and A. P. Alivisatos, *Science*, 2009, **324**, 1309–1312.
256 K. L. Jungjohann, J. E. Evans, J. A. Aguiar, I. Arslan and N. D. Browning, *Microsc. Microanal.*, 2012, **18**, 621–627.
257 J. M. Grogan and H. H. Bau, *J. Microelectromech. Syst.*, 2010, **19**, 885–894.
258 M. Gu, L. R. Parent, B. L. Mehdi, R. R. Unocic, M. T. McDowell, R. L. Sacci, W. Xu, J. G. Connell, P. Xu, P. Abellan, X. Chen, Y. Zhang, D. E. Perea, J. E. Evans, L. J. Lauhon, J.-G. Zhang, J. Liu, N. D. Browning, Y. Cui, I. Arslan and C.-M. Wang, *Nano Lett.*, 2013, **13**, 6106–6112.
259 R. Unocic, L. Adamczyk, N. Dudney, D. Alsem, N. Salmon and K. More, *Microsc. Microanal.*, 2011, **17**, 1564–1565.
260 C.-M. Wang, W. Xu, J. Liu, J.-G. Zhang, L. V Saraf, B. W. Arey, D. Choi, Z.-G. Yang, J. Xiao, S. Thevuthasan and D. R. Baer, *Nano Lett.*, 2011, **11**, 1874–1880.
261 E. R. White, S. B. Singer, V. Augustyn, W. A. Hubbard, M. Mecklenburg, B. Dunn and B. C. Regan, *ACS Nano*, 2012, **6**, 6308–6317.
262 N. de Jonge and F. M. Ross, *Nat. Nanotechnol.*, 2011, **6**, 695–704.
263 F. Wu and N. Yao, *Nano Energy*, 2015, **11**, 196–210.
264 M. E. Holtz, Y. Yu, D. Gunceler, J. Gao, R. Sundararaman, K. A. Schwarz, T. A. Arias, H. D. Abruña and D. A. Muller, *Nano Lett.*, 2014, **14**, 1453–1459.
265 C. M. Hill, D. A. Clayton and S. Pan, *Phys. Chem. Chem. Phys.*, 2013, **15**, 20797–20807.
266 S. E. Fosdick, M. J. Anderson, E. G. Nettleton and R. M. Crooks, *J. Am. Chem. Soc.*, 2013, **135**, 5994–5997.
267 W. Wang, Y. Yang, S. Wang, V. J. Nagaraj, Q. Liu, J. Wu and N. Tao, *Nat. Chem.*, 2012, **4**, 846–853.
268 Y. Wang, X. Shan, S. Wang, N. Tao, P.-Y. Blanchard, K. Hu and M. V. Mirkin, *Anal. Chem.*, 2016, **88**, 1547–1552.

269 R. A. Marcus and N. Sutin, *Biochim. Biophys. Acta, Rev. Bioenerg.*, 1985, **811**, 265–322.
270 S. S. Skourtis, D. H. Waldeck and D. N. Beratan, *Annu. Rev. Phys. Chem.*, 2010, **61**, 461–485.
271 D. N. Beratan, C. Liu, A. Migliore, N. F. Polizzi, S. S. Skourtis, P. Zhang and Y. Zhang, *Acc. Chem. Res.*, 2015, **48**, 474–481.
272 P. S. Singh and S. G. Lemay, *Anal. Chem.*, 2016, **88**, 5017–5027.
273 H. Ishikawa, K. Kwak, J. K. Chung, S. Kim and M. D. Fayer, *Proc. Natl. Acad. Sci. U. S. A.*, 2008, **105**, 8619–8624.
274 S. W. Fan, R. A. George, N. L. Haworth, L. L. Feng, J. Y. Liu and M. A. Wouters, *Protein Sci.*, 2009, **18**, 1745–1765.
275 C. Amatore and E. Maisonhaute, *Anal. Chem.*, 2005, **77**, 303A–311A.
276 R. Wen, B. Rahn and O. M. Magnussen, *Angew. Chem., Int. Ed.*, 2015, **54**, 6062–6066.
277 D. Momotenko, J. C. Byers, K. McKelvey, M. Kang and P. R. Unwin, *ACS Nano*, 2015, **9**, 8942–8952.
278 C. Amatore, E. Maisonhaute and G. Simonneau, *J. Electroanal. Chem.*, 2000, **486**, 141–155.
279 X.-S. Zhou, B.-W. Mao, C. Amatore, R. G. Compton, J.-L. Marignier, M. Mostafavi, J.-F. Nierengarten and E. Maisonhaute, *Chem. Commun.*, 2016, **52**, 251–263.
280 S. Guo, J. Hihath and N. Tao, *Nano Lett.*, 2011, **11**, 927–933.
281 H. D. Sikes, J. F. Smalley, S. P. Dudek, A. R. Cook, M. D. Newton, C. E. D. Chidsey and S. W. Feldberg, *Science*, 2001, **291**, 1519–1523.
282 K. Krischer, in *Advances in Electrochemical Science and Engineering*, Wiley-VCH Verlag GmbH & Co. KGaA, 2003, vol. 8, pp. 89–208.

Faraday Discussions

PAPER

Single-molecule electrochemistry in nanochannels: probing the time of first passage†

Shuo Kang, Ab F. Nieuwenhuis, Klaus Mathwig,‡ Dileep Mampallil,§ Zinaida A. Kostiuchenko and Serge G. Lemay*

Received 11th April 2016, Accepted 19th April 2016
DOI: 10.1039/c6fd00075d

The diffusive mass transport of individual redox molecules was probed experimentally in microfabricated nanogap electrodes. The residence times for molecules inside a well-defined detection volume were extracted and the resulting distribution was compared with quantitative analytical predictions from random-walk theory for the time of first passage. The results suggest that a small number of strongly adsorbing sites strongly influence mass transport at trace analyte levels.

Introduction

The ability to address and probe individual molecules has led to significant new insights on molecular transport and internal dynamics in fields ranging from molecular electronics to biophysics.[1,2] It is thus of considerable interest to extend the range of single-molecule analysis methods to other transduction mechanisms such as electrochemistry. While single-molecule detection by electrochemical means has indeed been reported,[3–7] efforts so far have focused almost exclusively on demonstrating the basic ability to resolve single molecules and have not allowed for deeper analysis. This is largely due to the experimental challenge inherent in measuring the extremely small signal levels associated with single-molecule faradaic processes.

Here we demonstrate for the first time the ability to statistically analyse the diffusive dynamics of individual molecules using an electrochemical method. We

MESA+ Institute for Nanotechnology, University of Twente, PO Box 217, 7500 AE Enschede, The Netherlands. E-mail: S.G.Lemay@utwente.nl

† In response to comments raised during the discussion of this manuscript, Eqn (2) and Fig. (3) of the manuscript have been altered from the preprint version. Eq. (2) was rewritten in its exact form instead of the earlier, approximate form and an error in the normalization of the theoretical curves in Fig. (3) was corrected.

‡ Present address: Groningen Research Institute of Pharmacy, University of Groningen, P.O. Box 196, 9700 AD Groningen, The Netherlands.

§ Present address: Indian Institute of Science Education & Research Tirupati, Mangalam P. O. Tirupati 517 507, India.

focus on the distribution of residence times for molecules randomly entering and leaving a nanofluidic cavity, an inherently single-molecule property. The residence time corresponds to a well-studied quantity in stochastic systems, the so-called first-passage time,[8] which appears in diverse fields including ecology,[9] biophysics,[10–12] finance,[13] diffusion-controlled chemical reactions,[14] *etc.*,[15] and which permits a quantitative comparison of the measurements with theory.

Experimental methods

Redox cycling in nanogap transducers offers the possibility to record extensive, high-quality single-molecule amperometric data.[6] The nanogap device is sketched in Fig. 1a. It consists of two electrodes with a length $L = 50$ μm and a width of 4 or 5 μm that are separated by a solution-filled gap of 40 nm. The device was microfabricated using optical lithography, as described earlier.[6] In brief, a stack of metal layers (Pt/Cr/Pt) was first deposited and patterned on a 4-inch Si wafer,

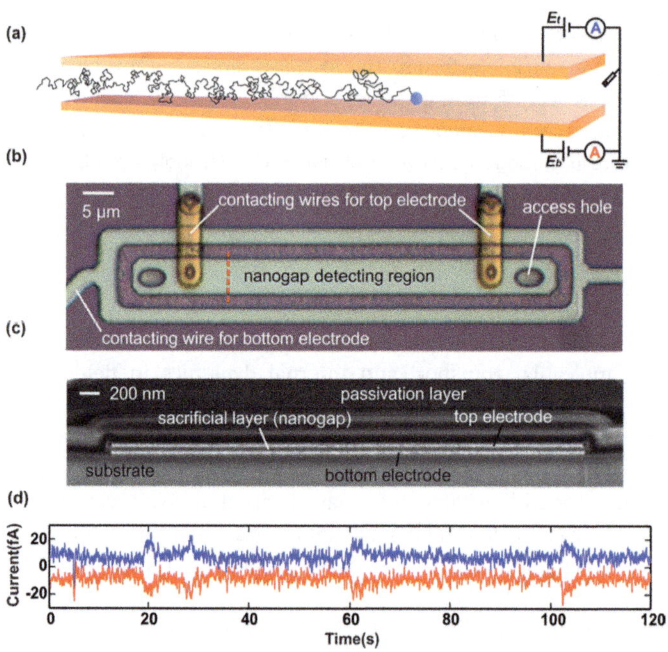

Fig. 1 (a) Schematic illustration of a nanogap device. (b) Optical microscope image (top view) of a device. The cyan rectangle in the center corresponds to the top electrode, below which lies the detection region. The remaining structures serve as electrical interconnects to the top and bottom electrodes (yellow and cyan lines, respectively). (c) Scanning electron microscopy image of the cross section of a device from a 52° viewing angle. To make this image the device was cut open using a focused ion beam along the line indicated by the dashed red line in (b). (d) Current-time traces simultaneously measured at the top (blue) and bottom (red) electrodes (width 5 μm) under redox-cycling conditions (10 pM FcTMABr in 0.1 M KCl aqueous solution, $E_t = 0.5$ V and $E_b = 0.35$ V). The traces have been offset vertically for clarity. Several single-molecule events with duration >1 second appear as abrupt changes in the currents compared to the baseline. These transients have opposite signs at the two electrodes as a result of redox cycling (oxidation and reduction at the top and bottom electrode, respectively).

followed by ion-beam etching (IBE) and subsequent wet etching to define the active nanogap detecting region. A final metal layer sandwiched between two passivation layers was then deposited and patterned to provide electrical connections to the top electrode. Finally, access holes for liquid were created by etching through the passivation and top electrode. The sacrificial layer was wet-etched *via* the access holes immediately before use, releasing the 40 nm high nanochannel. An optical microscope image of a complete device and a scanning electron microscope image of the nanogap cross section are shown in Fig. 1b and c, respectively.

The measurements were performed by interfacing the device with an external reservoir containing an aqueous solution with a concentration $C = 10$ pM of redox species and 0.1 M KCl as supporting electrolyte. The average number of analyte molecules in solution in the detection region was then $\langle N \rangle = CVN_A = 0.05$ for a device with a width of 4 μm, where $V = 8 \times 10^{-18}$ m^3 is the volume of the nanogap region and N_A is Avogadro's number. For a 5 μm wide device, $\langle N \rangle = 0.06$ under the same conditions. For such low occupancy, the fraction of the time that the device is occupied by two or more molecules at the same time is low (<0.2%) based on Poisson statistics.

For single-molecule experiments, the bottom and top electrodes were biased at reducing and oxidizing potentials, respectively. Under these redox-cycling conditions, each chemically reversible redox molecule present in the nanogap is repeatedly oxidized and reduced as it undergoes Brownian motion, generating faradaic currents of opposite polarity at the two electrodes. These reduction and oxidation currents were measured simultaneously; each time that a molecule randomly entered and subsequently exited the nanochannel, current steps with magnitudes of the order 10^{-14} A and opposite signs were observed in the measured currents. This is illustrated in Fig. 1d, which shows a time trace with four distinctly recognizable single-molecule events with durations of the order of a few seconds.

In order to achieve sufficient sensitivity to detect these extremely low currents, current-to-voltage converters with 10^{12} V A^{-1} trans-impedance gain and a response time of 350 ms (10–90% rise time) were employed (FEMTO DDPCA-300).[16] Such a slow response is essentially unavoidable due to the inverse relationship between gain and measurement bandwidth in high-gain amplifiers. Consequently, and as discussed further below, only events lasting longer than ~450 ms could be reliably resolved.

Results and discussion

Numerical simulations

To ascertain the feasibility of identifying single-molecule events from the experimental data and accurately determining their durations, we performed random walk simulations of individual molecular trajectories. We have previously shown that, due to the devices' highly anisotropic geometry, Brownian motion in nanogap transducers such as the ones employed here can be modelled as a discrete 1D random walk.[17] At each time step a diffusive molecule makes a random jump to the left or right with a step size $\Delta x = L/a$, where L is the length of the device and $a - 1$ ($\gg 1$) is the number of spatial grid points in the simulation. In order to ensure consistency with the macroscopic diffusion coefficient,

D, the size of each time step is taken as $\Delta t = \Delta x^2/2D$. We emphasize that this simple algorithm is sufficient to provide an accurate description of Brownian motion on length and time scales significantly larger than Δx and Δt, respectively, even when Δx and Δt are themselves much larger than the microscopic scales associated with individual molecular collisions in the liquid. In the simulation, the two grid points at the edges of the simulated domain correspond to positions adjacent to the two access holes of the device. Molecules that hop outside the simulated domain are considered to have exited the device through these holes, while molecules diffusing into the device are represented by a probability p that a new molecule appears at either end of the simulated domain at each time step. In the steady state, the average rates of entry and egress are equal and the average number of molecules inside the device, $\langle N \rangle$, is constant; the value of $\langle N \rangle$ can be tuned by adjusting the value of p. Interactions between molecules are neglected.

Fig. 2a shows a typical trace of the device occupancy *versus* time for an average occupancy $\langle N \rangle = 0.1$. Most events are very short, with only a minority of events having durations of the order of seconds. To simulate the role of the measurement electronics, the occupancy was first converted to a current as $I(t) = N(t)i_p$, where i_p is the redox-cycling current per molecule. White noise was then added and the resulting traces were convolved with a first-order impulse response (a simple exponential) with a time constant of $\tau = 350$ ms that corresponds to the experimental apparatus. The result for the trace in Fig. 2a is shown in Fig. 2b. Events with a duration of order 1 s reach (or even temporarily exceed, because of noise) the full single-molecule current, i_p. Such events exhibit a trapezoid shape, with a full width at half-maximum that agrees within ±10% with the duration of the underlying event in Fig. 2a. Longer events (not shown) instead exhibit a well-defined plateau, while events with durations of order $\tau = 450$ ms or shorter contribute to the apparent noise in the current but cannot be reliably identified. Occasionally two long events are juxtaposed and cannot be distinguished, as illustrated in Fig. 2 around 17 s. However, at this low average occupancy such occurrences are rare, as are occurrences of two long events overlapping for a short time. Multiple occupancy thus has minimal impact on the overall statistics. Taken together, these simulation

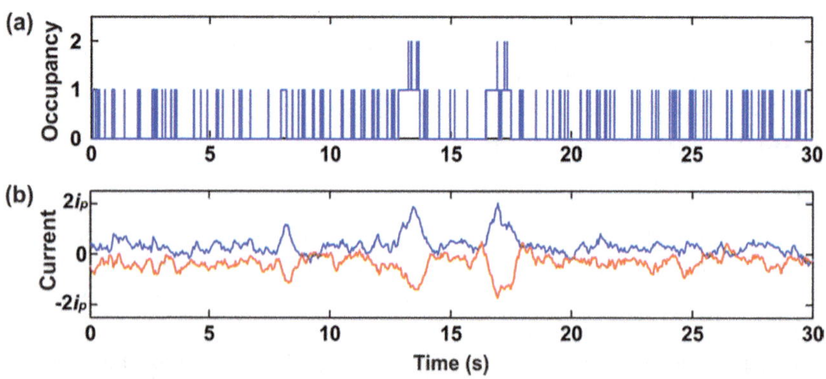

Fig. 2 (a) Simulated occupancy–time trace for an average occupancy $\langle N \rangle = 0.1$. (b) Corresponding simulated current–time trace including additive Gaussian noise (3 fA_{rms}) and filtered to account for the influence of the measurement electronics. The current per molecule, i_p, is assumed to be 15 fA.

results suggest that it is feasible to accurately identify single-molecule events longer than $\tau = 450$ ms and to determine their duration.

Single-molecule measurements

Extensive current–time traces were recorded with aqueous solutions of 10 pM ferrocenedimethanol (1Fc(MeOH)$_2$, $E^{\circ\prime} = 0.25$ V) and 10 pM ferricyanide $[Fe(CN)_6]^{3-}$, $E^{\circ\prime} = 0.23$ V) in 0.1 M KCl under redox cycling conditions ($E_t = 0.3$ V, $E_b = 0.2$ V for Fc(MeOH)$_2$ and $E_t = 0.3$ V, $E_b = 0.15$ V for $[Fe(CN)_6]^{3-}$) in 4 μm and 5 μm wide devices, respectively, as shown in Fig. 3a and c. Single-molecule events were identified and their durations were extracted from these amperometric data. Fig. 3b and d show the histograms of event durations extracted from 26 min and 20 min long traces measured under the same conditions as Fig. 3a and c, respectively. Events shorter than 450 ms were excluded from this analysis, yielding in total of 132 and 91 events for Fc(MeOH)$_2$ and $[Fc(CN)_6]^{3-}$, respectively. In both measurements, the histograms exhibit a sharply decreasing number of events with increasing event duration, with the majority of identifiable events having durations below 1 s.

The histograms in Fig. 3 encapsulate detailed quantitative information about the statistics of individual single-molecule trajectories. Furthermore, because mass transport is purely diffusive at the high supporting electrolyte concentrations employed here, the form of the histograms can be predicted quantitatively. This provides an unprecedentedly detailed test of electrochemical single-molecule data.

Time of first passage analysis

Within the discrete random-walk model, a single-molecule event as discussed in Fig. 3 corresponds to a molecule starting at one edge of the simulation domain and taking a number of random steps inside the device before making a final jump out either end of the simulation domain. While such an event can in principle be arbitrarily long (the molecule can hop left and right between two adjacent sites indefinitely, for example), intuition dictates that very long events should be scarce as the randomly hopping molecule is bound to hit one of the boundaries sooner or later. Conversely, very short events are expected to be common since a molecule that just entered the simulation domain need only take a few steps in the reverse direction before it exits again. This behaviour cannot be probed using measurements averaged over multiple molecules (by measuring, *e.g.*, the average current) since those signals by definition average over the contributions from multiple microscopic events. Conversely, a single event provides only very limited information. A meaningful comparison between experiment and theory can however be achieved by comparing the relative probability of the possible outcomes.

Thanks to their broad applicability in fields ranging from physics to ecology and economics, the statistics of random walks have been studied in detail. Furthermore, while numerical simulations such as those in Fig. 2 represent a useful tool for predicting the qualitative features of the data, quantitative comparison is facilitated by the use of exact analytical expressions. The event duration in our experiment and associated random-walk model corresponds to the classic problem of the so-called first-passage time for a particle undergoing a discrete random walk in a linear domain with absorbing boundaries.[18] The

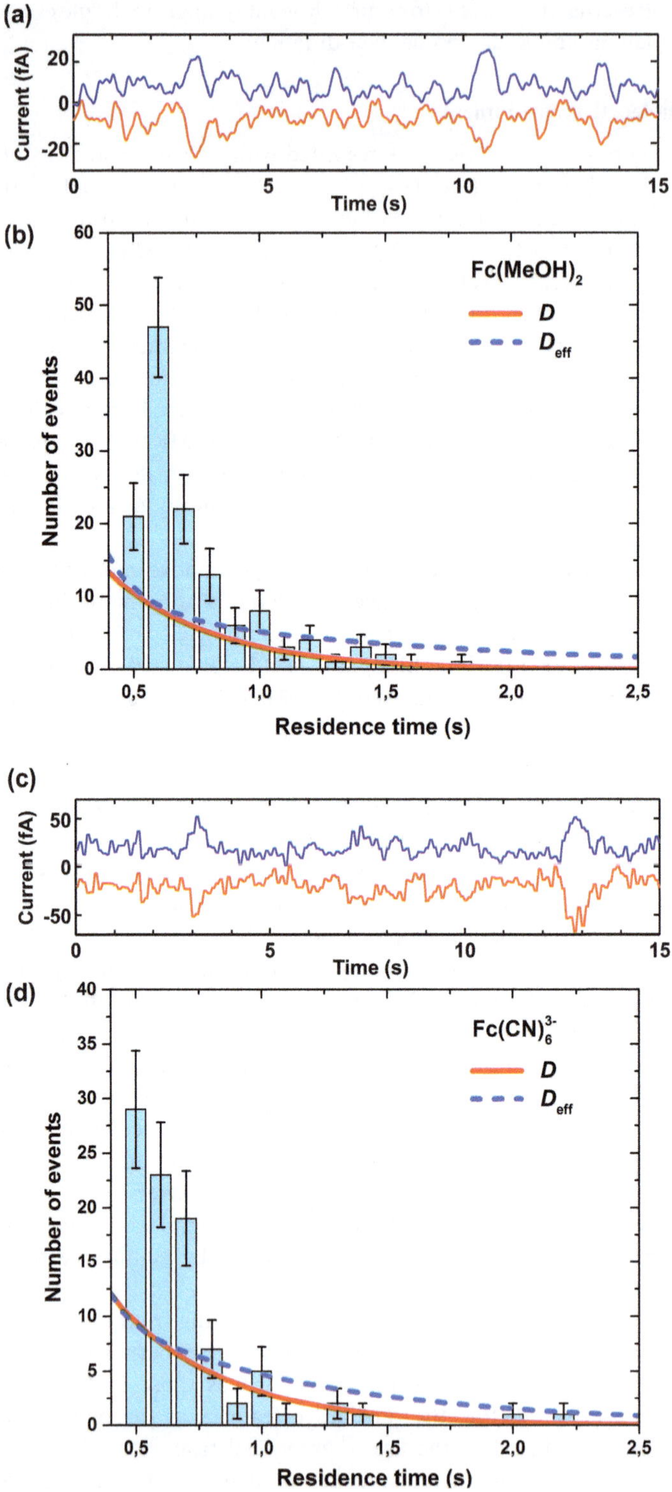
(a)
Current (fA)
20
0
-20
0
5
10
15
Time (s)
(b)
Number of events
Fc(MeOH)$_2$
D
D_{eff}
Residence time (s)
(c)
Current (fA)
50
0
-50
Time (s)
(d)
Number of events
$Fc(CN)_6^{3-}$
D
D_{eff}
Residence time (s)

probability of the particle exiting after taking exactly n steps is given by the analytic expression

$$P(n) = \frac{1}{a}\sum_{\mu=1}^{a-1}\cos^{n-1}\frac{\pi\mu}{a}\sin\frac{\pi\mu}{a}\sin\frac{\pi\mu z}{a} + \frac{1}{a}\sum_{\mu=1}^{a-1}\cos^{n-1}\frac{\pi\mu}{a}\sin\frac{\pi\mu}{a}\sin\frac{\pi\mu(a-z)}{a}. \quad (1)$$

Here a is the length of the simulation domain, as above, while z is the starting position of the particle ($1 \leq z \leq a-1$). We set $z = 1$ (or $z = a - 1$, which is equivalent from symmetry) to represent a molecule near either access hole immediately following entry of the device. The first term of the equation represents the probability that the particle exits from the same end as it entered and the second term from the opposite end. The first term diverges as $n^{-3/2}$ for small n; consequently, the most probable scenario for an event is for a molecule to enter the nanochannel and to exit again from the same end after only a few steps,[5,17] such that the residence time is very short and cannot be resolved experimentally. For large n, the distribution decays exponentially with n: in these relatively rare cases, the molecule penetrates further into the device, generating events that may be sufficiently long to be detected.

To compare theory with the experimental histogram data of Fig. 3, we renormalize eqn (1) to yield $M(t)$, the average number of events in a histogram bin of width γ and centred at time t:

$$M(t) = M_{\text{total}} \sum_{n=(t-\gamma/2)/\Delta t}^{(t+\gamma/2)/\Delta t} P(n). \quad (2)$$

Here Δt is the simulation time step as defined above and M_{total} is the total number of events that take place over the course of the measurement. The value of M_{total} is in turn obtained from $M_{\text{total}} = T\langle N\rangle/\tau_0$, where T is the total duration of the measurement, $\langle N\rangle$ is the average number of analyte molecules in the detection region, and τ_0 is the average duration of an event:

$$\tau_0 = \Delta t \sum_{n=1}^{\infty} nP(n). \quad (3)$$

These equations represent a quantitative prediction for the histogram, $M(t)$. The only inputs are experimental parameters, namely, the device length, L, the diffusion coefficient, D, the average occupancy, $\langle N\rangle$, and the duration of the measurement, T. Predictions based on the bulk diffusion coefficient ($D = 5.6 \times 10^{-10}$ m^2 s^{-1} for $Fc(MeOH)_2$ and $D = 6.0 \times 10^{-10}$ m^2 s^{-1} for $[Fe(CN)_6]^{3-}$, as determined from the diffusion-limited current at a 5 μm radius platinum ultramicroelectrode) and $\langle N\rangle$ based on the 10 pM bulk concentration are shown as red curves in Fig. 3b and d. In both cases the analytical calculation reproduces rather well the experimental data, with both the absolute number of

Fig. 3 (a, c) Representative current-time traces for 10 pM $Fc(MeOH)_2$ and $[Fe(CN)_6]^{3-}$ in 0.1 M KCl under redox cycling conditions ($E_t = 0.3$ V, $E_b = 0.2$ V, $E^{o\prime} = 0.25$ V and $E_t = 0.3$ V, $E_b = 0.15$ V, $E^{o\prime} = 0.23$ V, respectively). (b, d) Histograms (bin size $\gamma = 0.1$ s) of observed event durations and corresponding theoretical predictions for the experimental conditions of (a) and (c). Events shorter than 450 ms were excluded as they could not be resolved reliably.

events per bin and the decrease in the number of events with increasing duration being comparable between calculation and experiment apart from an excess of shorter events in the experiment.

This apparent good agreement may be fortuitous, however, as it contradicts our current understanding of microscopic diffusive dynamics in nanochannels. We have previously reported that the species studied here undergo some degree of reversible adsorption at the electrodes.[6,19,20] This is manifested by a suppression of the single-molecule current i_p such that the single-molecule current is multiplied by a factor χ_{ads} (with $0 \leq \chi_{ads} \leq 1$) compared to its ideal, diffusion-limited value,[3] $i_{p,ideal} = eD/z^2$, where $-e$ is the electron charge and z is the electrode spacing (in practice additional corrections are also included to take into account the electrode biases and the full three-dimensional geometry of the device, as per ref. 6). But reversible adsorption should also have a second consequence for mass transport, namely, that of decreasing the apparent longitudinal diffusion coefficient along the channel, D_{eff}, below its value for unhindered diffusion in bulk solution, D. These two adsorption-induced effects are directly coupled: for the simplest model in which adsorbed molecules are immobile on the electrode, $D_{eff} = \chi_{ads}D$.[6,19] We have previously observed this longitudinal behaviour in response time[21,22] and noise spectrum measurements.[17,19]

The amperometric data illustrated in Fig. 3a and c yield $\chi_{ads} = 0.3$ and 0.5 for $Fc(MeOH)_2$ and $[Fe(CN)_6]^{3-}$, respectively. Replacing D with the corresponding values of D_{eff} in eqn (1)–(3) yields the dashed green curves in Fig. 3b and d. Consistent with intuition, slower diffusion along the channel leads to a higher predicted rate of occurrence of long events for a given concentration (more precisely, the distribution gets stretched by a factor D/D_{eff} along the time axis and its amplitude is renormalized). Interestingly, however, these predictions do not agree well with the data, as significantly more long events are predicted than are observed in the experiment.

One possible interpretation of the observation that the histograms fit well with calculations using the bulk diffusion coefficient D but not with those using D_{eff} is that molecules adsorbed to the electrodes can diffuse along the channel without being slowed down substantially. In such a scenario the shuttling rate between the two electrodes is influenced by adsorption but the residence time inside the active region is not, as per the experimental findings. Surface diffusion of adsorbents has been studied in a variety of systems,[23–27] in particular using single-molecule fluorescence techniques. In some cases adsorbed molecules occupy specific binding sites but are able to move between binding sites,[24] qualitatively consistent with our observations. On the other hand, adsorption studies in nanogaps based on power spectral density (PSD) analysis[19] and transient chronoamperometric response[21,22] yield unambiguous evidence of an effective slowing down of longitudinal Brownian motion along the axis of the channel at μM and higher concentrations (five or more orders of magnitude higher than the concentration employed here). This would imply that facile hopping between binding sites is possible, but only at extremely low surface concentrations – a trend opposite to that commonly observed and described by the empirical Freundlich isotherm. Such an anomalous decrease of the surface diffusion coefficient with increasing surface concentration has been reported previously for some macromolecules and was attributed to adsorbate–adsorbate interactions.[28,29] Because the surface concentrations employed here correspond to a single molecule inside the device, however, it is implausible that such an interaction mechanism applies here.

An alternate interpretation is that D_{eff} is indeed the correct diffusion coefficient, in which case an additional mechanism must be invoked to explain the missing long events when compared to the theoretical predictions. Decay of the molecules on the scale of seconds would provide such a mechanism. This can however be ruled out on the basis that the species under study are stable, the oxidized form of ferrocene derivatives for example having a lifetime on the order of an hour.[30–33] Another, more plausible hypothesis is the existence of deep trap sites[34] at which molecules can remain adsorbed for several tenths of seconds or more, causing long events to be substituted by a succession of shorter events due to repeated binding/unbinding at these sites. Consequently a reduced number of long events compensated by additional short events is expected, consistent with the experimental data. Further basis for this hypothesis comes from earlier observations in nanogap devices that adsorption becomes more pronounced at low analyte concentrations.[19,21]

Conclusion

The first-passage times of individual redox molecule trajectories were probed using electrochemical nanogap devices and their statistical distribution was compared with analytical calculations. The inability to resolve the multitude of ultra-short events was circumvented by concentrating solely on long events, this possibility of analysing a subset of the data representing one of the main strengths of single-molecule techniques.

Analysis of the distribution of the time of first passage for individual events indicates a significant suppression of the number of long events compared to that expected for a purely diffusive random walk once slowing down due to reversible adsorption is taken into account. While this would be expected if surface-bound molecules undergo facile surface diffusion, this interpretation appears incompatible with measurements at higher concentrations. Instead, the data strongly suggest that single molecules can be trapped at rare, low-energy binding sites, such that long residence times inside the detection volume are divided into shorter periods of redox cycling separated by periods of inactivity on the surface.

These measurements represent the first time that diffusive mass transport has been characterized quantitatively with single-molecule resolution using an electrochemical method.

Acknowledgements

The authors gratefully acknowledge financial support from the Netherlands Organization for Scientific Research (NWO) and the European Research Council (ERC) under project number 278801.

References

1 N. G. Walter, C. Y. Huang, A. J. Manzo and M. A. Sobhy, *Nat. Med.*, 2008, **5**, 475–489.
2 D. R. Walt, *Anal. Chem.*, 2013, **85**, 1258–1263.
3 F. R. F. Fan and A. J. Bard, *Science*, 1995, **267**, 871–874.

4 P. Sun and M. V. Mirkin, *J. Am. Chem. Soc.*, 2008, **130**, 8241–8250.
5 M. A. G. Zevenbergen, P. S. Singh, E. D. Goluch, B. L. Wolfrum and S. G. Lemay, *Nano Lett.*, 2011, **11**, 2881–2886.
6 S. Kang, A. F. Nieuwenhuis, K. Mathwig, D. Mampallil and S. G. Lemay, *ACS Nano*, 2013, **7**, 10931–10937.
7 S. G. Lemay, S. Kang, K. Mathwig and P. S. Singh, *Acc. Chem. Res.*, 2013, **46**, 369–377.
8 S. Redner, *A Guide to First-passage Processes*, Cambridge University Press, Cambridge, UK, New York, 2001.
9 P. Fauchald and T. Tveraa, *Ecology*, 2003, **84**, 282–288.
10 I. Goychuk and P. Hanggi, *Proc. Natl. Acad. Sci. U. S. A.*, 2002, **99**, 3552–3556.
11 V. J. van Hijkoop, A. J. Dammers, K. Malek and M.-O. Coppens, *J. Chem. Phys.*, 2007, **127**, 085101.
12 V. Rajani, G. Carrero, D. E. Golan, G. de Vries and C. W. Cairo, *Biophys. J.*, 2011, **100**, 1463–1472.
13 R. Mcdonald and D. Siegel, *Quarterly Journal of Economics*, 1986, **101**, 707–727.
14 A. Szabo, K. Schulten and Z. Schulten, *J. Chem. Phys.*, 1980, **72**, 4350–4357.
15 L. Hofler and R. E. Gyurcsanyi, *Anal. Chim. Acta*, 2012, **722**, 119–126.
16 Datasheet, FEMTO DDPCA-300 Variable Gain Sub Femto Ampere Current Amplifier.
17 P. S. Singh, E. Katelhon, K. Mathwig, B. Wolfrum and S. G. Lemay, *ACS Nano*, 2012, **6**, 9662–9671.
18 W. Feller, *An Introduction to Probability Theory and its Applications*, Wiley, New York, 3rd edn, 1968.
19 M. A. G. Zevenbergen, P. S. Singh, E. D. Goluch, B. L. Wolfrum and S. G. Lemay, *Anal. Chem.*, 2009, **81**, 8203–8212.
20 P. S. Singh, H. S. M. Chan, S. Kang and S. G. Lemay, *J. Am. Chem. Soc.*, 2011, **133**, 18289–18295.
21 S. Kang, K. Mathwig and S. G. Lemay, *Lab Chip*, 2012, **12**, 1262–1267.
22 K. Mathwig and S. G. Lemay, *Electrochim. Acta*, 2013, **112**, 943–949.
23 R. D. Tilton, C. R. Robertson and A. P. Gast, *J. Colloid Interface Sci.*, 1990, **137**, 192–203.
24 R. Gomer, *Rep. Prog. Phys.*, 1990, **53**, 917–1002.
25 P. R. Unwin and A. J. Bard, *J. Phys. Chem.*, 1992, **96**, 5035–5045.
26 E. G. Seebauer and C. E. Allen, *Prog. Surf. Sci.*, 1995, **49**, 265–330.
27 S. Roy, J. M. Thomas, E. A. Holmes, J. T. Kellis, A. J. Poulose, C. R. Robertson and A. P. Gast, *Anal. Chem.*, 2005, **77**, 8146–8150.
28 V. Chan, D. J. Graves, P. Fortina and S. E. McKenzie, *Langmuir*, 1997, **13**, 320–329.
29 J. Zhao and S. Granick, *J. Am. Chem. Soc.*, 2004, **126**, 6242–6243.
30 M. Avron and N. Shavit, *Anal. Biochem.*, 1963, **6**, 549–554.
31 R. Prins, A. R. Korswagen and A. G. T. G. Kortbeek, *J. Organomet. Chem.*, 1972, **39**, 335–344.
32 J. Holecek, K. Handlir, J. Klikorka and N. Dinhbang, *Collect. Czech. Chem. Commun.*, 1979, **44**, 1379–1387.
33 Z. N. Gao, X. X. Han, H. Q. Yao, B. Liang and W. Y. Liu, *Anal. Bioanal. Chem.*, 2006, **385**, 1324–1329.
34 J. Lyklema, *Fundamentals of Interface and Colloid Science*, Academic Press, San Diego, 2000.

Faraday Discussions

PAPER

Electrochemistry at single molecule occupancy in nanopore-confined recessed ring-disk electrode arrays

Kaiyu Fu,[a] Donghoon Han,[b] Chaoxiong Ma[a] and Paul W. Bohn*[ab]

Received 30th March 2016, Accepted 4th April 2016
DOI: 10.1039/c6fd00062b

Electrochemical reactions at nanoscale structures possess unique characteristics, *e.g.* fast mass transport, high signal-to-noise ratio at low concentration, and insignificant ohmic losses even at low electrolyte concentrations. These properties motivate the fabrication of high density, laterally ordered arrays of nanopores, embedding vertically stacked metal–insulator–metal electrode structures and exhibiting precisely controlled pore size and interpore spacing for use in redox cycling. These nanoscale recessed ring-disk electrode (RRDE) arrays exhibit current amplification factors, AF_{RC}, as large as 55-fold with $Ru(NH_3)_6^{2/3+}$, indicative of capture efficiencies at the top and bottom electrodes, $\Phi_{t,b}$, exceeding 99%. Finite element simulations performed to investigate the concentration distribution of redox species and to assess operating characteristics are in excellent agreement with experiment. AF_{RC} increases as the pore diameter, at constant pore spacing, increases in the range 200–500 nm and as the pore spacing, at constant pore diameter, decreases in the range 1000–460 nm. Optimized nanoscale RRDE arrays exhibit a linear current response with concentration ranging from 0.1 μM to 10 mM and a small capacitive current with scan rate up to 100 V s^{-1}. At the lowest concentrations, the average pore occupancy is $\langle n \rangle \sim 0.13$ molecule establishing productive electrochemical signals at occupancies at and below the single molecule level in these nanoscale RRDE arrays.

Introduction

The rapid development of nanofabrication techniques has enabled electrochemical experiments with electrodes in the nanoscale regime.[1,2] Typically this has been realized by fabricating ultrasmall electrodes with critical dimensions below 100 nm that exhibit unique electrochemical properties, *e.g.* sigmoidal voltammetric response, negligible ohmic (iR) potential losses, and small intrinsic

[a]*Department of Chemistry and Biochemistry, University of Notre Dame, Notre Dame, IN 46556, USA. E-mail: pbohn@nd.edu*

[b]*Department of Chemical and Biomolecular Engineering, University of Notre Dame, Notre Dame, IN 46556, USA*

charging currents.[3,4] Nanoelectrodes have also been employed to advantage in investigations of mass transport and electron transfer with enhanced temporal and spatial resolution.[5,6] Many reports focus on the fabrication of single nanoelectrodes with characteristic sizes ranging from a few nanometers to >100 nm,[7–9] however, when compared to macroscopic electrodes, or even microelectrodes, the current is limited, and achieving device-to-device reproducibility is challenging. Nanoelectrode ensembles (NEEs)[10,11] and nanoelectrode arrays (NEAs)[12–14] have been fabricated to increase the measurable current while maintaining the large bandwidth benefits of nanoelectrodes.[15] Recently, we have explored massively parallel arrays of NEAs fabricated with dual electrodes embedded in nanopores and operated in redox cycling (RC) mode, either with or without supporting electrolyte.[16–21]

In RC mode, reversible electro-active species undergo successive oxidation and reduction reactions by cycling between two closely-spaced, independently-addressable electrodes, *i.e.* electroactive molecules that are oxidized at one electrode then move to the other electrode and are reduced, then they can move back to the anode and repeat the cycle.[22] Pioneering work on RC was accomplished with rotating ring-disc electrodes[23,24] and interdigitated array electrodes,[25–27] which used forced convection and the small size (5–50 μm, typically) of the interelectrode gap, respectively, to amplify current. Bard and coworkers[28–30] extended the electrode spacing concept in an elegant way to directly detect current resulting from redox cycling of single molecules in a small volume trapped between a surface and a scanning electrochemical microscopy tip. Improvements in nanofabrication technology have made it possible to fabricate more precise RC systems, *e.g.* nanofluidic channels with closely parallel electrodes,[31,32] nanogap electrodes enclosing attoliter volumes,[33] and nanopore-based arrays with recessed ring-disk electrodes[20] have all been designed for ultrasensitive electrochemical measurements. Nanopore-based electrodes constitute a special class of nanoscale RC systems, because redox species are confined and can become electrostatically trapped such that they react with both electrodes multiple times before diffusing out of the nanopore.

In order to produce nanopore-based RC systems with control over geometric parameters, such as pore size, density and interpore distance, advanced lithographic techniques, *e.g.*, electron-beam lithography (EBL) and focused ion beam (FIB) milling, have been employed. For example, Wolfrum and co-workers[34] reported an approach to high density nanopore electrode arrays by combining EBL and reaction ion etching (RIE). Recently, our group took advantage of FIB milling to fabricate high density ($\sim 2 \times 10^9$ cm^{-2}) dual-ring nanopore electrode arrays which exhibit enhanced electrochemical performance.[21] However, these techniques are restricted by the need for serial-write lithography processes, employ expensive facilities, are time-consuming, and produce low throughput, so they are not well-suited to fabricating high density patterns at large scale.

Nanosphere lithography (NSL), also known as colloidal lithography, is based on the natural self-assembly of colloidal particle templates on substrates and was introduced by Fischer and Zingsheim in 1981.[35] Since then, by using smaller size nanospheres[36,37] and well-designed coating methods,[38,39] nanostructured arrays with elements smaller than 100 nm have been achieved with long-range order over large distances. We have used this procedure to develop sub-micrometer arrays with dual electrodes, demonstrating excellent sensitivity and selectivity in voltammetric measurements.[17,19]

Here, we extend the geometry of NSL-templated recessed ring-disk electrode (RRDE) arrays to small spacings (<500 nm) and large pore densities (up to 10^9 pores per cm^2) and explore the dependence of voltammetric response on the geometry of the array, including pore size and inter-pore spacing, in order to produce structures capable of redox cycling under single molecule occupancy conditions. By varying the size of the nanospheres in the template and the exposure time to O_2 plasma etching, nanopore electrode arrays with a wide range of pore diameters (800 nm to 100 nm) and pore densities (10^8 to 10^9 pores per cm^2) have been achieved with good uniformity at large scale. The optimized RRDE arrays demonstrate high collection efficiency in RC mode, producing current amplification as large as 55-fold. At the lowest concentrations examined, this performance is sufficient to achieve electrochemical measurements at single molecule occupancy levels, *i.e.* signals are obtained from a parallel array of nanopores, each containing $n \leq 1$ molecules, with the single molecule cycling between the lower (disk) and upper (ring) electrodes. The dependence of voltammetric behavior on the potential of the generator and collector electrodes, and the size and spacing of the nanopores in RC mode has been investigated experimentally and compared with the results of finite element simulations, showing that both pore diameter and pore density are important factors that govern transport-limited electron transfer and the corresponding current amplification. These structures also exhibit a linear current response over a wide range of analyte concentration and exhibit remarkably small capacitive currents, even at very fast scan rates.

Experimental section

Chemicals and materials

Hexaammineruthenium chloride ($[Ru(NH_3)_6]Cl_3$), potassium chloride, sulfuric acid (95%), and hydrogen peroxide (30%) were obtained from Sigma-Aldrich and used as received. Deionized (DI) water was generated from a Milli-Q Gradient water purification system (Millipore). Photoresist AZ5214E (AZ Electronic Materials) and polydimethylsiloxane (PDMS) (Dow Corning) were used according to the manufacturers' specifications.

Device fabrication

High density RRDE arrays were fabricated by using photolithography, layer-by-layer deposition, nanosphere lithography (NSL), and multiple reactive ion etching (RIE), similar to previously reported procedures.[20] A glass slide was cleaned in piranha solution (3 : 1 sulfuric acid (95%) : hydrogen peroxide (30%) – **caution** – strong oxidizer, use with extreme care) for 1 h, rinsed with DI water, and then dried at 100 °C for 2 h. A negative photoresist (AZ5214E) was used to define the overall pattern of the bottom electrode on glass. A 95 nm thick Au film, with a 5 nm Cr adhesion layer, was deposited using electron-beam evaporation (UNIVEX 450B, Oerlikon). After metal lift-off, a 50 nm thick SiN_x layer was then deposited by plasma-enhanced chemical vapor deposition (PECVD 790, Plasma-Therm). Then the overall top electrode pattern was deposited perpendicular to and overlapping the bottom electrode using the same photolithography procedure. The area of top and bottom electrode layer overlap was 180 μm × 180 μm. Next, an additional 200 nm thick SiO_2 layer was deposited over the whole surface to passivate the top surface. The substrate was exposed briefly to an O_2 plasma (Drytek RF Plasma asher, 1500W) to make the surface hydrophilic and then was immediately stored in DI water. Then a highly ordered polystyrene (PS) nanosphere monolayer was prepared at a liquid–air interface and was transferred to the substrate by immersing it at ~45° through the interface. Once the substrate was dried in air, O_2 plasma etching (Plasmatherm 790, 100W) was used to reduce the nanosphere size – with longer exposure times producing smaller final nanospheres. Then a metal nanopore mask was produced by depositing a 10 nm thick Cr film on top of the SiO_2 layer and then removing the PS nanospheres by sonication in chloroform or acetone for 5 min. Afterwards, a square pattern (150 μm × 150 μm) at the center of the intersection area was defined by photolithography to act as an overall etch mask for subsequent dry etching. Finally, the RRDE arrays were produced by transferring the pattern through the template using multistep reactive ion etching (Plasmatherm 790). During etching, each upper layer acts as the etch mask for the layer below it, *e.g.*, the Cr layer protects the SiO_2 layer, the SiO_2 protects the Au layer, *etc.*

Modeling and calculations

Modeling calculations were performed using finite element simulation software (COMSOL Multiphysics 5.0). The electrochemical system treated here is a simple outer-sphere heterogeneous electron transfer reaction at the electrodes, $O + ne^- \rightleftharpoons R$, where n is the number of electrons transferred. The arrays were simulated in a 2D geometry, consisting of recessed disk electrodes, a 50 nm thick insulating layer, a 100 nm thick ring electrode layer, and a 200 nm thick top insulator layer. The domain above the pores was drawn sufficiently large to avoid interference from boundaries, and the mesh was refined both within the nanopores and in the region just above the pores to avoid misleading results. For the case of high overpotential applied to the electrodes, constant-concentration boundary conditions were applied based on the assumption of complete oxidation or complete reduction on the electrode surfaces. The various devices were characterized at steady-state by assuming that the diffusion coefficients of the reduced and oxidized forms are the same, leading to the result that the sum of oxidized (C_O) and reduced (C_R) forms is constant over the diffusion space and is equal to the sum of the bulk concentrations (C_{bulk}).

Electrochemical measurements

Cyclic voltammetry (CV) experiments were performed with a CH Instruments electrochemical analyzer (Model 842C) using a Pt wire and Ag/AgCl as auxiliary and reference electrodes, respectively. The top and bottom electrodes of the RRDE arrays were operated as the first and second working electrodes. All potentials are reported *vs.* Ag/AgCl at 300 K. Both the reference electrode and auxiliary electrode were immersed in a 200 μL solution inside a 8 mm diameter PDMS reservoir. In all CV measurements, the potential of the bottom electrodes was swept, and the top electrodes were either held at a constant potential (GC mode) or disconnected (non-GC mode).

Results and discussion

Fabrication of high density metal nanopore templates

A key part of our strategy to study the electrochemical behavior of single molecules involves the preparation of high density arrays of zeptoliter-volume nanopores with internal annular ring and disk electrodes. Previous studies from our laboratory[17,20] have demonstrated that the size of nanopore-embedded electrodes prepared by NSL is determined by the initial size of the polystyrene sphere and the duration and power of the subsequent O_2 plasma exposure. In order to produce elements with tunable size and spacing, here we utilized nanospheres with initial diameters ranging from 800 nm to 100 nm. Because the initial nanosphere diameter determines the site-to-site spacing in the final structure, starting with smaller diameter nanospheres produces higher density arrays. In addition, smaller initial size nanospheres can be processed using milder plasma treatment conditions to shrink the sphere size, thereby minimizing deformation of the nanospheres which can lead to non-uniform structures.

To test this fabrication scheme, nanospheres with initial diameters ranging from 800 nm, to 200 nm were first assembled on a glass slide and then exposed to an O_2 plasma in order to decrease the size of the nanospheres through isotropic etching.[37] Fig. 1 shows the dependence of the final nanosphere size on the duration of O_2 plasma etching. At shorter etching times, the average nanosphere size decreases monotonically with O_2 plasma exposure, and the resulting spheres exhibit smooth, spherical surfaces with the periodic structure of the nanosphere array being well preserved. At longer times ($t > 240$ s) the final nanosphere size becomes less sensitive to the etching time, and the surface of the resulting spheres is rougher. Based on these observations, the usable processing regime extends from the original diameter, d_0, to $\sim d_0/3$.

After shrinking the nanosphere size by O_2 plasma treatment, a 10 nm Cr layer was deposited uniformly over the etched nanosphere array. Then the nanospheres were removed by sonication in chloroform, revealing a Cr nanopore

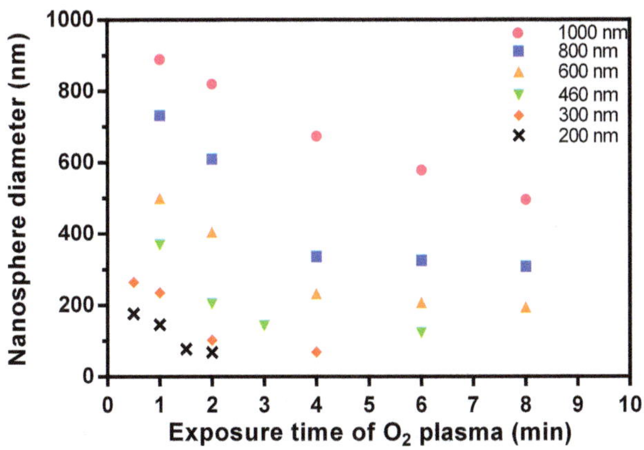

Fig. 1 The dependence of final nanosphere size on O_2 plasma exposure time for polystyrene nanospheres with initial diameters of 1000 nm (pink), 800 nm (blue), 600 nm (orange), 460 nm (green), 300 nm (red), and 200 nm (black).

mask. Cr was chosen as an etch mask based on its good adhesion to SiO_2 (passivating layer for the electrochemical device) and its selectivity in subsequent dry etching steps. Fig. 2 shows scanning electron microscope (SEM) images of nanopore arrays with a variety of pore sizes and spacings. The pore size could be adjusted from 800 nm to 50 nm, and the pore density could be controlled from low ($\sim 10^8$ cm^{-2}) to high density ($\sim 10^{10}$ cm^{-2}) by varying the O_2 plasma exposure conditions and the initial size of nanospheres, respectively. At the smallest initial nanosphere diameter, $d_0 = 100$ nm, the translational symmetry of the arrays was degraded relative to arrays produced from larger d_0 values. These images clearly demonstrate that NSL is a flexible, high-fidelity method to fabricate well-ordered, high-density nanopore arrays.

Fabrication of high density dual nanopore electrode arrays (NEAs)

Dual NEAs are nanopore arrays in which each nanopore has two electrodes, constructed in a metal–insulator–metal (MIM) structure in the vertical direction, as illustrated schematically in Fig. 3(A). First, the electrode pattern and the intersecting area of the two electrode layers, *i.e.* bottom disk electrode and top ring electrode, were defined by photolithography. The thicknesses of the metal and insulating layers were controlled by metal deposition and chemical vapor deposition, respectively. Then the metal nanopore template was fabricated on top of the intersecting electrodes. Finally, the dual NEAs were produced by dry etching to propagate the nanopore-templated pattern through to the bottom metal disk, at which the etch was stopped. Well ordered NEAs with tunable pore size and spacing were achieved at large scale, *viz.* Fig. 3(B). The cross-sectional SEM image, Fig. 3(C), shows the MIM structure in the NEA, each pore exhibiting

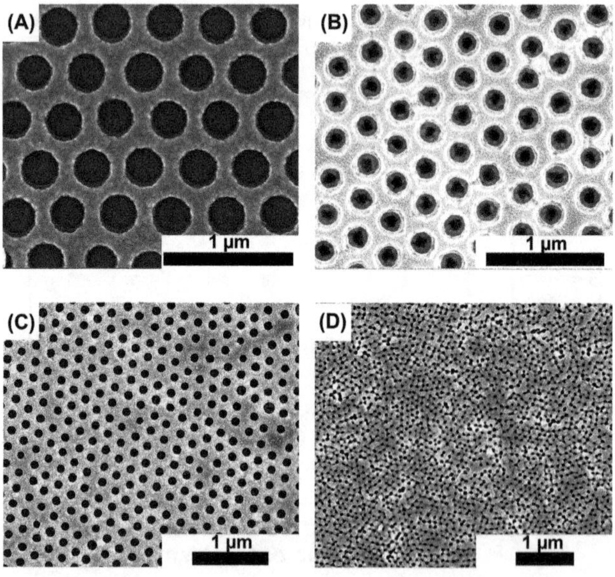

Fig. 2 SEM images of nanopore arrays with different pore diameters and densities. From (A) to (D), the initial sizes of nanosphere templates are 460 nm, 300 nm, 200 nm and 100 nm, respectively.

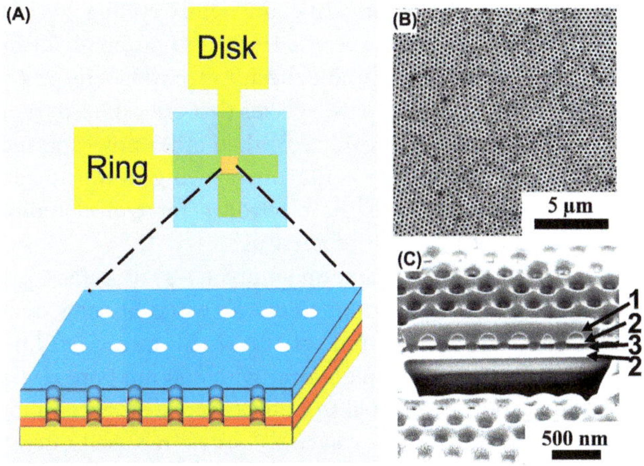

Fig. 3 (A) Schematic diagram showing the layout of a high density nanoscale RRDE array, with each nanopore consisting of an MIM structure in the vertical direction. (B) Plan view and (C) cross-sectional SEM images of the RRDE array. From top to bottom, the grey (1), white (2), black (3), and white (2) layers in (C) indicate SiO_2, Au, SiN_x, and Au layers, respectively.

from top to bottom: SiO_2 protection layer (grey), top Au ring electrode (white), SiN_x insulating layer (black) and bottom Au disk electrode (white). The average bottom and top diameters for this particular conical structure are *ca.* 90 nm and 150 nm, respectively.

Considering that the combined thickness of the top Au layer and SiN_x layer is ~150 nm, and that the nanopore shape is a right circular conical frustum, the effective volume of the ring-disk electrode region of each nanopore, *i.e.* the volume pertinent to the redox cycling reaction is $\sim 1.25 \times 10^{-18}$ L. Thus, the concentration corresponding to single molecule occupancy is ~1.3 μM. However, both the pore volume and the pore spacing are important for functional analysis of the electrochemical behavior of NEAs – the pore volume setting the single molecule occupancy criteria, and the pore spacing determining the ability to re-capture molecules that escape from a given pore in neighboring pores. The limitations on fabricating NEAs with high pore densities and small pore volumes by the NSL-templating strategy employed here arises due to the uniformity of the etching process when the top diameter of the nanopore is smaller than 100 nm. Additionally, the thickness of the top SiO_2 layer is important to achieve uniform shapes and small diameter nanopore arrays. In order to balance pore density against the requirement to operate with well-structured nanopores for electrochemical measurements, the smallest initial nanosphere size used to fabricate NEAs for electrochemical measurements was 300 nm, corresponding to the image in Fig. 2(B). This design rule produces arrays with pore densities, $\rho_p \sim 10^9$ cm^{-2}, which combined with an intersecting electrode area of 150 μm × 150 μm, leads to arrays with ~225 000 pores.

Current amplification in NEAs

The unique geometry of the dual NEA, *i.e.* two individually addressable and closely spaced working electrodes within an ultrasmall volume nanopore, is a nearly ideal geometry for investigating the redox cycling reaction of small numbers of reversible electroactive species. Here, the bottom (disk) and top (ring) electrodes were used as generator and collector electrodes, respectively, during redox cycling experiments. Accordingly, species that are oxidized (reduced) at the bottom electrode can be reduced (oxidized) when efficiently captured by the top electrode. Furthermore, given the ~100 nm spacing between generator and collector, diffusion times between the electrodes for typical redox species in aqueous electrolyte are of the order of a few μs.

Fig. 4(A) shows typical cyclic voltammograms (CVs) of $Ru(NH_3)_6^{2/3+}$ in 0.1 M KCl obtained by scanning the bottom electrode while floating (non-GC mode, black line) or holding the top electrode at +0.1 V (GC mode, red line) or −0.4 V (GC mode, blue line) *vs.* Ag/AgCl on an array with 400 nm pore size and 460 nm pore-to-pore spacing. When operated in non-GC mode at a scan rate of 100 mV s^{-1}, the nanoscale RRDE array behaves like a single recessed microelectrode of the same electrode size (150 μm × 150 μm). Under non-GC mode conditions, a characteristic transport-limited CV (inset of Fig. 4(A)) with substantial capacitive contribution was observed. This is likely due to the overlap of diffusion zones of neighboring nanopore electrodes.[40] Using a model developed by Bond *et al.*,[41] a current of ~57 nA was calculated for a single recessed microelectrode of the same electrode size and depth as the nanoscale RRDE array. This value agrees well with the 56 nA peak current observed experimentally. When the potential of the top electrode was fixed to either a relatively reducing (−0.4 V) or oxidizing (0.1 V) value for $Ru(NH_3)_6^{2/3+}$, the limiting current of the RRDE array in GC-mode was significantly enhanced. The amplification of measurable current from 60 nA (solid, black line) to 3.30 μA (solid, red line) is clear evidence that the analyte undergoes redox cycling. The amplification factor for redox cycling (AF_{RC}), calculated from the ratio of limiting current in GC mode to that in non-GC mode, is $AF_{RC} = 55$ under these conditions.

AF_{RC} is strongly related to the collection efficiency of the generator–collector system, which is defined as the ratio of the limiting current at the collector electrode to that at the generator electrode. In the structures studied here, collection efficiencies for the top electrode, Φ_t, as high as 99.4% ($d_0 = 400$ nm, 460 nm spacing, $E_{top} = +0.1$ V) were measured. If the fixed potential of the top electrode is switched from +0.1 V to −0.4 V *vs.* Ag/AgCl, the analyte oxidized at the bottom electrode is then reduced at the top electrode. Under these conditions, the limiting currents for top and bottom electrodes were 3.51 μA (dashed blue line) and 3.47 μA (solid blue line), respectively, which gives the collection efficiency of the bottom electrode as Φ_b = 98.9%. Based on the theory of Tabei and coworkers,[25] the number of redox cycles N_{RC} can be estimated from,

$$N_{RC} = \frac{1}{1 - \Phi_t \Phi_b} \tag{1}$$

For the nanoscale RRDE array described above, Φ_t and Φ_b were 99.4% and 98.9%, respectively, yielding $N_{RC} = 57.7$, in excellent agreement with the experimentally measured value of AF_{RC}. This value is substantially higher than previously reported AF_{RC} values obtained from lower density RRDE arrays,[19] an observation that can be attributed to the higher array density and smaller pore size, both of which work to improve the redox cycling efficiency.

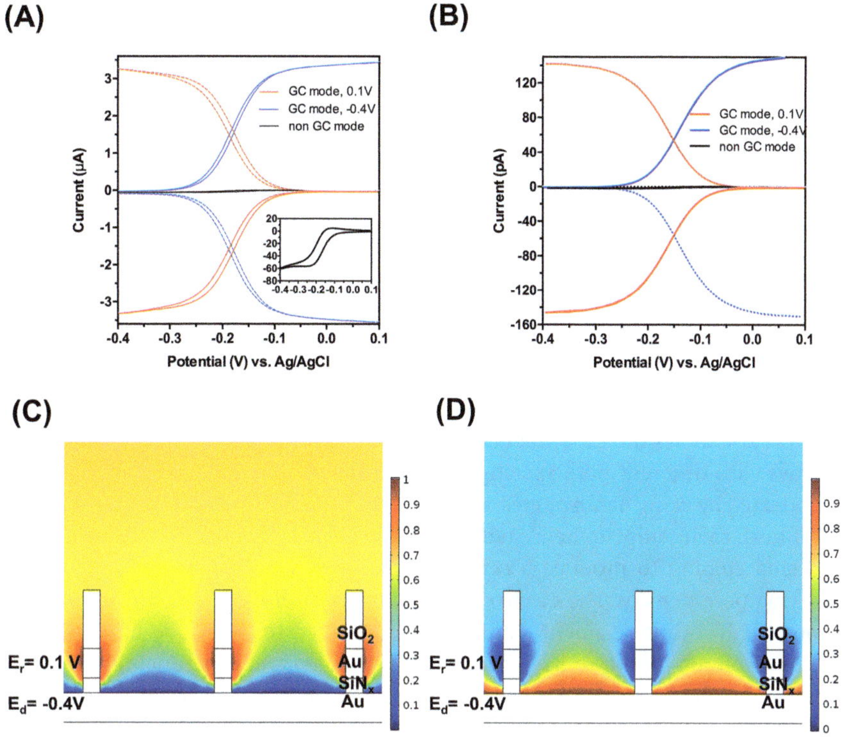

Fig. 4 (A) Cyclic voltammograms of 1 mM $Ru(NH_3)_6{}^{3+}$ on the nanoscale RRDE array (d_0 = 400 nm, 460 nm spacing) obtained by sweeping the bottom disk electrode while at the same time holding the top ring electrode at +0.1 V (red), −0.4 V (blue) or floating (black) in 0.1 M KCl. (Inset) magnified version of the CV obtained with the ring electrode floating. Current in nA. (B) Simulated CV response of an array of 10 nanopore electrodes under the same measurement conditions. In (A) and (B) dashed lines give current at the collector electrode while the generator electrode is swept. (C) and (D) Spatial distribution of $Ru(NH_3)_6{}^{3+}$ inside and outside the pore region taken from simulations of an array of 10 nanopore electrodes under conditions where the top ring electrode is held at an oxidizing (C) or reducing (D) potential. Scale bars in (C) and (D) are normalized to the maximum concentration.

Fig. 4(B) shows simulated CVs of nanoscale RRDE arrays with the same geometric parameters as the experimental structure, with the top electrode held at +0.1 V (red line), −0.4 V (blue line) or floating (black line). The simulated single-nanopore limiting current (~15 pA) is somewhat smaller than the experimental value (~30 pA) obtained through dividing the total limiting current (3.30 μA) by the number of nanopores (~112 500) on the array. The simulated amplification factor AF_{sim} = 54.8 calculated from the ratio of the simulated limiting currents in GC mode to that in non-GC mode agrees very well with the experimental AF_{RC} = 55. The calculated spatial distributions of redox species under different applied potentials are shown in Fig. 4(C) and (D), with the top ring electrode potential held at +0.1 V and −0.4 V *vs.* Ag/AgCl, respectively. With the top electrode set at an oxidizing potential, Fig. 4(C), $Ru(NH_3)_6{}^{2+}$ reduced at the bottom disk electrode is efficiently oxidized to $Ru(NH_3)_6{}^{3+}$ at the top ring electrode. Interestingly, the simulated spatial distributions reveal pockets of unreacted $Ru(NH_3)_6{}^{3+}$ near the surface of the insulator in Fig. 4(C). Conversely, when the top electrode is set to a negative potential (−0.4 V), Fig. 4(D), $Ru(NH_3)_6^{3+}$ produced at the bottom electrode is reduced at the top electrode.

Effect of pore size and spacing

The characteristic collision times of molecules with the electrodes inside the nanopores depend strongly on pore size. On the other hand, the recapture of escaped molecules by adjacent nanopore-embedded electrodes depends largely on pore density. Thus, the collection efficiency and the current amplification should depend strongly on the geometry of the array. Both experiments and simulations were conducted to understand the dependence of RC response on the nanopore geometry, especially the effect of pore size and pore density, on the efficiency of redox cycling.

Table 1 shows the experimental and simulated AF values for nanoscale RRDE arrays with different pore spacings, S_p. A series of nanoscale RRDE arrays were fabricated by using nanospheres with different initial diameters (800 nm, 600 nm, 460 nm), each template being reduced to the same sphere size (~200 nm) by O_2 plasma etching to produce a series of structures with constant pore size but varying pore density. The electrochemical behavior of these structures was then compared to finite element simulations performed to probe the effect of pore spacing on AF. As shown in Table 1, both simulated and experimental AF values increase with decreasing pore spacing. This is likely due to overlapping diffusion zones on high density nanoscale RRDE arrays, as is evident in the simulated concentration distributions in Fig. 4(C) and (D). At high pore densities the overlapping diffusion zones improve the collection efficiency by recapturing any molecules that escape from a neighboring nanopore.

The dependence of AF on pore size was also investigated on a series of nanoscale RRDE arrays of constant pore spacing but different pore diameters. These structures were prepared by starting from NSL templates with the same initial nanosphere diameter, but then exposing them to O_2 plasma for differing times to produce different final pore diameters. The simulated and experimental AF values of this series are summarized in Table 2. Both show that AF increases with increasing nanopore diameter, reflecting a larger net capture efficiency for larger pores, thus giving rise to larger AF_{RC}. At these small sizes, this likely reflects the effect of Φ_b, which increases at larger disk electrode sizes. This was confirmed by simulations which illustrate that Φ_b increases with increasing nanopore diameter (Table 2).

Linear response and scan rate dependence

In order to test the performance of the high density nanopore electrode arrays to small molecular loadings, the whole array was operated in GC mode to detect

Table 1 Experimental and simulated AF values as a function of pore spacing (S_p)

S_p	460	600	800	1000
AF_{exp}	38.1	35.0	32.1	20[a]
AF_{sim}	47.2	36.2	28.8	21.1

[a] Value is adopted from ref. 19.

Table 2 Experimental and simulated AF values and collection efficiencies as a function of pore diameter (D_p)

D_p	400	350	300	250	200
AF_{exp}	54.8	40.8	31.0	23.6	21.8
AF_{sim}	55.4	54.1	52.1	49.5	46.1
Φ_t	99.1%	99.3%	99.4%	99.6%	99.7%
Φ_b	98.7%	98.5%	98.3%	98.1%	97.7%

redox species of $Ru(NH_3)_6^{3+}$ at concentrations down to 100 nM. The limiting voltammetric current obtained from $Ru(NH_3)_6^{3+}$ as a function of concentration in GC mode is displayed in Fig. 5(A), illustrating that it is linear over 5 orders of magnitude in concentration. The current signals in these experiments were collected from a 150 μm × 150 μm nanopore electrode array with a (top) pore diameter and spacing of 150 nm and 300 nm, respectively. In this geometry, the conical frustum gives an average pore diameter ~ 150 nm at the top and ~90 nm at the bottom. Considering that the combined thickness of the top Au and SiN_x layers is 150 nm, the effective volume for redox cycling reaction is 1.25 aL. In order to obtain single molecule occupancy, the required concentration of analyte is ~1.3 μM. As shown in Fig. 5(A), the limiting current at 1 μM was ~0.4 nA, which means on average each nanopore produced ~1.8 fA current signal under these conditions (~225 000 pores on the entire array) corresponding to *ca.* 10^4 electron transfer events per second, a realistic value given an estimated diffusion time of 10 μs. This value is ~10× smaller than the value (20 fA) reported by Lemay and coworkers from single molecule diffusion inside an opposed-planar electrode nanofluidic electrochemical device.[32] We attribute this to the lower collection efficiency of the open geometry of nanopore electrodes compared to the closed structure used by Lemay's group.

Another advantage of high density nanopore electrode arrays is the small capacitive current in GC mode, as shown in Fig. 5(B). At high electrolyte concentrations, such as those used here, the magnitude of the charging current for nanoelectrodes is typically comparable to the faradaic current at low analyte concentrations. Here, the measured charging current at both the top and bottom

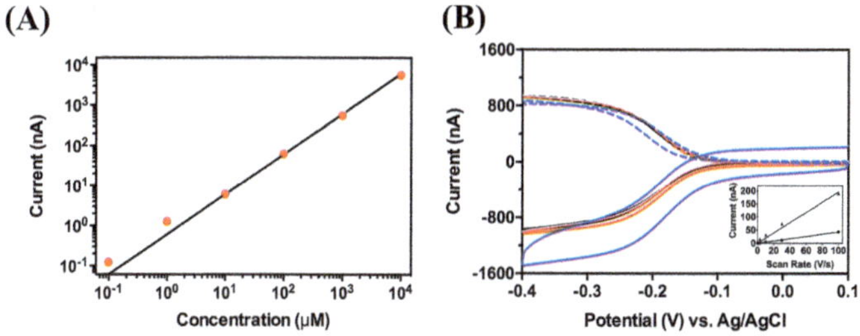

Fig. 5 (A) Limiting GC-mode currents for different concentrations of $Ru(NH_3)_6^{3+}$ with aqueous 0.1 M KCl on the nanoscale RRDE array (d_O = 150 nm, 300 nm spacing). (B) Cyclic voltammograms of 1 mM $Ru(NH_3)_6^{3+}$ solution with 0.1 M KCl as a function of scan rate. Bottom disk electrode is swept at 100 V s^{-1} (blue), 3 V s^{-1} (red) and 0.1 V s^{-1} (black); current acquired at ring (dashed) and disk (solid) electrodes. (Inset) capacitive current of ring (lower line) and disk (upper line) electrode as a function of scan rate.

electrode increases linearly with the scan rate ranging from 0.1 V s^{-1} to 100 V s^{-1}, *viz.* Fig. 5(B) inset. However, the capacitance of the whole array had little effect on the voltammetric signal when the scan rate was below 1 V s^{-1}. The smaller capacitive current (~0.2 nA) of the top electrode even at scan rates as high as 100 V s^{-1} indicates that these devices may serve as a suitable electrode platforms for ultrafast electrochemistry, as well as single molecule, measurements.

Conclusion

In this work we demonstrate a simple and robust approach to fabricate high density nanopore arrays with well-controlled dual-electrodes for ultrahigh efficiency redox cycling. Nanosphere lithography and O_2 plasma etching are clearly competent to fabricate nanopore arrays with controlled pore diameter and spacing over a wide range. Electrochemical reactions of redox couples operated in generator and collector mode produced greatly enhanced currents relative to non-GC operation, with collector electrode collection efficiencies as high as Φ_t = 99.4%. Both experiments and simulations illustrate the significance of pore size and spacing in determining high performance redox cycling.

The nanoscale RRDE arrays developed and studied here represent a new part of electrochemical phase space, where the small electrochemical detection volume (~1 aL) of each nanopore means that single molecule occupancy can be achieved at μM concentrations. In fact, at the lowest concentration at which a chemically significant electrochemical signal was measured (100 nM), the average occupancy was $\langle n \rangle \sim 0.13$. This measurement was enabled by a combination of high redox cycling collection efficiency at both bottom and top electrodes, $\Phi_{b,t} \sim 99\%$, and a massively parallel array of 2.25×10^5 pores giving rise to aggregate signals significantly above the noise level. Furthermore, these large area nanoscale RRDE arrays exhibit insignificant charging current at scan rates below 1 V s^{-1} and acceptable charging currents at scan rates as high as 100 V s^{-1}, suggesting their use for simultaneous ultrafast, ultrasensitive electrochemical measurements.

Acknowledgements

This work was supported by the National Science Foundation grant 1404744 (K. F. and C. M.) and the Defense Advanced Research Projects Agency FA8650-15-C-7546 (D. H.). Fabrication and structural characterization of the devices studied here were accomplished at the Notre Dame Nanofabrication Facility and the Notre Dame Integrated Imaging Facility, whose generous support is gratefully acknowledged.

References

1 D. W. M. Arrigan, *Analyst*, 2004, **129**(12), 1157–1165.
2 J. T. Cox and B. Zhang, *Annu. Rev. Anal. Chem.*, 2012, **5**(1), 253–272.
3 S. M. Oja, M. Wood and B. Zhang, *Anal. Chem.*, 2013, **85**(2), 473–486.
4 S. M. Oja, Y. Fan, C. M. Armstrong, P. Defnet and B. Zhang, *Anal. Chem.*, 2016, **88**(1), 414–430.
5 K. Dawson and A. O'Riordan, *Annu. Rev. Anal. Chem.*, 2014, **7**(1), 163–181.

6 R. W. Murray, *Chem. Rev.*, 2008, **108**(7), 2688–2720.
7 D. Krapf, M.-Y. Wu, R. M. M. Smeets, H. W. Zandbergen, C. Dekker and S. G. Lemay, *Nano Lett.*, 2006, **6**(1), 105–109.
8 G. Wang, B. Zhang, J. R. Wayment, J. M. Harris and H. S. White, *J. Am. Chem. Soc.*, 2006, **128**(23), 7679–7686.
9 A. P. Ivanov, K. J. Freedman, M. J. Kim, T. Albrecht and J. B. Edel, *ACS Nano*, 2014, **8**(2), 1940–1948.
10 W. Cheng, S. Dong and E. Wang, *Anal. Chem.*, 2002, **74**(15), 3599–3604.
11 V. P. Menon and C. R. Martin, *Anal. Chem.*, 1995, **67**(13), 1920–1928.
12 Y. H. Lanyon, G. De Marzi, Y. E. Watson, A. J. Quinn, J. P. Gleeson, G. Redmond and D. W. M. Arrigan, *Anal. Chem.*, 2007, **79**(8), 3048–3055.
13 M. E. Sandison and J. M. Cooper, *Lab Chip*, 2006, **6**(8), 1020–1025.
14 F. Santoro, S. Dasgupta, J. Schnitker, T. Auth, E. Neumann, G. Panaitov, G. Gompper and A. Offenhäusser, *ACS Nano*, 2014, **8**(7), 6713–6723.
15 C. N. LaFratta and D. R. Walt, *Chem. Rev.*, 2008, **108**(2), 614–637.
16 C. Ma, N. M. Contento, L. R. Gibson and P. W. Bohn, *Anal. Chem.*, 2013, **85**(20), 9882–9888.
17 C. Ma, W. Xu, W. R. A. Wichert and P. W. Bohn, *ACS Nano*, 2016, **10**, 3658–3664.
18 C. Ma, L. P. Zaino III and P. W. Bohn, *Chem. Sci.*, 2015, **6**(5), 3173–3179.
19 C. Ma, N. M. Contento and P. W. Bohn, *J. Am. Chem. Soc.*, 2014, **136**(20), 7225–7228.
20 C. Ma, N. M. Contento, L. R. Gibson and P. W. Bohn, *ACS Nano*, 2013, **7**(6), 5483–5490.
21 D. H. Han, L. P. Zaino III, K. Fu and P. W. Bohn, *J. Phys. Chem. C*, 2016, DOI: 10.1021/acs.jpcc.6b01287.
22 E. O. Barnes, G. E. M. Lewis, S. E. C. Dale, F. Marken and R. G. Compton, *Analyst*, 2012, **137**(5), 1068–1081.
23 A. Frumkin, L. Nekrasov, B. Levich and J. Ivanov, *J. Electroanal. Chem.*, 1959, **1**(1), 84–90.
24 S. Bruckenstein and B. Miller, *Acc. Chem. Res.*, 1977, **10**(2), 54–61.
25 O. Niwa, M. Morita and H. Tabei, *Anal. Chem.*, 1990, **62**(5), 447–452.
26 A. J. Bard, J. A. Crayston, G. P. Kittlesen, T. Varco Shea and M. S. Wrighton, *Anal. Chem.*, 1986, **58**(11), 2321–2331.
27 E. Goluch, B. Wolfrum, P. Singh, M. G. Zevenbergen and S. Lemay, *Anal. Bioanal. Chem.*, 2009, **394**(2), 447–456.
28 F.-R. F. Fan and A. J. Bard, *Science*, 1995, **267**(5199), 871–874.
29 A. J. Bard and F.-R. F. Fan, *Acc. Chem. Res.*, 1996, **29**(12), 572–578.
30 F.-R. F. Fan, J. Kwak and A. J. Bard, *J. Am. Chem. Soc.*, 1996, **118**(40), 9669–9675.
31 M. A. G. Zevenbergen, D. Krapf, M. R. Zuiddam and S. G. Lemay, *Nano Lett.*, 2006, **7**(2), 384–388.
32 M. A. G. Zevenbergen, P. S. Singh, E. D. Goluch, B. L. Wolfrum and S. G. Lemay, *Nano Lett.*, 2011, **11**(7), 2881–2886.
33 T. Li, L. Su, W. Hu, H. Dong, Y. Li and L. Mao, *Anal. Chem.*, 2010, **82**(4), 1521–1526.
34 M. Huske, R. Stockmann, A. Offenhausser and B. Wolfrum, *Nanoscale*, 2014, **6**(1), 589–598.
35 U. C. Fischer and H. P. Zingsheim, *J. Vac. Sci. Technol.*, 1981, **19**(4), 881–885.

36 A. Plettl, F. Enderle, M. Saitner, A. Manzke, C. Pfahler, S. Wiedemann and P. Ziemann, *Adv. Funct. Mater.*, 2009, **19**(20), 3279–3284.
37 L. Ji, Y.-F. Chang, B. Fowler, Y.-C. Chen, T.-M. Tsai, K.-C. Chang, M.-C. Chen, T.-C. Chang, S. M. Sze, E. T. Yu and J. C. Lee, *Nano Lett.*, 2014, **14**(2), 813–818.
38 G. D. Moon, T. I. Lee, B. Kim, G. Chae, J. Kim, S. Kim, J.-M. Myoung and U. Jeong, *ACS Nano*, 2011, **5**(11), 8600–8612.
39 A. V. Whitney, B. D. Myers and R. P. Van Duyne, *Nano Lett.*, 2004, **4**(8), 1507–1511.
40 A. Wahl, K. Dawson, J. MacHale, S. Barry, A. J. Quinn and A. O'Riordan, *Faraday Discuss.*, 2013, **164**(0), 377–390.
41 A. M. Bond, D. Luscombe, K. B. Oldham and C. G. Zoski, *J. Electroanal. Chem. Interfacial Electrochem.*, 1988, **249**(1), 1–14.

Faraday Discussions

PAPER

On the mechanism of electrochemical vesicle cytometry: chromaffin cell vesicles and liposomes†

Jelena Lovrić,‡[a] Neda Najafinobar,‡[a] Johan Dunevall,[a] Soodabeh Majdi,[b] Irina Svir,[c] Alexander Oleinick,[c] Christian Amatore[c] and Andrew G. Ewing*[ab]

Received 29th April 2016, Accepted 16th May 2016
DOI: 10.1039/c6fd00102e

The mechanism of mammalian vesicle rupture onto the surface of a polarized carbon fiber microelectrode during electrochemical vesicle cytometry is investigated. It appears that following adsorption to the surface of the polarized electrode, electroporation leads to the formation of a pore at the interface between a vesicle and the electrode and this is shown to be potential dependent. The chemical cargo is then released through this pore to be oxidized at the electrode surface. This makes it possible to quantify the contents as it restricts diffusion away from the electrode and coulometric oxidation takes place. Using a bottom up approach, lipid-only transmitter-loaded liposomes were used to mimic native vesicles and the rupture events occurred much faster in comparison with native vesicles. Liposomes with added peptide in the membrane result in rupture events with a lower duration than that of liposomes and faster in comparison to native vesicles. Diffusional models have been developed and suggest that the trend in pore size is dependent on soft nanoparticle size and diffusion of the content in the nanometer vesicle. In addition, it appears that proteins form a barrier for the membrane to reach the electrode and need to move out of the way to allow close contact and electroporation. The protein dense core in vesicles matrixes is also important in the dynamics of the events in that it significantly slows diffusion through the vesicle.

[a]*Chalmers University of Technology, Department of Chemistry and Chemical Engineering, Kemivägen 10, 412 96 Gothenburg, Sweden. E-mail: andrewe@chalmers.se*

[b]*University of Gothenburg, Department of Chemistry and Molecular Biology, Kemivägen 10, 412 96 Gothenburg, Sweden*

[c]*Ecole Normale Supérieure-PSL Research University, Département de Chimie, Sorbonne Universités-UPMC Univ. Paris 06, CNRS UMR 8640 PASTEUR, 24 rue Lhomond, 75005 Paris, France*

† Electronic supplementary information (ESI) available. See DOI: 10.1039/c6fd00102e

‡ Shared contribution.

Introduction

Chemical intercellular communication is based on a process of exocytosis, which involves the fusion of secretory transmitter vesicles with the plasma membrane and release of neurotransmitters into the extracellular space.[1–3] In neurons, exocytosis mostly involves clear synaptic vesicles. However, certain neurons and broad spectrum of non-neuronal cells, such as neuroendocrine cells, have another type of vesicles called large dense-core vesicles (LDCVs) that undergo the process of exocytosis as well.[1,2] Over the last 50 years, the mechanism of transmitter vesicle exocytosis has been one of the most important studied in neurochemistry and neurobiology. The traditional view of exocytosis involves an all-or-nothing mechanism during which vesicles completely fuse with the plasma membrane and release its entire cargo.[4,5] In contrast to the full fusion concept, extensive work has been done confirming the existence of partial release mechanisms where a fusion pore is formed between the vesicle and the cell membrane that does not fully dilate during exocytosis.[6,7] The modes of partial release include kiss-and-run exocytosis,[8,9] flickering fusion,[10] and extended kiss-and-run or open and closed exocytosis.[11]

Amperometry is a widely used technique to study exocytosis of single cells due to its high temporal resolution and capability to quantitatively measure the content of the electroactive transmitters released upon cell stimulation.[12,13] However, in order to obtain direct evidence for fractional exocytosis, it was necessary to probe the entire content of vesicles and to compare to the amount measured during exocytotic release upon stimulation. Electrochemical cytometry was developed to carry out amperometric detection of the entire content of single soft particles including liposomes and native vesicles after their separation by capillary electrophoresis and lysis on a cylindrical carbon fiber microelectrode.[14–16] More recently, this has been carried out without the separation step to probe single soft particles such as emulsion droplets, liposomes, and nanometer-sized LDCVs isolated from bovine adrenal glands by detecting their content as they impact an electrode.[17–20] This latter application, vesicle electrochemical cytometry, has allowed us to detect the cargo of these vesicles and compare this to the amount released during exocytosis. We have done this both with vesicle isolations from tissue and inside individual PC12 cells where vesicle content can be measured *in situ* in a live cell by inserting a flame-etched carbon-fiber microelectrode.[21]

Even though it is evident that much work has been done to quantify the chemical messengers inside mammalian vesicles, the mechanism of vesicle rupture on the electrode remains unclear. Extensive studies have been performed by various research groups in order to understand the adsorption and rupture pathways of liposomes, the simplified models of native vesicles. They have led many to conclude that the lipid composition of liposomes, adsorption substrate nature, the nature of the membrane bathing solution *etc.* have an important role in defining the adsorption–rupture mechanisms.[22–25] LDCVs are complex systems having a two compartment morphology with a protein rich core and surrounding solution called a halo.[26] Moreover, the membrane nature of LDCVs having transmembrane proteins such as ATPase, cytochrome b_{561}, catecholamine transporter and membrane bound proteins as dopamine β-hydroxylase, actinin and others,[27,28] makes the understanding of their rupture mechanism challenging.

In this paper, we investigate the mechanism of mammalian vesicle rupture on the surface of a polarized carbon fiber microelectrode during electrochemical vesicle cytometry.[20] In our approach, we measure the transmitter cargo of the vesicle by oxidation at the electrode surface. We consider that following adsorption to the surface of the polarized electrode, electroporation leads to formation of a pore at the interface between the liposome and the electrode and is potential dependent. The chemical cargo is then released through this pore to be oxidized at the electrode surface. Furthermore, we employed a bottom up approach to vesicle structure and used electrochemical vesicle cytometry to measure the rate of neurotransmitter oxidation upon pore opening. We used simple, lipid-only transmitter-loaded liposomes to mimic native vesicles and observed that the rupture events occur much faster in comparison with native vesicles. To understand if the proteins in isolated vesicles are responsible for the changes in rupture times, we employed liposomes with added peptide in the membrane and found that the duration of rupture events is slower than that of liposomes and faster in comparison to native vesicles. Diffusional models have been developed and suggest the trend in pore size is dependent on soft nanoparticle size and diffusion of the content inside the nanometer vesicle.

Experimental section

Reagents

1,2-Dioleoyl-*sn-glycero*-3-phosphocholine (DOPC, >99%), 1,2-dioleoyl-*sn-glycero*-3-phosphethanolamine (DOPE, >99%) and cholesterol (ovine wool) were purchased from Avanti Polar Lipids, USA. Fatty acid conjugated peptide, stearoyl octaarginine (STR8, 97.26%) was purchased from Life Tein, USA. Salts for buffer preparations were analytical grade and together with dopamine hydrochloride, chloroform (≥99.9%), methanol (≥99.9%) and protease inhibitors (Complete, EDTA-free) were obtained from Sigma-Aldrich, Sweden. All aqueous solutions were prepared using 18 MΩ cm^{-1} water from a Purelab Classic purification system (ELGA, Sweden).

Solutions

Lipid film hydration buffer. Dopamine hydrochloride solution concentration 150 mM and 277 mOsm kg^{-1} was prepared in 10 mM HEPES with pH 7.4, and was purged with argon (6.0, AGA Sweden) for 30 min prior to adding oxygen sensitive dopamine hydrochloride.

Liposome storage and running buffer. Amperometric measurements and nanoparticle tracking analysis (NTA) of liposome samples were performed in buffer with 10 mM HEPES, 142 mM NaCl, pH 7.4 and osmolality 277 mOsm kg^{-1}.

Homogenizing buffer. Chromaffin vesicles were isolated with buffer containing 184 mM NaCl, 10 mM KCl, 1 mM $MgSO_4$, 1 mM EDTA, 10 mM HEPES, protease inhibitors, with pH 7.4 and 380 mOsm kg^{-1} osmolality. This buffer was also used for NTA of chromaffin vesicles and as a storage buffer for amperometric measurements of potential dependence.

Chromaffin vesicle running buffer. All electrochemical experiments of chromaffin vesicles were performed in buffer with 300 mM sucrose, 10 mM KCl, 1 mM $MgSO_4$, 1 mM EDTA, 10 mM HEPES, protease inhibitors, with pH 7.4 and

380 mOsm kg^{-1} osmolality. This buffer was used as a vesicle storage buffer except in the case of amperometric measurements of potential dependence (see above).

Liposome preparation

Lipids were dissolved in chloroform, whereas STR8 was prepared in methanol.

Control liposomes. DOPC, DOPE and cholesterol were mixed in mass ratio 50 : 25 : 25 to obtain a concentration of 9.07 mM DOPC, 4.80 mM DOPE and 9.24 mM cholesterol before chloroform evaporation.

Peptide liposomes. We applied a modified method described in work by Iwasa *et al.*[29] Briefly, DOPC, DOPE, cholesterol and STR8 were mixed in mass ratio 49.3 : 24.6 : 24.6 : 1.5 to obtain concentration of 6.33 mM DOPC, 3.35 mM DOPE, 6.44 mM cholesterol and 99 μM STR8 before chloroform evaporation.

A thin lipid film was made by swirling the flask under argon and evaporating traces of organic solvent for 3 h under the vacuum. The lipid film was rehydrated with 1.5 ml of hydration buffer for 30 min under argon at room temperature. The rehydrated lipid mixture was exposed to five freeze–thaw cycles by placing it in liquid nitrogen and a warm water bath, alternately. The liposomal suspension was extruded 21 times with Avanti Mini-Extruder (Avanti Polar Lipids, Inc., USA) at room temperature over three 400 nm polycarbonate membranes (Whatman, USA). In order to exclude dopamine outside the liposomes, the liposomal suspension was 12 × diluted with liposome running buffer and centrifuged (Eppendorf, 5810R) at 10 000 × *g* and +4 °C for 20 min.

Comments on liposome preparation. Here we would like to emphasize our observation of success rate dependence of dopamine hydrochloride encapsulation efficiency and different batches of purchased lipids. Additionally, we also observed the aggregation of the peptide decorated liposomes over the course of time. Fig. SI-1† shows this phenomenon, where at approximately 105 min from sample preparation the aggregation was observed. Therefore, to exclude artifacts from the aggregation in electrochemical measurements and NTA, we used samples only approximately 1 hour from the time point when liposome samples were prepared.

Chromaffin vesicle isolation

Chromaffin vesicles were isolated according to the procedure developed in the Borges research group (University of La Laguna, Spain). Briefly, medulla tissue obtained from the central part of bovine adrenal glands was mechanically homogenized and with subsequent centrifugation steps, chromaffin vesicles were isolated.[20]

Nanoparticle tracking analysis (NTA)

Particle size and concentration measurements were performed with a NanoSight LM10 (Malvern Instruments Ltd, UK) equipped with a 488 nm (blue) laser and Hamamatsu C11440-50B/A11893-02 camera. The software used for capturing and analyzing the data was NTA 3.1. Experiments were performed at room temperature. Samples were diluted to obtain 40–100 particles per frame, where liposome samples were diluted with running buffer, whereas chromaffin vesicles were re-suspended in homogenizing buffer with NaCl. For each sample, 3 measurements consisting of 5 runs for 90 s were performed.

Electrochemical vesicle cytometry measurements

Electrochemical vesicle cytometry measurements were performed using amperometry at a carbon fiber microelectrode placed in liposome or vesicle suspension. The microelectrodes were prepared by aspirating a 33 μm carbon fiber into a glass capillary (o.d. 1.2 mm; i.d. 0.69 mm; no filament; Sutter Instrument Co., USA). A commercial micropipette puller (PE-21, Narishige, Japan) was used to pull the capillary and produce two carbon fiber-containing pipettes. In the following step a scalpel was used to cut the protruding carbon fiber close to the glass tip and then dipped into freshly made epoxy (EpoTek 301, Epoxy Technology, USA) for 3 min. The epoxy was allowed to cure and the electrodes were consequently polished at a 45° angle on a commercial micro grinder (EG-400, Narishige, Japan) and backfilled with 3 M KCl. All electrodes were tested prior to the experiments in 100 mM dopamine solution in PBS (pH 7.4) and only electrodes with stable I–E curves were used for experiments. For amperometric measurements, the electrode was kept at 700 mV *vs.* a Ag/AgCl reference electrode (made in house) using a commercial instrument (Axopatch 200B, Axon Instruments, USA). The signal was digitized at 10 kHz and filtered with an internal low pass Bessel filter at 2 kHz. The signal was displayed in real time (AxoScope 8.1, Axon Instruments, USA) and stored digitally. The collected data were analyzed with Igor Pro 6 (Version 6.3.4.1; WaveMetrics, Lake Oswego, OR) using an Igor Procedure File designed for analysis of quantal release from the group of David Sulzer.[30] Peaks were detected if exceeding a threshold of 3 times the RMS noise.[11]

Electrochemical vesicle cytometry of liposome systems. The liposomal pellet was re-suspended in 100 μl of liposome running buffer and 5 × diluted with the same buffer prior to the amperometric recordings which were done for 5 min at 37 °C.

Electrochemical chromaffin vesicle cytometry for investigation of potential dependence. The electrodes were dipped for 30 min at 4 °C in 10 × diluted stock suspension of chromaffin vesicles (stored in homogenizing buffer), and immediately transferred to chromaffin vesicle running buffer. The recordings were performed for 20 min at 37 °C.

Other electrochemical chromaffin vesicle cytometry experiments. Stock chromaffin vesicle suspension (in chromaffin vesicle running buffer) was 10 × diluted with chromaffin vesicle running buffer prior to the experiments. The recordings were done for 5 min at 37 °C.

Results and discussion

In electrochemical vesicle cytometry, the electroactive content of a soft nanoparticle is released and oxidized at the electrode surface resulting in a peak transient that can be attributed to the single vesicle fusion event.[18,20] The shape of the current peak reveals the kinetic information of the vesicle bursting events[6,7] whereas Faraday's law ($Q = nNF$) provides information about the amount of the transmitter released by each vesicle.[31,32] The quantity detected, the time course for the event to occur, and the number of events, are all parameters that can be used to evaluate the nature of the event at the electrode.

Electroporation as a mechanism of vesicle opening

A critical parameter in quantification of liposome contents in electrochemical vesicle cytometry is the opening of the vesicle and the direction of the pore at the electrode when it forms. We have considered that the vesicle might open *via* a mechanism involving electroporation. Application of 0.7 V to the electrode means that a field of 1.4×10^6 V cm^{-1} across the 5 nm membrane will form. This is clearly sufficient for electroporation.[33]

Increasing the electrode potential V should increase the probability of pore formation (*i.e.*, the number of events) due to the increased electrostatic energy (proportional to $CV^2/2$ where C is the capacitance of the membrane and V is the potential difference across the membrane) and, within the limits of the lateral extent of the electrostatic field, the pore size, and this could result in faster oxidation (pore size calculations are given in more detail in a later section). We examined the average time (peak half-width) for electrochemical vesicle cytometry experiments (Fig. 1A) and found that it varies distinctly from 0.5 to 0.7 V *vs.* the Ag/AgCl reference electrode. The difference from 0.7 to 0.9 V does not vary significantly, which might reflect a threshold for pore formation between 0.5 and 0.7 V corresponding to the electrostatic energy exceeding the energy of the pore edge. Importantly, the number of molecules observed per vesicle does not vary significantly at these potentials (Fig. 1B) demonstrating that the time effect is not from impartial electrolysis. Additionally, we have examined the number of events observed in these experiments and the height of the bar plots varies linearly for the events per time observed with potential, with more events at high potential (Fig. 1C).

We speculate that the proteins on the surface of the vesicle membrane inhibit the phospholipid bilayer from getting close enough to the electrode for electroporation, that membrane proteins need to move or migrate away from the contact area with the electrode to allow electroporation, and that electroporation is enhanced as expected at higher potentials. As we will demonstrate below, there are fewer events for the chromaffin vesicles compared to protein-free liposomes.

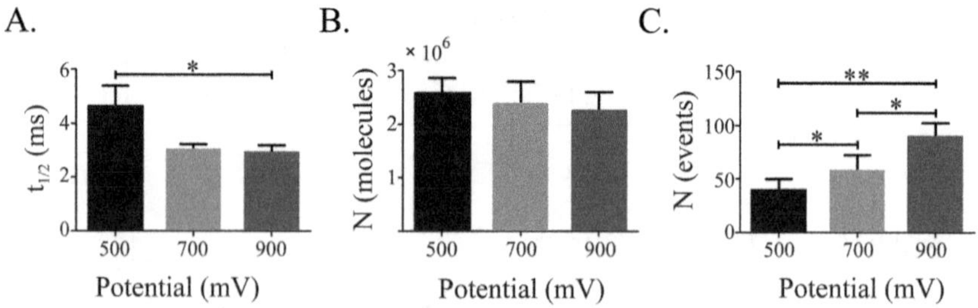

Fig. 1 The dependency of: (A) average $t_{1/2}$ (width of the peak at its half maximum), significance bar compares 500 mV to both 700 and 900; (B) number of molecules per vesicle; (C) number of events, *vs.* different potentials. *$p < 0.05$, **$p < 0.01$; two tailed paired student's t-test. Trace acquisition was as follows: 4 vesicle isolations, 10 traces per potential value; chromaffin cell vesicles were stored in homogenizing buffer and all experiments were run in chromaffin vesicle running buffer.

Dynamics of oxidation of vesicle content

Owing to the membrane complexity with different biomolecules, especially proteins and carbohydrates, and the presence of a protein dense core in the interior of epinephrine and norepinephrine rich chromaffin vesicles, we have applied synthetic vesicles, liposomes, as a first step to better understand the mechanism of mammalian vesicle rupture onto the electrode. Use of liposomes provides the advantage of having greater control over the system as their composition is relatively well defined and the protein dense core is absent. We have applied two types of liposome samples: control liposomes (without surface modification) and liposomes that are surface decorated with an octaarginine oligopeptide. Both liposome species were packed with dopamine.

In Fig. 2 we show the comparison between amperometric spikes for control liposomes and chromaffin vesicles. A significant difference is observed in the dynamics of these two different soft nanoparticles rupturing at the surface of the electrode. Liposomes result in significantly faster amperometric events compared to vesicles. Summaries of the time courses are shown in Table 1. The half-width of the peak ($t_{1/2}$) which summarizes the duration of the event, the rise time (t_{rise}) related to the time that takes for a pore to open, and the fall time (t_{fall}) needed for oxidation of the contents, are all significantly shorter for control liposomes. This phenomenon points to the simplicity of the membrane of the control liposomes, without a membrane protein component, but it is not clear how this would slow the dynamics of the oxidation event. It is possible that the protein content increases the viscosity of the membrane and slows down the electrostatic energy dissipation. It may also reduce the likelihood that an event occurs once the vesicle is adsorbed to the electrode and this is described below in more detail. For the change in the dynamics of these events, it seems more likely that it is the presence of the dense core in chromaffin vesicles that makes the vesicle more stable sitting at the electrode surface after adsorption and slows diffusion.

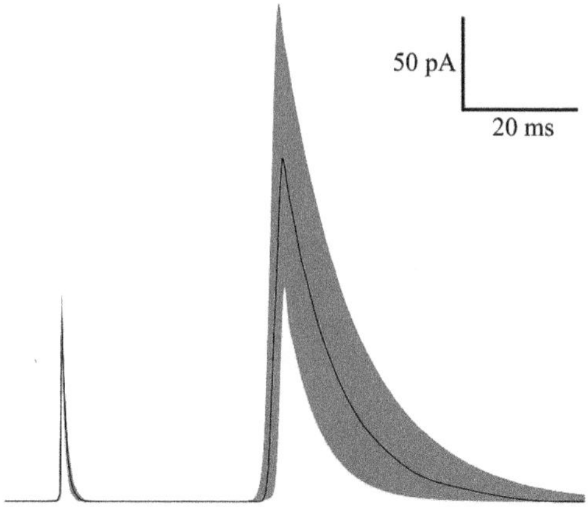

Fig. 2 The spike shape differences between liposome control (left) and chromaffin vesicles (right). Standard deviation is represented in grey. Chromaffin cell vesicles: storage and electrochemical cytometry were done in chromaffin vesicle running buffer.

Table 1 Summary of amperometric kinetic parameters for liposome systems and chromaffin vesicles[a]

Sample	t_{rise}[b] (ms)	$t_{1/2}$[c] (ms)	t_{fall}[d] (ms)
Liposome control	0.11 ± 0.03	0.57 ± 0.15	0.79 ± 0.19
Liposome with peptide	0.21 ± 0.09	1.18 ± 0.40	1.26 ± 0.42
Chromaffin vesicle	0.89 ± 0.32	7.20 ± 2.19	10.22 ± 3.76

[a] The data are presented as a mean ± standard deviation (SD). [b] Time from 25% to 75% of the front side of each peak signal. [c] Width of the peak at half maximum. [d] Time from 75% to 25% of the backside of each current transient.

Regulation of the frequency of vesicle opening events

We have estimated the percentage of soft nanoparticles from the bulk suspension that hit the electrode surface by applying a simplified diffusional theory as a means to look at the impact, sticking, and opening frequency for vesicles at the electrode surface. Here, we applied the linear form of the Cottrell eqn (1),

$$i = \frac{nFA\sqrt{D}C}{\sqrt{\pi}}\frac{1}{\sqrt{t}} \tag{1}$$

where i is the current, n is number of electrons, F is the Faraday constant, A is electrode area, C is the concentration of electroactive analyte, D is the diffusion coefficient and t is the time. In our approach, i is equivalent to the number of vesicles that rupture on the electrode surface during time t and thus represents the frequency of rupturing events. Therefore, we have omitted the n and F from the eqn (1) and the new equation becomes

$$\nu = \frac{A\sqrt{D}C}{\sqrt{\pi}}\frac{1}{\sqrt{t}} \tag{2}$$

where ν represents the frequency of rupturing events. We omit the radial diffusion terms of the equation as the diffusion distance, $d = (D \times t)^{1/2}$, is relatively small compared to the 33 μm size of the electrode due to the quite low diffusion coefficient for vesicles. We realize this is a simplification, but we additionally note that for the comparisons here, we are mainly interested in the depletion of vesicles and not the steady state behavior of diffusion. The use of this approach seems valid as the plot of ν *vs.* $t^{-1/2}$ is linear over the range of values used for these estimates (Fig. 3, right side).

Plots of frequency of amperometric events *vs.* acquisition time are shown in Fig. 3 (left side) for liposome models and chromaffin vesicles. We have calculated the diffusion coefficient for both liposomes and vesicles from diameter data obtained by NTA and by applying the Stokes–Einstein eqn (3),

$$D = \frac{k_B T}{6\pi\eta r} \tag{3}$$

where k_B is the Boltzmann constant, T is the absolute temperature, η is the viscosity of the media, and r is the hydrodynamic radius of the soft nanoparticle. Taking into account the value for the slope $\nu/t^{-1/2}$ from the plots for

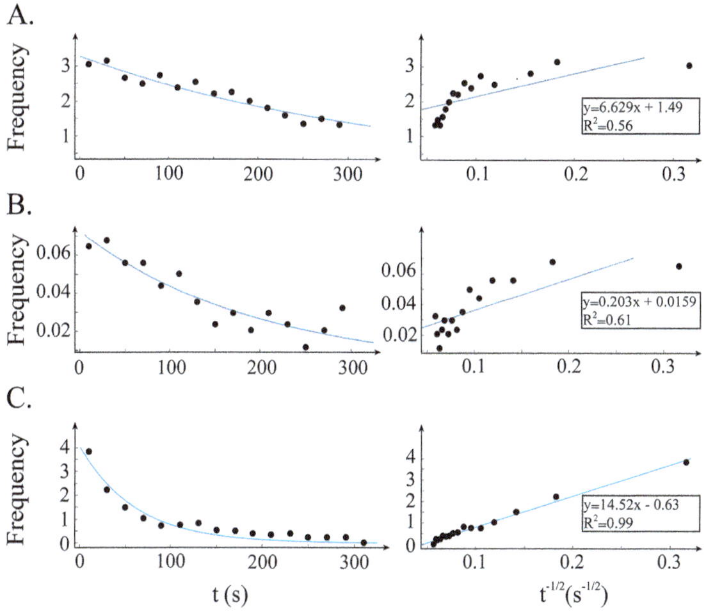

Fig. 3 Frequency *vs.* time (left) and frequency *vs.* time$^{-1/2}$ (right) for: (A) liposome control; (B) chromaffin vesicles; (C) liposomes with peptide. Chromaffin cell vesicles: storage and electrochemical cytometry was done in chromaffin vesicle running buffer.

vesicles and liposomes (Fig. 3, right side), soft nanoparticle diffusion coefficients, and surface area of the electrode, we have calculated the number of soft nanoparticles per millilitre (ml) that hit, stick, and rupture at the electrode surface. NTA of the liposomes and vesicles allows us to estimate the concentration of these soft nanoparticles in stock solution used for the impact cytometry experiment. By dividing the calculated concentration of particles obtained from the experimental data and using eqn (2) and concentration of particles in a bulk suspension obtained by NTA, we estimate that 1.5% of all control liposomes in bulk suspension hit the electrode during the 5 minutes of data acquisition. As it seems, all liposomes open once they stick, this gives us the means to estimate the percentage that stick upon impact. For vesicles that value was 0.077% for the event efficiency. Here it is clear that many vesicles stick on the surface, but do not get to the event stage, perhaps a space in the membrane by proteins moving away from the electrode contact area is not made on this time scale. This would probably prevent electroporation by maintaining a too large distance between the membrane and the electrode, hence a too small electrostatic field across the membrane. If we assume the vesicles have the same sticking efficiency as the liposomes (it is not clear this is correct, but can be used as a first approximation), then we can calculate that the fraction of vesicles opening relative to liposomes over 5 minutes time window is 5%. These data support the observation that vesicles rupture in a more difficult and complex manner, the phenomenon that might involve the complex protein nature of the vesicle membrane.

Building from the bottom up: liposomes with peptides

In order to support the hypothesis that the membrane protein component is important in regulating the vesicle rupture mechanism, we have added peptide to the liposome model to more closely approximate the native vesicle membrane. Here, a fatty acid chain conjugated oligopeptide, stearoyl octaarginine (STR8) was added to the liposome membrane. Amperometric events for peptide-decorated liposomes are compared to control liposomes and shown in Fig. 4. Slower rupture events are observed for peptide-decorated liposomes and an amperometric summary is given in Table 1 above.

The data for duration of release events for all three samples are summarized in a bar plot in Fig. 5. Significance testing shows that incorporating peptide into the membrane of the liposome slows down the process of rupturing in comparison to liposomes without peptide (control liposomes). This supports the assumption that proteins restrict the pore opening and consequently prolong the time for vesicles to rupture. However, even though adding peptide slows down the kinetics for liposome release events, though compared to chromaffin vesicles this process is still very fast. This is likely explained from the vesicle protein dense core, which both binds cationic messengers and restricts diffusion in the vesicle.[34,35] We conclude that the decoration of the membrane surface with peptide or protein affects the dynamics as well as the frequency of the vesicle cytometry events.

A model of vesicle opening in electrochemical vesicle cytometry

We have used the rate of the exponential fall for the amperometric events (time constant, τ) to evaluate the size of the pore needed for diffusion across the inside of these soft nanoparticles to the electrode surface.[6,7] The value of the diffusion in the vesicle interior is difficult to attain, but some estimates have recently been published[7] and in that work they calculated the size of the fusion pore that opens between a vesicle and a cell membrane as measured by an electrode in close vicinity to the cell membrane. This is a similar geometry to that used here, except

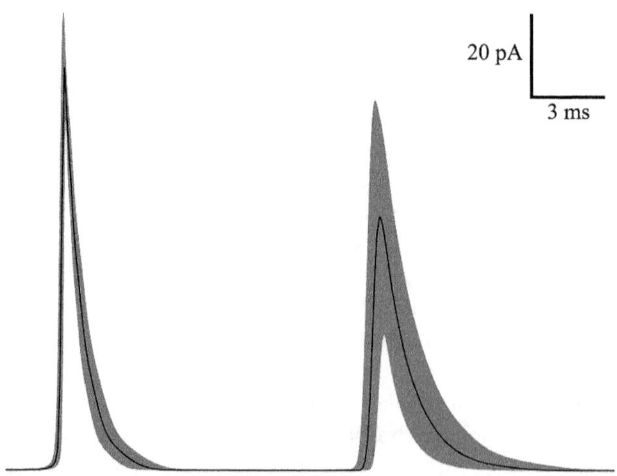

Fig. 4 The spike shape differences between liposomes control (left) and liposomes with peptide (right). Standard deviation is represented in grey.

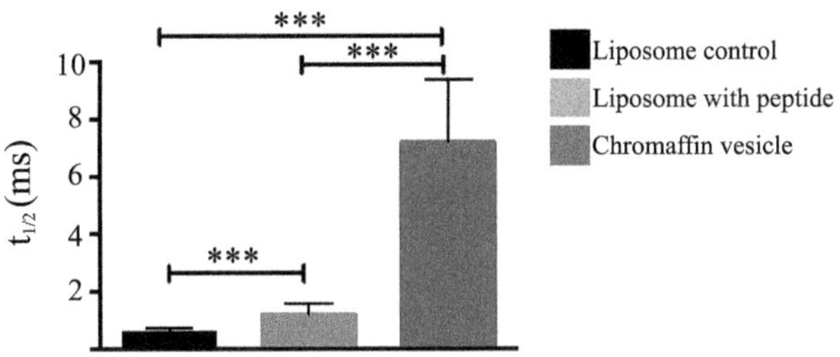

Fig. 5 The duration of events related to control liposomes (without peptide), liposomes with peptide and chromaffin vesicles; $*p < 0.05$, $**p < 0.01$ and $***p < 0.001$ using a two tailed paired student's t-test. Trace acquisition was as follows: control liposomes: 3 sample preparations, 16 traces in total; peptide liposomes: 4 sample preparations, 24 traces in total and chromaffin vesicles: 6 vesicle isolations, 18 traces in total. Chromaffin cell vesicles: storage and electrochemical cytometry were done in chromaffin vesicle running buffer.

we have an electrode with an adsorbed vesicle and so no membrane–membrane fusion occurs. To carry out these calculations, we first determined the time constant (τ) of the spike decay tail using an exponential model $i_{\text{spike-tail}} = A \times \exp(-t/\tau)$. Then $R_{\text{pore-max}}/R_{\text{ves}} = 1/(\text{Kappa} \times \tau)$, where Kappa $= D_{\text{ves}}/{R_{\text{ves}}}^2$ defines the diffusion rate of the molecules oxidized from inside the vesicle. $R_{\text{pore-max}}$ in absolute terms is given as $R_{\text{pore-max}} = {R_{\text{ves}}}^3/(D_{\text{ves}} \times \tau)$; however, since not all R_{ves} values are the same, the mean $R_{\text{pore-max}}$ value is obtained using the mean value for $R_{\text{pore-max}}/R_{\text{ves}}$ value times the mean R_{ves}. For chromaffin cells we used the experimental Kappa value (415 s^{-1}) determined previously.[7] For the liposomes, we assume that D_{ves} is close to the value in water solution (6×10^{-6} cm^2 s^{-1} for catecholamines). R_{ves} is the radius of the soft nanoparticle and was obtained with NTA. It is important to note that the Kappa value used for the cell vesicles assumes that they have the same structure as in the living cell for which the Kappa value used was determined considering that the initial fusion pore had a 1.2 nm radius for cell exocytosis fusion pores during the foot as previously determined in patch clamp studies.[36–38] If in the future, the community scales this estimate to lower radius values, then the size of the pore in the cytometry experiment might be somewhat smaller, being scaled down accordingly.

The pore diameters for the different systems are given in Table 2. The smallest size is for the liposome with the peptides added and the largest is for the vesicles. At first this seemed surprising as we fully expected the vesicles to have the

Table 2 Calculation of pore size opening from the time constants of spike decay branches

Sample	Vesicle/liposome diameter[a] (nm)	$\text{Diameter}_{\text{pore}}$ (nm)	Pore as % of vesicle/liposome diameter
Liposome control	328.3	18.4	5.6
Liposome with peptide	206.5	2.8	1.3
Chromaffin vesicle	339.8	96	28

[a] Diameters measured with nanoparticle tracking analysis (see Table SI-1 for STDs).

smallest pore size as their dynamics are the slowest experimentally. However, the significantly slower diffusion rate inside the matrix, owing to the dense core, appears to be the key feature here. This could be in fact predicted from recent work showing a slow and a fast phase for diffusion from exocytosis events and relating this to diffusion from the dense core and the halo or inter-protein spaces.[39] In the living cell, the pore opening and closing appears to be regulated by dynamin and actin.[40,41] In the isolated vesicles there are no intracellular proteins to restrict the pore from opening so it may open wider to release the internal pressure due to the constrained matrix swelling,[9,39] hence leading to a pore size (96 nm) that is approximately three times larger than that measured for events in living cells (*ca.* 30 nm).[7] The smaller pore values for liposomes appear to reflect the absence of any internal pressure from constrained matrix swelling. Thus, the size of the pore results only from equilibration of the electrostatic energy and the edge tension energy of the pore. For the peptide-decorated liposomes the pore might be smaller than for control liposomes owing to kinetic effects (slower viscous dissipation of the energy)[9,42,43] and/or due to the smaller electrostatic field imposed by a larger membrane–electrode distance imposed by the peptides.

Adsorbed vesicles have ruptured at a rate that is 1/20 that of the liposomes. We hypothesize that the proteins decorating the surface of the vesicle make a spatial barrier between the membrane and the electrode. This barrier reduces the potential field available for electroporation and inhibits this process. After some time, the proteins either diffuse or migrate in directions that open enough space for the membrane to wobble towards the electrode and when close enough, this will result in an electroporation pore opening and oxidation of the contents. This mechanism is depicted in Fig. 6. The peptides in the peptide-decorated liposomes lead to more events (Fig. 3C), and this we think is a result of fewer oligopeptide molecules in the membrane compared to the number of proteins in vesicles and, additionally, peptides probably enhance the sticking ability of the liposomes on the electrode. We plan to study this in more detail in future work.

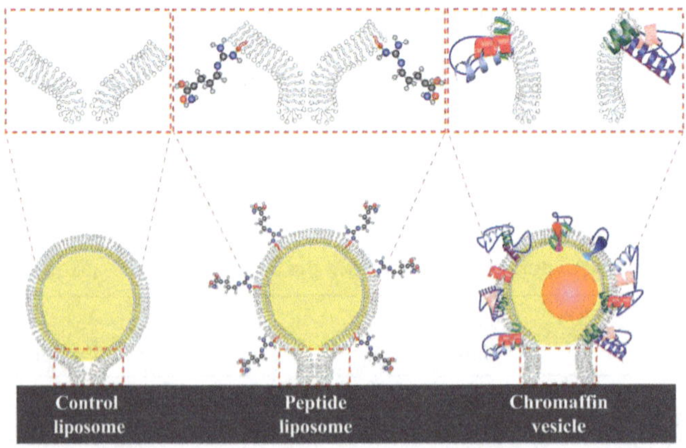

Fig. 6 Schematic illustration of liposome models and a chromaffin vesicle. In excerpts are shown pores in more detail for different soft nanoparticles. Peptides and proteins are not drawn to scale.

Summary and conclusions

The work reported here for electrochemical vesicle cytometry suggests that when membranous nanoparticles rupture on the electrode surface, this process involves electroporation to form a pore and the cargo diffuses through this pore and oxidizes at the electrode surface. Other work has suggested that vesicles opening on surfaces form pores at the point of highest membrane tension and this is usually at the side of the vesicle as it flattens towards the surface.[44,45] We are certainly not disputing this, but in the case of electrochemical vesicle cytometry, a pore is more likely to form at the membrane–electrode junction over the range of closer contact where the electrical field across the membrane is the highest, *i.e.*, where it is not yet fully screened by the double layer. This method allows us to quantify contents even at nanotip electrodes as it restricts diffusion away from the electrode and coulometric oxidation takes place. Proteins decorating the surface of the soft nanoparticles affect the pore size and the frequency of rupture. It appears that proteins form a barrier for the membrane to reach the electrode and need to move out of the way to allow close contact and electroporation. The protein dense core in vesicle matrices is also important in the dynamics of the events, in that it significantly slows diffusion through the vesicle[7] and while its constrained, swelling increases the membrane tension[9] resulting in the formation of pores of larger dimensions than for the same vesicles in living cells where this tension may be affected by dynamin and actin.[40,41]

Acknowledgements

We acknowledge Dalsjöfors Kött AB, Dalsjöfors, Sweden for donation of bovine adrenal glands for chromaffin vesicle experiments. The European Research Council (Advanced Grant), Knut and Alice Wallenberg Foundation, the Swedish Research Council (VR), and the National Institutes of Health are acknowledged for financial support. In Paris, this work was supported by CNRS, ENS and UPMC (UMR 8640).

References

1 R. D. Burgoyne and A. Morgan, *Physiol. Rev.*, 2003, **83**, 581–632.

2 S. An and D. Zenisek, *Curr. Opin. Neurobiol.*, 2004, **14**, 522–530.

3 J. Meldolesi and B. Ceccarelli, *Philos. Trans. R. Soc., B*, 1981, **296**, 55–65.

4 J. del Castillo and B. Katz, *J. Physiol.*, 1954, **124**, 560–573.

5 E. Neher and A. Marty, *Proc. Natl. Acad. Sci. U. S. A.*, 1982, **79**, 6712–6716.

6 C. Amatore, A. Oleinick and I. Svir, *ChemPhysChem*, 2010, **11**, 159–174.

7 A. Oleinick, F. Lemaitre, M. G. Collignon, I. Svir and C. Amatore, *Faraday Discuss. R. Soc. Chem.*, 2013, **164**, 33–55.

8 C. R. Artalejo, A. Elhamdani and H. C. Palfrey, *Curr. Biol.*, 1998, **8**, R62–R65.

9 C. Amatore, Y. Bouret, E. R. Travis and R. M. Wightman, *Biochimie*, 2000, **82**, 481–496.

10 R. G. W. Staal, E. V. Mosharov and D. Sulzer, *Nat. Neurosci.*, 2004, **7**, 341–346.

11 L. J. Mellander, R. Trouillon, M. I. Svensson and A. G. Ewing, *Sci. Rep.*, 2012, **2**, 907.

12 R. M. Wightman, J. A. Jankowski, R. T. Kennedy, K. T. Kawagoe, T. J. Schroeder, D. J. Leszczyszyn, J. A. Near, E. J. Diliberto and O. H. Viveros, *Proc. Natl. Acad. Sci. U. S. A.*, 1991, **88**, 10754–10758.
13 A.-S. Cans and A. G. Ewing, *J. Solid State Electrochem.*, 2011, **15**, 1437–1450.
14 D. M. Omiatek, M. F. Santillo, M. L. Heien and A. G. Ewing, *Anal. Chem.*, 2009, **81**, 2294–2302.
15 D. M. Omiatek, Y. Dong, M. L. Heien and A. G. Ewing, *ACS Chem. Neurosci.*, 2010, **1**, 234–245.
16 D. M. Omiatek, A. J. Bressler, A.-S. Cans, A. M. Andrews, M. L. Heien and A. G. Ewing, *Sci. Rep.*, 2013, **3**, 1447.
17 B.-K. Kim, A. Boika, J. Kim, J. E. Dick and A. J. Bard, *J. Am. Chem. Soc.*, 2014, **136**, 4849–4852.
18 W. Cheng and R. G. Compton, *Angew. Chem., Int. Ed.*, 2014, **53**, 13928–13930.
19 E. Lebègue, C. M. Anderson, J. E. Dick, L. J. Webb and A. J. Bard, *Langmuir*, 2015, **31**, 11734–11739.
20 J. Dunevall, H. Fathali, N. Najafinobar, J. Lovric, J. Wigström, A.-S. Cans and A. G. Ewing, *J. Am. Chem. Soc.*, 2015, **137**, 4344–4346.
21 X. Li, S. Majdi, J. Dunevall, H. Fathali and A. G. Ewing, *Angew. Chem., Int. Ed.*, 2015, **54**, 11978–11982.
22 R. Dimova, *Adv. Colloid Interface Sci.*, 2014, **208**, 225–234.
23 K. Dimitrievski and B. Kasemo, *Langmuir*, 2008, **24**, 4077–4091.
24 E. Reimhult, B. Kasemo and F. Höök, *Int. J. Mol. Sci.*, 2009, **10**, 1683–1696.
25 H. Schönherr, J. M. Johnson, P. Lenz, C. W. Frank and S. G. Boxer, *Langmuir*, 2004, **20**, 11600–11606.
26 T. L. Colliver, S. J. Pyott, M. Achalabun and A. G. Ewing, *J. Neurosci.*, 2000, **20**, 5276–5282.
27 H. Winkler, *Neuroscience*, 1976, **1**, 65–80.
28 R. G. Johnson, *Physiol. Rev.*, 1988, **68**, 232–307.
29 A. Iwasa, H. Akita, I. Khalil, K. Kogure, S. Futaki and H. Harashima, *Biochim. Biophys. Acta, Biomembr.*, 2006, **1758**, 713–720.
30 E. V. Mosharov and D. Sulzer, *Nat. Methods*, 2005, **2**, 651–658.
31 L. Mellander, A.-S. Cans and A. G. Ewing, *ChemPhysChem*, 2010, **11**, 2756–2763.
32 L. Tabares, M. Lindau and G. Alvarez de Toledo, *Biochem. Soc. Trans.*, 2003, **31**, 837–841.
33 K. A. DeBruin and W. Krassowska, *Biophys. J.*, 1999, **77**, 1213–1224.
34 K. B. Helle, R. K. Reed, K. E. Pihl and G. Serck-Hanssen, *Acta Physiol. Scand.*, 1985, **123**, 21–33.
35 M. S. Montesinos, J. D. Machado, M. Camacho, J. Diaz, Y. G. Morales, D. Alvarez de la Rosa, E. Carmona, A. Castañeyra, O. H. Viveros, D. T. O'Connor, S. K. Mahata and R. Borges, *J. Neurosci.*, 2008, **28**, 3350–3358.
36 E. Neher and B. Sakmann, *Nature*, 1976, **260**, 799–802.
37 A. E. Spruce, L. J. Breckenridge, A. K. Lee and W. Almers, *Neuron*, 1990, **4**, 643–654.
38 R. Fernández-Chacón and G. A. de Toledo, *FEBS Lett.*, 1995, **363**, 221–225.
39 A. Oleinick, R. Hu, B. Ren, Z.-Q. Tian, I. Svir and C. Amatore, *J. Electrochem. Soc.*, 2016, **163**, H3014–H3024.
40 R. Trouillon and A. G. Ewing, *ACS Chem. Biol.*, 2014, **9**, 812–820.
41 R. Trouillon and A. G. Ewing, *ChemPhysChem*, 2013, **14**, 2295–2301.

42 C. Amatore, Y. Bouret, E. R. Travis and R. M. Wightman, *Angew. Chem., Int. Ed.*, 2000, **39**, 1952–1955.

43 O. Sandre, L. Moreaux and F. Brochard-Wyart, *Proc. Natl. Acad. Sci.*, 1999, **96**, 10591–10596.

44 E. Reimhult, M. Zäch, F. Höök and B. Kasemo, *Langmuir*, 2006, **22**, 3313–3319.

45 J. Zimmerberg and L. V. Chernomordik, *Adv. Drug Delivery Rev.*, 1999, **38**, 197–205.

Faraday Discussions

PAPER

Membrane patches as ion channel probes for scanning ion conductance microscopy

Wenqing Shi,[a] Yuhan Zeng,[a] Lushan Zhou,[a] Yucheng Xiao,[b] Theodore R. Cummins*[b] and Lane A. Baker*[a]

Received 12th May 2016, Accepted 20th May 2016
DOI: 10.1039/c6fd00133e

We describe dual-barrel ion channel probes (ICPs), which consist of an open barrel and a barrel with a membrane patch directly excised from a donor cell. When incorporated with scanning ion conductance microscopy (SICM), the open barrel (SICM barrel) serves to measure the distance-dependent ion current for non-invasive imaging and positioning of the probe in the same fashion of traditional SICM. The second barrel with the membrane patch supports ion channels of interest and was used to investigate ion channel activities. To demonstrate robust probe control with the dual-barrel ICP-SICM probe and verify that the two barrels are independently addressable, current–distance characteristics (approach curves) were obtained with the SICM barrel and simultaneous, current–time (I–T) traces were recorded with the ICP barrel. To study the influence that the distance between ligand-gated ion channels (*i.e.*, large conductance Ca^{2+}-activated K^+ channels/BK channels) and the ligand source (*i.e.*, Ca^{2+} source) has on channel activations, ion channel activities were recorded at two fixed probe–substrate distances (D_{ps}) with the ICP barrel. The two fixed positions were determined from approach curves acquired with the SICM barrel. One position was defined as the "In-control" position, where the probe was in close proximity to the ligand source; the second position was defined as the "Far" position, where the probe was retracted far away from the ligand source. Our results confirm that channel activities increased dramatically with respect to both open channel probability and single channel current when the probe was near the ligand source, as opposed to when the probe was far away from the ligand source.

Introduction

Scanning ion conductance microscopy (SICM) is extremely versatile for high resolution, *in situ* imaging of delicate biological structures.[1–3] An oft cited

[a]Department of Chemistry, Indiana University, 800 E. Kirkwood Avenue, Bloomington, Indiana 47405, USA. E-mail: lanbaker@indiana.edu; Fax: +1 812 856 8300; Tel: +1 812 856 1873

[b]Department of Pharmacology and Toxicology, Indiana University-Purdue University Indianapolis, Stark Neurosciences Research Institute, 320 W. 15th St., Indianapolis, IN 46202, USA. E-mail: trcummin@iupui.edu; Tel: +1 317 278 9342

shortcoming of SICM is the inability to acquire chemical information, especially when compared to techniques like scanning electrochemical microscopy (SECM). A variety of approaches designed to overcome deficiencies of SICM for chemical measurement have been reported previously. For example, SICM has been coupled with SECM by incorporation of a carbon or metallic electrode at the probe tip to realize SICM–SECM.[4–10] In other approaches, chemical information has been collected in concert with SICM *via* secondary techniques, such as optical fluorescence.[11–13] SICM has also been combined with patch clamp measurements, namely the "smart patch clamp" reported by Korchev and co-workers,[14,15] in which the combination of SICM and patch clamp recording was accomplished through a nanopipette probe. First, a high-resolution topographic image of the cell was obtained by raster-scanning the nanopipette over the cell surface. Next, the pipette was guided to the area of interest *via* the high-resolution image, and patch clamp recordings were performed. With this technique, single channel recording from small cells and subcellular structures inaccessible in conventional patch clamp techniques was accomplished. In another report, Korchev *et al.* demonstrated mapping of ATP-regulated K^+ channels (K_{ATP} channels) with simultaneous topographic imaging on cardiomyocyte membrane,[16] where two pipettes were used; one for SICM topographical scanning, and a second to monitor the electrical response of the cell in whole-cell configuration. In these studies, the experiment was limited to only the target cell once the cell-attached or whole-cell configuration was established by the pipette. If data from multiple cells were desired, new patch clamp pipettes had to be used, which added uncertainty to the experiments. Moreover, the experiment was limited to ion channels that present within the target cell membranes.

Integrating ion channels at the probe tip is a logical step in the evolution of scanning ion conductance microscopy (SICM) for chemical/biochemical sensing. Recently, we described a method termed "Ion Channel Probe" (ICP)-SICM, in which the conductive probes are biological ion channels (in the initial report, alpha hemolysin (αHL)) incorporated within a black lipid membrane (BLM) at the pipette tip.[17] In the aforementioned work, feedback for SICM was demonstrated with ICPs, and preliminary results also demonstrated the possibility of using ICPs for topographic imaging. Although successful in imaging, this initial proof-of-concept experiment suffered several shortcomings. First, ion channels are directly utilized as current pathways in the SICM feedback loop, which may ultimately hamper the sensing utility of this design. Second, for the relatively large probes used in the initial report, the number of αHL channels required to generate suitable feedback was larger than ideal (in excess of 10). Third, the black lipid membranes suffer somewhat from stability issues (although there are methods to overcome this issue). Finally, the method of ion channel incorporation (direct insertion into a preformed BLM) at the pipette tip could, in some cases, limit selection of ion channels. Subsequent reports by White and coworkers reproduced these initial results, and added preliminary studies of imaging and detection of cyclodextrin with such probes at relatively low spatial resolution (compared to what is ultimately achievable with SICM).[18]

In this report, we describe a route to ICP-SICM that overcomes the major limitations of our initial efforts through two key advances. First, the ICP is constructed from a dual-barrel (or so-called "theta") pipette, where one barrel is unmodified and controls probe position and feedback, and the second barrel contains a bilayer/ion channels. Second, instead of forming BLMs at the probe tip, followed by ion channel insertion, membrane patches were extracted as inside-out or as outside-out patches directly from cells in a "sniffer patch" configuration.[19–21] The membrane patches exhibited excellent mechanical and chemical stability in our system, which allowed experimentation over the course of multiple hours. Use of intact cell patches obviates the need to reconstitute ion channels in synthetic bilayers and offers the opportunity to consider sensing applications that utilize fickle or delicate ion channels. Dual-barrel ICP-SICM probes can be constructed from a wide range of cells that contain native or transfected ion channels. Furthermore, when combined with SICM, the ICP can be positioned at desired locations over a sample. Importantly, as we demonstrate with this dual-barrel ICP-SICM configuration, the two barrels are controlled independently throughout the experiment. The SICM barrel serves only for probe–substrate distance control, while the ICP barrel was used only for sensing. As a result, ion current feedback and chemical/biochemical sensing at the pipette tip are largely decoupled.

Here we describe both outside-out (Nav1.5 channels[22]) and inside-out (BK channels[23]) membrane patches on ICPs. To demonstrate the feasibility of the dual-barrel ICP-SICM probe configuration as a two-in-one imaging/sensing platform, electrical characterization of the membrane-modified pipettes is presented. Current–time (I–T) traces recorded at the ICP barrel as a function of probe–substrate distance (D_{ps}) are reported in concert with feedback provided by the SICM barrel. A key advantage of the ICP route developed here is the ability to sense or detect molecules that are not necessarily electroactive or fluorescent, as the approach can be applied to any molecule with selective ion channel activity. ICPs are particularly exciting for applications in measurement of neurotransmitters such as glutamate (Glu), γ-aminobutyric acid (GABA) and acetylcholine (ACh), which otherwise require derivatization or special modification prior to electrochemical detection.[24] A proof-of-concept experiment was performed to demonstrate selective chemical sensing with ICPs and to verify the distance relationship between ligand-gated ion channels in an ICP and the ligand source. A dual-barrel ICP-SICM probe with a membrane patch containing BK channels was approached towards a Ca^{2+} source with positional feedback provided by the SICM barrel. I–T traces were measured with the ICP barrel at two fixed distances relative to the Ca^{2+} source, according to the approach curve. At $D_{ps} \approx 300$ nm, we observed significant ion channel activities (channels were in the OPEN state); at $D_{ps} > 10$ μm, most of the ion channels remained in the CLOSED state.

Combination of the precise probe position control offered by SICM and the powerful chemical sensitivity/selectivity provided by ion channels makes the dual-barrel ICP-SICM probe a promising candidate to study cell–cell interactions in biological systems with high spatial resolution. The possibility of measuring local release of neurotransmitters, signaling molecules and metabolites at subcellular scales will be possible through the ICP route described here.

Experimental

Chemicals and materials

All chemicals were used as received without further purification. Sodium chloride (VWR, Radnor, PA); potassium chloride (Macron Chemicals, Phillipsburg, NJ);

magnesium chloride hexahydrate (J. T. Baker, Phillipsburg, NJ); calcium chloride dihydrate (Sigma-Aldrich, St. Louis, MO); cesium chloride (Sigma-Aldrich); ethylenediaminetetraacetic acid (EDTA) (Sigma-Aldrich); ethylene glycol-bis(2-aminoethylether)-*N*,*N*,*N*′,*N*′-tetraacetic acid (EGTA) (Sigma-Aldrich) and 4-(2-hydroxyethyl)-1-piperazineethanesulfonic acid (HEPES) (Sigma-Aldrich) were used to prepare the recording buffers. Hydrochloric acid (EMD Millipore, Billerica, MA) and aqueous solutions of potassium hydroxide (Mallinckrodt, St. Louis, MO), sodium hydroxide (EMD Millipore, Billerica, MA) or cesium hydroxide (Sigma-Aldrich) were used for buffer pH adjustment. All aqueous solutions were prepared with deionized water (resistivity = 18 MΩ cm at 25 °C, EMD Millipore, Billerica, MA) and were filtered with 0.22 μm PVDF filter membranes (EMD Millipore, Billerica, MA) prior to use.

Probe fabrication

Dual-barrel theta pipettes were obtained by pulling borosilicate theta capillaries (1.5 mm O.D. (outer diameter), 1.17 mm I.D. (inner diameter), 0.165 mm septum, 10 cm length, Sutter Instruments, Novato, CA) with a P-2000 laser-based pipette puller (Sutter Instrument). Puller program parameters were as follows; heat: 415; filament: 4; velocity: 32; delay: 200; pull: 75. Pipettes were fire-polished (F-500 Fine Microforge System, Technical Products International, St. Louis, MO) to minimize capacitive artifacts and to facilitate gigaseal formation.[25] The size of each barrel after fire-polishing ranged from 400 to 800 nm, as characterized with scanning electron microscopy (SEM) and scanning transmission electron microscopy (STEM) images taken with a Quanta-FEG 600F scanning electron microscope (FEI, Hillsboro, OR), shown in Fig. 1a and b.

Cell preparation

Human embryonic kidney 293 (HEK 293) cells were grown under standard tissue culture conditions (5% CO_2 and 37 °C) in Dulbecco's Modified Eagle's Medium (DMEM-high glucose, D6429, Sigma-Aldrich, St. Louis, MO) supplemented with 10% (v/v) fetal bovine serum (F4135, Sigma-Aldrich) and 1% penicillin–streptomycin (P0781, Sigma-Aldrich). The stably transfected Nav1.5 cell line was prepared using the method described previously in Xiao *et al.*,[26] with cells used 2–3 days after seeding. Transient transfection of hSlo cDNA (for BK channels) constructs into HEK 293 cells was performed using the Magnetofection™ method with a commercially-available NeuroMag kit (OZ Biosciences, San Diego, CA) following manufacturer protocols. Neuromag (2.5 μL) was mixed with 2.5 μg of total plasmid cDNA (2 μg hSlo + 0.5 μg EGFP) in 200 μL of FBS-free DMEM. This solution was added to the cells after a 20 min incubation at room temperature. Next, the mixture of Neuromag-DNA and cells was placed on top of a magnetic plate, and incubated for an additional 20 min at 37 °C/5% CO_2. The following day, cells were detached and replated onto sterile 35 mm Petri dishes. Transfected cells were selected *via* fluorescence microscopy, where co-expression of EGFP (enhanced green fluorescent protein) was used as a marker of successful transfection of BK channels. BK-transfected cells were selected for patch clamp experiments 24 to 72 h after transfection. In Fig. 1c, a bright field image (200 μm × 200 μm) of HEK 293 cells 48 h after transient transfection is shown. The corresponding fluorescence image is shown in Fig. 1d. Two dashed circles were

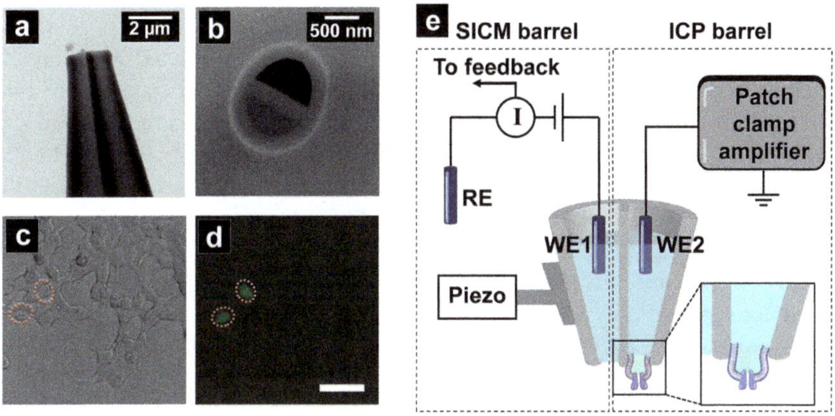

Fig. 1 (a) Scanning transmission electron micrograph (side-on view) of the dual-barrel theta pipette; (b) scanning electron micrograph of a theta pipette showing an end-on view; (c) bright field image of HEK 293 cells transient transfected with BK channels; (d) corresponding fluorescence image of (c), scale bar = 50 µm, red dashed circles were added to the cells that were successfully transfected. (e) Instrumental set-up for the dual-barrel ICP-SICM measurements.

added to highlight the position of EGFP expressing cells, which were successfully transfected.

Instrumentation for ICP-SICM measurements

To realize dual-barrel ICP-SICM experiments, a scanIC scanning ion conductance microscope (Ionscope Ltd., London, UK) equipped with an Axopatch 200 current amplifier (Molecular Devices, Sunnyvale, CA) was used to control probe position. A second current amplifier (PC505B, Warner Instruments, Hamden, CT) was used to record membrane patch formation and ion channel activity at the ICP barrel. A schematic of the dual-barrel ICP-SICM setup is depicted in Fig. 1e. The dual-barrel pipette was mounted vertically on a piezoelectric positioner and manipulated through SICM software. One of the two barrels, hereafter referred to as the SICM barrel, was back-inserted with a Ag/AgCl electrode (WE1) and connected to the headstage of the Axopatch amplifier. A second Ag/AgCl electrode (WE2) was back-inserted into the second barrel, and the PC505B amplifier was used to apply potential/record current at the second barrel, hereafter referred to as the ICP barrel. WE1 and WE2 shared a common reference electrode (RE, Ag/AgCl) in the bath solution, which was held at ground. In the final configuration of dual-barrel ICP-SICM experiments, the SICM barrel remained open, and ion current flow between WE1 and RE served as signal for feedback to control the probe–substrate distance during approach curves or imaging. The ICP barrel contained a membrane patch (inside-out membrane patch or outside-out membrane patch) with ion channels of interest, and the current from ion channel activities was recorded as a current–time trace. Currents were low-pass filtered at 1 kHz (−3 dB, 4-pole Bessel) and sampled at either 4 kHz (SICM barrel) or 25 kHz (ICP barrel). Data were acquired and analyzed with pClamp 10 data acquisition software (Molecular Devices, Sunnyvale, CA).

Electrophysiology recordings

For outside-out patch recordings from HEK 293 cells constructed with Nav1.5 channels, the pipette was filled with intracellular solution that consisted of 140 mM CsCl, 10 mM NaCl, 1.1 mM EGTA and 10 mM HEPES (pH was adjusted to 7.2 with CsOH). The external bath solution consisted of 140 mM NaCl, 5 mM KCl, 2 mM $CaCl_2$, 1 mM $MgCl_2$ and 10 mM HEPES (buffered at pH = 7.4 with NaOH).

For inside-out patch recordings from HEK 293 cells constructed with BK channels, the pipette solution consisted of 140 mM KCl, 5 mM NaCl, 10 mM HEPES and 1.1 mM EGTA (buffered at pH = 7.2 with KOH). The bath solution used consisted of 140 mM KCl, 5 mM NaCl, 2 mM $MgCl_2$ and 10 mM HEPES (buffered at pH = 7.4 with KOH). A micropipette (nominal diameter ≈ 2.5 μm) filled with 140 mM KCl, 5 mM NaCl, 10 mM HEPES and 1 mM $CaCl_2$ (buffered at pH = 7.4 with KOH) was used as a Ca^{2+} source for BK channel activation. The Ca^{2+} micropipette was mounted on an NMN-25 micromaniputalor (Narishige International USA, Long Island, NY) and was positioned directly at the substrate surface.

Results and discussion

Formation of dual-barrel ICP-SICM probes

To obtain a dual-barrel ICP-SICM probe with an open barrel and an outside-out membrane patch (from HEK293 cells transfected with Nav1.5 channels) barrel, a theta pipette was filled with intracellular solution (as detailed in the experimental section). Before submerging the pipette tip into the bath solution, a gentle positive pressure was applied to the side port of the electrode holder to prevent dirt/cellular debris from entering the pipette tip. After the pipette was immersed in the bath solution, junction potentials were cancelled to establish a zero baseline for each barrel. A square wave test pulse (V_{pp} = 5 mV, f = 60 Hz) was applied to each barrel and the corresponding currents were measured. The pipette resistance (R_p) was calculated according to Ohm's law. In Fig. 2a and b, R_p was calculated to be ~3.3 MΩ for both the SICM barrel and the ICP barrel. With the test pulse active, the pipette was lowered by the SICM Z stepper motor until the probe made contact with the target cell surface (indicated by a ~20% increase in R_p). In most cases, a gigaseal was formed at both barrels spontaneously after releasing the applied positive pressure; if a seal did not form immediately, light suction was applied to facilitate gigaseal formation. After gigaseal formation, the applied test pulse was increased to V_{pp} = 100 mV (f = 60 Hz), and the peak-to-peak leak current was measured to determine the seal resistance (R_{seal}) at each barrel. In Fig. 2c and d, R_{seal} was measured to be ~4.0 GΩ for the SICM barrel and ~2.5 GΩ for the ICP barrel. A brief pulse of suction was then applied to rupture the cell membrane to enter the whole-cell configuration (Fig. 2e and f), which is evident by an increase in the amplitude and duration of the capacitive charging peak, compared to that in the gigaseal configuration (Fig. 2c and d). In the gigaseal configuration, only the capacitance of the pipette and the membrane patch contributed to the total capacitive charging peak. However in the whole-cell configuration, the capacitance of the entire cell membrane contributed to the capacitive component. To form a membrane patch on both barrels, the pipette was retracted and moved away from the cell. Successful outside-out patch

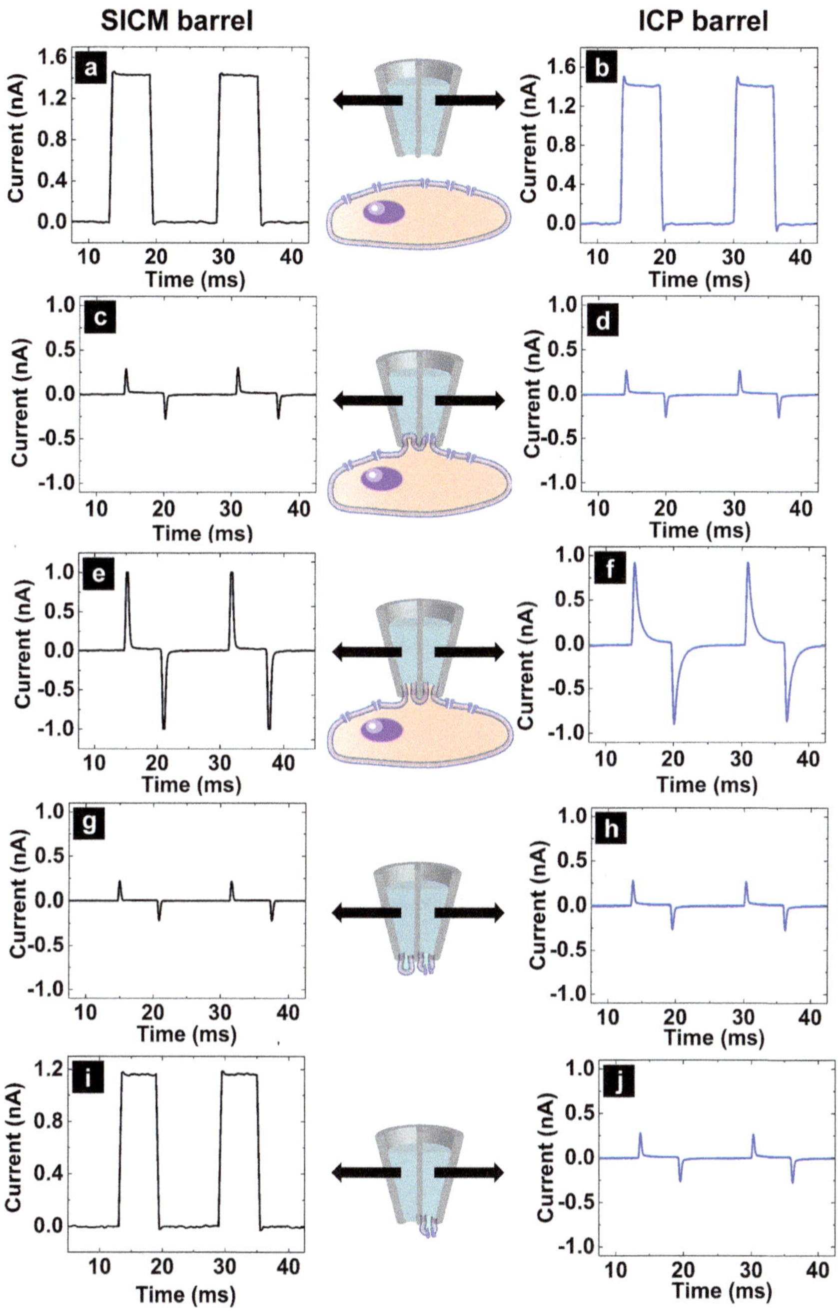

Fig. 2 Steps involved in the formation of a dual-barrel ICP-SICM probe, where both the SICM barrel and ICP barrel were open, (a) and (b), respectively; gigaseal configuration on both the SICM barrel (c) and the ICP barrel (d); whole-cell configuration on both the SICM barrel (e) and the ICP barrel (f); outside-out membrane patch on both the SICM barrel (g) and the ICP barrel (h); and when the SICM barrel was open (i) and ICP barrel contained an outside-out membrane patch (j).

formation was indicated by a significant decrease in both the amplitude and the duration of the capacitive charging current (Fig. 2g and h), as now the capacitive charging peak originated only from the membrane patch, which had a much smaller surface area compared to that of the entire cell. To break the membrane patch at the SICM barrel, electrically-induced breakdown of the cell membrane was achieved by application of increasing potential across the membrane patch. A 10 ms step of 0 mV was applied to the SICM barrel, followed by sequential 50 ms steps with amplitudes ranging from 0 to 1000 mV (in 5 mV increments), after which another 10 ms holding step of 0 mV was applied. In Fig. 3, the potential steps (left) and the corresponding current responses (right) are shown. For the case shown, the membrane patch remained intact until ~380 mV (Fig. 3a and b). At 385 mV, the membrane patch became unstable (Fig. 3c) and eventually at 390 mV, complete breakdown of the membrane patch was observed (Fig. 3d). Importantly, breakdown of the membrane patch was only observed at the barrel to which a potential was applied. As a result, a dual-barrel ICP-SICM probe with an open barrel (SICM barrel) and a barrel with an outside-out membrane patch (ICP barrel) was obtained (Fig. 2i and j). Of note, R_p at the SICM barrel in Fig. 2i was calculated to be ~4.2 MΩ, slightly larger than the 3.3 MΩ calculated in Fig. 2a. Increase in R_p is attributed to remaining cellular debris at the SICM barrel opening after the membrane patch was broken.

To construct a dual-barrel ICP-SICM probe with an inside-out patch at the ICP barrel, instead of retracting the pipette after whole-cell configuration as was done with the outside-out patch formation, the pipette was withdrawn and moved away from the cell after gigaseal formation.[19]

Approach curves

In dual-barrel ICP-SICM, probe–substrate distance is controlled *via* feedback provided by the ion current flow through the SICM barrel. As a constant potential is applied, ion current is at a maximum value when the probe is far away from the surface, and remains unaffected by the D_{ps}. As the probe approaches towards the surface, ion current decreases, owing to hindered ion flow in the probe–substrate gap. The D_{ps}-dependent current is used to control the pipette position along the Z axis. In Fig. 4a and b, a bare dual-barrel pipette (both the SICM barrel and the ICP barrel were open) was used. For the SICM barrel, a current *vs.* D_{ps} curve (Fig. 4a) was obtained; while for the ICP barrel, an I–T trace (Fig. 4b) was recorded (*i.e.*, the approach curve from the SICM barrel and the I–T trace recorded from the ICP barrel were collected simultaneously). The potentials of WE1 (SICM barrel) and WE2 (ICP barrel) were set at 5 mV. In both cases, the ion current decreased as the tip approached the substrate surface. D_{ps} equal zero was determined by finding the position at which the derivative of the current–distance response changed abruptly, which we take as the position at which the tip touches the surface. In Fig. 4c and d, approach curve and I–T trace obtained from a dual-barrel ICP-SICM probe are shown. Potentials applied to WE1 (SICM barrel) and WE2 (ICP barrel) were 5 mV and −20 mV, respectively. Ion current of the SICM barrel (Fig. 4c) exhibited the same distance-dependent behavior, and decreased with decreasing D_{ps}. For the ICP barrel, the current was approximately zero and remained unchanged throughout the approach, as the barrel opening was covered with an outside-out membrane patch. Of note, for this specific membrane patch,

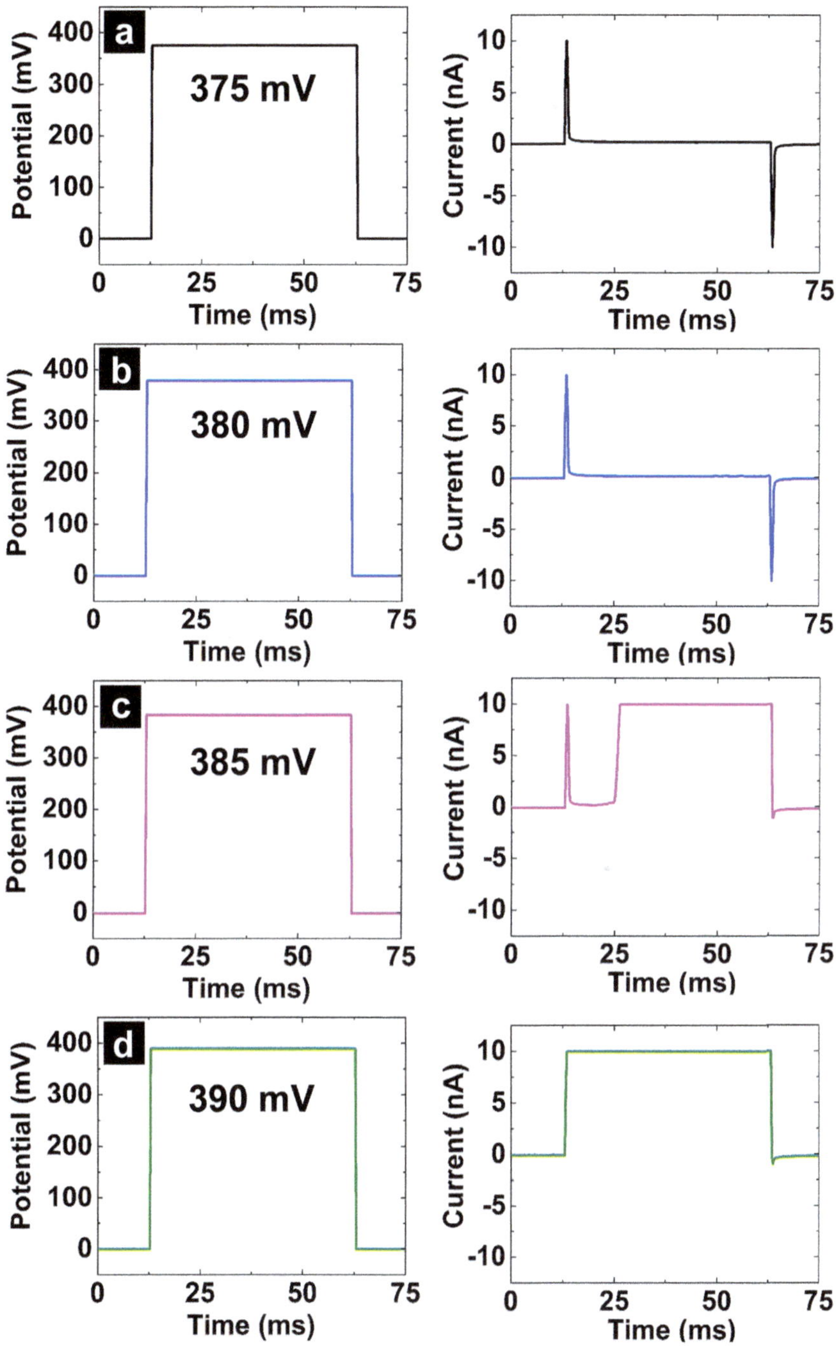

Fig. 3 Representative potential steps (left) and corresponding current responses (right) during electrically-induced breakdown of the membrane patch at the SICM barrel. Membrane patch remained intact at 375 mV (a) and 380 mV (b); and became unstable at 385 mV (c); and complete breakdown of the membrane patch occurred at 390 mV (d).

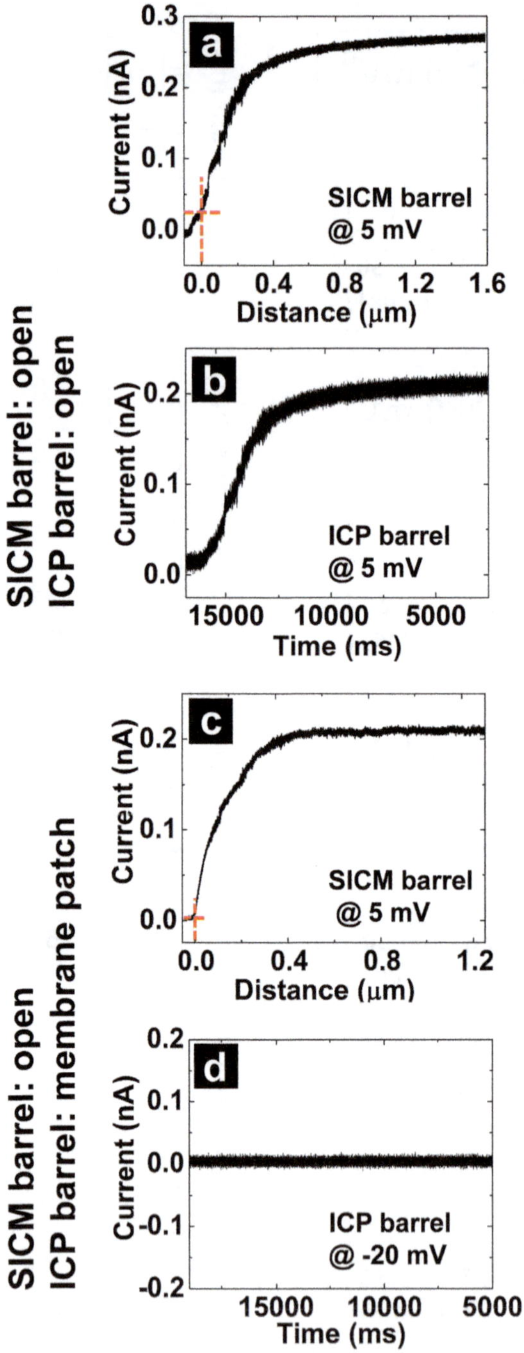

Fig. 4 Approach curve (a) and current–time (I–T) trace (b) recorded with a bare dual-barrel pipette, both barrels were biased at 5 mV. Approach curve (c) and I–T trace (d) recorded with an ICP-SICM probe, SICM barrel bias was 5 mV, ICP barrel bias was −20 mV.

although it is possible that Nav1.5 ion channels might be present, the holding potential here was set at only −20 mV, at which the channel opening activities are unlikely. Herein, our experiments further confirm that the potentials/currents at the two barrels can be controlled/monitored separately. The feedback offered by the SICM barrel enabled precise probe position control.

Distance-dependent ion channel activities

To evaluate the significance of the distance between the ligand-gated ion channels and the ligand source on channel activation, ion channel activities were recorded at two fixed probe–substrate distances. A dual-barrel ICP-SICM probe with an inside-out membrane patch containing BK channels was used. In Fig. 5a, a cartoon of the probe approaching a Ca^{2+} source is shown. Two positions were defined in Fig. 5a. The position "Far" refers to the position when the probe is more than 10 μm away from the surface, at which the ion current is at the maximum value. In typical SICM operations, the probe is kept at a position where the current drops to ~98% of the maximum current value, and this position is referred to as the "In-control" position. In Fig. 5b, an approach curve taken with the ICP-SICM probe is shown, and the "In-control" point was labeled with a red arrow, at which the D_{ps} was ~300 nm (as measured from subsequent approach curves to the surface). The probe was then held at the "In-control" point, at which the distance between the probe and the Ca^{2+} source was around 300 nm. An important point to note in this case is that the Ca^{2+} source was near but not directly under the ICP tip. Next, the SICM barrel was manually disconnected to minimize the noise in the system for single channel recordings. An I–T trace was then recorded with the ICP barrel (Fig. 6a). The bias applied to the ICP barrel was 60 mV, and single channel currents of ~8 pA were measured. From Fig. 6a, the BK channel was mostly in the OPEN state when the probe was in close proximity to the Ca^{2+} source, a feature commonly observed for BK channels under significant stimulation.[27–30] Next, the probe was retracted and held at the "Far" position and another I–T trace was taken with the ICP barrel (Fig. 6b). The applied bias was kept at 60 mV. In Fig. 6b, the BK channel remained in the CLOSED state for the majority of the time. Another observation was that the single channel current was ~5 pA. In the experiments described here, both channel activation frequency and single channel current increased with decreasing distance between the probe and the Ca^{2+} source. BK channel activation is dependent on both the applied potential and the Ca^{2+} concentration at the internal face of the membrane.[31] If we compare between the I–T trace taken at the "In-control" position and that taken at the "Far" position, the applied potential was kept constant. Thus the difference in the single channel current magnitude was most likely due to the difference in Ca^{2+} concentration. When the probe was operating close to the target surface, the Ca^{2+} source, BK channels experience higher Ca^{2+} concentration, which facilitates channel activation. On the other hand, when the probe was far away from the Ca^{2+} source, the Ca^{2+} concentration experienced by the channels was very low, and thus both the single channel current and channel open probability decreased dramatically. The importance of distance control in this case cannot be overstated. The advantage of utilizing ICP-SICM to enable sniffer patches can be demonstrated if we consider the release of molecules from a point source under purely diffusional control. As described previously, the time (t) dependent

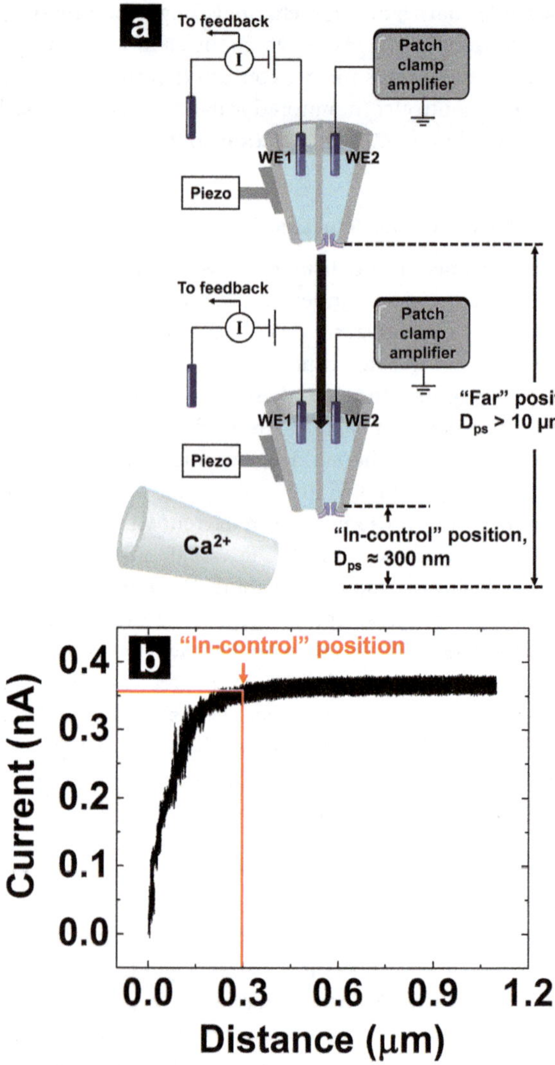

Fig. 5 (a) Cartoon illustration of a dual-barrel ICP-SICM probe approaching a Ca^{2+} source, two positions were defined in the graph; "In-control" position and "Far" position. (b) Approach curve taken with the SICM barrel of a double-barrel ICP-SICM probe toward a Ca^{2+} source, applied bias = 20 mV. The "In-control" position was labelled with a red arrow for clarification.

concentration ($C(t)$) at a fixed radial distance (r) from a source can be determined from eqn (1).[32] If we consider the case of the release of 1000 Ca^{2+} ions (M, 1.66×10^{-21} moles), with a diffusion coefficient (D) of 7.92×10^{-6} cm^2 s^{-1},[33] the concentration at a fixed position (plotted as 300 nm, 500 nm and 1000 nm from the surface) is shown in Fig. 7.

$$C(t) = \frac{M}{8(\pi Dt)^{3/2}} e^{(-r^2/4Dt)} \quad (1)$$

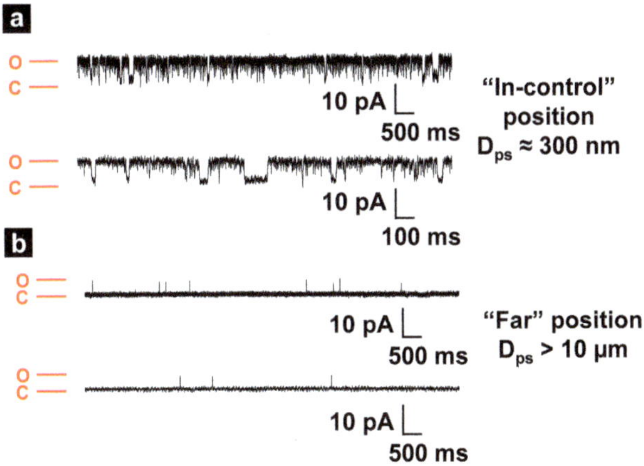

Fig. 6 (a) I–T trace recorded with the ICP barrel at "In-control" position, $D_{ps} \approx 300$ nm, applied potential = 60 mV; (b) I–T trace recorded with the ICP barrel at "Far" position, $D_{ps} > 10$ µm, applied potential = 60 mV. C = CLOSED state, O = OPEN state.

From this plot, the difference in concentration at a sensor (in this case the ICP) held at a fixed distance from the point source is observed to be a strong function of position. Initially, at time zero, there is minimal concentration, followed by pronounced region of maximum concentration, after which position-dependent concentration decays as the analyte diffuses into the bulk solution. Not only is the position a critical factor for obtaining local concentration enhancement, but the time response is dramatically increased for closely-positioned probes. In comparison to the experimental results presented here, the Ca^{2+} source is clearly not a point source in relation to the ICP tip, but the general phenomenon – a marked

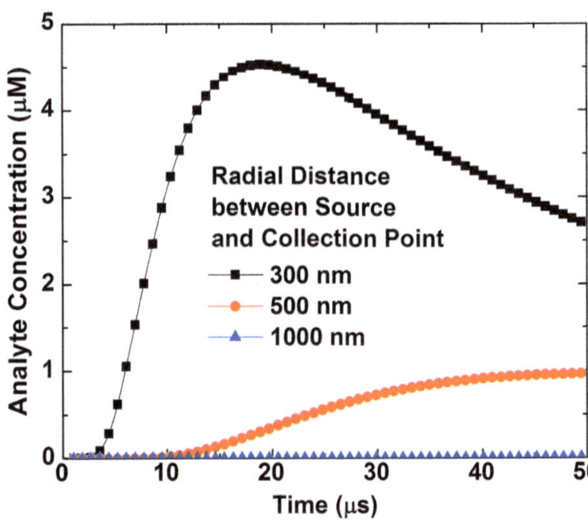

Fig. 7 Plot of concentration emitted as a function of time at a fixed radial position from the point source (eqn (1)).

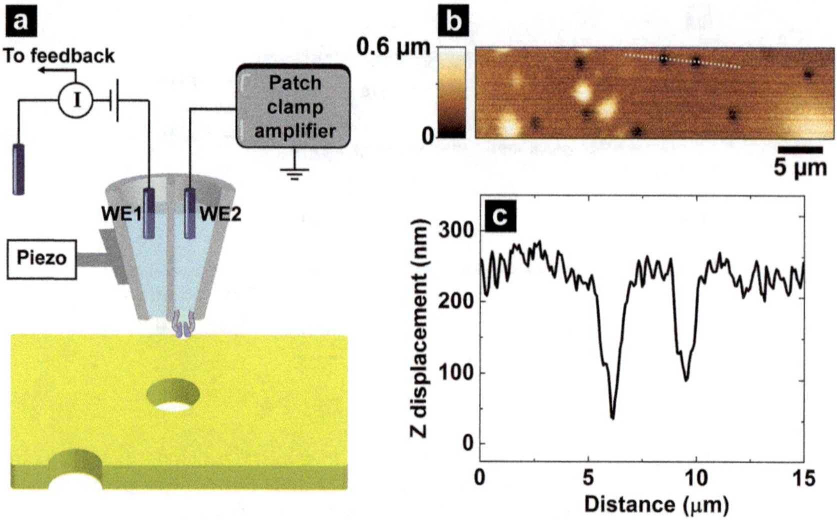

Fig. 8 (a) Cartoon illustration of a dual-barrel ICP-SICM probe imaging a porous polyimide (PI) membrane; (b) image of a PI membrane with pores taken with a dual-barrel ICP containing one open barrel for imaging and a second patched barrel. The ICP can obtain robust feedback, allowing topographic features such as the pores of this membrane to be visualized. (c) A line scan shows the topographic response of the ICP as it traverses two pores in the membrane, the position of this line scan is indicated by the dashed line in (b).

increase in concentration and time response for closely positioned ICPs – present in this approximation remains valid.

Results presented here clearly demonstrate the importance of having independent and accurate distance control when studying ion channel activities. First, channel activation events depend strongly on the distance between ion channels and the ligand source, and thus robust probe control which enables constant D_{ps} provided by SICM is key. Second, ion channel activities and the SICM feedback can be decoupled with the dual-barrel approach, enhancing the utility of ICPs for chemical analysis.

Conclusion

In conclusion, we presented a novel method of constructing a dual-barrel ICP-SICM probe *via* patch clamp techniques. Dual-barrel ICP-SICM probes consist of an open barrel used for SICM feedback and an ICP barrel used for ion channel activity recording. The ICP barrel was constructed by directly excising a membrane patch containing ion channels of interest from a donor cell. Robust probe position control was obtained with the SICM barrel, as demonstrated with approach curves. Influence of distance between a ligand-gated ion channel and the ligand source on channel activation was investigated by fixed position measurements.

This report opens the door to future studies for quantification with ICP probes. However, a number of important issues remain, which we are presently investigating. First, determining the number of channels present in membrane patches is an important question to consider, although calibrating the ICP to known concentrations may prove a route to normalize the activity of a given probe. Second, the nature of different patch configurations – *i.e.*, do the patches reside inside of the pipette tip or outside, and if so, to what extent – are important questions to consider for determining the SICM response of the open barrel. Such characterization is quite challenging, but previous studies on larger pipettes have examined flash-frozen membrane patches at a pipette tip to determine the exact patch geometry,[34] an approach which could be tractable here. Third, taking dynamic images, as opposed to fixed position measurements, with ICPs in a scanning mode is the next critical step in development. We can demonstrate that dual-barrel ICP probes can be used in preliminary imaging experiments (Fig. 8). However, for sake of quantitation, influence of the SICM barrel on distribution of ions at the tip of the pipette is non-trivial. Previous studies have shown that the narrow gap between the pipette tip and surface in SICM has interesting properties dependent on the charge of surfaces present and ions in solution.[35–37] That is to say, the case presented above (Fig. 7) for diffusion from a point source omits key features (explicitly the electric field at the SICM barrel) that will be relevant in ICP-SICM imaging applications, and these features require significant further investigation.

Compared to previous attempts of adding chemical specificity to SICM measurements, especially those which have relied on stochastic insertion of αHLs into black lipid bilayers, the method described here effectively eliminates the αHL reconstitution step. This aspect cannot be overlooked, as functional ion channels can require additional membrane proteins/components to function properly, and the approach taken here affords the opportunity to make use of more complicated biochemical systems. Ultimately, membrane patches can be formed from cells containing a variety of ion channels, and the resultant ICPs can be utilized to study release of signaling molecules in a wide range of cell- and tissue-based systems. Platforms that mimic cell–cell interactions can also be created to enable almost endless combinations of ICP-SICM investigations.

Acknowledgements

Alicia Friedman and Beth Yuill are acknowledged for assistance in manuscript preparation. Indiana University and the National Science Foundation (1507341) are acknowledged for funding. Dr. Jianghui Hou (Washington University in St. Loius) is greatly appreciated for the use of PC505B current amplifier. Electronic Instrument Services and Mechanical Instrument Services at Indiana University are gratefully acknowledged for assistance with instrumentation.

References

1 P. Hansma, B. Drake, O. Marti, S. Gould and C. Prater, *Science*, 1989, **243**, 641–643.
2 C.-C. Chen, Y. Zhou and L. A. Baker, *Annu. Rev. Anal. Chem.*, 2012, **5**, 207–228.
3 C.-C. Chen, Y. Zhou, C. A. Morris, J. Hou and L. A. Baker, *Anal. Chem.*, 2013, **85**, 3621–3628.

4 Y. Takahashi, A. I. Shevchuk, P. Novak, Y. Zhang, N. Ebejer, J. V. Macpherson, P. R. Unwin, A. J. Pollard, D. Roy and C. A. Clifford, *Angew. Chem., Int. Ed.*, 2011, **50**, 9638–9642.
5 R. Thakar, A. E. Weber, C. A. Morris and L. A. Baker, *Analyst*, 2013, **138**, 5973–5982.
6 C. A. Morris, C.-C. Chen and L. A. Baker, *Analyst*, 2012, **137**, 2933–2938.
7 K. McKelvey, B. P. Nadappuram, P. Actis, Y. Takahashi, Y. E. Korchev, T. Matsue, C. Robinson and P. R. Unwin, *Anal. Chem.*, 2013, **85**, 7519–7526.
8 Y. Takahashi, A. I. Shevchuk, P. Novak, Y. Murakami, H. Shiku, Y. E. Korchev and T. Matsue, *J. Am. Chem. Soc.*, 2010, **132**, 10118–10126.
9 D. J. Comstock, J. W. Elam, M. J. Pellin and M. C. Hersam, *Anal. Chem.*, 2010, **82**, 1270–1276.
10 W. Shi and L. A. Baker, *RSC Adv.*, 2015, **5**, 99284–99290.
11 J. Gorelik, A. Shevchuk, M. Ramalho, M. Elliott, C. Lei, C. F. Higgins, M. J. Lab, D. Klenerman, N. Krauzewicz and Y. Korchev, *Proc. Natl. Acad. Sci. U. S. A.*, 2002, **99**, 16018–16023.
12 D. Klenerman, A. Shevchuk, P. Novak, Y. E. Korchev and S. J. Davis, *Philos. Trans. R. Soc., B*, 2013, **368**.
13 A. I. Shevchuk, J. Gorelik, S. E. Harding, M. J. Lab, D. Klenerman and Y. E. Korchev, *Biophys. J.*, 2001, **81**, 1759–1764.
14 J. Gorelik, Y. Gu, H. A. Spohr, A. I. Shevchuk, M. J. Lab, S. E. Harding, C. R. Edwards, M. Whitaker, G. W. Moss and D. C. Benton, *Biophys. J.*, 2002, **83**, 3296–3303.
15 A. Bhargava, X. Lin, P. Novak, K. Mehta, Y. Korchev, M. Delmar and J. Gorelik, *Circ. Res.*, 2013, **112**, 1112–1120.
16 Y. E. Korchev, Y. A. Negulyaev, C. R. Edwards, I. Vodyanoy and M. J. Lab, *Nat. Cell Biol.*, 2000, **2**, 616–619.
17 Y. Zhou, L. K. Bright, W. Shi, C. A. Aspinwall and L. A. Baker, *Langmuir*, 2014, **30**, 15351–15355.
18 F. C. Macazo and R. J. White, *J. Am. Chem. Soc.*, 2016, **138**, 2793–2801.
19 O. P. Hamill, A. Marty, E. Neher, B. Sakmann and F. Sigworth, *Pflügers Archiv*, 1981, **391**, 85–100.
20 T. G. J. Allen, *Trends Neurosci.*, 1997, **20**, 192–197.
21 R. H. Kramer, *Neuron*, 1990, **4**, 335–341.
22 N. Klugbauer, L. Lacinova, V. Flockerzi and F. Hofmann, *EMBO J.*, 1995, **14**, 1084–1090.
23 A. Marty, *Nature*, 1981, **291**, 497–500.
24 W. A. Jacobs, *J. Chromatogr. A*, 1987, **392**, 435–441.
25 B. Sakmann and E. Neher, *Annu. Rev. Physiol.*, 1984, **46**, 455–472.
26 Y. Xiao, J.-P. Bingham, W. Zhu, E. Moczydlowski, S. Liang and T. R. Cummins, *J. Biol. Chem.*, 2008, **283**, 27300–27313.
27 C. Solaro, M. Prakriya, J. Ding and C. Lingle, *J. Neurosci.*, 1995, **15**, 6110–6123.
28 S. Ghatta, D. Nimmagadda, X. Xu and S. T. O'Rourke, *Pharmacol. Ther.*, 2006, **110**, 103–116.
29 S. Grissmer, R. S. Lewis and M. D. Cahalan, *J. Gen. Physiol.*, 1992, **99**, 63–84.
30 A. Blatz and K. Magleby, *Trends Neurosci.*, 1987, **10**, 463–467.
31 D. P. McCobb, N. Fowler, T. Featherstone, C. Lingle, M. Saito, J. Krause and L. Salkoff, *Am. J. Physiol.: Heart Circ. Physiol.*, 1995, **269**, H767–H777.
32 J. Crank, *The Mathematics of Diffusion*, Oxford university press, 1979.

33 M. M. Mbogoro, M. E. Snowden, M. A. Edwards, M. Peruffo and P. R. Unwin, *J. Phys. Chem. C*, 2011, **115**, 10147–10154.

34 A. Ruknudin, M. Song and F. Sachs, *J. Cell Biol.*, 1991, **112**, 125–134.

35 N. Sa and L. A. Baker, *J. Am. Chem. Soc.*, 2011, **133**, 10398–10401.

36 N. Sa, W.-J. Lan, W. Shi and L. A. Baker, *ACS Nano*, 2013, **7**, 11272–11282.

37 T. Plett, W. Shi, Y. Zeng, W. Mann, I. Vlassiouk, L. A. Baker and Z. S. Siwy, *Nanoscale*, 2015, **7**, 19080–19091.

Faraday Discussions

PAPER

Voltammetry and molecular assembly of G-quadruplex DNAzyme on single-crystal Au(111)-electrode surfaces – hemin as an electrochemical intercalator

Ling Zhang, Jens Ulstrup* and Jingdong Zhang*

Received 25th April 2016, Accepted 27th May 2016
DOI: 10.1039/c6fd00091f

DNA quadruplexes (qs) are a class of "non-canonical" oligonucleotides (OGNs) composed of stacked guanine (G) quartets stabilized by specific cations. Metal porphyrins selectively bind to G-qs complexes to form what is known as DNAzyme, which can exhibit peroxidase and other catalytic activity similar to heme group metalloenzymes. In the present study we investigate the electrochemical properties and the structure of DNAzyme monolayers on single-crystal Au(111)-electrode surfaces using cyclic voltammetry and scanning tunnelling microscopy under electrochemical potential control (*in situ* STM). The target DNAzyme is formed from a single-strand OGN with 12 guanines and iron(III) porphyrin IX (hemin), and assembles on Au(111) through a mercapto alkyl linker. The DNAzyme monolayers exhibit a strong pair of redox peaks at 0.0 V (NHE) at pH 7 in acetate buffer, shifted positively by about 50 mV compared to free hemin weakly physisorbed on the Au(111)-electrode surface. The voltammetric hemin signal of DNAzyme is enhanced 15 times compared with that of hemin adsorbed directly on the Au(111)-electrode surface. This is indicative of both the formation of a close to dense DNAzyme monolayer and that hemin is strongly bound to the immobilized 12G-qs in well-defined orientation favorable for interfacial ET with a rate constant of 6.0 ± 0.4 s^{-1}. This is supported by *in situ* STM which discloses single-molecule G-quartet structures with a size of 1.6 ± 0.2 nm.

1. Introduction

DNA-based single- and double-strand (ss and ds) oligonucleotides (OGNs) immobilized on solid surfaces are focus systems in molecular diagnostics and other chemical and biological sensing.[1–4] Long-range electron transfer (ET) mechanisms in redox metalloprotein systems have become understood in increasing detail over the last couple of decades.[5–8] The more recent discovery of long-range photo-induced ET in ds OGNs has triggered strong interest in the fundamental mechanisms of DNA-based electron and hole conductivity.[9–16]

Department of Chemistry, Technical University of Denmark, Building 207, Kemitorvet, DK-2800 Kgs. Lyngby, Denmark. E-mail: jz@kemi.dtu.dk; ju@kemi.dtu.dk

Electron or hole "hopping", with full vibrational relaxation between each single-charge hop has emerged as an attractive mechanism for photo-induced OGN electronic conductivity. This notion is, however, faced with challenges in interfacial electrochemical ET in redox marked surface immobilized OGN molecules.[9,11–13,17–20] Much lower energies than in photochemical ET are involved here, questioning the feasibility of actual electron or hole generation (reduction and oxidation) in the nucleobase pairs in the electrochemical potential ranges that are accessible. It was also found that in addition to the faradaic ET processes between the electrode surface and intercalated redox probes,[12,13,17–20] potential induced surface structural changes in the OGN adlayer on atomically planar Au(111)-electrode surfaces can induce capacitive voltammetric signals in the presence of multiply charged counter ions (spermidine)[20] that stabilize the ds form.[21,22] These peaks can resemble the faradaic signals of the ET probes in redox marked OGNs otherwise forwarded as fingerprints of long-range electrochemical ET along the axis of the ds OGNs.[11,12,17–19] Central issues regarding the mechanisms of interfacial electrochemical ET of redox probes intercalated or covalently bound to immobilized target OGNs are therefore open.

We report here a study of electrochemical OGN conductivity using a different approach with the following perspectives:

• Our molecular target is a "non-canonical" OGN form, *viz.* a guanine-rich OGN quadruplex (G-qs) instead of the ds OGN. G-qs structures can be viewed as stacking of a variable number of G-quartets folding along ss guanine-rich strands.[23–26] G-qs form spontaneously in aqueous biological buffers in the presence of specific cations, potassium or sodium ions in particular, Fig. 1A.[26] The G-qs structures resemble the telomere ends in the chromosomes.[23–28] Their natural biological functions are largely unknown, but G-qs peptide binding and recognition has attracted interest in peptide sensing, therapy and other biotechnology.[26–28]

• The rigid G-qs structure compared with ds OGNs offers a perspective for facile long-range ET as reflected in the good conductivity of G-qs structures.[29–32] As for spermidine induced ds OGN structural tightening, G-qs structures might also show capacitive voltammetry that reflects reorientation of the rigid G-qs structure at the well-defined Au(111)-electrode surfaces.

• Studies of the electrochemical behavior of G-qs immobilized on polycrystalline Au-electrode surfaces have been reported.[33] These have, for example been brought to distinguish between ds and qs structures based on surface accessibility of small mobile redox molecules such as $[Ru(NH_3)_6]^{2+/3+}$ and

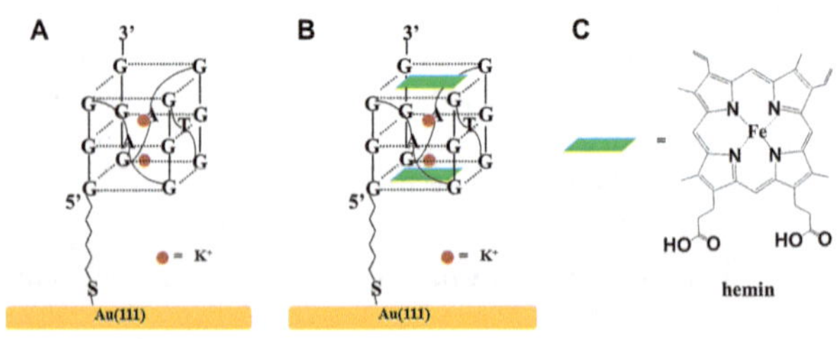

Fig. 1 (A) Scheme of G-qs structure stabilized by K$^+$ ions. (B) Scheme of G-qs DNAzyme with terminally bound hemin molecules. (C) Molecular structure of hemin.

$[Fe(CN)_6]^{4-/3-}$. In the present report, we use instead firstly atomically planar single-crystal Au(111)-electrodes onto which the target G-qs structures are linked *via* a protected thiol linker and a chain of six methylene groups. Use of this configuration enables, secondly structural mapping of the immobilized G-qs molecules by electrochemical scanning tunnelling microscopy (*in situ* STM) to single-molecule resolution.

• Finally, recent reports have pointed to the feasibility of both structural and functional single-molecule mapping of complex molecules including both redox metalloproteins and redox marked DNA-based molecules, by scanning tunnelling current/overpotential spectroscopy under electrochemical potential control (*in situ* STS).[34,35] These studies include the use of a redox probe intercalated or covalently bound to the OGN frame and electron exchange between the redox probe and the electrode.[36–39] Other recent studies have addressed single-biomolecule conductivity *in situ* using direct conductivity approaches.[39] The G-qs structure can be combined with iron(III) porphyrin IX (hemin) through π–π stacking between the porphyrin and the guanine quartet plane.[26,40,41] This "hybrid", known as DNAzyme, catalyzes the reduction of dioxygen or hydrogen peroxide similarly to the peroxidase enzyme class. The hemin molecule is believed to be terminally bound to either of the two terminal quartet planes of the G-qs, Fig. 1A and B. In the present work hemin is thus a redox probe that could potentially monitor long-range electrochemical ET through the G-qs molecules.

We address first as a reference structural and electrochemical properties of hemin itself adsorbed on bare Au(111)-electrode surfaces using a combination of cyclic voltammetry and *in situ* STM to single-molecule resolution. This is fraught with some challenge as hemin is adsorbed much more weakly on Au(111)-electrode surfaces in biological pH-ranges than on, say highly oriented pyrolytic graphite (HOPG) surfaces. We explore next the interfacial electrochemical structure and dynamics of the shortest, ss 15-base OGN, 5′-SH-$(CH_2)_6$-GGG A GGG T GGG A GGG (MW 5147.3, T30695),[26] folded into the G-qs structures in the presence of potassium ions, and attached to the Au(111)-electrode surface *via* a protected -S-$(CH_2)_6$- linker. The 12G target is addressed both in pure form and with terminally bound hemin, using the same combination of CV and *in situ* STM as for hemin itself. Notable differences between the adlayers of pure and hemin marked G-qs (DNAzyme) are observed. Intriguingly, both the electrochemical hemin signal and the UV-vis absorption spectra of the G-qs DNAzyme in the hemin Soret band region are much stronger than for hemin itself, indicative that a robust organized and dense 12G-qs/hemin adlayer has formed.

Based on *in situ* STM, it is shown that the single 12G-qs molecular entities can organize into monolayers on the Au(111)-electrode surface possibly *via* hydrogen bond interactions among the guanine quartets. The domains are composed of quartets and octagons with intermolecular distance of about 1.6 ± 0.2 nm. The organization follows, intriguingly, adsorption patterns of single-G-quartets of guanine molecules on Au(111)-surfaces at 350 K in ultra-high vacuum (UHV).[42] Approaches such as these to G-qs/hemin monolayers have not been, however, reported before.

2. Experimental

2.1 Chemicals and preparation

Chloroprotoporphyrin IX iron(III) (hemin, porcine, $C_{34}H_{32}ClFeN_4O_4$, ≥ 98.0%), sodium hydroxide (NaOH, 99.99%), sodium acetate (CH_3COONa, 99+%), potassium acetate (CH_3COOK, for molecular biology, ≥ 99.0%), were all from Sigma-Aldrich and used without further purification. Hemin solution was prepared by dissolving 1 mM hemin in 10 mM sodium hydroxide and diluting to 2.5 μM with sodium acetate buffer at pH 7. Millipore water (Milli-Q, 18.2 MΩ cm) was used throughout. The purified 12G oligonucleotide (5′-SH-$(CH_2)_6$-GGG A GGG T GGG A GGG, MW 5147.3) was synthesized using commercially available building blocks at the Nucleic Acid Center, University of Southern Denmark.

The 12G samples were prepared in 0.51 μM stock solution in 5 mM sodium acetate solution at pH 7.2 and stored at −20 °C. Before use, the solution was annealed at 88 °C for 10 min to ensure complete melting and then cooled slowly to room temperature.[26,40]

2.2 Preparation of G-quadruplex DNAzyme monolayers on Au(111) electrode surfaces

Single-crystal Au(111) bead electrodes (*ca.* Ø 3 mm) were prepared as described and checked by cyclic voltammetry in 0.1 M H_2SO_4.[43–47] The Au(111) bead electrodes were cleaned prior to use by H_2 flame annealing and quenched in Millipore water saturated with H_2.

To form the G-qs quadruplex structures, 0.51 μM 12G stock solution was mixed with 10 μM potassium acetate at room temperature (21 °C). After 45 min, the G-qs formed were incubated with 0.25 μM hemin solution for one hour leading to binding or intercalation of hemin molecules in 12G-qs G-qs/hemin complexes, here denoted as DNAzyme. After quenching in Millipore water, the Au(111)-electrodes were soaked in DNAzyme solution for 1 hour, rinsed with a mixture of 5 mM sodium and 10 μM potassium acetate, and then used in the cyclic voltammetry and *in situ* STM.

2.3 Preparation of hemin monolayers on Au(111)-electrode surfaces

Self-assembled monolayers of hemin molecules were prepared by immersing the freshly cleaned single-crystal Au(111)-electrodes in 0.25 μM or higher hemin solutions in 5 mM sodium acetate buffer, pH 7.2 for 1 hour. The modified electrodes were gently rinsed and kept in the same buffer for further use.

2.4 Instrumentation

The Au(111)-electrodes for cyclic voltammetry were used in the hanging meniscus mode. Cyclic voltammetry was performed using an Autolab potentiostat/galvanostat system (Eco Chemie, The Netherlands) controlled by the NOVA 2.0 software. An electrochemical setup including a cell and three-electrode configuration was employed as described.[43–47] A bright platinum wire was used as counter electrode and a freshly prepared reversible hydrogen electrode (RHE) exposed to the same supporting electrolyte and bubbled with pure H_2 as the reference electrode. The RHE was calibrated against the saturated calomel electrode (SCE) after each experiment. The electrolyte was deoxygenated by bubbling purified argon (Chrompack, 5 N) into the solution for one hour, and an argon atmosphere maintained throughout the experiments. UV-visible absorbance spectra were recorded using an Agilent G1103A Spectrophotometer.

A Pico SPM instrument (Molecular Imaging Co., U.S.A) was used for *in situ* STM. Home-made STM Tips were prepared by etching either tungsten or Pt/Ir (80/20) wire and coated with Apiezon wax.[47] The *in situ* STM cell was an in-house-built three-electrode PTFE cell with two Pt wires as reference and counter electrodes. The Au(111)-electrode was the substrate. The potential of the Au(111)-electrode substrate and of the tungsten or Pt/Ir tips was controlled in the supporting electrolyte by a bipotentiostat.

3. Results and discussion

3.1 Voltammetry of hemin monolayers on Au(111)-electrode surfaces

Hemin adsorbed on the Au(111)-electrode surface was found to display a complex behavior that reflects both protonation equilibria and dimer/oligomer formation on the surfaces in different pH-ranges. We shall communicate these observations elsewhere. Presently we focus on hemin surface behavior on a bare Au(111)-electrode surface in the same aqueous buffer, pH $\approx$ 7 as the stability range of surface immobilized G-qs to be probed by bound hemin.

Without the hemin monolayer, acetate adsorbs on the Au(111)-electrode surface, giving a set of strong and composite adsorption/desorption peaks in the range 0.1–0.6 V (*vs.* NHE), Fig. 2A.[44] Hemin adsorption causes the acetate adsorption peaks largely to disappear, and a pair of peaks with a midpoint potential around −0.05 V (NHE) appears instead, Fig. 2B. This peak is fraught

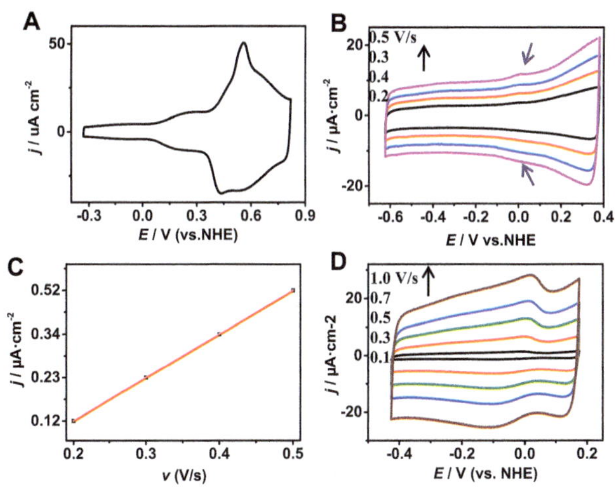

Fig. 2 (A) Cyclic voltammetry (CV) of Au(111)-electrode in 5 mM sodium acetate solution, pH 7.3. Scan rate, 0.2 V s^{-1}. (B) CV of hemin modified Au(111)-electrode at different scan rates. 5 mM sodium acetate, pH 7.1. The hemin adlayer was prepared from 0.25 μM hemin solution, 5 mM sodium acetate, pH ~ 7. The arrows inserted at ≈0.0 V point to the (small) hemin faradaic peaks. (C) Anodic peak current *versus* the scan rate. (D) CV of hemin modified Au(111)-electrode prepared from 0.5 μM hemin solution at different scan rates.

with a high capacitive background but is conspicuous enough that the dependence of the peak height, i_p on the scan rate ν could be determined. The i_p/ν correlation is clearly linear up to about 0.5 V s^{-1}, Fig. 2C, identifying the voltammetric behavior as surface confined.

The coverage of the hemin on the surface depends on the hemin concentration in the soaking solution. Taking the process to be a one-ET transition, the surface coverage was found to be about $(6.0 \pm 1.4) \times 10^{-13}$ mol cm^{-2} from the peak area when the hemin adlayer was prepared from 0.25 μM hemin solution. If the adlayer is prepared from 0.5 μM hemin solution, significantly stronger peak currents appear, Fig. 2D, giving now a coverage of hemin molecules of $(5.4 \pm 0.4) \times 10^{-12}$ mol cm^{-2}. Assuming a dense monolayer of a 1×1 nm^2 molecular hemin entity, the coverage is 1.7×10^{-10} mol cm^{-2}. If all the hemin molecules in 1 mL solution are adsorbed on a Au(111)-electrode with a diameter of 3.0 mm and form a monolayer, this would correspond to a hemin concentration of 0.012 μM. As hemin molecules are involved in an adsorption equilibrium between the electrode and solution, a concentration of 0.25–0.5 μM hemin appears insufficient to form a full monolayer on the Au(111) surface.

The anodic and cathodic peak separation increases from about 60 mV at the lowest scan rates to about 113 mV at higher scan rates approaching 1 V s^{-1}, Fig. 2D. At still higher scan rates the hemin monolayer seemed to loosen from the surface and the peak current dependence on the scan rate was replaced by a $i_p/\nu^{1/2}$ dependence (data not shown). The peak separation dependence on the scan rate up to 1 V s^{-1} enabled estimating the standard interfacial electrochemical ET rate constant, k_s using Laviron's method, giving the value of $k_s \approx 15 \pm 2$ s^{-1}. This value can be compared with the value found from hemin stacked onto G-qs, see Section 3.3.

3.2 *In situ* STM of hemin monolayers on Au(111)-electrode surfaces

In situ STM of the adlayer structure of hemin on the Au(111)-electrode surface in sodium hydroxide solution was recorded next. A full self-assembled monolayer (SAM) of hemin molecules was obtained by soaking the Au(111) electrode in 10 μM hemin aqueous solution at pH 12. This coverage is much higher than the

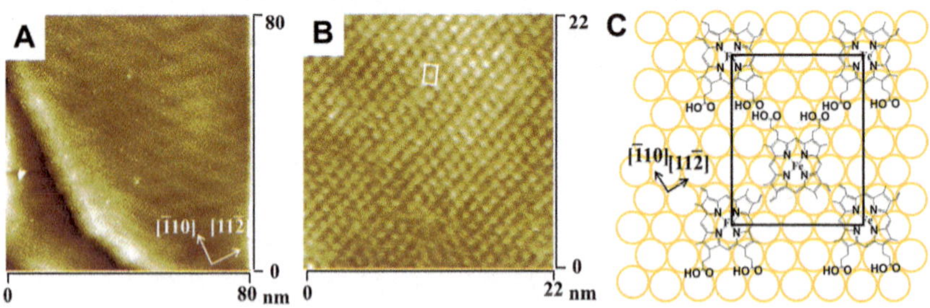

Fig. 3 (A) and (B) *In situ* STM images of hemin monolayer on the single-crystal Au(111)-electrode in 10 mM sodium hydroxide, pH 12. The electrode modified in 10 μM hemin in 10 mM sodium hydroxide solution. Working electrode potential $E_{work} = 0.2$ V (*vs.* NHE), tunnelling current $I_{tunnel} = 0.04$ nA. Bias voltage, $V_{bias} = E_{tip} - E_{work} = 0.26$ V (A), $V_{bias} = -0.14$ V (B). (C) Proposed model for iron porphyrin monolayer formed on Au(111) surface superimposed with a $c(4 \times 3\sqrt{3}) - 2$ unit cell.

coverage obtained by soaking in 0.25 μM hemin solution around pH 7. As shown in Fig. 3A, ordered lines with 120° angle and 6.4 nm spacing on the atomically flat terrace were observed. These lines are attributed to the Au(111) herringbone ($\sqrt{3} \times 22$) reconstruction lines on Au(111) and indicates clearly that the hemin molecules are physisorbed on the Au(111) surface.

Ordered arrays of hemin molecules are thus assembled on the reconstructed Au(111) surface. A close-up image is shown in Fig. 3B. Detailed distance and angle analysis shows that one row of the adlayers is parallel to one of the atomic rows of the Au(111)-surface with a periodic distance of 1.2 ± 0.1 nm. The other row is perpendicular to the atomic row of the substrate underneath with the periodic distance 1.6 ± 0.2 nm. The surface structure is therefore assigned to a $c(4 \times 3\sqrt{3}) - 2$ unit cell containing two hemin molecules. The unit cell is marked in Fig. 3B and a likely surface binding mode is proposed in Fig. 3C. The hemin plane is parallel to the Au(111)-surface probably supported by the –OH axial ligand adsorbed on the surface. Both hydrophobic and hydrophilic interactions act among the hemin molecules in the monolayer. The hemin molecule at the center is moreover mirror symmetric relative to the molecules at the corners of the rectangle, and the pairs are stabilized by both hydrogen bonds and hydrophobic interactions between the hemin side groups.[48–51] The unit cell is close to the $c(5 \times 6\sqrt{3}) - 2$ unit cell for Co(II)phthalocyanine on Au(111) reported by Itaya and associates[52] but differs from the unit cells of octaethylporphyrin iron(III) chloride ($a \times b$ = 1.23 nm × 1.4 nm with an angle of 66°) on Au(111)[53] and hemin (1.22 nm × 1.34 nm with 68°) on HOPG.[48] These differences are most likely caused by different lateral interactions among the molecules and between the molecules and the electrode substrates.[51] Very different packing lattices in cysteine monolayers on Au(111)- and Au(110)-electrode surfaces in solution and vacuum have also been found.[47,54]

3.3 Cyclic voltammetry of 12G-qs and DNAzyme SAMs

Our qs target, 12G-qs is the smallest G-qs with only three G-quartets. The DNAzyme is formed when hemin molecules are bound onto the 12G-qs ends by π–π interaction or intercalated into the 12G-qs structures. As shown in Fig. 4A, the Soret absorption peak at 404 nm from the constructed DNAzyme is 20 times higher than the peak for the same concentration of free hemin in solution and furthermore red-shifted by 2 nm compared to that of free hemin molecules. Both spectral observations indicate that there is strong electronic interaction between hemin and the terminal quartet(s) in spite of the non-covalent nature of this interaction. The OGN absorption peak at 254 nm is maintained after the binding of hemin.

The DNAzyme was investigated by cyclic voltammetry (CV). Hemin free 12G-qs or the DNAzyme complex was adsorbed on the Au(111)-electrode surface through the protected SH-group. Fig. 4B and C show CVs of hemin-free 12G-qs and the 12G-qs/hemin complex (DNAzyme), respectively. The CVs of hemin free 12G-qs show a broad featureless capacitive background in the potential range −0.4 V to +0.1 V. There is, however, also a conspicuous pair of peaks at ≈−0.4 V for the anodic peak and ≈−0.55 V for the cathodic peak (midpoint potential −0.48 V). Indications of less well-defined peaks around +0.2 V are not likely to be associated with hemin but either with the adsorption of acetate or reconstruction lift of the

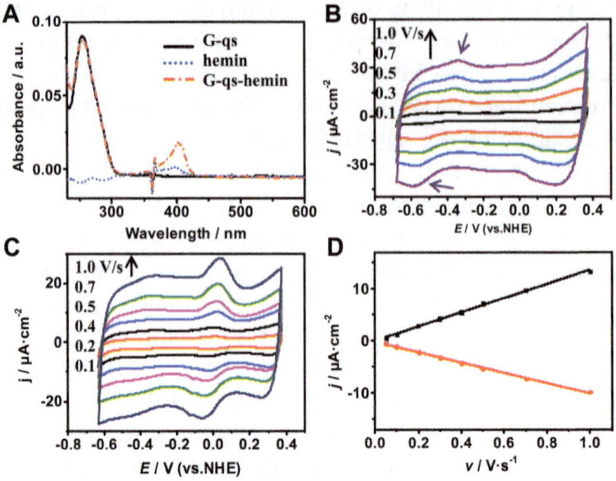

Fig. 4 (A) UV-vis spectra of 0.5 μM 12G-qs, 0.5 μM 12G-qs with 0.25 μM hemin, and of 0.25 μM hemin solutions in 5 mM sodium acetate buffer with 10 μM K^+ ions at pH 7. (B) CVs of 12G-qs modified Au(111)-electrode in 5 mM sodium acetate, pH 7.2 at different scan rates. Tilted arrows show the voltammetric peaks discussed in the text. (C) CVs of DNAzyme modified Au(111)-electrode in 5 mM sodium acetate, pH 7.2 at different scan rates. (D) Anodic and cathodic peak currents *versus* scan rates.

Au(111)-electrode surface. The origin of the −0.4 and −0.55 V peaks is not clear. Reductive and oxidative desorption/re-adsorption cannot be excluded, but both the midpoint potential and shape of this low-potential pair resemble those for the capacitive peaks observed for 20-base pair ds ONS when the triply charged counter ion spermidine is present.[20] Spermidine stabilizes the ds form,[20–22] and the peak – specific to base pair matching – was assigned to potential induced surface structural reorganization in the double-strand DNA adlayer which was further tentatively supported by *in situ* STM in the peak potential range.[34] The −0.4/−0.55 V 12G-qs pair could have a similar origin, *i.e.* arise from surface reorientation or order/disorder transition in the SAM of the compact 12G-qs molecules. *In situ* STM imaging studies to explore the surface dynamics further are in progress.

Fig. 4C shows CVs of the DNAzyme, *i.e.* hemin bound in the 12G-qs/hemin complex. In order to avoid having free hemin in the solution the complex was always prepared from a 2 : 1 G-qs/hemin mole ratio, leaving statistically one bound hemin molecule per 12G-qs/hemin complex. The most striking observation now is a strong signal at ≈ 0.0 V (*vs.* NHE), shifted positively by about 50 mV compared to the midpoint potential of hemin physisorbed on the Au(111)-electrode surface. This is strongly indicative of favorable interfacial ET of hemin bound to 12G-qs, possibly through the G-quadruplex structures, and further of high 12G-qs coverage on the Au(111)-electrode surface. The peak current of the strongly 12G-qs bound hemin is thus increased about 15 times compared to physisorbed hemin on the Au(111)-electrode surface. The total amount of hemin bound to surface immobilized 12G-qs after π–π binding or intercalation is $(2.5 \pm 0.4) \times 10^{-11}$ mol cm^{-2}, which is 40 times the amount of physisorbed hemin in Fig. 2B. Both anodic and cathodic peak currents depend linearly on the scan rates, corresponding to surface behavior of the DNAzyme monolayer, Fig. 5D. The anodic peak widths are in the 103–135 mV range at scan rates from 0.1 to 1.0 V s^{-1}. This can be compared with the theoretical value (91 mV) for reversible one-electron transfer and supports that as an average, there is one bound hemin molecule per 12G-qs structural entity. The amount of the DNAzyme obtained from the CVs, $(2.5 \pm 0.4) \times 10^{-11}$ mol cm^{-2} is not much lower than for a dense monolayer of 12G-qs/hemin, which is $(6.6 \pm 0.7) \times 10^{-11}$ mol cm^{-2} obtained from the STM results, Section 3.4. The standard interfacial electron transfer rate constant k_s is 6.0 ± 0.4 s^{-1}, which is comparable even to that of the hemin monolayer itself adsorbed directly on the Au(111)-electrode surface (15 ± 2 s^{-1}, Section 3.1).

3.4 *In situ* STM of 12G-qs DNAzyme SAMs on Au(111)-surfaces

The DNAzyme adlayer assembled on the Au(111)-electrode surface was further characterized by *in situ* STM, Fig. 5. These data are somewhat preliminary, and further work is in progress. The potential applied to the sample is 0.49 V, which is the open circuit potential and the hemin redox center bound to 12G-qs is in the

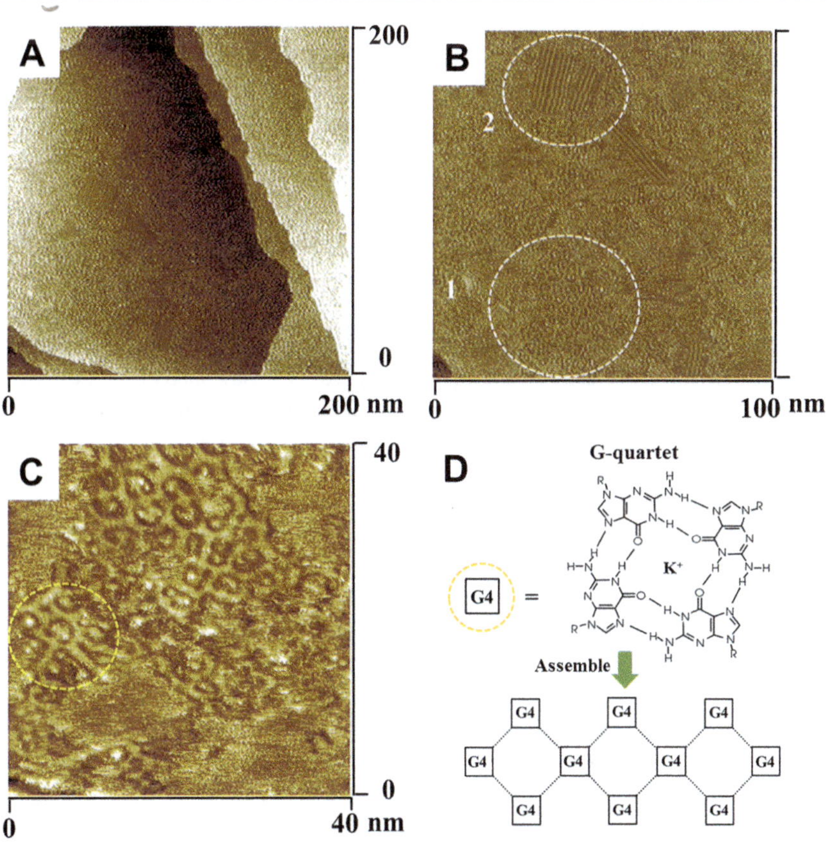

Fig. 5 (A)–(C) *In situ* STM images of DNAzyme adlayer on Au(111)-electrode surface in 5 mM sodium acetate at pH 7.0. $E_w = 0.49$ V (*vs.* NHE), $V_{bias} = -0.48$ V. $I_{tunn} = 0.09$ nA. (D) As a comparison, the structure of the planar G-quartet and the octagonal superstructure of the G-quadruplex SAMs formed on the Au(111)-electrode surface.

oxidized form Fe(III). Our present objective is thus to achieve single-molecule imaging, but tunnelling current/overpotential and bias voltage spectroscopy will be addressed in further research. Ordered domains are seen already over a large scan area of 200 × 200 nm^2. The reconstruction lines of the Au(111)-surface are lifted in the presence of the DNAzyme adlayer due to formation of S–Au bonds, Fig. 5A. Zooming in to 100 × 100 nm^2, two kinds of domains become apparent. Domain 1 shows network-like structures, while Domain 2 shows linear arrays (Fig. 5B). Smaller bright structures, possibly liberated hemin molecules are also apparent. The fine-structure in Domain 1 resolved to 40 × 40 nm^2 shows both quadratic molecular scale and larger octagonal entities (Fig. 5C). The quadratic entities could represent the folded 12G-qs box composed of three vertically stacked planar G-quartets. The octagonal units would then be larger structural entities formed through hydrogen bonds stretching out from the quartets, Fig. 5D and further mediated *via* networks of surface bound water molecules. Similar quartet structures were discovered by the assembly of guanine molecules directly on Au(111) surface in ultra-high vacuum (UHV) as imaged by STM at 350 K.[42] The size of the G-quartet with the four guanine molecules in UHV is 0.67 nm. The size of the 12G-qs in aqueous solution presently studied is larger and measured to be 1.6 ± 0.2 nm. The larger side length is most likely caused by water molecules engaged in hydrogen bonding stretching out from the 12G-qs boxes.

Domain 2 is composed of parallel lines (Fig. 5B). The origin of these structures is presently more elusive, but the lines could possibly suggest that unfolded single OGN strands are present and could align *via* hydrogen bonding between the unfolded strands.

4. Conclusion

The electronic conductivity of redox metalloproteins and metalloenzymes,[34,36–39,55,56] and double strand DNA-based molecules[11–13,16–19,57,58] immobilized at electrochemical interfaces have been explored comprehensively. Electronic conductivity of "non-canonical" DNA forms, quadruplexes in particular, have also been addressed, but such reports have only begun to appear much more recently,[27–33] and with little if any direct approach to interfacial electrochemical ET of the quadruplex targets themselves. As noted, perspectives of addressing the electrochemistry of the quadruplexes offer other perspectives: (a) the box-like qs structure, Fig. 1 offers a mechanistic perspective for electronic conductivity through a three-dimensional DNA-based structures more rigid than ds DNA; (b) the rigid "solid" structure may also hold a perspective for mapping potential induced structural reorientation of the surface immobilized qs DNA adlayers on the atomically planar Au(111)-electrode surfaces similar to what was reported for 20-base ds OGNs bound to spermidine[20] and for other biomolecules such as the amino acid homocysteine[59] and other complex molecules[60] as well as the nucleobases themselves[61]; (c) surface immobilized G-quadruplexes finally offer perspectives for ultra-sensitive detection of metal ions,[40–42] complex biomolecules such as peptides,[23–30] and perhaps other analytes.

In the present study we have addressed both the interfacial electrochemical behavior and single-molecule structural *in situ* STM imaging of the smallest three-quartet stacked 12G-qs immobilized on single-crystal Au(111)-electrode surfaces. The planar macrocyclic Fe(II)/(III) complex hemin terminally bound non-covalently to the quadruplex molecule, or possibly intercalated in the 12G-qs structure is exploited as a redox probe for interfacial electrochemical electron transfer. Biological buffers with pH around 7 were the reaction media.

Hemin itself was found to adsorb directly on the Au(111)-electrode surface and showed complex electrochemistry depending on pH, soaking concentration *etc.*, these results will be communicated elsewhere. A single monolayer voltammetric redox peak ascribed to the Fe(II)/(III) transition appears, however, at −0.05 V (NHE) at pH 7. Soaking at hemin concentrations in the sub-μM range gives a coverage of only a few per cent but soaking at high (10 μM) hemin concentration leads to a dense and highly ordered monolayer with a $c(4 \times 3\sqrt{3}) - 2$ surface lattice as clearly mapped by *in situ* STM.

12G-qs and the 12G-qs/hemin complex both adsorb on the Au(111)-electrode surface. Voltammetry of hemin-free 12G-qs shows a broad featureless capacitive background but also a conspicuous pair of voltammetric peaks at the significantly negative potentials, −0.4 to −0.55 V (*vs.* NHE). This pair resembles the capacitive peak reported earlier for 20-base pair DNA OGNs with bound spermidine.[20] The triply charged spermidinium cation increases both the structural stability and rigidity of the ds DNA backbone. In this respect the 12G-qs appears to resemble spermidinium stabilized ds DNA, and we tentatively assign the −0.4/−0.55 V peaks to potential induced 12G-qs adlayer reorientation. We have not, however, presently reached the same level of support for 12G-qs structural reorganization as for spermidinium bound ds 20-base pair OGNs.

Intriguingly, the hemin voltammetric peak intensity, sharpness, and stability all increase strongly compared with hemin physisorbed directly on the Au(111)-electrode surface, when hemin is bound to 12G-qs on the Au(111)-electrode surface. Close to full monolayer coverage of hemin bound 12G-qs emerges from the CVs, testifying to a close to full monolayer coverage of hemin-bound 12G-qs. The well-defined peaks enable estimating the interfacial electrochemical ET rate constant as $6.0 \pm 0.4\ s^{-1}$. This value is high, and as the terminal hemin binding site can be at either end of the 12G-qs target molecule, the standard rate constant can reflect ET of hemin bound either at the adjacent or the remote end of vertically bound or tilted 12G-qs. The electronic charge flow could thus reflect ET through the 12G-qs box, but this conclusion is not presently definite.

A dense monolayer of 12G-qs/hemin complex was supported by *in situ* STM. Domains of G-quartets/octagons ordered collectively possibly *via* H-bonded water molecules are clearly visible, Fig. 5. Formation of ordered adlayers is, however, highly sensitive to the preparation conditions, quality of the electrodes, and other controlling factors. Domains of linear structures, suggested to be single-strand unfolded 12G-qs molecules are thus visible together with the quartet structures in the same scan area, Fig. 5.

In conclusion, 12G-qs can be immobilized on well-defined Au(111)-electrode surfaces. A capacitive voltammetric peak at negative potential is indicative of potential induced structural surface reorientation in the adlayer similar to what was reported for spermidine bound 20-base pair ds OGN[20] although not as solidly supported. Hemin bound to surface immobilized 12G-qs structures displays well-defined faradaic voltammetry with much stronger and better defined hemin-based peaks than for hemin directly physisorbed on the Au(111)-surface and with fast interfacial electrochemical electron transfer.

The peak area reflects that a dense monolayer of 12G-qs/hemin DNAzyme has formed, supported by *in situ* STM imaging. Even though they are still voltammetrically highly active, these attractive looking structures are, however, fragile and at risk of decomposition into both unfolded but still laterally organized 12G single strands and less ordered surface adlayers of mixed single-entity structures and a disordered background. Work towards optimization of the imaging conditions is in progress. This work will address both tunnelling current/overpotential and current/bias voltage spectroscopy and also target qs structures with two hemin molecules bound. The latter approach would give a clearer view of long-range ET through the 12G-qs structure either by ET directly between the remote hemin group and the electrode surface or by sequential ET that would also involve electron exchange between the two bound hemin molecules.

Acknowledgements

LZ acknowledges the award of a H.C. Ørsted/Marie Curie Co-fund grant. Financial support from The Danish Council for Independent Research for the YDUN project (DFF 4093-00297) and the Lundbeck Foundation for project R141-2013-13273 to JZ is gratefully acknowledged. We wish to thank Professor Jesper Wengel at the Nucleic Acid Center, University of Southern Denmark for providing the thiolated target 12G strand.

References

1 Electrochemistry of Nucleic Acids and Proteins, *Towards Electrochemical Sensors for Genomics and Proteomics*, ed. E. Paleček, F. Scheller and J. Wang, Elsevier, Amsterdam, 2005.
2 A. Sassolas, B. D. Leca-Bouvier and J. L. Blum, *Chem. Rev.*, 2008, **108**, 109–139.
3 *Chapters in: DNA Conjugates and Sensors*, ed. K. R. Fox and T. Brown, RSC Publishing, London, 2012.
4 *Electrochemical DNA Biosensors*, ed. M. S. Ozsoz, Pan Stanford Publishing, Singapore, 2012.
5 H. B. Gray and J. R. Winkler, *Proc. Natl. Acad. Sci. U. S. A.*, 2005, **102**, 3534–3539.
6 J. J. Warren, M. E. Ener, A. Vlček Jr, J. R. Winkler and H. B Gray, *Coord. Chem. Rev.*, 2012, **256**, 2478–2487.
7 C. C. Moser, J. L. R. Anderson and P. L. Dutton, *Biochim. Biophys. Acta*, 2010, **1797**, 1573–1586.
8 J. R. Winkler, A. R. Dunn, C. R. Hess and H. B. Gray, in *Bioinorganic Electrochemistry*, ed. O. Hammerich and J. Ulstrup, Springer, Dordrecht, 2008, pp.1–23.
9 A. M. Kuznetsov and J. Ulstrup, in *Bioinorganic Electrochemistry*, ed. O. Hammerich and J. Ulstrup, Springer, Dordrecht, 2008, pp.161–205.
10 N. J. Turro and J. K. Barton, *J. Biol. Inorg. Chem.*, 1998, **3**, 201–209.
11 M. A. O'Neil and J. K Barton, in *Charge Transfer in DNA: From Mechanism to Application*, ed. H.-A. Wagenknecht, Wiley-VCH, Weinheim, 2004, pp. 27–75.
12 J. C. Genereux and J. K. Barton, *Chem. Rev.*, 2010, **110**, 1642–1661.
13 E. L. S. Wong, E. Chow and J. Gooding, *Langmuir*, 2005, **21**, 6957–6965.
14 J. Jortner, M. Bixon, T. Langenbacher and M. E. Michel-Beyerle, *Proc. Natl. Acad. Sci. U. S. A.*, 1998, **95**, 12759–12765.

15 M. Bixon, B. Giese, S. Wessely, T. Langenbacher, M. E. Michel-Beyerle and J. Jortner, *Proc. Natl. Acad. Sci. U. S. A.*, 1999, **96**, 11713–11716.

16 *Chapters in: Charge Transfer in DNA: From Mechanism to Application*, ed. H.-A. Wagenknecht, Wiley-VCH, Weinheim, 2004.

17 M. G. Hill and S. O. Kelley, in *Bioinorganic Electrochemistry*, ed. O. Hammerich and J. Ulstrup, Springer, Dordrecht, 2008, pp.129–160.

18 C. G. Pheenwy and J. K. Barton, *Langmuir*, 2012, **28**, 7063–7070, and references there.

19 N. B. Muren, E. D. Olmon and J. K. Barton, *Phys. Chem. Chem. Phys.*, 2012, **14**, 13754–13771.

20 P. Salvatore, K. K. Karlsen, A. G. Hansen, J. Zhang, R. J. Nichols and J. Ulstrup, *J. Am. Chem. Soc.*, 2012, **134**, 19092–19098.

21 A. A. Kornyshev and S. Leikin, *Phys. Rev. Lett.*, 2001, **86**, 3666–3669.

22 A. A. Kornyshev, D. J. Lee, S. Leikin and A. Wynveen, *Rev. Mod. Phys.*, 2007, **79**, 943–996.

23 S. Müller, S. Kumari, R. Rodriguez and S. Balasubramanian, *Nat. Chem.*, 2010, **2**, 1095–1098.

24 A. N. Lane, J. B. Chaires, R. D. Gray and J. O. Trent, *Nucleic Acids Res.*, 2008, **36**, 5482–5515.

25 D. Koirala, S. Dhakal, B. Ashbridge, Y. Sannohe, R. Rodriguez, H. Sugiyama, S. Balasubramanian and H. Mao, *Nat. Chem.*, 2011, **3**, 782–787.

26 T. Li, L. Shi, E. Wang and S. Dong, *Chem.–Eur. J.*, 2009, **15**, 1036–1042.

27 B. Ruttkay-Nedecky, J. Kudr, L. Nejdi, D. Maskova, R. Kizek and V. Adam, *Molecules*, 2013, **18**, 14760–14779.

28 O. Doluca, J. M. Withers and V. V. Filichev, *Chem. Rev.*, 2013, **113**, 3044–3083.

29 S. Liu, S. H. Weisbrod, Z. Tang, A. Marx, E. Scheer and A. Erbe, *Angew. Chem., Int. Ed.*, 2010, **49**, 3313–3316.

30 S. Liu, X. Zhang, W. Luo, Z. Wang, X. Guo, M. L. Sterigerwald and X. Fang, *Angew. Chem., Int. Ed.*, 2010, **50**, 2496–2502.

31 J. Choi, J. Park, A. Tanaka, M. J. Park, Y. J. Jang, M. Fujitsuka, S. K. Kim and T. Majima, *Angew. Chem., Int. Ed.*, 2013, **52**, 1134–1138.

32 C. J. Lech, A. T. Phan, M. E. Michel-Beyerle and A. A. Voityuk, *J. Phys. Chem.*, 2013, **117**, 9851–9856.

33 A. de Rache, T. Doneux and C. Buess-Hermann, *Anal. Chem.*, 2014, **86**, 8057–8065.

34 J. Zhang, Q. Chi, A. G. Hansen, P. S. Jensen, P. Salvatore and J. Ulstrup, *FEBS Lett.*, 2012, **586**, 526–535.

35 A. G. Hansen, P. Salvatore, K. K. Karlsen, R. J. Nichols, J. Wengel and J. Ulstrup, *Phys. Chem. Chem. Phys.*, 2013, **15**, 776–786.

36 J. Zhang, A. M. Kuznetsov, I. G. Medvedev, Q. Chi, T. Albrecht, P. S. Jensen and J. Ulstrup, *Chem. Rev.*, 2008, **108**, 2737–2791.

37 E. A. Della Pia, Q. Chi, J. E. Macdonald, J. Ulstrup, D. D. Jones and M. Elliott, *Nanoscale*, 2012, **4**, 7106–7113.

38 B. Bonanni, L. Andolfi, A. R. Bizarri and S. Cannistraro, *J. Phys. Chem. B*, 2007, **111**, 5062–5075.

39 J. M. Artés, Y. Li, J. Qi, M. P. Anantram and J. Hihath, *Nat. Commun.*, 2015, **6**, 8870, DOI: 10.1038/ncomms9870.

40 T. Li, E. Wang and S. Dong, *J. Am. Chem. Soc.*, 2009, **131**, 15082–15083.

41 G. Pelossof, R. Tel-Vered and I. Willner, *Anal. Chem.*, 2012, **84**, 3703–3709.

42 C. Zhang, L. Xie, L. Wang, H. Kong, Q. Tan and W. Xu, *J. Am. Chem. Soc.*, 2015, **137**, 11795–11800.
43 A. Hamelin, *J. Electroanal. Chem.*, 1996, **407**, 1–11.
44 Q. Chi, J. Zhang, E. P. Friis, J. E. T. Andersen and J. Ulstrup, *Electrochem. Commun.*, 1999, **1**, 91–96.
45 Q. Chi, J. Zhang, J. U. Nielsen, E. P. Friis, I. Chorkendorff, G. W. Canters, J. E. T. Andersen and J. Ulstrup, *J. Am. Chem. Soc.*, 2000, **122**, 4047–4055.
46 J. Zhang, A. Bilič, J. R. Reimers, N. S. Hush and J. Ulstrup, *J. Phys. Chem. B*, 2005, **109**, 15355–15367.
47 J. Zhang, Q. Chi, J. E. T. Andersen, E. P. Friis, J. U. Nielsen and J. Ulstrup, *Langmuir*, 2000, **16**, 7229–7237.
48 N. J. Tao, G. Cardenas, F. Cunha and Z. Shi, *Langmuir*, 1995, **11**, 4445–4448.
49 S. Yoshimoto, A. Tada and K. Itaya, *J. Phys. Chem. B*, 2004, **108**, 5171–5174.
50 S. Yoshimoto, N. Yokoo, T. Fukuda, N. Kobayashi and K. Itaya, *Chem. Commun.*, 2006, 500–502.
51 J. R. Reimers, D. Panduwinata, J. Visser, Y. Chin, C. Tang, L. Goerick, M. J. Ford, M. Sintic, T.-J. Sum, M. J. J. Coenen, B. L. M. Henriksen, J. A. A. W. Elemens, N. S. Hush and M. J. Crossley, *Proc. Natl. Acad. Sci. U. S. A.*, 2015, **112**, E6101–E6110.
52 S. Yoshimoto, A. Tada, K. Suto and K. Itaya, *J. Phys. Chem. B*, 2003, **107**, 5836–5843.
53 S. Yoshimoto, A. Tada and K. Itaya, *J. Phys. Chem. B*, 2004, **108**, 5171–5174.
54 J. Zhang, A. C. Welinder, Q. Chi and J. Ulstrup, *Phys. Chem. Chem. Phys.*, 2011, **13**, 5526–5545.
55 J. Zhang, A. C. Welinder, A. G. Hansen, H. E. M. Christensen and J. Ulstrup, *J. Phys. Chem. B*, 2003, **104**, 12480–12484.
56 V. Climent, J. Zhang, E. P. Friis, L. H. Østergaard and J. Ulstrup, *J. Phys. Chem. C*, 2012, **116**, 1232–1243.
57 T. Oshiro and M. Maeda, *Chem. Commun.*, 2010, **46**, 2581–2583.
58 T. Nishino and P. T. Bui, *Chem. Commun.*, 2013, **49**, 3437–3439.
59 J. Zhang, A. Demetriou, A. C. Welinder, T. Albrecht, R. J. Nichols and J. Ulstrup, *Chem. Phys.*, 2005, **319**, 210–221.
60 G. J. Su, H. M. Zhang, L. J. Wan, C. L. Bai and T. Wandlowski, *J. Phys. Chem. B*, 2004, **108**, 1931–1937.
61 M. H. Holzle, T. Wandlowski and D. M. Kolb, *J. Electroanal. Chem.*, 1995, **394**, 271–275.

Faraday Discussions

PAPER

Evidence for a hopping mechanism in metal|single molecule|metal junctions involving conjugated metal–terpyridyl complexes; potential-dependent conductances of complexes $[M(pyterpy)_2]^{2+}$ (M = Co and Fe; pyterpy = 4′-(pyridin-4-yl)-2,2′:6′,2″-terpyridine) in ionic liquid†‡

Sarah Chappell,[a] Carly Brooke,[a] Richard J. Nichols,[a] Laurence J. Kershaw Cook,[b] Malcolm Halcrow,[b] Jens Ulstrup[c] and Simon J. Higgins*[a]

Received 15th April 2016, Accepted 25th April 2016
DOI: 10.1039/c6fd00080k

Extensive studies of various families of conjugated molecules in metal|molecule|metal junctions suggest that the mechanism of conductance is usually tunnelling for molecular lengths < *ca.* 4 nm, and that for longer molecules, coherence is lost as a hopping element becomes more significant. In this work we present evidence that, for a family of conjugated, redox-active metal complexes, hopping may be a significant factor for even the shortest molecule studied (*ca.* 1 nm between contact atoms). The length dependence of conductance for two series of such complexes which differ essentially in the number of conjugated 1,4-C_6H_4- rings in the structures has been studied, and it is found that the junction conductances vary linearly with molecular length, consistent with a hopping mechanism, whereas there is significant deviation from linearity in plots of log(conductance) *vs.* length that would be characteristic of tunnelling, and the slopes of the log(conductance)–length plots are much smaller than expected for an oligophenyl system. Moreover, the conductances of molecular junctions involving the redox-active molecules, $[M(pyterpy)_2]^{2+/3+}$ (M = Co, Fe) have been studied as a function of electrochemical potential in ionic liquid electrolyte, and the conductance–overpotential relationship is found to fit well with the Kuznetsov–Ulstrup relationship, which is essentially a hopping description.

[a]*Department of Chemistry, University of Liverpool, Crown Street, Liverpool L69 7ZD, UK. E-mail: shiggins@liverpool.ac.uk*
[b]*School of Chemistry, University of Leeds, Leeds LS2 9JT, UK*
[c]*Department of Chemistry, Technical University of Denmark, Kemitorvet, DK-2800 Kgs. Lyngby, Denmark*

† Electronic supplementary information (ESI) available. See DOI: 10.1039/c6fd00080k

‡ Data supporting this study has been archived; see DOI: 10.17638/datacat.liverpool.ac.uk/186

Introduction

The development of techniques that allow the electrical properties of molecules to be investigated in metal|molecule|metal 2-terminal junctions down to the single-molecule level has stimulated renewed interest in truly 'molecular' electronics in the last decade.[1-3] Early work focused mainly on testing different molecular backbones and metal–molecule contact groups to maximise molecular junction conductance.[4] For molecules that conduct by a tunnelling mechanism, the tunnelling current I_T decays exponentially with molecular length according to the equation $I_T = Ae^{-\beta n}$ where the factor A is largely governed by the nature of the contact, β is called the decay constant and is mainly determined by the nature of the molecular backbone, and n can be expressed either as the number of repeat units in oligomers of varying length, or simply as the length of the molecule (in which case the units of β are $(\text{length})^{-1}$). Since the Fermi energy of the metal contacts must lie between the HOMO and LUMO levels of the molecule, it follows that in general, the lower the HOMO–LUMO separation, the more slowly the molecular conductance falls with length (the smaller is β). The lowest values of β found for purely organic, non-redox-active molecules so far have been with oligo-carbynes ($\beta = 0.6 \pm 0.3$ nm^{-1}).[5] For some conjugated redox-active molecules, such as 'fused-tape' oligoporphyrins (0.19 ± 0.01 nm^{-1})[6] and oligoviologens (0.06 ± 0.004 nm^{-1}),[7] even lower values have been found, but in some of these cases, it is difficult to distinguish between an exponential decay of conductance with length involving a very low β (tunnelling), and a linear decay with length expected for a 'hopping' mechanism.

There is considerable interest in molecules with electrical properties that may be modulated by an external influence, for instance by electrostatic 'gating' from a third electrode,[8,9] by irradiation with light,[10,11] or by a chemical binding event.[12,13] In the longer term, this will be important for molecular-scale devices or sensors. We have had a long-standing interest in the control of single molecule junction conductance using electrochemical gating.[14,15] The use of a bipotentiostat and an electrolyte medium offers a way of overcoming the technical challenges inherent in fabricating a 3-terminal nanoscale device in which a molecule is in close enough proximity to the electrostatic gate to be switched by its electric field.[16] In strong electrolytes the Debye screening length can be extremely short, so although the counter and reference electrodes are physically far from the molecular bridge, any charge on the molecular bridge is screened by charge in solution at a distance much shorter than can usually be achieved in a solid state device. The result is a good and, significantly, a reproducible gate field. Using this approach, the junction conductances involving a structurally-diverse range of organic redox molecules (*e.g.* bis(thiaalkyl)viologens,[14,17,18] bis(thiahexyl)pyrrolo-tetrathiafulvalene (6PTTF6),[15,17] oligoanilines[19,20] and perylenetetracarboxylicdiimides[21,22]) have been studied as a function of electrochemical potential. Two distinct forms of behaviour were observed. In one case, the conductance changed monotonically as the redox process was traversed over a rather broad voltage range (denoted 'off-on' behaviour).[14,21,22] More rarely, the conductance rose to a peak value as the half-wave potential was approached, then fell again beyond it ('off-on-off' behaviour).[15,17,19] Attempts were made to model this theoretically, but it was not clear what factors were at work in determining the form of behaviour observed for a particular molecule or medium employed.[23,24]

We, and others, have previously shown that ionic liquids (*e.g.* 1-butyl-3-methylimidazolium triflate or bis(trifluoromethyl-sulfonyl)imide salts, BMIM-OTf and BMIM-TFSI respectively) are useful media for single molecule conductance studies under potential control.[15] Their negligible vapour pressure makes them ideally suited for use with STM instrumentation and they have significantly wider potential windows than aqueous electrolytes. We have recently demonstrated that the conductance-potential behaviour observed for a small family of viologen redox molecules depends not upon the molecular structure, but rather, upon the degree of gate coupling mediated by the electrolyte, denoted in the well-known Kuznetsov–Ulstrup model by the gate coupling parameter ξ.[25] In aqueous electrolytes, the viologens displayed 'off-on' behaviour (as observed in earlier work[14,17]) and ξ was 0.2, whereas in ionic liquids, they displayed 'off-on-off' behaviour and ξ was 1.

Although the great majority of single molecule junction studies has to date involved organic molecules, the incorporation of metal centres into 'molecular wires' could offer significant potential advantages, for example in providing additional frontier orbital energies better-matched to the metal contact Fermi energies (in concert with reversible redox activity),[26,27] or stable paramagnetic centres with a potential role in molecular spintronics.[28] Two families of metal complexes in particular have often been put forward as candidates for molecular electronics (although actual metal|molecule|metal device studies on these systems remain sparse). The first is metal–alkynyl organometallic complexes, including multimetallic examples with bridging oligophenyleneethynylene (OPE) or oligocarbyne ligands.[29,30] These could provide very long, rigid-rod molecules, readily functionalised with terminal contact groups which are straightforward to characterise by NMR spectroscopy since they are diamagnetic.[31–33] Foci of interest with these molecules include comparison of their electrical properties with non-metal-containing OPE molecules to probe the influence of the metal(s),[34,35] and the transition between tunnelling and hopping conductance mechanisms as a function of length.[36] The second family is derivatives of the ligand 2,2′:6′,2″-terpyridine (terpy), which have an extensive coordination chemistry; the ligand structure is amenable to modification (for example to incorporate potential contact groups for junction formation), their metal complexes often have reversible redox activity and they can form relatively inert paramagnetic metal complexes amenable to junction formation experiments. For instance, oligomeric complexes of 'back-to-back' binucleating derivatives of terpy such as ligands **B** and **C** (Scheme 1) have been built up as monolayers on electrode surfaces by alternating reactions with excess ligand followed by excess metal ion (typically Co(II) or Fe(II)), and their redox behaviour has been characterised.[37–39] The electrical properties of some of these monolayers have been determined as a function of molecular length by using a macroscopic mercury electrode to make contact with the top of the monolayer. Remarkably low β values were claimed in this work, and these were metal-dependent (for Fe(II), 0.28 nm^{-1} and for Co(II), 0.01 nm^{-1}). This was ascribed to the operation of an intramolecular hopping mechanism.[37] Single-molecule conductance studies involving metal–terpy complexes are a very recent development,[27] although thiolated $[Co(terpy)_2]^{n+}$ derivatives were early subjects of

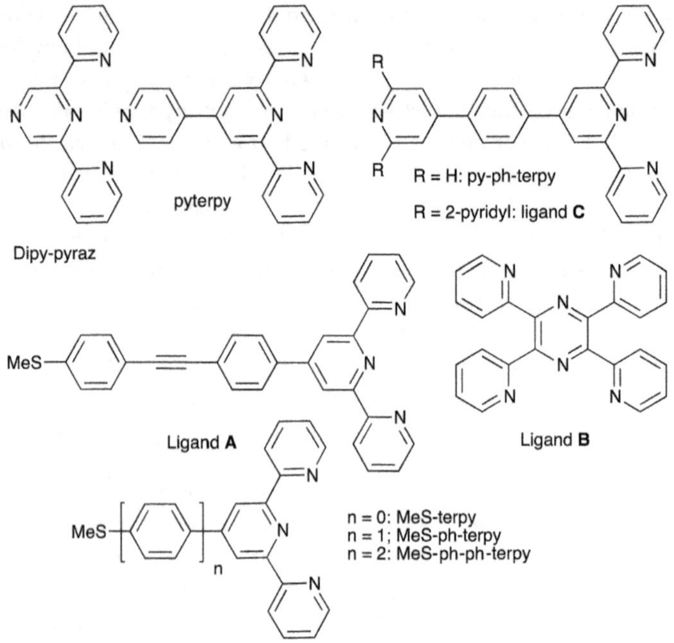

Scheme 1 Ligands discussed in this work.

a low-temperature Coulomb blockade and Kondo effect study using break junctions made by electromigration.[8] Using pre-synthesised mono- and dimetallic complexes of ruthenium(II), with combinations of ligands such as **A** (Scheme 1) and the back-to-back terpy derivative ligand **B** to afford thiomethyl-terminated complexes for junction formation, Davidson *et al.* found a β value of 1.5 nm^{-1} for this system, although unfortunately, synthetic limitations (ligand scrambling) precluded the selective syntheses of trimetallic or longer examples.[27]

Complexes of the ligand 4′-(pyridin-4-yl)-2,2′:6′,2″-terpyridine (pyterpy)[40] have been extensively used in supramolecular chemistry, for instance in polyheterometallic complexes in which octahedral $[M(pyterpy)_2]^{n+}$ complexes are coupled through subsequently-induced metal–pyridin-4-yl coordination,[41] in 'metalloviologen' compounds in which the pendant pyridine-4-yls are quaternised,[42] and in self-assembled monolayers where they are used to bind to a metal surface.[39] In this contribution, we report single molecule conductance studies on complexes $[M(pyterpy)_2]^{n+}$ ($n = 2$; M = Co, Fe, Ru: $n = 3$; M = Cr) and of some derivatives of pyterpy with varying molecular length, using the pendant pyridine-4-yl units as binding sites for gold|terpyridyl complex|gold junctions. For $[M(pyterpy)_2]^{2+}$ (M = Co and Fe), we have examined the potential-dependent conductance of these complexes over their M^{II}/M^{III} redox waves in an ionic liquid medium. To test the effect of changing the binding group, we have also examined the length dependence of conductance for a related family of methylthioether-contacted terpy derivatives.

Results and discussion

Synthesis and characterisation of the complexes

The ligand pyterpy[43] and its complexes [M(pyterpy)$_2$](PF$_6$)$_2$ (M = Co,[44] Fe[40] or Ru[45]) were synthesised by literature methods, or adaptations thereof. In addition, [Co(pyterpy)$_2$](PF$_6$)$_3$ was prepared by Br_2 oxidation of the Co(II) complex to allow characterisation by NMR spectroscopy. The complex [Cr(pyterpy)$_2$](PF$_6$)$_3$ was prepared from $CrCl_2$ and pyterpy, followed by metathesis with NH_4PF_6 and oxidation with $AgPF_6$, a method previously used to prepare [Cr(terpy)$_2$](PF$_6$)$_3$.[46] However, attempts to obtain a mass spectrum of this complex using the electrospray method failed, and we were unable to obtain microanalyses owing to the high percentage of fluorine in the complex. Characterisation was therefore restricted to a solution UV-visible spectrum, which showed similar charge transfer bands to those observed for $[Cr(terpy)_2]^{3+}$, and cyclic voltammetry; the complex showed a reversible 3+/2+ redox wave at −0.54 V *vs.* Fc/Fc$^+$ (CH_3CN/0.1 M Bu_4NBF_4), slightly anodic of the corresponding process for $[Cr(terpy)_2]^{2+/3+}$ at −0.63 V (ref. 46) as expected for the somewhat more electron-deficient ligand. The 'extended pyterpy' ligand 4′-(4-(pyridin-4-yl)phenyl)-2,2′:6′,2″-terpyridine (py-ph-terpy) was prepared by the route shown in Scheme 2; Suzuki coupling between 4-formylphenylboronic acid and 4-bromopyridine hydrochloride yielded 4-(pyridin-4-yl)benzaldehyde in 47% yield. Subsequent reaction with two equivalents of 2-acetylpyridine followed in one pot by oxidative ring closure with ammonium acetate, afforded the target molecule in a yield of 34%. This compound has previously been prepared in similar yield by Suzuki coupling of 4′-(4-bromophenyl)-2,2′:6′,2″-terpyridine with pyridin-4-ylboronic acid.[47]

A shorter analogue of pyterpy, 2,6-di(pyridin-2-yl)pyrazine (dipy-pyraz; Scheme 1), was also prepared as described previously.[48] Ruthenium(II) complexes of dipy-pyraz and of py-ph-terpy, were prepared as hexafluorophosphate salts, from the ligand and '$RuCl_3 \cdot 3H_2O$' in hot ethylene glycol followed by metathesis with NH_4PF_6 (Experimental). Both the ligands and their Ru(II) complexes were characterized by ^{1}H NMR spectroscopy, electronic spectroscopy, microanalyses and mass spectrometry before single molecule junction experiments.

The ligands MeSterpy and MeS-ph-terpy (Scheme 1) were prepared using literature methods,[49,50] as were their Ru(II) complexes (as hexafluorophosphate salts).[50] The new extended analogue 4′-(4′-(methylthio)-[1,1′-biphenyl]-4-yl)-2,2′:6′,2″-terpyridine (MeS-ph-ph-terpy) was prepared (albeit in low yield) by Suzuki coupling of 4′-(4-bromophenyl)-2,2′:6′,2″-terpyridine[51] with (4-(methylthio)

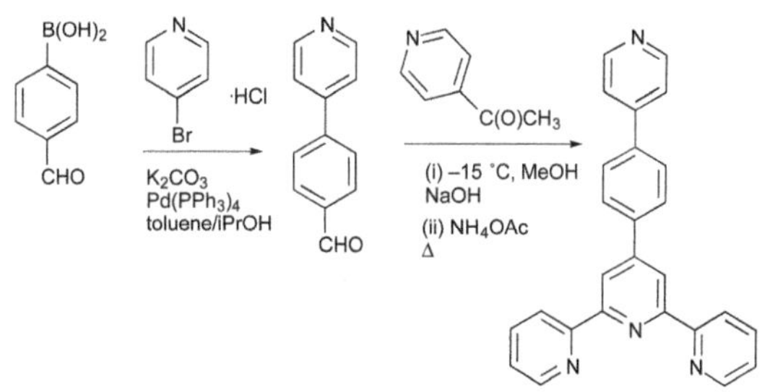

Scheme 2 Synthesis of py-ph-terpy.

phenyl)boronic acid. The corresponding Ru(II) complex [Ru(MeS-ph-ph-terpy)$_2$](PF$_6$)$_2$ was then prepared using the same conditions as for the other thiomethyl-terpy derivatives.

Single molecule junction conductance measurements in ambient conditions

To measure the molecular junction conductances, we used the $I(s)$ technique developed by Haiss *et al.*[4,14] In this technique, contact between the Au STM tip and the surface is avoided. The tip is moved to a fixed distance above the Au substrate using the set point current, the feedback loop is disconnected and the tip is then retracted at constant speed while the tunnelling current I_s is monitored. Submonolayer coverages of molecules were previously deposited on the substrates by briefly dipping them in dilute solutions of the appropriate molecule (see Experimental for details). The formation of a metal|molecule|metal junction is indicated by the presence in the current–distance (I_s–s) plot of a plateau. Pyridyl contact groups have been shown to result in multiple conductance values for the same molecule depending upon the precise experimental conditions used.[5,52] In this work, we chose to use the conditions that we previously found to result in the lowest of the three conductance values seen for 4,4′-bipyridine.[5] In brief preliminary experiments with 4,4′-bipyridine as a model compound, the value found here (Table 1) was close to that found previously ($1.34 \pm 0.17 \times 10^{-4} G_0$).[5] The 'hit rate' (the proportion of I–s retraction experiments resulting in a plateau) was approximately 10% with the set point current I_0 (40 nA) and bias V (0.6 V) used here.

We then measured the conductances of junctions with each of the metal-pyterpy complexes, again measured in ambient with the same bias and I_0 as for 4,4′-bipyridine. Fig. 1 shows a typical data set, that of [Fe(pyterpy)$_2$](PF$_6$)$_2$ (similar plots for the other complexes are presented in the ESI†). The hit rate for the complexes was somewhat lower than for 4,4′-bipyridine, approximately 5%. The conductance values for [M(pyterpy)$_2$](PF$_6$)$_2$ (M = Fe, Ru and Co) are the same within experimental uncertainty, whereas the value for [Cr(pyterpy)$_2$](PF$_6$)$_3$ is somewhat lower (Table 1). Whereas both the Fe(II) and Ru(II) complexes are low-spin d^6, [Co(pyterpy)$_2$]$^{2+}$ is d^7 and, like some other Co(II)–terpy derivatives, it exhibits spin-crossover behaviour.[53] The Jahn–Teller-distorted, low-spin (^{2}E) state predominates at low temperature (110 K).[53,54] The form of the temperature-dependent paramagnetism in the solid state is anion- and solvate-dependent; for

Table 1 Conductance values for all molecules measured in ambient conditions

Compound	Conductance (G_0)	Expt. break-off (theoretical) (nm)
4,4′-Bipyridine	$(1.0 \pm 0.3) \times 10^{-4}$	1.34 ± 0.24 (1.1)
[Fe(pyterpy)$_2$](PF$_6$)$_2$	$(4.4 \pm 1.3) \times 10^{-5}$	1.9 ± 0.3 (2.2)
[Co(pyterpy)$_2$](PF$_6$)$_2$	$(4.5 \pm 1.4) \times 10^{-5}$	1.95 ± 0.25 (2.2)
[Cr(pyterpy)$_2$](PF$_6$)$_2$	$(3.3 \pm 0.6) \times 10^{-5}$	1.92 ± 0.35 (2.2)
[Ru(dipy-pyraz)$_2$](PF$_6$)$_2$	$(8.6 \pm 1.9) \times 10^{-5}$	2.1 ± 0.4 (1.4)
[Ru(pyterpy)$_2$](PF$_6$)$_2$	$(4.8 \pm 1.2) \times 10^{-5}$	2.04 ± 0.27 (2.2)
[Ru(py-ph-terpy)$_2$](PF$_6$)$_2$	$(1.4 \pm 0.3) \times 10^{-5}$	3.0 ± 0.4 (3.1)
[Ru(MeSterpy)$_2$](PF$_6$)$_2$	$(9.6 \pm 2.3) \times 10^{-5}$	2.5 ± 0.5 (1.6)
[Ru(MeS-ph-terpy)$_2$](PF$_6$)$_2$	$(5.2 \pm 1.9) \times 10^{-5}$	2.2 ± 0.4 (2.4)
[Ru(MeS-ph-ph-terpy)$_2$](PF$_6$)$_2$	$(1.5 \pm 0.5) \times 10^{-5}$	3.2 ± 0.3 (3.3)

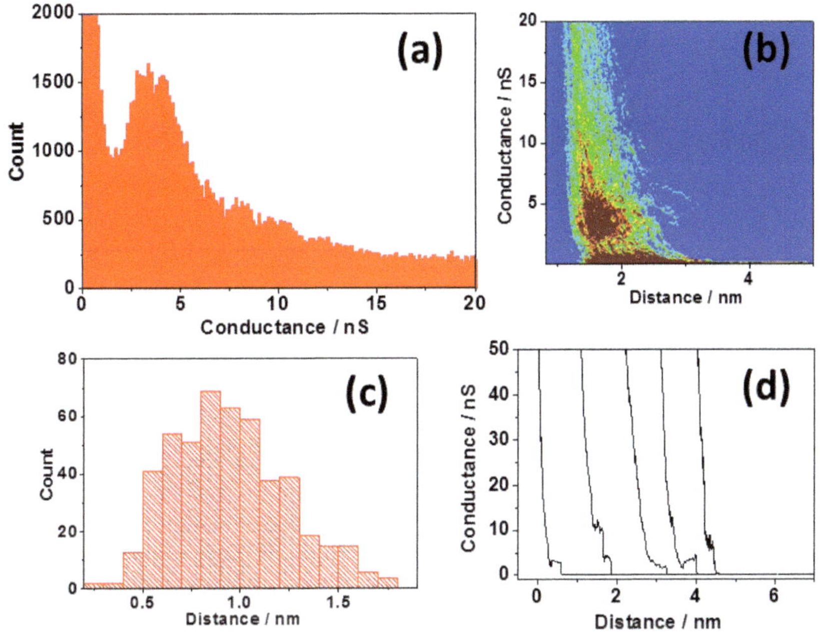

Fig. 1 *I*(*s*) data recorded for $[Fe(pyterpy)_2](PF_6)_2$ in ambient conditions; $I_0 = 40$ nA, $V_{bias} = 0.6$ V. (a) One-dimensional histogram of the 510 *I*(*s*) traces collected that showed plateau(s), (b) two-dimensional representation of conductance *vs.* corrected break-off distance for all plateau(s)-containing traces, (c) histogram of uncorrected break-off distances for all plateau(s)-containing traces, (d) some representative plateau(s)-containing *I*–*s* traces.

the unsolvated PF_6 salt, 56% of the complex is in the high-spin state at room temperature.[53] The Co(II)/Co(III) redox potential for this complex is −0.04 V *vs.* Fc/Fc^+ (q.v.) and it is possible that on Au(111) in air, it becomes oxidised to Co(III); the Co(III) complex is also low-spin d^6, which would mean that all three complexes in the conductance measurements are structurally- and electronically-similar, accounting for the similar conductance values. Attempts to determine the conductance of junctions starting with $[Co(pyterpy)_2](PF_6)_3$ were unsuccessful; it appears that this complex does not significantly adsorb on gold from those solvents in which it is soluble. It is interesting in this respect that the d^3 Cr(III) complex is the only example in this group with a significantly different junction conductance.

The conductances of the Ru(II) complexes with the pyterpy analogues, $[Ru(dipy\text{-}pyraz)_2]^{2+}$ and $[Ru(py\text{-}ph\text{-}terpy)_2]^{2+}$ were similarly determined in ambient (Table 1). The longest complex necessitated the use of a smaller I_0, 20 nA. Plots of conductance *vs.* molecular length and of log(conductance) *vs.* molecular length, with experimental uncertainties denoted as error bars, are shown in Fig. 2. It should be noted that the plot of conductance *vs.* length is clearly linear, while the plot of log(conductance) *vs.* length deviates significantly from linearity. This may be preliminary evidence that a hopping-type mechanism is operating with these molecules, as suggested earlier for oligomeric metal complexes with 'back to back' terpy ligands.[37] A straight line of best fit for the log plot yields a β value of

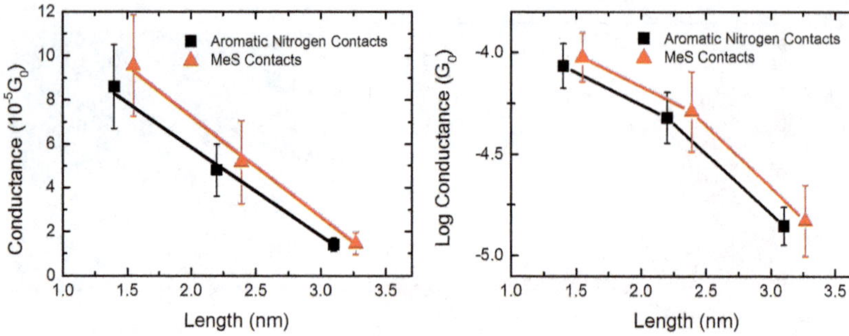

Fig. 2 Conductance *vs.* length (left) and log(conductance) *vs.* length (right), for the ruthenium(II) complexes with aromatic nitrogen contacts, and with MeS contacts.

1.06 nm^{-1}. This low β value can be compared with that previously found in STM break junction experiments, of 4.3 nm^{-1} for both amine- and carbon-contacted oligophenyls.[55] The fact that β is much smaller than that for the purely organic systems, in spite of the fact that in both cases the molecules are related by the successive extension of a polyphenylene-type chain, again suggests a hopping-type mechanism may be operative for this series of ruthenium complexes.

To obtain more data on this issue, we extended the work to encompass a related series of thioether-contacted terpy derivatives (Scheme 1; Table 1), conveniently available by straightforward synthetic routes. The Ru(II) complexes of these ligands all gave good results in *I*(*s*) experiments under ambient conditions, producing a slightly higher hit rate (10–15%) than the terpy-contacted molecules with otherwise identical experimental parameters. The conductance values are given in Table 1, and Fig. 2 shows both linear and semi-log plots of conductance *vs.* molecular length. Again, it is clear that the linear fit is appreciably better than the semilogarithmic one, consistent with the notion of a hopping mechanism in this system. The linear best fit of the semilogarithmic plot gives a β value of 1.09 nm^{-1}, very similar to the value for the pyridyl-contacted series.

In all of these experiments, we also determined the mean break-off distances (the height *s* at which the junction breaks down), using a published method[56] (described in the ESI†), and found that in most cases it was consistent with the full length of the molecules as calculated using molecular mechanics (Spartan14© version 1.1.8), implying that the conductance is characteristic of the fully extended molecules for this set of complexes. Very short molecules, such as 4,4′-bipyridine, $[Ru(dipy\text{-}pyraz)_2]^{2+}$ and $[Ru(MeSterpy)_2]^{2+}$ in this work, often give apparent break-off distances longer than the full molecular length with this method; this could be due to significant deviation from linearity in d(ln *I*)/d(*s*) at very short tip-sample separations in ambient conditions.

Single molecule junction conductance measurements on $[M(pyterpy)_2](PF_6)_2$ (M = Co, Fe) under potential control

We chose 1-(*n*-butyl)-3-methylimidazolium bis(trifluoromethylsulfonyl)imide (BMIM-TFSI) as the medium for studying the conductances of $[M(pyterpy)_2](PF_6)_2$ (M = Co, Fe) under electrochemical potential control. First, we examined the

M(II)/M(III) redox waves of the complexes in BMIM-TFSI as a function of scan rate (Fig. 3).

The Fe(II)/Fe(III) process ($E_{1/2}$ = +0.78 V *vs.* Fc/Fc$^+$) is chemically reversible and the peak to peak separation increases only from 60 mV at 0.02 V s^{-1} to 75 mV at 1.0 V s^{-1}. The Co(II)/Co(III) process ($E_{1/2}$ = −0.04 V *vs.* Fc/Fc$^+$) is also chemically reversible, although clearly with slower electron transfer kinetics since the peak-to-peak separation increases from 80 mV at 0.02 V s^{-1} to 240 mV at 1.0 V s^{-1}. This is as expected since related low spin d^6/d^5 couples generally have small metal–ligand bond length differences and hence, low inner sphere reorganisation energies, while the d^7/low-spin d^6 process will involve a significant Co–N bond length change irrespective of whether the Co(II) state is high- or low-spin.[57]

Given the long timescale for collection of sufficient *I*(*s*) data to compile meaningful conductance histograms at many different electrode potentials, the fact that both redox forms are chemically stable on the experimental timescale in the ionic liquid, as demonstrated here, is important. *I*(*s*) data and corresponding conductance histograms were therefore collected for both complexes as a function of electrochemical potential across the M(II)/M(III) redox waves. The results are summarised in Fig. 4 (histograms for individual experiments at each potential are presented in the ESI†). It is clear from the results that in ionic liquid electrolyte, both the Fe and Co complexes show 'off-on-off' behaviour, as observed earlier for 6PTTF6 in BMIM-TFSI across both of its redox waves (neutral/radical cation, and radical cation/dication),[15] and for viologens across the radical cation/dication redox wave.[25]

As part of an earlier study of the relationship between molecular conductances and electrochemical rate constants for structurally-varied redox-active molecules, Zhou *et al.* studied the conductance of junctions of the complex [Os(py-terpy)$_2$](PF$_6$)$_2$ in aqueous 1 M NaClO$_4$ as a function of potential, using a technique very similar to the one employed here.[58] In their experiments, this complex showed 'on-off' conductance switching; the conductance changed sigmoidally from 23.1 × 10^{-5}G_0 (17.8 nS) in the Os(II) redox state to 2.70 × 10^{-5}G_0 (2.1 nS) in the Os(III) state. In contrast, we find that in the ionic liquid, the isoelectronic Fe complex switches from 2.70 × 10^{-5}G_0 (2.1 nS) in the 'off' state to *ca.* 8.4 × 10^{-5}G_0 (6.5 nS) in the 'on' state, then back to 2.70 × 10^{-5}G_0 (2.1 nS) as the complex is fully oxidised to Fe(III). Similarly, the Co complex switches from 4.5 × 10^{-5}G_0 (3.5 nS) in the 'off' state, to *ca.* 10.0 × 10^{-5}G_0 (7.7 nS) in the 'on' state, and back to 4.5 ×

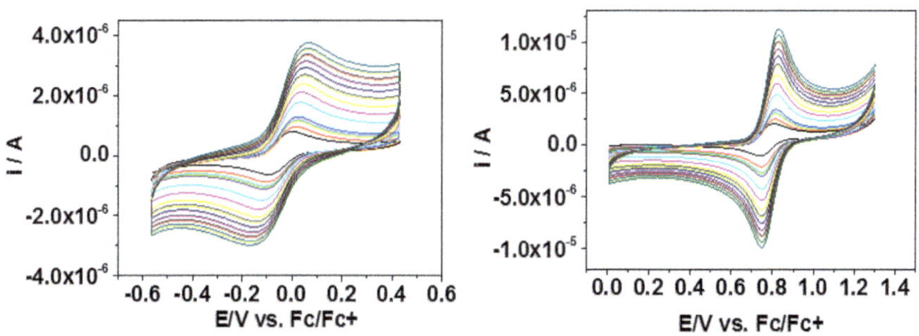

Fig. 3 Cyclic voltammograms of (left) [Co(pyterpy)$_2$](PF$_6$)$_2$ and (right) [Fe(pyterpy)$_2$](PF$_6$)$_2$ in BMIM-TFSI, scan rates 0.02, 0.05, 0.08, then 0.1–1.0 V s^{-1} in 100 mV intervals.

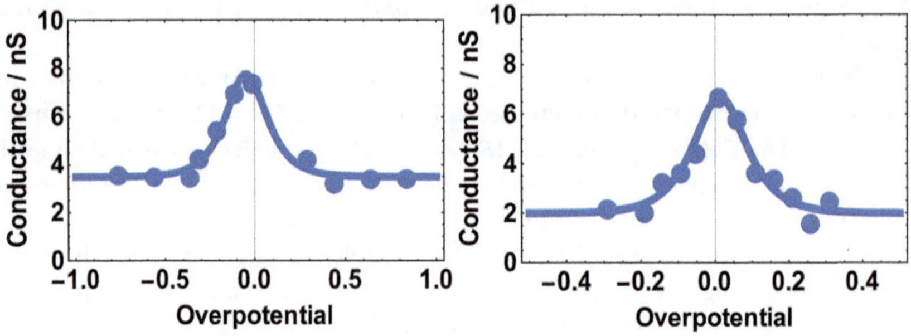

Fig. 4 Conductance–overpotential plots for $[Co(pyterpy)_2]^{2+/3+}$ (left) and $[Fe(pyterpy)_2]^{2+/3+}$ (right) in BMIM TFSI. Data at each potential shown as points; the line corresponds to the fitting to the Kuznetsov–Ulstrup model, q.v.

$10^{-5}G_0$ (3.5 nS) as the complex is fully oxidised to Co(III). This is similar to the situation for viologens, which show 'off-on-off' behaviour in ionic liquid but 'off-on' behaviour in aqueous electrolytes.

We have modelled the electrochemical single model conductance switching data for $[Co(pyterpy)_2]^{2+/3+}$ with the Kuznetsov–Ulstrup 2-step charge transfer model. This model is illustrated in Fig. 5. As shown in this figure the molecule bridges between the two electrodes (the gold substrate surface and gold STM tip), with the metal redox centre tethered in the middle of this nano-gap. Hole transfer occurs through this redox centre in two electrochemical steps in what can be described as an electrochemical charge-hopping model. In this model the hole tunnels from the left electrode to the metal centre following pre-organisation of the molecule and its environment so that charge transfer can follow a Frank–Condon-type transition. The molecule and its environment then relax, with the

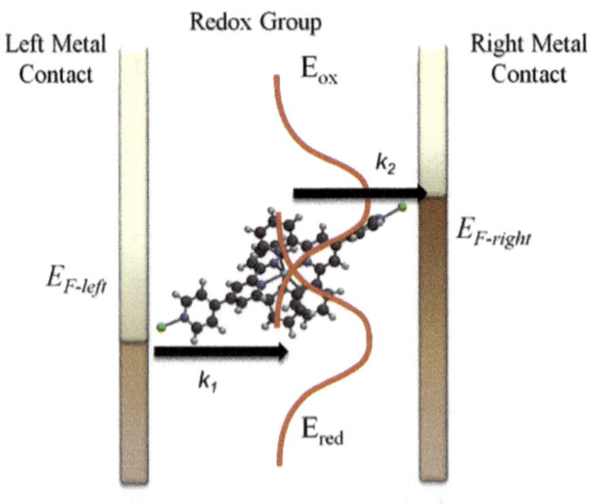

Fig. 5 An illustration of the Kuznetsov–Ulstrup model showing $[M(pyterpy)_2]^{2+/3+}$ attached between two gold electrode, labelled left and right respectively. The redox states of the molecule are illustrated in red with charge transfer occurring through these through two sequential electron transfer processes labelled k_1 and k_2.

hole losing coherence during this relaxation. The hole then tunnels from the metal centre to the right electrode. In the adiabatic limit, which is used here in the modelling, the relaxation prior to the second hole transfer step is partial. Following the first hole transfer the M(II) centre is oxidised and then subsequently reduced following charge transfer to other electrode. In the limit of strong electronic interactions between the redox centre and the electrodes the steady state current flow through the molecular junction is given by:

$$i = 2en\frac{k_1k_2}{k_1 + k_2}$$

where n represents the number of holes or electrons that can transfer in a "cascade" while the redox centre and its environment relaxes, and e is the electronic charge. k_1 and k_2 are electrochemical rate constant expressions.

Ref. 59 describes the resulting equation which models the enhanced current flowing across the molecular junction (j_{enh}) following the two-step adiabatic model of Kuznetsov and Ulstrup described above.

$$j_{\text{enh}} \approx j_0 \exp(-\lambda/4k_\text{B}T)\frac{\exp(e|V_{\text{bias}}|/4k_\text{B}T)}{\cosh\left(\dfrac{e(0.5-\gamma)V_{\text{bias}} - e\xi\eta}{2k_\text{B}T}\right)}$$

In this equation k_B is the Boltzmann constant, V_{bias} the bias voltage, T the temperature and e the charge on an electron. The overpotential is represented as η while γ is a modelling parameter representing the fraction of the bias voltage dropped at the redox site. The total reorganization energy is λ, which includes both inner and outer-sphere contributions. The fraction of the electrochemical potential experienced at the redox site is ξ. An expression for the pre-factor, j_0, can be found in ref. 59:

$$j_0 = en\omega_{\text{eff}}/2\pi$$

with

$$n \approx eV_{\text{bias}}\left(\frac{1}{2\kappa_\text{L}\rho_\text{L}} + \frac{1}{\kappa_\text{R}\rho_\text{R}}\right)^{-1}$$

The additional terms here are as follows: κ is the electron transmission coefficient, ρ is the density of electronic states in the metal electrodes near the Fermi level and ω_{eff} is the effective nuclear vibrational frequencies. The subscripts L and R refer to the left and right electrodes, respectively. Using these equations a numerical form for the enhanced current flowing across the molecular junction (j_{enh}) can be obtained and single molecule conductance data *vs.* electrochemical potential can then be quantitatively analysed with respect to this model.

Fig. 4 shows the results of such analyses for $[M(pyterpy)_2]^{2+/3+}$ (M = Fe, Co). The points in this plot show the experimental single molecule conductance data while the line is the fitting to the Kuznetsov–Ulstrup 2-step model in the adiabatic limit. The fitting parameters for the cobalt complex are $\lambda = 0.80$ eV, $\xi = 0.5$ and $\gamma = 0.40$, while those for the iron complex are $\lambda = 0.77$ eV, $\xi = 0.8$ and $\gamma = 0.55$. The reorganization energies determined here are considerably less than previously found for a viologen single molecule bridge (1.3 eV) also determined in ionic liquid electrolytes.[25] The viologen single molecule bridge used in the latter work was called "6V6", where "V" refers to the viologen (bipyridinium) core which was attached to gold contacts at both ends through $-(CH_2)_6S-$ groups (denoted here as "6"). The reorganisation energies of the pyterpy complexes obtained here are also considerably less than a similar analysis for both the first and second redox transitions for a 6PTTF6 bridge in ionic liquid environment, which gave values of $\lambda = \sim 1.2$ eV (6PTTF6 refers to the redox-active pyrrolo-tetrathiafulvalene (PTTF) moiety attached to gold contacts at both ends through $-(CH_2)_6S-$ groups). It is also noteworthy that in the 6V6 and 6PTTF6 examples $\xi = \sim 1$ was obtained. This value corresponds to the full electrochemical potential being experienced at the redox centre, indicative of the short Debye screening lengths of the ionic liquids and effective structuring of the ionic liquid in the nano-gap junction. In the case of the pyterpy studied here, ξ is lower indicating that the screening is not so effective. This may be due to the more "voluminous" redox centre with the metal ion redox centre being surrounded by a bulky ligand shell.

Conclusions

In a previous $I(s)$ study of four related mono- and bi-metallic MeS-contacted Ru(II)-terpy complexes (including $[Ru(MeS\text{-}ph\text{-}terpy)_2]^{2+}$), it was found that a semi-logarithmic plot of junction conductance *vs.* molecular length was approximately linear, with a decay constant β of 1.5 nm^{-1}, and a tunnelling mechanism was therefore suggested.[27] However, these four molecules did not comprise a truly homologous series. In the present study, we have examined two homologous series, one with pyridyl and the other with MeS contacts, and both show a better fit of conductance *vs.* length than log(conductance) *vs.* length. Both also show a very low β for the semilogarithmic plots, of <1.1 nm^{-1}, whereas for a purely organic oligophenylene series, one would expect β to be 4.3 nm^{-1}.[55] This is evidence that a hopping-type mechanism may be more appropriate for molecules of the type studied here. This is further supported by the observation that the conductance–overpotential relationships for the conductances of $[M(pyterpy)_2]^{2+/3+}$ under electrochemical potential control fit the Kuznetsov–Ulstrup relationship (essentially, a hopping-based description) well.

A further examination of the conductances of junctions involving these molecules as a function of temperature may shed more light on the mechanism, although in this respect, it is worth noting that although we found evidence for a temperature dependence of the conductance of some oligoporphyrins, this could be accommodated within a tunnelling-based theoretical description, owing to relatively small thermal effects on the transmittance functions of these highly conjugated molecules having a significant effect upon their conductances.[60]

Experimental section

Synthesis and characterisation of ligands and complexes

Reagents were purchased from Sigma-Aldrich Chemical Company and used as received, except where otherwise stated. Proton and ^{13}C NMR spectra were recorded using a Bruker Avance 400 MHz spectrometer and were referenced to internal TMS. Mass spectra were recorded using a Carlo Erba 8000 Trio-1000 quadrupole spectrometer using electrospray, electron or chemical ionization, as appropriate. CHN microanalyses were recorded using an isoprime elemental analyser (we were unable to record microanalytical data for samples containing >20% fluorine with this analyser).

The ligand pyterpy[43] and its complexes $[M(pyterpy)_2](PF_6)_2$ (M = Co,[44] Fe[40] or Ru[45]) were synthesised by literature methods and the characterising data obtained was consistent with the literature. The ligands MeSterpy[49] and MeS-ph-terpy[50] and their Ru(II) complexes[50] were similarly prepared by literature methods or adaptations thereof.

Bis(4′-(pyridin-4-yl)-2,2′:6′,2″-terpyridine)chromium(III) hexafluorophosphate, $[Cr(pyterpy)_2](PF_6)_3$. To degassed water (30 cm^3) was added 4′-(pyridin-4-yl)-2,2′:6′,2″-terpyridine (0.67 g, 2.19 mmol) and $CrCl_2$ (0.135 g, 1.10 mmol). The solution was stirred at room temperature for 20 h during which time the colour changed from light blue to dark brown. A degassed solution of aqueous NH_4PF_6 (5 mL; 0.44 M, 2.20 mmol) was added and a precipitate formed. The solid was filtered using a Schlenk stick, then washed with 50 cm^3 of degassed water, dried under vacuum, extracted with the minimum volume of dry CH_3CN (*ca.* 2 cm^3) and filtered. To this solution was added $AgPF_6$ (0.030 g, 0.12 mmol) in CH_3CN (1 cm^3) over 20 s with stirring. After 15 minutes the CH_3CN was removed under vacuum to afford a brown residue. The product was taken up in CH_3CN and filtered through Celite, and the solvent was removed under vacuum. The residue was triturated and then washed with THF, filtered and dried under vacuum to give the title compound as an orange solid (0.45 g, 37%). UV/vis (CH_3CN, 1.2×10^{-5} mol dm^{-3}) λ_{max}/nm ($\varepsilon_{max}/10^3$ dm^3 mol^{-1} cm^{-1}): 368 (21), 318 (sh), 282 (89.2). Cyclic voltammetry (0.1 M Bu_4NBF_4/CH_3CN): −0.54 V *vs.* Fc/Fc^+, reversible.

4′-(4-(Pyridin-4-yl)phenyl)-2,2′:6′,2″-terpyridine (py-ph-terpy; see Scheme 2)

(a) 4-(Pyridin-4-yl)benzaldehyde. This compound has previously been prepared in somewhat higher yield using a cyclopalladated species as catalyst.[61] (4-Formylphenyl)boronic acid (0.16 g, 1.06 mmol) and 2 M aqueous K_2CO_3 (2.6 cm^3) were added to a solution of 4-bromopyridine hydrochloride (0.2 g, 1.34 mmol) in toluene : i-PrOH (50 : 50, 4 cm^3). After degassing the solution for 15 min using Ar, $Pd(PPh_3)_4$ (0.036 g, 0.030 mmol) was added and the mixture was refluxed for 7 h under Ar. After this time the reaction mixture was diluted with water and extracted with EtOAc. The combined organic layers were dried with $MgSO_4$ and the solvent was evaporated. Recrystallisation from EtOAc yielded the title compound as white needles (0.116 g, 47%); mp: 85–88 °C; ^{1}H NMR (400 MHz, $CDCl_3$): 10.11 (s, 1H, C*H*O), 8.75 (dd, *J* 4.6, 1.6 Hz, 2H), 8.03 (dd, *J* 6.8, 1.7 Hz, 2H), 7.83 (dd, *J* 6.8, 1.6 Hz, 2H), 7.65 (dd, *J* 4.65, 1.6 Hz, 2H); ^{13}C NMR (100 MHz, $CDCl_3$): 191.47 (*C*=O), 148.9, 147.0, 143.3, 136.8, 130.5, 127.9, 122.3; MS (CI^+ NH_3): *m*/*z*: 184.3.

(b) Reaction with 2-acetylpyridine. A solution of 4-(pyridin-4-yl)benzaldehyde (0.05 g, 0.273 mmol) in methanol (4 mL) was cooled to −15 °C. Solutions of 2-acetylpyridine (0.066 g, 0.54 mmol) in MeOH (2 mL) and 20% aq. NaOH (4 mL) were added simultaneously over a period of 20 minutes. The reaction mixture was then further stirred for 3 h at −10 °C then allowed to reach room temperature, and was filtered. The filtrate was treated with NH_4OAc (0.1 g, 1.297 mmol) and heated to reflux for 4 h. The resulting suspension was filtered and the solid washed several times with hot water and hot ethanol. The title compound was recovered as a white solid (0.036 g, 34%); mp: 220–225 °C. ^{1}H NMR (400 MHz, $CDCl_3$): 8.82 (s, 2H), 8.76 (s, 2H), 8.68 (m, 4H), 8.09 (d, *J* 8.3 Hz, 2H), 7.92 (td, *J* 7.7,

1.6 Hz, 2H), 7.84 (d, J 8.3 Hz, 2H), 7.75 (d, J 6.2 Hz, 2H), 7.41–7.38 (m, 2H). ^{13}C NMR (100 MHz, $CDCl_3$): 156.1, 156.0, 149.2, 149.1, 149.0, 139.6, 138.1, 137.0, 128.2, 127.6, 124.0, 121.9, 121.5, 118.8. MS (EI$^+$): m/z: 386.5. UV/vis (CH_3CN, 1.2×10^{-5} mol dm^{-3}) λ_{max}/nm ($\varepsilon_{max}/10^3$ dm^3 mol^{-1} cm^{-1}): 285 (209.4).

Bis(4′-(4-(pyridin-4-yl)phenyl)-2,2′:6′,2″-terpyridine)ruthenium(II) hexafluorophosphate, [Ru(py-ph-terpy)$_2$](PF$_6$)$_3$. A solution of '$RuCl_3 \cdot 3H_2O$' (0.019 g, 0.120 mmol) and 4′-(4-(pyridin-4-yl)phenyl)-2,2′:6′,2″-terpyridine (0.092 g, 0.240 mmol) in ethane-1,2-diol (10 cm^3) was heated to reflux for 3 h. The solution was allowed to cool, and water (10 cm^3) and an excess of methanolic NH_4PF_6 were added. The resulting dark brown precipitate was collected on Celite by filtration and then redissolved in CH_3CN. The CH_3CN was reduced to minimum volume *in vacuo*, and purification was by flash column chromatography (SiO_2, eluting with CH_3CN : sat. aq. KNO_3 : water, 7 : 1 : 0.5). The first major orange fraction was collected, an excess of methanolic $NH_4[PF_6]$ was added, and the solution was evaporated to low volume to precipitate [Ru(py-ph-terpy)$_2$](PF$_6$)$_2$. The compound was filtered and dried in air to yield the title compound as a red solid (0.048 g, 46%). ^{1}H NMR (CD_3CN), (400 MHz): 9.08 (s, 4H), 8.77 (dd, J 4.7, 1.5 Hz, 4H), 8.68 (d, J 7.95 Hz, 4H), 8.38 (d, J 8.45 Hz, 4H), 8.18 (d, J 8.45 Hz, 4H), 7.99–7.94 (m, 8H), 7.45 (d, J 5.5 Hz, 4H), 7.22–7.94 (m, 4H). MS (ES$^+$, CH_3OH): m/z: 873 [M − H − 2PF$_6$]$^+$, 1019 [M − PF$_6$]$^+$. UV/vis (CH_3CN, 1.2×10^{-5} mol dm^{-3}) λ_{max}/nm ($\varepsilon_{max}/10^3$ dm^3 mol^{-1} cm^{-1}): 493 (24.3), 310 (60.8), 287 (sh).

Bis(2,6-di(pyridin-2-yl)pyrazine)ruthenium(II) hexafluorophosphate, [Ru(dipy-pyraz)$_2$](PF$_6$)$_2$. A solution of '$RuCl_3 \cdot 3H_2O$' (0.027 g, 0.105 mmol) and 2,6-di(pyridin-2-yl)pyrazine (0.05 g, 0.214 mmol) in ethane-1,2-diol (10 cm^3) was heated to reflux for 3 h. The solution was allowed to cool, and water (10 cm^3) and an excess of methanolic NH_4PF_6 were added. The resulting dark brown precipitate was collected on Celite by filtration and then extracted into CH_3CN. The CH_3CN was evaporated to minimum volume, and purification was by flash column chromatography (SiO_2, eluting with CH_3CN : sat. aq. KNO_3 : water, 7 : 1 : 0.5). The first major orange fraction was collected, an excess of methanolic $NH_4[PF_6]$ was added, and the solution was reduced in volume to precipitate [Ru(2,6-di(pyridin-2-yl) pyrazine)$_2$][PF$_6$]$_2$. The compound was filtered and dried in air to yield the title compound as a red solid (0.04 g, 47%). ^{1}H NMR (CD_3CN): 9.86 (s, 1H), 8.62 (d, J 8 Hz, 1H), 7.99 (td, J 8 Hz, 1.3 Hz, 1H), 7.36 (d, J 5.51 Hz, 1H), 7.20 (td, J 5.66, 1.13 Hz, 1H); ^{13}C NMR (100 MHz, CD_3CN): 153.76, 150.44, 144.39, 139.41, 128.56, 125.62; MS (ES+, CH_3OH): m/z: 569 [M − H − 2PF$_6$], 715 [M − PF$_6$]; UV/vis (CH_3CN, 1.2×10^{-5} mol dm^{-3}) λ_{max}/nm ($\varepsilon_{max}/10^3$ dm^3 mol^{-1} cm^{-1}): 469 (9.5), 340 (sh), 308 (26.8), 269 (23.5).

4′-(4′-(Methylthio)-[1,1′-biphenyl]-4-yl)-2,2′:6′,2″-terpyridine, MeS-ph-ph-terpy. To a solution of 4′-(4-bromophenyl)-2,2′:6′,2″-terpyridine (0.23 g, 0.59 mmol) in toluene : i-PrOH (50 : 50, 4 cm^3) was added (4-(methylthio)phenyl)boronic acid (0.115 g, 0.61 mmol) and 2 M aqueous K_2CO_3 (2.6 cm^3). After degassing the solution for 15 min with Ar, $Pd(PPh_3)_4$ (0.016 g, 0.014 mmol) was added and the mixture was refluxed for 7 h under Ar. After this time the reaction mixture was diluted with water and extracted several times with EtOAc. The combined organic layers were dried with $MgSO_4$ and the solvent was evaporated. Recrystallisation from EtOAc yielded the title compound as white needles (0.025 g, 0.058 mmol, 10%); mp: 190–195 °C; ^{1}H NMR ($CDCl_3$), (400 MHz): 8.82 (s, 2H), 8.71 (m, J 7.7 Hz, 4H), 7.99 (d, J 8.3 Hz, 2H), 7.92 (t, J 7.7 Hz, 2H), 7.65 (d, J 8.3 Hz, 2H), 7.53 (d, J

8.4 Hz, 2H), 7.37 (t, J 5.6 Hz, 2H), 7.29 (d, J 8.4 Hz, 2H), 2.47 (s, 3H). $^{13}C\{^1H\}$ NMR (101 MHz, $CDCl_3$): 148.4, 141.3, 138.2, 137.0, 127.9, 127.4, 127.3, 127.0, 124.1, 121.9, 119.2, 14.8. MS (CI^+ CH_4): m/z [M + H] 432; UV/vis (CH_3CN, 9×10^{-6} mol dm^{-3}) λ_{max}/nm ($\varepsilon_{max}/10^3$ dm^3 mol^{-1} cm^{-1}): 279 (163.2).

Bis(4′-(methylthio)-2,2′:6′,2″-terpyridine)ruthenium(II) hexafluorophosphate, [Ru(MeSterpy)$_2$](PF$_6$)$_2$. A solution of '$RuCl_3 \cdot 3H_2O$' (0.19 g, 1.20 mmol) and 4′-(methylthio)-2,2′:6′,2″-terpyridine (0.66 g, 2.4 mmol) in ethane-1,2-diol (100 cm^3) was heated at reflux for 12 h. The solution was allowed to cool, and water (10 cm^3) and an excess of methanolic NH_4PF_6 were added. The resulting dark brown precipitate was collected on Celite by filtration and then redissolved in CH_3CN. The CH_3CN was evaporated to minimum volume, and purification was by flash column chromatography (SiO_2, eluting with CH_3CN : sat. aq. KNO_3 : water, 7 : 1 : 0.5). The first major orange fraction was collected, an excess of methanolic NH_4PF_6 were added, and the solution reduced in volume to precipitate [Ru(4′-(methylthio)-2,2′:6′,2″-terpyridine)$_2$][PF$_6$]$_2$. The compound was filtered and dried in air to yield the title compound as a red solid (0.21 g, 20%). ^1H NMR (CD_3CN, 400 MHz): 8.54 (s, 4H), 8.50 (d, J 8.0 Hz, 4H), 7.90 (m, J 9.1, 1.3 Hz, 4H), 7.38 (d, J 5.0 Hz, 4H), 7.16–7.14 (m, 4H), 2.93 (s, SMe, 6H). $^{13}C\{^1H\}$ NMR (CD_3CN, 101 MHz): 158.4, 155.1, 153.1, 152.3, 138.4, 128.1, 125.0, 120.2, 14.8. MS (ES^+, CH_3OH): m/z: 805.0 [M − PF_6]. UV/vis (CH_3CN, 9×10^{-6} mol dm^{-3}) λ_{max}/nm ($\varepsilon_{max}/10^3$ dm^3 mol^{-1} cm^{-1}): 491 (11.2), 301 (sh), 281 (115.6). UV/vis data are consistent with those for the chloride salt, prepared previously.[62]

Bis(4′-(4′-(methylthio)-[1,1′-biphenyl]-4-yl)-2,2′:6′,2″-terpyridine)ruthenium(II) hexafluorophosphate, [Ru(MeS-ph-ph-terpy)$_2$](PF$_6$)$_2$. A solution of '$RuCl_3 \cdot 3H_2O$' (0.007 g, 0.03 mmol) and 4′-(4′-(methylthio)-[1,1′-biphenyl]-4-yl)-2,2′:6′,2″-terpyridine (0.025 g, 0.06 mmol) in ethane-1,2-diol (2.5 cm^3) was heated to reflux for 3 h. The solution was allowed to cool, and water (2.5 cm^3) and an excess of methanolic $NH_4[PF_6]$ were added. The resulting dark brown precipitate was collected on Celite by filtration and then redissolved in CH_3CN. The CH_3CN was reduced to minimum volume, and purification was by flash column chromatography (SiO_2, eluting with CH_3CN : sat. aq. KNO_3 : water, 7 : 1 : 0.5). The first major orange fraction was collected, an excess of methanolic $NH_4[PF_6]$ was added, and the solution was reduced in volume to precipitate [Ru(MeS-ph-ph-terpy)$_2$](PF$_6$)$_2$. The compound was filtered and dried in air to yield a red solid (0.012 g, 33%). ^1H NMR (CD_3CN, 400 MHz): 9.07 (s, 4H), 8.68 (d, J 8.0 Hz, 3H), 8.34 (d, J 8.4 Hz, 3H), 8.10 (d, J 8.1 Hz, 4H), 8.03 (d, J 8.4 Hz, 4H), 7.96 (t, J 7.3 Hz, 6H), 7.85 (d, J 8.2 Hz, 4H), 7.45 (m, 4H), 7.19 (t, J 6.1 Hz, 4H), 2.76 (s, SCH_3, 6H). MS (ES^+, CH_3OH): m/z: 995 (M − HPF_6 − PF_6 + MeOH), 1141 (M − PF_6 + MeOH). UV/vis (CH_3CN, 9×10^{-6} mol dm^{-3}) λ_{max}/nm ($\varepsilon_{max}/10^3$ dm^3 mol^{-1} cm^{-1}): 494 (9.9), 311 (38.9), 287 (sh).

STM-based molecular junction conductances

First, molecules were adsorbed onto flame-annealed gold-on-glass samples from a 5×10^{-5} M CH_2Cl_2 solution of the appropriate complex. A freshly-cut Au STM tip was held above the surface at a fixed set point current (40 nA except for the longest molecules for which 20 nA was used) and bias voltage (+0.6 V) and then withdrawn at a speed of 40 nm s^{-1} while maintaining a constant x–y position. A current–vertical distance ($I(s)$) curve was collected during tip withdrawal. When molecular junction formation occurs, I decreases exponentially with s but is larger, at a given s, than in cases where no junction forms, and I typically then settles at a plateau value as s is increased, before abruptly decaying to zero on junction breakdown. Many such $I(s)$ traces are recorded. Only those that contain plateaus (*ca.* 500 per data set) were collected together in histogram plots to determine junction conductance values. The occurrence of plateaus in many such $I(s)$ curves results in peaks in the histograms. The percentage of scans showing evidence of junction formation was usually in the range 5–10% (*ca.* 15% for the MeS-contacted molecules). Further details of data collection and processing, and of the procedure for estimating s_0, are in the ESI.†

Acknowledgements

We thank Dr Andrea Vezzoli for recording some NMR spectra and the EPSRC for funding (grants EP/C011511/1, EP/H035184/1 and EP/H015639/1).

References

1 J. P. Bergfield and M. A. Ratner, *Phys. Status Solidi B*, 2013, **250**, 2249–2266.
2 J. C. Cuevas and E. Scheer, *Molecular Electronics: an Introduction to Theory and Experiment*, World Scientific, Singapore, 2010.
3 R. J. Nichols and S. J. Higgins, in *Ann. Rev. Anal. Chem.*, ed. R. G. Cooks and J. E. Pemberton, 2015, vol. 8, pp. 389–417.
4 R. J. Nichols, W. Haiss, S. J. Higgins, E. Leary, S. Martin and D. Bethell, *Phys. Chem. Chem. Phys.*, 2010, **12**, 2801–2815.
5 C. Wang, A. S. Batsanov, M. R. Bryce, S. Martin, R. J. Nichols, S. J. Higgins, V. M. Garcia-Suarez and C. J. Lambert, *J. Am. Chem. Soc.*, 2009, **131**, 15647–15654.
6 G. Sedghi, L. J. Esdaile, H. L. Anderson, S. Martin, D. Bethell, S. J. Higgins and R. J. Nichols, *Adv. Mater.*, 2012, **24**, 653–657.
7 V. Kolivoska, M. Valasek, M. Gal, R. Sokolova, J. Bulickova, L. Pospisil, G. Meszaros and M. Hromadova, *J. Phys. Chem. Lett.*, 2013, **4**, 589–595.
8 J. Park, A. N. Pasupathy, J. I. Goldsmith, C. Chang, Y. Yaish, J. R. Petta, M. Rinkoski, J. P. Sethna, H. D. Abruna, P. L. McEuen and D. C. Ralph, *Nature*, 2002, **417**, 722–725.
9 S. Kubatkin, A. Danilov, M. Hjort, J. Cornil, J.-L. Brédas, N. Stuhr-Hansen, P. Hedegård and T. Bjørnholm, *Nature*, 2003, **425**, 698–701.
10 S. Martin, W. Haiss, S. J. Higgins and R. J. Nichols, *Nano Lett.*, 2010, **10**, 2019–2023.
11 A. S. Kumar, T. Ye, T. Takami, B.-C. Yu, A. K. Flatt, J. M. Tour and P. S. Weiss, *Nano Lett.*, 2008, **8**, 1644–1648.
12 E. Leary, H. Hobenreich, S. J. Higgins, H. van Zalinge, W. Haiss, R. J. Nichols, C. M. Finch, I. Grace, C. J. Lambert, R. McGrath and J. Smerdon, *Phys. Rev. Lett.*, 2009, **102**, 086801, 4 pp.
13 A. Vezzoli, I. Grace, C. Brooke, K. Wang, C. J. Lambert, B. Xu, R. J. Nichols and S. J. Higgins, *Nanoscale*, 2015, **7**, 18949–18955.
14 W. Haiss, H. van Zalinge, S. J. Higgins, D. Bethell, H. Höbenreich, D. J. Schiffrin and R. J. Nichols, *J. Am. Chem. Soc.*, 2003, **125**, 15294–15295.
15 N. J. Kay, S. J. Higgins, J. O. Jeppesen, E. Leary, J. Lycoops, J. Ulstrup and R. J. Nichols, *J. Am. Chem. Soc.*, 2012, **134**, 16817–16826.

16 M. Poot, E. Osorio, K. O'Neill, J. M. Thijssen, D. Vanmaekelbergh, C. A. van Walree, L. W. Jenneskens and H. S. J. van der Zant, *Nano Lett.*, 2006, **6**, 1031–1035.

17 E. Leary, S. J. Higgins, H. van Zalinge, W. Haiss, R. J. Nichols, S. Nygaard, J. O. Jeppesen and J. Ulstrup, *J. Am. Chem. Soc.*, 2008, **130**, 12204–12205.

18 Z. H. Li, I. Pobelov, B. Han, T. Wandlowski, A. Blaszczyk and M. Mayor, *Nanotechnology*, 2007, **18**, 044018.

19 F. Chen, J. JHe, C. Nuckolls, T. Roberts, J. E. Klare and S. Lindsay, *Nano Lett.*, 2005, **5**, 503–506.

20 J. He, F. Chen, S. Lindsay and C. Nuckolls, *Appl. Phys. Lett.*, 2007, **90**, 072112.

21 X. Li, J. Hihath, F. Chen, T. Masuda, L. Zang and N. Tao, *J. Am. Chem. Soc.*, 2007, **129**, 11535–11542.

22 C. Li, V. Stepanenko, M.-J. Lin, W. Hong, F. Wuerthner and T. Wandlowski, *Phys. Status Solidi B*, 2013, **250**, 2458–2467.

23 J. Zhang, Q. Chi, A. M. Kuznetsov, A. G. Hansen, H. Wackerbarth, H. E. M. Christensen, J. E. T. Andersen and J. Ulstrup, *J. Phys. Chem. B*, 2002, **106**, 1131–1152.

24 W. Haiss, T. Albrecht, H. van Zalinge, S. J. Higgins, D. Bethell, H. Hobenreich, D. J. Schiffrin, R. J. Nichols, A. M. Kuznetsov, J. Zhang, Q. Chi and J. Ulstrup, *J. Phys. Chem. B*, 2007, **111**, 6703–6712.

25 H. M. Osorio, S. Catarelli, P. Cea, J. B. G. Gluyas, F. Hartl, S. J. Higgins, E. Leary, P. J. Low, S. Martin, R. J. Nichos, J. Tory, J. Ulstrup, A. Vezzoli, D. C. Milan and Q. Zeng, *J. Am. Chem. Soc.*, 2015, **137**, 14319–14328.

26 F. Lissel, F. Schwarz, O. Blacque, H. Riel, E. Loertscher, K. Venkatesan and H. Berke, *J. Am. Chem. Soc.*, 2014, **136**, 14560–14569.

27 R. Davidson, J.-H. Liang, D. C. Milan, B.-W. Mao, R. J. Nichols, S. J. Higgins, D. S. Yufit, A. Beeby and P. J. Low, *Inorg. Chem.*, 2015, **54**, 5487–5494.

28 A. C. Aragones, D. Aravena, J. I. Cerda, Z. Acis-Castillo, H. Li, J. Antonio Real, F. Sanz, J. Hihath, E. Ruiz and I. Diez-Perez, *Nano Lett.*, 2016, **16**, 218–226.

29 L. Luo, S. H. Choi and C. D. Frisbie, *Chem. Mater.*, 2011, **23**, 631–645.

30 S. Rigaut, *Dalton Trans.*, 2013, **42**, 15859–15863.

31 S. J. Higgins, R. Nichols, S. Martin, P. Cea, H. S. J. van der Zant, M. M. Richter and P. J. Low, *Organometallics*, 2011, **30**, 7–12.

32 G. Grelaud, N. Gauthier, Y. Luo, F. Paul, B. Fabre, F. Barriere, S. Ababou-Girard, T. Roisnel and M. G. Humphrey, *J. Phys. Chem. C*, 2014, **118**, 3680–3695.

33 A. Mulas, Y.-M. Hervault, L. Norel, S. Rigaut and C. Lagrost, *ChemElectroChem*, 2015, **2**, 1799–1805.

34 L. Luo, A. Benameur, P. Brignou, S. H. Choi, S. Rigaut and C. D. Frisbie, *J. Phys. Chem. C*, 2011, **115**, 19955–19961.

35 S. Marques-Gonzalez, D. S. Yufit, J. A. K. Howard, S. Martin, H. M. Osorio, V. M. Garcia-Suarez, R. J. Nichols, S. J. Higgins, P. Cea and P. J. Low, *Dalton Trans.*, 2013, **42**, 338–341.

36 B. Kim, J. M. Beebe, C. Olivier, S. Rigaut, D. Touchard, J. G. Kushmerick, X. Y. Zhu and C. D. Frisbie, *J. Phys. Chem. C*, 2007, **111**, 7521–7526.

37 N. Tuccitto, V. Ferri, M. Cavazzini, S. Quici, G. Zhavnerko, A. Licciardello and M. A. Rampi, *Nat. Mater.*, 2009, **8**, 41–46.

38 Y. Nishimori, K. Kanaizuka, T. Kurita, T. Nagatsu, Y. Segawa, F. Toshimitsu, S. Muratsugu, M. Utsuno, S. Kume, M. Murata and H. Nishihara, *Chem.–Asian J.*, 2009, **4**, 1361–1367.

39 E. Figgemeier, L. Merz, B. A. Hermann, Y. C. Zimmermann, C. E. Housecroft, H. J. Guntherodt and E. C. Constable, *J. Phys. Chem. B*, 2003, **107**, 1157–1162.

40 E. C. Constable and A. Thompson, *J. Chem. Soc., Dalton Trans.*, 1992, 2947–2950, DOI: 10.1039/dt9920002947.

41 S. S. Sun and A. J. Lees, *Inorg. Chem.*, 2001, **40**, 3154–3160.

42 E. C. Constable, C. E. Housecroft, M. Neuburger, D. Phillips, P. R. Raithby, E. Schofield, E. Sparr, D. A. Tocher, M. Zehnder and Y. Zimmermann, *J. Chem. Soc., Dalton Trans.*, 2000, 2219–2228, DOI: 10.1039/b000940g.

43 L. Persaud and G. Barbiero, *Can. J. Chem.*, 1991, **69**, 315–321.

44 R. Indumathy, S. Radhika, M. Kanthimathi, T. Weyhermuller and B. U. Nair, *J. Inorg. Biochem.*, 2007, **101**, 434–443.

45 E. C. Constable and A. M. W. Cargill-Thompson, *J. Chem. Soc., Dalton Trans.*, 1994, 1409–1418, DOI: 10.1039/dt9940001409.

46 C. C. Scarborough, K. M. Lancaster, S. DeBeer, T. Weyhermueller, S. Sproules and K. Wieghardt, *Inorg. Chem.*, 2012, **51**, 3718–3732.

47 W. Goodall, K. Wild, K. J. Arm and J. A. G. Williams, *J. Chem. Soc., Perkin Trans. 2*, 2002, 1669–1681, DOI: 10.1039/b205330f.

48 L. J. K. Cook, F. Tuna and M. A. Halcrow, *Dalton Trans.*, 2013, **42**, 2254–2265.

49 K. T. Potts, P. Ralli, G. Theodoridis and P. Winslow, *Org. Synth.*, 1986, **64**, 189.

50 E. C. Constable, C. E. Housecroft, E. Medlycott, M. Neuburger, F. Reinders, S. Reymann and S. Schaffner, *Inorg. Chem. Commun.*, 2008, **11**, 518–520.

51 P. Korall, A. Borje, P. Norrby and B. Åkermark, *Acta Chem. Scand.*, 1997, **51**, 760–766.

52 S. Y. Quek, M. Kamenetska, M. L. Steigerwald, H. J. Choi, S. G. Louie, M. S. Hybertsen, J. B. Neaton and L. Venkataraman, *Nat. Nanotechnol.*, 2009, **4**, 230–234.

53 X. Zhang, H. Xie, M. Ballesteros-Rivas, Z.-X. Wang and K. R. Dunbar, *J. Mater. Chem. C*, 2015, **3**, 9292–9298.

54 X. Zhang, Z.-X. Wang, H. Xie, M.-X. Li, T. J. Woods and K. R. Dunbar, *Chem. Sci.*, 2016, **7**, 1569–1574.

55 W. Chen, J. R. Widawsky, H. Vazquez, S. T. Schneebeli, M. S. Hybertsen, R. Breslow and L. Venkataraman, *J. Am. Chem. Soc.*, 2011, **133**, 17160–17163.

56 G. Sedghi, K. Sawada, L. J. Esdaile, M. Hoffmann, H. L. Anderson, D. Bethell, W. Haiss, S. J. Higgins and R. J. Nichols, *J. Am. Chem. Soc.*, 2008, **130**, 8582–8583.

57 N. Sutin, B. S. Brunschwig, C. Creutz and J. R. Winkler, *Pure Appl. Chem.*, 1988, **60**, 1817–1830.

58 X.-S. Zhou, L. Liu, P. Fortgang, A.-S. Lefevre, A. Serra-Muns, N. Raouafi, C. Amatore, B.-W. Mao, E. Maisonhaute and B. Schoellhorn, *J. Am. Chem. Soc.*, 2011, **133**, 7509–7516.

59 J. Zhang, A. M. Kuznetsov, I. G. Medvedev, Q. Chi, T. Albrecht, P. S. Jensen and J. Ulstrup, *Chem. Rev.*, 2008, **108**, 2737–2791.

60 G. Sedghi, V. M. Garcia-Suarez, L. J. Esdaile, H. L. Anderson, C. J. Lambert, S. Martin, D. Bethell, S. J. Higgins, M. Elliott, N. Bennett, J. E. Macdonald and R. J. Nichols, *Nat. Nanotechnol.*, 2011, **6**, 517–523.

61 Z. C. Xiong, N. D. Wang, M. J. Dai, A. Li, J. H. Chen and Z. Yang, *Org. Lett.*, 2004, **6**, 3337–3340.

62 K. T. Potts, D. A. Usifer, A. Guadalupe and H. D. Abruña, *J. Am. Chem. Soc.*, 1987, **109**, 3961–3967.

Faraday Discussions

PAPER

Single molecular catalysis of a redox enzyme on nanoelectrodes

Lianhuan Han, Wei Wang, Jacques Nsabimana, Jia-Wei Yan, Bin Ren and Dongping Zhan*

Received 30th March 2016, Accepted 28th April 2016
DOI: 10.1039/c6fd00061d

Due to a high turnover coefficient, redox enzymes can serve as current amplifiers which make it possible to explore their catalytic mechanism by electrochemistry at the level of single molecules. On modified nanoelectrodes, the voltammetric behavior of a horseradish peroxidase (HRP) catalyzed hydroperoxide reduction no longer presents a continuous current response, but a staircase current response. Furthermore, single catalytic incidents were captured through a collision mode at a constant potential, from which the turnover number of HRP can be figured out statistically. In addition, the catalytic behavior is dynamic which may be caused by the orientation status of HRP on the surface of the electrode. This modified nanoelectrode methodology provides an electrochemical approach to investigate the single-molecule catalysis of redox enzymes.

Introduction

Single-molecule detection is the ultimate goal of analytical chemistry, which is essential for lowering the limit of detection, fabricating molecular devices, understanding chemical and biological processes, *etc.*[1] Aside from fluorescence spectroscopy or electrical conductance measured by scanning probe microscopy, it is difficult for electrochemistry since even the best instrument is incapable of detecting the Faraday current resulting from one or a few electron transfers. For a one-electron transfer reaction, a current of 1 pA means $\sim 6.0 \times 10^7$ molecules are involved per second. So, special current amplifiers are needed to probe the electrochemistry at the single-molecule level.[2]

The first electrochemical detection was reported by Bard's group, in which single molecules was trapped in a 10 nm-thick gap between a conductive substrate and a nanoelectrode (NE) tip in a scanning electrochemical microscopy (SECM) system.[3] The protocol was followed by Mirkin's group through inserting a slightly recessed platinum NE into a mercury pool.[4] Recently, Lemay's group employed nanofluidics as the thin-layer cell[5] and Unwin's group reported their

State Key Laboratory of Physical Chemistry of Solid Surfaces, Collaborative Innovation Center of Chemistry for Energy Materials (iChEM), Department of Chemistry, College of Chemistry and Chemical Engineering, Xiamen University, Xiamen 361005, China

design by employing a four-barrel glass nanopipette as the tip in a scanning electrochemical cell microscopy (SECCM) system.[6] In the nanometre-scale thin layer, the current is amplified by the to-and-fro diffusion of single redox molecules therein.[7] Once coupled with spectroscopic technologies, the thin-layer technique is expected to be a powerful tool to uncover the reaction mechanism of single molecules, especially with respect to the orientation of adsorbates, changes in molecular structure, and also unstable reaction intermediates.[8]

Alternatively, the Faraday current can be amplified by redox reactions rather than a diffusion loop. Since 2007 Bard's group has established a collision mode where the electrochemistry of single particles or droplets can be evaluated.[9] Recently, the blocking effect of single protein and DNA macromolecules was also employed to estimate their size.[9d,10] A current amplified by a few enzyme molecules was first reported by Lemay *et al.*, where the turnover number was obtained on a [NiFe]-hydrogenase film-modified nanoelectrode.[11] When preparing this paper, we found the newly published work of Sekretaryova *et al.* where they investigated the catalytic ORR incident of a single laccase molecule, which has a thinner insulate shell and offers a direct electron transfer between its active site and the bare gold ultramicroelectrode.[12] Due to either the isolation caused by hydrophobic amino acid residues or the denaturation of the redox enzymes, direct charge transfer between the bare metal electrode and the redox enzyme was hindered.[13] The modified electrode methodology may be more helpful in the studies of enzyme catalysis at the single molecular level. Moreover, modified NEs show discrete behaviour in cyclic voltammetry and dynamic current response in chronoamperometry, which may give more detailed information on single-molecule enzyme catalysis and their individual diversity.

Experimental

Egg lecithin, octyltriethoxysilane, ethyl alcohol (EtOH), acetonitrile (CH_3CN), sulfuric acid (H_2SO_4), hydrogen peroxide (H_2O_2), potassium chloride (KCl), potassium dihydrogen phosphate (KH_2PO_4), and dipotassium hydrogen phosphate (K_2HPO_4) were provided by Sinopharm Co., China. Horseradish hydroperoxidase (HRP) and ferrocenemethanol (FcMeOH) were purchased from Sigma. All aqueous solutions are prepared with deionized water (18.2 MΩ cm, Milli-Q, Millipore Co.).

The fabrication of gold (Au) NEs was performed as described previously. An annealed 25 μm Au wire (Goodfellow) was pulled into a borosilicate capillary (Drummond; 1.0 mm o.d., 0.2 mm i.d.) under vacuum with the help of a Sutter P-2000/G laser pipette puller. The pulled electrodes were polished on 50 nm lapping tape under microscopic video control and washed with distilled water. The effective radius of an electrode was evaluated using steady-state voltammetry, and only electrodes producing well-shaped sigmoidal voltammograms in 1 mM FcMeOH at a scan rate of 50 V s^{-1} were used for experiments.

The Au NEs were cleaned in freshly prepared H_2SO_4/H_2O_2 (v/v 4 : 1) solution and in deionized H_2O and pure CH_3CN alternatingly several times. The cleaned Au NEs were then immersed in 1% octyltriethoxysilane/EtOH solution for 12 hours to make the surface hydrophobic, dipped in egg lecithin solution for 10 minutes and drawn back and dried in the air. As a result, a standing bilayer lipid membrane (BLM) formed spontaneously on the surface, which can provide a mimic environment for the HRP molecules and ensure the direct communication between HRP and the Au NEs. The concentration of H_2O_2 was as high as 20 mM to exclude the mass transfer problem. If a perfect octyltriethoxysilane-supported BLM is formed on the surface of nanoelectrode, the direct reduction of H_2O_2 will be hindered as described in the followed section. All the experiments were performed with an Autolab workstation with a low-current-detection module (Metrohm Autolab). An Ag/AgCl wire was adopted as both the reference and counter electrode.

Results and discussion

Curve 1 in Fig. 1 shows the typical cyclic voltammogram of the BLM-modified Au NE (radius: 12 nm) in a solution with 20 mM H_2O_2, 100 mM NaCl and 25 mM PBS (pH: 7.4), which indicates that the hydrophobic layer of octyltriethoxysilane can block H_2O_2 reduction effectively. After being immersed in a 0.5 U mL^{-1} HRP solution for 30 min, the HRP-BLM modified Au NE was rinsed in deionized H_2O for several times and then transferred into the same solution. The voltammetric behavior was observed to be much different from those obtained on the conventional macroelectrode: during the cathodic scan the current showed a sharp step at −280 mV and then increased discretely; during the anodic scan the current decreased in a staircase way and no sharp steps were observed.

Considering the mass transfer of H_2O_2, the limiting reduction current on a bare 12 nm-radius Au NE is estimated to be ~100 pA ($i = 4nFDaC$), much higher than the observed current. The result shows the octyltriethoxysilane layer makes the Au NE hydrophobic enough to block the direct reduction of H_2O_2 on the BLM-modified Au NE. The average size of HRP molecules are reported to be 2 nm by 3 nm by 3 nm. Thus, a 12 nm-radius Au NE can accommodate 15 HRP molecules if the coverage is supposed to be 30%. Since the electrode size is so small and the loaded HRP molecules are countable, the redox behaviour shouldn't be controlled by the mass transfer of H_2O_2, but the HRP catalytic reaction or the electron transfer process between HRP and the modified Au NE, *i.e.*, the "electrocatalysis" of single HRP molecules. Moreover, the current fluctuations during the potential

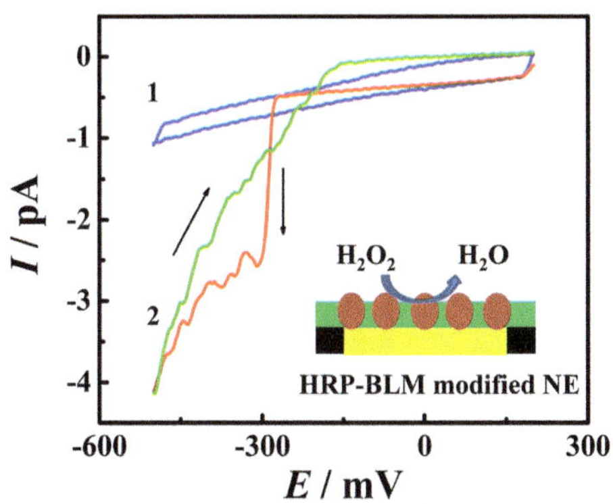

Fig. 1 Cyclic voltammograms of the BLM-modified Au NE (curve 1) and the HRP-BLM-modified Au NE (curve 2) in a solution of 20 mM H_2O_2, 100 mM NaCl and 25 mM PBS (pH: 7.4). The scanning rate is 50 mV s^{-1} and the radius of the Au NE is 12 nm.

scan predict that the electrocatalytic properties of single HRP molecules might be dynamic themselves and have individual diversity from each other.

Chronoamperometric experiments were performed to detect the collision incidents of electrocatalytic H_2O_2 reduction by single HRP molecules. To obtain the transient current response of single incidents, the sample frequency was 1 kHz and the potential was held at −450 mV. Fig. 2 shows the dynamic current response of four HRP molecules (see the lines to guide the eye). After injecting 0.1 U mL^{-1} of HRP (final concentration), the initial current changes ($0 < t < 30$ s) might be caused by either the first collision incident of HRP molecules or just by the perturbation of the electric double layer. However, at the moment of 30 s, there is a clear current step which lasts for almost 80 s and should be the HRP catalysed H_2O_2 reduction. From 110 s to 130 s, it seems that one HRP leaves and then comes back, or a new HRP adheres to the electrode. The plateau current is turbulent (~50 fA) and much higher than the inherent noise of the instrument in cyclic voltammetry (~10 fA). The current fluctuations might be caused by the dynamic activity of individual HRP molecules.

A series of parallel experiments were performed to capture the collision incident of single HRP molecules (Fig. 3). The step current that is higher than 3 times the perturbation of plateau current (*i.e.*, >150 fA) will be considered as a single incident. The log-normal statistic current is derived as (0.23 ± 0.05) pA, from which the turnover number can be derived as $(6.9 \pm 1.5) \times 10^5$ s^{-1}. Actually, in some specific collision incidents the calculated turnover number can reach as high as $\sim 10^6$ s^{-1}. The results show that the BLM-modified Au NEs can provide good electric communication between the HRP molecules and the electrode and improve the turnover number.[12]

The individual diversity is the basic characteristics of single molecular catalysis of redox enzymes in the collision mode. If one HRP molecule is adhered, the electrode should have a diameter of 3.0 nm or lower. However, after polishing, the laser-pulled NEs are usually much larger than that. Since a collision incident is random, sometimes it is difficult to judge whether it is caused by one molecule or the turbulence of a few molecules. The complexity is shown as the inserts in

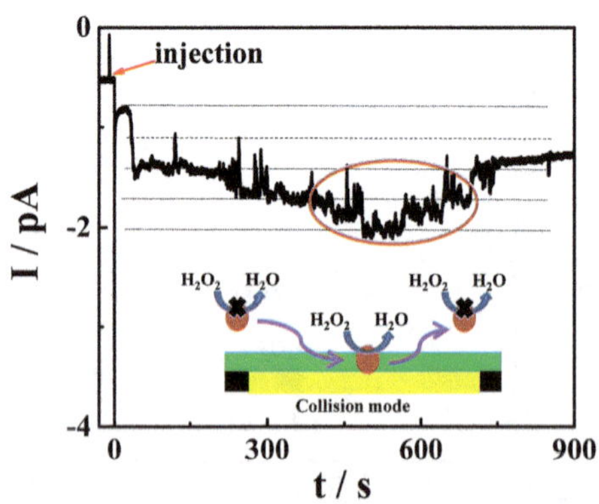

Fig. 2 Chronoamperometric response of the 12 nm-radius Au NE in solution containing 0.1 U mL^{-1} HRP, 20 mM H_2O_2, 100 mM NaCl and 25 mM PBS (pH: 7.4). The Au NE was biased at −450 mV *vs.* Ag/AgCl wire, and the sample frequency is 1 kHz. The complexity shown in the red circle area shows the dynamic activity of single HRP molecules.

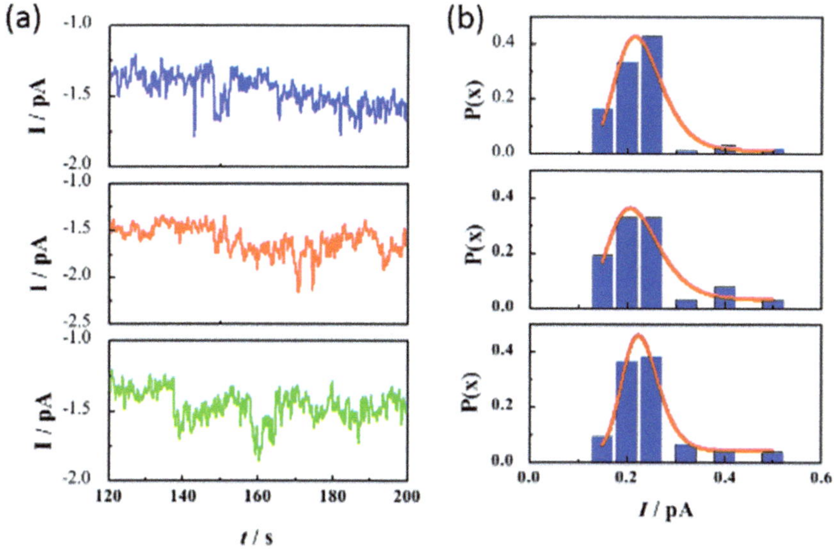

Fig. 3 (a) A group of chronoamperometric experiments of the 12 nm-radius Au NE in the solution containing 0.1 U mL^{-1} HRP, 20 mM H_2O_2, 100 mM NaCl and 25 mM PBS (pH: 7.4), and (b) the log-normal statistical analysis of the probability and step currents based on the data shown in (a) respectively.

Fig. 4c–e: Fig. 4b looks like two independent incidents which give a very uniform current pulse. In Fig. 4c the first incident gives a perfect current plateau, however, both of the incidents show an activated process of a few seconds. Fig. 4d tells of dynamic activity, and in Fig. 4e the last incident undergoes a passivation process. Despite the individual diversity of the collision incidents, the average turnover number is calculated as $(6.2 \pm 1.9) \times 10^5\ s^{-1}$ from the integrated charge in unit time. The results are in accordance with those obtained from the statistic results of the step current. To clarify individual diversity, future work might be: (1)

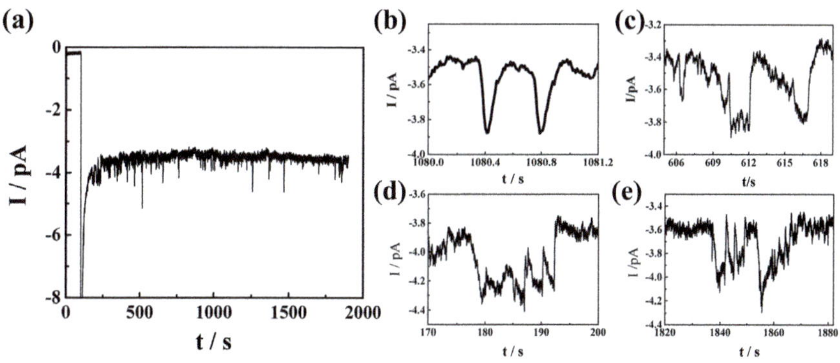

Fig. 4 Individual diversity of single enzyme catalysis: (a) chronoamperometric curve of the 12 nm-radius Au NE in solution containing 0.1 U mL^{-1} HRP, 20 mM H_2O_2, 100 mM NaCl and 25 mM PBS (pH: 7.4). The Au NE was biased at −450 mV *vs.* Ag/AgCl wire. (b–e) selected details of several incidents recorded in (a).

employing a 3.0 nm-diameter or less nanoelectrode to hold only one enzyme molecule; (2) coupling with high-resolution spectroscopy to reveal how the "orientation" of enzyme molecules in the BLM affects their catalytic activity. This is crucial for both molecular biology and single molecular sensing.

Conclusions

The electrocatalytic reduction of H_2O_2 through single molecules of horseradish hydroperoxidase was studied on a bilayer lipid membrane-modified gold nanoelectrode. Because of the small electrode size, countable HRP molecules and excess H_2O_2, the rate-determining step was ensured as the electrocatalytic process of HRP to H_2O_2. The cyclic voltammetric behaviour was no longer continuous but a coulomb staircase. A collision mode was adopted to record the single-molecule catalytic current. The turnover number of HRP was derived to be as high as $\sim 10^6$ s^{-1}. The individual diversity of single HRP molecules were observed, which showed the complexity of single molecular catalysis. The methodology is applicable to many redox enzymes and should be valuable in understanding their roles in biological processes, chemical analysis and bioenergy devices.

Acknowledgements

W. Wang and L. Han contributed equally this paper. The authors declare no competing financial interest. Financial support from the National Basic Research Program of China (2012CB932902), the National Science Foundation of China (91323303, 21327002, and 21321062), and the NSF-NSFC Collaborative Program in Chemistry (21061120456) are appreciated.

Notes and references

1 (*a*) J. Zhang, A. M. Kuznetsov, I. G. Medvedev, Q. Chi, T. Albrecht, P. S. Jensen and J. Ulstrup, *Chem. Rev.*, 2008, **108**, 2737–2791; (*b*) X.-S. Zhou, L. Liu, P. Fortgang, A.-S. Lefevre, A. Serra-Muns, N. Raouafi, C. Amatore, B.-W. Mao, E. Maisonhaute and B. Schöllhorn, *J. Am. Chem. Soc.*, 2011, **133**, 7509–7516; (*c*) E. J. Schelter, F. Karadas, C. Avendano, A. V. Prosvirin, W. Wernsdorfer and K. R. Dunbar, *J. Am. Chem. Soc.*, 2007, **129**, 8139–8149; (*d*) R. Bilewicz and M. Majda, *J. Am. Chem. Soc.*, 1991, **113**, 5464–5466; (*e*) R. Bilewicz, T. Sawaguchi, R. V. Chamberlain and M. Majda, *Langmuir*, 1995, **11**, 2256–2266.

2 P. Sun and M. V. Mirkin, *Anal. Chem.*, 2006, **78**, 6526–6534.

3 (*a*) F. R. F. Fan and A. J. Bard, *Science*, 1995, **267**, 871–874; (*b*) A. J. Bard and F.-R. F. Fan, *Acc. Chem. Res.*, 1996, **29**, 572–578; (*c*) F.-R. F. Fan, J. Kwak and A. J. Bard, *J. Am. Chem. Soc.*, 1996, **118**, 9669–9675; (*d*) F. R. F. Fan and A. J. Bard, *Science*, 1997, **277**, 1791–1793.

4 P. Sun and M. V. Mirkin, *J. Am. Chem. Soc.*, 2008, **130**, 8241–8250.

5 (*a*) M. A. G. Zevenbergen, P. S. Singh, E. D. Goluch, B. L. Wolfrum and S. G. Lemay, *Nano Lett.*, 2011, **11**, 2881–2886; (*b*) P. S. Singh, E. Kätelhön, K. Mathwig, B. Wolfrum and S. G. Lemay, *ACS Nano*, 2012, **6**, 9662–9671; (*c*) S. Kang, A. F. Nieuwenhuis, K. Mathwig, D. Mampallil and S. G. Lemay, *ACS Nano*, 2013, 7, 10931–10937.

6 J. C. Byers, B. Paulose Nadappuram, D. Perry, K. McKelvey, A. W. Colburn and P. R. Unwin, *Anal. Chem.*, 2015, **87**, 10450–10456.

7 R. J. White and H. S. White, *Langmuir*, 2008, **24**, 2850–2855.

8 (*a*) D. Kurouski, M. Mattei and R. P. Van Duyne, *Nano Lett.*, 2015, **15**, 7956–7962; (*b*) S. G. Lemay, S. Kang, K. Mathwig and P. S. Singh, *Acc. Chem. Res.*, 2013, **46**, 369–377; (*c*) S. Zaleski, M. F. Cardinal, J. M. Klingsporn and R. P. Van Duyne, *J. Phys. Chem. C*, 2015, **119**, 28226–28234; (*d*) A. Lauchner, A. E. Schlather, A. Manjavacas, Y. Cui, M. J. McClain, G. J. Stec, F. J. García de Abajo, P. Nordlander and N. J. Halas, *Nano Lett.*, 2015, **15**, 6208–6214; (*e*) E. Cortés, P. G. Etchegoin, E. C. Le Ru, A. Fainstein, M. E. Vela and R. C. Salvarezza, *J. Am. Chem. Soc.*, 2010, **132**, 18034–18037; (*f*) J. M. Artés, I. Díez-Pérez, F. Sanz and P. Gorostiza, *ACSNano*, 2011, **5**, 2060–2066; (*g*) E. Cortés, P. G. Etchegoin, E. C. Le Ru, A. Fainstein, M. E. Vela and R. C. Salvarezza, *J. Am. Chem. Soc.*, 2013, **135**, 2809–2815.

9 (*a*) X. Xiao and A. J. Bard, *J. Am. Chem. Soc.*, 2007, **129**, 9610–9612; (*b*) A. J. Bard, *ACS Nano*, 2008, **2**, 2437–2440; (*c*) X. Xiao, F.-R. F. Fan, J. Zhou and A. J. Bard, *J. Am. Chem. Soc.*, 2008, **130**, 16669–16677; (*d*) J. E. Dick, C. Renault and A. J. Bard, *J. Am. Chem. Soc.*, 2015, **137**, 8376–8379.

10 (*a*) S. M. Oja, Y. Fan, C. M. Armstrong, P. Defnet and B. Zhang, *Anal. Chem.*, 2016, **88**, 414–430; (*b*) K. Mathwig, Q. Chi, S. G. Lemay and L. Rassaei, *ChemPhysChem*, 2016, **17**, 452–457.

11 F. J. M. Hoeben, F. S. Meijer, C. Dekker, S. P. J. Albracht, H. A. Heering and S. G. Lemay, *ACS Nano*, 2008, **2**, 2497–2504.

12 A. N. Sekretaryova, M. Y. Vagin, A. P. F. Turner and M. Eriksson, *J. Am. Chem. Soc.*, 2016, **138**, 2504–2507.

13 L. J. C. Jeuken, S. D. Connell, P. J. F. Henderson, R. B. Gennis, S. D. Evans and R. J. Bushby, *J. Am. Chem. Soc.*, 2006, **128**, 1711–1716.

Faraday Discussions

DISCUSSIONS

From single cells to single molecules: general discussion

Justin Gooding, Olaf Magnussen, David Fermin, Richard Crooks, Frederic Kanoufi, Wolfgang Schuhmann, Richard Nichols, Wolfgang Schmickler, Nongjian Tao, Shengli Chen, Paolo Actis, Ashley Page, Kristina Tschulik, Sanli Faez, Martin Edwards, Robert Johnson, Wojciech Nogala, Christine Kranz, Michael Eikerling, Patrick Unwin, Bradley Thomas, Venkateshkumar Prabhakaran, Jan Clausmeyer, Kylie Vincent, Marc Koper, Zhongqun Tian, Andy Mount, Mario A. Alpuche-Aviles, Henry White, Andrew Ewing, Simon Higgins, Lane Baker, Dongping Zhan, Jens Ulstrup, Paul W. Bohn and Serge Lemay

DOI: 10.1039/c6fd90066f

Mario A. Alpuche-Aviles opened a general discussion of the papers by Serge Lemay: The concentrations have to be very low and one worries about background issues and the effect of contaminants. Why do they have to be this low? Could these experiments be performed at higher concentrations?

Serge Lemay replied: We would love to be able to work at higher concentrations! The concentration is currently set by the need to work under conditions where there is less than one molecule on average present inside the device. This is needed because the finite time resolution of femtoampere-level electronics leads to blurring of single-molecule signals if they overlap significantly in time.[1] The device volume, in particular the channel width and height, are in turn mostly limited by our fabrication approach. While it would be possible to make shorter devices, the expected residence times would be reduced, which would in turn make the experiment more challenging.

1 M. A. G. Zevenbergen, P. S. Singh, E. D. Goluch, B. L. Wolfrum and S. G. Lemay, *Nano Lett.*, 2011, **11**, 2881–2886.

Nongjian Tao commented: In the design of your nanochannels and measurements, how do you select or optimize the size of the electrodes, the spacing between the top and bottom electrodes, concentration of analytes and other parameters?

Serge Lemay answered: The device dimensions are largely limited by microfabrication constraints. The width of the channels is set by the resolution of the photolithography method that we employ. Their height is set by the thickness of a sacrificial chromium layer that is etched away prior to measurements; while this can be controlled accurately, going below 40 nm tends to cause a much lower yield as imperfections near the edges and access holes cause short circuits between the electrodes. The length of the devices is very flexible (we've made devices with electrodes ranging from 10 microns to a millimeter) but for single-molecule experiments it was selected to yield longitudinal diffusion times ($L^2/2D$) on the scale of seconds (unfortunately such long events are missing in our experimental observations, as discussed in detail in the paper). The concentration is selected based on the device volume to yield less than one molecule in the device on average. Smaller volumes would simplify experiments as they would allow working at higher concentrations; this is an aspect that we are currently trying to address.

Olaf Magnussen commented: Did you assume in your theoretical treatment of the distribution times that these are given by the time of first return to the starting point in a 1D random walk? If yes, this may be problematic, since the same particle could reenter the channel directly (or after a short excursion of a few nanometers into the entrance hole) again, leading to a second event that is not statistically independent. Experimentally these events could not be resolved and would contribute to effectively longer events. This may significantly alter your statistics.

Serge Lemay replied: This is an excellent point and this behavior was indeed an important factor in our first published work on single-molecule detection.[1] In that case, the nanogap electrodes were imbedded away from the entrance of a one-dimensional channel, and it was likely that a molecule undergoing one-dimensional diffusion near the edge of the detection region would enter and exit repeatedly this region as suggested.

In the work presented here, a different geometry was employed in which the access holes into the redox-cycling region punch through the top electrode.[2] As a result, molecules coming back to the access holes after spending some time in the device will with a high likelihood diffuse back into solution. More technically, this corresponds to radiative boundary conditions at the inlets of the device, which yield a very low probability of exiting molecules returning into the device. We therefore do not expect that this behavior has a significant impact on the statistics described here.

We further note that 'composite events', in which a molecule repeatedly enters and leaves the detection region so fast that the train of events is interpreted as a single event, would lead to an excess of long events. The experiment however shows precisely the opposite tendency.

1 M. A. G. Zevenbergen, P. S. Singh, E. D. Goluch, B. L. Wolfrum and S. G. Lemay, *Nano Lett.*, 2011 **11**, 2881–2886.
2 S. Kang, K. Mathwig and S. G. Lemay, *Lab Chip*, 2012, **12**, 1262–1267.

Zhongqun Tian asked: Can you introduce or combine a micro-/nano-droplet technique with your micro-/nano channel system to make this new method even more powerful? If the electrolyte solution in the channel can be separated into many droplet with controllable droplet size and number of reactive species/

particle. In addition, the flow rate of the droplet and moving direction can be controlled to move into or leave the electrode pair region.

Serge Lemay responded: Thank you for this suggestion. It is certainly possible to combine our devices with droplet microfluidics since the materials and fabrication processes are compatible. This would make for very exciting experiments!

Wolfgang Schmickler remarked: These are elegant experiments. We believe that random walk theory holds for such processes, so the most interesting observations are deviations from this theory. You mentioned that traps may play a role. Could you please explain this in greater detail?

Serge Lemay replied: The distribution of residence times expected from random-walk theory is well defined and does not include any unknown parameters. Which diffusion coefficient to use, however, is subject to debate. It is our belief, based primarily on the interpretation of noise spectra at high concentrations[1–3] and of single-molecule current levels at low concentrations[4,5] that the effective diffusion coefficient is reduced due to reversible adsorption. Applying the same logic to single-molecule experiments, we would expect to see more 'long' events than we do in practice. This suggests the existence of sites in which the redox molecules can be trapped for an extended period of time.

1 M. A. G. Zevenbergen, P. S. Singh, E. D. Goluch, B. L. Wolfrum and S. G. Lemay, *Analytical Chemistry*, 2009,**81**, 8203–8212.
2 P. S. Singh, H.-S. Chan, S. Kang and S .G. Lemay,*J. Am. Chem. Soc.*, 2011, **133**, 18289–18295.
3 D. Mampallil, K. Mathwig, S. Kang and S. G. Lemay,*J. Phys. Chem. Lett.* , 2014, **5**, 636–640.
4 M. A. G. Zevenbergen, P. S. Singh, E. D. Goluch, B. L. Wolfrum and S. G. Lemay, *Nano Lett.*, 2011, **11**, 2881–2886.
5 S. Kang, A. F. Nieuwenhuis, K. Mathwig, D. Mampallil and S. G. Lemay, *ACS Nano*, 2013, **7**, 10931–10937.

Kristina Tschulik asked: The histogram in **Fig. 2b** shows a maximum residence time of 450 ms, while for a purely diffusional scenario the expected first passage maxima would be expected to be at lower values. Is the number of data points in the histogram statistically meaningful and if so does this observation hint towards the adsorption of $Fc(MeOH)_2$ at the electrode?

Serge Lemay replied: The appearance of a maximum in **Fig. 3b** is unexpected in any simple mechanism characterized by, for example, a single release rate. More likely our analysis is missing some of the shorter events with a duration around 0.5 s due to noise.

Sanli Faez said: To circumvent the lack of resolution for ultra-short events, you have based the analysis on events of a certain length or longer. If three or more electrodes were used instead of two, then correlative measurements could also resolve short paths. What is the prospect of performing these experiments with more than two electrodes?

Serge Lemay responded: This is absolutely correct and we have fabricated devices incorporating two or four electrode pairs for this purpose. The fabrication strategy employed for the devices in the present article cannot be applied to such devices, however, due to geometry constraints. As a result, these devices have a relatively large 'dead volume' on either side of the active region, which decreases the single-molecule signal and introduces additional noise. We have not yet succeeded in obtaining single-molecule data with these devices.

Patrick Unwin commented: If you sum the residence time histograms in **Fig. 3**, you end up with a time-averaged fractional occupancy which is very close to what you would expect based on the concentration of redox species employed (between 0.05 and 0.06). However, events less than 450 ms are thrown away. This therefore suggests that the average occupancy is higher than might be expected. Is this additional evidence for adsorption on the electrodes? Theoretically, what is the total time of the $<$ 450 ms events that are lost compared to those observed?

Serge Lemay responded: This was an acute observation, and after checking the calculation of the renormalization factor we have found an error. The curves were indeed inaccurately amplified. This will be corrected in the final manuscript and a footnote stating this change will be added.

Henry White remarked: This question regards adsorption of molecules in your thin-layer electrochemical cells. Have your ever put a molecule in one of these cells that does not adsorb? It appears to me that the observed 'adsorption' response may not be a chemical process but perhaps results from molecule entrapped in physical crevices, *etc.* Our laboratory has also observed a response in similarly constructed thin layer cells that suggests that adsorption of $Ru(NH)_6{}^{3+}$ on positively charged Pt electrodes.[1]

1 Q. Chen, K. McKelvey, M. A. Edwards and H. S. White, *J. Phys. Chem. C*, 2016, **120**, 17251–17260.

Serge Lemay answered: We have never seen a response that did not exhibit at least some degree of what we interpret as adsorption, either in the fluctuation spectrum or the time-dependent response. While this can be mitigated by temperature, potential or electrode modification, we have not found conditions that eliminate it completely. This is also true of $Ru(NH)_6{}^{3+}$, consistent with the observations of Chen *et al.* The question as to whether entrapment could be responsible is a fair one, but so far we have no direct evidence for this possibility: while we have not imaged devices post-measurement, metal films on which a sacrificial layer has been deposited and then etched using the same protocol as in actual devices did not exhibit additional roughness.

1 M. A. G. Zevenbergen, B. L. Wolfrum, E. D. Goluch, P. S. Singh and S. G. Lemay, *J. Am. Chem. Soc.*, 2009, **131**, 11471–11477.

David Fermin asked: Your analysis does not appear to include the possibility of ion pairing which is expected to be quite significant in the case of $Fe(CN)_6{}^{3-}$. Can you comment on whether this process can have an effect on the unexpected first-passage time statistics observed experimentally?

Serge Lemay responded: Ion pairing in the bulk would influence the effective diffusion coefficient of the species, but this is taken into account as we use diffusion coefficients measured in the same electrolytes using ultra-microelectrodes (UMEs). At the surface, however, there is evidence that the nature of the anion greatly influences the degree of adsorption, to the point where it follows the Hofmeister series.[1] Whether ion pairing is the correct theoretical framework to describe these species-dependent effects however remains unclear.

1 D. Mampallil, K. Mathwig, S. Kang and S. G. Lemay, *J. Phy. Chem. Lett.*, 20145, 636–640.

Andy Mount asked: If reversible adsorption were the process which leads to the observed value of D, D_{eff}, being lower in **Fig. 3** than the accepted value of D measured in solution, might one not expect a dependence of D_{eff} on the applied electrode potential (within the constraints that this potential should always be sufficient to ensure mass transport limited reaction of the redox species at the electrode), as the potential would be expected to affect the adsorption equilibrium position and the rates of the adsorption and desorption processes? Has this been observed?

Serge Lemay answered: We have indeed observed a dependence of the effective diffusion coefficient in measurements at high concentrations.[1,2] We did not perform a systematic study at single-molecule level, however. This is certainly a worthwhile experiment to attempt in the future.

1 P. S. Singh, H.-S. Chan, S. Kang and S. G. Lemay, *J. Am. Chem. Soc.*, 2011, **133**, 18289–18295.
2 S. Kang, K. Mathwig and S. G. Lemay, *Lab Chip*, **12**, 1262–1267.

Wolfgang Schmickler asked: I am surprised that the two redox couples, which are chemically quite different and also carry different charges, behave in such a similar way. I supsect that the trapping is not chemical, but a physical effect. Do you agree?

Serge Lemay responded: Yes, this is the most natural hypothesis. Note however that the apparent level of adsorption as revealed by noise analysis also depends both on the nature of the anion[1] and on modifications to the electrode surface.[2]

1 D. Mampallil, K. Mathwig, S. Kang and S. G. Lemay, *J. Phys. Chem. Lett.*, 2014, **5**, 636–640.
2 P. S. Singh, H.-S. Chan, S. Kang and S. G. Lemay, *J. Am. Chem. Soc.*, 2011, **133**, 18289–18295.

Richard Crooks asked: The electrolyte solution in your experiments is not degassed, and therefore the concentration of oxygen is about 8 orders of magnitude higher than the redox probe that you use for measurements. Accordingly, it seems likely that bimolecular reactions between the redox probe and oxygen will have a significant effect on measurements and hence conclusions from the experiments. Thoughts?

Serge Lemay responded: This is a very good point. Because of the small volumes involved, degassing in our setup would require building a gas-tight cell and we have not done this yet. We therefore cannot rule out interactions between our simple redox molecules and oxygen.

Richard Nichols said: Could you clarify any cleaning procedures for the nanochannels, this would appear to be not so straightforward as for planer disc gold or platinum electrodes which can be polished or flame annealed. 'Aggressive' oxidation/reduction (for example at high scan rates) have been used for *in situ* electrode cleaning. These can also dramatically alter surface micro/nano structure and increase surface roughness. Would such be suitable for an *in situ* cleaning? This may in itself be useful for increasing potential 'traps' and further addressing the issue of dwell time for the redox species on the surface. Also for macro electrodes and solution free redox couples the peak to peak separation in cyclic voltammograms is often useful for assessing the state of the surface and the reversibility of the electron transfer process. I realise that this could not be directly applied to such small cavity channels but are there any other diagnostic tests to assess the reversibility in nanochannels?

Serge Lemay answered: It is absolutely the case that a limitation of microfabricated devices is the inability to apply standard electrode polishing procedures. As an alternative, we routinely employ cyclic voltammetry in sulfuric acid to clean the electrodes following etching of the chromium sacrificial layer to create the nanochannel. Repeating this procedure after prolonged measurements (a few hours) yields non-ideal responses for the first scan, indicating that the electrode surface degrades over the course of the measurements. As an alternative to peak-to-peak separation for diagnosing electrode quality, mass transport in nanogap electrodes is also efficient enough to observe non-ideal, Butler–Volmer-like cyclic voltammograms even for very fast couples.[1] The apparent heterogeneous rate constant observed in such experiments does tend to decay over time on the scale of hours, but this can only be checked after the conclusion of single-molecule experiments since the introduction of sufficient concentrations for standard cyclic voltammetry make it impossible to return to ultra-low concentrations afterward.

1 M. A. G. Zevenbergen, B. L. Wolfrum, E. D. Goluch, P. S. Singh and S. G. Lemay, *J. Am. Chem. Soc.*, 2009, **131**, 11471–11477.

Frederic Kanoufi communicated: Based on the discussion, your device may behave as a nanocolumn of adsorption chromatography. Ferrocene has likely high affinity for the channel walls. One could get inspiration from separation science strategies: using a non-adsorbing compound, coating the channels with short-chain thiols to decrease adsorption and prefer partitioning, using a flow to optimize the solute residence time, using an optical readout such as fluorescence single-molecule analysis…

Serge Lemay communicated in reply: Thank you for the suggestions. We have indeed already partially implemented some of these approaches. First, we did observe a decrease in the level of adsorption at high concentrations upon modifying the electrodes with organothiols.[1] We have not yet attempted to combine this with single-molecule experiments, however. Second, we have built devices allowing convection of the sample along the channel, but it turns out that the PDMS material that we employed introduces too much contamination to allow sub-nanomolar measurements. Rectifying this problem is currently one of our main goals.

1 P. S. Singh, H.-S. Chan, S. Kang and S. G. Lemay, *J. Am. Chem. Soc.*, 2011, **133**, 18289–18295.

Sanli Faez opened a general discussion of the papers by Paul Bohn: You have mentioned that you could not image all nanopores together during the experiment, which explains why some of the electrically detected signal is not captured in the optical channel. What is the main limitation for measuring on a single nanopore? How about increasing the optical field of view to include the whole array of pores? What limits that?

Paul W. Bohn replied: The design of the nanopore array was undertaken to optimize two somewhat countervailing factors. In order to maximize electrochemical efficiency, we need small pore-to-pore spacing, which improves capture efficiency of molecules desorbing from a given nanopore by a neighboring nanopore. On the other hand, larger pore-to-pore spacings are required to allow the observation of a single nanopore by optical microscopy. The particular structure on which the published results were obtained was designed as a compromise between these two constraints. The pore spacing is small enough to yield good electrochemical capture efficiency but at the expense of not being able to resolve an individual pore.

Christine Kranz asked: All electrochemical data shown are obtained with a outer sphere redox species (hexaammineruthenium chloride), however most of relevant analytes in electroanalytical applications exhibit inner sphere behavior. Hence, how is the performance of the nanopore-confined recessed ring-disk electrode arrays for such applications? What is the effect of adsorption on the long-term performance?

Paul W. Bohn answered: We have conducted numerous additional experiments addressing the application of nanopore arrays to electrochemical detection, using, for example, $Ru(CN)_6^{3/4-}$, anthraquinone-2-sulfonic acid, ferrocene/ferrocenium, dopamine and ascorbic acid. These experiments establish that nanopore geometry is a key factor determining the magnitude of the electric double layer effect, as is the role played by the magnitude and sign of the charge of the redox-active species. Redox pairs exhibiting charges of different magnitude and sign show completely different response upon decreasing the supporting electrolyte concentration. Thus, the combined effects of ion accumulation and ion migration in these nanopore arrays overwhelm distinctions between inner and outer sphere redox processes.

Andy Mount asked: In a previous Faraday Discussions paper,[1] we reported significant changes in the voltammograms observed at our Microsquare Nanoband Edge Electrode (MNEE) array systems on reducing the ionic strength from hundereds of mM to tens of mM when utilising bands of 50 nm width. An additional current was observed in the diffusion-limited region of the voltammogram for two different redox species, ferrocenecarboxylate and hexaammineruthenium(III), which we attributed to the additional contribution from transport of these redox species due to migration. Have you observed similar effects in your dual electrode systems at lower ionic strengths and how does/would this complicate generator/collector analysis in your dual band systems?

1 I. Schmueser, A. J. Walton, J. G. Terry, H. L. Woodvine, N. J. Freeman and A. R. Mount, *Faraday Discuss.*, 2013, **164**, 295–314.

Paul W. Bohn answered: Yes, this is a very interesting phenomenon which we have also observed in our nanopore electrode arrays.[1,2] In fact, we are currently preparing a manuscript which clearly shows that the geometry of NEAs and the magnitude and sign of the charge on the redox species strongly affect both ion accumulation (permselectivity) and ion migration contributions to enhanced currents in NEAs. Our results confirm the screening of negatively charged species, in contrast to the accumulation of cations. Quantitative modeling and experiments in somewhat larger nanopores (d $\geq$ 500 nm) has shown that at the lowest concentrations of redox species (1 μM), the resulting current enhancements can be assigned roughly 80% to ion accumulation and 20% to ion migration. Furthermore, the combination of ion accumulation and unscreened ion migration at low ionic strength provides additional sensitivity when coupled to redox cycling effect in our generator/collector system. In optimal cases redox cycling gives a ~50-fold current enhancement, while working without supporting electrolyte produces an additional factor of ~50–60, resulting in overall current enhancements as large as 3000 in favorable cases.

1 C. Ma, N. M. Contento and P. W. Bohn, *J. Am. Chem. Soc.*, 2014, **136**, 7225–7228.
2 C. Ma, W. Xu, W. R. A. Wichert, and P. W. Bohn, *ACS Nano*, 2016, **10**, 3658–3664.

Frederic Kanoufi remarked: I am wondering what is the sensitivity of your electrode array system, in terms of current detection/noise or highest frequency achievable. Actually the μA currents, provided by the large number of nanoelectrodes may not allow the use of a very fast acquisition rate which could be necessary for single entity detection.

Paul W. Bohn replied: The large currents referenced (μA range), which are exhibited in **Fig. 4** arise from two sources. One is the large number of nanoelectrodes (~10^5) in the array, another is due to the high concentration (1 mM) of redox species used. We operated under these conditions to illustrate the fundamental nature of the NEAs in GC and non-GC mode operation (sigmoidal I–V curves with little capacitance). For single entity detection, the concentration for single molecule occupancy in each nanopore is *ca.*1 μM. Under these conditions with ~10^5 nanopores we observe limiting currents of ~0.3 nA. Certainly, the large number of nanopores precludes observation of stochastic signals corresponding to single electron transfer events. Thus, in our more recent experiments, we work with smaller array sizes and at much lower concentration (1 nM, where the average electrochemical pore occupancy is 0.0011). Under these conditions, we observe highly correlated current fluctuations between the generator and collector electrodes. In addition, we observe smaller charging currents for the top collector electrode (**Fig. 5(B)**), which makes it possible to operate the experiment at scan rates up to 100 V s^{-1}.

Richard Crooks opened a general discussion of the papers by Andrew Ewing: In your model of the vesicle contacting the electrode surface, followed by a pore opening and subsequent rapid redox chemistry of the contents, how are the changes in the charge on the redox molecules compensated by the supporting electrolyte? I ask because it seems like there is no way for the redox molecule and supporting electrolyte to interact across the bilayer membrane.

Andrew Ewing replied: At least for cell vesicles and decorated liposomes we think that proteins on the vesicle make space between the remaining membrane after pore opening to allow protons to diffuse out as they are oxidised and ions to diffuse in. The oxidation reaction is at the electrode where this would all occur, so I think it makes sense. We are working on this and will send you a manuscript when we finalize!

Wolfgang Schuhmann asked: You had to apply a quite high potential to oxidise dopamine. Does this potential change the adhesion of the vesicle (or the proteins inside the bilayer membrane) and is hence influencing the opening of the pore?

What did you mean with 'a non-electrified interface'? Each interface in an electrolyte is adapting to a certain potential.

Andrew Ewing answered: For dopamine amperometric detection it is standard to use about 0.7 V *vs.* AgAgCl so as to be on the wave plateau. This is not a very high potential. But, with the vesicles, 0.5 to 0.7 V across a 5 nm membrane is about in the range of field needed to electroporate a membrane. That is a main point of what we are stating here. We think the membrane is electroplated when the lipid bilayer gets close enough to the membrane to get to this field strength. The diffusional evidence suggests that the potential does not influence the adhesion of the vesicle to the electrode.

I cannot find where I used 'non-electrified' Otherwise, I agree.

Justin Gooding asked: The model of the membrane pores in the different liposomes and vesicles suggests that the size of the pore is related to membrane tension. What do you feel the role of the membrane viscosity is? Do you think adding cholesterol to make the membrane less fluid is worth exploring?

Andrew Ewing responded: Excellent idea. We have not explored membrane tension in this regard, but it has been discussed. It is difficult with vesicles to manipulate, but the cholesterol idea is a great one to try.

Jan Clausmeyer commented: There is a marked difference in the opening behavior between artificial liposomes decorated with peptides and chromaffin vesicles which carry a complex mixture of proteins. Is there already any information on how the nature of the proteins might affect the interaction with the electrode surface? How was the peptide selected? Does the electrode material matter for the vesicle opening mechanism, too?

Andrew Ewing responded: We only looked at the one peptide so far. In the paper we conclude though that the main difference is probably from the presence of the protein dense core in the vesicles that will inhibit diffusion. The electrode material needs to be studied more.

Patrick Unwin asked: I'd like to support the arguments made by Henry White about the location of the pore opening. Even if the pore opens at the top of the impacting vesicle, the area of the collector electrode is enormous and there is a strong concentration gradient towards it because it acts as sink. I would have thought that the collection efficiency would be reasonably close to 1 and that the timescale and profile would be of the order that seen in the paper. Have you done the simulations for this case?

Andrew Ewing answered: I agree with the arguments for diffusion alone, but we have other arguments for the pore opening at the electrode *via* electroporation including the potential dependence which matches electroporation, simulations with the nano tip, the peptide result, we see pre-spike feet, *etc.* I will not say more here as we are preparing a separate manuscript and will send it to you later. But, fundamentally, on the diffusion argument you are correct and I was incorrect, except maybe the overall time, which we will address in the new manuscript.

Martin Edwards communicated: The numerical simulations presented below complement the calculations of characteristic diffusion times as discussed by Profs. Unwin and White and support that suggestion that the time for the current to fall upon vesicle opening and the amount collected is not sufficient on its own to suggest that the vesicle opens on the bottom.

The simulated flux-time curve presented (Fig. 1) is for the 'worst case' scenario where the vesicle opens with a pore on top. Despite this, essentially all of the contents is recovered by the electrode within a relatively short time (integral of flux equals molar content of entire vesicle = vesicle volume × concentration). It is interesting to note that the timescale is comparable with that shown in **Fig. 3** in the paper.

Also shown is a concentration plot taken at 2 ms (Fig. 2). The flux lines shown in white show how all of the flux from the vesicle is collected at the electrode.

Simulation details: Comsol Multiphysics 5.2a, diffusion coefficient 5×10^{-6} $cm^2\ s^{-1}$, vesicle diameter 330 nm, pore opening diameter 33 nm radius, vesicle initial concentration 80 mM. Note: changes to simulation parameters will change the exact shape of the curve, but the collection efficiency is expected to be comparable.

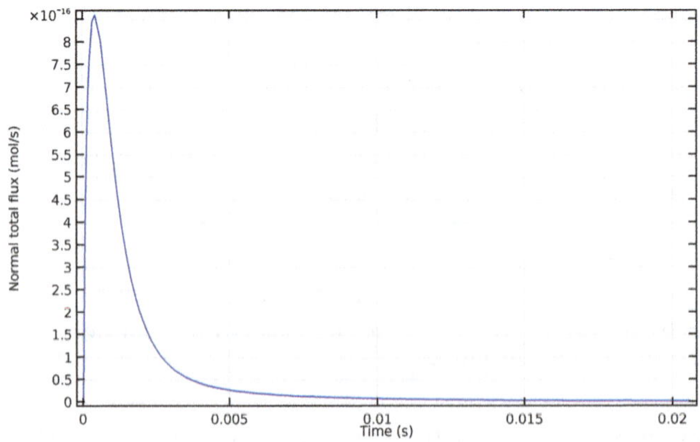

Fig. 1 Flux *vs.* time for release from a pore at the top of a vesicle.

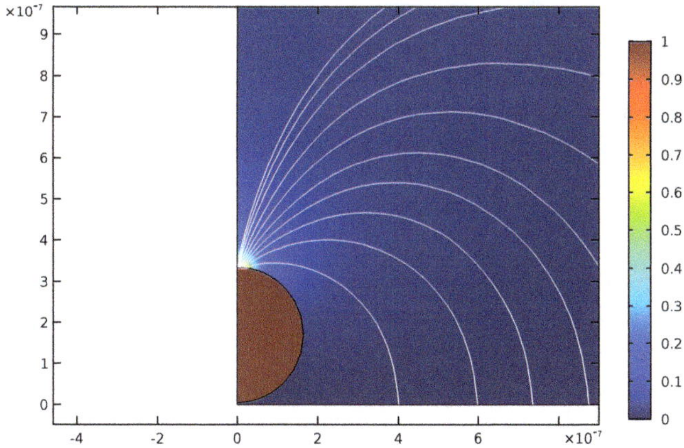

Fig. 2 Concentration of the molecule released from the vesicle (colour) and flux lines (white). Taken 2 ms after the start of simulations.

Andrew Ewing communicated in reply: I am surprised you publish this now, and we have discussed this in detail. It looks like the sim has a slower tail and it does not explain the differences between the peptide liposomes and the liposomes, which for a pore at the top should be the same. It also does not explain the potential dependence, the consistency with nano tip electrodes in and out of cells, and especially the presence of the pre-spike foot.

Christine Kranz opened the discussion of the paper by Lane Baker: I wondered about the set up for the experiment sketched in **Fig. 5a**. The positioning of the probe at the Ca^{2+} source seems to be rather cumbersome. Why wasn't the transport of Ca^{2+} through pores (as shown in **Fig. 8**) not used to demonstrate the functionality of the ICP-SICM probe?

Also what are the white spots visible in **Fig. 8b**?

Lane Baker replied: The setup is cumbersome, but quick. To be honest, we were rushing to meet the deadline for this submission, and the setup in **Fig. 5a** was much quicker to implement. As you point out, the setup in **Fig. 8** is a much better way to quantitate what is happening, and really demonstrate diffusion (or transport) measurements, and we have done that now also. In **Fig. 8b** the white spots are debris adsorbed to this particular membrane surface. We have since collected images absent of such features.

Christine Kranz added: In your paper you state that an increase in R_p is attributed to remaining cellular debris at the SICM barrel opening after the membrane patch was broken. Could you comment on whether such debris is affecting the imaging performance and in general if clogging of the second 'open' channel is an issue? Could you comment on success rate (some statistical data) on gigaseal formation on such dual barrels?

Lane Baker replied: The pipettes do drift some with time, but they generally never clog completely, and this is generally the case for SICM tips. The tips aren't usually capable of making multiple consecutive patches. So far the imaging performance isn't such a big deal, but we don't have a lot of data on that yet (no studies of resolution). The gigaseal formation is ~50% of the time and a channel is present another ~50% of the time (these are rough numbers), but a fair degree of this is (1) user experience and (2) protein expression level in the cell lines. Expression level is related to transfection efficiency and time, as right of now we use transient expression, where cells stop expressing eventually.

Martin Edwards communicated: Scanning ion conductance microscopy is typically used in a distance modulation mode.[1] Using a lock-in amplifier to detect an alternating current gives a much more stable feedback mode, especially if there is drift in the current or one is working with smaller currents such as is inevitable with smaller probes.[2,3] Can you operate these membrane patch probes in distance modulated mode? I wonder whether the bilayers would be stable under mechanical stresses; would there be any artefacts due to, for example, capacitive currents due to fluid forces causing changes in the membrane area?

1 C.-C. Chen, Y. Zhou and L. A. Baker, *Annu. Rev. Anal. Chem.*, 2012, **5**, 207–228.
2 A. I. Shevchuk, J. Gorelik, S. E. Harding, M. J. Lab, D. Klenerman and Y. E. Korchev, *Biophys. J.*, 2001, **81**, 1759–1764.
3 M. A. Edwards, C. G. Williams, A. L. Whitworth and P. R. Unwin, *Anal. Chem.*, 2009, **81**, 4482–4492

Lane Baker answered: The probe configuration we show here is indeed stable with distance modulation. The last figure in our paper here is taken in AC mode, with feedback from the open barrel of the pipette. The patch remained intact throughout the course of scanning this image (a capacitive membrane test was performed before and after the image). This might give some clues as to where the patch is actually located on the probe, as if the patch were inside of the pipette it might be more stable under close probe surface distances that can be attained in with feedback imaging routines.

Richard Crooks asked: Your technique is based on detection through an ion channel immobilized on the capillary by patch clamping. How reproducible are the ion channels once attached to the capillary, what are the effects of irreproducibility on the conclusions that can be drawn from the experiment, and is there a non-electrochemical method to characterize the ion channel and its environment?

Lane Baker responded: Each probe is a unique tool, meaning that there is variability in the pipette, the patch and ion channel(s) in the probe. With that said, the conductance of the ion channels don't change a lot, but the number could change. Pipette geometry has an effect, although that may be a lesser effect for some aspects of sensing (because the ion channel is so small relative to the tip opening). With that in mind, each probe has to be calibrated if quantitative results are desired. Electron microscopy (especially STEM) is useful for characterizing the pipette. Previous work on larger pipettes has used TEM to look at patches in pipettes. These are hard experiments, but something on our radar.

Henry White asked: Ion channels frequently diffuse rapidly in lipid bilayers. I was wondering how this diffusion would affect imaging resolution where the ion channel is in a relatively large lipid patch suspended across the orifice of a pipette?

Lane Baker answered: Yes, this is definitely true, channels will diffuse. There is a difference here relative to synthetic bilayers, as 'real' bilayers from cells have components (*e.g.* cholesterol) that change diffusion properties, but it surely won't stop diffusion completely. Because we are imaging with the open barrel, the topographic resolution probably wont suffer much, but the 'chemical' resolution from the ion channel will be influenced. This also will depend on if the patch is an inside-out or outside-out patch, as the membrane area available to the channel(s) is likely different also, although we still have no real structural information on the geometry of the patched barrel. We have some plans to try and measure diffusion within the patch by holding the tip at a fixed distance and seeing if we can used diffusion of the channel to stochastically map the area under the patched barrel. That may or may not help define the extent to which diffusion influences this technique.

Paolo Actis commented: To achieve high resolution imaging there is a need for very sharp nanopipettes but it is very difficult if not impossible to excise a membrane from a sharp nanopipette. Novak *et al.*[1] proposed a method to precisely break the nanopipette tip to solve this issue. How are you planning to address the spatial resolution/selectivity issue?

1 P. Novak, J. Gorelik, U. Vivekananda, A. I. Shevchuk, Y. S. Ermolyuk, R. J. Bailey, A. J. Bushby, G. W. J. Moss, D. A. Rusakov, D. Klenerman, D. M. Kullmann, K. E. Volynski and Yuri E. Korchev, *Neuron*, 2013, **79**, 1067–1077.

Lane Baker replied: While SICM is surely capable of high resolution imaging, I'm not sure we will go after that aspect with these probes. The probes have fairly strict geometric requirements to allow the dual barrel approach shown here, and the *in situ* breaking method of Novak might be hard to implement here. We have decent resolution for finding larger locations at cell/tissue interfaces, and that is what we are targeting first. It is really interesting to think about using smaller probes in the future, we just aren't there right now.

Martin Edwards asked: Due to the way you produce the probe I assume that the number of pores you end up with is a stochastic quantity. Your data appear to show a single channel. How may pores do you typically get in your probe? And do you do anything to control this number?

Lane Baker answered: So far we haven't done anything to control the number, in general the protein expression level seems to be variable. We have looked at populations of channels in different patches and typically get 0, 1 or 2 open conductance states.

Zhongqun Tian opened a general discussion of the paper by Paul Bohn: I think the most important subject in the field of single entity electrochemistry could be the single active sites, *e.g.*, single adsorption/reaction sites. It may be necessary to note that single-atom catalysts is becoming a new frontier in heterogeneous catalysis, focusing on single atoms anchored to metal oxide/metal surfaces. These reaction active sites are monodispersed at the surface and show very high activity and selectivity.[1,2] More importantly, two reports just published directly relate to electrocatalysis.[3,4] The next extremely hard target could be to characterize and thus understand structurally well-defined sites. Since the 1970s electrochemists have tried to use structurally well-defined single crystal surfaces to mimic the active sites because the atomic structure is the same periodically across the whole surface, which is easily characterized and simulated theoretically. However, people have realized that these single crystal surfaces cannot represent the whole active sites in real electrochemical systems well, *e.g.*, in electrocatalysis, electroplating, corrosion, because the single site includes the support/substrate. Preparing single entity active sites then well characterizing them one by one are a high challenge target. Vibrational spectroscopy may be a good diagnostic tool, such as tip-enhanced Raman spectroscopy (TERS). Could I have your comment on this?

1 X.-F. Yang, A. Wang, B. Qiao, J. Li, J. Liu and T. Zhang, *Acc. Chem. Res.*, 2013, **46**, 1740–1748.
2 J. D. A. Pelletier and J. M. Basset, **Acc. Chem. Res.**, 2013, **49**, 664–677.
3 P. Yin, T. Yao, Y. Wu, L. Zheng, Y. Lin, W. Liu, H. Ju, J. Zhu, X. Hong, Z. Deng, G. Zhou and S. Wei, Y. Li, *Angew. Chem. Int. Ed.*, 2016, **55**, 10800–10805.
4 S. Yang, J. Kim, Y. J. Tak, A. Soon and H. Lee, *Angew. Chem. Int. Ed.*, 2016, **55**, 2058–2062.

Paul W. Bohn replied: We agree that the preparation, characterization, and study of single-atom reactive sites is a 'holy grail' problem in (electro)chemical dynamics. Ideally, one could also study the reaction events one-at-a-time. These experiments are, as you appreciate, exceptionally difficult. From the outset, one may distinguish between two limiting systems: (a) those in which the special atomic sites are placed with structural regularity – and are thus relatively straightforward to find, but difficult to synthesize/fabricate; and (b) those in which the atomic sites are randomly placed – thus making the structures relatively easy to fabricate but the atomic sites difficult to locate for single-entity studies. We also agree that, realizing the caveats above, high spatial resolution vibrational spectroscopy (such as TERS) would be an excellent tool for such studies provided that the sensitivity is sufficient for the observation of the extremely limited number of reactants and products. In this regard, the construction of artificial nanostructures (not our nanopores necessarily, but other nanostructures allowing the volume to be focused on the single-atom site of interest) may provide a significant advantage.

Ashley Page directed a general discussion to the paper by Lane Baker: In the paper you deliberately over-express certain ion channels under the hope that a given membrane patch contains an individual ion channel of the desired type (Ca^{2+} gated K^+ channels/BK channels *etc.*). How often and to what extent do you see a convolution of the signal from native ion channels and how do you rule this out as a contributing factor in your experiments?

Lane Baker responded: We are using HEK cells that don't have a lot of other high conductance channels. But for sure, some are there. Because we know what the conductance and the shape of the current–voltage response for the channels we've chosen to look at, we can make sure the probe responds in a manner that strongly suggests what we have at the tip. One advantage we have here is – so far – we aren't doing any new biology, so we have decades of results from real patch-clamp measurements to use for reference.

Nongjian Tao opened a general discussion of the paper by Andrew Ewing: It would be great to visualize individual particles (or vesicles) as they land and react on the electrode with an optical imaging technique. Of course, there are always important details that are impossible to 'visualize' experimentally. That's where theory (and your model) can play an important role.

Andrew Ewing responded: Thank you. We will think about it.

Richard Crooks opened a general discussion of the introductory lecture by NJ Tao: What is the source of the signal in the SPR image for *Cyt C* and why is the background signal so low?

Nongjian Tao replied: Both traditional and plasmonic cyclic voltammograms (CVs) show redox peaks of *Cyt C* adsorbed on an electrode, but the peaks in the plasmonic CV are more pronounced than those in the traditional CV. This is because the plasmonic image measures the optical polarizability difference between the reduced and oxidized molecules. In the case of proteins, *e.g.*, *Cyt C*, the difference is large because of the conformational change in the protein associated with oxidation and reduction, which leads to relatively large redox peaks in the plasmonic CV. The background current in the plasmonic CV arises from double layer charging (like the traditional CV), which is small because of surface passivation by a mercaptopropionic acid layer.

Patrick Unwin opened a general discussion of the paper by Paul Bohn: You presented some new data where the anti-correlated 'step changes' in the current of the generator and collector electrodes seemed quite long, of the order of a second or more. Given the high density of elements in the array and the concentration of molecules in solution, are discrete changes in the population of molecules in the device (loss/removal of a molecule) on that timecale reasonable? Have you run simulations for these experiments?

Paul W. Bohn responded: We are currently conducting detailed simulations of the behavior of these nanopore arrays, but do not yet have complete results. However, the observed times are reasonable considering a simple model based on freely diffusing molecules. The nanopore electrode arrays were patterned in a 20 μm × 20 μm square array with an inter-pore (center-to-center) separation of 250 nm, giving an 80 × 80 array. The key feature affecting transport dynamics is that when a molecule escapes a nanopore, it is likely to be captured be a neighboring nanopore. Thus, in this simple picture a molecule is only lost when it escapes the entire array which would require a series of capture-hop-capture events. Diffusion over 10 μm would be of the order $t = x^2/2D = (10\ \mu\text{m})^2/(2 \times 10^{-6}\ \text{cm}^2\ \text{s}^{-1}) = 500$ ms for all of the inter-pore diffusion events. When the molecule resides in a pore, the average duration time in single nanopore is about ~10 ms. Thus, ~40 captures would be required to reach the edge of the array and escape, giving rise to an additional 400 ms. Thus, within the assumptions and simplifications of this picture, capture times of the order 1 s, as observed experimentally, are reasonable.

Bradley Thomas asked: In response to two different questions posed to you by other delegates you have said you never perform post-experiment analysis of your material. This would clearly be a worthwhile pursuit given the questions it would enable you to answer, so is there a fundamental reason for not doing so this given the breadth of techniques available?

Paul W. Bohn responded: There is no fundamental reason (beyond time and cost) that would prevent us from examining the nanopore arrays in cross-sectional SEM imaging *ex post facto*, and we appreciate the excellent suggestion. However, it is important to understand the limitations of such an experiment. First, the cross-sectional images will at most be able to reveal the structure of a few nanopores located proximal to one another. We could possibly get around this by sequentially moving and re-sectioning the device in the FIB. The other issue with respect to the optically-interrogated nanopores is that although we can identify the locale from which the optical data are acquired, it is difficult to pinpoint these pores exactly, so sectioning to reveal the state of these pores after the experiment would involve some uncertainty.

Michael Eikerling opened a general discussion of the paper by Andrew Ewing: **Fig. 6** in your article shows the different surface-functionalized vesicles considered in your study. The figure is used in an attempt to provide a mechanistic explanation of the observed rupture process. What role does the surface tension of the vesicle (that varies with chemical composition and surface structure) play in the process? Can this effect be explored systematically by varying the size of the vesicle and the chemical structure of chemisorbed species (using simpler surfactants)?

Andrew Ewing responded: I would guess that tension will play a role in the size of the pore that opens. I think this is a good idea to test the vesicle sizes and perhaps osmolality to look at this effect. I suspect, however, that the tension is the smaller affect *versus* the wettability and pore restrictions that are likely more from the functionality of the surface. Experiments like you propose will be needed to test this I think.

Patrick Unwin opened a discussion of all three papers by communicating: With small scale electrodes, the issue of contamination becomes increasingly problematic and I wonder whether each of you could comment on the significance of such effects in the experiments and systems you have described:

(a) For Andrew's studies on vesicle cytometry, it is well known that the oxidation of dopamine at high concentrations leads to the fouling of carbon electrodes[1] and as many electrochemical impacts are recorded in these studies and the 'electrochemical footprint' will be a bit larger than the projected area of the vesicle, does this lead to a deterioration of the electrode response over time? Could such effects contribute to the decay plots of frequency *vs.* time shown in **Fig. 3**?

(b) For Paul's studies on recessed ring-disk electrodes, how stable is the voltammetric response over time and how long can experiments be run?

(c) Could Lane make some comment on the advantages and disadvantages of SECM *vs.* SICM tips in terms of long term use and fouling effects?

1 A. N. Patel, S. Tan, T. S. Miller, J. V. Macpherson and P. R. Unwin, *Anal. Chem.*, 2013, **85**, 11755–11764.

Andrew Ewing responded: Actually, the projected area is defined by an area that is based on a twice the diameter of the vesicle. But, the vesicles are so small compared to the electrode we see no decrease over time during these experiments. Considering the overpotential and that we are doing amperometry, this might not be surprising. Also, this might also argue for the pore being at the electrode–vesicle interface. If this is the only place for messengers to escape, with high over potential, they will be trapped and oxidised at the small region of the electrode. If they are coming from the top, then the area affected by the response is much larger and could be fouled for future events. I think, however, this is less likely.

Paul W. Bohn responded: If the recessed ring-disk electrodes are characterized in full supporting electrolyte solution, the electrochemical signal is very stable – at least 100 CV scans (each scan around 10 s). In addition, the measurements are qualitatively reproducible across different sample conditions (*i.e.* draining and refilling the nanopore array) even over periods as long as a few months. This is quite fortunate, because the fabrication and conditioning of the arrays is non-trivial.

Lane Baker replied: For the question of SECM *versus* SICM tips for electrode fouling, there are two significant points. First, ease of preparation of SICM tips makes electrode fouling a less significant issue because less time is invested in the preparation of each individual tip (relative to SECM tips). Second, SICM pipettes have an advantage that they don't rely on faradaic electron transfer at the smallest, most crucial part of the probe, so they seem to be more resistant to fouling. SICM tips do change subtly in terms of ion currents when used for extended periods, but they can be used for extended periods in solutions of proteins without failure. SICM tips do clog at times, but not really to the extent one might imagine. I think once a passivating layer of whatever is in solution adsorbs to the glass wall, that likely mitigates further adsorption.

Venkateshkumar Prabhakaran opened a general discussion of the paper by Paul Bohn communicating: The diffusion of redox molecules in the nano pore array electrodes is also one of the important factor that determines performance, efficiency and sensitivity of redox cycling. Could we use electrochemical impedance spectroscopy in this study to estimate diffusion parameters and correlate to performance data?

Paul W. Bohn communicated in reply: Impedance spectroscopy is definitely possible, given the low noise floor ($\leq$ 20 fA) we routinely observe in our measurements. Although this paper and our previous papers on electrochemistry in nanopore arrays are focused on voltammetry, the diffusive transport of redox species can induce a Warburg impedance, from which (in principle) information could be obtained regarding the diffusion of oxidant and reductant. We appreciate the question, because it may give rise to a route to measure hindered diffusion – a characteristic behavior of molecules in confined environments, such as our nanopores.

Richard Nichols opened a general discussion of the paper by Jens Ulstrup: When hemin is attached to the surface through the thiolated quadraplex the voltammetry is much sharper, while when it is physisorbed it is much less distinct. What is the reason for this? What is your model for the changes?

Jens Ulstrup replied: This observation was in fact somewhat unexpected. Physisorption around pH 7 is weak and gives generally poorly ordered adlayers. Both the voltammetry and the order as disclosed by *in situ* STM are much better at high (10–12) pH. Hemin marking of the quadruplex must, however, be around pH 7 to maintain stability of the quadruplex. The way we look at it, is that hemin is bound terminally to the quadruplex molecules by π–π stacking in well-defined configurations and that the quadruplex in turn is bound to the surface in coverages higher that hemin binding directly to the Au(111)-electrode surface. In addition the strong signals might then also originate in facile electron transfer through the (not so long) quadruplex molecule.

Nongjian Tao commented: Is there a way to characterize your DNA quadruplex and redox-modified quadruplex structures with a structural technique, *e.g.*, X-Ray?

Jens Ulstrup answered: The single-strand oligonucleotide used is not a commercial product but prepared and purified by the Nucleic Acid Center at University of Southern Denmark. We are keen to characterize the quadruplex structure further and have initiated steps first to use CD spectroscopy. I do not know how easy it would be to crystallize the samples, but NMR spectroscopy might be an alternative option.

Olaf Magnussen asked: You observed 2 different structures in your STM studies of the DNAzyme. Could you give the characteristic dimensions of those, *i.e.*, the line spacing and mesh width, respectively? How do these compare those to the size of 12G-qs and the DNAzyme? Do you see both structures also in the absence of hemine? Could hemin bind to to two neighboring 12G-qs and thus assist the formation of the observed small patches of aggregates in a more disordered matrix?

Jens Ulstrup responded: The mesh width was measured to be 1.6 ± 0.2 nm, which involves hydrogen bonding between two adjacent guanine bases. We can therefore say that the side length of the G-quartet is 1.6 ± 0.2 nm. The line spacing of Domain 2 in **Fig. 5B** was measured to be 1.5 ± 0.1 nm, which corresponds to the intermolecular distance between the two adjacent single strands of unfolded 12G-qs.

Both structures would most likely form also without hemin, but we are still working towards *in situ* STM visualization of hemin-free 12G-qs.

Hemin binding to two neighboring 12G-qs might be possible and would, presumably, then extend the linear structural units in Domain 2. This might also lead to small patches of aggregates, but there is no immediate evidence in the images that this is what is seen.

Robert Johnson said: At negative potentials *circa* −0.8 V *vs.* SCE, double-stranded DNA at the electrode surfaces can 'unwind' or 'denature' into its constituent single stranded components. See, for example, Johnson *et al.*[1] and Palecek.[2] A G-quadraplex is held together by Hoogsteen hydrogen bonding, and so I suspect when immobilized at the electrode it may be possible to unravel the quadraplex as the potential is scanned negative. Do you observe denaturation and/or structural changes to the quadraplex at negative potentials?

1 R. P Johnson, A. M. Fleming, Q. Jin, C. J. Burrows and H. S. White, *Biophys. J.*, 2014, **107**, 924–931.
2 E. Palecek, *Collect. Czech. Chem. Commun.*, 1974, **39**, 3449–3460.

Jens Ulstrup answered: The G-quadruplex voltammograms were reversible in the sense that successive scanning gave the same voltammetric features, with no indications of irreversible denaturation, but we deliberately stayed away from very negative potentials. This was to avoid interference from reductive desorption of the linking Au–S bond, which we have seen previously to pose challenges around neutral pH. We agree that we must be aware of G-quadruplex unwinding at very negative potentials.

Kylie Vincent commented: The voltammograms you show in **Fig. 2** show Au electrodes after hemin adsorption from dilute solution, but the STM measurements (**Fig. 3**) were made on electrodes prepared from much more concentrated hemin. Have you looked at the voltammetry of electrodes prepared from concentrated hemin solition? Can you be sure that the STM image reflects a single monolayer of hemin, rather than stacks of hemin?

Jens Ulstrup answered: The pH of the preparation solution for the hemin modified electrodes in **Fig. 2** and **3** are different. The hemin modified electrode and the quadruplex voltammetry in **Fig. 2** was prepared from 0.2 or 0.5 μM hemin in 5 mM sodium acetate at pH just above 7. The hemin modified electrode in **Fig. 3** (STM of hemin) was prepared from 10 μM hemin in 10 mM sodium hydroxide, pH 12. Both hemin solutions were diluted from 1 mM hemin stock solution prepared by dissolving hemin powder in sodium hydroxide at pH 12.

The voltammetry of the hemin modified Au(111) electrode in **Fig. 3** was recorded also at pH 12 and exhibits surface-controlled behaviour. The coverage of the hemin monolayer calculated from the voltammetric peak area is $(4.3 \pm 0.6) \times 10^{-11}$ mol cm^{-2}. The STM image in Figure 3 shows a highly ordered dense monolayer of hemin with a coverage of 1.7×10^{-10} mol cm^{-2}. Stacks of hemin molecules would have to be highly uniform to accord with such images. We are therefore sure that the STM images reflect a single dense monolayer of hemin and the voltammetry a high-coverage sub-monolayer.

Patrick Unwin asked: Have you tried to produce an adsorption isotherm of hemin on gold? I noted that only limited concentrations have been studied and that when the concentration was doubled from 0.25 μM to 0.5 μM the amount of hemin adsorbed went up by about an order of magnitude, which would appear to suggest strong interactions between adsorbed hemin molecules.

Jens Ulstrup replied: Yes, this strong increase suggests strong interactions between the hemin molecules in the adsorbed layers. Strong interactions are also implied by the *in situ* STM images of the dense monolayers and may arise from hydrogen bonds between partially deprotonated propionate groups of the two hemin molecules in neutral solution. We have not produced a proper adsorption isotherm but I am sure that such an isotherm would offer interesting information regarding parameters of the interaction between the adsorbed hemin molecules.

Richard Nichols remarked: This question concerns **Fig. 2** and the cyclic voltammogram from the surface adsorption with the low concentration of hemin. Would square wave voltammetry help here in enhancing these peaks? Also is reductive desorption a possibility for characterising the adsorbed thiolated quadraplex species? Reductive desorption has often been found to be very useful for characterising adsorbed thiol species. Would the thiolate reductively desorb as a solution free species?

Jens Ulstrup replied: To the first part of the question, yes square wave (or differential pulse) voltammetry would be expected almost certainly to help improve the voltammetric sensitivity at low hemin coverage of the electrode and is therefore worth trying. To the second part, yes given that there would be conspicuous reductive desorption signals, such signals would offer useful additional characterization. We did this previously with some success for single- and double-strand DNA where we could determine the surface coverage and also distinguish between Faradaic and capacitive contributions to the reductive desorption signals. To the third part, since the G-quadruplex part is electrostatically highly charged, the molecule would most likely desorb as a solution free species.

Kylie Vincent asked: You mention that 'DNAzyme' activity has been demonstrated for the hemin–DNA G-quadruplex structures. Can you make use of this electrocatalytic activity to further characterise electron transfer rates through the DNA structure?

Jens Ulstrup answered: It is frequently (though not necessarily always) observed that redox metalloenzymes give no voltammetric signals in the absence of their substrate but strong electrocatalytic signals when substrate is added. We have seen this also at the single-molecule scale for the copper enzymes laccase[1] and nitrite reductase[2] using *in situ* STM. If no substrate is present (dioxygen or nitrite), no *in situ* STM contrasts of the protein molecules are seen, but when substrate is present strong molecular scale contrasts appear, as if a conduction channel through the protein is triggered on substrate binding. This effect also seems to be accompanied by notable 'swelling' of the enzyme (nitrite reductase) as disclosed by electrochemical AFM.[3]

Substrate binding to hemin bound to the G-quadruplexes may trigger both electronic and molecular structural changes in hemin. Whether this would transmit into the quadruplex structure is hard to predict but something that we are in the process of looking into.

1 V. Climent, J. Zhang, E. P. Friis, L. H. Østergaard and J. Ulstrup, *J. Phys. Chem. C*, 2012, **116**, 1232–1243.
2 J. Zhang, A. C. Welinder, A. G. Hansen, H. E. M. Christensen and J. Ulstrup, *J. Phys. Chem. B*, 2003, **107**, 12480–12484.
3 X. Hao, . Zhang, H. E. M. Christensen, H. Wang and J. Ulstrup, *ChemPhysChem*, 2012, **13**, 2919–2924).

Sanli Faez opened a general discussion of the paper by Simon Higgins: Could measuring conductance *vs.* temperature serve as an extra indicator of the hopping mechanism that is suggested? Why are the conductance measurements as a function of length (*e.g.* in **Fig. 2**) limited to only 3 lengths? What is the limiting factor?

Simon Higgins answered: Certainly the measurement of conductance as a function of temperature could be useful additional evidence for hopping *vs.* tunnelling. However, it is worth introducing a note of caution here. In the case of conjugated oligoporphyrin molecular wires, we have found that although the conductance of some of these families of molecules is indeed temperature-dependent, DFT combined with non-equilibrium Green's function-based transport calculations established that this was not inconsistent with a tunnelling mechanism.[1]

The limitation here is simply one of synthetic chemistry. As oligoaryls get longer, their solubility generally decreases exponentially. The longest ligands used in this study were already at the limit of solubility to be useful as reagents for making their desired metal complexes. At the cost of greater synthetic complexity, it would be feasible to attach long alkyl chains to one or more of the aryl rings in longer analogues, to increase their solubility. But as this would also add another variable, namely the presence or absence of alkyl substituents, we decided that the time costs outweighed the likely benefits, and instead, we chose to make the alternative family of thioether-contacted molecules for comparison with the pyridyl-contacted examples.

1 G. Sedghi,V. M. García-Suárez, L. J. Esdaile, H. L. Anderson, C. J. Lambert, S. Martín, D. Bethell, S. J. Higgins, M. Elliott,N. Bennett, J. E. Macdonald and R. J. Nichols, *Nature Nanotech.*, 2011, **6**, 517–523.

Wolfgang Schmickler opened a general discussion of the paper by Jens Ulstrup: I was surprised to see that changing the solvent resulted in a completely different current–potential curve, from a Gaussian-like peak to a sigmoidal shape. In your model, changing the solvent changes the energy of reorganization, and perhaps the part of the potential that drops off between the electrode and the conducting center. This can deform the current–potential curve somewhat, but cannot result in a completely different form. I suggest that a sigmoidal form indicates that the electrode is blocked in a certain potential range.

Jens Ulstrup responded: The sigmoidal current–potential curve of the viologen in aqueous solution could be reproduced by a model resting on 'gated' electron transfer, *i.e.* pre-organization of soft twisting of the two halves of the molecule from twisted to coplanar orientation leading to faster electron transfer. The current–potential curve would assume a Gaussian-like shape if twisting is prevented or no longer soft in the ionic liquid. We are in the process of addressing this issue further.

Kylie Vincent opened a general discussion of the paper by Simon Higgins: Could you comment on why the hit rate for the complexes is lower than that seen for 4,4'-bipyridine, and why the hit rate is higher for the thioether complexes compared to the complexes with N-donor ligands?

Simon Higgins replied: We are not sure about the reasons for the lower hit rate for the complexes, but it could be connected with the fact that 4,4'-bipyridine (under ambient conditions) is neutral, whereas the complexes are charged. Probably the thioether group has a higher affinity for gold surfaces than a pyridyl group and this accounts for the higher hit rate.

Patrick Unwin said: Could you give us an idea of how easy or difficult these measurements are for small molecules? What's the success rate? Is there an art to doing these experiments?

Simon Higgins answered: I suppose there is always an art to making measurements! In this case, I would say that it is easier to learn to make the measurements than it is to gain the experience to know when the experiment is working well before the data are analysed. It is worth noting that I regularly supervise undergraduate MChem students who both synthesise molecules, and then undertake metal | molecule | metal junction conductance measurements on them, in a project lasting one academic year where the student spends 50% of their time on project work. In that case, though, the measurements would be performed in ambient conditions, and the experiments in the paper, performed in an electrolyte environment under electrochemical potential control, are more tricky. It takes a good PhD or postdoc a few weeks of effort to learn to do those measurements reliably.

Nongjian Tao remarked: You describe two different behaviours in the conductance *vs.* potential plots. These data are from conductance histogram analysis. It might be interesting to trap a molecule between the STM tip and substrate, and then sweep the potential to see the potential dependence of the conductance for the molecule. This may allow you see things that are missing in the histogram analysis.

Simon Higgins answered: In fact, we did try to take I/V traces, because we hoped to measure the transition voltage of 6V6 under electrochemical control using Fowler–Nordheim type plots, to see whether these varied as a function of oxidation state. However, it proved impossible to collect meaningful data over the required range of gate voltages; the preamp saturated easily, perhaps due to a faradaic current following cracking of the wax coating at high bias, and at high overvoltages the I/V plots were unfortunately irreproducible.

David Fermin asked: In the presentation, you concluded that the conductance dependence on the overpotential is determined by the environment (electrolyte) rather than the electronic structure of the molecule itself. The model used in your studies contains two parameters associated with this behaviour, the reorganisation energy ξ and the fraction of the potential experienced by the redox centre (λ). Could you describe how these two parameters compare in aqueous electrolyte and ionic liquids?

Simon Higgins replied: The molecules I discussed in the paper were only studied in ionic liquid, so I cannot comment on the relative values of λ and ξ in ionic liquid and aqueous electrolyte for them. The viologen molecule '6V6' showed the very broad 'off-on' behaviour in aqueous electrolyte but later, we found it showed a much sharper 'off-on-off' conductance–overpotential relation in ionic liquid, and on fitting these curves, we found the aqueous electrolyte behaviour was best fitted with values of $\lambda = 0.83$ eV, $\xi = 0.2$, whereas in ionic liquid the values were $\lambda = 1.3$ eV and $\xi = 1$.[1]

1 H. M. Osorio, S. Catarelli, P. Cea, J. B. G. Gluyas, F. Hartl, S. J. Higgins, E. Leary, P. J. Low, S. Martín, R. J. Nichols, J. Tory, J. Ulstrup, A. Vezzoli, D. C. Milan and Q. Zeng, *J. Am. Chem. Soc.*, 2015, **137**, 14319–14328.

Olaf Magnussen asked: The pronounced differences between the behaviour in aqueous electrolytes and ionic liquids are intriguing. In the paper you assign those to the short Debye screening lengths of the ionic liquids and effective structuring of the ionic liquid in the nano-gap junction. However, the difference in the screening by a typical double layer at 0.1 M concentration and an IL, which typically consists of rather large ions, should not be that different. The concrete interaction of the conducting molecule with the molecular components of the liquid may be more relevant for these effects.

Simon Higgins responded: It may well be the combination of the use of an ionic liquid and the very small nano gap involved in our experiments that contributes to the very significant observed differences. The layered structuring of ionic liquids at interfaces may be particularly effective at screening in the nano-gap junction. We have seen more effective charge screening (through the parameter zeta in the KU equation) for both the viologen (6V6) and pyrrole-tetrathiafulvalene (6PTTF6) systems (in the latter case, across both of its 2 successive redox state transitions). Ionic liquids have been shown to build clathrate cage structures around molecules in solution. If such were present in our systems this may also promote charge screening. Moreover, the nano-gap dimensions could create partially overlapping double layers, which may be more effectively mitigated in the 'charge dense' ionic liquids. Of course, ionic liquids do not have solvation shells to lose.

Kristina Tschulik remarked: Ionic liquids often contain impurities (like chloride or water) that remain after the synthesis procedure. Do you expect that such impurities might play a crucial role in your measurements?

Simon Higgins replied: We dry the ionic liquids as thoroughly as possible by heating in high vacuum overnight, and the ionic liquid is added to the dried electrochemical cell under argon; the STM measurements are taken in a controlled atmosphere chamber under argon. We also examine the potential 'window' of each batch after drying, using cyclic voltammetry, and have not seen any evidence for Cl^- oxidation in the batches we have bought.

Wojciech Nogala opened a general discussion of the papers by Dongping Zhan: It is clearly seen that a bilayer lipid membrane (BLM) effectively blocks hydrogen peroxide reduction on the BLM modified gold nanoelectrode (**Fig. 1**, curve 1). After injection of horseradish peroxidase (HRP) into the cell, a few hundred femtoampere current steps that are observed (**Fig. 2**), which are ascribed to hydrogen peroxide electroreduction catalyzed by single HRP molecules are unusually high. As calculated in the paper, a corresponding average turnover number of 620 000 s^{-1} is enormously higher than other literature reported turnover numbers for HRP. This suggest that the observed current steps are possibly caused by phenomenon other than direct wiring of active HRP molecules to the nanoelectrode. One possible mechanism is that the HRP molecules enter the bilayer lipid membrane creating channels for H_2O_2, which is directly reduced on the negatively polarized gold surface.

Another issue is the frequency of the current step appearance. Is this comparable with expected collision frequency of HRP molecules with the nanoelectrode?

Dongping Zhan responded: We don't know exactly what happens here and we prefer to believe it is caused by the electric double layer charging because it is always there and difficult to recover even after a long experiment. We don't think the other steps are caused by the leakage of H_2O_2 during the orientation of HRP in the lipid bilayer, because it is not random but in good regularity, from which we calculated the turnover number in an acceptable error range. The turnover number is much higher because we think it is closest to the practical bio-environment of single HRP electrocatalysis because of the bio-mimic membrane. By using a nanoelectrode, the collision frequency will be much lower than that oberved in the microelectrode. Adopting nanoelectrodes is good for decreasing the collision frequency and helpful to judge whether the collision is caused by one single molecule or a few molecules.

Kylie Vincent commented: How do the turnover frequencies you measure for HRP from the contact experiments compare to turnover frequencies measured for HRP films at macroelectrodes? Admittedly mass transport may be limited at macroelectrodes, but can be significantly improved at a rotating disk electrode. Have you made comparable experiments at a macroelectrode? It would be useful to include a cyclic voltammogram recorded at a macroelectrode modified with lipid bilayer and HRP for comparison. This would also enable comparison of the onset potential for catalysis to that observed in **Fig. 1**. It would also be useful to show control experiments using deactivated or inhibited horseradish peroxidase to confirm that the electrocatalytic currents rely upon active molecules of HRP contacting the electrode.

Dongping Zhan replied: Actually, we calculate the turnover frequency from the most probable current steps or the integrated charge of single collision incidents, which are obtained from the chronopotentiometry. Since the concentration of H_2O_2 is as high as 20 mM and the number of HRP is countable, mass transfer is not the problem. The current step shows the limit of the catalytic capacitor of HRP. There are many reports on the direct electrochemistry of HRP on a macro-electrode. That is why we didn't do the experiments. It is a good suggestion to use an inhibitor to stop the catalytic experiment, which we didn't pay attention to before.

Wolfgang Schuhmann said: You had to apply a very low potential at the bilayer-modified electrode to see the assumed direct electrochemistry of horseradish peroxidase. However, it is known that the catalytic cycle of horseradish peroxidase is through an iron(IV) oxo radical cation at very high potentials. Hence, you need an overpotential of more than 1 V to transfer the electrons with the enzyme. Why did you have to apply such a huge overpotential?

Dongping Zhan replied: Although the two reversible peaks were observed in the direct electrochemistry of HRP,[1] the reduction of H_2O_2 catalyzed by HRP occurs at a potential region of (−0.1 V, −0.3 V) *vs.* the AgCl reference electrode.[2] During the reduction of hydrogen peroxide, we think the Fe(III) is reduced on the electrode which needs only a small overpotential.

1 J. Li and S. Dong, *J. Electroanal. Chem*, 1997, **431**, 19–22.
2 X. Liu, Y. Huang, L. Shang, X. Wang, H. Xiao and G. Li, *Bioelectrochemistry*, 2006, **68**, 98–104.

Wojciech Nogala said: According to **Fig. 1** in the paper (curve 2), H_2O_2 reduction in the studied system does not occur at electrode potentials higher (more positive) than *ca.* −0.2 V *vs.* Ag|AgCl. Direct electron transfer HRP-catalyzed electroreduction of H_2O_2 occurs at much higher potentials.[1] This suggests that H_2O_2 reduction in the studied system occurs directly on gold and the biocatalytic activity of HRP might play no role in the electrode process.

1 F. Chekin, L. Gorton and I. Tapsobea, *Anal. Bioanal. Chem.*, 2015, **407**, 439–446.

Dongping Zhan replied: This issue shows the importance of the reference electrode. However, in the experiment, to simplify the measurements, an Ag/AgCl wire is well adopted as the reference in bioelectrochemistry. Different publications have reported different ‘experiment’ potential. Thus, the calibration of the potential is very important. As shown in **Fig. 1**, the lipid bilayer can block the reduction of H_2O_2 very well. It is in the presence of the HRP that the current step is observed and the background becomes much smaller. We think HRP should play a role.

Kylie Vincent asked: It is not uncommon for redox enzymes to show higher catalytic rates on an electrode compared to their behaviour in solution assays due to effective direct electron transfer to the immobilised enzyme. However, the turnover frequencies for HRP that you measure seem unusually high. Can you comment?

HRP is not necessarily a poor choice of enzyme for these experiments, but it will be useful to carry out future experiments on other redox enzymes that are able to engage in direct electron transfer in order to see whether rate enhancements are observed more generally.

Dongping Zhan replied: We chose HRP necause we have some experience with HRP. It is a good suggestion to try other enzymes. The high turnover number might benefit from the bio-mimic membrane of the lipid bilayer.

Richard Nichols opened the discussion of the papers by Dongping Zhan and Jens Ulstrup: All three papers in this session have the commonality of electron transfer through molecular species. It would be useful to have a comparison of the electron transfer distances for each paper. The most complex system in this respect seems to be the paper on single molecule catalysis at redox enzymes where more than one layer is implicit, with a silane layer, a bilayer membrane and then the HRP docking and presumably orienting to this upper surface. Some comments and comparisons between the electron transfer distance would be interesting.

Dongping Zhan answered: The size of HRP is about 2 nm by 3 nm by 3 nm. And we think the thickness of the silane supported lipid bilayer is at the same scale. If the orientation of HRP in the lipid layer is correct, the charge transfer distance is about 1.0 to 1.5 nm. This is a reasonable distance in electrochemistry for out-of-sphere reactions. By the way, we have no evidence whether the memebrane is multiscale or whether there is internal communication between the adsorbed HRP molecules.

Jens Ulstrup replied: The quadruplex box length in the first of the three papers is about 0.7 nm if the distances between the G-quartets are roughly the same as between the base pairs in double-strand DNA. The length of the 6C-sulfur linker is about 1 nm. Assuming that the quadruplex box is vertically oriented or slightly tilted on the Au(111)-electrode surface, the electron transfer distance would then be about 1 and 2 nm for hemin bound at the end adjacent and remote from the electrode surface, respectively. The apparent conductance from **Fig. 5** is 0.18 nS or $2.3 \times 10^{-6} \times G_0$. The distance dependence could in principle be addressed by comparison with G-quadruplexes of longer box lengths.

The total length of the target transition metal complexes with aromatic ligands and linking groups in the second of the three papers varies from about 1.5 nm to about 3.2 nm, *cf.* **Scheme 1** and **Fig. 2** in the paper, *i.e.* in a certain sense in the same range as for the quadruplex. Notably, the aromatic linkers give both more than an order of magnitude higher conductance $(1.5–9.6) \times 10^{-5} \times G_0$ and a very weak distance dependence reflecting the two-step hopping mechanism of these molecules and the aromatic nature of the linking groups. The electron transfer distance in the multi-layer catalytic system addressed in the third of the papers is probably longer but harder to assess. The three layers on the modified Au-nanoelectrode are first octyltriethoxysilane, secondly egg lecitine, and third the enzyme horseradish peroxidase (HRP). The former layer thickness would be about 1.5 nm. Uncoiled lecithin would be about 3 nm long, but the degree of folding on the surface is not clear. The spatial extension of HRP is about 6 nm but it is not clear to what extent the enzyme is embedded in the lecithin layer. Overall, however, the composite multi-layer system appears to correspond to a significantly longer electron transfer distance than in the quadruplex and transition metal systems.

Kylie Vincent remarked: In different ways, the ordering of molecules on a gold surface is important in all of the three papers we are discussing here. Jens Ulstrup has identified different types of ordered domains in the hemin–DNA G quadruplex structures on gold using STM. Dongping Zhan is relying on ordering of a lipid bilayer membrane on a gold electrode to prevent access of peroxide directly to the electrode. Simon Higgins' experiments rely on the complexes aligning perpendicular to the surface, lying between the Au STM tip and the Au surface. This question is really addressed to Simon: to what extent do the complexes really align vertically between the tip and the surface, and does the distribution of break-off distances observed in your experiments result from a range of tilted orientations of the complexes? How does this complicate the interpretation of your results?

Simon Higgins replied: I didn't have time to exemplify in my presentation that we measure the break-off distance (the distance at which the metal | molecule | metal junctions break down), as well as the conductance of the junctions. The procedure has been described in our previous papers. This break-off distance should correspond with the length of the molecule. This does not depend upon the molecules being vertically oriented before junction formation because the molecules are lifted up as the tip–substrate distance is increased. The molecules may begin by lying flat on the surface, or may migrate into the tip–substrate gap as the junction forms. The molecules can either be adsorbed in a sub-monolayer state on the surface, or may be present in solution.

David Fermin opened a general discussion of the papers by Dongping Zhan: It remains unclear to me the nature of the so-called HRP activation which leads to the unusual voltammetric behaviour in **Fig. 1**. Could you elaborate on the origin of the step increase in the cathodic current at −0.3 V in the forward scan as well as the hysteresis and current loop observed in the reversed scan? You also mentioned that hydrogen peroxide reduction is effectively blocked at the modified nanoelectrode in the absence of HRP. However, this does not appear to be the case as there is a clear negative background current over large portion of the potential window.

Dongping Zhan responded: This is indeed a one-cycle voltammogram, in which the potential is held at +0.2 V for 6 s and then scanned negatively. This phenomenon is very reproduciable and we will provide a similar figure obtained with a different electrode in our revised manuscript. We didn't do the experiment beginning at −0.3V because we want to observe the HRP catalytic reaction from inactive status to active status. It is a good suggestion that we may try in the future. However, the collision incidents were recorded at a constant potential at −0.45 V, where the HRP is always active.

Actually, the hydrogen peroxide is blocked sufficiently by the lipid bilayer. Or else, the current will be 20-fold higher than the current observed in **Fig. 1**. We can observe the cathodic current from the blue curve in **Fig. 1**. It is the dramatic difference that makes the case for HRP catalysis convincing.

If the modofied nanoelectrode is kept polarized to a more negative potential, I am afraid that the bilayer will be detached.

Christine Kranz commented: The experiments were performed at gold nanoelectrodes exhibiting radii as small as 12 nm. The limiting current was calculated based on the equation for disc-shaped microelectrodes. However, considering the reported dimensions herein and the fact that no TEM images were recorded to evaluate the actual size and geometry of the electrode, the authors should comment on the uncertainty that is introduced for the presented data related to an erroneous shape and/or size of the electrode. Besides the electrochemical characterization, which complementary techniques have been used to evaluate the actual size and geometry of the nanoelectrode?

Secondly, Figure 1 shows the scheme of an HPR-BLM modified nanoelectrode. How was ensured that the enzyme is integrated in the BLM layer and not just adsorbed and stacked at the bilayer? How would this affect the turnover rate, and again, what is the influence on the accuracy of the reported values?

Dongping Zhan responded: As indicated in a response to a previous question, we accept that nanoelectrodes are controversial because they are too small to be characterized, easily broken and unstable during the CV. Frankly, we don't know the exact state of the nanoelectrode, we just judge its quality by ultra-high scanning voltammetry and calculate its apparent diameter using the steady-state limiting current. We think the nanoelectrode should be OK unless it is abused at extreme potentials either in the oxidation region or in the hydrogen evolution region.

We don't have evidence on the orientation state of HRP in the lipid bilayer. We believe that the turnover rate is dependent on the orientation from the individual diversity of single collision incidents. In future, *in situ* coupling techniques should be developed to obtain this information.

Patrick Unwin remarked: I would like to point out that measuring the size of nanoscale electrodes by voltammetry alone is somewhat ambiguous. One has to assume a planar electrode geometry, but the electrode could be recessed and (much) larger, or protruding and smaller. Better is to do approach curve measurements, as demonstrated, for example, by Takahashi *et al.*[1] Better still is to use transmission electron microscopy (TEM), which is very powerful for examining the geometry of nanopipettes, as demonstrated by Baker's group,[2] and which we have used to produce detailed mass transport models for nanopipettes.[3] TEM can also be used for the types of electrodes (filled nanopipettes) described here, as shown for example in **Fig. 1** in a recent paper by Minkyung Kang *et al.* on nanoscale electrochemical imaging.[4] In fact, Macpherson and coworkers were using TEM more than a decade ago to characterise nanoscale electrodes integrated into AFM tips.[5]

1 Y. Takahashi, A. I. Shevchuk, P. Novak, B. Babakinejad, J. V. Macpherson, P. R. Unwin, H. Shiku, J. Gorelik, D. Klenerman, Y. E. Korchev and T. Matsue, *Proc. Natl. Acad. Sci. U. S. A.*, 2012, **109**, 11540–11545.
2 L. Zhou, Y. Zhou and L. A. Baker, *Electrochem. Soc. Interface*, 2014, **23**, 47–52.
3 D. Perry, D. Momotenko, R. A. Lazenby, M. Kang, and P. R. Unwin, *Anal. Chem.*, 2016, **88**, 5523–5530.
4 M. Kang, D. Momotenko, A. Page, D. Perry and P. R. Unwin, *Langmuir*, 2016, **32**, 7993–8008.
5 D. P. Burt, N. R. Wilson, J. M. Weaver, P. S. Dobson and J. V. Macpherson, *Nano Lett.*, 2005, **5**, 639–643.

Wolfgang Schuhmann commented: This comment is related to the comment by Pat Unwin concerning the easy characterisation of nanotips using TEM. Unlike in the case of capillaries, in the case of nanoelectrodes, charging by touching the electrode may destroy the electrode tip (see paper by S. Amemiya *et al.*).[1] Hence, frequent insertion of a nanoelectrode into a TEM may not always be so easy for in-depth characterisation of the tip.

1 N. Nioradze, R. Chen, J. Kim, M. Shen, P. Santhosh and S. Amemiya, *Anal. Chem.*, 2013, **85**, 6198–6202.

Dongping Zhan replied: Nanoelectrodes are controversial because they are difficult to prepare and characterize, fragile and easily broken, and the metal nanowire is unstable during the experiments. However, if nanoelectrodes are used in a proper way, it can provide convincing information over a certain period. Even if there are some defects, it depends on what kind of electrochemical reaction is being investigated. For example, leakage is the main problem for nanoelectrodes. However, it may have little effect on experiments under ultrahigh scanning rate because, due to the high resistance and capacitance in the ultra-narrow leakage layer and, thus, the high time constant, it has no time to response the ultra-high voltage scanning ($> 100\ V\ s^{-1}$) or ultra-fast voltage pulse (> 1 GHz) . We don't know what exactly happens in SEM experiments. The strong electron beam may destroy the nanoelectrode directly, especially a very small one. As we know, focused electron beam is a very important microfabrication method.

Shengli Chen opened a general discussion of the papers by Simon Higgins: In Prof. Higgins' paper, the electron transfer mechanism has been distinguished into tunneling and hopping mechanisms respectively. From my understanding, the hopping electron transfer should be also through a tunneling process which occurs over relatively large distance and requires orbital overlapping. So, what's the intrinsic difference between tunneling and hopping transfer here?

Simon Higgins replied: Yes, in the Kuznetsov–Ulstrup model, the electron has to tunnel from the electrode to the redox site in the molecule, and then again from the redox site to the other electrode. This is distinct from the situation where the frontier orbitals are far removed in energy from the Fermi energies of the electrodes (*e.g.* for molecules like alkanedithiols, or for short conjugated molecules), where the electron can then only get from one electrode to the other by tunnelling across the molecule. The first case is incoherent to a degree depending upon the extent to which the molecule can relax after the first tunnelling process; the second case is coherent.

Venkateshkumar Prabhakaran communicated: Ionic liquids are interesting and promising electrolytes in many electrochemical technologies as most of them do not need solvents and so avoid many side reactions caused because of them. An interesting property of ionic liquid is that it has differential double charging and formation of the solvated clusters with active species of interest during forward and reverse scanning due to the asymmetry of ion sizes. These properties makes them very difficult to model the interfaces comprising ionic liquid. Dissimilar double layer charging could significantly change the redox activity of the conjugated molecules. Alexei Kornyshev *et al.* have published a series of papers on the double layer charging of ionic liquids. My question is if this property has been considered during studies, and whether these properties have been studied separately both in oxidation and reduction scans.

Simon Higgins communicated in reply: In our measurements, we hold the potentials of the two working electrodes (tip and substrate) fixed while collecting data; these are not cycling experiments. However, with a suitably long-lived junction, perhaps using strongly-bonded contacts like thiolates (Ar–S–Au) or direct Au–C bonds rather than 4-pyridyl groups, it may be possible to measure junction conductance while cycling the electrode potentials with a fixed bias, hoping that the junction lives long enough, although we haven't tried this.

Faraday Discussions

PAPER

In situ video-STM studies of the mechanisms and dynamics of electrochemical bismuth nanostructure formation on Au†

H. Matsushima,‡[a] S.-W. Lin,[b] S. Morin[b] and O. M. Magnussen*[a]

Received 20th April 2016, Accepted 9th May 2016

DOI: 10.1039/c6fd00086j

The microscopic mechanisms of Bi electrodeposition on Au(111) and Au(100) electrodes in the overpotential regime were studied by *in situ* scanning tunneling microscopy with high spatial and temporal resolution. Atomic resolution images of the needle-like Bi(110) deposits formed on Au(111) reveal the central influence of covalent Bi–Bi bonds on the deposit morphology. In the straight steps along the needle edges the Bi atoms are interlinked by these bonds, whereas at the needle tip and at kinks along the needle edges dangling bonds exist, explaining the rapid structural fluctuations at these sites. For ultrathin Bi deposits on Au(100) a more open atomic arrangement was found within the surface plane, which was tentatively assigned to an epitaxially stabilised Bi(111) film. Furthermore, well-defined nanowires, consisting of zigzag chains of Bi surface atoms, were observed on this surface.

Introduction

Quasi one-dimensional nanostructures, such as nanowires, nanotubes and nanoribbons, have unique physical and chemical properties and are of substantial interest for nano-electronics and sensor applications.[1] Electrodeposition has established itself as a popular technique for the formation of such nanostructures.[2] On the one hand, nanowire growth can be achieved by templates, for example by electrodeposition into porous membranes[3–5] or by heterogeneous nucleation and growth at one-dimensional defects, such as steps.[6–9] On the other hand, intrinsic anisotropies and kinetic effects can result in the spontaneous formation of one-dimensional nanostructures. The latter includes the well-known

[a]*Christian-Albrechts-Universität zu Kiel, Olshausenstr. 40, D-24098 Kiel, Germany. E-mail: magnussen@physik.uni-kiel.de*

[b]*York University, 4700 Keele Street, Toronto, Canada*

† Electronic supplementary information (ESI) available: STM videos showing the dynamics of needle-like Bi islands. See DOI: 10.1039/c6fd00086j

‡ New address: Faculty of Engineering, Hokkaido University, Sapporo, Japan.

electrochemical growth of whiskers and dendrites[10] as well as the lateral epitaxial growth of nanowires and strongly anisotropic crystallites on electrode surfaces.[11–15] Here we investigate the microscopic growth mechanisms of such structures and the atomic-scale growth dynamics for the case of electrodeposited bismuth on Au electrodes.

Bi thin films and nanostructures have been receiving much attention in recent years owning to the unusual surface electronic structure of Bi.[16,17] Bi exhibits strong spin–orbit coupling, making it of great interest to the field of topological insulators, and has electronic surface states that cross the Fermi level. Phenomena such as surface superconductivity, enhanced thermopower, and high electronic mobility have been observed, which are still under debate. As a consequence the behaviour of Bi nanowires and nanoribbons also strongly deviates from that of the bulk material and changes as a function of the nanostructure diameter and type.[18–22] Furthermore, thin Bi films for potential applications in magnetic sensors,[23,24] thermoelectric materials[17,25] and spin filters[26,27] depend strongly on the integrity of the microstructure.[28] Bi thin films prepared by electrochemical deposition have been studied in recent years because of this method's ability to provide a better control of the thin film's microstructure, simple processing, such as room temperature operation at atmospheric pressure, and low costs. Electrochemical growth of crystalline Bi thin films has been demonstrated on Au(111)[13,29] and GaAs(111),[30] resulting in surface morphologies with needle-like and pyramidal islands, respectively. These deposits can be converted into highly ordered single-crystal Bi films by annealing. Similar needle-like morphologies were also found for Bi deposition from the gas phase on Ag(111),[31] Si(111)[32,33] and graphite.[34,35]

Detailed studies of the initial stage of Bi electrodeposition on Au(111) and Au(100) substrates were performed using *in situ* STM and *in situ* surface X-ray scattering measurements.[13,36–41] Most of these studies focused on Bi underpotential deposition (UPD), revealing a similar sequence of Bi monolayer phases on both surfaces. With decreasing potential first commensurate Bi superstructures were found, followed by uniaxially-incommensurate quasi-hexagonal adlayers with high packing density. As shown by X-ray diffraction studies, bulk Bi deposition on Au(111) occurs in the form of an epitaxial film, in which the Bi(110) plane§ is oriented parallel to the Au surface (see Fig. 1).[29] An *in situ* STM study of the Bi growth behaviour was reported by Jeffrey *et al.*[13] At low overpotential ($\approx -0.08\ V_{SCE}$) the growth of these films was found to start with the formation of an isotropic Bi bilayer on the terraces, followed by the formation of nuclei at step edges and defect sites. These nuclei develop into highly anisotropic needle-like structures which are oriented with the long axis parallel to the Au surface and which grow quite rapidly. This needle growth mode is observed up to many hundreds of layers, resulting in a Bi film of increasing roughness. However, a clear mechanism of the nucleation and growth of these needles is still lacking, because of the high growth rate and the resulting difficulties in following this process using a conventional *in situ* STM.

§ We employ throughout the paper the rhombohedral unit cell for crystallographic notation. In the discussion of literature where a hexagonal unit cell was employed, we have converted the notation for clarity. Specifically, we note that (110) and (111) in the rhombohedral system corresponds in the hexagonal notation to (012) and (001), respectively.

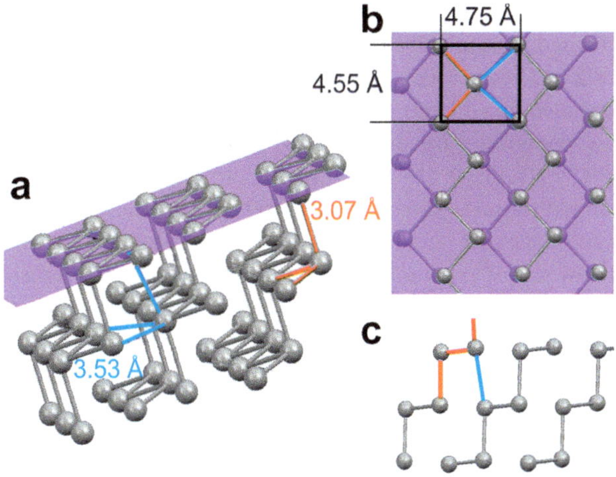

Fig. 1 (a) Structure of bulk bismuth, indicating short covalent (red) and long non-covalent (blue) bonding distances between neighbouring atoms. The shaded plane is the (110) plane. (b) In-plane structure and (c) side view of the (110) planes.

Such atomic-scale studies of fast dynamic processes at electrochemical interfaces have become feasible using *in situ* video STM, performed in constant height mode at scan rates of 10 to 30 images per second.[42] In particular, video STM has been employed in the past for studies of electrochemical crystal growth and dissolution processes[43–45] and of the dynamics of quasi one-dimensional nanostructures on electrode surfaces.[46–49] Bi deposition on Au single crystal electrodes in $HClO_4$ solution is well suited for such video STM studies because (i) the ClO_4^- anions are not specifically adsorbed on Bi and hence do not affect the Bi growth and (ii) the Bi electrodeposit exhibits a well-defined crystalline structure with extended terraces. In this paper we describe the structure and dynamics of Bi growth on Au(111) and Au(100). We especially focus on the origin of the pronounced anisotropy in this deposition process, manifesting in the formation of Bi needles and nanowires. The dynamic behaviour of kinks along the Bi needle edges is analysed quantitatively and interpreted taking the Bi lattice structure into account.

Experimental

The study was performed using the home-built *in situ* electrochemical video STM of the Kiel group.[42] All images presented in this work were recorded at a rate of 10–20 images per second in constant height mode. STM tips were etched from W-wire and coated with ethylene–vinyl acetate polymer, leaving the very end exposed. The tip and sample were independently kept under potentiostatic control.

The electrochemical cell consisted of a Kel-f cell with a charged palladium hydride electrode (Pd–H) as a reference, a Pt wire as a counter electrode and a Au(111) or Au(100) single crystal (orientation $\leq \pm 0.5°$, Monocrystal Co.) as a working electrode. The Pd–H electrode was charged prior to the experiment in a dilute sulfuric acid solution and exhibited a stable potential (*ca.* −0.20 V, *vs.* SCE) in the bismuth containing electrolytes. All the cell potentials reported in this work are given with respect to the saturated calomel electrode (SCE). The Au crystal was cleaned in suprapure sulfuric acid and rinsed in ultrapure water (Elga Lab water, 18.2 MΩ resistance). The crystal was then flame annealed for several minutes and cooled down in air before being mounted in the video STM cell. The Bi electrolytes, containing 0.5 or 1 mM Bi^{3+}, were prepared by dissolving Bi_2O_3 (Alfa Aesar, 99.9995%) in 0.1 M $HClO_4$ (ultrapure grade, Seastar Chemicals Inc.) prepared using ultrapure water. In all experiments, the initial potential was chosen between 0.05 and −0.08 V, *i.e.* at a value more positive than the corresponding Bi^{3+}/Bi Nernst potential, to avoid Bi bulk deposition. Subsequently, the potential was stepped down to more negative values in the overpotential deposition (OPD) regime to investigate the growth of bulk Bi.

Results and discussion

Bismuth growth on Au(111)

In situ video STM experiments performed on Au(111) electrodes in the Bi OPD potential regime reveal the same needle growth behaviour observed using the conventional STM. When the cell potential is stepped to higher overpotentials (<−0.10 V), rapid formation of Bi multilayer deposits commences. Because the surface roughness in this stage of the growth process is initially high, the (constant-height mode) video-rate imaging is impeded and high-resolution images of the deposit cannot be obtained. However, 1 to 2 minutes after the onset of Bi bulk deposition the growth rate slows down and the characteristic Bi needle structure can be observed with atomic resolution (Fig. 2a). The Bi surface lattice is described by a pseudo-square structure with a prominent maximum at the corners of the unit cell and a weaker maximum slightly of center (shifted to the left of the unit cell center in the image in Fig. 2a). This structure is in full

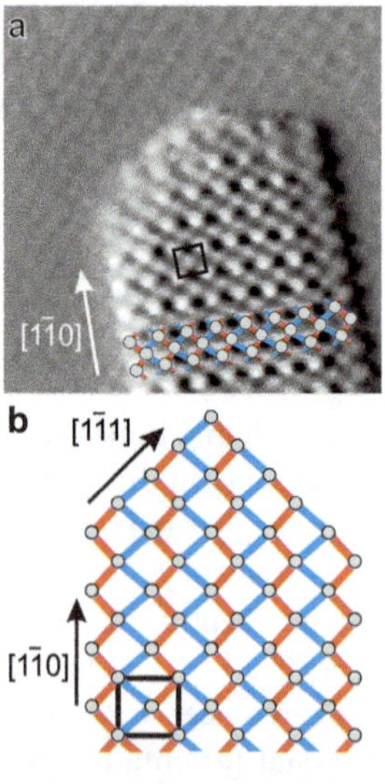

Fig. 2 (a) High-resolution *in situ* video STM image of the needle-like Bi deposit structure obtained on Au(111) at −0.19 V (46 × 46 Å^2). (b) Schematic model of the atomic structure of the Bi needles.

agreement with the arrangement of Bi atoms in the Bi(110) plane (Fig. 1b). Within this plane parallel running zigzag chains of covalently bound Bi atoms are located. Each surface atom thus exhibits two covalent bonds of 3.07 Å length (indicated in red) to atoms within the chain and two longer non-covalent bonds with 3.53 Å bonding distance (indicated in blue) to atoms in the neighbouring chain. Very similar UHV-STM images were obtained on needle-like Bi deposits on Ag(111) and assigned to an identical structure.[31] In some *in situ* video STM images obtained at larger bias only one of the atoms in the pseudo-square unit cell was resolved. Those observations strongly resemble high-resolution *in situ* STM images obtained in the previous study of Bi on Au(111) by Jeffrey *et al.*[13] as well as in UHV-STM experiments of Bi(110) films on Si(111)-(7 × 7).[32]

The characteristic needle-like growth morphology can be clearly correlated with the anisotropic atomic structure of the Bi deposit. The Bi needles are oriented perfectly parallel to the $[1\bar{1}0]$ direction of the Bi lattice, *i.e.* along the direction of the covalently bound chains. Apart from a few kinks during needle growth and dissolution (see below), the needle edges along this direction are perfectly straight. Interestingly, each needle consists of an odd number of Bi rows, *i.e.* the needles are $(n + 1/2)$ unit cells wide (with n being a natural number). As illustrated schematically in Fig. 2b, these observations can be rationalised by an atomic arrangement where the needles are on both edges terminated by Bi chains. In this type of structure all Bi atoms at the needle edges are energetically stabilized by two covalent bonds to neighbouring edge atoms.

The needle-like Bi islands shown as examples in Fig. 2a grow on top of a smooth Bi(110) terrace (the pseudo-square unit cell of the latter surface is only faintly visible due to the much higher tunnelling distance in these constant height images). They are of monolayer height, which is in contrast to the needle-like Bi islands observed in UHV-STM studies of gas phase deposited Bi films.[31–35] In the latter case the step heights typically corresponded to multiples of bilayers. This was attributed to a bilayer arrangement as in black phosphorous, where the out-of-plane dangling bonds of the Bi atoms pair up. Depending on the substrate the planes in the black phosphorous structure were found to be either flat or buckled. Our experimental observations suggest a bulk like (110) structure of the Bi deposit on Au(111) electrodes rather than a flat or buckled black phosphorous phase: first, no preference for films with an even number of Bi layers is found. Second, the two atoms in the Bi(110) surface unit cell appear significantly different in the STM images whereas no such difference in contrast was observed in high-resolution UHV-STM studies.[31,32,34,50] For a bulk-like (110) film the two atoms in the unit cell are chemically non-equivalent, which could explain this difference in contrast. Specifically, the atoms in the center of the unit cell of Fig. 1b exhibit dangling bonds towards the electrolyte while those in the unit cell corners would have covalent bonds to the underlying Bi layer. This structural difference of the Bi(110) film in our study and under vacuum conditions may be caused by the termination of the dangling bonds by adsorbed species from the electrolyte. In view of the negative deposition potential the most likely candidate is adsorbed hydrogen, which readily binds to these sites.[32,51–53] Evidence for the adsorption of hydrogen species, especially on Bi surfaces exhibiting covalent bonds, was also obtained in studies of hydrogen evolution.[54] In contrast, the presence of other species, such as adsorbed anions or OH, can be ruled out in this potential regime.[55]

The dynamics of the Bi islands reflects their anisotropic structure. While the long, atomically smooth steps along the $[1\bar{1}0]$ direction are very stable, the tips of the needle-like islands exhibit rough and highly mobile steps. In atomic resolution *in situ* video STM measurements these tips often appear blurry and change from image to image (see Fig. 3a and video 1 in the ESI†), indicating fast attachment and detachment of Bi adatoms. This instability of the step edge atoms at the needle tips can be explained within the structural model proposed above. Because the zigzag chains are terminated at the tips, these atoms have a lower number of covalent bonds to neighbouring Bi atoms than the atoms at the $[1\bar{1}0]$

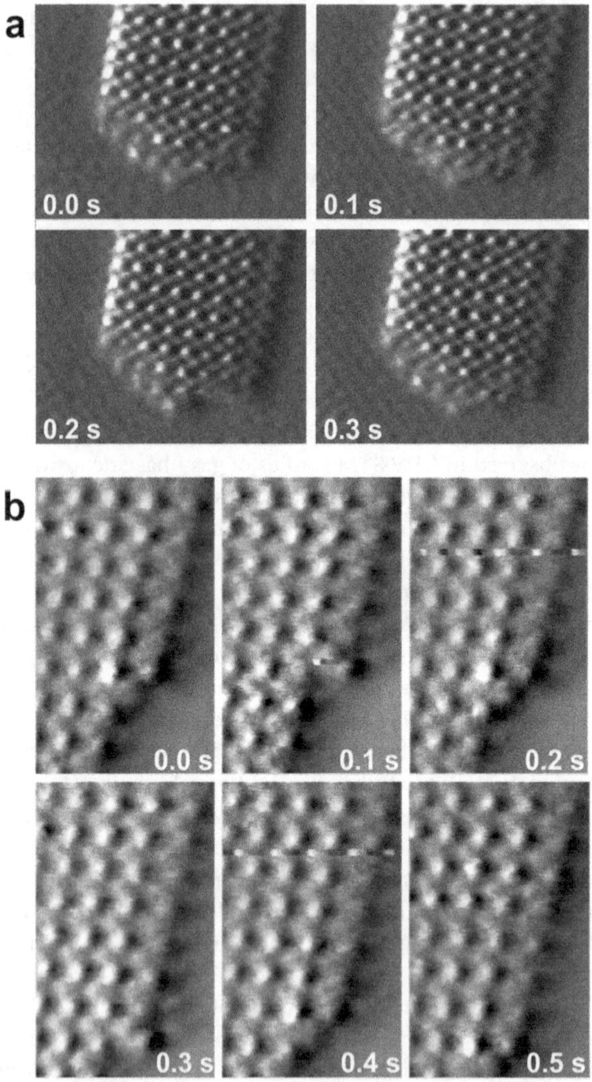

Fig. 3 Sequences of STM images of Bi deposits on Au(111), taken from videos recorded at 10 Hz. (a) Fluctuations at the tip of a needle-like island, obtained at −0.19 V (62 × 51 $Å^2$). (b) Kink propagation along the edge of a needle-like island, recorded at −0.40 V (25 × 38 $Å^2$).

oriented steps (see Fig. 2b). The weak tendency of the steps at the needle tips to orient roughly along the [1$\bar{1}$1] orientation suggests that the step edge atoms can also be partly stabilized by the non-covalent bonds; however, this effect is much less pronounced than that of the covalent coordination.

The same tendencies manifest in video STM studies of the motion of kinks along the straight [1$\bar{1}$0] oriented steps. These studies reveal that the kink motion proceeds *via* the addition or removal of two rows of Bi atoms in a zigzag fashion, *i.e. via* one-dimensional growth or dissolution of a chain of covalently bound Bi surface atoms along the island edge (see Fig. 3b and video 2†). Three types of kinks were observed in the STM videos, which are illustrated in Fig. 4: typically, kinks are of the type denominated as A or B in Fig. 4 and propagate along the step edge by alternating between these two states. Examples of these types of kinks can be observed *e.g.* in Fig. 3b. In more rare cases steps of the type C are observed in the images. Also in this case kink propagation occurs *via* subsequent changes between the type C and the type B state.

The microscopic growth rates found in these experiments are much lower than those which would be expected at the rather high overpotentials employed. We attribute this to geometric shielding by the STM tip, which is a well-known effect

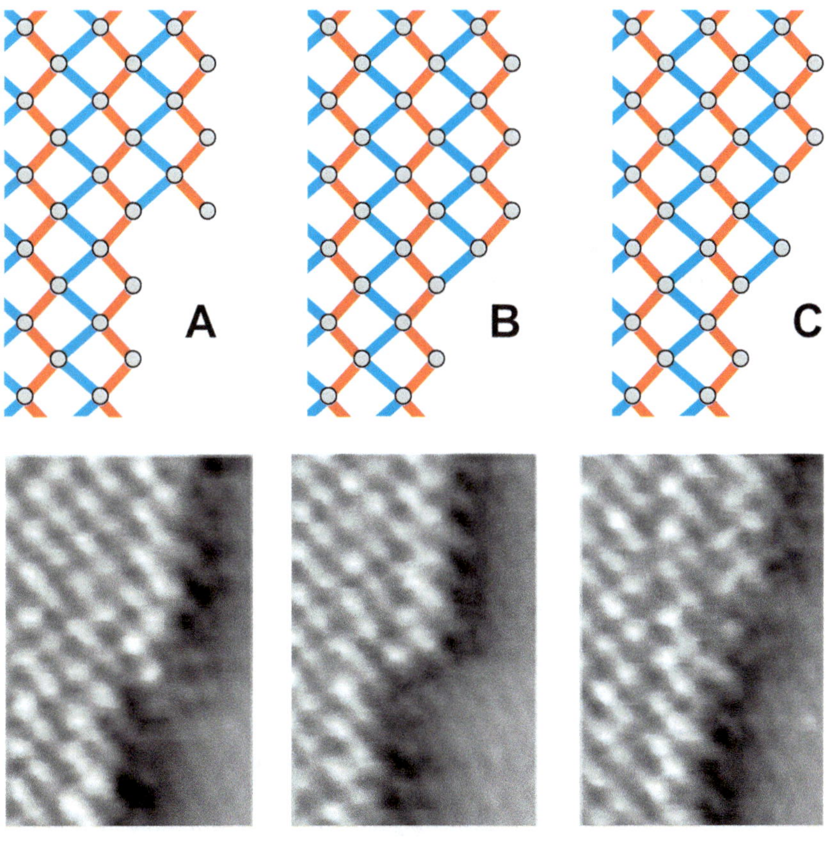

Fig. 4 Different types of kinks observed along the edges of Bi needles on Au(111) and corresponding examples, taken from an STM video.

in *in situ* STM studies of electrodeposition processes.[56] Nevertheless, the video STM observations of the structural fluctuations at the kinks provide interesting additional information on the Bi deposits. Because these fluctuations occur on the hundred millisecond time scale, the different kink configurations can be sampled with good statistics. According to our data, the different types of kinks are observed with very different probabilities. Based on an analysis of 750 kinks, we find the type B kinks with 81% probability, type A kinks with 16% probability and type C kinks with only 3% probability. Within the experimental errors no clear dependence on potential was found in the range −0.10 to −0.64 V. These statistical data reflect the different probabilities p_j (j = A, B, C) for a kink to be in one of the three observed states, *i.e.* the different residence times of the kinks to be in one of these configurations. Because the studies were performed at low kink propagation rates and thus are close to equilibrium, these statistical data allow the calculation of the relative energy difference ΔE_{ij} between these kinks *via* the Boltzmann equation $\Delta E_{ij} = -k_B T \ln(p_i/p_j)$. Using the statistical data obtained in the video STM measurements (performed at $T = 298$ K) one obtains an increase in energy of $\Delta E_{AB} = 42 \pm 3$ meV of the type A and of $\Delta E_{CB} = 86 \pm 7$ meV of the type C kinks as compared to the type B kinks, respectively.

We interpret these results in terms of the different coordination of the Bi atoms at the kink in the three different states. In the type A kinks the Bi atoms at the kink are missing two non-covalent bonds to neighbouring atoms as compared to the type B kinks; type C kinks are even lacking two covalent bonds. Taken this into account, the energy differences at a first glance seem smaller than expected from the Bi cohesive energy (2.16 eV per atom[57]) or the Bi–Bi bond energy (2.08 eV (ref. 58)). However, the energy balance may be strongly affected by the presence of coadsorbates. Recent density functional theory calculations for two-dimensional Bi nanoribbons with similar covalently bound Bi chains along the edges found strong modification of the geometric and electronic structure by hydrogen adsorption at the dangling bonds of the terminating Bi atoms.[51–53] Hydrogen adsorption energies in the range of 2.2 eV were reported.[52] Similar behaviour may be expected in the electrochemical environment and may significantly reduce the kink energies.

Bismuth growth on Au(100)

Similar *in situ* video STM studies of Bi electrodeposition were performed on Au(100) electrodes. In contrast to the structure of the Bi UPD adlayer on Au(100), which has been characterized in detail by STM and X-ray surface diffraction,[36,40] the initial stages of Bi bulk electrodeposition have not been reported before. High-resolution video STM images recorded in the overpotential range (Fig. 5a) find a lattice structure that clearly differs from that of the UPD layer or of the Bi(110) film observed on Au(111). In particular, these Bi deposits exhibit a much more open atomic structure, which is incompatible with Bi(110) or the c($p \times 2$) Bi UPD layer. The observed surface structure is described by a rectangular unit cell with one prominent and one weaker maximum that form alternating rows. Upon assigning both of these to Bi surface atoms, a (pseudo-) hexagonal arrangement results. The interatomic distances are in good agreement with those in an atomic layer of the (111) plane of bulk Bi (shown in the model in Fig. 5b). In this plane the Bi atoms form hexagonally ordered bilayers, interlinked by covalent bonds. This

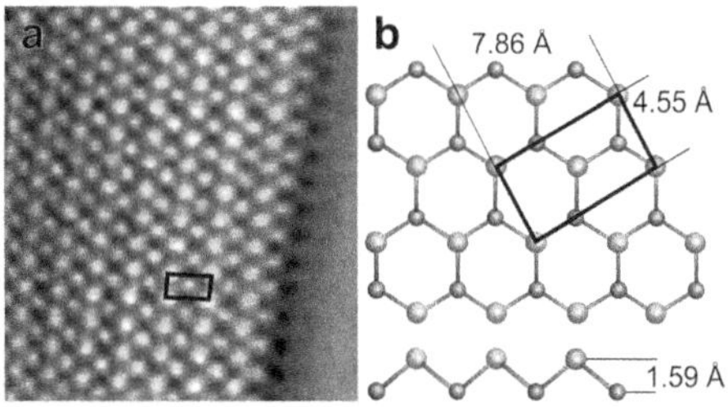

Fig. 5 (a) High-resolution *in situ* video STM image of the Bi deposit structure obtained on Au(100) at −0.17 V (75 × 60 Å^2). (b) Model of the atomic structure of the Bi(111) plane, showing the first (light grey) and second (dark grey) Bi layer. In the bottom part a side view of the bilayer arrangement in this plane is shown.

assignment assumes that only the topmost Bi(111) layer is visible in the STM images, which is likely in view of the 1.59 Å spacing to the underlying second layer (see Fig. 5b, bottom).

Bi(111) is the natural cleavage plane of bismuth and under UHV conditions the preferred orientation in epitaxial growth.[16] The surface atoms in Bi(111) have a lower number of non-covalent long bonds to neighbouring atoms as compared to those in the Bi(110) surface, but – in contrast to the latter – do not exhibit dangling bonds. For thick epitaxial Bi films formed by electrodeposition on Au(100) a (110) orientation has been found.[59] However, Bi(110) and Bi(111) films are similar in energy and structural transitions between them have been found in ultrathin epitaxial films.[32,60] It therefore seems possible that a Bi(111) structure is epitaxially stabilized in the very thin films studied in this work (typically consisting of only a few layers).

Although a Bi(111) film would account for the observed inter-atomic spacings of the surface atoms, it cannot explain the row-like modulation of the structure. The latter would not be expected for true Bi(111) films and was not observed in previous STM studies of (111) oriented Bi films on Si(111).[32,60] Most likely, these rows result from epitaxy to the underlying Au substrate lattice. This is supported by observations of the deposit morphology on larger length scales, which show that the rows in different Bi islands are arranged at well-defined angles of 90° with respect to each other (see below). Interestingly, the 7.86 Å spacing between next-nearest neighbour atoms of the topmost Bi(111) layer differs by only 3.7% from $2a_{Au}$ (with $a_{Au} = 4.08$ Å being the Au lattice constant). Furthermore, the Bi spacing in the perpendicular direction is only 11.5% larger than a_{Au}. Minor deformation of the Bi(111) bilayer lattice thus allows it to transform into a structure that is fully commensurate by the Au(100) lattice and can be described by a $(\sqrt{2} \times 2\sqrt{2})$R45° superstructure. In this structure half of the second layer Bi atoms would occupy the energetically preferred fourfold-hollow sites while the remaining half would reside on top sites, explaining the vertical modulation. The length of the covalent Bi–Bi bonds can be preserved in this arrangement by a small vertical relaxation of the bilayer distance; the bond angles likewise will only be slightly distorted. Such a commensurate structure may significantly reduce the Bi–Au interface energy and by this stabilize ultrathin Bi(111) films. For thicker deposits the epitaxial stress associated with this structure should drive a structural transition, apparently resulting in a Bi(110) film.

Observations of the growth morphology and dynamics support that the ultrathin Bi deposits on Au(100) differ structurally from the Bi(110) film found on Au(111). First, the observed Bi islands are rectangular in shape rather than needle-like. Similar to the latter they exhibit relatively straight edges which are preferentially oriented along the direction of the atomic rows. Typically, the islands are terminated by one of the rows that appear to be of lower apparent height in the STM images (see Fig. 5a). However, compared to the Bi deposits on Au(111) these steps usually exhibit a much larger number of kinks that often are multiple rows wide.

The typical structure and dynamic behaviour of these kinks is illustrated in Fig. 6. Kinks in the step are usually one unit cell (or a multiple thereof) wide, *i.e.* consisting of two Bi rows. They are predominantly terminated by an atomically smooth edge (see *e.g.* Fig. 6, 0.0 s) or an additional atom in the kink corner (Fig. 6, 0.1 s), resembling the B and C type kinks illustrated in Fig. 4. However, the kinks propagate in a much more irregular fashion along the step edges than for Bi on Au(111). In particular, single rows that extend several atoms away from the kink are often formed transiently during island growth or dissolution (Fig. 6, 0.2–0.3 s). Such structures were never observed in videos of kink propagation along the edges of the needle-like Bi(110) islands, where they would correspond to atoms that are bound to the step only by long non-covalent bonds (see model of C type kinks in Fig. 4) and thus are energetically very unfavourable. In contrast, these observations can be easily reconciled with a Bi(111) deposit. Because all covalent bonds of the Bi surface atoms in the (111) plane are to atoms in the underlying second layer, the different kink configurations should be similar in energy.

Fig. 6 Sequences of STM images of Bi deposits on Au(100), taken from a video recorded at 10 Hz and a potential of −0.17 V. The images show kink propagation along the island edge (75 × 70 Å^2).

A difference between the high and the low atomic rows could in this case only come from the different epitaxial relationship with respect to the substrate lattice, which should be a more minor effect.

In addition to these two-dimensional islands spontaneous formation of truly one-dimensional nanowires was observed for Bi deposition on the Au(100) surface (Fig. 7). These nano-wires are of well-defined width and can be several tens of nanometers long. They grow predominantly away from the edges of the Bi islands, following the directions of the atomic rows (see Fig. 7c). Furthermore, structural fluctuations are observed in which Bi islands are converted into nanowires and *vice versa*. Often the nanowires form arrays with well-defined spacings that are exactly

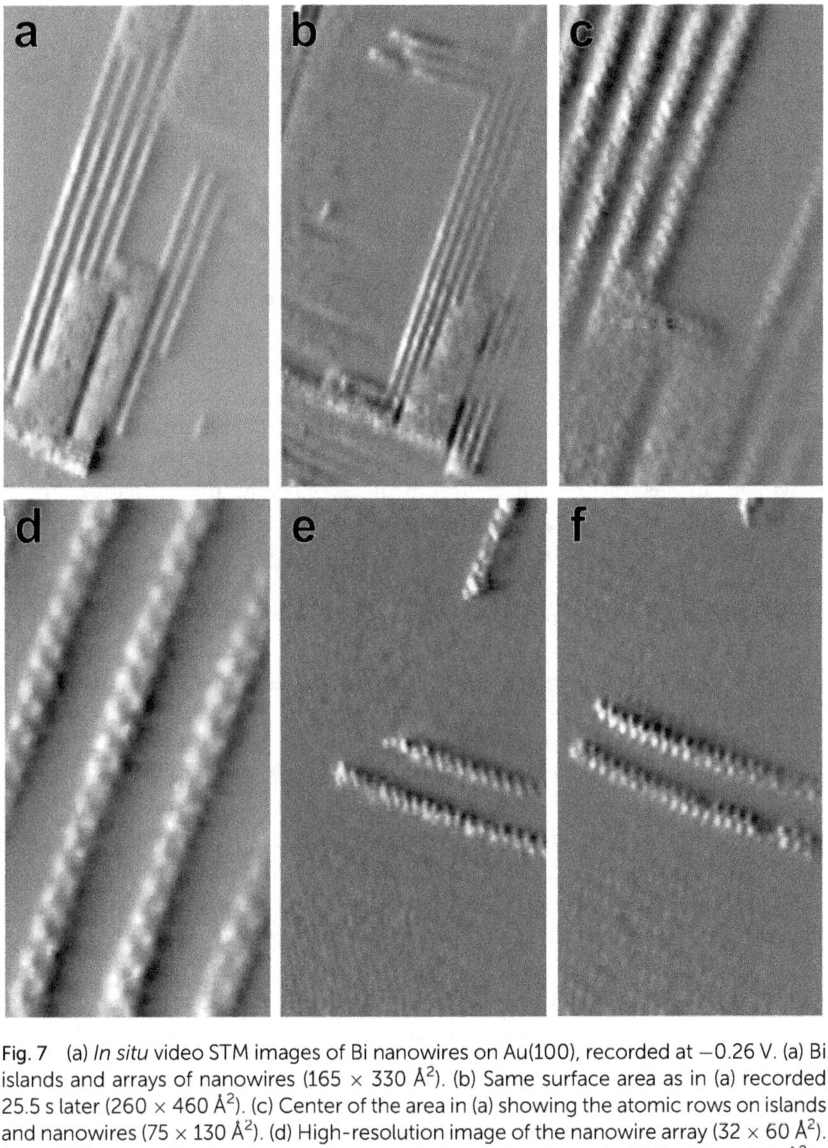

Fig. 7 (a) *In situ* video STM images of Bi nanowires on Au(100), recorded at −0.26 V. (a) Bi islands and arrays of nanowires (165 × 330 Å^2). (b) Same surface area as in (a) recorded 25.5 s later (260 × 460 Å^2). (c) Center of the area in (a) showing the atomic rows on islands and nanowires (75 × 130 Å^2). (d) High-resolution image of the nanowire array (32 × 60 Å^2). (e and f) Images of isolated nanowires recorded at a time interval of 1.0 s (85 × 170 Å^2).

five times the distance between the atomic rows in the Bi islands (Fig. 7a–d). In more rare cases freestanding nanowires are also observed (Fig. 7e and f). The nanowires slowly grow with time, both in length and number, as can be seen in the image in Fig. 7b, recorded 25.5 s after the image in Fig. 7a in the same surface area. In addition, rapid fluctuations are found, especially for isolated free-standing nanowires. However, these fluctuations may be (at least partly) induced by the STM tip.

High-resolution images of the nanowires (Fig. 7d–f) reveal that they consist of a zigzag chain of atoms. As visible in the images in which the underlying Bi film is atomically resolved (Fig. 7e and f), these chains are commensurate to the Bi lattice. That is, the nanowires are exactly one of the rectangular unit cells of the Bi film wide and are oriented in the direction of the shorter unit cell vector. In areas where the nanowires are 90° rotated (*e.g.* in the upper right corner of Fig. 7e and f) the atomic rows in the underlying Bi substrate also run in the perpendicular direction. These observations indicate that the nanowires grow epitaxially on the Bi bulk deposit and, most likely, are structurally related to the latter. However, it is currently unclear what stabilises the unique one-dimensional morphology of these structures.

Conclusions

As demonstrated by these *in situ* video STM studies, Bi electrodeposition exhibits a rich microscopic growth behaviour, reflecting its unique material properties. A central aspect governing the deposit morphology and dynamic is the presence of short bonds with pronounced covalent character. These bonds have already been found to determine the epitaxial growth of ultrathin Bi films under UHV conditions as well as to strongly influence the electronic surface and edge states of one-dimensional Bi nanostructures – a topic of great current interest. Our observations indicate that the presence of covalent Bi–Bi bonds also governs epitaxial growth in an electrochemical environment. Specifically, the formation of the needle-like Bi deposits on Au(111) found in previous STM studies can be clearly rationalised by the structural anisotropy in the deposit caused by these bonds. The latter interlink the atoms in the Bi(110) surface in the form of one-dimensional chains that define the needle direction. Growth and dynamic fluctuations occur predominantly at the end of these chains.

While thick Bi electrodeposits grow in the form of a Bi(110) film on Au(111) as well as on Au(100), video STM investigations of ultrathin deposits reveal a very different structure and growth morphology. As indicated by these data, the structure of ultrathin Bi electrodeposits may be strongly affected by epitaxial effects. On the basis of our structural observations we suggested the formation of a Bi(111) like film on Au(100), which is commensurate to the square substrate lattice. Independent confirmation of this deposit structure, *e.g.* by surface X-ray diffraction studies, would be highly desirable. Particularly interesting is the surprising observation of truly one-dimensional Bi nanowires on this surface. Structures of this type have not been reported in previous studies of Bi deposits, including deposits formed under vacuum conditions. Although their origin is currently unclear, they represent unique nanostructures with potentially interesting novel properties. In particular, one-dimensional Bi nanostructures are of great current interest in the field of topological insulators. The epitaxial electrodeposition of Bi may thus offer attractive new possibilities for the formation of such structures.

Acknowledgements

We gratefully acknowledge financial support by the Deutsche Forschungsgemeinschaft *via* MA 1618/11 and MA 1618/15, NSERC for a Discovery Grant and a Canada Research Chair (S. M.), a fellowship for S. M. by the Alexander von Humboldt-Stiftung, and field trip funds from York University for S.-H. L.

Notes and references

1 Y. Xia, P. Yang, Y. Sun, Y. Wu, B. Mayers, B. Gates, Y. Yin, F. Kim and H. Yan, *Adv. Mater.*, 2003, **15**, 353.
2 G. Staikov, in *Electrocrystallization in nanotechnology*, Wiley-VCH, Weinheim, 2007.
3 T. R. Kline, M. Tian, J. Wang, A. Sen, M. W. H. Chan and T. E. Mallouk, *Inorg. Chem.*, 2006, **45**, 7555.
4 H. Schloerb, V. Haehnel, M. S. Khatri, A. Srivastav, A. Kumar, L. Schultz and S. Faehler, *Phys. Status Solidi B*, 2010, **247**, 2364.
5 U. S. Mohanty, *J. Appl. Electrochem.*, 2011, **41**, 257.
6 R. M. Penner, *J. Phys. Chem. B*, 2002, **106**, 3339.
7 R. M. Penner, *Mod. Aspects Electrochem.*, 2009, **45**, 175.
8 S. Morin, A. Lachenwitzer, O. M. Magnussen and R. J. Behm, *Phys. Rev. Lett.*, 1999, **83**, 5066.
9 X. Xiao, M. Nielinger and H. Baltruschat, *Electrochim. Acta*, 2003, **48**, 3093.
10 J. K. Chang, C. M. Wu and I. W. Sun, *J. Mater. Chem.*, 2010, **20**, 3729.
11 F. Möller, O. M. Magnussen and R. J. Behm, *Phys. Rev. Lett.*, 1996, **77**, 3165.
12 S. Strbac, O. M. Magnussen and R. J. Behm, *Phys. Rev. Lett.*, 1999, **83**, 3246.
13 C. A. Jeffrey, D. A. Harrington and S. Morin, *Surf. Sci.*, 2002, **512**, L367.
14 S. Takakusagi, K. Kitamura and K. Uosaki, *J. Phys. Chem. C*, 2008, **112**, 3073.
15 R. Wen, G. B. Pan and L. J. Wan, *J. Am. Chem. Soc.*, 2008, **130**, 12123.
16 P. Hofmann, *Prog. Surf. Sci.*, 2006, **81**, 191.
17 T. E. Huber, K. Owusu, S. Johnson, A. Nikolaeva, L. Konopko, R. C. Johnson and M. J. Graf, *J. Appl. Phys.*, 2012, **111**, 043709.
18 L. Li, Y. Yang, X. Fang, M. Kong, G. Li and L. Zhang, *Solid State Commun.*, 2007, **141**, 492.
19 M. Tian, J. Wang, Q. Zhang, N. Kumar, T. E. Mallouk and M. H. W. Chan, *Nano Lett.*, 2009, **9**, 3196.
20 K. Lee, S. Lee, S. N. Holmes, J. Ham, W. Lee and C. H. W. Barnes, *Phys. Rev. B: Condens. Matter Mater. Phys.*, 2010, **82**, 245310.
21 J. Kim, S. Lee, Y. M. Brovman, M. G. Kim, P. Kim and W. Lee, *Appl. Phys. Lett.*, 2014, **104**, 043105.
22 M. Tian, J. Wang, W. Ning, T. E. Mallouk and M. H. W. Chan, *Nano Lett.*, 2015, **15**, 1487.
23 F. Y. Yang, K. Liu, K. M. Hong, D. H. Reich, P. C. Searson and C. L. Chien, *Science*, 1999, **284**, 1335.
24 F. Y. Yang, K. Liu, C. L. Chien and P. C. Searson, *Phys. Rev. Lett.*, 1999, **82**, 3328.
25 Z. Zhang, X. Sun, M. S. Dresselhaus, J. Y. Ying and J. Heremans, *Phys. Rev. B: Condens. Matter Mater. Phys.*, 2000, **61**, 4850.
26 Y. Koroteev, G. Bihlmayer, J. E. Gayone, E. V. Chulkov, S. Blügel, P. M. Echenique and P. Hofmann, *Phys. Rev. Lett.*, 2004, **93**, 046403.

27 T. Hirahara, T. Nagao, I. Matsuda, G. Bihlmayer, E. V. Chulkov, Y. Koroteev, P. M. Echenique, M. Saito and S. Hasegawa, *Phys. Rev. Lett.*, 2006, **97**, 146803.
28 B. O'Brien, M. Plaza, L. Y. Zhu, L. Perez, C. L. Chien and P. C. Searson, *J. Phys. Chem. C*, 2008, **112**, 12018.
29 C. A. Jeffrey, S. H. Zheng, E. Bohannan, D. A. Harrington and S. Morin, *Surf. Sci.*, 2005, **600**, 95.
30 Z. L. Bao and K. L. Kavanagh, *Appl. Phys. Lett.*, 2006, **88**, 063708.
31 H. L. Zhang, W. Chen, X. S. Wang, J. Yuhara and A. T. S. Wee, *Appl. Surf. Sci.*, 2009, **256**, 460.
32 T. Nagao, J. T. Sadowski, M. Saito, S. Yaginuma, Y. Fujikawa, T. Kogure, T. Ohno, Y. Hasegawa, S. Hasegawa and T. Sakurai, *Phys. Rev. Lett.*, 2004, **93**, 105501.
33 S. Yaginuma, T. Nagao, J. T. Sadowski, M. Saito, K. Nagaoka, Y. Fujikawa, T. Sakurai and T. Nakayama, *Surf. Sci.*, 2007, **601**, 3593.
34 P. J. Kowalczyk, O. Mahapatra, D. N. McCarthy, W. Kozlowski, Z. Klusek and S. A. Brown, *Surf. Sci.*, 2011, **605**, 659.
35 P. J. Kowalczyk, O. Mahapatra, S. A. Brown, G. Bian, X. Wang and T. C. Chiang, *Nano Lett.*, 2013, **13**, 43.
36 S. H. Zheng, K. Krug, F. Golks, D. Kaminski, S. Morin and O. M. Magnussen, *J. Electroanal. Chem.*, 2010, **649**, 189.
37 C.-H. Chen and A. A. Gewirth, *J. Am. Chem. Soc.*, 1992, **114**, 5439.
38 C. Chen, K. D. Kepler, A. A. Gewirth, B. M. Ocko and J. Wang, *J. Phys. Chem.*, 1993, **97**, 7290.
39 K. Tamura, B. M. Ocko, J. X. Wang and R. R. Adzic, *J. Phys. Chem. B*, 2002, **106**, 3896.
40 M. Hara, Y. Nagahara, S. Yoshimoto, J. Inukai and K. Itaya, *J. Electrochem. Soc.*, 2004, **151**, E92.
41 M. Nakamura, N. Sato, N. Hoshi and O. Sakata, *Langmuir*, 2010, **26**, 4590.
42 L. Zitzler, B. Gleich, O. M. Magnussen and R. J. Behm, *Proceedings of the Electrochemical Society*, 2000, vol. 99–28, p. 29.
43 O. M. Magnussen, L. Zitzler, B. Gleich, M. R. Vogt and R. J. Behm, *Electrochim. Acta*, 2001, **46**, 3725.
44 O. M. Magnussen, W. Polewska, L. Zitzler and R. J. Behm, *Faraday Discuss.*, 2002, **121**, 43.
45 W. Polewska, R. J. Behm and O. M. Magnussen, *Electrochim. Acta*, 2003, **48**, 2915.
46 M. Labayen and O. M. Magnussen, *Surf. Sci.*, 2004, **573**, 128.
47 M. Labayen, C. Haak and O. M. Magnussen, *Phys. Rev. B: Condens. Matter Mater. Phys.*, 2005, **71**, 241409–241411.
48 H. Matsushima, A. Taranovskyy, C. Haak, Y. Gründer and O. M. Magnussen, *J. Am. Chem. Soc.*, 2009, **131**, 10362.
49 H. Matsushima, C. Haak, A. Taranovskyy, Y. Gründer and O. M. Magnussen, *Phys. Chem. Chem. Phys.*, 2010, **12**, 13992.
50 Y. Lu, W. Xu, M. Zeng, G. Yao, L. Shen, M. Yang, Z. Luo, F. Pan, K. Wu, T. Das, P. He, J. Jiang, J. Martin, Y. P. Feng, H. Lin and X. S. Wang, *Nano Lett.*, 2015, **15**, 80.
51 Z. F. Wang, L. Chen and F. Liu, *Nano Lett.*, 2014, **14**, 2879.
52 L. Chen, G. Cui, P. Zhang, X. Wang, H. Liu and D. Wang, *Phys. Chem. Chem. Phys.*, 2014, **16**, 17206.

53 D. Wang, L. Chen, H. Liu, X. Wang, G. Cui, P. Zhang, D. Zhao and S. Ji, *Phys. Chem. Chem. Phys.*, 2015, **17**, 3577.
54 K. Lust, E. Perkson and E. Lust, *Russ. J. Electrochem.*, 2000, **36**, 1257.
55 X. Li and A. A. Gewirth, *J. Am. Chem. Soc.*, 2003, **125**, 7086.
56 O. Skylar, T. H. Treutler, N. Vlachopoulos and G. Wittstock, *Surf. Sci.*, 2005, **597**, 181.
57 C. Kittel, in *Introduction to solid state physics*, Wiley, NY, 3rd edn, 1968.
58 D. R. Lide, in *CRC handbook of chemistry and physics*, CRC Press, Boca Raton, 73th edn, 2000.
59 S. H. Zheng, PhD thesis, York University, 2011.
60 T. Nagao, S. Yaginuma, M. Saito, T. Kogure, J. T. Sadowski, T. Ohno, S. Hasegawa and T. Sakurai, *Surf. Sci.*, 2005, **590**, L247.

Faraday Discussions

PAPER

1/f^2 noise in bistable electrocatalytic reactions on mesoscale electrodes†

S. Bozdech,[a] K. Krischer,*[b] D. A. Crespo-Yapur,[b] E. Savinova[a] and A. Bonnefont*[c]

Received 3rd May 2016, Accepted 18th May 2016
DOI: 10.1039/c6fd00115g

The formation of a self-organized spatial domain during current-controlled CO oxidation, a kinetically bistable reaction, is investigated experimentally and by deterministic simulations as a function of the electrode size and of the supporting electrolyte concentration. Decreasing the microelectrode size leads to the suppression of the spatial instability at the electrode and thus stabilizes the S-NDR branch of the reaction. The critical microelectrode size capable of supporting sustained domain formation is shown to be strongly affected by the sulfuric acid concentration, the characteristic time of the positive feedback loop increasing with the sulfate concentration. Furthermore, we demonstrate that for microelectrode diameters close to the instability threshold, small amplitude electrochemical potential fluctuations appear in the system. These potential fluctuations cannot be captured by deterministic mathematical models and are attributed to a strong enhancement of molecular fluctuations or intrinsic noise in the vicinity of the spatial instability. Analysis of the electrochemical noise revealed a 1/f^2 frequency dependence and several common features with neuronal shot noise.

Introduction

Spatio-temporal instabilities may occur in a large number of catalytic and electrocatalytic reactions.[1–6] Depending on the nonlinearity of the reaction kinetics and on the nature of the spatial coupling, the emergence of self-organized states, such as stationary domains, Turing patterns, chemical waves, or spatio-temporal chaos have been frequently observed under nonequilibrium conditions. These self-organization processes are well understood for large systems, when the dynamics of the system behaves deterministically and the phenomena can be

[a]Institut de Chimie et Procédés pour l'Energie, l'Environnement et la Santé, UMR 7515 of CNRS-Université de Strasbourg-ECPM, 25 rue Becquerel, 67087 Strasbourg, France

[b]Nonequilibrium Chemical Physics, Technical University of Munich, James-Franck-Str. 1, 85748 Garching, Germany. E-mail: krischer@tum.de

[c]Institut de Chimie de Strasbourg, UMR 7177, CNRS-Université de Strasbourg, 4 rue Blaise Pascal, 67000 Strasbourg, France. E-mail: bonnefont@unistra.fr

† Electronic supplementary information (ESI) available. See DOI: 10.1039/c6fd00115g

described with reaction-diffusion-type models. However, when going to the meso- or nanoscale, the performance of the system might be drastically influenced by intrinsic noise, such as molecular fluctuations, due to its interaction with the positive and negative feedback loops in the reaction network.[7,8]

In contrast to nonlinear behavior in macroscopic systems, *i.e.* deterministic systems, the understanding of dynamic instabilities in the presence of noise is still very limited and only relatively few theoretical and experimental studies of self-organization processes in small catalytic or electrocatalytic systems have been carried out. These studies can be subdivided in two groups: in nanoscale systems, *e.g.* on catalytic nanoparticles, the diffusion length (or more generally the characteristic interaction length) of the active species, *e.g.* an adsorbed intermediate, is much larger than the lateral extension of the system such that spatial instabilities are suppressed and intrinsic noise acts on the reaction dynamics only. In mesoscale systems, noise also affects the spatial self-organization or instabilities.

In most studies, catalytic CO oxidation served as a model system.[9–14] Gas-phase CO oxidation on Pd or Pt(111) surfaces exhibits a kinetic bistability which manifests itself by the co-existence of two stable stationary homogeneous states for the same set of parameters: the so-called passive state with high CO coverage and low reaction rate and the active state, corresponding to a high rate of CO_2 production and low CO coverage. In the study of gas-phase CO oxidation on Pd nanoparticles with molecular beam experiments, Johanek *et al.* demonstrated that the kinetic bistability of the system vanishes when the particle size decreases.[10] This experimental finding was attributed to the occurrence of frequent fluctuation-induced kinetic transitions between the passive and the active states.[10,11] Such fluctuation-induced transitions between active and passive branches have also been studied using field electron microscopy in the case of CO oxidation on a Pt field emitter tip.[12–14] Another set of experimental and theoretical studies focused on the alteration of kinetic oscillations on nm-sized catalyst particles through the presence of molecular fluctuations.[15–19]

The presence of intrinsic or external noise was also shown to influence the spatial propagation of fronts in bistable systems when the diffusion length of the active species becomes smaller than the size of the system. In the case of bistable CO oxidation on Pd(111), it was experimentally demonstrated that the intrinsic noise of the system is sufficient to induce some transitions between the two active and passive states.[20] In contrast, an external noise had to be applied to induce such transitions when the reaction is performed on Ir(111) surfaces.[21] Other noise-induced patterns observed during gas-phase CO oxidation on Pt surfaces at intermediate pressures are so-called "raindrop" patterns, where small reactive islands randomly occur on an otherwise CO covered surface and collapse again after a short time.[22–24]

As for electrochemical systems, studies on the impact of noise on self-organization processes are practically lacking altogether. In the case of nanoelectrodes, it was theoretically predicted that fluctuations of the electrode potential result in significantly enhanced reaction kinetics in comparison to macroscopic electrodes, and this was independent of whether the system exhibits a dynamical instability or not.[25–28] Furthermore, reversible adsorption of redox species was reported to break ergodicity, which, in turn, goes along with power law fluctuations.[29] In comparison with the above mentioned heterogeneous gas-phase reactions, experimental studies on individual nanoelectrodes are impeded by the rapid transport of even the smallest impurities in the electrolyte to the nanoelectrodes, hindering investigation under constant reaction conditions for a sufficiently long period of time. A further experimental difficulty is the measurement of the tiny currents with a level of measurement noise far beyond the intrinsic or molecular noise of the electrochemical reaction.

In this article, we study the impact of noise on a bistable mesoscale electrochemical system that exhibits a spatial instability. More precisely, we investigate the dynamic behaviour of a single Pt microelectrode during an electrochemical bistable surface reaction, the CO electrooxidation on Pt, as a function of the microelectrode diameter and the concentration of the supporting electrolyte. The CO electrooxidation on Pt is a prominent example of an S-shaped Negative Differential Resistance (S-NDR) system in electrochemistry,[30] with an active and a passive, or a high and a low current state, very similar to the gas-phase system. The S-shaped current–potential curve results from a positive feedback loop in the reaction network that is caused by the interplay of the competitive Langmuir–Hinshelwood mechanism for CO oxidation and the mass transport of CO in the solution. The bistable potential window is readily measured in cyclic voltammograms with moderate to slow scan rate. Under current-controlled conditions the system exhibits a spatial instability: when the applied current value lies in the intermediate current range between the passive and the active states, the electrode does not attain a uniform state with intermediate current density but it splits into two domains which adopt states on the passive and the active branch, respectively.[31–33] The domains remain stationary in time and the potential adjusts to the equistability potential which lies approximately in the middle of the bistable region. A change in current density leads to a change of the ratio of active and passive domains at constant electrode potential. The system resembles in some aspects an equilibrium first order phase transition system, where at the equistability point of the two phases, which is obtained by Maxwell's construction, one phase grows on the expense of the other one.

In the following work, experimental and simulated current potential curves obtained on Pt microelectrodes of various diameters will be presented and discussed.

Experimental

The Pt microdisk electrodes were purchased from Metrohm Autolab and consisted of small diameter Pt wire (either 10, 25 or 50 μm-diameter, reference numbers PT10, PT25, PT50 respectively) sealed into a glass enclosure in such a way that only the cross section of the wire is exposed to the electrolyte solution. Before each experiment, the microelectrodes were first polished to a mirror finish by using successively 1 (Reflex LDP 1μ) and 0.25 μm (Reflex LDP 1/4μ) diamond polishing suspension and then abundantly rinsed with ultrapure water (18.2 MΩ cm and total organic carbon TOC < 3 ppb). For comparison with the behaviour on macroscopic electrodes, a 3 mm diameter polycrystalline Pt disk (Metrohm Autolab) was used.

The electrochemical experiments were performed in a three-electrode cylindrical glass cell that was cleaned in a 1 : 1 mixture of concentrated $H_2SO_4 + H_2O_2$, followed by thorough rinsing in ultrapure water. The working electrode was the Pt microelectrode, a large-surface-area Pt wire placed far away from the microelectrode was used as counter-electrode and the reference electrode was a RHE. These three electrodes were connected to a Gamry ref. 600 potentiostat/galvanostat capable of measuring low currents. The sulfuric acid supporting electrolytes were prepared from 96 wt% H_2SO_4 (Merck suprapure) and ultrapure water. Before starting the electrochemical experiments in saturated CO (4.7 N) electrolytes, the microelectrodes were cleaned by many potential cycles between 0.07 and 1.32 V *vs.* RHE in N_2 (4.7 N) saturated supporting electrolyte until stable cyclic voltammograms (CVs) were obtained. Typical CVs obtained with the various microelectrodes are displayed in Fig. S1.†

Simulations

To gain further insight into the influence of the microelectrode diameter on the spatial instability, finite element simulations were performed using the software Freefem++.[34] The objective of this work was to investigate qualitatively the influence of the diameter of the microelectrode on the occurrence of the domain formation during current-controlled CO electrooxidation in the framework of a deterministic model. For the sake of simplification, a two-dimensional (2D) geometry was assumed for the simulations of the current–potential curves. The Pt microelectrode is modelled (see the geometry of the simulations in Fig. 3) as a hemicircle Γ of radius r and centered at $x = x_0$, placed in the middle of a planar substrate and surrounded by the aqueous electrolyte. The mass transport of CO in solution is described by Fick's second law in two dimensions:

$$\frac{\partial C_{CO}}{\partial t} = D_{CO}\left(\frac{\partial^2 C_{CO}}{\partial x^2} + \frac{\partial^2 C_{CO}}{\partial y^2}\right) \tag{1}$$

where x and y are the spatial coordinate parallel and perpendicular to the substrate of the microelectrode, respectively, C_{CO} the concentration of CO dissolved in solution and D_{CO} the diffusion coefficient of CO. Eqn (1) is solved on a two-dimensional spatial mesh with the following boundary conditions:

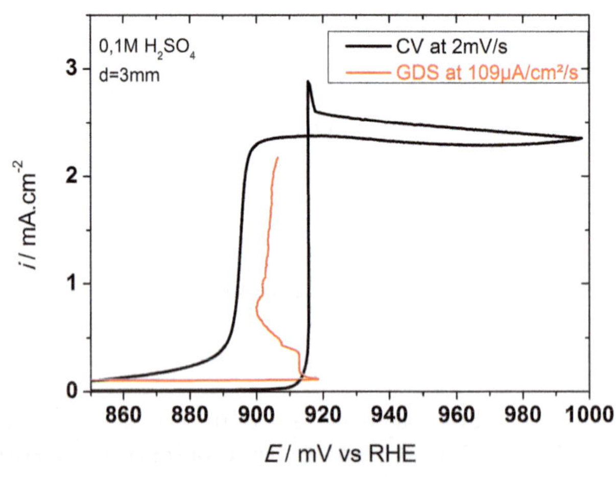

Fig. 1 Cyclic voltammogram measured at 2 mV s^{-1} (CV, black curve) and galvanodynamic scan (GDS, red curve) measured at 109 μA cm^{-2} s^{-1} on a 3 mm diameter Pt rotating disk electrode (2500 rpm) in CO saturated 0.1 M H_2SO_4 electrolyte.

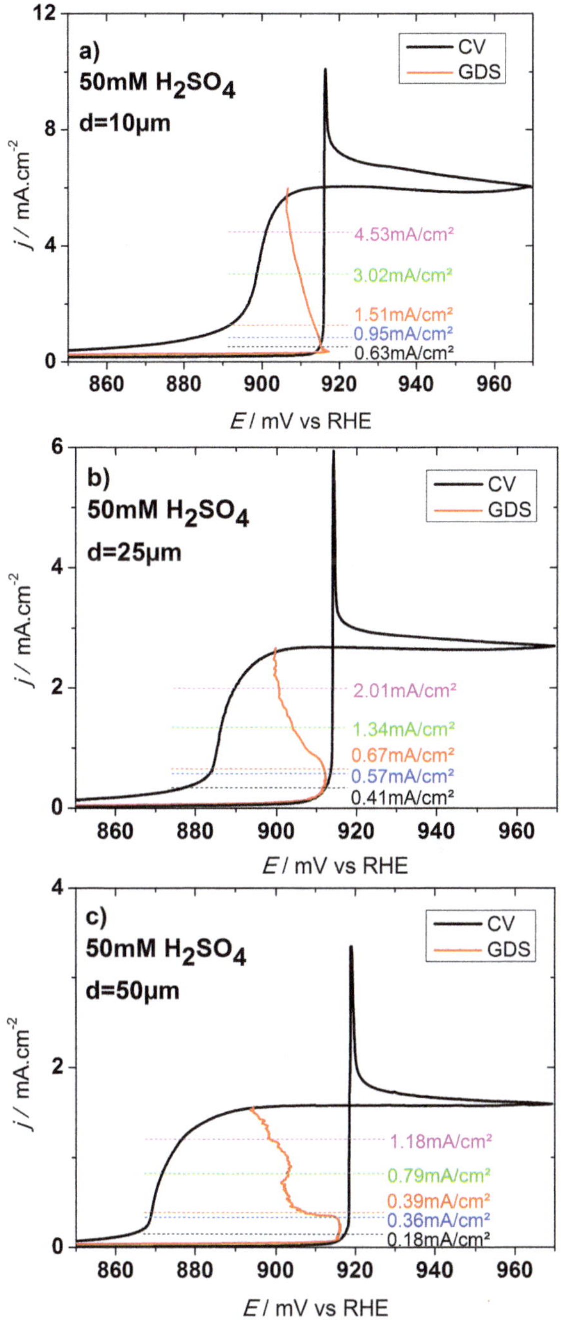

Fig. 2 Cyclic voltammogram at 2 mV s^{-1} (CV, black curves) and galvanodynamic scan (GDS, red curves) on a 10 μm (a), 25 μm (b) and 50 μm (c) diameter Pt disk microelectrode in CO saturated 50 mM H_2SO_4 electrolyte.

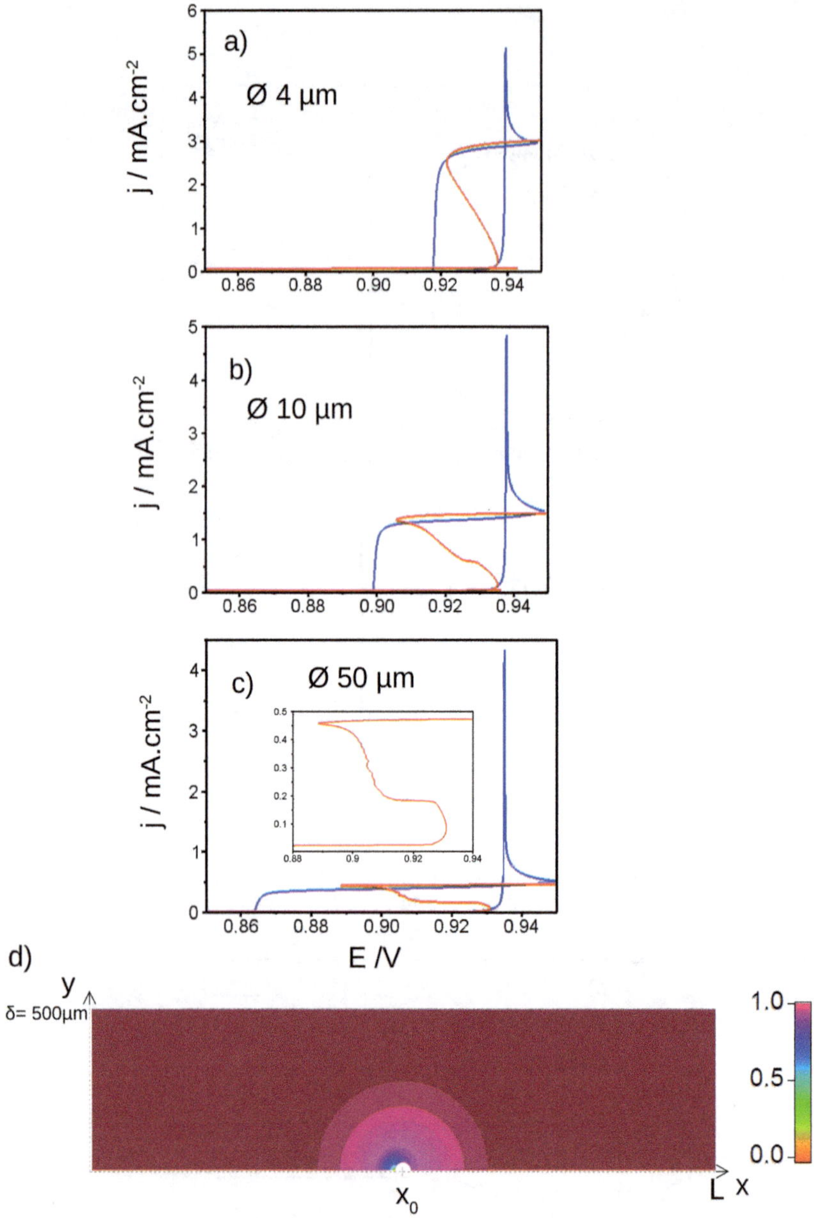

Fig. 3 Simulated cyclic voltammograms at 2 mV s^{-1} (blue curves) and galvanodynamic scans (red curves) during CO electroxidation on a 4 μm (a), 10 μm (b) and 50 μm (c) diameter hemicircle microelectrode. The inset displays a close-up of the galvanodynamic scan of the 50 μm microelectrode. (d) Simulated CO concentration profile around the 50 μm diameter hemicircle electrode at $j = 0.19$ mA cm^{-2}.

(i) In the bulk solution at $y = \delta$, $C_{CO} = C^{b}_{CO}$, C^{b}_{CO} being the CO concentration in CO saturated aqueous electrolyte;

(ii) On the non-reactive substrate, at $0 \le x < x_0 - r$ and at $x_0 + r < x \le L$,

$$\frac{\partial C_{CO}}{\partial y} = 0 \tag{2}$$

(iii) At the borders of the solution, $x = 0$ and $x = L$,

$$\frac{\partial C_{CO}}{\partial x} = 0 \tag{3}$$

(iv) At the hemi-circular microelectrode Γ, at $(x - x_0)^2 + y^2 = r^2$,

$$\frac{\partial C_{CO}}{\partial n} = k_a C_{CO} S_t (1 - \theta_{CO}) - k_d S_t \theta_{CO} \tag{4}$$

here n is the coordinate along the normal vector to the microelectrode surface, θ_{CO} is the adsorbed CO coverage on the microelectrode, k_a and k_d are respectively the CO adsorption and desorption rate constants and S_t the mole number of active sites per unit surface area of the microelectrode. The kinetics of the electrochemical oxidation of CO at the Pt microelectrode is described using a modified version of the model discussed by Koper *et al.*[30] where the surface coverage of adsorbed oxygenated species has been adiabatically eliminated.[35,36] The temporal evolution of adsorbed CO coverage, θ_{CO} is given by:

$$\frac{\partial \theta_{CO}}{\partial t} = k_a C_{CO}(1 - \theta_{CO}) - k_d \theta_{CO} - k_r \theta_{CO}(1 - \theta_{CO}) \exp\left(\frac{2\alpha FE}{RT}\right) \tag{5}$$

where k_r is the CO reaction rate constant. The total faradaic current density i_F at the Pt microelectrode is obtained by:

$$i_F = 2\int_\Gamma F S_t k_r \theta_{CO}(1 - \theta_{CO}) \exp\left(\frac{2\alpha FE}{RT}\right) ds \tag{6}$$

where E is the microelectrode potential. Under controlled current conditions, the temporal evolution of the microelectrode potential is given by the charge balance equation:

$$C\frac{\partial E}{\partial t} = i_T - i_F \tag{7}$$

Table 1 List of parameters used in simulations

Parameter	Meaning	Value
C_b	Bulk CO concentration	10^{-3} M
S_t	Surface active site density	2.2×10^{-9} mol cm^{-2} or 1.1×10^{-9} mol cm^{-2}
α	Charge transfer coefficient	0.5
$k_{CO,a}$	CO adsorption rate constant	5×10^7 cm^3 s^{-1} mol^{-1}
$k_{CO,d}$	CO desorption rate constant	5×10^{-3} s^{-1}
k_r	Reaction rate constant	8×10^{-15} s^{-1}
D	CO diffusion coefficient	1.5×10^{-5} cm^2 s^{-1}
C	Specific double layer capacitance	200 μF cm^{-2}
F	Faraday constant	96 485 C mol^{-1}
R	Gas constant	8.314 J K^{-1} mol^{-1}
T	Temperature	300 K

here, C is the specific microelectrode capacitance and i_T the total applied current density. The meaning of the symbols and the values of the parameters entering eqn (1)–(7) are given in Table 1.

Results and discussion

The typical bistable behaviour obtained on a Pt rotating disk electrode in CO saturated 0.1 M H_2SO_4 solutions can be seen in the cyclic voltammogram displayed in Fig. 1, which is in good agreement with the ones reported in ref. 30, 37 and 38. For an applied potential value in the bistability region (from 0.895 to *ca.* 0.915 V), the Pt electrode can be either in the passive, low current state, which is covered by adsorbed CO, or in the active state corresponding to the diffusion limited current for CO oxidation. The autocatalytic electrooxidation of CO manifests itself in a sharp ignition current peak at the first saddle node (sn1) potential at 0.915 V, while at the second saddle node (sn2) potential at 0.9 V, the autocatalytic readsorption of CO occurs. The corresponding galvanodynamic curve obtained for a positive current scan on the macroscopic Pt electrode is also plotted in Fig. 1. As mentioned in the introduction, the S-NDR branch of CO electrooxidation is not observed on a macroscopic electrode. Upon increase of the applied current, the electrode potential first increases to the ignition potential of the reaction. Further current increase causes a decrease of the electrode potential until, at an applied current density of *ca.* 0.5 mA cm^{-2}, the equistability potential is reached, which lies approximately in the middle of the bistability window. Then, the electrode potential remains almost independent of the applied current until the current density is close to the one of the active branch. As revealed in previous studies using spatially resolved infrared spectroscopy and simulations,[31–33] the reason for the latter behaviour is the formation of self-organized domains on the macroscopic Pt electrode surface. At the equistability potential, one part of the electrode is fully covered by CO, while the other part is almost free of CO, the surface of the active domain increasing with the applied current value. As already mentioned, this self-organized domain formation presents some analogies with equilibrium first order phase transitions in van der Waals fluids.

To explore the influence of the electrode size on the occurrence of this kinetic phase transition, the electrooxidation of CO from a CO-saturated solution was studied on Pt microelectrodes of 10, 25 and 50 μm diameter. In Fig. 2, the current potential-curves obtained with these three electrodes in 50 mM H_2SO_4 are plotted. Overall, the CVs in Fig. 2 exhibit similar features as the ones observed on macroscopic electrodes. The limiting current density in the CVs increases from *ca.* 1.5 to 6 mA cm^{-2} when the size of the microelectrode decreases to 10 μm in diameter, which is a manifestation of the well-known mass transport enhancement to microelectrodes. Furthermore, the width of the bistability window decreases also with the microelectrode diameter, which illustrates the role of the mass transport of CO in the bistable kinetics of CO electrooxidation.[30]

In contrast to the CVs, the galvanodynamic scans differ considerably compared to the one of the macroscopic electrode. For all three electrode sizes, the microelectrode potential tends to decrease during the positive current scan in the intermediate current range between the passive and the active branches. However, while the potential decreases smoothly in case of the 10 μm diameter microelectrode (Fig. 2a), and still traces the negative differential resistance current branch in case of the 25 μm diameter microelectrode, though with minor fluctuations (Fig. 2b), a pronounced potential drop, similar to the one described above for the macroscopic electrode, is observed at a current density of *ca.* 0.5 mA cm^{-2} with the 50 μm-diameter microelectrode (Fig. 2c). The accessibility of the S-NDR branch over the entire current range between active and passive states for the 10 and 25 μm diameter microelectrode evidences that the sizes of these microelectrodes are below the critical threshold for the spatial instability in 50 mM H_2SO_4. The somewhat noisy potential response in the case of the 25 μm diameter microelectrode is an indication of the close proximity of the domain bifurcation in parameter space. In contrast, the sudden potential drop at intermediate current density in case of the 50 μm electrode is a clear indication that the parameter values were beyond the spatial instability and a patterned state spontaneously formed. Yet, in other features the potential response differs from the one of the macroscopic electrode: for increasing applied current the potential still tends to decrease, and it exhibits an enhanced noise level.

To corroborate our interpretation of the critical electrode size for domain formation during CO electrooxidation, deterministic finite element simulations were performed for a hemicircular microelectrode geometry. The mathematical model takes into account the competitive Langmuir–Hinshelwood mechanism for CO electrooxidation combined with the diffusive mass transport of CO at the microelectrode solved in a two-dimensional geometry for the sake of simplicity. The simulated current–potential curves for 4, 10, and 50 μm hemicircle diameters are displayed in Fig. 3. In fact, this deterministic model is capable to capture qualitatively most of the experimentally observed features. The width of the hysteresis region in the simulated CVs increases with the diameter of the microelectrode, which is in good agreement with the experimental results. The simulated galvanodynamic scans confirmed also the disappearance of the spatial instability when the diameter of the hemicircle shrinks. Thus, the numerical simulations predict that the S-NDR branch of the reaction can be accessed for a diameter of 4 μm over its whole current range (Fig. 3a), while for 10 and 50 μm, the spontaneous formation of stationary spatial domains manifests itself by a sharp decrease of the potential at a current density of *ca.* 0.6 and 0.15 mA cm^{-2}, respectively (Fig. 3b and c). Furthermore, in contrast to macroscopic electrodes where the potential remains at the equistability value once the spatial domains emerged, a negative potential slope is still persistent in the galvanodynamic scan on microelectrodes despite of the domain formation. In fact, the simulations reveal that this remarkable feature, observed both in experiments and simulations, can be attributed to the interplay of the inhomogeneous reaction rate at the microelectrode with the particular CO concentration field established in the solution surrounding the microelectrode.

The normalized CO concentration profile in the solution is depicted in Fig. 3d for the 50 μm diameter hemicircle for an applied current density of 0.19 mA cm^{-2}. An inhomogeneous distribution of CO concentration spontaneously develops in the vicinity of the electrode surface, only one part of the microelectrode being active with a high CO oxidation rate. Indeed the passive and active domains are strongly coupled through the diffusion layer of the microelectrode, as apparent in the simulated concentration profile displayed in Fig. 3d. It should be noted that a small amplitude modulation of the potential is present on the simulated galvanodynamic current–potential curves in Fig. 3c. However, below we will show that the experimentally observed potential fluctuations are independent of the scan rate of the current, *i.e.* they also exist in stationary experiments. In contrast, the simulated potential modulation completely disappears when the applied current is kept constant. This potential modulation, observed only for the 50 μm diameter hemicircle, can be tentatively attributed to the influence of the slow dynamics of the diffusion layer rearrangement in response to the relatively fast galvanodynamic scan. Therefore, the experimentally observed potential noise on the S-NDR branch of the 50 μm-diameter microelectrode could not be captured by the deterministic simulations confirming its stochastic nature.

In addition to the microelectrode diameter, the supporting electrolyte concentration was found to have a strong influence on the emergence of the nonequilibrium phase transition during CO electrooxidation.[39] This can be seen in Fig. 4, where the CVs and the galvanodynamic scans of a 10 μm diameter disk

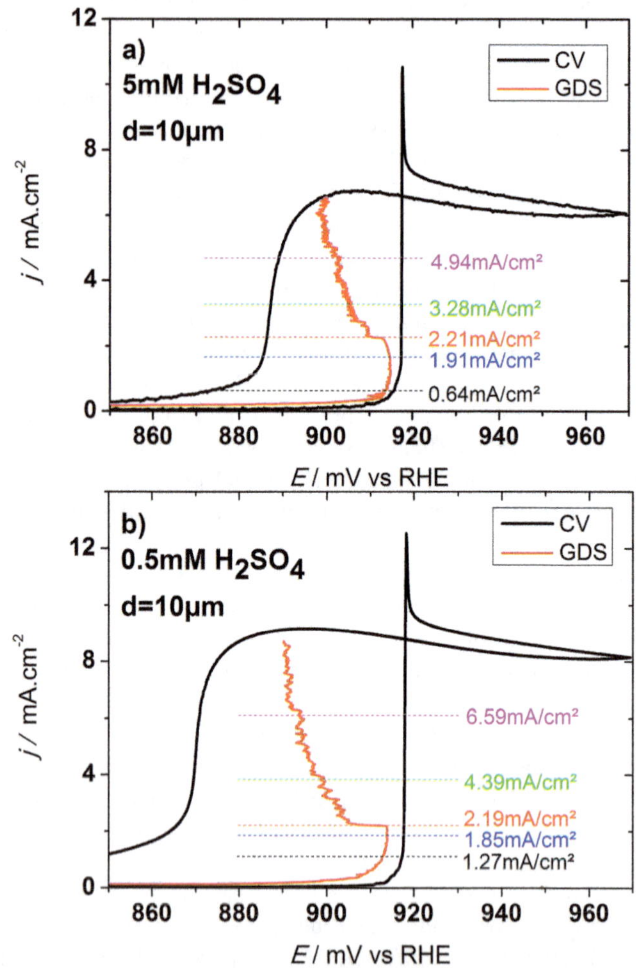

Fig. 4 Cyclic voltammogram at 2 mV s^{-1} (CV, black curves) and galvanodynamic scan (GDS, red curves) on a 10 μm diameter Pt disk microelectrode in CO saturated 5 mM (a) and 0.5 mM (b) H_2SO_4 electrolyte.

microelectrodes in CO saturated 5 and 0.5 mM H_2SO_4 solution are plotted. At lower H_2SO_4 concentration, the width of the bistability window and the value of the limiting current plateau in the CVs are larger. Similar trends were reported by Angelucci *et al.* in CO oxidation on Pt(111) electrodes in 0.1 M $HClO_4$, upon addition of a small amount of H_2SO_4.[40] These features indicate that the positive feedback loop in CO electrooxidation is strongly affected by the H_2SO_4 concentration in solution. The increase of the bistability window is mostly caused by a negative shift of the sn2 potential, where the active state ceases to exist, and is a consequence of the higher limiting current. Interestingly, both galvanodynamic curves in Fig. 4a and b exhibit the pronounced potential decrease associated to the transition to the patterned state at the onset of the S-NDR branch. This reveals that the decrease of the H_2SO_4 concentration affects the characteristic time of the autocatalytic process in such a way that the 10 μm microelectrode is now above the threshold for domain formation. Indeed, the critical size for domain formation is expected to be close to twice the width of the interfacial region between the domains, which is proportional to $(D_{CO}/k)^{0.5}$ where D_{CO} is the CO diffusion coefficient in solution and k is characteristic rate of the autocatalytic process. It is also worth noting that the electrode potential becomes particularly noisy in the presence of the spatial instability.

The mechanism by which the CO electrooxidation autocatalysis is slowed down at high sulfuric acid concentration is still unclear. One possible explanation could be that the kinetics of the reaction and thus the autocatalytic process are influenced by the local pH of the solution. To validate this hypothesis, the proton mass transport in the electrochemical cell as well as its involvement in the reaction kinetics has to be explicitly taken into account in the simulations. While the first can be done easily, any implementation of how the proton concentration affects the reaction rate (other than by the trivial Nernst shift) would be pure speculation at this point. However, there is also another tentative explanation that can be easily tested numerically. It is based on the assumption that adsorption of (bi)sulfate species leads to a partial blocking of Pt active sites, which translates to a decrease of the parameter S_t in the simulations. The simulated current–potential curves assuming a lower S_t value for a 10 μm diameter hemicircle electrode are plotted in Fig. 5. The simulations reproduce qualitatively the features observed in the experiments: the limiting current as well as the bistability window decrease with the density of the active surface sites. This can be attributed to the lower rate of CO consumption in the diffusion layer arising from the decreased Pt active site density. The corresponding galvanodynamic curves exhibit both a smooth decrease of the potential in the S-NDR branch and reveal that the 10 μm diameter hemicircle becomes too small to sustain domain formation when S_t is decreased. The results of the simulations seem to support the crucial influence of the sulfate concentration on both the characteristic time of the autocatalytic CO electrooxidation process and on the shift in the critical size required for pattern formation.

Further insight on the dynamics in the phase transition region can be gained by recording and analysing long times series under stationary, *i.e.* galvanostatic, conditions for different electrode sizes and H_2SO_4 concentrations (Fig. 6 and 7). For each time series, the potential value was sampled at a frequency of 200 Hz and recorded for 20 minutes. The colored dashed lines in Fig. 2 and 4 indicate the

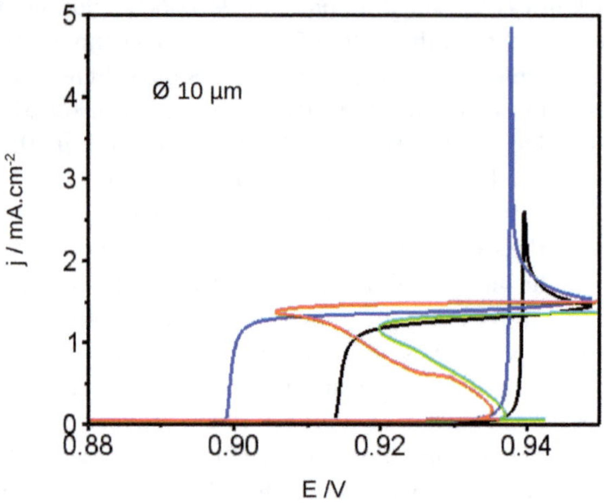

Fig. 5 Simulated cyclic voltammograms (blue and black curves) at 2 mV s^{-1} and galvanodynamic scans (red and green curves) during CO electroxidation on a 10 µm diameter hemicircle microelectrode for $S_t = 2.2 \times 10^{-9}$ mol cm^{-2} (blue and red curves) and $S_t = 1.1 \times 10^{-9}$ mol cm^{-2} (black and green curves).

current densities at which the time series were recorded. The corresponding power spectral densities $S(f)$ can be written as:

$$S(f) = \frac{(\delta t)}{N}\left(\sum_{n=0}^{N-1} V(n)\mathrm{e}^{(-\mathrm{i}2\pi fn)}\right)\left(\sum_{n=0}^{N-1} V(n)\mathrm{e}^{(\mathrm{i}2\pi fn)}\right) \quad (8)$$

where N is the number of points in the time series, δt the time interval between two measurements, f the frequency and V the microelectrode potential minus its average value. $S(f)$ was calculated using the Welch method, averaging periodograms calculated over 40.96 s and applying a Hanning window function.

Independent of the applied current density, the time series exhibit hardly any potential fluctuations when the system is too small to support domain formation, as in the case of 10 and 25 µm electrodes and 50 mM H_2SO_4 supporting electrolyte (Fig. 6a and c). They show a slow positive potential drift, which becomes more pronounced with increasing current density (green and magenta curves), and is more distinct for the smaller electrode. Similar potential drifts were also reported for macroscopic Pt electrodes[41,42] in oscillatory methanol and formic acid oxidation. The potential drift manifests itself by a contribution below 0.03 Hz in the power spectra densities, *i.e.* below the smallest frequency shown in the corresponding power spectral density plots and are of no further interest in our context. For all current densities the power spectra density is almost constant between 0.05 and 30 Hz (Fig. 6b and d). This is characteristic for white noise, and the spectral density plots in Fig. 6b and d can be seen as a signature of the background noise of the experimental setup.

For microelectrodes able to support sustained domain formation, the potential time series exhibit to the largest part strong fluctuations (Fig. 6e, 7a and c). Only those on the passive stable state that are sufficiently far from the spatial instability have the same low noise level as the measurements discussed above

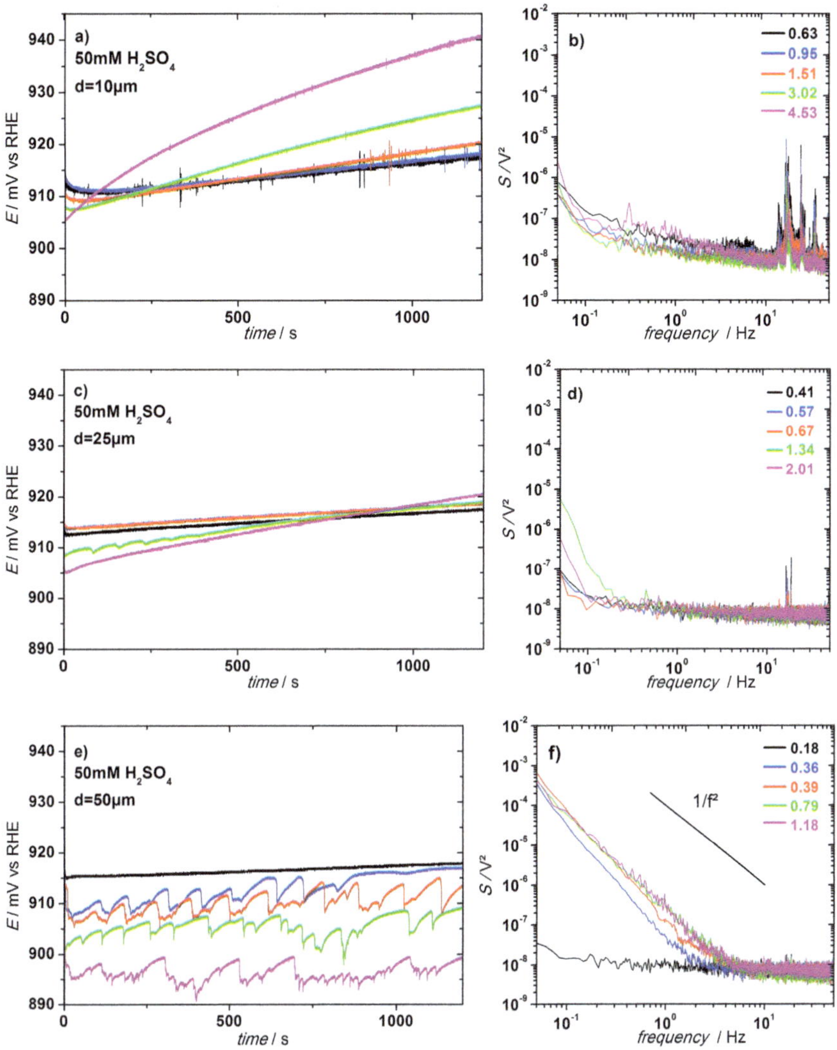

Fig. 6 Potential times series (a, c and e) and corresponding power spectra (b, d and f) recorded under various constant current densities as indicated in Fig. 2 for a 10 µm (a and b), 25 µm (b and c) and 50 µm (d and e) diameter Pt disk microelectrode in 50 mM CO saturated H_2SO_4 electrolyte.

(black curves in Fig. 6e and 7a). The fluctuating time series have diverse characteristics, and on a phenomenological level, the noisy potential time series can be coarsely subdivided into three groups: group 1 exhibits irregular potential spikes composed of fast transitions towards lower potentials followed by a slow relaxation back to the original value (*e.g.* all colored time series in Fig. 6e). Group 2 jumps erratically between two states, a noisy state at lower potentials and a low-noise state at higher potentials (*e.g.* red time series in Fig. 7c). Group 3 is found at the lowest potentials and exhibits rather rapid low amplitude fluctuations (*e.g.* green and magenta time series in Fig. 7a). However, independent of the

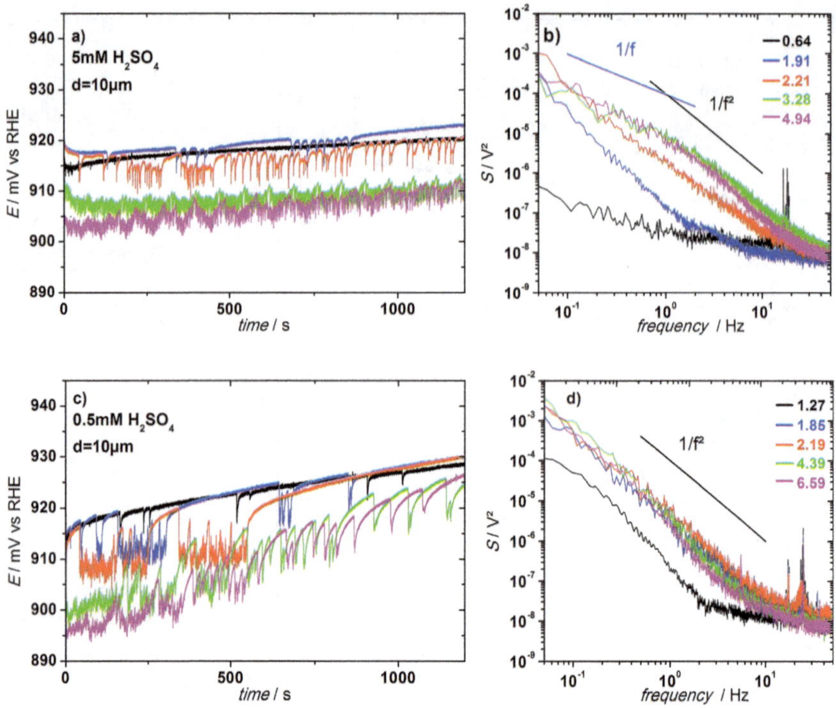

Fig. 7 Potential times series (a and c) and corresponding power spectra (b and d) recorded under various constant current densities (indicated in the figure) for a 10 μm diameter Pt disk microelectrode in CO saturated 5 mM (a and b) and 0.5 mM (c and d) H_2SO_4 electrolyte.

characteristics no predominant frequency could be detected in the power spectral densities plots. Rather, the latter display power law dependence in wide ranges of frequencies, in most cases with an exponent of about −2.

In the literature, there are a few examples for a similar $1/f^2$ dependence. Based on a mathematical analysis, Steyer *et al.* predicted that $1/f^2$ noise might arise from a coalescence-growth process,[43] the corresponding time series exhibiting a spike-like decrease followed by a slow relaxation, very similar to our group 1 fluctuating time series. Such a coalescence-induced $1/f^2$ noise was indeed found in recent experiments of galvanostatic hydrogen evolution on Pt electrodes under the influence of a magnetic field.[44] However, in our case this mechanism seems to be unlikely since it would imply the formation of a large number of growing and coalescing CO domains, while we are operating close to the critical size of domain formation. $1/f^2$ noise behaviour was also found during the recording of local field potential by using intracranial electrodes in a variety of cerebral regions.[45–47] Inspired by models developed for electronic shot noise, two analytical models of such 'neuronal shot noise' were proposed by Milstein *et al.*[45] The first model considers statistically independent spiking neurons undergoing slow decay of a dendrosynaptic extracellular field, while in the second model, $1/f^2$ noise arises from temporally correlated cells generating up-down states. Interestingly, the noisy time series in these two models resemble our fluctuating potential time series in group 1 and 2, respectively. Stochastic mathematical modeling, which would allow a deeper comparison of neuronal shot noise, up-down noise and our groups 1 and 2 noise, is beyond the scope of this paper. However, an interpretation of the possible dynamic features of CO electrooxidation which might cause the different types of strongly enhanced noise reveal that they in fact share common characteristics with the neuronal systems.

Type 1 noise exhibits sharp spikes towards lower potential from a state that is itself not noisy. According to our discussion above, this state is likely to be a uniform state of high CO coverage which is at the edge of becoming unstable with respect to domain formation. Small intrinsic noise in this state might thus ignite locally the autocatalytic reaction of CO thereby forming reactive or active spots with low CO coverage. Under galvanostatic conditions, the enhanced current triggers instantaneously a decrease in the potential. The formed domain pattern is however not stable, and diffusion controlled CO adsorption brings the system back to the original state. Comparing type 1 time series of the 50 and the 10 μm electrode reveals that the characteristic time of the slow relaxation process increases with the electrode size which supports the interpretation that the slow rearrangement of the CO concentration is determined by diffusion. Thus, from a dynamic point of view, type 1 current spikes could arise due to the interaction of low level intrinsic noise with the fast and autocatalytic reaction of adsorbed CO and slow recovery of the original, in the deterministic limit weakly stable state due to diffusion. In other words, the interaction between small intrinsic noise and the nonlinear reaction network gives rise to stochastic, multi-timescale spike events. From this point of view, CO electrooxidation seems to be similar to the above mentioned neural model.

In the type 2 fluctuating time series, the potential jumps back and forth between two states, the passive, CO-covered state, and the patterned domain state. This suggests that at these parameter values both states are linearly stable in the deterministic system, a low noise level being sufficient to trigger transitions between the two states. In so far, this system can be seen as an up-down system, which, as mentioned above, is also known to produce $1/f^2$ noise. However, in contrast to the described up-down system, our type 2 potential time series exhibit fast low-amplitude fluctuations when the system is in the patterned state. We are thus facing a more complicated situation. In fact, the portions of the time series in which the system is in the patterned state strongly resemble type 3 fluctuations and have most likely the same origin.

As for the origin of type 3 noise, it is the one which is least clear and we can only speculate about it. Visualizing a domain pattern, it is composed of three parts, the reactive, nearly bare domain, the CO-covered passive domain and an interfacial region in between, in which the CO-coverage continuously changes from a high value in the passive state to a low value in the active state. Since the sizes of our electrodes are close to the critical size of domain formation, the interfacial region amounts also to a considerable fraction of the electrode. It seems to be intuitive that intrinsic noise in this interfacial region can more easily trigger a local transition to the active or the passive state, respectively, from an intermediate CO coverage than a transition from the active to the passive state or *vice versa*. Our current hypothesis is thus that this fast, low amplitude noise in the domain state is caused by local transitions to one of the two stable states and the subsequent restoration of the interface, *i.e.* the dynamical stable domain structure.

Conclusions

We have investigated the dynamics of a pattern forming system on microelectrodes with a lateral extension close to the critical size for pattern formation. For the system we chose the electrochemical oxidation of CO on Pt electrodes, which exhibits bistability under potential control, the most basic manifestation of complexity in a nonlinear system. Under galvanostatic conditions, the uniform electrode surface might become unstable and the electrode splits into two domains, each domain adopting one of the two locally stable states. In some respect, the formation of the domains, the most simple spatial instability in a nonlinear system, can be seen as the nonequilibrium variant of an equilibrium phase transition of first order.

The use of microelectrodes gave not only important insights into the electrochemical mechanism of CO oxidation but brought about also general findings on the impact of noise in bistable, pattern forming systems. As for the former, we found that the critical microelectrode size able to sustain domain formation is strongly affected by the sulfuric acid concentration, the characteristic time of the positive feedback loop increasing with the sulfate concentration. Furthermore, the current density of the active state decreased with increasing sulfate concentration. Both facts could be reproduced in simulations when assuming that an increasing sulfate concentration reduces the effective number of active Pt sites.

From the dynamical point of view, we could demonstrate that in systems with a spatial domain size close to the instability threshold, molecular or intrinsic noise is strongly enhanced due to its interaction with the nonlinear reaction kinetics. Analysis of the electrochemical noise revealed a $1/f^2$ frequency dependence and several common features with neuronal shot noise. A deeper understanding of these fluctuations requires the development of advanced statistical approaches. The enhancement of molecular fluctuations by non-linear reaction kinetics on individual elements is expected to play a crucial role in the global properties of macroscopic systems when several elements are coupled.[39,48–50]

Acknowledgements

Financial support from the International Center for Frontier Research in Chemistry, from the Labex "Chimie des Systèmes Complexes" (University of Strasbourg) and from the Bayerisch-Französisches Hochschulzentrum (BFHZ) is gratefully acknowledged.

References

1 K. Krischer, in *Principles of Temporal and Spatial Pattern Formation in Electrochemical Systems, Modern Aspects of Electrochemistry*, ed. B. E. Conway, J. O. Bockris and R. E. White, Kluwer Academic Publishers, Boston, 2002. pp. 1–142, available from: http://link.springer.com/10.1007/0-306-46916-2_1.

2 R. Imbihl and G. Ertl, Oscillatory Kinetics in Heterogeneous Catalysis, *Chem. Rev.*, 1995 May, **95**(3), 697–733.

3 G. Ertl, Reactions at Surfaces: From Atoms to Complexity (Nobel Lecture), *Angew. Chem., Int. Ed.*, 2008, **47**(19), 3524–3535.

4 A. S. Mikhailov and K. Showalter, Control of waves, patterns and turbulence in chemical systems, *Phys. Rep.*, 2006, **425**(2–3), 79–194.
5 D. Luss and M. Sheintuch, Spatiotemporal patterns in catalytic systems, *Catal. Today*, 2005, **105**(2), 254–274.
6 G. Ertl and H.-H. Rotermund, Spatiotemporal pattern formation in reactions at surfaces, *Curr. Opin. Solid State Mater. Sci.*, 1996, **1**(5), 617–621.
7 M. Hildebrand and A. S. Mikhailov, Mesoscopic Modeling in the Kinetic Theory of Adsorbates, *J. Phys. Chem.*, 1996, **100**(49), 19089–19101.
8 D.-J. Liu and J. W. Evans, Realistic multisite lattice-gas modeling and KMC simulation of catalytic surface reactions: Kinetics and multiscale spatial behavior for CO-oxidation on metal (100) surfaces, *Prog. Surf. Sci.*, 2013, **88**, 393–521.
9 M. Berdau, G. G. Yelenin, A. Karpowicz, M. Ehsasi, K. Christmann and J. H. Block, Macroscopic and mesoscopic characterization of a bistable reaction system: CO oxidation on Pt(111) surface, *J. Chem. Phys.*, 1999, **110**(23), 11551.
10 V. Johanek, Fluctuations and Bistabilities on Catalyst Nanoparticles, *Science*, 2004, **304**(5677), 1639–1644.
11 M. Pineda, R. Imbihl, L. Schimansky-Geier and C. Zülicke, Theoretical analysis of internal fluctuations and bistability in CO oxidation on nanoscale surfaces, *J. Chem. Phys.*, 2006, **124**(4), 044701.
12 Y. Suchorski, J. Beben, E. W. James, J. W. Evans and R. Imbihl, Fluctuation-Induced Transitions in a Bistable Surface Reaction: Catalytic CO Oxidation on a Pt Field Emitter Tip, *Phys. Rev. Lett.*, 1999, **82**(9), 1907–1910.
13 Y. Suchorski, J. Beben, R. Imbihl, E. W. James, D.-J. Liu and J. W. Evans, Fluctuations and critical phenomena in catalytic CO oxidation on nanoscale Pt facets, *Phys. Rev. B: Condens. Matter Mater. Phys.*, 2001, **63**, 165417.
14 R. Imbihl, Fluctuations in catalytic surface reactions, *New J. Phys.*, 2003, **5**, 62.
15 Y. Gong, Z. Hou and H. Xin, Optimal Particle Size for Reaction Rate Oscillation in CO Oxidation on Nanometer-Sized Palladium Particles, *J. Phys. Chem. B*, 2004, **108**(46), 17796–17799.
16 V. P. Zhdanov and B. Kasemo, Fluctuations in kinetic oscillations on nm-sized catalyst particles, *Surf. Sci.*, 2005, **588**(1–3), L220–L226.
17 C. Barroo, Y. De Decker, T. Visart de Bocarmé and P. Gaspard, Fluctuating Dynamics of Nanoscale Chemical Oscillations: Theory and Experiments, *J. Phys. Chem. Lett.*, 2015, **6**(12), 2189–2193.
18 J.-S. McEwen, P. Gaspard, T. V. de Bocarme and N. Kruse, Nanometric chemical clocks, *Proc. Natl. Acad. Sci. U. S. A.*, 2009, **106**(9), 3006–3010.
19 P. Grosfils, P. Gaspard and T. Visart de Bocarmé, The role of fluctuations in bistability and oscillations during the H_2 + O_2 reaction on nanosized rhodium crystals, *J. Chem. Phys.*, 2015, **143**(6), 064705.
20 S. Wehner, S. Karpitschka, Y. Burkov, D. Schmeisser, J. Küppers and H. R. Brand, Stochastic aspects of pattern formation during the catalytic oxidation of CO on Pd(111) surfaces, *Phys. D*, 2010, **239**(11), 746–751.
21 J. Cisternas, S. Wehner and O. Descalzi, CO oxidation on Ir(111) surfaces under large non-Gaussian noise, *J. Chem. Phys.*, 2012, **137**(6), 064105.
22 J. Starke, C. Reichert, M. Eiswirth, H. H. Rotermund and G. Ertl, Fluctuation-induced pattern formation in a surface reaction, *Europhys. Lett.*, 2006, **73**(6), 820–825.

23 M. Pineda, L. Schimansky-Geier and R. Imbihl, Fluctuation-induced phase transition in a spatially extended model for catalytic CO oxidation, *Phys. Rev. E: Stat., Nonlinear, Soft Matter Phys.*, 2007, **75**, 061107.

24 M. Pineda, R. Imbihl and L. Schimansky-Geier, Effects of surface size on minimalistic stochastic models for the catalytic CO oxidation, *Phys. A*, 2010, **389**(6), 1178–1188.

25 V. García-Morales and K. Krischer, Kinetic enhancement in nanoscale electrochemical systems caused by non-normal distributions of the electrode potential, *J. Chem. Phys.*, 2011, **134**(24), 244512.

26 V. Garcia-Morales and K. Krischer, Fluctuation enhanced electrochemical reaction rates at the nanoscale, *Proc. Natl. Acad. Sci. U. S. A.*, 2010, **107**(10), 4528–4532.

27 V. Garcia-Morales and K. Krischer, Superstatistics in nanoscale electrochemical systems, *Proc. Natl. Acad. Sci. U. S. A.*, 2011, **108**(49), 19535–19539.

28 V. P. Zhdanov and B. Kasemo, Role of the field fluctuations in electrochemical reactions, *Appl. Surf. Sci.*, 2003, **219**(3–4), 256–263.

29 D. Krapf, Nonergodicity in nanoscale electrodes, *Phys. Chem. Chem. Phys.*, 2013, **15**, 459–465.

30 M. T. M. Koper, T. J. Schmidt and N. M. Marković, Ross PN. Potential Oscillations and S-Shaped Polarization Curve in the Continuous Electro-oxidation of CO on Platinum Single-crystal Electrodes, *J. Phys. Chem. B*, 2001, **105**(35), 8381–8386.

31 R. Morschl, J. Bolten, A. Bonnefont and K. Krischer, Pattern Formation during CO Electrooxidation on Thin Pt Films Studied with Spatially Resolved Infrared Absorption Spectroscopy, *J. Phys. Chem. C*, 2008, **112**(26), 9548–9551.

32 A. Bonnefont, R. Morschl, P. Bauer and K. Krischer, Electrochemical impedance spectroscopy of patterned steady states on electrode surfaces, *Electrochim. Acta*, 2009, **55**(2), 410–415.

33 P. R. Bauer, A. Bonnefont and K. Krischer, Spatially Resolved ATR-FTIRS Study of the Formation of Macroscopic Domains and Microislands during CO Electrooxidation on Pt, *ChemPhysChem*, 2010, **11**(13), 3002–3010.

34 F. Hecht, New development in freefem++, *J. Numer. Math.*, 2012, **20**(3–4), 251–266.

35 D. Zhang, O. Deutschmann, Y. E. Seidel and R. J. Behm, Interaction of Mass Transport and Reaction Kinetics during Electrocatalytic CO Oxidation in a Thin-Layer Flow Cell, *J. Phys. Chem. C*, 2011, **115**(2), 468–478.

36 S. Malkhandi, P. R. Bauer, A. Bonnefont and K. Krischer, Mechanistic aspects of oscillations during CO electrooxidation on Pt in the presence of anions: Experiments and simulations, *Catal. Today*, 2013, **202**, 144–153.

37 H. A. Gasteiger, N. M. Markovic and P. N. Ross, H_2 and CO Electrooxidation on Well-Characterized Pt, Ru, and Pt–Ru. 1. Rotating Disk Electrode Studies of the Pure Gases Including Temperature Effects, *J. Phys. Chem.*, 1995, **99**(20), 8290–8301.

38 S. Malkhandi, A. Bonnefont and K. Krischer, Dynamic instabilities during the continuous electro-oxidation of CO on poly- and single crystalline Pt electrodes, *Surf. Sci.*, 2009, **603**(10–12), 1646–1651.

39 D. A. Crespo-Yapur, A. Bonnefont, R. Schuster, K. Krischer and E. R. Savinova, Sequential Activation and Oscillations of Globally Coupled Microelectrodes during a Bistable Reaction, *ChemElectroChem*, 2014, **1**(6), 1046–1056.

40 C. A. Angelucci, F. C. Nart, E. Herrero and J. M. Feliu, Anion re-adsorption and displacement at platinum single crystal electrodes in CO-containing solutions, *Electrochem. Commun.*, 2007, **9**(5), 1113–1119.

41 N. Perini, B. C. Batista, A. C. D. Angelo, I. R. Epstein and H. Varela, Long-Lasting Oscillations in the Electro-Oxidation of Formic Acid on PtSn Intermetallic Surfaces, *ChemPhysChem*, 2014, **15**(9), 1753–1760.

42 R. Nagao, E. Sitta and H. Varela, Stabilizing Nonstationary Electrochemical Time Series, *J. Phys. Chem. C*, 2010, **114**(50), 22262–22268.

43 A. Steyer, P. Guenoun and D. Beysens, Coalescence-induced $1/f^2$ noise, *Phys. Rev. Lett.*, 1992, **68**(12), 1869–1871.

44 Z. Diao, P. A. Dunne, G. Zangari and J. M. D. Coey, Electrochemical noise analysis of the effects of a magnetic field on cathodic hydrogen evolution, *Electrochem. Commun.*, 2009, **11**(4), 740–743.

45 J. Milstein, F. Mormann, I. Fried, C. Koch and L. M. de la Prida, Neuronal Shot Noise and Brownian $1/f^2$ Behavior in the Local Field Potential, *PLoS One*, 2009, **4**(2), e4338.

46 B. J. He, J. M. Zempel, A. Z. Snyder and M. E. Raichle, The Temporal Structures and Functional Significance of Scale-free Brain Activity, *Neuron*, 2010, **66**(3), 353–369.

47 G. Baranauskas, E. Maggiolini, A. Vato, G. Angotzi, A. Bonfanti, G. Zambra, *et al.*, Origins of $1/f^2$ scaling in the power spectrum of intracortical local field potential, *J. Neurophysiol.*, 2012, **107**(3), 984–994.

48 D. A. Crespo-Yapur, A. Bonnefont, R. Schuster, K. Krischer and E. R. Savinova, Cooperative Behaviour of Pt Microelectrodes during CO Bulk Electrooxidation, *ChemPhysChem*, 2013, **14**(6), 1117–1121.

49 N. Masuda, Y. Kawamura and H. Kori, Collective fluctuations in networks of noisy components, *New J. Phys.*, 2010, **12**(9), 093007.

50 Y. Kawamura, H. Nakao, K. Arai, H. Kori and Y. Kuramoto, Collective Phase Sensitivity, *Phys. Rev. Lett.*, 2008, **101**(2), 024101.

Faraday Discussions

PAPER

Anisotropic etching of rhodium and gold as the onset of nanoparticle formation by cathodic corrosion

Thomas J. P. Hersbach,[a] Vladislav A. Mints,[a] Federico Calle-Vallejo,[a] Alexei I. Yanson[b] and Marc T. M. Koper*[a]

Received 13th April 2016, Accepted 25th April 2016
DOI: 10.1039/c6fd00078a

Cathodic corrosion is a phenomenon in which negatively polarized metal electrodes are degraded by cathodic etching and nanoparticle formation. Though these changes are dramatic and sometimes even visible by eye, the exact mechanisms underlying cathodic corrosion are still unclear. This work aims to improve the understanding of cathodic corrosion by studying its onset on rhodium and gold electrodes, which are subjected to various constant cathodic potentials in 10 M NaOH. After this polarization, the electrodes are studied using cyclic voltammetry and scanning electron microscopy, allowing a corrosion onset potential of −1.3 V *vs.* NHE for rhodium and −1.6 V *vs.* NHE for gold to be defined. The mildness of the potentials on both metals suggests that cathodic corrosion is less extreme and more ubiquitous than expected. Furthermore, we are able to observe well-defined rectangular etch pits on rhodium. Combined with rhodium cyclic voltammetry, this indicates a strong preference for forming (100) sites during corrosion. In contrast, a (111) preference is indicated on gold by voltammetry and the presence of well-oriented quasi-octahedral nanoparticles. This different etching behavior is suggested to be caused by preferential adsorption of sodium ions to surface defects, as is confirmed by density functional theory calculations.

1 Introduction

All metals undergo oxidation under sufficiently anodic (positive) polarization, as can readily be deduced from the electrochemical series.[1] This oxidation lies at the basis of a variety of processes, such as the formation of catalytically active metal oxides,[2] reorganization of the metallic surface[3] and corrosion and dissolution of material.[4] In contrast, metals are generally considered to be chemically stable under cathodic (negative) polarization; reduction of metals is not listed in the electrochemical series, Pourbaix diagrams generally consider uncharged metallic

[a]*Leiden Institute of Chemistry, Leiden University, PO Box 9502, 2300 RA Leiden, The Netherlands. E-mail: m.koper@lic.leidenuniv.nl*
[b]*Cosine Measurement Systems, Oosteinde 36, 2361 HE Warmond, The Netherlands*

species to be most stable at low potentials,[5] and metals are commonly polarized cathodically to prevent anodic corrosion.[6]

However, cathodic potentials are able to induce remarkable changes that are similar to those observed anodically, such as the generation of catalytically active undercoordinated sites[7] and reorganization of the surface.[8,9] In fact, it is even possible to degrade metallic electrodes in a process called cathodic corrosion.[10–12] This process etches metallic electrodes and is capable of forming nanoparticles by applying cathodic potentials, possibly by generating anionic metal intermediates. Though this process has typically been studied using strongly negative potentials,[13] a recent study demonstrated that the cathodic corrosion of platinum starts at the relatively mild potential of −0.4 V *versus* the reversible hydrogen electrode (RHE) in 10 M NaOH.[14] Furthermore, this study showed a strong preference for the creation of (100) sites during corrosion. In order to gain more insight into cathodic corrosion, it is important to explore the onset potential and etching preference of other metals. However, a key factor in determining these properties for platinum was the ability to create a reproducible electrode surface by flame annealing using a standard butane–oxygen burner,[15] which is not possible for all metals. Two of these metals are rhodium and gold: rhodium is rapidly covered by its oxide during annealing,[16] whereas gold electrodes of the required size melt almost immediately after reaching their glowing temperature.

In this study, we circumvent this problem by replacing the annealing step by a chemical cleaning step in which organic contaminants are removed from the electrodes using a 3 : 1 mixture of sulfuric acid and hydrogen peroxide. This procedure yields a highly polycrystalline, yet clean and reproducible electrochemical response for rhodium. A reproducible gold surface can also be created if repeated cyclic voltammograms (CVs) are performed in dilute sulfuric acid after chemical cleaning. After this preparation procedure, both rhodium and gold can be studied using the protocol developed earlier:[14] electrodes are characterized using cyclic voltammetry in 0.1 M H_2SO_4, after which they are treated cathodically in 10 M NaOH. Subsequent cyclic voltammetry in H_2SO_4 will then reveal any changes in the electrode surface, allowing the tentative cathodic corrosion onset potentials to be pinpointed and detecting anisotropic etching preferences for both metals. Further study using scanning electron microscopy (SEM) reveals well-defined etch pits on rhodium and crystallites on gold, which are in excellent agreement with the electrochemically observed etching anisotropy. This etching anisotropy is explored further using density functional theory (DFT), which suggests that the etching preference arises from the specific adsorption of sodium ions to surface defects.

These results demonstrate that cathodic corrosion can occur as readily as anodic corrosion, since both rhodium and gold etch anisotropically at mild cathodic potentials. Furthermore, this study provides a protocol for investigating cathodic corrosion on other metals that cannot be flame annealed, which is a crucial step in expanding cathodic corrosion studies.

2 Experimental

2.1 Electrochemical characterization and DC polarization

Electrochemical experiments were performed with an Autolab PGSTAT12. All water used in this study was demineralized and ultrafiltered by a Millipore MilliQ system (resistivity > 18.2 MΩ cm, TOC < 5 ppb) before use. Working electrolytes were deoxygenated by purging argon (Linde, 6.0 purity) for at least 30 minutes prior to starting experiments. Deoxygenation was maintained by flowing argon over the solution during experiments.

Short lengths of Au (Materials Research Corporation, März Purity; ϕ = 0.125 mm) and Rh (Mateck, 99.9%; $\phi = 0.125$ mm) wire were used as working electrodes. Each working electrode was rinsed with water, dipped in a 3 : 1 mixture of H_2SO_4 and H_2O_2 to remove organic contamination and rinsed again before transfer into a standard 3-electrode cell filled with 0.1 M H_2SO_4 (Merck, Ultrapur). This cell contained an Au or Pt spiral counter electrode for Au and Rh experiments, respectively. A reversible hydrogen electrode (RHE) was used as the reference electrode and connected to an Au or Pt wire using a 4.7 μF capacitor in order to filter noise during voltammetry. After careful immersion of the working electrode, which was controlled using a micrometer screw, gold electrodes were treated electrochemically by running 200 cyclic voltammetry cycles between 0 and 1.75 V *vs.* RHE at a scan rate of 1 V s^{-1}, in order to obtain a stable cyclic voltammogram (CV).[32] Such a cycling procedure was not necessary for rhodium. Next, the working electrodes were characterized by measuring four CVs at a scan rate of 50 mV s^{-1}.

Following characterization, the working electrodes were transferred to a home-made fluorinated ethylene propylene cell containing 10 M NaOH (Fluka, Trace-select), a Ti counter electrode and a HydroFlex RHE (Gaskatel). After immersion of the working electrodes, a constant cathodic potential was applied for 60 s. This cathodic potential was not IR corrected, but ohmic drop estimates will be included in the main text when relevant. Following polarization, working electrodes were removed from the electrolyte under potential control, rinsed and transferred back to the H_2SO_4 cell. In this cell, the electrodes were re-characterized using cyclic voltammetry. Finally, the working electrodes were taken out of the cell, rinsed and stored for later analysis using scanning electron microscopy.

2.2 Scanning electron microscopy

Scanning electron micrographs were measured with a FEI NOVA NanoSEM 200 microscope, using an acceleration voltage of 5 kV and a beam current of 0.9 nA.

2.3 Density functional theory

Density Functional Theory (DFT) calculations were performed using VASP,[17] the PBE exchange-correlation functional[18] and the projector augmented-wave (PAW) method.[19] The (111), (100), (211) and (553) surfaces contained four metal layers: the two topmost ones and the adsorbates were free to relax in all directions, while the two bottommost layers were fixed at the optimized bulk interatomic distances. The relaxations were performed with the conjugate-gradient scheme and a cut-off of 450 eV for the plane-wave basis set, until the maximum force on any relaxed atom was below 0.01 eV Å^{-1}. $6 \times 6 \times 1$ *k*-point meshes were used for the 2×2 (111) surfaces; $6 \times 8 \times 1$ *k*-point meshes were used for the 3×2 (100) surfaces; $6 \times 4 \times 1$ *k*-point meshes were used for the 2-atom-wide (211) surface; and $5 \times 3 \times 1$ *k*-point meshes were used for the 2-atom-wide (553) surfaces. The distance between periodically repeated slabs was larger than 14 Å in all cases and dipole corrections were applied. For the calculations, $k_BT = 0.2$ eV was used, and the energies were extrapolated to $T = 0$ K. Na was calculated in cubic boxes of 15 Å × 15 Å × 15 Å, using a gamma point distribution and an electronic temperature of 0.001 eV. The following reaction was used to assess the adsorption energies of Na:

$$* + Na_{(g)} \rightarrow *Na$$

where * is a free adsorption site. Thus, the adsorption energies were calculated as:

$$\Delta E_{Na} = E_{*Na} - E_{*} - E_{Na}$$

3 Results and discussion

In an attempt to formulate coherent corrosion models for rhodium and gold, each metal will be discussed separately. First, we will discuss the electrochemical characterization of rhodium, which is followed by the corresponding scanning electron microscopy. These observations will then be combined into a consistent corrosion model. Gold will be discussed in the same manner. Finally, we will draw parallels and highlight important differences between the corrosion characteristics of both materials, while also comparing them to platinum.[14] The most notable difference, which is the different etching preference of these metals, will be explained with the aid of density functional theory calculations.

3.1 Rhodium

3.1.1 Electrochemical characterization. Rhodium electrodes are characterized using cyclic voltammetry. By measuring cyclic voltammograms in H_2SO_4 before and after the application of a constant cathodic potential in NaOH, we are able to detect changes in the electrode surface. Rhodium voltammograms feature two regions of particular interest. The first region lies above approximately 0.4 V *vs.* RHE and corresponds to the formation and reduction of surface rhodium oxide.[20] Though the shape of this region depends on the presence of various surface facets, the lack of sharp peaks impairs assessment of the relative distribution of surface sites.

This distribution can be assessed more reliably using the second CV region of interest, which lies below 0.3 V *vs.* RHE. Cathodic peaks in this region are attributed to a combination of hydrogen adsorption and (bi)sulfate desorption, whereas anodic peaks correspond to hydrogen desorption and (bi)sulfate adsorption. The location of these peaks relates directly to the distribution of surface sites. For this study, we will focus on the anodic peak, which is located respectively at 0.122 V, 0.157 V and 0.120 V *vs.* RHE for (111), (100) and (110) surfaces.[20]

Cyclic voltammograms of rhodium before and after cathodic polarization are displayed in Fig. 1. As can be seen in panel a, several small changes are visible after cathodic polarization at −0.3 V *vs.* RHE. Firstly, a small shoulder develops on the anodic and cathodic peaks between 0.15 and 0.19 V *vs.* RHE. Secondly, the main cathodic and anodic peaks are slightly more well-defined after cathodic treatment. Though the exact reason for these changes is unclear, they appear consistently after polarization at and above −0.3 V *vs.* RHE. Finally, it is important

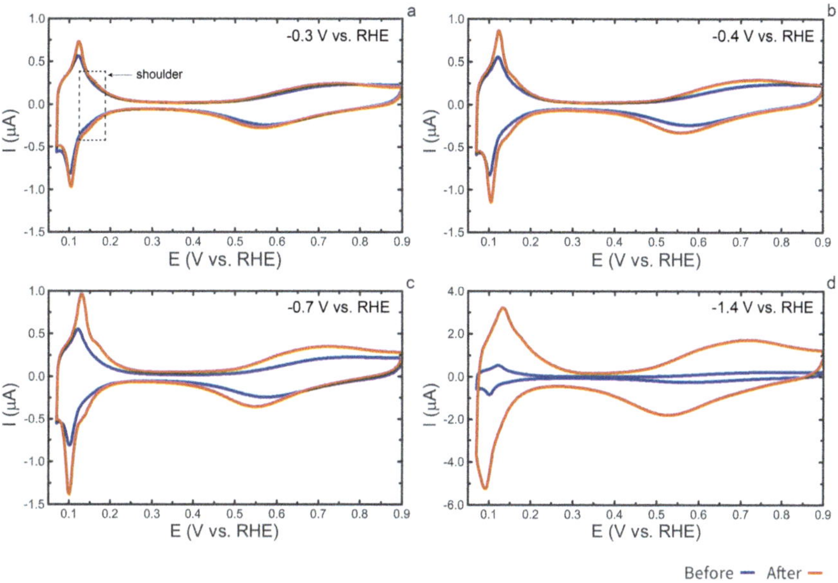

Fig. 1 Cyclic voltammograms of rhodium electrodes before (blue trace) and after (red trace) cathodic polarization in 10 M NaOH at (a) −0.3 V *vs.* RHE, (b) −0.4 V *vs.* RHE, (c) −0.7 V *vs.* RHE and (d) −1.4 V *vs.* RHE. Voltammograms were recorded in 0.1 M H_2SO_4, at a scan rate of 50 mV s^{-1}.

to note that polarization at these potentials does not shift the location of the anodic peak by more than 1 mV (Fig. 2b): it lies at approximately 0.121 V *vs.* RHE before and after treatment in NaOH. This position indicates a high abundance of (111) and (110) sites.

More prominent changes are visible after polarization at −0.4 V *vs.* RHE (Fig. 1b): the shoulder between 0.15 and 0.19 V *vs.* RHE is more pronounced and the charges corresponding to all oxidative and reductive surface reactions have increased. These changes, which become more prominent if the cathodic

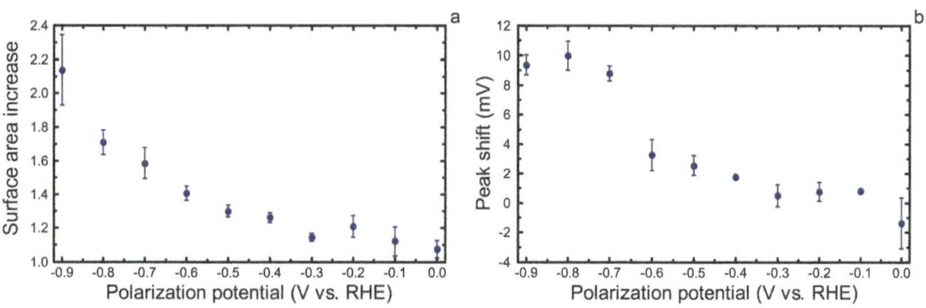

Fig. 2 (a) Relative rhodium surface area increase after cathodic polarization *vs.* polarization potential, as determined from the charge increase of the hydrogen desorption region. (b) Shift of the hydrogen desorption peak position after cathodic polarization *vs.* polarization potential. Each data point is the average of 3 measurements and error bars represent one standard deviation. If no error bar is visible, it overlaps with its corresponding data point.

potential is lowered to −0.5 and −0.6 V *vs.* RHE, indicate an increase in the electrode area. This area increase is displayed as a function of the polarization potential in Fig. 2a. Furthermore, the hydrogen desorption peak shifts positively by approximately 2 mV. This shift increases slightly as the cathodic potential is decreased to −0.5 and −0.6 V *vs.* RHE (Fig. 2b), which points towards a small increase in the relative abundance of (100) sites. Thus, cyclic voltammetry suggests that, after polarization at −0.4 V *vs.* RHE, the electrode is roughened by a process that favors the creation of (100) sites.

While the relative increase of (100) sites seems to develop only slightly until −0.6 V *vs.* RHE, a bigger increase in the amount of these sites is visible after treatment at −0.7 V *vs.* RHE (Fig. 1d).† This is apparent from the position of the hydrogen desorption peak, which shifts positively by roughly 9 mV, as is also visible in Fig. 2b. This shift remains reasonably constant upon further lowering of the cathodic polarization potential. In contrast, the electrode area after polarization continues to increase monotonously (Fig. 2a).

Though no other remarkable changes can be detected electrochemically, an important and surprising observation can be made visually: a black streak slowly extends from the electrode when a potential of −0.9 V *vs.* RHE‡ or lower is applied, which indicates detachment of nanoparticles from the electrode. This leads to severe roughening of the electrode, which is readily apparent from the voltammogram after treatment at −1.4 V *vs.* RHE:§ the increase in hydrogen desorption charge suggests that the electrode has 7.6 times more surface area, while the broadening of this peak is likely caused by the formation of nanoparticles.

3.1.2 Scanning electron microscopy. More information regarding the cathodic corrosion of rhodium can be gained by studying the polarized electrodes using scanning electron microscopy (SEM). Scanning electron micrographs of electrodes treated at −0.7, −0.8 and −1.2 V *vs.* RHE are presented in Fig. 3. These wire electrodes were mounted such that they are aligned vertically in the micrograph within a 5-degree error.

Several features stand out in the micrographs of electrodes treated at or above −0.7 V *vs.* RHE (Fig. 3a and d). Firstly, the electrodes exhibit a high degree of polycrystallinity, since grain boundaries are visible at high magnifications. This polycrystallinity is also clearly visible in Fig. 3 as a variety of spots that appear to be darker than others, since grains with a significantly different orientation have a different contrast in SEM.[21,22] The degree of polycrystallinity is higher than that observed on platinum,[14] which is likely due to the platinum wires being flame-annealed before treatment. Such an annealing step should allow for the merging of smaller grains. Secondly, the electrodes typically exhibit pits and ridges, such as those presented in Fig. 3d. These ridges are not related to cathodic corrosion, because they are visible after polarization at all potentials above −0.7 V *vs.* RHE and on wires that were imaged without prior cathodic polarization. Thus, they are

† It should be noted that −0.7 V *vs.* RHE is the least negative applied potential for rhodium where ohmic drop effects start approaching the magnitude of the 0.1 V potential steps: an ohmic drop of approximately 0.05 V is present here.

‡ Ohmic drop causes the 'real' electrode potential to be −0.8 V *vs.* RHE.

§ Fully correcting for an estimated 240 mV ohmic drop and rounding to the nearest 100 mV leads to a 'real' electrode potential of −1.2 V *vs.* RHE.

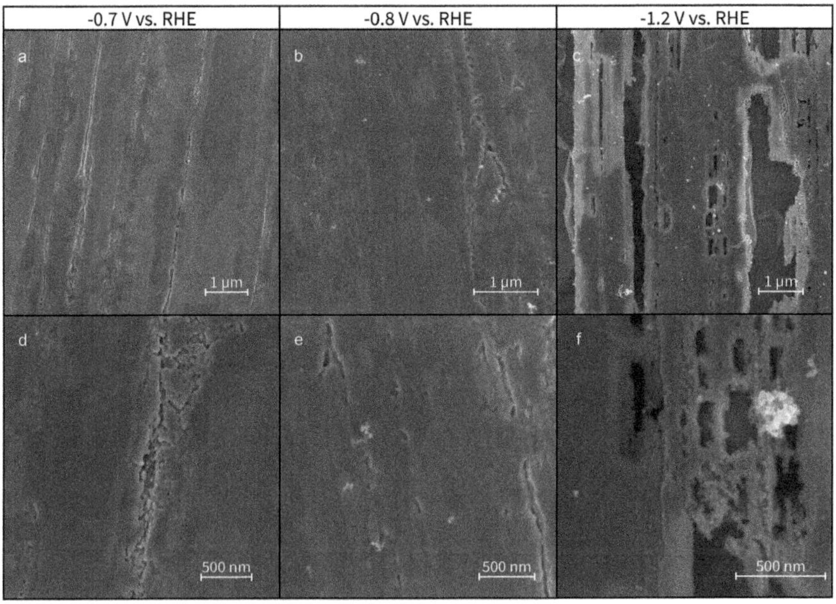

Fig. 3 Scanning electron micrographs of rhodium electrodes treated at (a and d) −0.7 V *vs.* RHE, (b and e) −0.8 V *vs.* RHE and (c and f) −1.2 V *vs.* RHE.

likely defects that were created during electrode preparation and not removed by an annealing step. Therefore, they do not seem to be related to cathodic corrosion, indicating that no signs of corrosion are detectable by SEM on wires that were treated at potentials of −0.7 V *vs.* RHE and higher.

In contrast, the effects of cathodic corrosion are ubiquitous on electrodes polarized at −0.8 V *vs.* RHE¶ or lower. As can be seen in Fig. 3b and e, nanoparticle clusters of varying lengths are visible after cathodic treatment. These are the earliest sign of cathodic corrosion on rhodium. Upon decreasing the polarization potential to −0.9 V *vs.* RHE, rectangular etch pits such as those in Fig. 3c and f start to develop. Though the borders of these pits are generally decorated with nanoparticles, their sides are offset by 90 degrees and the pit walls descend straight down. These etch pits can therefore be considered rectangles in terms of their outline and side-wall orientation, though their floor does not necessarily conform to this rectangular shape. Finally, it is interesting to note that these rectangles are always oriented such that their long sides are parallel to the wire direction.

3.1.3 Model. Combining the electrochemical and microscopic data, it is possible to formulate a preliminary model for the cathodic corrosion of rhodium. This process seems to start at approximately −0.4 V *vs.* RHE, since cyclic voltammetry indicates roughening of the electrode and the formation of (100) sites after polarization at this potential. More severe corrosion takes place below −0.7 V *vs.* RHE, as is indicated by a 9 mV shift of the hydrogen desorption peak in voltammetry. Though any changes in the surface are too small to be detected by

¶ The electrode potential is −0.7 V *vs.* RHE after full ohmic drop correction.

SEM after polarization at or above −0.7 V *vs.* RHE, nanoparticles can be seen after treatment at −0.8 V *vs.* RHE. These particles are accompanied by rectangular etch pits that are oriented along the wire if the potential is decreased to −0.9 V *vs.* RHE or lower. At this potential, dispersion of nanoparticles into the working solution is visible.

Though the voltammetry before corrosion suggests that these etch pits are created in (111) or (110) surfaces, the shape of these pits rules out (111) surfaces; creating a (100)-type hole in a (111)-type surface should create triangular etch pits.[14] In contrast, creating a (100)-type hole in a (110)-type surface will form a rectangular hole with walls that are parallel to the [110] direction. These walls have a (100) surface orientation, as is visualized in the model etch pit in Fig. 4. Though the real cathodic corrosion etching kinetics are likely more complex than the formation of simple rectangles in an ideal (110) surface, the first-order approximation presented in Fig. 4 is able to explain important corrosion features by assuming the preferential formation of (100) sites in a (110)-type surface.

We will highlight two of these features. First of all, the model clarifies why the etch pits are rectangular and not square, since electrochemistry indicates that (100) sites are formed preferentially. Such preferential (100) site formation implies that the corresponding (100) pit walls should elongate at a higher rate than the (110) walls. This leads to the formation of a rectangular etch pit. Secondly, the model explains the orientation of the etch pits along the electrode direction, since it is likely that the grains in which they are formed are not oriented randomly due to the electrode production process.[23] Instead, it is reasonable to assume that the [110] crystal axes are aligned with the wire direction: the voltammetry displayed in Fig. 1 suggests a (110) surface preference, while the shape of grains on non-corroded electrodes in Fig. 3 indicates a non-random

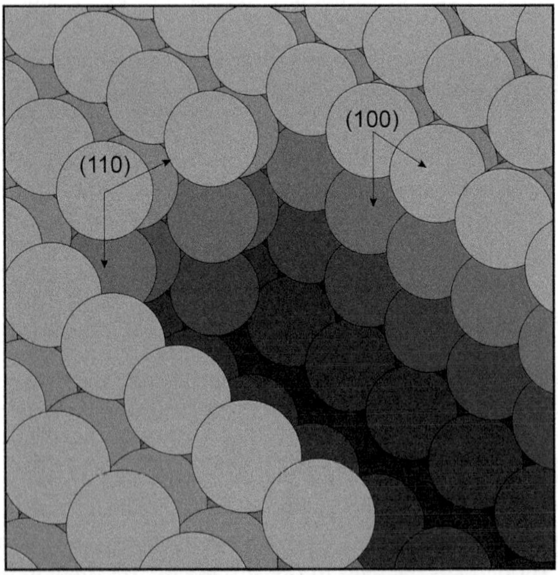

Fig. 4 Model (110) surface with a rectangular etch pit. The bottom left and top right walls of the etch pit have a (100) orientation, whereas the top left side and bottom have a (110) configuration.

grain orientation. Based on the surface mainly consisting of (110) grains with a fiber-like orientation along the wire, the model in Fig. 4 would then dictate that the etch pits should be aligned such that their long (100) walls lie along the wire direction.

3.2 Gold

3.2.1 Electrochemical characterization. In contrast with rhodium, gold does not exhibit any appreciable hydrogen adsorption in its voltammogram. Instead, its oxide formation region shows sensitivity towards the orientation of the surface.[24] CVs of gold electrodes before and after polarization at various potentials are shown in Fig. 5.

Fig. 5a shows the typical behavior of gold electrodes when polarized at −0.6 V *vs.* RHE or higher. No significant changes are induced by these cathodic potentials; the peak and shoulder in the voltammogram around 1.4 and 1.5 V *vs.* RHE indicate that the electrodes exhibit features corresponding to (110) and (100) surfaces,[24] whereas a sharp feature corresponding to (111) sites around 1.59 V *vs.* RHE is absent before and after cathodic treatment. In addition, the electrode surface area remains reasonably constant after cathodic polarization.

In contrast, a minor increase in surface area is indicated by a small increase in the charges associated with oxide formation and reduction after polarization at −0.8 V *vs.* RHE. However, this increase in area is not as reliable in defining the corrosion onset potential as it is for rhodium, since it is relatively small and is sometimes even observed for electrodes that were polarized above −0.6 V *vs.* RHE. A more reliable descriptor is the development of a distinct shoulder at 1.59 V *vs.*

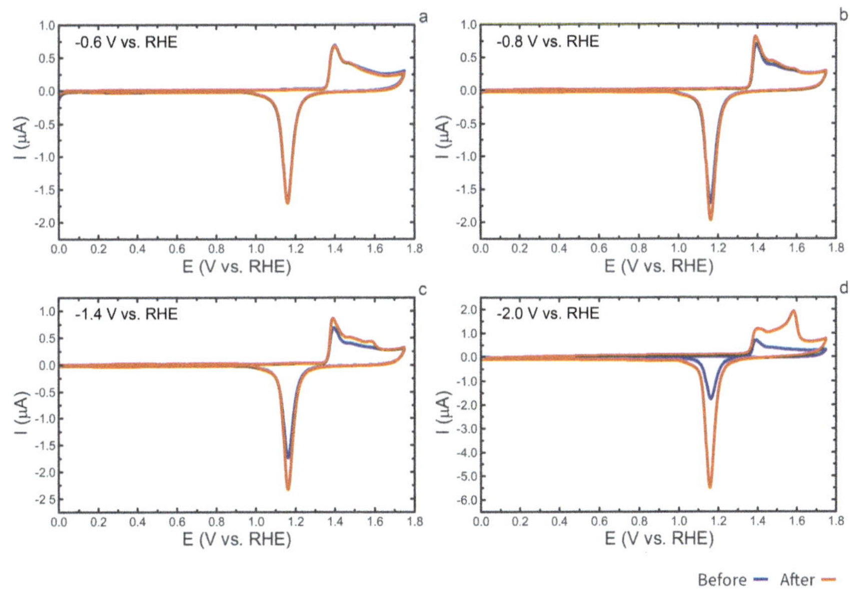

Fig. 5 Cyclic voltammograms of gold electrodes before (blue trace) and after (red trace) cathodic polarization in 10 M NaOH at (a) −0.6 V *vs.* RHE, (b) −0.8 V *vs.* RHE, (c) −1.4 V *vs.* RHE and (d) −2.0 V *vs.* RHE. Voltammograms were recorded in 0.1 M H_2SO_4, at a scan rate of 50 mV s^{-1}.

RHE, corresponding to the formation of (111) sites. This development is subtle yet reproducible, so that closer inspection of the oxide formation region allows us to reveal the start of this (111) feature formation.

Such an analysis is performed in Fig. 6, which displays voltammograms of the oxide formation region of electrodes treated between −0.5 and −0.8 V *vs.* RHE. These voltammograms have been normalized by using the charge of the oxide reduction peak to determine the electrode surface area, assuming a specific charge of 390 μA cm^{-2}.[25] This normalization approach emphasizes relative changes in the orientation of the electrode surface. Fig. 6 does not show a (111) peak for electrodes treated at −0.5 and −0.6 V *vs.* RHE; both voltammograms run parallel at 1.59 V *vs.* RHE. However, a (111) shoulder develops in this region after polarization at −0.7 and −0.8 V *vs.* RHE. This indicates an increase in the relative amount of (111) sites after treatment at these potentials.

The amount of (111) sites steadily increases upon lowering the cathodic treatment potential, as is visible in Fig. 5c.‖ Though the current in all regions of the oxide formation region has increased, the (111) peak at 1.59 V *vs.* RHE is clearly visible. This peak continues developing as the treatment potential is lowered, which is visible in Fig. 5d; a well-defined (111) peak is visible after corrosion and the increase in charge of the oxide reduction peak indicates a factor of 2.7 area increase.

3.2.2 Scanning electron microscopy. Scanning electron microscopy (Fig. 7) allows for further analysis of the corrosion of gold. After polarization at −0.9 V *vs.* RHE or above (Fig. 7a and d), the presence of many small crystal grains indicates a similar degree of polycrystallinity as of rhodium. Another similarity with rhodium is the orientation of the grains, which show a fiber-like arrangement along the direction of the electrode. Finally, ridges are visible between some grain boundaries. As was the case for rhodium, the presence of these features can be attributed to the inability to flame anneal electrodes before experiments. No other

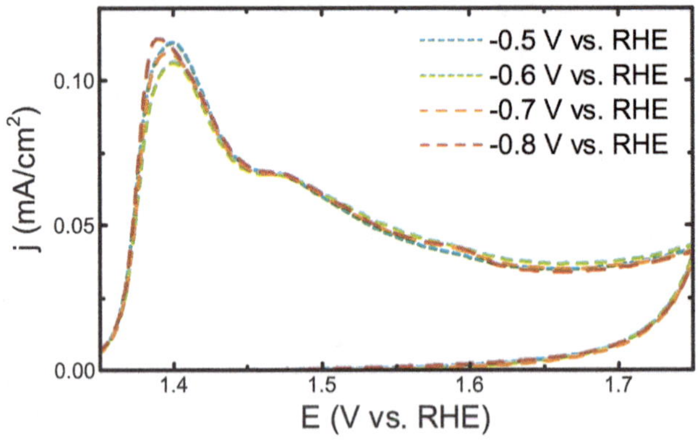

Fig. 6 Voltammograms of the gold oxide formation region after polarization at −0.5 V, −0.6 V, −0.7 V and −0.8 V *vs.* RHE.

‖ For the gold experiments, the ohmic drop becomes significant at −1.0 V *vs.* RHE, where it shifts the potential by approximately 0.06 V. This causes the 'real' electrode potentials to be −1.2 and −1.5 V *vs.* RHE in panel c and d of Fig. 5, respectively.

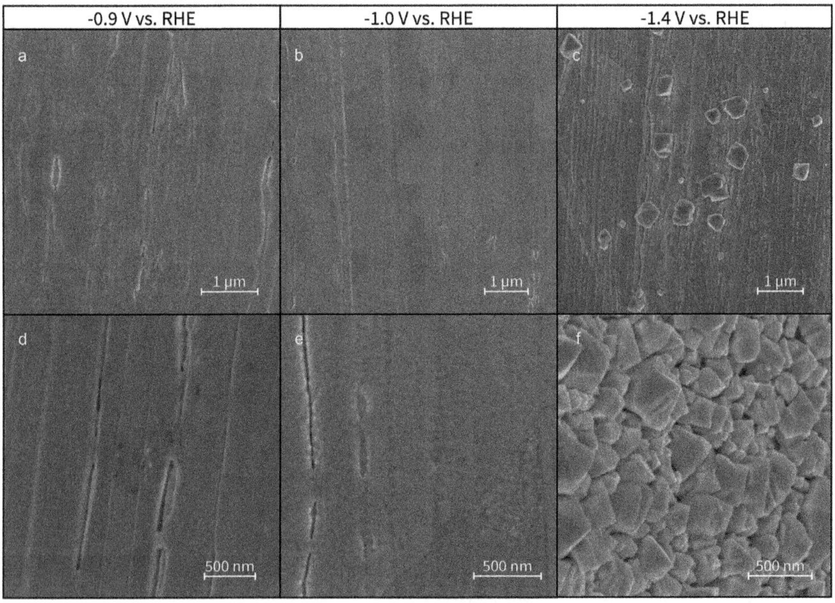

Fig. 7 Scanning electron micrographs of gold electrodes treated at (a and d) −0.9 V *vs.* RHE, (b and e) −1.0 V *vs.* RHE and (c and f) −1.4 V *vs.* RHE.

remarkable features are present on the electrodes, indicating that no signs of corrosion are observable by SEM after polarization at and above −0.9 V *vs.* RHE.

When lowering the treatment potential to −1.0 V *vs.* RHE, no signs of corrosion can be seen at lower magnifications (Fig. 7b). However, such signs are visible upon closer inspection of the crystal grains and ridges between them (Fig. 7e). Along these ridges and on these grains, bright spots appear in the micrograph. These spots can be attributed to material budding out of the electrode. This budding becomes more pronounced as the cathodic potential is lowered. Furthermore, etch lines develop in the direction of the electrode, as is partly visible in Fig. 7c. These lines indicate etching of gold, whereas the bright spots indicate redeposition of gold.

Along with etch lines, Fig. 7c displays the formation of crystallites which are several hundreds of nanometers in size. These particles are first observable after polarization at −1.3 V *vs.* RHE** and predominantly contain 60° and 120° angles. The amount of observable particles increases as the cathodic potential is lowered. Their prevalence also increases towards the tip of the electrode, which is the furthest immersed point during polarization. This is readily visible when comparing panel c and f in Fig. 7, since both micrographs were taken at different locations on the same electrode.

3.2.3 Model. These SEM and electrochemical observations lead to the following model of cathodic gold corrosion, which electrochemistry indicates to start at −0.7 V *vs.* RHE. This process favors the creation of (111) sites and initially leads to a minor increase in electrode surface area. Because this increase is small,

** −1.1 V *vs.* RHE after 100% IR correction.

no signs of corrosion are detectable by SEM until electrodes are polarized at or below −1.0 V *vs.* RHE. After polarization at these potentials, corrosion is initially visible as budding of material from the electrode surface. Corrosion becomes more pronounced as the potential is lowered to −1.3 V *vs.* RHE, after which particles are visible.

The shape of these particles strongly suggests their surface to have the (111) orientation, as can be visualized using the model in Fig. 8. This model displays an octahedral gold nanoparticle, which exclusively exposes (111) surface sites. Though this is an ideal, high-symmetry model particle, it allows for the rationalization of the shape of the observed particles; the angles that should arise from the formation of (111) sites are immediately visible, since the model exclusively contains 60° angles. 120° angles can also be obtained, by simply merging two model particles through a shared surface. In fact, merging model particles of various sizes or selectively extending one of the sides of a model particle allows for a relatively satisfying description of the particles in Fig. 7.

3.3 Comparison

After formulating models for the corrosion of rhodium and gold, comparing corrosion on these metals yields further insights.

Firstly, it is vital to notice the similar corrosion onset potentials for both rhodium and gold. Upon conversion to the normal hydrogen electrode (NHE) scale, these corrosion onset potentials respectively are −1.3 and −1.6 V *vs.* NHE for rhodium and gold, which is similar to the −1.3 V *vs.* NHE found for platinum.[14] Such potentials are relatively mild, especially in alkaline media; indeed, they are not too different from the potentials applied in reactions such as CO_2 reduction.[26] This supports the line of reasoning that cathodic corrosion might, in fact, be just as detrimental to metal electrodes as the much better known phenomenon of anodic corrosion.

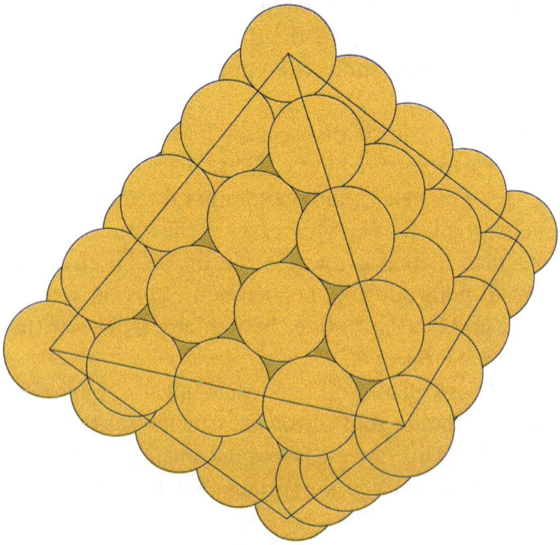

Fig. 8 Model octahedral (111) nanoparticle.

An important difference between both metals is their surface orientation preference. Gold favors the creation of (111) sites, whereas rhodium shows a preference for (100) sites. A similar (100) preference was observed for platinum and suggested to be caused by specific adsorption of sodium cations during cathodic polarization.[14] This hypothesis is explored further in Fig. 9, which displays the DFT-calculated adsorption energy of sodium on the (111), (100), (211) and (553) facets of Rh, Pt and Au. The (211) and (553) surfaces both contain (111) terraces, but possess different step sites: (211) contains 3-atom-long (111) terraces and (100) steps (denoted [3(111) × (100)]), while (553) contains 5-atom-long (111) terraces and (111) steps (denoted [5(111) × (111)]). The adsorption on (211) and (533) surfaces was evaluated at these different steps. From Fig. 9, it is visible that sodium adsorbs most strongly to platinum, followed respectively by rhodium and gold. This order is in agreement with the trends observed in scaling relations.[27,28] More importantly, Fig. 9 demonstrates that, for a given metal, sodium prefers to adsorb on open sites with lower coordination numbers; similar behaviour has previously been observed for oxygen- and hydrogen-containing adsorbates.[29,30] Such an adsorption preference for steps indicates that adsorbate-induced anisotropies in cathodic corrosion should be caused by preferential sodium adsorption on steps or defect sites. Importantly, sodium adsorption is more favorable on (100)-type sites on rhodium and platinum, while it shows a small preference for (111)-type sites on gold. Thus, the DFT-calculated sodium adsorption energies are able to explain why platinum and rhodium preferentially expose (100) facets, while gold favors (111) facets: these sites are selectively stabilized by sodium adsorption. The minuteness of this stabilization on gold also indicates why the anisotropy at the onset of corrosion is less pronounced than for the other metals.

A second distinction between rhodium and gold is the size of their etch features and formed nanoparticles: a comparison of Fig. 3 and 7 demonstrates that rhodium displays much larger anisotropic etch features than gold, whereas the comparison of rhodium and gold nanoparticles in Fig. 10 demonstrates that gold forms larger anisotropic nanoparticles. This can tentatively be explained by the differing mobility

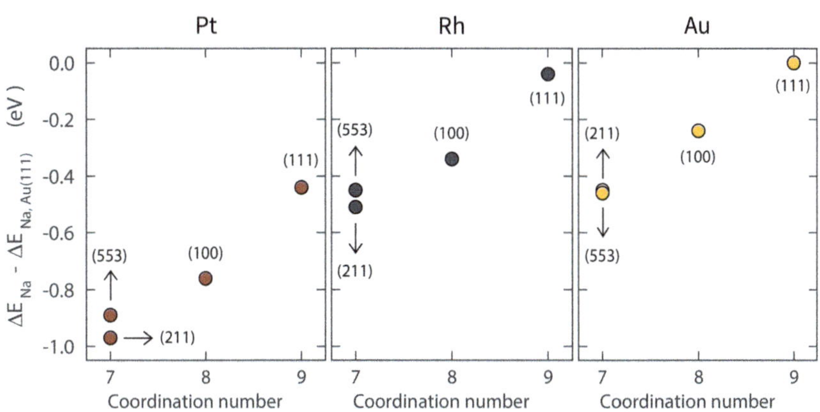

Fig. 9 Adsorption energies of Na on various flat and stepped surfaces of Au, Pt and Rh, as a function of the coordination number of the adsorption sites. Energies are reported with respect to the adsorption of Na on Au(111).

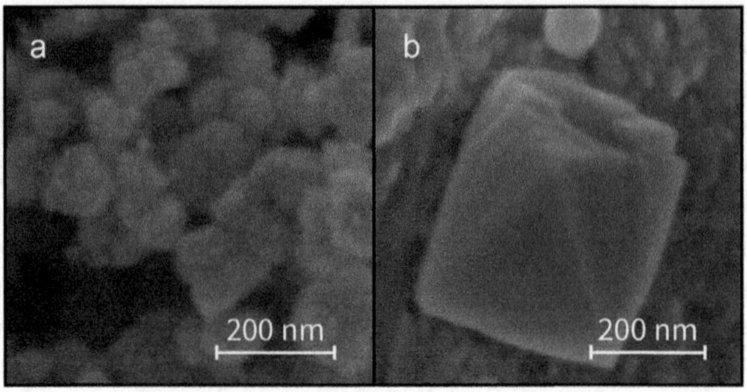

Fig. 10 Scanning electrode micrographs of nanoparticles on (a) a rhodium electrode treated at −1.3 V *vs.* RHE and (b) a gold electrode treated at −1.4 V *vs.* RHE.

of surface atoms of the two metals. Since gold has a lower melting point and sublimation energy,[31] it is presumably more mobile than rhodium. This would allow gold to reconstruct more easily in order to minimize its surface energy, thereby counteracting the formation of etching features. In contrast, etching features in rhodium are lifted less easily by long-range reordering, causing the formation of large anisotropic etch pits. Similarly, the lower mobility of rhodium would lead to the formation of smaller nanoparticles with a high surface area such as those in Fig. 10a, whereas gold particles can merge and reconstruct more easily to form larger particles with less exposed surface area (Fig. 10b). Thus, this difference in mobility explains both the differences observed in SEM and the electrochemically observed higher surface area increase for rhodium, as compared to gold.

4 Conclusions

In this work, we have studied the onset of cathodic corrosion on rhodium and gold. In doing so, we found tentative corrosion onset potentials of −1.3 V *vs.* NHE for rhodium and −1.6 V *vs.* NHE for gold in 10 M NaOH. The mildness of these potentials suggests that cathodic corrosion can be at least as effective as anodic corrosion in inducing surface changes in metal electrodes.

Additionally, we were able to observe different crystallographic corrosion preferences for both metals: rhodium favors the formation of (100) sites, whereas gold preferentially forms (111) sites. This anisotropy is expressed in the formation of well-oriented rectangular etch pits on rhodium and quasi-octahedral nanoparticles on gold. The difference in anisotropy is attributed to the different adsorption of electrolyte cations on both metals, whereas the qualitative size differences in corrosion features are likely related to their different surface atom mobility.

Not only are these conclusions essential towards understanding cathodic corrosion, but the experimental procedure leading to these conclusions also provides a useful framework for studying cathodic corrosion on metals that cannot be flame-annealed.

Acknowledgements

FCV received funding from NWO, Veni project 722.014.009. We thank NCF with support from NWO for use of their supercomputers.

References

1 P. Vanýsek, *CRC Handbook of Chemistry and Physics*, CRC Press, Boca Raton, FL, 96th edn, 2015, ch. 5, pp. 5-80–5-89.

2 H.-Y. Su, Y. Gorlin, I. C. Man, F. Calle-Vallejo, J. K. Nørskov, T. F. Jaramillo and J. Rossmeisl, *Phys. Chem. Chem. Phys.*, 2012, **14**, 14010–14022.

3 A. M. Gómez-Marín and J. M. Feliu, *Electrochim. Acta*, 2012, **82**, 558–569.

4 S. Cherevko, A. R. Zeradjanin, A. A. Topalov, N. Kulyk, I. Katsounaros and K. J. J. Mayrhofer, *ChemCatChem*, 2014, **6**, 2219–2223.

5 M. Pourbaix, *Atlas of electrochemical equilibria in aqueous solutions*, National Association of Corrosion Engineers, 2nd edn, 1974, p. 644.

6 W. v. Baeckmann, W. Schwenk and W. Prinz, *Handbook of Cathodic Corrosion Protection*, Elsevier, 3rd edn, 1997, p. 568.

7 V. Díaz and C. F. Zinola, *J. Colloid Interface Sci.*, 2007, **313**, 232–247.

8 D. Kolb, *Prog. Surf. Sci*, 1996, **51**, 109–173.

9 Y.-G. Kim, J. H. Baricuatro, A. Javier, J. M. Gregoire and M. P. Soriaga, *Langmuir*, 2014, **30**, 15053–15056.

10 F. Haber, *Trans. Am. Electrochem. Soc.*, 1902, **2**, 189–196.

11 B. N. Kabanov, I. I. Astakhov and I. G. Kiseleva, *Electrochim. Acta*, 1979, **24**, 167–171.

12 A. I. Yanson, P. Rodriguez, N. Garcia-Araez, R. V. Mom, F. D. Tichelaar and M. T. M. Koper, *Angew. Chem., Int. Ed.*, 2011, **50**, 6346–6350.

13 Y. I. Yanson and A. I. Yanson, *Low Temp. Phys.*, 2013, **39**, 304–311.

14 T. J. P. Hersbach, A. I. Yanson and M. T. M. Koper, *Nat. Commun.*, 2016, 7, 12653.

15 J. Clavilier and J. P. Chauvineau, *J. Electroanal. Chem. Interfacial Electrochem.*, 1979, **100**, 461–472.

16 L. A. Carol and G. S. Mann, *Oxid. Met.*, 1990, **34**, 1–12.

17 G. Kresse and J. Furthmüller, *Phys. Rev. B: Condens. Matter Mater. Phys.*, 1996, **54**, 11169–11186.

18 J. P. Perdew, K. Burke and M. Ernzerhof, *Phys. Rev. Lett.*, 1996, **77**, 3865–3868.

19 G. Kresse and D. Joubert, *Phys. Rev. B: Condens. Matter Mater. Phys.*, 1999, **59**, 1758–1775.

20 Q. Xu, U. Linke, R. Bujak and T. Wandlowski, *Electrochim. Acta*, 2009, **54**, 5509–5521.

21 R. Zauter, F. Petry, M. Bayerlein, C. Sommer, H.-J. Christ and H. Mughrabi, *Philos. Mag. A*, 1992, **66**, 425–436.

22 S. Canovic, T. Jonsson and M. Halvarsson, *Int. J. Mod. Phys.: Conf. Ser.*, 2008, **126**, 012054.

23 I. L. Dillamore and W. T. Roberts, *Metall. Rev.*, 1965, **10**, 271–380.

24 A. Hamelin, *J. Electroanal. Chem.*, 1996, **407**, 1–11.

25 S. Trasatti and O. A. Petrii, *Pure Appl. Chem.*, 1991, **63**, 711–734.

26 Y. Hori, *Modern Aspects of Electrochemistry*, Springer, New York, 42nd edn, 2008, ch. 3, pp. 89–189.

27 F. Calle-Vallejo, J. I. Martínez, J. M. García-Lastra, J. Rossmeisl and M. T. M. Koper, *Phys. Rev. Lett.*, 2012, **108**, 116103.

28 F. Calle-Vallejo, D. Loffreda, M. T. M. Koper and P. Sautet, *Nat. Chem.*, 2015, 7, 403–410.

29 F. Calle-Vallejo, J. Tymoczko, V. Colic, Q. H. Vu, M. D. Pohl, K. Morgenstern, D. Loffreda, P. Sautet, W. Schuhmann and A. S. Bandarenka, *Science*, 2015, **350**, 185–189.
30 J. Tymoczko, F. Calle-Vallejo, W. Schuhmann and A. S. Bandarenka, *Nat. Commun.*, 2016, 7, 10990.
31 W. M. Haynes, *CRC Handbook of Chemistry and Physics*, CRC Press, Boca Raton, FL, 96th edn, 2015, p. 2677.
32 C. Jeyabharati, P. Ahrens, U. Hasse and F. Scholz, *Journal of Solid State Electrochemistry*, 2016, DOI: 10.1007/s10008-016-3228-1.

Faraday Discussions

PAPER

Electrochemistry of single nanobubbles. Estimating the critical size of bubble-forming nuclei for gas-evolving electrode reactions

Sean R. German,[ab] Martin A. Edwards,[a] Qianjin Chen,[a] Yuwen Liu,[a] Long Luo[c] and Henry S. White*[a]

Received 28th April 2016, Accepted 10th May 2016
DOI: 10.1039/c6fd00099a

In this article, we address the fundamental question: "What is the critical size of a single cluster of gas molecules that grows and becomes a stable (or continuously growing) gas bubble during gas evolving reactions?" Electrochemical reactions that produce dissolved gas molecules are ubiquitous in electrochemical technologies, *e.g.*, water electrolysis, photoelectrochemistry, chlorine production, corrosion, and often lead to the formation of gaseous bubbles. Herein, we demonstrate that electrochemical measurements of the dissolved gas concentration, at the instant prior to nucleation of an individual nanobubble of H_2, N_2, or O_2 at a Pt nanodisk electrode, can be analyzed using classical thermodynamic relationships (Henry's law and the Young–Laplace equation – including non-ideal corrections) to provide an estimate of the size of the gas bubble nucleus that grows into a stable bubble. We further demonstrate that this critical nucleus size is independent of the radius of the Pt nanodisk employed (<100 nm radius), and weakly dependent on the nature of the gas. For example, the measured critical surface concentration of H_2 of ~0.23 M at the instant of bubble formation corresponds to a critical H_2 nucleus that has a radius of ~3.6 nm, an internal pressure of ~350 atm, and contains ~1700 H_2 molecules. The data are consistent with stochastic fluctuations in the density of dissolved gas, at or near the Pt/solution interface, controlling the rate of bubble nucleation. We discuss the growth of the nucleus as a diffusion-limited process and how that process is affected by proximity to an electrode producing ~10^{11} gas molecules per second. Our study demonstrates the advantages of studying a single-entity, *i.e.*, an individual nanobubble, in understanding and quantifying complex physicochemical phenomena.

[a]*University of Utah, Department of Chemistry, 315 S 1400 E Salt Lake City, Utah 84112, USA. E-mail: white@chem.utah.edu*

[b]*Revalesio Corporation, 1200 East D Street, Tacoma, Washington 98421, USA*

[c]*The University of Texas at Austin, Department of Chemistry and the Texas Materials Institute, 105 East 24th Street, Stop A5300, Austin, Texas 78712, USA*

Introduction

The spontaneous phase transformation required to initiate bubble formation is a highly activated process requiring extreme conditions. Classical nucleation theory (CNT) describes the activation barrier to create a bubble nucleus in terms of the cohesive force of the liquid, while assuming a critical nucleus is large enough that it can be described by bulk thermodynamics.[1–4] CNT applies equally to single component systems, where a vapor bubble is generated by single component boiling,[5,6] as well as binary mixtures, where a dissolved gas phase separates into a bubble;[7–10] these two systems differ predominantly in their growth/dissolution dynamics. Vapor bubbles in single component systems are at all instances surrounded by their comprising material and are often violent in their dynamics, such as with cavitation.[5] Diffusional gradients control the growth of bubbles formed from dissolved gas. In this work, we report on nucleation of individual gas bubbles from a supersaturated binary mixture engendered by electrogeneration of dissolved gas at Pt nanodisk electrodes.

The free energy of formation of a gas bubble nucleus in solution, ΔG_{tot}, is the sum of the energy cost of creating the new interface and that gained through liberation of dissolved gas into the bubble volume. Fig. 1 (top) schematically presents the classic description of ΔG_{tot} as a function of the bubble's radius, r_{nb}. The free energy attributable to the gas/solution interface, $\Delta G_{surf} = 4\pi\gamma {r_{nb}}^2$ (dashed line, where γ is surface energy of the gas/solution interface), is proportional to the area of the bubble's interface. The bulk component term, ΔG_{bulk}, (dotted line, $(4\pi/3)\Delta G_V {r_{nb}}^3$) is proportional to the volume of the bubble and ΔG_V, the energy difference between the dissolved and gaseous state of the molecules in that volume. ΔG_{tot} is the sum of these two components,

$$\Delta G_{tot} = 4\pi\gamma {r_{nb}}^2 + (4\pi/3)\Delta G_V {r_{nb}}^3 \tag{1}$$

Eqn (1) indicates that ΔG_{tot} initially increases as a function of r_{nb} before reaching a maximum, E_a, at a critical radius, r_{crit}. The implication of this free energy maximum is that a bubble of radius greater than r_{crit} is energetically favored to continue to grow, whereas bubbles with radii less than r_{crit} are inclined to shrink. However, since bubbles of critical size necessarily arise from the growth of sub-critical nuclei, their formation relies upon relatively improbable fluctuations along the free energy barrier.

A bubble of critical radius is both in mechanical and chemical equilibrium with the dissolved gas. The Young–Laplace equation (eqn (2)) describes the pressure difference across the gas/solution interface, $\Delta P_{Y-L} = P_{internal} - P_{ambient}$, where $P_{internal}$ and $P_{ambient}$ are the total pressure within the bubble and the ambient pressure, respectively. For a bubble of radius r_{crit}

$$\Delta P_{Y-L} = 2\gamma/r_{crit}. \tag{2}$$

Large positive values of ΔP_{Y-L} tend to drive gas molecules out of sub-critical bubbles into the solution, resulting in bubble shrinkage. The bubble's internal hydrostatic pressure ($\Delta P_{Y-L} + P_{ambient}$) can be countered when the partial pressure of the dissolved gas, P_{gas}, is equal to the bubble's internal pressure. For any

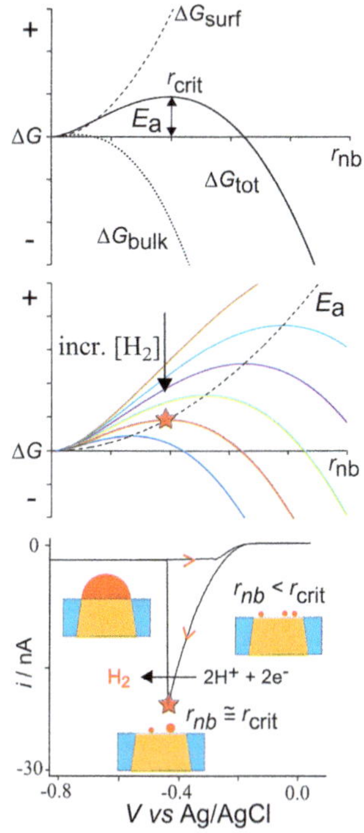

Fig. 1 (Top) Plot of total free energy of formation of a gas bubble nucleus, $\Delta G_{\text{tot}} = \Delta G_{\text{surf}} + \Delta G_{\text{bulk}}$, *vs.* the radius of the nanobubble, r_{nb}. The maximum in ΔG_{tot} corresponds to the critical nanobubble radius, r_{crit}, with an activation energy barrier of E_{a}. For a bubble nucleus of radius greater than r_{crit}, the continued growth of the nucleus into a stable bubble is energetically favorable. (Middle) Plots of ΔG_{tot} *vs.* r_{nb} at different concentrations of dissolved H_2 (*i.e.* different supersaturations). Lower values of $[H_2]$ are associated with higher values of E_{a}, (indicated by the dashed line), reducing the probability of bubble formation. Increasing $[H_2]$ decreases the activation energy such that thermal fluctuations may nucleate a bubble of critical radius, as depicted by the red star for an arbitrary H_2 concentration. (Bottom) Cyclic voltammogram for H^+ reduction in 0.5 M H_2SO_4 at a 33 nm radius Pt nanodisk electrode (500 mV s^{-1}). The peak (red star) corresponds to a phase change as a nucleus grows into a stable nanobubble covering the electrode. Inlaid schematics show the spontaneous formation of nuclei near the surface of the electrode. At a sufficiently high rate of H_2 electrogeneration and, thus, supersaturation at the electrode/electrolyte interface, a nucleus of critical size may exist, which grows into a bubble covering the electrode. Note: while the schematic shows nucleation as a homogeneous process occurring just above the electrode, we also consider the possibility of heterogeneous nucleation on the electrode surface (see Results and discussion).

specific supersaturation of dissolved gas, $P_{\text{gas}} - P_{\text{ambient}}$, there is a corresponding E_{a} required for formation of a continuously growing bubble. As shown in the middle plot of Fig. 1, a critical bubble size is associated uniquely with a specific E_{a} and supersaturation. As qualitatively illustrated in this figure for H_2 bubble formation, the values of E_{a} at low supersaturations are sufficiently large such that nuclei are unlikely to reach r_{crit} and grow into stable bubbles. Conversely, large supersaturations of H_2 serve to reduce r_{crit} such that thermal fluctuations of the magnitude of E_{a} may occur at an appreciable rate.

We recently reported on the formation of single bubbles of H_2, N_2, and O_2 at the surface of Pt nanoelectrodes that are created by electrogeneration of large gas supersaturations at the electrode/electrolyte interface.[11–15] The lower frame of Fig. 1 shows an example cyclic voltammogram for the formation of a single H_2 bubble. In this experiment, H^+ reduction at a 33 nm radius Pt nanoelectrode in 0.5 M H_2SO_4 is used to drive H_2 production. On the initial sweep of the *i*–*E* curve, the current for H_2 electrogeneration increases when the electrode potential, *E*, is scanned to potentials negative of the formal potential for H^+ reduction $\left(E^0_{H^+/H_2} = -0.23 \text{ V } \textit{vs.}\text{Ag/AgCl}\right)$. As the potential is scanned to more negative values, the current and rate of H_2 electrogeneration continue to increase nearly exponentially until there is an abrupt drop in the current to a non-zero residual current that is nearly independent of *E*. The abrupt drop defines the peak current, i^p_{nb}, that coincides with the formation of a stable nanobubble on the surface of the electrode.[12] The bubble, which covers the electrode surface, significantly diminishes the rate of H^+ reduction at the electrode, the latter occurring along the circumference of the bubble at the Pt/gas/electrolyte interface.[11] In this specific voltammogram, the formation of a single bubble occurs at $\sim$−0.4 V, as indicated by the sudden drop in current magnitude from i^p_{nb} $\sim$20 nA to $\sim$3 nA. The bubble is only stable as long as the potential is held sufficiently negative of $\sim$−0.3 V, such that the constant production of H_2 at the electrode circumference replenishes the H_2 that dissolves from the bubble into the solution. As the potential is swept back towards positive potentials, the current for H^+ reduction decays and the bubble rapidly dissolves.[15] Note, throughout this work, our analysis does not require knowledge of the precise mechanism of the gas evolving electrode reactions, only that such reactions produce gas as described by their stoichiometric reactions, *e.g.*, $2H^+ + 2e^- \rightarrow H_2$.

At the foot of the voltammetric wave, the rate of H_2 generation is low, and thus the amount of dissolved H_2 is also correspondingly low. Based on the above description, we expect that the transient H_2 bubble nuclei, schematically shown in the insets of Fig. 1, are smaller than the corresponding r_{crit}, and thus are unstable and dissolve rapidly. As the current is swept to more negative potentials, the concentration (*i.e.*, supersaturation) of dissolved H_2 in the vicinity of the electrode increases and r_{crit} decreases until the probability of a bubble nucleus of radius approaching or exceeding r_{crit} becomes sufficiently large. At this electrode potential, continued growth of a nucleus leading to a stable bubble is thermodynamically favorable. This critical point coincides with the red star on each of the graphs. While the thermodynamics suggest that the bubble will continue to grow *ad infinitum*, its growth is self-limited at the Pt disk edges, which, as noted above, acts as the source of H_2 for its growth.[11] A stable bubble is established in a dynamic steady state when the rate of H_2 dissolution and electrogeneration are equal.

Herein, we show that the value of i^p_{nb}, recorded at the instant prior to formation of a single bubble, is proportional to the electrode size, with a constant of proportionality that depends on the identity of the gas. Our voltammetric results indicate a sharp potential-dependent threshold in nucleation rates, from immeasurably slow to extremely fast. We demonstrate that the peak current corresponds to critical supersaturations for H_2, N_2, and O_2, above which a stable phase transformation immediately proceeds. The critical supersaturation is a direct measure of the chemical potential of gas within the critical bubble nucleus, allowing calculation of the critical nucleus pressure and radius, as well as the number of constituent gas molecules contained within the nucleus.

Results and discussion

Fig. 2 shows typical voltammograms for electrochemical reactions at Pt nanoelectrodes that correspond to the formation of stable individual nanobubbles of H_2, N_2, and O_2. Each voltammogram displays a similar shape to that for H_2 nanobubble formation shown in Fig. 1; however, the values of potentials and currents are different for the different gases. In each case we expect the bubble to form from a critically sized nucleus, although the critical radius for each gas will differ due to the distinct chemical properties of the gases. The peak current i^p_{nb} just prior to bubble formation is labeled on each voltammogram. In all cases, i^p_{nb} is reproducible between cycles and is nearly independent of the scan rate up to $\sim$10 V s^{-1}.

The reproducibility of i^p_{nb} and insensitivity to scan rate suggests a very sharp threshold in the electrode potential over which the probability of a nucleus having radius greater than r_{crit} changes from essentially zero to unity, and bubble formation immediately proceeds. If the electrode potential threshold was more gradual, one would expect significant variability in i^p_{nb}. Similarly, if the formation of a bubble nucleus was a slow event on the voltammetric timescale, one might expect an increase in the peak current at increasing scan rates.

The rate of critical bubble formation is a function of the activation energy (indicated in Fig. 1) by the Arrhenius equation $J = Z\exp(-E_a/kT)$. For homogeneous nucleation, J is a steep function of γ and supersaturation ($P_{gas} - P_{ambient}$)[7]

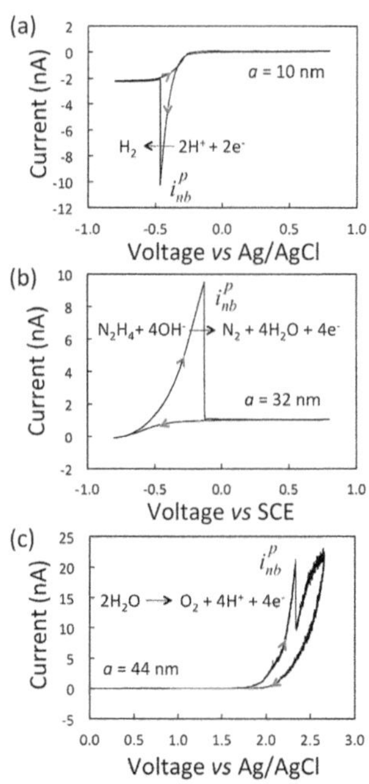

Fig. 2 Voltammograms corresponding to H_2, N_2, and O_2 bubble formation at Pt nanoelectrodes. The gas-evolving electrode reactions are noted on the figure. The electrode radii, *a*, indicated on each plot were determined from the steady-state diffusion-limited current for ferrocene oxidation in acetonitrile. The aqueous solutions contained: (a) 0.5 M H_2SO_4, (b) 1 M N_2H_4, and (c) 0.25 M H_2SO_4.

$$J = Z \exp\left(\frac{-16\pi\gamma^3}{3kT(P_{\text{gas}} - P_{\text{ambient}})^2}\right) \quad (3)$$

where J (m^{-3} s^{-1}) is the rate of critical bubble formation, the exponential prefactor Z is treated as a constant, k is Boltzmann's constant, and T is temperature. For nucleation of bubbles from dissolved gas in solution, there is wide disagreement between experiment and theory, with nucleation often occurring at supersaturation levels that are orders of magnitude lower than predicted by eqn (3).[10,16–19] The literature presents a variety of explanations for the observed discrepancy, ranging from new thermodynamic interpretations of the surface tension of small nuclei,[6,20] to impinging cosmic rays,[21] to pre-existing stable nuclei.[16] Our current data does not allow speculation about nucleation rates aside from the observation that the rate appears to vary from extremely slow to very fast over a small voltage range (~5 mV), corresponding to a small change in the dissolved gas concentration.

A majority of the studies of nucleation in the literature similarly report a supersaturation value at which J becomes appreciably fast. There are three types of approaches that have been employed to measure the critical supersaturation at which bubble formation is observed: homogeneous gas-producing chemical reaction in bulk solution,[18,20,22,23] release of pressure after equilibrium is reached at elevated pressure[24,25] and electrochemical gas generation.[17,26] Lubetkin has compiled the results of a large number of these reports and points out that these measurements often reflect kinetic limitations and not true thermodynamic values.[27]

In the context of nucleation rates, i^{p}_{nb} represents a point in the J *vs.* supersaturation curve where bubble formation occurs quickly relative to the rate of change in gas supersaturation. For instance, at a 1 V s^{-1} scan rate, bubble nucleation reproducibly occurs within an ~5 mV range. This voltage range, ΔV, is related to the characteristic time of the liquid-to-gas phase transition (τ) and the voltammetric scan rate, ν, by the relationship, $\tau = \Delta V/\nu$. In our experiments, at least one critical bubble forms at that dissolved gas supersaturation in the volume near the electrode at least every ~5 ms, corresponding to a minimum nucleation frequency of 200 s^{-1}. Of course, the nucleation frequency may be much larger than this estimate of the minimum value.

Fig. 3 shows the dependence of i^{p}_{nb} for N_2 bubble formation on the radius of the Pt nanodisk, a. The upper part of the figure shows voltammograms recorded in aqueous 1 M N_2H_4 at Pt nanoelectrodes of 10, 23, 47, and 80 nm radius, and at a scan rate of 200 mV s^{-1}. It is apparent from these data that i^{p}_{nb} increases for N_2 bubble formation as a function of the electrode size; similar voltammetric behavior is observed from H_2 and O_2 bubble formation. The lower part of Fig. 3 shows that i^{p}_{nb} is a linear function of a for the nucleation of all three gases ($R^2 > 0.97$ in all cases). The solid lines in this plot represent least squares fits to the data that pass through the origin with slopes of 0.81 ± 0.02 nA nm^{-1}, 0.48 ± 0.03 nA nm^{-1} and 0.31 ± 0.01 nA nm^{-1} for H_2, O_2 and N_2, respectively.

The steady-state current at a nanodisk electrode, while typically written in terms of the bulk concentration of the reactants, can also be written as a function of the surface concentration of electrogenerated dissolved gas, C_{surf}. Assuming

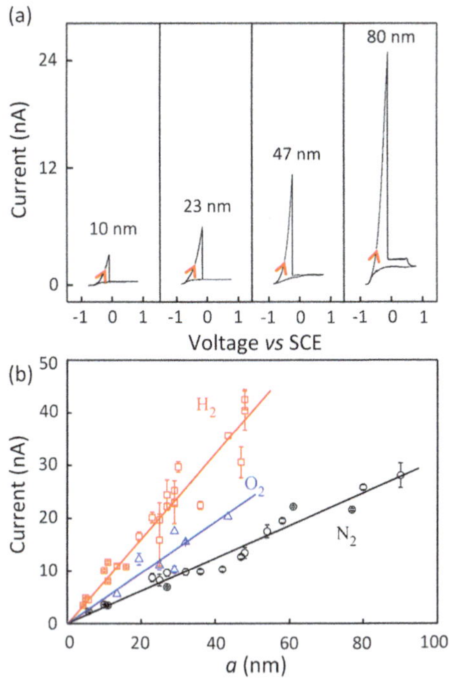

Fig. 3 Top: Voltammetry corresponding to the formation of stable N_2 bubbles at Pt nanoelectrodes of varying radii (labeled) in a 1 M N_2H_4 aqueous solution. Bottom: Plot of the i_{nb}^{p} as a function of the electrode radius, a, for H_2, N_2, and O_2. Similar data were originally presented in ref. 12–14.

zero concentration of the gas in bulk solution, the relation between current and C_{surf} is given by:[28]

$$i = 4naFDC_{surf} \tag{4}$$

where D is the diffusion coefficient of the gas in solution and n the number of electrons transferred in the reaction ($n = 2$ for H_2, 4 for N_2, and 4 for O_2). Rearranging this equation, and setting i to i_{nb}^{p} allows one to obtain the critical concentration of dissolved gas at the electrode surface at the moment that a dynamically stable bubble is nucleated.

$$C_{crit} = i_{nb}^{p}/4naFD \tag{5}$$

Because the plots of i_{nb}^{p} *vs.* a in Fig. 3 are linear, we conclude that C_{crit} is independent of the electrode size and can be calculated from eqn (5). Values of C_{crit} for H_2 (0.23 M), N_2 (0.09 M) and O_2 (0.14 M) are tabulated in Table 1, with errors determined from the variance in linear fits and uncertainty in the value of D (see Appendix C for details). The large supersaturation values associated with i_{nb}^{p} for each gas justify the use of eqn (4) and (5), which are based on the assumption that the bulk dissolved gas concentration is negligibly small.

A number of properties of the critical bubble nucleus, also listed in Table 1, are derivable from the values of C_{crit}. In computing these properties, we note that

Table 1 Parameters describing the critical nuclei for H_2, N_2, and O_2 bubble generation

	H_2		H_2 + surfactant		N_2		O_2	
	Ideal	Corrected	Ideal	Corrected	Ideal	Corrected	Ideal	Corrected
C_{crit} (M)	0.23 ± 0.02		0.14 ± 0.02		0.09 ± 0.02		0.14 ± 0.01	
ΔP_{Y-L} (atm)	290 ± 30[a]	350 ± 30[b]	180 ± 30	210 ± 30	140 ± 30	160 ± 40	100 ± 10	120 ± 10
r_{crit} (nm)	4.9 ± 0.6[c]	3.6 ± 0.3[d]	4.1 ± 0.8	3.5 ± 0.6[e]	10 ± 3	7.9 ± 1.6	14 ± 2	10 ± 1
n_{crit}[f]	3500 ± 900	1700 ± 300	1300 ± 500	900 ± 300	15 000 ± 9000	8100 ± 2500	28 000 ± 9000	12 000 ± 3000

[a] Pressure difference calculated directly from Henry's law at 25 °C and 1 atm in pure water ($\Delta P_{Y-L} = C_{crit}/K_H - 1$). [b] Pressure difference corrected using experimental gas solubility at high pressure in pure water and accounting for electrolyte reduction of solubility, see Appendix C, eqn (A6). [c] Calculated using the ideal ΔP_{Y-L} and surface tension of the air/solution interface at 1 atm. [d] Calculated using corrected ΔP_{Y-L} and modification of the interfacial tension by surface adsorption of gas at high pressures, see Appendix C. [e] Calculated using the surface tension of the air/solution interface at 1 atm, see Appendix C. [f] Calculated from r_{crit}, ΔP_{Y-L} and the ideal gas law (eqn (7)) assuming a bubble of spherical geometry.

although the overall electrochemical system is not present in a true equilibrium state, local equilibrium is assumed to be maintained between the solution containing dissolved gas and the gas bubble. The Laplace pressure of a critical nucleus is directly related to C_{crit} by the partial pressure of the dissolved gas ($P_{gas} = \Delta P_{Y-L} + P_{ambient}$). This relationship is given in a rudimentary way using Henry's law ($K_H P_{gas} = C$), which upon substitution of $P_{gas} = \Delta P_{Y-L} + P_{ambient}$, and rearrangement, yields:

$$\Delta P_{Y-L} = (C_{crit}/K_H) - P_{ambient}. \quad (6)$$

K_H is equal to 0.78 mM atm^{-1} for H_2 in pure water at 25 °C.[29] Thus, for the measured $C_{crit} = 0.23$ M, eqn (6) yields $\Delta P_{Y-L} = 290$ atm. This Laplace pressure is a thermodynamic quantity, which is related through the surface tension, γ, to the bubble radius of curvature, *i.e.*, the critical bubble radius, by eqn (2). Using the macroscopic surface tension of 0.5 M H_2SO_4 at ambient temperature and pressure (0.072 N m^{-1})[30] as a simple estimation of γ, we calculate the value of $r_{crit} = 4.9$ nm. From this we compute the number of molecules, n_{crit}, in the critical nucleus using the ideal gas law, assuming a spherical nucleus of volume $(4\pi/3)r_{crit}^3$.

$$n_{crit} = (\Delta P_{Y-L} + P_{ambient})4\pi r_{crit}^3/3RT \quad (7)$$

For the 4.9 nm radius critical H_2 nucleus, formed at 1 atm and 298 K, the total internal pressure is 291 atm, and the critical nucleus contains ~3500 H_2 molecules.

The thermodynamic relations used above provide a straightforward means to estimate parameters describing the critical bubble nucleus and are listed as ideal values in Table 1. Better estimates of ΔP_{Y-L}, r_{crit}, and n_{crit} can be obtained by consideration of known non-idealities at high pressures. The corrected values in Table 1 are based on reported measurements of gas solubility in pure water at high pressure, the reduction of gas solubility by electrolyte at ambient pressures (*i.e.*, 'salting out'), and reduction of surface tension by gas adsorption at high pressure. These corrections are detailed in Appendix C. Another consideration is that the surface tension is a function of the interfacial radius of curvature for very small bubbles.[31] The size at which the effect becomes important is highly debated. However, experimental studies of capillary condensation have shown the surface tension of water and cyclohexane droplets as small as ~5 nm to be very nearly equal to the bulk value.[32,33] Thus, the possible dependence of γ on bubble radius is ignored here.

It is expected from classical nucleation theory that a solution/gas interface with lower surface tension will have a significantly lower value of ΔG_{surf} and hence a lower value of E_a for the spontaneous formation of a critical nucleus, Fig. 1. However, both Rubin *et al.*[18] and Hemmingsen[24] observed only marginally lower critical values for N_2 supersaturation induced by a chemical reaction, and for argon supersaturation induced by pressure release, respectively, in the presence of surfactant, and argued that the surfactant concentration in the bulk was too low to affect homogeneous nucleation. To test this idea, we investigated the effect of surfactant on the H_2 system. Fig. 4 demonstrates the influence of adding cetyltrimethylammonium bromide (CTAB) to the solution on the value of i_{nb}^p. The voltammograms show the formation of a H_2 bubble in 0.5 M H_2SO_4 in the

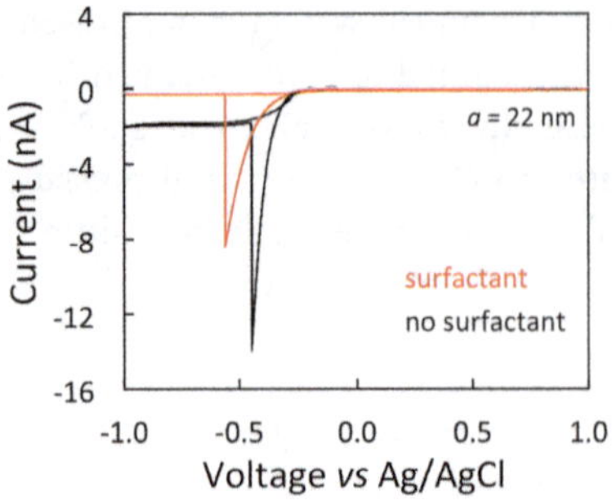

Fig. 4 i–V responses from a 22 nm radius Pt electrode in 0.5 M H_2SO_4 with (red) and without 0.2 mg mL^{-1} CTAB surfactant (black). Scan rate 100 mV s^{-1}.

presence and absence of 0.2 mg mL^{-1} CTAB, recorded at the same 22 nm radius electrode. We observe a reduction of ~40% of i^{p}_{nb} in the presence of surfactant, which is reproducible with different electrodes, indicating that nucleation proceeds at a lower critical supersaturation of H_2. This finding indicates that a critical bubble nucleus in the presence of CTAB has a lower internal pressure, suggesting that the lifetime of a sub-critical radius nucleus, as it grows to critical size, is long enough for surfactant to adsorb to its surface. While both gas and surfactant adsorption reduce surface tension, it has been shown that the two components are not additive. Rather, the two components are in competition for available surface area.[34] Thus, for the CTAB data, in contrast to the other experiments, we do not account for gas adsorption in the corrected value of r_{crit} reported in Table 1.

Both ideal and corrected values of ΔP_{Y-L}, r_{crit}, and n_{crit} for O_2 and N_2 bubble nucleation were computed as above and are listed in Table 1. The supersaturation pressure values corresponding to a nucleation event range between 100 and 350 atm and are dependent upon gas type. These values are as large as the largest reported values listed by Lubetkin,[27] and the magnitude of supersaturation and ordering by gas type $O_2 < N_2$ (H_2 not studied) agree approximately with Hemmingsen,[24] who visually observed bubble formation after release of pressure to ambient conditions. In comparison to the decompression method, our results are significantly more precise and the electrochemical method is simpler. Our measured critical supersaturation values are an order of magnitude larger than other reports of electrochemically induced supersaturation. Dapkus and Sides[17] attributed the relative ease of nucleation in their electrochemical system to a reduction of surface tension by the large electric field within the double layer at the electrode surface. Here we do not speculate on electric field effects; however, in the following section, we do consider how bubble nucleation might be influenced by proximity to a gas-evolving surface.

The results presented in Table 1 indicate at least a few thousand molecules need to spontaneously assemble to form a critical bubble with r_{crit} between ~4 and ~10 nm. In the case of H_2, this process occurs at a C_{crit} of ~0.23 M. If we consider a homogeneous solution containing 0.23 M H_2, the requisite number of molecules to form a critical bubble (n_{crit} = 1700) is found within a spherical volume of solution having a radius of 14 nm. A bubble nucleus will deplete the surrounding solution of dissolved gas as it grows, relying upon diffusion of individual H_2 molecules to the growing bubble. Random diffusion of thousands of molecules to a growing bubble within this volume is an unlikely event. However, in our experiment, the Pt nanoelectrodes can support an immense rate of H_2 generation due to the high catalytic rate for H^+ reduction at Pt, as well as the large convergent diffusional flux of H^+ to the nanoscale electrode. For instance, a typical 16 nA peak current at a 20 nm radius electrode is equivalent to the production of $\sim 10^{11}$ H_2 molecules per second, corresponding to the generation of the requisite $\sim 10^3$ molecules, in 3 nanoseconds, for a H_2 bubble of radius equal to r_{crit}. Thus a growing sub-critical bubble nucleus near the Pt nanoelectrode will not deplete the local concentration of dissolved gas, and its growth will thus not likely be transport limited.

We performed steady-state finite element simulations to gain further insight into the flux and concentration distributions of H_2 around a bubble nucleus at an electrode. Mass transport of H_2 in the solution and H_2 transfer exchange at the bubble–solution interface were modeled in these simulations (detailed description in Appendix B). The results are shown in Fig. 5 for three bubble nuclei of radii that are, respectively, smaller, larger and nearly equal to the value of $r_{crit} = 4.9$ nm reported in Table 1. In each simulation, the bubble is situated 1 nm above a 20 nm radius electrode that generates a uniform surface flux of H_2 equivalent to the current experimentally observed at i_{nb}^{p} (16.2 nA). The concentration of H_2 is represented by color (red = high concentration, blue = low concentration); its value at the surface of the bubble is fixed to that given by Henry's law and the bubble's internal pressure (eqn (6)). The streamlines represent H_2 flux, from which we observe that the H_2 generated at the electrode underneath the bubble diffuses into the bubble, whereas the H_2 generated at the electrode far from the bubble

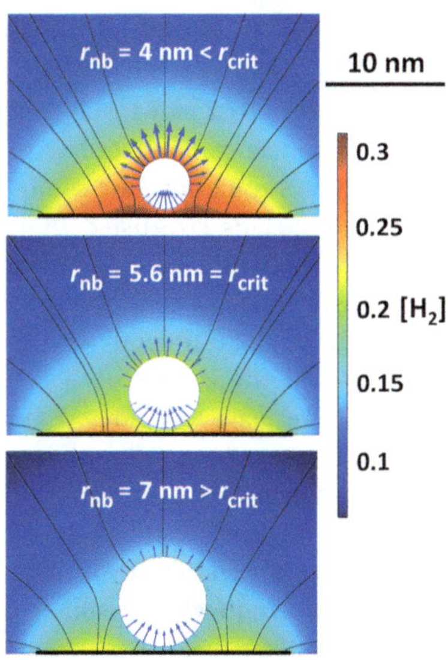

Fig. 5 Finite element simulations of 4, 5.6, and 7 nm-radius bubble nuclei positioned one nanometer above a 20 nm radius Pt electrode. Streamlines represent H_2 flow away from the electrode and vectors show the flux across the bubble surface.

diffuses into bulk solution. The arrows at the bubble–solution interface represent H_2 flux vectors across the interface; the results in Fig. 5 indicate that H_2 enters the bottom of the bubble that is close to the electrode surface, and exits through the top surface of the bubble. The magnitudes of these fluxes are quite different for the different sized nuclei. The smallest bubble nucleus ($r_{nb} = 4$ nm), which has the highest internal pressure, has a larger H_2 flux out of the top of the nucleus and a smaller flux into its lower side. Overall, this gives a net outward flow of H_2, and hence this sub-critical nucleus will tend to collapse. In contrast, for the largest bubble ($r_{nb} = 7$ nm), the lower internal pressure generates a lower surface concentration and the net flux of H_2 is into the bubble; *i.e.*, this is a growing bubble. The net inward flux (7.1 fmol s^{-1} integrated over the bubble surface) is such that enough gas enters the bubble to grow it to a hemisphere equal to the size of the electrode in ~7 μs. The 5.6 nm bubble corresponds to a size where the H_2 fluxes into and out of the bubble, through the lower and upper surfaces, are nearly perfectly balanced; however, any perturbation would cause the bubble to either grow or shrink. Note that this simulated bubble size is just slightly larger than $r_{crit} = 4.9$ nm reported in Table 1 for H_2.

For simplicity, the lower edge of the bubble nucleus in Fig. 5 was fixed 1 nm from the electrode center in the simulations; however, moving the nucleus horizontally or vertically only subtly affects the calculated values and not the qualitative conclusion that the proximity of the electrode provides a sufficient supply of the gas to support the rapid growth of the bubble. Moreover, simulations with the different gases show similar flux distributions, although the exact concentrations/critical sizes depend on the gas. As the simulations were performed at steady state, they do not capture the time dependence; however, they do reflect the expected direction and magnitude of the local fluxes (NB: the use of steady-state simulations accounts for the anomalously high concentration observed around the 4 nm bubble, which, rather than constantly generating H_2 while maintaining its size, would actually be shrinking).

The primary conclusion we draw from the simulations is that the close proximity of the bubble nucleus to the electrode/electrolyte interface greatly reduces kinetic limitations to bubble growth that may result from improbable stochastic fluctuation in the dissolved gas concentration required to reach r_{crit}. However, we also note that the results of these simulations are semi-quantitatively consistent with the values of the critical nucleus size listed in Table 1. Specifically, the simulations indicate that bubbles with $r_{nb} < r_{crit}$ (~5 nm for H_2) will dissolve, while bubbles with $r_{nb} > r_{crit}$ will continue to grow. The apparent consistency of the simulated kinetic phenomena (*i.e.*, diffusion) with the conclusions based on thermodynamic relationships is highly satisfactory.

In the above discussions, we have considered homogeneous nucleation occurring in the solution. However, as suggested by the above simulations, nucleation might be a pseudo-homogeneous process in which bubble growth is a surface mediated process whereby the constituent gas is delivered over a short diffusive distance. Heterogeneous nucleation involving gas adsorption on the Pt surface is also a possible mechanism, in which a sphere cap nucleus is formed at the electrode surface. Our computed values of r_{crit}, which began with the experimentally measured C_{crit}, did not require knowledge of the bubble geometry, other than assuming a surface curvature that is spherical. Hence the critical radius for nucleation we calculate is independent of mechanism. However, for heterogeneous nucleation, the bubble surface area and volume are defined by the contact angle of the sphere cap on the Pt surface. The C_{crit} for heterogeneous nucleation may be lower compared to homogeneous nucleation because fewer molecules are required for a smaller sphere cap shaped bubble on the electrode surface, and the energy cost for creating a smaller bubble surface is lower. If heterogeneous nucleation is occurring in our system, surface adsorption of the bubble nucleus to the electrode would likely need to be directed by hydrophobic surface contaminants. Surface roughness would also be variable from electrode to electrode and consequently the critical concentration, and thus values of observed values of i_{nb}^{p}, would be expected to be less reproducible using different electrodes than our results suggest. However, we cannot discount this possibility.

Conclusions

In this report, we have demonstrated that the study of single gas bubbles at Pt nanodisk electrodes allows precise quantitative evaluation of key parameters describing bubble nucleation. Specifically, analysis of the voltammetric peak currents at the instant preceding stable bubble formation provides the size, internal pressure, and number of gas molecules of the critical size nucleus that grows into a stable bubble. We find that the critical nuclei for H_2, O_2 and N_2 bubbles have a radius between 4 and 10 nm, an internal pressure of 100 to 350 atm, and contain between 2000 and 10 000 molecules, with the variation arising from a number of physicochemical parameters of the gas (*e.g.*, diffusion coefficient, gas solubility, surface tension, *etc.*). The origin of the largest uncertainty in these values arises from the uncertainty in the gas diffusivity in electrolyte solution, as well as potential unknown limitations of applying thermodynamic relationships at nanoscale length scales for single entities (*i.e.*, individual gas nanobubbles).

Our current investigations do not shed significant insight into the frequency of transient bubble formation and collapse prior to the formation of a nucleus with radius > r_{crit} that grows into a dynamically stable bubble. The voltammetric data place a lower bound of $\sim$200 s^{-1} for nuclei generation when the electrode is held at a potential in immediate vicinity of the nucleation potential. We are currently exploring measurements on single bubble experiments to estimate nucleation rates for gas bubbles, and these studies will be communicated in a future report.

Appendix A: experimental methods

Nanoelectrodes were fabricated as previously reported.[35] Briefly, the end of a 25 μm Pt wire was sharpened by etching in 6 M sodium cyanide, the wire attached to a tungsten rod, and the sharp end sealed in a glass capillary in a H_2/air flame. The glass was then polished on silicon carbide sandpaper (400/1200 grit Buehler) until a Pt nanodisk was exposed, as indicated by an electronic feedback circuit. The electrochemically-apparent electrode radius, a, was determined by measurement of the steady-state diffusion-limited current for ferrocene oxidation in an acetonitrile solution containing 0.1 M tetrabutylammonium hexafluorophosphate ($TBAPF_6$). Radii were calculated using[28]

$$i_{lim} = 4nFD_{Fc}C^*a$$

where n is the number of electrons transferred ($=1$ for Fc), F is Faraday's constant ($=96\,485$ C mol^{-1}), and D_{Fc} (2.5×10^5 cm^2 s^{-1})[36] and C^* are the diffusion coefficient and bulk concentration of ferrocene (5 mM), respectively.

Sulfuric acid, hydrazine, and CTAB (Sigma Aldrich) were used as received. Ferrocene was purified by sublimation. A Dagan Cornerstone Chem-Clamp and a Pine RDE4 (used as a waveform generator) were interfaced with a PCI data acquisition card (National Instruments) to perform the voltammetric experiments.

Appendix B: description of finite element simulations

The concentrations and fluxes of H_2 presented in Fig. 5 were calculated by numerically solving the equations describing H_2 diffusion and H_2 transfer at the bubble–solution interface. Boundary conditions were chosen to match those of experiments with a 20 nm radius electrode in 0.5 M H_2SO_4 at the experimentally determined i^p_{nb} (=16.2 nA) and for a range of bubble nuclei sizes. The lower edge of the bubble was placed 1 nm above the nanoelectrode center, as shown in Fig. 6.

Mass transfer was described by the steady-state diffusion equation (eqn (A1)) within the axially symmetric domain shown in Fig. 6.

$$0 = D_{H_2}\nabla^2 c \tag{A1}$$

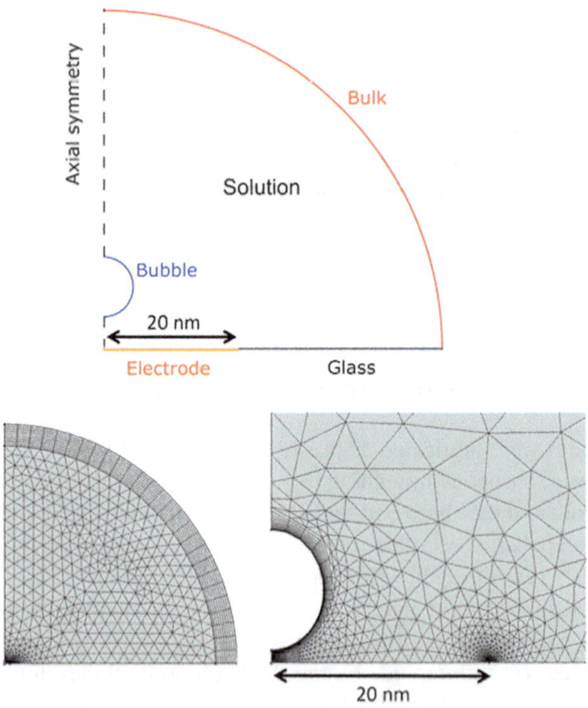

Fig. 6 Geometry (not to scale) and mesh for finite element simulations of a bubble nucleus above a nanoelectrode.

here, D_{H_2} ($=4.5 \times 10^{-5}$ cm^2 s^{-1}) and c are the diffusion coefficient and concentration of H_2, respectively. The concentration at the bulk boundary was taken to match that of the experimental conditions, that is

$$c_{H_2} = 0 \text{ M} \tag{A2}$$

There is no flux of H_2 into, or out of, the glass surrounding the nanoelectrode, as described by

$$0 = D_{H_2} \nabla c \cdot \vec{n} \tag{A3}$$

where $\vec{n}$ is the inward pointing unit normal to the surface.

On the surface of the bubble, we assume that the gas transfer kinetics are infinitely fast, *i.e.*, always at equilibrium. We use the Young–Laplace equation (eqn (6)) to describe the internal pressure and Henry's law to determine its equilibrium concentration. The bubble surface concentration of H_2 is thus described by:

$$C_{H_2} = k_H \left(P_{Ext} + \frac{2\gamma}{r_{nb}} \right). \tag{A4}$$

The boundary condition on the electrode was matched to the experimentally determined current ($i_{nb}^{p} = 16.2$ nA, for $r_{elec} = 20$ nm) such that the H_2 flux is distributed evenly across the electrode surface, which is a reasonable assumption for a kinetically limited reaction, *i.e.* at moderate electrode overpotentials.

$$\frac{-i_{nb}^{p}}{2F\pi a^2} = D_{H_2} \nabla c \cdot \vec{n} \tag{A5}$$

Solution of eqn (A1) was achieved using the commercial finite element package COMSOL Multiphysics (version 5.2). The mesh, which was heavily refined on the boundary between the glass and the Pt and the bubble surface, is shown in Fig. 6.

The "infinite elements" coordinate transform was used to map the outer region (quadrilateral mesh in Fig. 6) to infinity to rule out any influence of a finitely sized domain of simulation.

Appendix C: calculation of parameters for the critical nuclei in Table 1

Table 2 lists diffusion coefficients for H_2, N_2, and O_2 in water from different literature reports. In evaluation of C_{crit} from the i_{nb}^{p} *vs.* a slopes in Fig. 3, the diffusion coefficients for each gas were averaged and the extrema were used to estimate the uncertainty in D.

The ideal values in Table 1 in the main text are calculated using straightforward classical thermodynamic relationships (see main text). Henry's law is the most well studied reference point for gas solubility. At 25 °C and 1 atm, K_H is 0.78, 0.69, and 1.39 mM for H_2, N_2, and O_2 in pure water, respectively.[29,40,41] However, calculation of the values using Henry's law implicitly assumes that gas solubility increases linearly with increasing pressure. Empirical measurements of gas

Table 2 List of diffusion coefficients found in the literature for each gas type

	H_2	N_2	O_2
Diffusivity (10^{-5} cm^2 s^{-1}) at 25 °C (ref. 37)	4.50	1.88	2.10
Diffusivity (10^{-5} cm^2 s^{-1}) at 20 °C (ref. 38)	4.25	2.00	2.30
Diffusivity (10^{-5} cm^2 s^{-1}) at 20 °C (ref. 39)	5.0	2.6	2.3
Average ± extremum (10^{-5} cm^2 s^{-1})	4.58 ± 0.33	2.16 ± 0.44	2.23 ± 0.13

solubility in pure water at extreme pressures are available in the literature and report significant deviation from the linear approximation.[29,42,43] Fig. 7 compares the linear approximation with the empirical data. The pressure at which the equilibrium solubility matches our measured C_{crit} is a more accurate partial pressure, which we denote by $P_{gas}(C_{crit})$.

From the data presented in Fig. 7 we obtain the pressure in equilibrium with the critical concentration for H_2 ($C_{crit} = 0.23$ M), as $P_{gas}(C_{crit}) = 310$ atm, slightly larger than the ideal value, 290 atm, calculated with the Henry's law (linear) approximation. An additional correction concerns the electrolyte's effect on dissolved gas solubility ("salting out"). As no measurements of this effect are available at high pressure, we assume that the fractional change is the same at all pressures. Thus, this is included through multiplying $P_{gas}(C_{crit})$ by the ratio of equilibrium solubility at 1 atm in pure water and with electrolyte present. The corrected value for $\Delta P_{Y\text{-}L}$, which is reported in Table 1 as 'corrected' is given by

$$\Delta P_{Y\text{-}L} = P_{gas}(C_{crit})K_H(\text{water})/K_H(\text{electrolyte}) - P_{ambient} \tag{A6}$$

For instance, the Henry's law constant for H_2 in 0.5 M H_2SO_4 is 89% of the value in pure water (both at 1 atm);[44] consequently, our measured C_{crit} represents a higher $\Delta P_{Y\text{-}L}$ of 350 atm. Experimental studies of gas solubility in the electrolyte solutions employed in this study would allow us to further refine these approximations.

Additionally, it is not expected that the surface tension of a bubble nucleus is equal to that of the air/solution interface at atmospheric pressure. At high

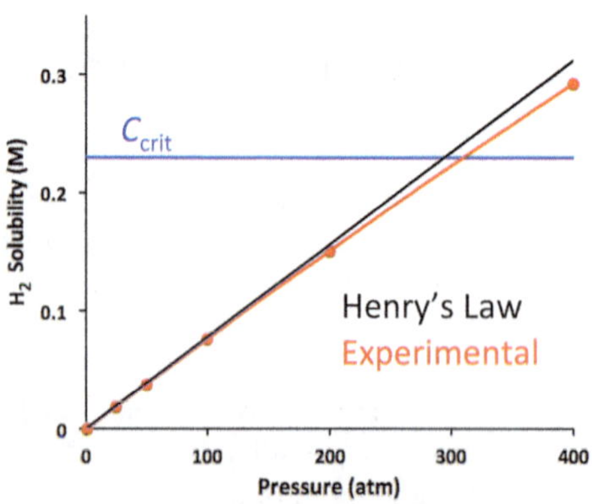

Fig. 7 Plot of experimental values (red) of H_2 solubility at high pressures from Wiebe *et al.*[29] Compared with Henry's law linear approximation.

pressures, gases are known to adsorb to the gas/solution interface, reducing the surface tension to different extents according to their polarizability. Depending on the gas and the pressure within the critical nucleus, the reduction in surface tension can be significant and would proportionally lower the estimated r_{crit}. For the gases in this study, the order of increasing positive gas adsorption to the interface has been empirically measured as $H_2 < O_2 < N_2$.[45] Measurements of the surface tension of gas/water interfaces were reported up to 80 atm. Polynomial fits to the data are provided in the form $\gamma = \gamma_0 + BP + CP^2$ where γ_0 (dyn cm^{-1}) is the surface tension at 1 atm, P is the applied pressure, and B and C are the fit parameters. Values for B and C were reported as −0.0250 and 0, −0.0835 and 0.000194, and −0.0779 and 0.000104 for H_2, N_2, and O_2, respectively. Values of 72 and 73 dyn cm^{-1} were used for γ_0 for the H_2SO_4 and N_2H_4 solutions, respectively,[30,46] giving corrected values for γ of 63, 65, and 64 dyn cm^{-1} for H_2, N_2, and O_2, respectively. Note, as stated in the main text, as the influence of the surfactant (CTAB) cannot easily be combined with that of the gas altering the surface tension, we used the surface tension at 1 atm in all cases ($\gamma = 37$ dyn cm^{-1}).[47] From these surface tensions and the corrected pressure we derived the corrected values of r_{crit} displayed in Table 1, from which the number of molecules in the critical nucleus, n_{crit}, were deduced using the ideal gas law (eqn (7), with the corrected pressure).

Conflict of interest

The authors declare the following competing financial interest: Sean German is an employee of Revalesio Corporation.

Acknowledgements

We thank Stephen Feldberg for helpful discussions. This work was funded by the Office of Naval Research Award Number N000141211021.

References

1 M. Volmer and A. Weber, *Z. Phys. Chem.*, 1926, **119**, 277.
2 R. Becker and W. Doring, *Ann. Phys.*, 1935, **24**, 719.
3 J. Frenkel, *Kinetic Theory of Liquids*, Dover, New York, 1955.
4 Y. B. Zeldovich, *Acta Physicochim. URSS*, 1943, **18**, 1.
5 M. S. Plesset and A. Prosperetti, *Annu. Rev. Fluid Mech.*, 1977, **9**, 145.
6 V. K. Shen and P. G. Debenedetti, *J. Chem. Phys.*, 2003, **118**, 768.
7 V. Talanquer and D. W. Oxtoby, *J. Chem. Phys.*, 1995, **102**, 2156.
8 S. Goldman, *J. Phys. Chem. B*, 2008, **112**, 16701.
9 M. Blander and J. L. Katz, *AIChE J.*, 1975, **21**, 833.
10 C. Ward, A. Balakrishnan and F. C. Hooper, *Trans. ASME*, 1970, **92**, 695.
11 L. Luo and H. S. White, *Langmuir*, 2013, **29**, 11169.
12 Q. Chen, L. Luo, H. Faraji, S. W. Feldberg and H. S. White, *J. Phys. Chem. Lett.*, 2014, **5**, 3539.
13 Q. Chen, H. S. Wiedenroth, S. R. German and H. S. White, *J. Am. Chem. Soc.*, 2015, **137**, 12064.
14 Q. Chen, L. Luo and H. S. White, *Langmuir*, 2015, **31**, 4573.

15 S. R. German, Q. Chen, M. A. Edwards and H. S. White, *J. Electrochem. Soc.*, 2016, **163**, 3160.
16 E. N. Harvey, D. K. Barnes, W. D. McElroy, A. H. Whiteley and D. C. Pease, *J. Am. Chem. Soc.*, 1945, **67**, 156.
17 K. V. Dapkus and P. J. Sides, *J. Colloid Interface Sci.*, 1986, **111**, 133.
18 M. B. Rubin and R. M. Noyes, *J. Phys. Chem.*, 1992, **96**, 993.
19 P. M. Wilt, *J. Colloid Interface Sci.*, 1986, **112**, 530.
20 P. G. Bowers, K. Bar-Eli and R. M. Noyes, *J. Chem. Soc., Faraday Trans.*, 1996, **92**, 2843.
21 D. Sette and F. Wanderlingh, *Phys. Rev.*, 1962, **125**, 409.
22 Y. Mori, K. Hijikata and T. Nagatani, *Int. J. Heat Mass Transfer*, 1976, **19**, 1153.
23 M. J. Hey, A. M. Hilton and R. D. Bee, *Food Chem.*, 1994, **51**, 349.
24 E. A. Hemmingsen, *J. Appl. Phys.*, 1975, **46**, 213.
25 S. Jones, G. Evans and K. Galvin, *Adv. Colloid Interface Sci.*, 1999, **80**, 27.
26 D. E. Westerheide and J. W. Westwater, *AIChE J.*, 1961, **7**, 357.
27 S. Lubetkin, *Langmuir*, 2003, **19**, 2575.
28 A. J. Bard and L. R. Faulkner *Electrochemical Methods: Fundamentals and Applications*, John Wiley and Sons, New York, 2001.
29 R. Wiebe, V. L. Gaddy and C. Heins Jr, *Ind. Eng. Chem.*, 1932, **24**, 823.
30 R. M. Suggitt, P. M. Aziz and F. Wetmore, *J. Am. Chem. Soc.*, 1949, **71**, 676.
31 R. C. Tolman, *J. Chem. Phys.*, 1949, **17**, 333.
32 M. M. Kohonen and H. K. Christenson, *Langmuir*, 2000, **16**, 7285.
33 L. R. Fisher and J. N. Israelachvili, *Nature*, 1979, **277**, 548.
34 C. Jho and A. D. King Jr., *J. Colloid Interface Sci.*, 1979, **69**, 529.
35 B. Zhang, J. Galusha, P. G. Shiozawa, G. Wang, A. J. Bergren, R. M. Jones, R. J. White, E. N. Ervin, C. C. Cauley and H. S. White, *Anal. Chem.*, 2007, **79**, 4778.
36 T. Kuwana, D. E. Bublitz and G. Hoh, *J. Am. Chem. Soc.*, 1960, **82**, 5811.
37 E. L. Cussler, *Diffusion: Mass Transfer in Fluid Systems*, Cambridge University Press, New York, 1997.
38 P. T. H. M. Verhallen, L. J. P. Oomen, A. J. J. M. v. d. Elsen and J. Kruger, *Chem. Eng. Sci.*, 1984, **39**, 1535.
39 D. L. Wise and G. Houghton, *Chem. Eng. Sci.*, 1966, **21**, 999.
40 R. Battino, T. R. Rettich and T. Tominaga, *J. Phys. Chem. Ref. Data*, 1983, **12**, 163.
41 R. Battino, T. R. Rettich and T. Tominaga, *J. Phys. Chem. Ref. Data*, 1984, **13**, 563.
42 V. I. Baranenko, V. S. Sysoev, L. N. Falkovskii, V. S. Kirov, I. Piontkovskii and A. N. Musienko, *At. Energy*, 1990, **68**, 162.
43 M. Geng and Z. Duan, *Geochim. Cosmochim. Acta*, 2010, **74**, 5631.
44 P. Ruetschi and R. F. Amlie, *J. Phys. Chem.*, 1966, **70**, 718.
45 R. Massoudi and A. D. King Jr, *J. Phys. Chem.*, 1974, **78**, 2262.
46 N. B. Baker and E. C. Gilbert, *J. Am. Chem. Soc.*, 1940, **62**, 2479.
47 Z. Adamczyk, G. Para and P. Warszynski, *Langmuir*, 1999, **15**, 8383.

Faraday Discussions

PAPER

The oxygen reduction reaction at the three-phase boundary: nanoelectrodes modified with Ag nanoclusters

Jan Clausmeyer,* Alexander Botz, Denis Öhl and Wolfgang Schuhmann*

Received 28th April 2016, Accepted 18th May 2016
DOI: 10.1039/c6fd00101g

Silver nanoclusters are deposited on bifunctional Θ-shaped nanoelectrodes consisting of a carbon nanoelectrode combined with a hollow nanopipette. The Θ-nanoelectrodes are used as model systems to study interfacial mass transport in gas diffusion electrodes and in particular oxygen-depolarized cathodes (ODC) for the oxygen reduction reaction (ORR) in chlor-alkali electrolysers. By local delivery of O_2 gas to the electroactive Ag nanoclusters through the adjacent nanopipette, enhanced currents for the ORR at the Ag nanoparticles are recorded which are not accountable when considering the low solubility and slow diffusion of O_2 in highly alkaline media. Instead, local oversaturation of O_2 leads to current enhancement at the Ag nanoclusters. Due to the intrinsic high mass transport rates at the nanometric electrodes accompanied by local delivery of reactants, the method generally allows to study electrochemical reactions at single nanoparticles beyond the limitations induced by slow diffusion and low reactant concentration. Kinetic and mechanistic information, for instance derived from Tafel slopes, can be obtained from kinetic regimes not accessible to standard techniques.

Introduction

Electrocatalytic reactions involving gaseous reactants or products play a crucial role in electrochemical energy conversion and the industrial synthesis of bulk chemicals. For instance, chlorine is produced *via* chlor-alkali electrolysis with a global annual production volume of $\sim10^7$ tons of Cl_2. Electrolytic chlorine production is one of the most energy-intensive processes in the world.[1] In strongly alkaline solution, chlorine is evolved anodically, while the cathodic reaction is classically the generation of hydrogen under concomitant formation of NaOH. Alternatively, oxygen-depolarized cathodes (ODC) are employed to reduce the overall cell voltage and thus increase the energy efficiency.[2] In chlor-alkali

Analytical Chemistry – Center for Electrochemical Sciences (CES), Ruhr-Universität Bochum, Universitätsstrasse 150, D-44780, Bochum, Germany. E-mail: jan.clausmeyer@rub.de; wolfgang.schuhmann@rub.de

electrolysers, ODCs provide high current densities despite the very low solubility and slow diffusivity of oxygen which limits the availability of O_2 for the electrochemical reaction.[3,4] It is surprising how the low O_2 concentrations in highly concentrated NaOH solutions can sustain industrially relevant current densities of some kA m^{-2}. Moreover, the high concentration of electrolyte ties water to the solvation shell of solute molecules and thus decreases the abundance of free H_2O molecules. Thus, to ensure efficient mass transport with the gas phase as well as with the bulk solution, ODCs require gas diffusion membranes in intimate contact with the highly porous cathode material. The reaction only takes place at significant rates at the three-phase boundary between the catalyst material, the alkaline electrolyte and the O_2-rich gas phase. The design of numerous composite gas diffusion electrodes has led to a large number of publications and patents.[1,2] Nevertheless, mass and charge transport phenomena at the critical three-phase boundary is yet poorly understood. This might be due to the difficulty in probing the catalytically active micro- and nano-domains experimentally.

Herein, we present a nanoelectrochemical technique to study interfacial charge and mass transfer processes representative of the conditions present in oxygen-depolarized cathodes. Our approach is based on the use of bifunctional carbon nanoelectrodes. The oxygen reduction reaction (ORR) is investigated at nanometric Ag clusters adjacent to a nanometric source of gaseous O_2. Carbon nanoelectrodes were shown to be powerful tools to probe the electrochemical characteristics of micro- and nanometric entities.[5–8] They can be functionalized in manifold ways to comprise electro-chemical sensors[9] to access the chemical microenvironment[10,11] and probe the interior of single living cells.[12] Moreover, we recently investigated the electrochemistry of single O_2-evolving nanoparticles.[13] The extraordinarily fast diffusion allows the study of electrochemical reactions at high current densities without mass transport limitation.[14] The methodological approach in this study increases the effective mass transport rate even further. By applying the gaseous reactants locally to the nanometric electrode in question, the kinetics of fast electrochemical reactions can be studied, even at low solubility and/or low diffusivity of the reactant.

Experimental

Bifunctional nanoprobes and single barrel carbon nanoelectrodes were fabricated adapting a previously published procedure.[12,15] Θ-shaped quartz glass capillaries ($ø_{outer}$: 1.2 mm, wall thickness: 0.15 mm; Sutter Instruments) or regular quartz capillaries ($ø_{outer}$: 1.2 mm, $ø_{inner}$: 0.9 mm; Sutter Instruments) were pulled in a P-2000 capillary puller (Sutter Instruments) to nanopipettes with only a few nanometres orifice. To fill only one of the two barrels with pyrolytic carbon, one barrel was blocked with blu-tack modelling clay (Bostik). The capillary was connected to a butane/propane source (CV 300 plus, Campingaz) and heated in a butane flame while inserted in a second capillary flushed with Argon. After pyrolysis the clay-filled part was cut off.

Electrochemical measurements at single-barrel nanoelectrodes were recorded either using a ModuLab (Solartron) potentiostat, whereas the double barrel nanoelectrodes were investigated utilizing either a Jaissle PGU-100 (IPS) or an Axopatch 200B (Axon Instruments). All measurements were performed in a two-electrode setup *versus* a Ag/AgCl/3 M KCl reference/counter electrode.

The dimensions and quality of nanoelectrodes were evaluated from steady-state voltammograms in 5 mM $[Ru(NH_3)_6]Cl_3$ (Sigma Aldrich) in 100 mM KCl (J.T. Baker). Only electrodes showing diffusion-limited $[Ru(NH_3)_6]^{3+}$ reduction currents smaller than 10 pA were used in this study. Silver was electrochemically deposited on the carbon nanoelectrode by polarization to typically 0.1 V *vs.* Ag/AgCl/3 M KCl in 5 mM $AgNO_3$ (Sigma Aldrich) for different electrolysis times. The amount of silver deposited on the nanoelectrodes was determined by stripping voltammetry in the Ag deposition solution (0.1 M $AgNO_3$) or in 0.1 M $NaClO_4$ (Sigma Aldrich) by sweeping the potential (scan rate 25 mV s^{-1}) from 0.1 to 0.7 V *vs.* Ag/AgCl/3 M KCl.

Oxygen reduction measurements were performed in a closed container filled with NaOH solution (VWR) of various concentrations. A potential ramp (scan rate 10 or 25 mV s^{-1}) from typically 0 to −0.7 V *vs.* Ag/AgCl/3 M KCl was applied to the Ag-modified electrode. Oxygen was delivered through PVC tubing which was connected to the back end of the bifunctional Θ-capillary. A thin copper wire ($\o$ = 150 μm) was used to contact the carbon electrode (Fig. 4). Pure O_2 (≥99.5%; Air Liquide) was delivered from a gas cylinder by adjusting the pressure using a GDH 200-14 digital manometer (Greisinger). Pressure values are reported as overpressures with respect to atmospheric pressure, *i.e.* atmospheric pressure corresponds to 0 bar. Before each experiment all tubing was flushed with O_2 to remove air. To saturate solutions globally with oxygen, the solution was purged with O_2 at atmospheric pressure for 15 min prior to the ORR measurements in the respective measurement container.

SEM images were taken after Ag deposition but before systematic study of the ORR in alkaline solutions using a Quanta 3D ESEM (FEI) at 20 kV in the high vacuum mode. The electrodes were not coated with a metal layer. Electrostatic charging of the carbon-filled nanopipettes was minimized by attaching the pipettes to the metal holder thoroughly with conductive carbon tape. No change of the activity for the ORR was observed when comparing voltammograms in globally O_2-saturated solution before and after the SEM measurement.

Results and discussion

Nanoelectrodes are suitable analytical tools to rationalize the relationship between the structure of nanoparticles and their electrochemical behaviour and catalytic activity. In non-ensemble measurements, single nanoparticles are investigated to exclude the detrimental effect of polydispersity in particle population for the interpretation of activity studies. Nanoelectrodes can be functionalized with pre-synthesised single particles[16–18] or serve as the template for deposition of single particles.[13,14,19,20] At the same time, nanopipettes allow for pressure- or voltage-driven highly localised delivery of reagents.[21–24] Here we use carbon nanoelectrodes, combined with a nanopipette to locally apply gaseous O_2 to an adjacent catalyst cluster at which the ORR is taking place. Double-barrel Θ-shaped quartz capillaries are pulled to comprise dual nanopipettes. Then pyrolytic carbon is selectively deposited in one of the barrels to yield a carbon nanoelectrode neighbouring a hollow capillary orifice.[15,25,26] It was shown previously that the size of the nanoprobes can be tuned by changing parameters of the laser-assisted pulling process, whereas diffusion-limited electrochemical steady-state currents allow a fair assessment of electrode size and quality.[9,11,12,27]

Nanoclusters of Ag are electrochemically deposited on carbon nanoelectrodes. Ag is commonly used as the active catalyst in ODCs where it compares favourably with Pt. At the cost of somewhat higher overpotentials for the ORR, Ag has a lower price and higher stability in electrolysers than Pt.[1,28] While the activity of Ag at milder alkaline pH value is worse than that of Pt,[29,30] in particular at high pH values over 14, typical of industrial electrolysis conditions, Ag fares well as an ORR catalyst.[31]

To test the electrodeposition of Ag on nanoelectrodes,[19] Ag was first deposited on single-barrel carbon nanoelectrodes[9] from solutions of $AgNO_3$ at a potential of 0.1 V *vs.* Ag/AgCl/3 M KCl. The amount of Ag deposited is determined by means of stripping voltammetry by integrating the peak for the anodic dissolution of Ag to Ag^+. Fig. 1a shows stripping peaks obtained after Ag deposition for varying deposition times. The size of the Ag nanoclusters can be estimated from the charge, assuming that the clusters have approximately hemispherical shape. The estimated cluster radii increase linearly with the deposition time (Fig. 1b). The relation between cluster size and deposition time varies between different individual nanoelectrodes but is consistent for a particular given electrode.

For bifunctional nanoelectrodes, different characteristic shapes for the stripping voltammograms are observed. In some cases, stripping peaks are very narrow suggesting a single well-defined and electrically well connected Ag nanoparticle (Fig. 2a). In other examples broad peaks are followed by a general increase of the current at higher potentials, consistent with a Ag cluster whose dissolution is hampered by substantial electrical resistance (Fig. 2b and c). Some stripping voltammograms show multiple superimposed peaks hinting at Ag particles with slightly varying redox potential (Fig. 2d). These findings reflect some degree of heterogeneity in the geometry and size of the electrodeposited Ag nanoclusters caused by poor control during the electrodeposition potentially imposed by Ag deposition on the glass walls or the growth of a Ag-particle chain. For the bifunctional nanoelectrodes, no consistent correlation between electrodeposition time and the shape and charge of the stripping peaks was observed.

We investigated the presence, shape and size of Ag clusters by scanning electron microscopy (SEM). Some small clusters are not clearly identifiable by SEM (Fig. 3a), despite the presence of a cluster was proved by stripping

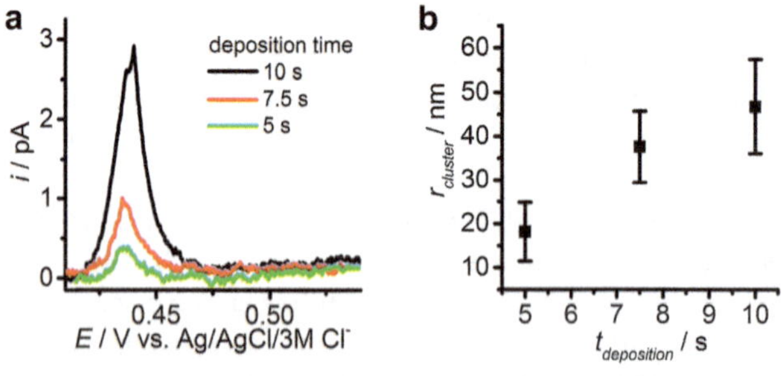

Fig. 1 Deposition of Ag on nanoelectrodes affords nanometric Ag clusters. The cluster size is determined by stripping voltammetry showing a silver dissolution peak (a). The estimated cluster radii assuming hemispherical geometry increase with increasing Ag deposition time (b). Each data point is obtained from at least three consecutive deposition–stripping cycles on the same nanoelectrode.

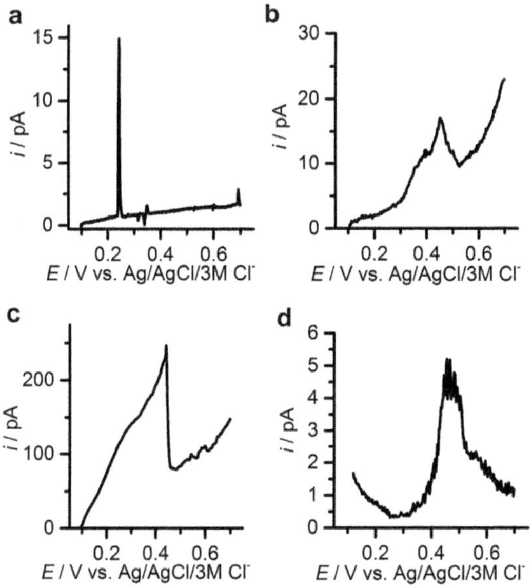

Fig. 2 Voltammograms for the stripping of Ag from bifunctional nanoelectrodes show different characteristic shapes.

voltammetry. Some nanoelectrodes exhibit multiple, separated Ag clusters adsorbed to the glass insulator in proximity of the tip of the nanoelectrode (Fig. 3b). One might conclude that the Ag nanoparticles grown on the side of the electrode are due to insufficient insulation by the quartz glass sheath. In contrast, the comparison of SEM images obtained from a nanoelectrode directly after Ag deposition and the same nanoelectrode after stripping of the Ag cluster show that only the cluster directly in contact with the active carbon disk is removed, while the neighbouring ones remain attached to the sheath of the electrode. Hence, the particles are electrically connected only *via* the active carbon electrode.

The Ag-modified bifunctional nanoprobes are model systems to study the electrocatalytic oxygen reduction at the three-phase boundary between the Ag catalyst, alkaline electrolyte and the gas phase. Pressurized O_2 is applied through the open barrel and by this dosed in a controlled manner to the adjacent Ag nanoelectrode (Fig. 4). Only a small volume of electrolyte solution separates the

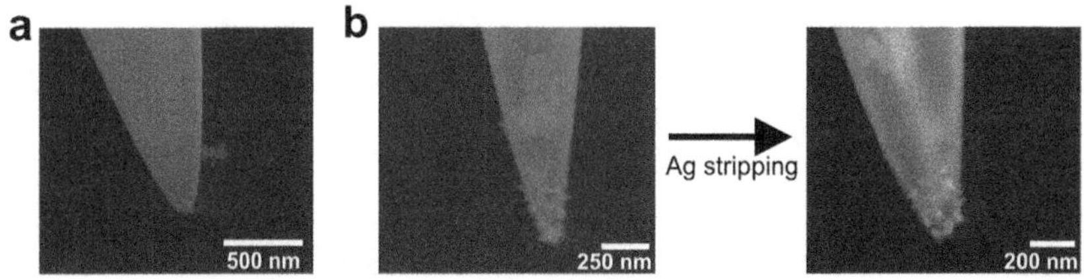

Fig. 3 SEM images show the size and shape of deposited Ag nanoclusters. Image of a bifunctional electrode modified with a Ag nanocluster identified by stripping voltammetry after SEM (a). Images before and after stripping of an individual nanoelectrode (b).

nanometric O_2 source from the Ag cluster. This configuration resembles the situation in gas diffusion electrodes where porous channels for O_2 are adjacent to particles of the dispersed catalyst material.

The high viscosity of the electrolyte solution and the low solubility of oxygen lead to diffusion coefficients of around 10^{-7} cm^2 s^{-1} and low maximum concentrations of 40 μM for O_2 in 10 M NaOH.[4] Accordingly, when studying the ORR at a nanometric Ag cluster, low cathodic currents are observed for the ORR when the solution is globally saturated with O_2 (Fig. 5a). However, upon applying oxygen overpressure to the capillary, O_2 is locally delivered from its orifice, leading to an O_2 oversaturation in the direct proximity of the Ag nanocluster. Consequently, higher ORR currents are recorded with increasing pressure. In the locally oversaturated environment, currents exceed the ones recorded in a globally saturated solution by many times, indicated by the quantity i/i_{sat}, the current at a given pressure normalized by the current at atmospheric pressure (Fig. 5b). On a bare carbon nanoelectrode which was not modified with silver, absolute currents are by far smaller than the ones recorded at Ag clusters and the current enhancement during local application of O_2 is negligibly small (data not shown). Interestingly, when applying high pressure no departure of gas bubbles was observed. The Ag-modified electrodes do not show periodic modulations of the ORR currents as it would be expected for the formation and departure of bubbles.[32,33] No gas bubble evolution was detected using optical microscopy with a 320-fold magnification. Instead, the results suggest a stable interface between the gas and liquid phase close to the orifice of the capillary with oxygen crossing the phase boundary to create a nanometric diffusion layer for O_2. The results suggest the formation of a stable nanobubble, which was previously described for the hydrogen evolution reaction at nanoelectrodes.[34] Notably, the steady-state voltammograms for the ORR do not show current plateaus as it would be expected for mass transport-limited reaction rates. The fast mass transport characteristic of nanoelectrodes is further enhanced by the local O_2 source. Hence, even at high overpotentials, kinetic parameters of the electrochemical reaction can be studied. Fig. 5c shows a Tafel plot for the ORR while locally applying O_2 at a pressure of 2.5 bar. The Tafel slope extracted in the low overpotential region between 0.9 V and 1.0 V *vs.* RHE amounts to −85 mV dec^{-1} and thus matches the ones reported in the literature for the ORR at silver electrodes in alkaline media.[31,35] However, in

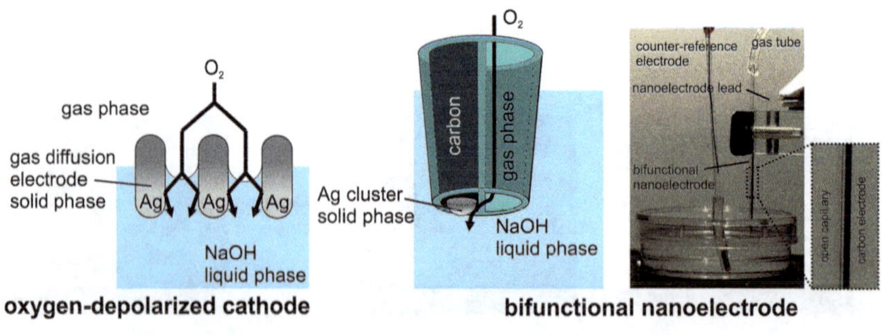

Fig. 4 The critical three-phase boundary found in gas diffusion electrodes is simulated using bifunctional nanoelectrodes to apply gaseous oxygen locally while studying the ORR at the neighbouring Ag nanocluster.

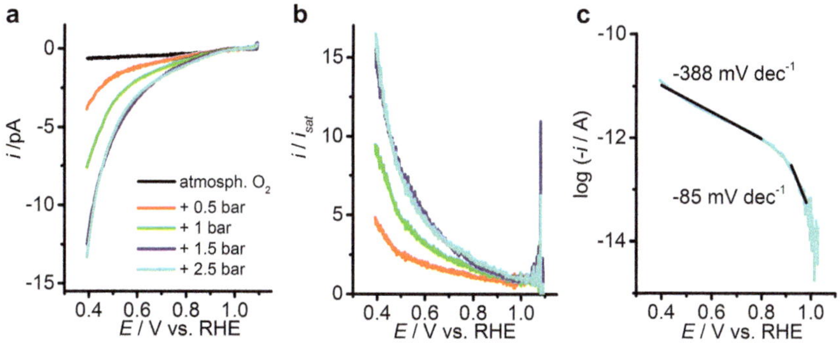

Fig. 5 Local application of O_2 oversaturates the electrolyte near the three-phase boundary. Steady-state voltammograms recorded at a Ag nanocluster in 10 M NaOH while applying varying O_2 pressure through the open barrel of a bifunctional nanoelectrode (a). ORR currents normalized by the current recorded in globally O_2 saturated solution (b). Tafel plot for the ORR at an O_2 overpressure of 2.5 bar (c).

contrast to previous results obtained using classical methods, the kinetic region extends far into the high overpotential region. The Tafel plot shows linear behaviour until potentials as low as 0.4 V *vs.* RHE with a high, yet measurable Tafel slope of -388 mV dec^{-1}.

When varying the NaOH concentration of the electrolyte solution, the absolute ORR currents recorded at an individual Ag cluster decrease with increasing NaOH concentration due to the increasingly low solubility and slow diffusion of O_2 (Fig. 6a). The steady-state voltammograms show currents in the order 0.1 M > 5 M > 10 M NaOH. In all cases when applying O_2 pressure, the ORR currents increase with respect to the situation with atmospheric pressure in the capillary. However, the extent of local oversaturation depends on the NaOH concentration. The normalized quantity i/i_{sat} allows to compare the results from different Ag-modified nanoelectrodes whose Ag clusters have different sizes. Fig. 6b compares the

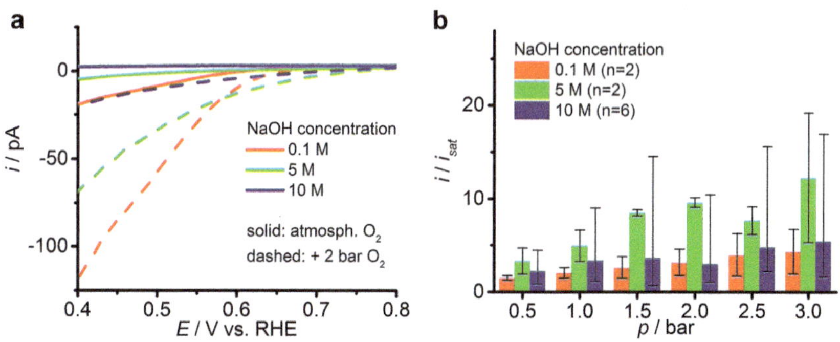

Fig. 6 Local O_2 oversaturation depends on the pressure and concentration of NaOH. Absolute ORR currents decrease with increasing NaOH concentration. Steady-state voltammograms at one individual Ag-modified bifunctional nanoelectrode recorded in different solutions (a). Currents of different electrodes at $E = 0.45$ V *vs.* RHE normalized by the corresponding current in globally O_2-saturated solution vary with O_2 pressure and NaOH concentration (b). Error bars show the smallest and largest recorded values.

ORR currents at 0.45 V *vs.* RHE during local O_2 application, normalized by the currents at globally O_2-saturated solution as a function of pressure and electrolyte composition. Higher O_2 pressure generally results in larger oversaturation, consistent with a larger flux of O_2 to the active catalyst. The highest current enhancement is achieved at a NaOH concentration of 5 M. The current during local application of O_2 is on average up to 20 times larger compared to globally saturated solution. At lower NaOH concentration as well as at higher concentration, local O_2 oversaturation only induces a current enhancement of around 5. This finding can not only be explained by the slower diffusion at high alkalinity. Other factors governing the local oversaturation must be taken into account as well. For instance, the surface tension and thus the capillary action is expected to depend on the pH value and ionic strength of the solution. This affects the shape and position of the gas–liquid interface in nanometric channels and thus the distance between the electroactive catalyst cluster and the local source of O_2.

Conclusions

The bifunctional nanoelectrodes used in this study carry an electrodeposited Ag nanocluster on one side and exhibit a hollow nanocapillary for the local delivery of reagents on the other. We use these nanoelectrochemical tools to mimic the situation in gas diffusion electrodes and in particular oxygen-depolarized cathodes with silver as the active ORR catalyst. Our model suggests that the high current densities observed for the ORR in chlor-alkali electrolysers can only be explained by local oversaturation of O_2 in the direct vicinity of microscopic gas channels. The technique presented herein helps to understand how O_2 reaches the electroactive catalyst surface despite its low solubility and slow diffusion in the highly alkaline solution. The O_2-detecting probe for the evaluation of O_2 transport at the three-phase boundary needs to be very close to the O_2 source so that both diffusion layers overlap. Here we use a nanoelectrode that accommodates a Ag nanocluster adjacent to the nanometric O_2 source. More thorough investigation is necessary to describe the flux of O_2 quantitatively and thus predict the ORR currents at the nanoelectrodes *a priori*. This also requires improved control and knowledge of the geometry of the deposited electroactive catalyst clusters. Future efforts will aim on achieving better control over the electrodeposition. Apart from its utility for the case study seeking to rationalise oxygen-depolarized cathodes, the presented method is generic and applicable to the fundamental study of electrochemical reactions. The particular experimental configuration constitutes a nanoscale electrochemical reactor including a controllable reactant supply and a catalyst stage for small catalyst amounts down to single nanoparticles.[13,14,16–20] Mass transport, already being extremely fast due to the nanoscale dimensions, can be further accelerated by local delivery of reactants. In particular, observing the ORR at single catalyst entities is difficult because of the low solubility of O_2 in aqueous media. The limitation is overcome by local delivery and oversaturation of O_2. This strategy allows the access of new regimes to extract kinetic and mechanistic information for catalytic reactions whose kinetics are hidden behind the "mass-transport wall". Here, we observe that the kinetic region for the ORR at Ag electrodes extends to exceptionally high overpotentials. The technique might be adapted to other catalyst materials and electrochemical reactions involving gaseous and non-gaseous reactants.

Acknowledgements

The authors are grateful to the Deutsche Forschungsgemeinschaft in the framework of the Cluster of Excellence "Resolv" (EXC1069) and the Forschergruppe (FOR 2397-1; SCHU 929/15-1) "Multi-scale analysis of complex three-phase systems: oxygen reduction on gas-diffusion electrodes in aqueous electrolyte".

References

1 I. Moussallem, J. Jörissen, U. Kunz, S. Pinnow and T. Turek, *J. Appl. Electrochem.*, 2008, **38**, 1177–1194.

2 E. Yeager and P. Bindra, *Chem. Ing. Tech.*, 1980, **52**, 384–391.

3 D. Tromans, *Ind. Eng. Chem. Res.*, 2000, **39**, 805–812.

4 C. Zhang, F.-R. F. Fan and A. J. Bard, *J. Am. Chem. Soc.*, 2009, **131**, 177–181.

5 D. W. M. Arrigan, *Analyst*, 2004, **129**, 1157–1165.

6 T. Li and W. Hu, *Nanoscale*, 2011, **3**, 166–176.

7 S. M. Oja, Y. Fan, C. M. Armstrong, P. Defnet and B. Zhang, *Anal. Chem.*, 2016, **88**, 414–430.

8 J. Clausmeyer and W. Schuhmann, *TrAC, Trends Anal. Chem.*, 2016, **79**, 46–59.

9 J. Clausmeyer, P. Actis, A. López Córdoba, Y. Korchev and W. Schuhmann, *Electrochem. Commun.*, 2014, **40**, 28–30.

10 T. Matsue, *Anal. Sci.*, 2013, **29**, 171–179.

11 Y. Zhang, J. Clausmeyer, B. Babakinejad, A. López Córdoba, T. Ali, A. Shevchuk, Y. Takahashi, P. Novak, C. Edwards, M. Lab, S. Gopal, C. Chiappini, U. Anand, L. Magnani, R. C. Coombes, J. Gorelik, T. Matsue, W. Schuhmann, D. Klenerman, E. V. Sviderskaya and Y. Korchev, *ACS Nano*, 2016, **10**, 3214–3221.

12 P. Actis, S. Tokar, J. Clausmeyer, B. Babakinejad, S. Mikhaleva, R. Cornut, Y. Takahashi, A. López Córdoba, P. Novak, A. I. Shevchuck, J. A. Dougan, S. G. Kazarian, P. V. Gorelkin, A. S. Erofeev, I. V. Yaminsky, P. R. Unwin, W. Schuhmann, D. Klenerman, D. A. Rusakov, E. V. Sviderskaya and Y. E. Korchev, *ACS Nano*, 2014, **8**, 875–884.

13 J. Clausmeyer, J. Masa, E. Ventosa, D. Öhl and W. Schuhmann, *Chem. Commun.*, 2016, **52**, 2408–2411.

14 S. Chen and A. Kucernak, *J. Phys. Chem. B*, 2004, **108**, 3262–3276.

15 Y. Takahashi, A. I. Shevchuk, P. Novak, Y. Zhang, N. Ebejer, J. V. Macpherson, P. R. Unwin, A. J. Pollard, D. Roy, C. A. Clifford, H. Shiku, T. Matsue, D. Klenerman and Y. E. Korchev, *Angew. Chem., Int. Ed.*, 2011, **50**, 9638–9642.

16 J. Lakbub, A. Pouliwe, A. Kamasah, C. Yang and P. Sun, *Electroanalysis*, 2011, **23**, 2270–2274.

17 Y. Li, J. T. Cox and B. Zhang, *J. Am. Chem. Soc.*, 2010, **132**, 3047–3054.

18 Y. Yu, Y. Gao, K. Hu, P.-Y. Blanchard, J.-M. Noël, T. Nareshkumar, K. L. Phani, G. Friedman, Y. Gogotsi and M. V. Mirkin, *ChemElectroChem*, 2015, **2**, 58–63.

19 J. Velmurugan, J.-M. Noël, W. Nogala and M. V. Mirkin, *Chem. Sci.*, 2012, **3**, 3307.

20 P. Sun, F. Li, C. Yang, T. Sun, I. Kady, B. Hunt and J. Zhuang, *J. Phys. Chem. C*, 2013, **117**, 6120–6125.

21 B. Babakinejad, P. Jönsson, A. López Córdoba, P. Actis, P. Novak, Y. Takahashi, A. Shevchuk, U. Anand, P. Anand, A. Drews, A. Ferrer-Montiel, D. Klenerman and Y. E. Korchev, *Anal. Chem.*, 2013, **85**, 9333–9342.

22 J. Hong, J. B. Edel and A. J. deMello, *Drug Discovery Today*, 2009, **14**, 134–146.

23 L. Ying, A. Bruckbauer, D. Zhou, J. Gorelik, A. Shevchuk, M. Lab, Y. Korchev and D. Klenerman, *Phys. Chem. Chem. Phys.*, 2005, **7**, 2859.

24 L. Hirt, R. R. Grüter, T. Berthelot, R. Cornut, J. Vörös and T. Zambelli, *RSC Adv.*, 2015, **5**, 84517–84522.

25 M. A. O'Connell and A. J. Wain, *Anal. Chem.*, 2014, **86**, 12100–12107.

26 B. Paulose Nadappuram, K. McKelvey, R. Al Botros, A. W. Colburn and P. R. Unwin, *Anal. Chem.*, 2013, **85**, 8070–8074.

27 K. McKelvey, B. Paulose Nadappuram, P. Actis, Y. Takahashi, Y. E. Korchev, T. Matsue, C. Robinson and P. R. Unwin, *Anal. Chem.*, 2013, **85**, 7519–7526.

28 J. S. Spendelow and A. Wieckowski, *Phys. Chem. Chem. Phys.*, 2007, **9**, 2654.

29 C. Coutanceau, L. Demarconnay, C. Lamy and J.-M. Léger, *J. Power Sources*, 2006, **156**, 14–19.

30 F. H. B. Lima, J. Zhang, M. H. Shao, K. Sasaki, M. B. Vukmirovic, E. A. Ticianelli and R. R. Adzic, *J. Phys. Chem. C*, 2007, **111**, 404–410.

31 M. Chatenet, L. Genies-Bultel, M. Aurousseau, R. Durand and F. Andolfatto, *J. Appl. Electrochem.*, 2002, **32**, 1131–1140.

32 A. R. Zeradjanin, E. Ventosa, A. S. Bondarenko and W. Schuhmann, *ChemSusChem*, 2012, **5**, 1905–1911.

33 R. A. Rincón, A. Battistel, E. Ventosa, X. Chen, M. Nebel and W. Schuhmann, *ChemSusChem*, 2015, **8**, 560–566.

34 Q. Chen, L. Luo, H. Faraji, S. W. Feldberg and H. S. White, *J. Phys. Chem. Lett.*, 2014, **5**, 3539–3544.

35 B. B. Blizanac, P. N. Ross and N. M. Marković, *J. Phys. Chem. B*, 2006, **110**, 4735–4741.

Faraday Discussions

PAPER

A theoretical consideration of ion size effects on the electric double layer and voltammetry of nanometer-sized disk electrodes

Yu Gao, Yuwen Liu and Shengli Chen*

Received 21st April 2016, Accepted 16th May 2016
DOI: 10.1039/c6fd00087h

Considering that an electric-double-layer (EDL) structure may significantly impact on the mass transport and charge transfer kinetics at the interfaces of nanometer-sized electrodes, while EDL structures could be altered by the finite sizes of electrolyte and redox ions, the possible effects of ion sizes on EDL structures and voltammetric responses of nanometer-sized disk (nanodisk) electrodes are investigated. Modified Boltzmann and Nernst–Planck (NP) equations, which include the influence of the finite ion volumes, are combined with the Poisson equation and modified Butler–Volmer equation to gain knowledge on how the finite sizes of ions and the nanometer sizes of electrodes may couple with each other to affect the structures and reactivities of a nanoscale electrochemical interface. Two typical ion radii, 0.38 nm and 0.68 nm, which could represent the sizes of the commonly used aqueous electrolyte ions (*e.g.*, the solvated K^+) and the organic electrolyte ions (*e.g.*, the solvated TEA^+) respectively, are considered. The finite size of ions can result in decreased screening of electrode charges, therefore magnifying EDL effects on the ion transport and the electron transfer at electrochemical interfaces. This finite size effect of ions becomes more pronounced for larger ions and at smaller electrodes as the electrode radii is larger than 10 nm. For electrodes with radii smaller than 10 nm, however, the ion size effect may be less pronounced with decreasing the electrode size. This can be explained in terms of the increased edge effect of disk electrodes at nanometer scales, which could relax the ion crowding at/near the outer Helmholtz plane. The conditions and situations under which the ion sizes may have a significant effect on the voltammetry of electrodes are discussed.

Introduction

Understanding the structures and processes at nanoscale electrochemical interfaces is of significance in diverse areas such as energy conversion, application of

Key Laboratory of Analytical Chemistry for Biology and Medicine (Ministry of Education), Hubei Key Laboratory of Electrochemical Power Sources, Department of Chemistry, Wuhan University, Wuhan 430072, China. E-mail: slchen@whu.edu.cn; Fax: +86-27-68754693; Tel: +86-27-68754693

scanning probe microscopes in electrolyte solutions, electrochemical fabrication and characterization of nanomaterials, and so on.[1,2] Nanometer-sized electrodes (nanoelectrodes) provide model systems to investigate electrochemistry at nanoscale interfaces. When the electroactive sizes of electrodes or the structural units of electrodes are reduced to the nanoscale so that they are comparable with the characteristic lengths of electric-double-layers (EDLs), some distinct interfacial features which are inappreciable at macro/microscale electrodes may arise, one of which is the significantly enhanced EDL effects on the charge transfer kinetics and transport dynamics even in the presence of an excess of supporting electrolytes. Such EDL effects make the electrode size impact on the electrochemical responses not only quantitatively but also qualitatively. During the past two decades, there have been numerous theoretical and experimental studies demonstrating distinct EDL effects at nanometer-size electrodes.[1–10]

In previous theoretical studies,[1,3,4] EDL effects on the electrochemistry of nanoelectrodes have been mostly modelled with continuum equations which neglect the size (volume) of electrolyte ions, for examples, the conventional Boltzmann equation for ion distribution at equilibrium interfaces and the Nernst–Planck equation for ion transport at a dynamic interface. Whether these continuum equations are applicable or not remains a question at present. Krapf *et al.*[6] have observed a strong nonlinear dependence of the limiting current on the concentration of electroactive ions at hemispherical electrodes smaller than 10 nm in diameter, which seemed to challenge the validity of continuum equations at such small electrodes. Experimental observations by Sun *et al.*[7] have shown an approximately linear dependence of the limiting current on the concentration of reactants at disk electrodes smaller than 10 nm, which deviates from the prediction of diffusion-based mass transport equation, but can be reasonably explained with the continuum Poisson–Nernst–Planck equations.

It is known that the sizes of ions would visibly impact the EDL structure when the concentrations of electrolyte or redox ions are considerably high or their sizes are considerably large, so that significant crowding of ions in the region near the electrode surface occurs.[11–16] One of the recent strategies in developing high performance energy and electronic systems is to use media that have high electrochemical windows, for example organic solvents,[17] ionic liquids[18–20] and molten salts.[21] Large-size cations such as tetra-alkylammonium (TEA^+), imidazolium and phosphonium, and anions such as bis(trifluoromethane sulfonyl) imide ($TFSI^-$) and $CF_3SO_3^-$ are the common components of ionic liquids or organic electrolytes; while ionic liquids and molten salts are typical concentrated media. Recently, highly concentrated aqueous LiTFSI solution (>20 M) has been shown to enable a potential window as high as 3.0 V.[22]

The application of nanoscopic electrode materials and the concentrated and large-size electrolytes in modern energy technologies make it necessary to investigate ion size effects on the structures and reactivity of nanoscale electrochemical interfaces. This could provide knowledge for designing new energy systems and/or improving current systems. As known from previous studies, electron transfer and ion transport processes would be more pronouncedly impacted by EDL structures at nanometer-sized electrodes,[1–8,10] while ions with large sizes can significantly impact EDL structures.[11–16] Therefore, the size of electrolyte ions should also impact the reactivity at nanoscale electrochemical interfaces.

In this paper, using nanometer-sized disk (nanodisk) electrodes as models, we study how the nanometre size effect of electrodes and the finite size effect of ions can be coupled to affect EDL structure and reactivity at the nanoscale electrochemical interface. The approach by Bikerman *et al.*[11,23–26] for treating the finite volume effect of ions is adopted to modify the conventional Boltzmann and Nernst–Planck (NP) equations, which, together with the Poisson equation and the modified Butler–Volmer equations, are used to gain knowledge on how the finite size effect of ions may be altered by nanometer sized electrodes. It is shown that the ion size effect become more pronounced for larger ions and at smaller electrodes as the electrode radii are larger than 10 nm. For electrodes with radii smaller than 10 nm, however, the ion size effect becomes less pronounced at smaller electrodes due to the increased edge effect of disk electrodes at nanometer scales, which could relax the ion crowding at/near the outer Helmholtz plane.

Model and simulation

We consider the following outer-sphere one-electron reduction reaction in solution containing a symmetric electrolyte M^+N^- at a nanodisk electrode with an electroactive radius of r_0 which is embedded in an insulation sheath (*e.g.*, glass) of infinite thickness.

$$O^x + e^- \leftrightarrow R^{x-1} \tag{1}$$

The interface at the nanodisk electrodes is geometrically modelled with a cylindrical coordinate system of axial symmetry containing radial coordinate r and normal coordinate z (Fig. 1). The center of the electrode surface is set as the origin of the coordinate system. Since we are interested mainly in how the ion size effect could be different at nanometer-sized electrodes, the complexities from other factors are ignored for simplicity. Therefore, it is assumed that the compact EDL forms on the surface of the insulation sheath as well as the electrode, with the same thickness μ; no free charge resides in the compact EDL, that is, none of ions in the system chemically adsorb on the electrode surface; the compact EDL can be split into two regions, namely, the outer Helmholtz layer (OHL) and inner

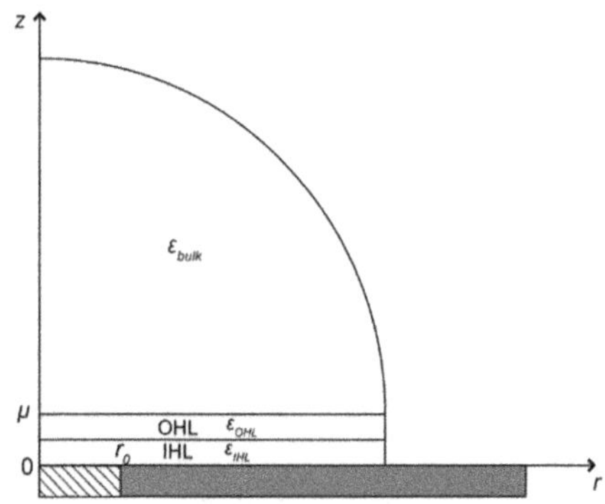

Fig. 1 Schematic illustration of the geometric model used for the nanodisk electrode interface.

Helmholtz layer (IHL), which are composed of closely packed solvent molecules and solvated ions respectively;[4] the relative dielectric constant of the media in the IHL and OHL ($\varepsilon_{\rm IHP}$ and $\varepsilon_{\rm OHP}$) are different from that in the solution due to the dielectric saturation effect under the strong electric field in the EDL;[4] all the ions have the same plane of closest approach at the outer boundary of the OHL, the-so-called outer Helmholtz plane (OHP); the O^x and R^{x-1} have the same diffusion coefficient ($D = 10^{-9}$ m^2 s^{-1}); and all ions have the same radius ($r_{\rm ion}$).

The finite size effect of electrolyte ions was first considered for the equilibrium EDL in the case of a symmetric electrolyte by Bikerman in the early 1940s.[11] Based on the lattice gas models which consider the ions as discrete cells of certain volumes, the finite size of ions result in a modified entropic contribution of ion concentration to the Helmholtz free energy.[11–13] These treatments have also been extended to more general cases, for examples, for unsymmetrical electrolytes[23] and EDL dynamics.[24] Various treatments to the finite size effects of ions can be generalized by adding an excess term to the chemical potential of the corresponding ions, $\mu_i^{\rm ex}$, that is[23–26]

$$\mu_i = RT\ln c_i + z_iF\phi + \mu_i^{\rm ex} \tag{2}$$

in which z_i, c_i and μ_i are the charge, local concentration and chemical potential of the $i^{\rm th}$ ion, ϕ is the local electrostatic potential, and R, T and F have their usual meanings. The excess chemical potential can be expressed as

$$\mu^{\rm ex} = -k_{\rm B}T\ln(1 - \Phi) \tag{3}$$

in which Φ is the volume fraction of all ions (anions + cations) in the solution, which can be expressed as

$$\Phi = \sum c_i/c_{\rm max} \tag{4}$$

where $c_{\rm max} = 1/8{r_{\rm ion}}^3N_{\rm A}$ is the maximum concentration the ions can reach, and $N_{\rm A}$ is the Avogadro constant. Based on eqn (2)–(4), a modified Boltzmann equation for equilibrium ion distribution can be obtained by using the condition that the chemical potential is uniform at the interface and in the solution. That is:[11–13]

$$c_i = \frac{c_i^0 {\rm e}^{-(z_iF\phi/RT)}}{1 + 2c_i^0\sinh^2(F\phi/RT)/c_{\rm max}} \tag{5}$$

in which c_i^0 is the bulk concentration of the ion. This is also called the Fermi distribution equation, which will reduce to the conventional Boltzmann distribution equation as the value of Φ approaches zero.

The chemical potential expression of eqn (2) can also be used to treat the finite size effect on the ion transport at the interface where redox reactions take place, by modifying the conventional Nernst–Planck (NP) equation. For steady-state ion transport in a cylindrical coordinate system, we use

$$\left(\frac{\partial c_i}{\partial t}\right) = -\frac{1}{r}\nabla(rJ_i) = -\frac{1}{r}\nabla\left(\frac{D_ic_i}{RT}r\nabla\mu_i\right) = 0 \tag{6}$$

in which J_i and D_i refer to the flux and diffusion coefficient of the $i^{\rm th}$ ion. In a cylindrical coordinate system, the gradient operator $\nabla = (\partial/\partial r + \partial/\partial z + 1/r)$. Combination of eqn (2)–(4) and (6) leads to

$$\nabla D_i\left(r\nabla c_i + \frac{z_i F c_i}{RT} r\nabla\phi + \frac{c_i}{c_{\max}(1-\Phi)} r\nabla \sum_i c_i\right) = 0. \tag{7}$$

Eqn (7) will reduce to the conventional NP equation as the value of Φ approaches zero. To get the equilibrium distribution or the transport rate of an ion, eqn (5) or (7) has to be solved together with the Poisson equation which has the following form in a cylindrical coordinate system

$$\rho = \frac{1}{r}\varepsilon_0 \nabla(-\varepsilon r \nabla\phi) \tag{8}$$

where ρ refers to the local charge density, ε and ε_0 are the relative dielectric constant and the permittivity of a vacuum.

Eqn (5) and (7) should be solved in the domain of $z > \mu$, where $\mu = l_{\text{IHP}} + l_{\text{OHP}}$ is the thickness of the compact region of EDL, and l_{IHP} and l_{OHP} are the thickness of the IHL and OHL respectively. The Poisson equation (eqn (8)) should be solved in the domain of $z > 0$. To solve these equations, the following boundary conditions are used:

$$r = \infty \text{ or } z = \infty\text{: } \phi = 0,\ \nabla\phi = 0 \text{ and } c_i = c_i^0 \tag{9}$$

$$-r_0 \leq r \leq r_0 \text{ and } z = 0\text{: } \phi = E \tag{10}$$

$$-r_0 \leq r \leq r_0 \text{ and } z = \mu\text{: } \phi = \phi_{\text{OHP}} \tag{11}$$

$$-r_0 \leq r \leq r_0 \text{ and } z = \mu : J_i = -m_j \frac{i}{nF} \tag{12}$$

$$r > r_0 \text{ (or } r < -r_0\text{) and } z = \mu\text{: } J_i = 0 \tag{13}$$

where E refers to the electrode potential, ϕ_{OHP} is the electrostatic potential at OHP, i is the current density on the electrode surface, m_j is an integer constant taking values +1, −1 and 0 for species R^{x-1}, O^x, and the inert electrolyte ions respectively. The reduction current is defined as positive here. For simplicity, we use the Butler–Volmer (BV) model to describe the kinetics of a charge transfer reaction at an electrode surface, that is

$$\frac{i}{F} = k^0\left\{c_{\text{O}^x} \exp\left(\frac{-F(E - E^0 - \phi_{\text{OHP}})}{2RT}\right) - c_{\text{R}^{x-1}} \exp\left(\frac{F(E - E^0 - \phi_{\text{OHP}})}{2RT}\right)\right\} \tag{14}$$

in which k^0 and E^0 refer to the standard rate contestant and the standard equilibrium potential of reaction (1). Eqn (14) is a modified form of the conventional BV equation to include the effect of EDL on the electron transfer kinetics, specifically, the term "$E - E^0$" in the normal BV equation is replaced by "$E - E^0 - \phi_{\text{OHP}}$".

For comparison, we also consider the simple situation when the mass transport and electron transfer processes are not affected by the EDL structure. In this case, the steady-state diffusion equation including the ion size effect is used to describe the mass transport of redox species, that is

$$\nabla D_i\left(r\nabla c_i + \frac{c_i}{c_{\max}(1-\Phi)}r\nabla\sum_i c_i\right) = 0 \tag{15}$$

and the term "$E - E^0 - \phi_{OHP}$" in eqn (14) should be changed to "$E - E^0$". Similarly, eqn (15) will reduce to the conventional diffusion equation when Φ is zero.

The equations are solved numerically using the finite-element-method package in Comsol Multiphysics. In the computations, the bulk concentrations of O^x, R^{x-1}, M^+ and N^- are 0.01, 0, 0.1 and 0.1 mol L^{-1} respectively; the diffusion coefficients for all ions are all 1×10^{-5} cm^2 s^{-1}; the dielectric constants are 6, 40, and 78.3 respectively in the IHL, OHL and region outside the compact double layer; the thicknesses of IHL and OHL are 0.3 nm and 0.29 nm respectively; the standard rate constant is 1.0 cm s^{-1}; the standard electrode potential is 0 V with respect to the potential of zero charge (PZC); and $T = 298.15$ K.

Results and discussion

Effect of electrolyte ion sizes on the equilibrium EDL structures at nanodisk electrodes

We mainly consider two sizes of electrolyte ions, 0.38 and 0.68 nm, which represent the sizes of commonly used aqueous electrolyte ions, *e.g.* the solvated K^+, and organic electrolyte ions, *e.g.* solvated TEA^+. Since the concentration of the redox species are very low compared with the electrolyte ions, we ignore the size effect of the redox species on the EDL structure here.

Fig. 2 shows the electrostatic potential profiles along the *z*-axis away from the OHP (the diffuse part of the interface) at $r = 0$ as the electrode potentials are -0.4 and -1.0 V respectively at electrodes with 10 and 2 nm radii. For electrodes with larger radii, the calculated electrostatic potential profiles (not shown) are very

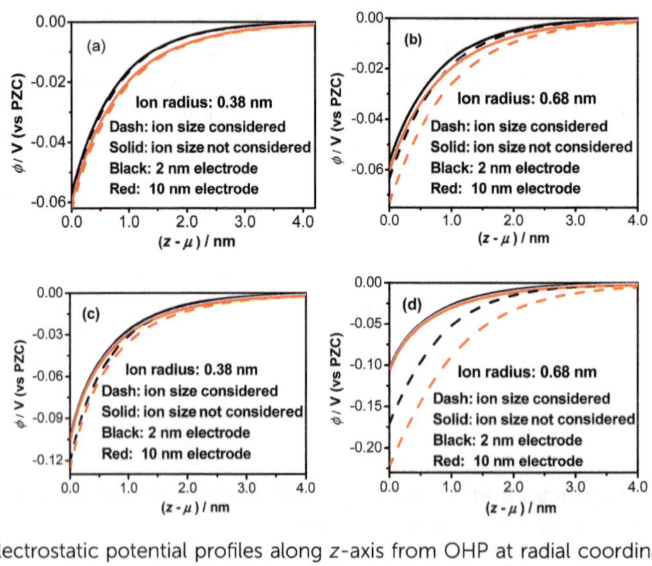

Fig. 2 Electrostatic potential profiles along *z*-axis from OHP at radial coordinate $r = 0$. Electrode potential: (a and b) -0.4 V; (c and d) -1.0 V. Solid and dashed lines: electrolyte ion sizes considered and not considered, respectively.

close to those at the 10 nm electrode, regardless of the electrolyte ion sizes. This is because the EDL structures at electrodes larger than 10 nm in radii are negligibly impacted with electrode sizes, as is known from previous studies.[1,4] However, the influence of the EDL structure on the mass transport dynamics and charge transfer kinetics would vary significantly with the electrode sizes as the electrode radii are below 1 μm; a more pronounced EDL effect should be seen at smaller electrodes, owing to the increased overlap between the EDLs and mass transport layers.[1,4] This also will be seen in the next section.

Only negative electrode potentials are considered because we consider the reduction reaction (1) in this study. It is found that the 0.38 nm size of electrolyte ions have a nearly invisible effect on the potential profiles in EDL (Fig. 2a and c) and the voltammetric responses of electrodes (shown in next section) are positive to −0.4 V. For 0.68 nm ion size, the electrostatic potential profiles are obviously affected by the finite size effect of the ion (Fig. 2b and d). The finite size of ions makes the electrostatic potentials at the OHP and that in the entire diffuse EDL (dashed line in the figures) higher in magnitude than those predicted by the conventional Poisson–Boltzmann theory which treats the electrolyte ions as point charges (solid lines in the figures). The point charge treatment, *i.e.*, ignoring the finite volume of ions, could overestimate the concentration of counter ions at and near the OHP, thus over-screen the electrode charge. This can be seen in Fig. 3. When the finite sizes of electrolyte ions are considered according to the modified Poisson–Boltzmann equations (eqn (5) and (8)), the calculated concentration of the counter ion (M^+ here) becomes apparently lower at OHP and in the diffuse EDL. The larger the ion size, the lower the concentration of counter ions, as given by the modified Poisson–Boltzmann equations.

An interesting phenomenon seen from Fig. 2 and 3 is that the effect of ion size becomes less pronounced at the 2 nm electrode than that at the 10 nm electrode.

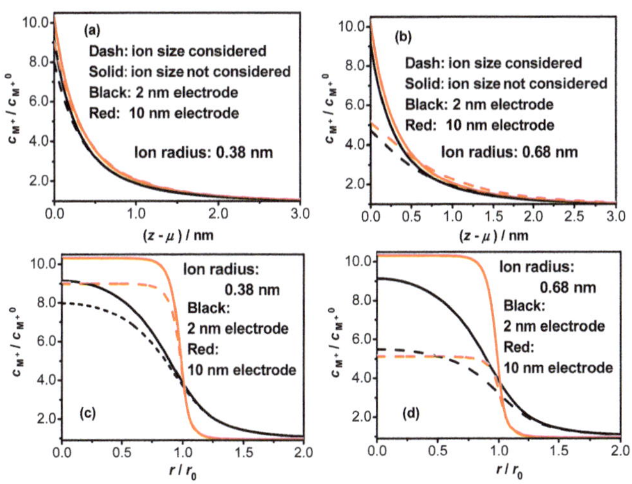

Fig. 3 Concentration profiles of M^+ ion along z-axis from OHP at radial coordinate $r = 0$ (a and b) and that along the r-axis at the OHP ($z = \mu$) (c and d) at electrode potential of −0.4 V. The concentration values are normalized by the corresponding value in the bulk solution. The other details are the same as those in Fig. 2.

The electrostatic potentials and counter ion concentrations calculated by the modified Poisson–Boltzmann equations (dashed lines) show relatively smaller deviation from those predicted by the conventional Poisson–Boltzmann equations (solid lines) at the 2 nm electrode (black lines) than that at the 10 nm electrode (red lines). It is somewhat counter intuitive that the smaller electrode exhibits a lesser size effect than the larger electrode. However, it becomes understandable from the radial concentration profiles of counter ions shown in Fig. 3c and d. One can see that the concentration profiles at the 2 nm electrode extend significantly to the region out of the electrode surface ($r > r_0$), which means that the crowding of the counter ions near the OHP can be relaxed at very small electrodes due to such a pronounced edge effect. The edge effect at small electrodes offers a large space to accommodate the counter ions that approach the OHP. Although the concentration of the counter ion at a given r coordinate value may be lower at the 2 nm electrode than that at the 10 nm electrode, a higher average concentration value can be obtained at the 2 nm electrode when first integrating the ion concentration over all r coordinates and then normalizing the integrated value with the electrode radius (or area). Thus, the electrode charge can be more pronouncedly screened by the counter ions at smaller electrode due to the electrode edge effect.

As shown in Fig. 4, the radial electrostatic potential profiles also exhibit a significant edge effect at the 2 nm electrode, similar to the ion concentration. Again, one can find that the 2 nm electrode exhibits a less pronounced ion size effect on the radial profile of the electrostatic potential.

Effect of finite ion sizes on the voltammetric responses of nanodisk electrodes

Fig. 5 shows the steady-state voltammetric responses of disk electrodes with radii from 1 μm to 2 nm for reaction (1) with $x = -1$ (anion reduction), calculated by solving the modified Poisson–Nernst–Planck (PNP) equations together with eqn (14) (considering both the EDL effect and the ion size effect), the conventional PNP equations together with eqn (14) (considering the EDL effect only), and the diffusion-based transport equations (eqn (15)) together with the normal BV equation (considering neither the EDL effect nor the ion size effect). For simplicity, we have assumed that all ions in the system have the same size when solving the modified PNP equations. It is noted that the voltammetric responses have been normalized by the diffusion limiting current (i_{dL}) on the same electrodes. We have found that the voltammetric responses for electrodes of various

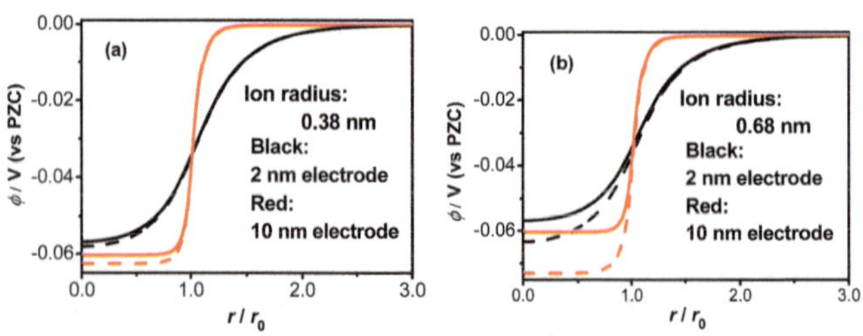

Fig. 4 Electrostatic potential profiles along the r-axis at the OHP ($z = \mu$) at an electrode potential of −0.4 V. The other details are the same as those in Fig. 2.

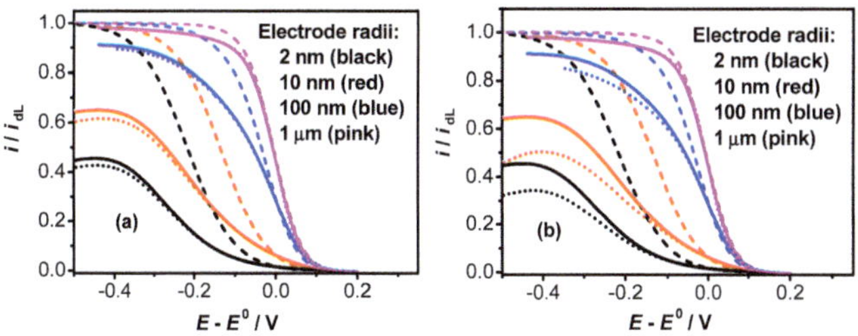

Fig. 5 Steady-state voltammetric responses of disk electrodes with different radii (distinguished by the coloured lines) for reduction of an anion with −1 charge, calculated with model including both the EDL and ion size effects (dotted lines), only the EDL effect (solid lines), and neither the EDL nor the ion size effect (dashed lines). The ion radii are (a) 0.38 nm and (b) 0.68 nm.

sizes obtained by solving eqn (15) are nearly identical to those obtained from the conventional diffusion equation without considering the ion size effect. This is because the ion size affects the voltammetric responses mainly by altering the EDL structure. When the EDL effect is trivial or it is not considered, the effect of the ion size on the voltammetric responses would be negligible.

The EDL effect decreases the current for the reduction of the anion reactant (comparing the solid and dash lines in Fig. 5). This is due to the mass transport of the reactant ions toward the OHP and the electron transfer kinetics between the electrode and the reactant ions at the OHP are both inhibited by the electrostatic potentials in the diffuse EDL.[1,3,4] As has been shown above, the finite size of ions would increase the magnitude of the electrostatic potential in the diffuse EDL due to the reduced screening of the electrode charge. Therefore, the ion size effect should result in further decline of the anion reduction current. This can be clearly seen in Fig. 5b for an ion with 0.68 nm radius. For an ion with 0.38 nm radius, a visible influence of ion size on the voltammetric responses is seen only at the electrode around 10 nm as the electrode potential is below −0.4 V (Fig. 5a).

The voltammetric responses of the 1 μm electrode show a slight EDL effect in the limiting current region (comparing the solid to the dotted lines), indicating that only the mass transport rates of the reactant ions are slightly affected by the diffuse EDL at such a large electrode. For electrodes with radii 100 nm and smaller, the voltammetric responses are significantly altered by the EDL effect, in both the limiting current and the kinetic regions, indicating that both the mass transport of ions and electron transfer kinetics are strongly affected by the EDL structures. When only the EDL effect is considered, one can see a simple variation trend that a more pronounced EDL effect occurs at smaller electrodes.

When the effect of the ion sizes is correlated with the electrode sizes, one can find different variation trends from Fig. 5. As is already known, the ion size effect will result in a further current diminishment in addition to the normal EDL effect (comparing the dotted and solid lines). For electrodes with radii between 1 μm and 10 nm, this current diminishment increases in magnitude with decreasing electrode radius, which is in consistency with the enhanced EDL effect. Upon further reducing the electrode radius from 10 nm, for example, to 2 nm, the current diminishment becomes smaller, although the ordinary EDL effect at the 2 nm electrode remains larger than that at the 10 nm electrode. The reasons for this should be the same as that discussed for the equilibrium EDL. That is, the pronounced edge effect at electrodes smaller than 10 nm can relax the ion crowding caused by the finite volumes.

We also have investigated the reduction of the +1-valent cation and the results are shown in Fig. 6. For cation reduction, the electrostatic potentials at the OHP and in the diffuse EDL would accelerate the mass transport of the reactant, while that at the OHP could inhibit the electron transfer kinetics. In most cases, the overall EDL effect should result in increased rates for cation reduction, but the magnitude of the current increase would be lower than the EDL-induced current decrease for anion reduction at electrodes of the same sizes. This can be seen through comparison of Fig. 6 with 5. Similar to the anion reduction, the EDL effect also becomes more pronounced at smaller electrodes for cation reduction.

One may expect that the EDL-induced current increase would be enhanced by the finite size effect of ions because of the less screened electrostatic interaction, as that seen for the anion reduction in Fig. 5. By comparing the dotted and solid lines in Fig. 6, however, it can be found that the current increase actually becomes less pronounced as the ion size effect is considered. This should be due to that the finite volumes of the redox ions make the actual concentration increase of cation reactant near the OHP lower than that estimated from the electrostatic interaction based on the continuum model. To manifest this effect more clearly, we compare the voltammetric responses of the 2 nm and 10 nm electrodes obtained by considering the sizes of both the electrolyte ions and redox ions and that by considering the sizes of the electrolyte ions but neglecting the sizes of the redox ions. As shown in Fig. 7, the current for cation reduction does show further increase when only the sizes of the electrolyte ions are considered (comparing the dotted and solid lines), which is consistent with the expectation of a magnitude increase of electrostatic potential in the diffuse EDL due to a lessened screening effect of electrolyte ions to the electrode charges. When the size effect of the redox ions is also considered, the current shows considerable diminishment (dashed lines), and the diminishment magnitude is larger at the 2 nm electrode. These results indicate that the EDL effect on the voltammetric responses for cation reduction would be significantly overestimated when ignoring the finite sizes of the redox ions. At very small electrodes (<10 nm radii), the electrode edge effect further weakens the EDL effect.

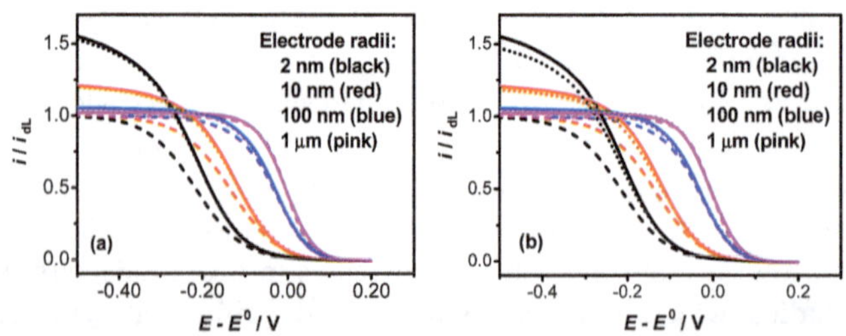

Fig. 6 Steady-state voltammetric responses of disk electrodes with different radii (distinguished by the line colours) for reduction of a cation with +1 charge. The ion radii are (a) 0.38 nm and (b) 0.68 nm respectively. The other details are the same as those in Fig. 5.

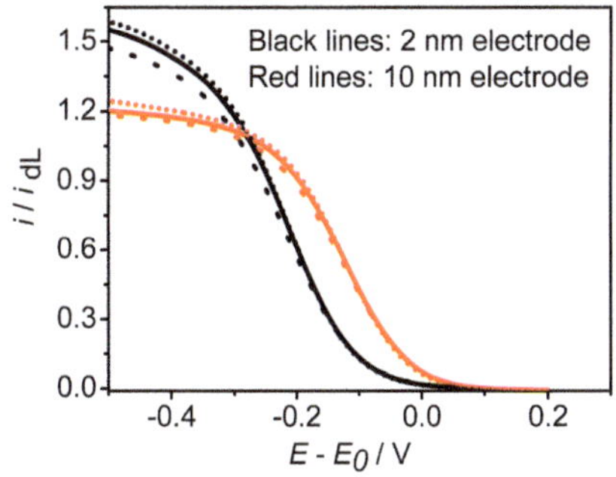

Fig. 7 Steady-state voltammetric responses of 10 and 2 nm electrodes for reduction of a cation with +1 charge, calculated with the model considering the EDL effect and no ion size effect (solid lines), the model considering the EDL effect and the size effect of the electrolyte ions (dotted lines), and the model considering the EDL effect and the size effects of both the electrolyte and redox ions (dotted lines). The ion radii are 0.68 nm.

Further discussion

The two ion radii considered here, namely 0.38 nm and 0.68 nm, could represent the sizes of commonly used aqueous electrolyte ions (*e.g.* solvated K^+) and organic electrolyte ions (*e.g.* solvated TEA^+), respectively. The concentration of 0.1 M considered here is also commonly employed for the supporting electrolyte in electrochemical measurements. For ions with 0.38 nm radius, as can be seen in Fig. 5a and 6a, the current deviation magnitude due to the ion size effect is actually very small, which could be within the errors of usual measurements. This seems to suggest that the effect of ion sizes on the voltammetric responses of nanoelectrodes in a usual aqueous system can be ignored. Therefore, one can simply use the conventional PNP equations to model the mass transport of redox ions. For 0.68 nm radius ions, the ion size could visibly affect the voltammetric responses of electrodes with radii smaller than 100 nm for anion reduction (Fig. 5b) and smaller than 10 nm for cation reduction (Fig. 6b); and in other cases, the ion size effects would be also negligible.

The magnitude of the ion size effects could be increased at some conditions that are not considered here. It is known from previous studies that the EDL effect would become more pronounced for reactions with lower k^0 values and those with E^0 values more deviating from the potential of zero charge (PZC).[1,3,4] In current study, we have considered the reactions with relatively facile kinetics ($k^0 = 1$ cm s^{-1}). For reactions with smaller k^0, one should expect a more pronounced ion size effect. We also have assumed that $E^0 = 0$ (*vs.* PZC) in the current study. Usually, the standard electrode potential of a reaction should deviate from the PZC, which makes the electrode surface relatively heavily charged. This can also result in pronounced ion crowding and EDL effects, and therefore increase the magnitude of the ion size effects. Finally, the electrochemical systems considered in the current study consist of a relatively low concentration of electrolyte (0.1 mol L^{-1}) in solvents. For systems dominated with electrolyte, for example, ionic liquids, molten salts and water-in-salts,[22] the ion size effects should become much more pronounced because the screening of electrode charges by solvent are absent. We are currently studying the ion size effect on EDL structure and reactivity at nanoelectrodes under these conditions and situations.

Conclusions

Modified Boltzmann and Nernst–Planck (NP) equations have been combined with the Poisson and modified Butler–Volmer equations to investigate the finite size effects of ions on the EDL structures and voltammetric responses of nanodisk electrodes. The results have shown that the ion size effect would become more pronounced for larger ions and at smaller electrodes as the electrode radii is larger than 10 nm. For electrodes with radii smaller than 10 nm, however, the ion size effect becomes less pronounced due to the increased edge effect of disk electrodes at nanometer scales, which could relax the ion crowding at/near the OHP. The effect of ion sizes on the voltammetric responses of nanoelectrodes in a usual aqueous electrochemical system may be ignored. Therefore, one can use conventional PNP equations to model the mass transport of redox ions. For 0.68 nm radius ion, which may represent some cations in organic electrolyte, the ion size may visibly affect the voltammetric response of electrodes with radii smaller than 100 nm for anion reduction and smaller than 10 nm for cation reduction. The EDL effect on the voltammetric responses for cation reduction would be significantly overestimated when ignoring the finite sizes of the redox ions; while it will be underestimated for anion reduction.

Acknowledgements

This work is supported by the Ministry of Science and Technology of China under the National Basic Research Program (2012CB932800) and the National Natural Science Foundation of China (21173162).

Notes and references

1 S. L. Chen, Y. W. Liu and J. X. Chen, *Chem. Soc. Rev.*, 2014, **43**, 5372–5386.
2 S. L. Chen and Y. W. Liu, *Phys. Chem. Chem. Phys.*, 2014, **16**, 635–652.
3 (*a*) J. D. Norton, H. S. White and S. W. Feldberg, *J. Phys. Chem.*, 1990, **94**, 6772–6780; (*b*) C. P. Smith and H. S. White, *Anal. Chem.*, 1993, **65**, 3343–3353; (*c*) E. J. F. Dickinson and R. G. Compton, *J. Phys. Chem. C*, 2009, **113**, 17585–17589.
4 (*a*) R. He, S. L. Chen, F. Yang and B. L. Wu, *J. Phys. Chem. B*, 2006, **110**, 3262–3270; (*b*) Y. W. Liu and S. L. Chen, *J. Phys. Chem. C*, 2012, **116**, 13594–13602; (*c*) Y. W. Liu, R. He, Q. F. Zhang and S. L. Chen, *J. Phys. Chem. C*, 2010, **114**, 10812–10822.
5 (*a*) J. J. Watkins, J. Y. Chen, H. S. White, H. D. Abruña, E. Maisonhaute and C. Amatore, *Anal. Chem.*, 2003, **75**, 3962–3971; (*b*) J. J. Watkins and H. S. White, *Langmuir*, 2004, **20**, 5474–5483; (*c*) J. L. Conyers and H. S. White, *Anal. Chem.*, 2000, **72**, 4441–4446.
6 D. Krapf, B. M. Quinn, M. Y. Wu, H. W. Zandbergen, C. Dekker and S. G. Lemay, *Nano Lett.*, 2006, **6**, 2531–2535.
7 Y. Sun, Y. W. Liu, Z. X. Liang, L. Xiong, A. L. Wang and S. L. Chen, *J. Phys. Chem. C*, 2009, **113**, 9878–9883.

8 (*a*) J. T. Cox and B. Zhang, *Annu. Rev. Anal. Chem.*, 2012, **5**, 253–272; (*b*) S. M. Oja, M. Wood and B. Zhang, *Anal. Chem.*, **85**, 473–486.

9 C. J. Slevin, N. J. Gray, J. V. Macpherson, M. A. Webb and P. R. Unwin, *Electrochem. Commun.*, 1999, **1**, 282–288.

10 B. Zhang, L. X. Fan, H. W. Zhong, Y. W. Liu and S. L. Chen, *J. Am. Chem. Soc.*, 2013, **135**, 10073–10080.

11 J. J. Bikerman, *Philos. Mag.*, 1942, **33**(220), 384–397.

12 I. Borukhov, D. Andelman and H. Orland, *Phys. Rev. Lett.*, 1997, **79**, 435–438.

13 A. A. Kornyshev, *J. Phys. Chem. B*, 2007, **111**, 5545–5557.

14 M. S. Kilic, M. Z. Bazant and A. Ajdari, *Phys. Rev. E: Stat., Nonlinear, Soft Matter Phys.*, 2007, **75**(2–1), 021503/1–021503/11.

15 P. H. R. Alijó, F. W. Tavares and E. C. Biscaia Jr, *Colloids Surf., A*, 2012, **412**, 29–35.

16 C. Liu and B. Eisenberg, *J. Phys. Chem. B*, 2012, **116**, 11422–11441.

17 W. Wang and S. Vince, Energy storage Redox flow batteries go organic, *Nat. Chem.*, 2016, **8**(3), 204–206.

18 J. Mindemark and L. Edman, *J. Mater. Chem. C*, 2016, **4**, 420–432.

19 C. C. Qi, Y. X. Hua, C. Y. Xu, J. Li, Q. B. Zhang and Y. Dylan Li, *General Review*, 2014, **14**(4), 694–707.

20 S. Kondrat, P. Wu, R. Qiao and A. A. Kornyshev, *Nat. Mater.*, 2014, **13**(4), 387–393.

21 V. Giordani, D. Tozier, H. J. Tan, C. M. Burke, B. M. Gallant, J. Uddin, J. R. Greer, B. D. McCloskey, G. V. Chase and D. Addison, *J. Am. Chem. Soc.*, 2016, **138**(8), 2656–2663.

22 L. M. Suo, O. Borodin, T. Gao, M. Olguin, J. Ho, X. L. Fan, C. Luo, C. S. Wang and K. Xu, *Science*, 2015, **350**(6263), 938–943.

23 H. N. Wang, A. Thiele and L. Pilon, *J. Phys. Chem. C*, 2013, **117**(36), 18286–18297.

24 S. Lamperski, C. W. Outhwaite and L. B. Bhuiyan, *Mol. Phys.*, 1996, **87**(5), 1049–1061.

25 M. Z. Bazant, M. S. Kilic, B. D. Storey and A. Ajdari, *Adv. Colloid Interface Sci.*, 2009, **152**, 48–88.

26 R. P. Buck, *J. Membr. Sci.*, 1984, **17**, 1–62.

Faraday Discussions

DISCUSSIONS

Reactions at the nanoscale: general discussion

Thom Hersbach, Julie MacPherson, Olaf Magnussen, Richard Crooks, Simon Higgins, David Fermin, Frederic Kanoufi, Wolfgang Schuhmann, Richard Nichols, Sushanta Mitra, Wolfgang Schmickler, Kristina Tschulik, Philip Bartlett, Sanli Faez, Wojciech Nogala, Michael Eikerling, Christine Kranz, Patrick Unwin, Marc Koper, Serge Lemay, Andrew Mount, Andrew Ewing, Zhongqun Tian, Henry White, Shengli Chen, Jan Clausmeyer and Katharina Krischer

DOI: 10.1039/C6FD90067D

Richard Nichols opened a general discussion of the paper by Olaf Magnussen: You state that perchlorate is not absorbed on bismuth. What is the evidence for this? For example is this apparent from the voltammetry?

Olaf Magnussen answered: Previous studies of Bi UPD on Au(111) and Au(100) have observed identical adlayer structures in perchloric and sulfuric acid solution and concluded that both anions are not specifically adsorbed or at least have no decisive influence on the UPD structure. For this reason, also no major anion influence is expected at the more negative potentials of the bulk deposition regime.

Wolfgang Schmickler asked: The bismuth ion is triply charged; it must have an energy of solvation of about 50 eV. So it is extremely difficult to shed this solvation sphere. How is this ion deposited? Does it form a complex with anions?

Olaf Magnussen replied: The chemistry of Bi complexes in aqueous solution is very rich. However, in the strongly acidic solution used in our experiments simple aquo complexes of Bi^{3+} with (six) water ligands should be the predominant species. RDE measurements of the charge transfer kinetics[1] report a stepwise charge transfer, but details on the ion transfer process are not understood to my knowledge.

1 I. Valsiūnas, L. Gudavičiūtė, V. Kapočius and A. Steponavičius, *Chemija*, 2006, **17**, 11.

Patrick Unwin asked: Are you able to determine kinetics from measurements at the nanoscale and do these scale up and relate to macroscopic measurements? Do you have a feeling for how big the inhibition of mass transport by the tip is?

Olaf Magnussen answered: The inhibition of Bi deposition is significant, approximately one to two orders of magnitude. For this reason we did not analyse the Bi growth rates quantitatively. However, for the quantitative analysis of the probability of a kink to be in one of the three different states this does not pose a problem, because the mass transport required for the latter is negligible (a change from a type B to a type A or a type C kink only requires addition of a single atom).

Marc Koper inquired: Can you say anything from the images about the nucleation of these structures?

Olaf Magnussen replied: As mentioned in the paper, nucleation of the Bi nanochains on Au(100) occurs predominantly at the edges of Bi islands, but can occasionally also be observed in the center of extended Bi terraces.

Kristina Tschulik asked: You occasionally observed the formation of Bi nanowires of well-defined orientation with respect to the substrate. Are these wires re-formed at the same position when you cycle the potential to sequentially dissolve and re-deposit them? If so this could hint towards a preferred nucleation of these wires at specific sites, *e.g.* point defects.

Olaf Magnussen answered: All Bi nanowires as well as the nanochains on Au(100) are epitaxial to the underlying substrate. Nucleation of these wires was indeed found to occur preferentially at surface defects, such as steps or domain boundaries in the Bi UPD layer (see also ref. 13 and 59 of the manuscript).

Zhongqun Tian remarked: You have demonstrated the power of this technique in terms of spatial and time resolution. I wonder how far can it move forward, especially regarding the sensitivity. The *in situ* monitoring of deposition growth is surely important, but for almost all practical applications in electroplating, additives are used to improve the smoothness and metal quality. Would it be possible to observe these or some additive adsorption during the deposition process?

Olaf Magnussen replied: Indeed, such insights are in principle obtainable. In previous video STM studies we have investigated the surface dynamic behaviour of anionic as well as small organic molecules, such as methyl thiolate, including the interaction of these species with Cu adatoms and steps.[1–3] However, similar studies of larger organic adsorbates are challenging at present. In this case low tunneling currents of ≤100 pA are required for STM imaging, in order to avoid significant modifications of the surface dynamics by the STM tip. In our current video STM the high bandwidth of 600 kHz effectively limits the average tunneling current to values of ≥1 nA. Lowering this value should be possible, but will require significant improvements in the design of the STM head and preamplifier.

1 Y.-C. Yang, A. Taranovskyy and O. M. Magnussen, *Angew. Chem. Int. Ed.*, 2012, **51**, 1966.
2 Y.-C. Yang, A. Taranovskyy and O. M. Magnussen, *Langmuir*, 2012, **48**, 14143.
3 Y.-C. Yang and O. M. Magnussen, *J. Electroanal. Chem.*, 2014, **716**, 80.

Serge Lemay opened a general discussion of the paper by Katharina Krischer: The power spectrum for a stochastic signal consisting of uncorrelated exponentially decaying pulses has the form $S(f) = S_0/(1+(2\pi f/a)^2)$, where a is the decay constant of the exponential. This universally yields a $1/f^2$ decay in $S(f)$ at high enough frequencies; the main piece of quantitative information that can be extracted from the spectrum is the value of a, which determines the frequency below which the spectrum deviates from the inverse square law. With this in mind, it seems that the observation of a nearly $1/f^2$ spectrum in the data presented here simply reflects the shape of the individual impulses, and that the more interesting dynamics may occur on longer time scales. However, the corresponding frequency range (below 10^{-1} Hz) is not studied in the manuscript. What are the prospects for obtaining sufficiently long time traces so that these ultra-low frequencies can also be analyzed?

Katharina Krischer answered: This is a very good point. If you look carefully at the time series in the manuscript, you observe that they exhibit a slow drift. This unfortunately restricts the measuring time, and for this reason we refrained from evaluating the spectra at lower frequency in more detail. Still, we observed that the power spectra do not exhibit any preferred frequency. Hence, it is unlikely that any of the three types of noisy or spiky time series are related to an underlying oscillatory dynamics. We argue that instead the fluctuations are induced by molecular noise that is amplified to a mesoscopic or macroscopic level. In our view, this is a quite remarkable feature. However, it is also clear that further studies, in particular stochastic simulations, are necessary to understand the dynamic features of the system better.

Wolfgang Schmickler commented: The critical size of 25 μm is quite large on an atomic scale. Can it be related to any other characteristic length in your system?

Katharina Krischer replied: Yes, the characteristic domain radius depends on an interaction between the autocatalytic surface reaction and diffusion of CO in front of the electrode. From the theory of nonlinear dynamics it is known that the critical radius of a circular domain can be estimated as $\frac{D}{\sqrt{Dk}}$, where D is the diffusion constant and k is the characteristic rate constant of the autocatalytic reaction. In our case, the 'autocatalytic reaction' is really a network of reactions and is not related in a simple manner to an individual rate constant entering the model. With the given D and k of approximately 1 s^{-1}, which seems to be a realistic value, a rough estimation of the critical radius of a domain amounts to about 30 microns. Note that this is only a rough estimate, and for microelectrodes the critical size depends also on the size of the microelectrode since diffusion becomes the more effective the smaller the electrode is. This trend is captured correctly in our simulations. Note that in the ms we estimate the critical size of the nucleus from the width of the interfacial region of a planar interface between a CO covered and a bare electrode, which is proportional to $\sqrt{\frac{D}{k}}$. This gives the same estimation.

Marc Koper asked: What would happen if instead of one microelectrode you would have an array of microelectrodes? Would you see phase separation as on one macro-sized electrode? Would you expect any spatial or temporal self-organisation?

Katharina Krischer replied: This is an interesting question. When one performs a slow galvanodynamic scan with an array of microelectrodes, one electrode after the other is ignited, leading thus at intermediate current densities to a partitioning of the array into active electrodes and CO covered passive ones. One of those electrodes might also exhibit an intra-electrode phase separation as in the present example. For an intermediate, constant current density, the response might be even more complex: under some conditions, inter-electrode switching is observed, such that some electrodes exhibit more or less regular transitions from an active state to a passive state and back. Furthermore, especially at relatively high current densities, *i.e.* when nearly all electrodes are activated, a Hopf bifurcation might occur resulting in true limit cycle behavior. Note that these oscillations are only due to the global coupling of the array of bistable electrodes. Hence, there is no feedback in the reaction mechanism that causes the oscillatory behavior. In fact, before we did the present study of the intra-electrode dynamics we studied the inter-electrode dynamics of an array of four microelectrodes. Some of what I explained can be found in our publications.[1,2] Finally, I should note that all dynamics I discussed refers to an array configuration and to simulations where the global coupling through the galvanostatic mode is dominating. When the electrodes are placed closer together, diffusional coupling between the electrodes also occurs. I would speculate that the dynamics will still be dominated by the global coupling, but there are so far no experiments or simulations that support this hypothesis. Hence, it is possible that the spatio-temporal behavior becomes even more complex.

1 D. A. Crespo-Yapur, A. Bonnefont, R. Schuster, K. Krischer and E. R. Savinova, *ChemPhysChem*, 2013, **14**(6), 1117–1121.
2 D. A. Crespo-Yapur, A. Bonnefont, R. Schuster, K. Krischer and E. R. Savinova, *ChemElectroChem*, 2014, **1**(6), 1046–1056.

Marc Koper said: How do you prepare your micro-electrodes, and how do you verify their cleanliness? We find that Pt ultramicro-electrodes tend to be too contaminated to study very structure sensitive electrocatalytic reactions.[1]

1 L. Jacobse, S. J. Raaijman and M. T. M. Koper, *Phys. Chem. Chem. Phys.*, 2016, **18**, 28451–28457.

Katharina Krischer answered: The preparation of the microelectrodes is certainly the most critical point in the experiments. Therefore, let me describe it in detail. However, before doing so I would like to note that in CO electrooxidation experiments the strong adsorption of CO on the electrode surface protects the surface from other impurities such that the problem here might be somewhat less severe than with other electrocatalytic reactions. As for the preparation of the electrode: The microelectrode surface is first polished on an adhesive polishing disk (PRESI, reference: 00162160) with a mixture of diamond suspension REFLEX LDP 1μ (PRESI, reference: 00242610) and Lubricant REFLEX LUB (water based lubricant, PRESI, reference: 00242700). The polishing is done by describing a figure 8 with the microelectrode, after 10 cycles, the electrode is rotated by 90° and the procedure is repeated again until the electrode comes back to its starting position. At the end the microelectrode is rinsed under a flow of ultrapure water for 1 minute.

The same procedure is repeated with a mixture of diamond suspension REFLEX LDP 1/4μ (PRESI, reference: 00242600) and Lubricant REFLEX LUB (water based lubricant, PRESI, reference: 00242700). After rinsing with water, the electrode is put into a small beaker (10 mL) containing the electrolyte for 5 minutes, then the electrode is taken out of the beaker, rinsed with electrolyte (50 mL), and again rinsed with a flow of ultrapure water for 1 minute.

The complete set of tools (beakers, pipettes, flasks, cells, corks…) used to prepare or contain electrolyte is put once a week overnight into a mixture of 1 : 1 sulfuric acid (96% Merck) and hydrogen peroxide (32% VWR), then rinsed with water (8 times) and subsequently filled with water for 4 h and rinsed again 8 times with ultrapure water. All the material is filled with water when it is not in use to minimise contamination. Before each use the material is rinsed again 8 times with ultrapure water.

Finally the microelectrode is inserted into our cylindrical glass cell that has been previously filled with electrolyte purged with N_2 (4.7 N). A first set of 3 cyclic voltammograms (50 mV s^{-1}; starting at 1 V going to 1.32 V then going down to 0.07 V and finishing at 1 V) is done to see the current range of the CV. If it fits with the expected values (for a 10 μm diameter electrode the current values are between 0.3 and −0.3 nA) then the electrochemical cleaning is started, if not, the electrode is again polished. The electrochemical cleaning consists of 250 cycles at 1 V s^{-1}; starting from 1 V going to 1.32 V then going down to 0.07 V and finishing at 1 V. The final shape of the CV is checked with 3 more cycles at 50 mV s^{-1}, where we carefully investigate the shapes of the H-upd peaks. If the peaks are well defined, then the electrolyte is saturated with CO, if not the electrode is polished again. In the CO saturated electrolyte, 20 cyclic voltammograms (50 mV s^{-1}; starting at 0.07 V going to 1.00 V then going down to 0.07 V and finishing at 0.07 V) are done to observe a stable CO diffusion limiting current and a sharp ignition peak. If the CVs are not stable, the electrode is sent back to polishing, if the CVs are stable, then the actual experiments are started.

Frederic Kanoufi asked: You are mentioning that the process is ruled by a characteristic length of the order of 10 micrometers, characteristic of an autocatalytic process whose apparent kinetics is then of the order of 10^3 s^{-1}. It is very nice that you proposed a deterministic simulation of the process, even though in 2D only. The simulation qualitatively reproduces the experimental bistability. Looking at the numbers you are using for this simulation, it appears that the desorption rate that you have chosen is of the order of 10^3 s^{-1}. I wonder if it is actually this desorption step which is rate determining. Is it maybe fortuitious? I wonder if this is fortuitious or if the same simulations can be obtained by changing k_{des} (for example to 1 order of magnitude higher)? Moreover, one has wondered if the bistability could be due to edge effects. I wonder if you could detect edge effects from your simulation (the edge of your insulating–UME boundary, although in 2D). We have indeed previously demonstrated that heterogeneous surface coverage could be detected both optically and by simulation at microelectrodes (edge *vs* center).[1] Would local monitoring (optically for example) be possible or helpful in your experimental situation?

1 S. Munteanu, J. P. Roger, Y. Fedala, F. Amiot, C. Combellas, G. Tessier and F. Kanoufi, *Faraday Discuss.*, 2013, **164**, 241–258.

Katharina Krischer responded: Desorption does not play a role for the size of the critical nucleus. It is only taken into account to prevent total poisoning of the surface, *i.e.* a CO surface coverage of 1, which would not allow any reaction to occur. Since in experiments, CO oxidation occurs also when the CO adlayer is saturated, we allowed for some desorption of adsorbed CO molecules. Alternatively, we could have assumed that the maximum CO coverage is less than 1, *e.g.* 0.99, and omitted CO desorption completely. This still gives rise to a bistability. The characteristic length is determined by an interaction of reaction kinetics and diffusional coupling. The reaction rate in front of the bare domain is diffusion limited, and thus the CO concentration in front of the electrode is very small. In contrast, the reaction rate on the CO covered patch is negligibly small; the CO concentration in front of a CO covered surface is thus close to the bulk CO concentration. Consequently, CO molecules adjacent to the electrode diffuse from the CO covered domain to the bare surface, where they may adsorb and react. This is the principal mechanism that adjusts the width of the interface between a CO covered and a reactive domain, and also the critical nucleus. Note that compared to diffusion in the electrolyte, surface diffusion of adsorbed CO is much slower. Therefore, we believe that it does not contribute significantly to the interface width or the critical size of a nucleus. Hence, as answered to Prof. Schmickler's question a rough estimate of the radius of a critical nucleus is given by a combination of the diffusion coefficient and the reaction constant of CO oxidation. As for the second part of your question: our simulations assumed a semicircular electrode surrounded by electrolyte. Hence, the simulation does not take into account the UME–insulator transition. Still, the simulations reproduce the spontaneous formation of domains on the electrode surface. Therefore, they suggest that the domain formation is intrinsic to the dynamics and not induced by the edges. On the other hand, in the experiments the edges impose inhomogeneous parameter distributions and might have an influence where the domains nucleate. It would certainly be interesting to study the local CO coverage on the UME. However, optical imaging does not give any contrast. IR imaging would give the contrast, but it brings about further experimental difficulties. The only IR technique I am aware of that allows for spatial resolution is SEIRAS measurements,[1] and here the spatial resolution is on the order of some 10 microns, and thus not sufficient in our case.

1 P. R. Bauer, A. Bonnefont and K. Krischer, *Sci. Rep.*, 2015, **5**, 16312.

Philip Bartlett remarked: The current distribution at the microelectrode can be non-uniform and at the edge there is a change in the double layer between that over the platinum and that over the glass insulation. As the microelectrode becomes smaller these edge effects might be expected to become more significant. Do you see any influence from the boundary between the platinum disc and the glass insulation in your experiments?

Katharina Krischer answered: *A priori*, it is not possible to see in the experiments whether the metal electrode–insulator transition has an impact on the phase transition we observe. However, in our simulations we assume a hemispherical metal electrode surrounded by electrolyte (except for the base plane). Hence, in the simulations we do not have inhomogeneities due to mixed boundaries. Since the inhomogeneous phase transition and the critical size of the nucleus as well as its dependence on the density of active surface sites are all well captured by the simulations, it is not necessary to consider the influence of the non-uniformities at the edge of the microelectrode to explain these phenomena. However, I could imagine that edge effects determine where a domain nucleates. It would certainly be interesting to do simulations with the mixed boundaries and compare them with those of the hemispherical electrode.

Kristina Tschulik commented: You mentioned that you observe an increase in the frequency with time. Do you assume that the origin of this effect is the co-adsorption of impurities from the electrolyte or that this is caused by the repeating phase transition in the course of the experiment? Both causes might be distinguished from each other by immersing the electrode in the electrolyte for several minutes prior to starting the experiment. If an increased frequency is observed then the first scenario seems likely, otherwise the second would likely be the origin.

Do you expect any influence of surface charge of the insulation sheath on the phase transition? If so the application of an electrophoretic paint instead of a glass sheath might be tested usefully.

Katharina Krischer responded: I would assume that the increase in frequency of, *e.g.*, the shot-noise-like spikes are due to a slow contamination of the electrode surface. But it would certainly be interesting to test it in the way you suggested, thanks for this comment. Concerning the influence of surface charge of the insulation sheath on the phase transition, I expect it to be of minor importance. It might determine where the bare electrode phase nucleates, but I do not expect any effect on the qualitative observations we report in the paper.

Richard Crooks asked: To what extent do you think solution impurities affect the types of measurements you make? I understand that you take precautions in this regard before experiments begin (*e.g.*, cycling the electrode until the signature of a clean Pt surface is observed), but no matter how much effort one puts into purifying water, there is still organic material present that can affect sensitive experiments like yours if it adsorbs to the surface AFTER initial electrode preparation. Could you comment on this possibility and the effect it might have on your observations?

Katharina Krischer answered: This is a very important aspect. It is clear that there will be always some contamination during the measurement. If you look, *e.g.*, at the time series measurements in our paper (Fig. 7 and 6), you observe a slow increase in potential with time. We attribute this increase to a slow and minor contamination of the surface (though it might also be due to the formation of Pt oxides with a higher valency that have a lower reactivity towards CO oxidation). The potential increase tends to be less pronounced the lower the current is, *i.e.* the larger the CO coverage is, which makes sense. However, looking at the figures it also becomes clear that the qualitative effects are not affected by the contamination: the critical electrode size clearly depends on the sulfuric acid concentration (compare *e.g.* the time series in Fig. 6a and Fig. 7a,b); the potential transients become noisy as soon as phase separation occurs, and so on. The main impact of contamination that I see on our measurements is that we are restricted in the length of the time series we can record. It would certainly be desirable to study the power spectra at lower frequencies since the different types of noise should have a different signature in the low frequency tail. But this is not possible due to the influence of minor contaminations.

Marc Koper asked: In relation to Dick Crooks' comment about electrolyte impurities: what would happen if you add small amounts of chloride to the electrolyte?

Katharina Krischer responded: In fact, the addition of chloride introduces a negative feedback into the reaction network rendering the local (or lumped) dynamics oscillatory.[1,2] Of course, whether oscillations are observed or not depends on the parameters of the system, most importantly the chloride concentration. We have just started to study the dynamics of CO oxidation in the presence of chloride on microelectrodes. The behavior is rather complex and we are still in the middle of our studies. But as a first, rough description I would say that for very low chloride concentrations one still observes the formation of two phases and corresponding fluctuations. At higher chloride concentrations, in addition to the random behavior, characteristic frequencies show up. Also the low frequency boundary up to which the $1/f^2$ scaling is observed is shifted to higher frequencies so that in this case it seems to be possible to evaluate it. But for the moment it is too early to make any further statements.

1 S. Malkhandi, A. Bonnefont and K. Krischer, *Surf. Sci.*, 2009, **603**(10–12), 1646–1651.
2 S. Malkhandi, P. R. Bauer, A. Bonnefont and K. Krischer, *Catal. Today*, 2013; **202**, 144–153.

Julie MacPherson opened a general discussion of the paper by Thom Hersbach: Given the importance of cation adsorption in the processes described, what is the role of cation size on the adsorption process, *e.g.* Li^+ versus Cs^+? Have you looked at anything other than Na^+? What do the DFT adsorption energy models predict?

Thom Hersbach responded: We have not explored any other cations in DFT. However, we are currently working on experiments that involve other cations, which show that the corrosion behavior can change when varying the cation. For example, initial experiments indicate that the corrosion preference of gold can be moved towards (100) when lithium is present in the solution.

Wolfgang Schuhmann asked: You used the term "onset potential" which is a physico-chemical undefined expression of a potential at which a reaction starts. Could you explain the meaning in your context? Which better expression would you propose?

Thom Hersbach replied: In our context, the onset potential is simply the earliest potential at which we can detect changes in the electrode surface. In other words, it is the least negative potential at which we can detect cathodic corrosion. The determination of this potential is limited by the accuracy of our equipment and our ability to reliably identify differences between voltammograms. This is the reason why we defined the onset potential with a 0.1 V accuracy and did not choose to use a more well-defined expression such as the standard equilibrium potential. But perhaps the most important limitation is that determining such a formally well-defined potential requires a lot more general understanding of cathodic corrosion than we have right now.

Wolfgang Schmickler remarked: During the 1980s the group of Arvia published a series of papers in which they showed that the surface of platinum can be modified by rapid cycling. Depending on the cycling rates and the potential range, they obtained different surface structures. Can your work be related to that of Arvia?

Thom Hersbach responded: Our work can certainly be related to that of the group of Arvia, in the sense that they were able to use an electrochemical method to roughen and preferentially orient an electrode. These electrodes even exhibit features that we observe on our electrodes, such as particles and well-defined etch pits. Arvia's approach differs from ours in the procedure and underlying chemistry: typically, a cycling procedure is used in which potentials above 0 V *vs.* RHE are applied. Since the upper potential limit is typically high, the underlying chemical processes are attributed to formation and reduction of various platinum oxide and hydroxide phases. In contrast, we modify our electrodes by applying constant potentials below 0 V *vs.* RHE. The underlying chemistry is thus strictly cathodic in our case, which is an important difference with Arvia's work.

Michael Eikerling asked: Coming back to the question asked before by Wolfgang Schuhmann, could you determine a proper equilibrium potential of the process instead of referring to a vaguely defined "onset potential"? What is limiting you to find the equilibrium potential?

Thom Hersbach responded: There are two important factors that prevent us from defining us an equilibrium potential, which is actually something we addressed in our publication on the cathodic corrosion of platinum.[1] First of all, one needs to know the identity of all involved reaction species. Currently, the exact nature of the corrosion intermediates is still one of the largest open questions regarding cathodic corrosion.

Secondly, after identifying all reaction species, one would need to accurately determine their concentrations. The main reaction intermediate seems to be so unstable that it decomposes instantly. This makes its concentration ill-defined and hard to measure, since one cannot simply wait until equilibrium establishes and then determine the concentration of all species.

Thus, a standard equilibrium potential cannot be determined until we learn more about these two key issues.

1 T. J. P. Hersbach, A. I. Yanson and Marc T. M. Koper, *Nat. Commun.*, 2016, 7, 12653.

Marc Koper commented: As a clarification: during the cathodic treatment, there is a fixed negative potential, and at no moment there would be any surface oxidation. The voltammetry shown is taken in sulfuric acid after the cathodic treatment in alkaline solution, and is used to characterize the surface structure obtained from the cathodic treatment.

Richard Crooks questioned: With regard to cathodic corrosion, could the corroding species be single ion species such as Au^-?

Thom Hersbach responded: We typically describe the corrosion intermediates as anions, without specifying the exact nature of these anions. Though we do not know the exact nature of these intermediates, we like to think of them negatively charged metal clusters. Such clusters might be able to stabilize the negative charge amongst the atoms of which they are composed, much like post-transition metal polyanions found in Zintl phases. By doing so, they might be more stable than single ion species.

However, single ion species are still valid candidates for the unknown reaction intermediate. In fact, the existence of single transition metal ions such as Au^- and Ag^- has been reported in non-aqueous solvents.[1,2] Similarly, Pt^{2-} can be prepared in air- and moisture-free environments.[3] So even though we simply describe the intermediates as unknown anions and like to think of them as clusters, we cannot definitively point towards one species to be the mysterious anionic intermediate; metastable single ion species are certainly candidates we are considering.

1 W. J. Peer and J. J. Lagowski, *J. Am. Chem. Soc.*, 1978, **100**, 6260–6261.
2 N. E. Tran and J. J. Lagowski, *Inorg. Chem.*, 2001, **40**, 1067–1068.
3 A. Karpov, J. Nuss, U. Wedig and M. Jansen, *Angew. Chem., Int. Ed.*, 2003, **42**, 4818–4821.

Frederic Kanoufi said: Indeed the formation Au^- anions has been demonstrated by Allen Bard in the late 1970s.[1] However this anion is formed at very negative potentials (close to that of solvated electron formation). Would you suggest that it could be responsible for Au cathodic corrosion in your alkaline aqueous solution?

Regarding cathodic corrosion of metals in non-aqueous solvents, the corrosion of Pt seems to proceed via the generation of negative Pt species as proposed by Simonet .[2] The role of cations (intercalation products) has also been evidenced. How would this compare to the corrosion of Au or Rh?

Then Pt corrosion was also evidenced in an aqueous environment; the role of reactive oxygen species, generated by reduction of oxygen, is of particular importance.[3] Does O_2 have any influence on your Au or Rh cathodic corrosion?

1 T. H. Teherani, W. J. Peer, J. J. Lagowski and A. J. Bard, *J. Am. Chem. Soc.*, 1978, **100**, 7768–7770.
2 J. Ghilane, C. Lagrost, M. Guilloux-Viry, J. Simonet, M. Delamar, C. Mangeney and P. Hapiot, *J. Phys. Chem. C*, 2007, **111**, 5701–5707.
3 J.-M. Noël, Y. Yu and Michael V. Mirkin, *Langmuir*, 2013, **29**, 1346–1350.

Thom Hersbach answered: The potential for the formation of gold anions in ammonia is an important consideration concerning the identification of the cathodic corrosion intermediate. At a first glance, the standard potential of 2.3 V *vs.* $Ag/AgNO_3$ that Bard *et al.* report does not seem that far off from the onset potential of −1.6 V *vs.* NHE that we found for gold corrosion. In fact, this onset potential is undoubtedly more positive than the true equilibrium that underlies cathodic corrosion, due to the low concentration of intermediate species that likely is present.

Still, it would be better to calculate the standard electrode potential for the formation of Au^- in water. Unfortunately, calculating this potential using classical thermodynamic formulas is challenging without running simulations; it is difficult to accurately calculate the solvation energy of an Au^- from known thermodynamic values. So, while the equilibrium potential of Au^- is a valid concern, we currently do not understand our system well enough to determine this value in order to definitively rule out or confirm Au^- as an intermediate.

Your second question regards the intercalation of cations in the bulk metal. Though these modifications do occur at potentials similar to our cathodic corrosion potentials, it seems counterintuitive for this to be the reason behind cathodic corrosion; for corrosion to occur through this mechanism, it would require incorporation and subsequent leaching of a cation at the same potential. Additionally, the shape of the reported iono-platinic phases seems irregular and reduction of these phases is reported to recover the original platinum structure.[1] These observations contrast with the anisotropic shape and stability of the features created by cathodic corrosion. So though the same initial reduction of metal may underlie cathodic corrosion and the formation of these iono-platinic phases, I think it is unlikely that these phenomena have similar reaction mechanisms.

The final question relates to the role of oxygen in our system. We purge our electrolytes with 6.0 purity argon before all experiments for at least 30 minutes. Therefore, there should be no influence of oxygen in our corrosion experiments.

1 J. Ghilane, M. Guilloux-Viry, C. Lagrost, P. Hapiot and J. Simonet, *J. Phys. Chem. B.*, 2005, **109**, 14925–14931.

Simon Higgins asked: I wonder if it would be possible to use cathodic corrosion as a synthesis tool, by incorporating a suitable ligand (*e.g.* a phosphite or CO) into the electrolyte to coordinate to very low-valent metal centres, although perhaps also it would be necessary to change the solvent. If it's not possible to do the latter, one candidate might be cyanide; in the 1960's there was a lot of work on using cyanide as a ligand for surprisingly low oxidation state metal complexes in liquid ammonia, for example $[Ni(CN)_4]^{4-}$ and even (if I remember correctly) $[Ti(CN)_4]^{4-}$. Could you comment on this from your experience of the phenomenon?

Thom Hersbach responded: That sounds like an interesting strategy. Though I do not have any experience with creating metal complexes by using cathodic corrosion, I can comment on several requirements and possible pitfalls related to this strategy.

First of all, the success of this procedure will strongly depend on the nature of the cathodic corrosion intermediate; one will only be able to obtain single metal complexes if the intermediate species is a single ion, as opposed to a polyanion. Secondly, the charge of the metal center during coordination of the ligands is important. Negatively charged intermediates will likely repel negatively charged ligands, making such ligands unsuitable for capture of the metastable corrosion intermediate. However, it might be possible to capture single metal atoms in the stage between reduction of the anion and clustering of the atoms into nanoparticles. This rules out solvents like liquid ammonia, since they will likely stabilize the negative oxidation state of the metal.

Finally, the ligand will have to be stable at the reductive potentials that are related to cathodic corrosion. Kinetic stability might be sufficient, such that one could choose a ligand for which the reduction is catalyzed poorly on the metal that is to be captured.

Taking all of this into account and assuming I were to attempt a one-step synthesis, I would work in a CO-saturated water solution if the cathodic corrosion intermediate species are indeed single ions. Water, on the one hand, will facilitate oxidation of the intermediate to the zerovalent state. CO, on the other hand, is uncharged and should be able to approach the intermediate anion. This might improve the chances of capturing the ion before or immediately after its reduction.

Wolfgang Schmickler remarked: Since the cations play an important role, would it not be useful to investigate systematically the alkali ions, starting from Li and going down the periodic table?

Thom Hersbach replied: This would indeed be very useful and this is a project that we are currently working on.

Olaf Magnussen asked: Does the electrode microstructure, in particular grain boundaries, play an important role in this process? A related cathodic corrosion phenomenon is hydrogen embrittlement and here grain boundaries play a important role. Studies of Pt single crystals may help to provide insight on the mechanism, in particular on the role of undercoordinated metal sites.

Thom Hersbach replied: The electrode microstructure does seem to play a role in the corrosion process, since the corrosion features seem to be more pronounced around grain boundaries. However, these features are still visible all over the electrode, which is even the case on platinum.[1] Platinum can be flame annealed before the corrosion experiments and thus has significantly less grain boundaries than the gold and rhodium electrodes used in this study. So though corrosion seems to be easier around grain boundaries, these boundaries may not be as important for cathodic corrosion as they are in hydrogen embrittlement, where they facilitate diffusion of hydrogen into the bulk metal.

Nonetheless, single crystal studies would generate important insights into cathodic corrosion. At this point we have not attempted these studies yet; the experiments may lead to the destruction of the single crystals, which are quite expensive.

1 T. J. P. Hersbach, A. I. Yanson and Marc T. M. Koper, *Nat. Commun.*, 2016, 7, 12653.

Zhongqun Tian commented: The mechanism of the cathodic corrosion and the interfacial components are very complicated and it seems to be necessary to get the surface information at the molecular level. However, due to severe hydrogen bubble evolution from the probed surface, most surface spectroscopic methods cannot be applied. The most suitable way could be the use of ATR configuration-based optical measurements. The gold film electrode can be used for plasmon-enhanced IR and Raman spectroscopy. In addition to the fundamental aspect, what is the advantage in practical applications for cathodic corrosion in comparison with the traditional means, *e.g.*, anodic processes for roughening Au and Pt electrodes?

Thom Hersbach answered: If the cathodic corrosion intermediates are IR- or Raman-active, these spectroscopy techniques could be quite valuable. Regarding your second question: in my opinion, the main benefit of cathodic corrosion would be the surface selectivity. In case of the platinum electrodes we studied earlier, we are able to remove almost all (110)-type sites. A similarly pronounced selectivity might be possible for gold, rhodium and other metals if the right conditions are found. In that sense, cathodic corrosion might prove to be a versatile surface modification technique.

Frederic Kanoufi addressed Thom Hersbach and Olaf Magnussen: Continuing the comment of Professor Tian regarding optical *in situ* monitoring, the dynamics of the cathodic corrosion of Au and Rh seems rather slow and would fit the image acquisition rate of *in-situ* STM described in Magnussen's talk. Would such *in-situ* STM imaging be possible?

Thom Hersbach replied: We have considered doing *in-situ* STM during cathodic corrosion. STM would provide valuable information, but *in-situ* STM is currently hindered by experimental complexities that are introduced when working in an STM cell. For example, one would have to switch between 10 M NaOH and 0.5 M H_2SO_4 in the same cell, which may introduce contamination and safety issues.[1] We have therefore not attempted STM while using the approach that we have used so far to study corrosion.

One could consider doing true *in operando* STM while cathodically corroding the electrode, but this is difficult due to the large amounts of hydrogen that are formed during cathodic corrosion. This hydrogen will be reduced on the STM tip and interfere with the measurements. So, though STM would be quite interesting, challenges in the experimental design have so far prohibited us from carrying out such experiments.

1 T. J. P. Hersbach, A. I. Yanson and M. T. M. Koper, *Nat. Commun.*, 2016, 7, 12653.

Olaf Magnussen replied: Because cathodic corrosion occurs deep in the hydrogen evolution regime, *in situ* STM studies of this process might be difficult. First, this reaction might lead to very unstable imaging conditions and, second, hydrogen bubbles may be trapped between tip and sample, resulting in blocking access of the electrolyte to the imaged surface area.

David Fermin asked: Is it possible to use mass spectrometry in order to detect the intermediate species generated during cathodic corrosion, prior to the formation of nanoparticles? Perhaps in a configuration similar to differential electrochemical mass spectrometry.

Thom Hersbach responded: So far, the main challenge in identifying this intermediate species has been its instability. Mass spectrometry might alleviate this problem if the intermediate species is stable in the vacuum under which the spectrometer operates. Proper experimental design will be vital in such an experiment, since one would not want the intermediate to decompose before entering the mass spectrometry system. Additionally, transport of the intermediate species through the membrane and into the mass spectrometer is likely not trivial. However, mass spectrometry might be an interesting tool for analyzing the intermediate species *in-situ*, if such a setup is made compatible with the corrosion process.

David Fermin opened a general discussion of the papers by Olaf Magnussen, Katharina Krischer and Thom Hersbach: All these papers deal with phase formation and dissolution processes. What is the role of water and ions in these processes? It is clear that cations play a role in cathodic corrosion, while anions contribute to the size of the CO phases at the micro-electrode. Is water a spectator?

Thom Hersbach answered: Water is certainly not a spectator species during cathodic corrosion. In fact, it might be a special enabler of the process because it possesses two important properties. Firstly, water seems to allow for the bulk metallic electrode to be reduced and dissolved into a layer in the solution that is practically water-free because of the vigorous hydrogen evolution that takes place during cathodic corrosion.[1] Secondly, water is able to re-oxidize these reduced metal clusters when they encounter free water. Thus, water allows for electrode material to dissolve and then precipitate. Solvents that only possess the first property, such as ammonia, will simply dissolve stable metal anions like Au^{-}. In contrast, solvents that lack the first property will not be able to dissolve the electrode at all.

1 A. I. Yanson, P. V. Antonov, P. Rodriguez and M. T. M. Koper, *Electrochim. Acta*, 2016, **112**, 913–918.

Katharina Krischer responded: This is a very good question. I believe that at least in sulfate containing aqueous electrolyte water is more than a spectator since it influences the adsorption of sulfate ions, which, as you also mentioned, plays a strong role for the critical domain size. To really answer this question, one would need to have a much better understanding of sulfate adsorption under reaction conditions, and thus of its interaction with the adsorption of CO and OH on the Pt surface, than we have at present. I am not sure whether our experimental or technical methods are advanced enough yet to study these aspects on an atomic scale.

Olaf Magnussen replied: Understanding the role of water in electrochemical processes is still a central challenge in interfacial electrochemistry. The group of Schmickler has recently shown by theoretical studies of metal electrodeposition that the compatibility of the ion hydration shells with the structure of interfacial water has tremendeous impact on the ion transfer kinetics.[1] Similar effects likely are relevant for many other electrochemical processes and need to be explored further.

1 L. M. C. Pinto, E. Spohr, P. Quaino, E. Santos and W. Schmickler, *Angew. Chem. Int. Ed.*, 2013, **52**, 7883.

Richard Nichols addressed Thom Hersbach: Going back to David Fermin's question about water: for your DFT calculations of sodium adsorption on the surface do you include any remaining solvated water with the adsorbed sodium ions? This would seem to be important since it is generally taken that alkali metal cations adsorb with at least some and maybe quite a lot of the water hydration intact, with the energetics of removal of solvated water being high.

Also there are papers on the electrodissolution of Pd, Rh and Pt electrodes which have analysed and detected metal dissolution into the electrolyte during potentiodynamic procedures. The metal ion concentration accumulating in the solution has been analysed by widely available atomic absorption spectrophotometery.[1] For your procedure do you envisage metal dissolution as complexes in solution to be a possibility?

1 R. Schumacher, W. Helbig, I. Haß, M. Wünsche and H. Meyer, *J. Electroanal. Chem.*, 1993, **354**(1–2), 59–70.

Thom Hersbach responded: Our DFT calculations do not include any solvated water. We are aware that solvated water can be an important factor during these calculations, but think it is unlikely that the explicit inclusion of remaining solvated water will affect the relative differences that we found. Regarding your second question: metastable metal complexes in solution, such as metal hydrides, are the final species that we think might be intermediates in cathodic corrosion. However, these complexes generally possess a positively charged metal atom and would therefore require the bulk metal to be oxidized in order to form. Because we operate under strongly reductive conditions, we think that such metal complexes might be less likely intermediates than anionic intermediates. *In-situ* studies shall have to shed light on this situation.

Richard Crooks addressed Olaf Magnussen: You didn't mention the possibility of tip-induced effects in your presentation. Back in the early 90s when my group worked in this field this was a serious problem, but perhaps things have changed. Are you able to rule out tip effects in the results you presented today, and if so how do you do that?

Olaf Magnussen replied: Tip effects are indeed an issue and have to be considered in every *in situ* STM experiment. Different types of such tip effects exist: First, the presence of the tip can block access of electrolyte species to the surface, which is in particular a problem for faradaic processes. As discussed in our paper, this leads to a significant reduction of the deposition rate underneath the tip, which is a well known phenomenon. Second, the tip may directly induce changes of the surface structure and dynamics. We typically perform experiments at different tunneling parameters to assess this influence. Because the surface is imaged in constant height mode in our video STM experiments, *i.e.* at a fixed average tip-sample distance, this is in particular problematic for freestanding high structural features on a planar surface, where this distance can become very small. An example are the Bi nanochains on Au(100), where the dynamic behaviour indeed may be significantly influenced by the tip and thus was not analysed quantitatively. In contrast, the kink dynamics along Bi steps is determined by processes on the lower terrace, where the tip-surface distance is large, and consequently is not noticeably affected. We note that only rather strong tip effects can influence these measurements of the dynamic fluctuations, since the tip is located only a short fraction of the time directly above the kink site.

Patrick Unwin commented: I'd like to pick up again on the point about what the tip does locally. There is a related AFM literature on crystal growth and dissolution which we have recently analysed.[1] For such nanoscale measurements, there is a tendency to focus on areas of the surface that are close to perfect, with just a few step edges, but when one deduces the kinetics from these measurements, they are typically one or more orders of magnitude slower than one might expect from macroscopic measurements. This is mainly due to flux processes outside the AFM scan area (region of interest) affecting the region of interest. The surfaces we tend to look at are very heterogeneous spatially, as are the interfacial fluxes, and as concentration boundary layers tend to extend by tens of microns (at least) above a surface, there is considerable diffusional cross-talk between neighbouring areas of a surface that needs to be taken into account to make kinetic measurements by scanned probe microscopy quantitative. Areas outside the scanned area tend to be rougher and more active and so have a substantial influence on the region of interest.

1. M. Peruffo, M. M. Mbogoro, M. Adobes-Vidal and P. R. Unwin, *J. Phys. Chem. C*, 2016, **120**(22), 12100–12112.

Olaf Magnussen answered: In order to monitor details at individual defects, such as crystal steps, high resolution STM/AFM requires that these defects are spaced at sufficient distances from each other. For a comparison of the microscopic and the macroscopic growth rate the ion transfer mechanism should play a role: If ion transfer occurs at defects ("direct deposition") one would have to normalize by the defect density, which may be much higher on real electrode surfaces. If ion transfer proceeds *via* the formation of adatoms on the surface, subsequent nucleation and growth of islands of the deposit will occur and the local deposition rate will be independent of the initial defect density. In both cases, strong deviations between microscopic and macroscopic growth may occur, if mass transport in solution plays a significant role, because of the mentioned cross-talk between different electrode areas and the effect of the tip. This is a particularly serious problem for deposition processes with fast ion transfer, *e.g.* Cu electrodeposition.[1] In contrast, for processes with strongly inhibited ion transfer kinetics, such as Ni, Co, and Fe deposition, the microscopically observed deposition rates are often similar to those determined macroscopically.

1 O. Skylar, T. H. Treutler, N. Vlachopoulos and G. Wittstock, *Surf. Sci.*, 2005, **597**, 181.

David Fermin asked Olaf Magnussen: In your images, Bi nucleation and growth occur over a Bi monolayer on both Au surface orientations. Is this a general process? One would expect that the UPD will be unstable and reconstruct into 3D nuclei at potentials more negative than the Bi reduction potential (overpotential region). Can the Bi UPD layer be lifted upon positive polarization, or does it go through a spontaneous irreversible adsorption?

Olaf Magnussen replied: It is difficult to judge on the basis of the STM experiments whether the structure of the Bi UPD layer is maintained underneath the Bi bulk deposit or whether this layer is restructured into a bulk-like lattice. This could in principle be clarified by surface X-ray diffraction methods, but corresponding data for Bi on Au(111) and Au(100) has not been published so far. However, that a bulk deposit can grow epitaxially on top of a UPD layer, whose structure is maintained underneath, has been obeserved in other systems, *e.g.* for Tl on Ag(100).[1] Regarding the second question, the Bi OPD as well as UPD is highly reversible. Positive polarization results in fast dissolution/desorption of Bi from the Au surface.

1 J. X. Wang, R. R. Adžić, O. M. Magnussen and B. M. Ocko, *Surf. Sci.*, 1995, **344**, 111.

Julie MacPherson returned to discussion of the paper by Thom Hersbach: Why choose 10M sodium hydroxide as the electrolyte for all these studies, it seems an extreme medium?

Thom Hersbach replied: Indeed, 10 molar is quite a high concentration. Though lower concentrations also allow for cathodic corrosion to take place, the corrosion process is most pronounced at these high concentrations. Since the changes caused by cathodic corrosion are more apparent, using such a high concentration thus facilitates their detection. This in turn allows for a more reliable definition of the onset potential.

It is worth mentioning that we are currently working on experiments in less concentrated electrolytes and the first data indicate that, for the systems we have tested so far, the onset potential does not shift negatively by more than 100 mV on the RHE scale.

Marc Koper added: The reason for using (strongly) alkaline media is partially historical, as this is the medium in which Pt wires are etched to make STM tips, and this is how we (re-)discovered the phenomenon of cathodic corrosion. Note, however, that during cathodic corrosion, strong hydrogen evolution generates a condition of high alkalinity near the electrode surface. Therefore, the exact bulk pH does not matter for the phenomenon,[1] and the local surface condition is always highly alkaline.

1 A. I. Yanson, P. Rodriguez, N. Garcia-Araez, R. V. Mom, F. D. Tichelaar and M. T. M. Koper, *Angew. Chem. Int. Ed.*, 2011, **50**, 6346.

Michael Eikerling opened a general discussion of the paper by Henry White: In your article as well as your presentation, you suggest that bubbles nucleate in the bulk solution, away from the catalyst surface. What evidence do you have for bulk nucleation as compared to surface-catalyzed heterogeneous nucleation (which seems the more likely option to me)? Looking at the sensitivity of the nucleation kinetics to the surface structure of the catalyst should provide valuable insight in this regard. Has this been explored experimentally?

Henry White responded: We would expect that differing roughness and/or hydrophobicity of the electrode surface should have dramatic effects upon the critical concentration if it were a surface catalyzed process. We are not able to explicitly control the structure of the electrode surface in these experiments. Therefore, the reproducibility of the critical concentration for nucleation electrode to electrode is surprising and is the main reason we suggest the nucleation we observe is homogeneous. Ultimately, we do not know whether nucleation in this case is surface catalyzed or not.

Olaf Magnussen commented: I would be interested in the case of heterogeneous bubble nucleation at the electrode surface. Because the gas concentration is highest directly at the surface, this should be considered. I agree with the authors that the general form of the equations and the analysis will be the same. Also in this case the free energy contribution of the volume will scale as r^3 and that of the bubble surface as r^2. However, the prefactors will depend on the wetting angle θ. Specifically, the latter term will have a contribution from the electrode–gas interface with a surface energy term given by the difference of the solid–gas and the solid–liquid interface tension, which can be calculated by the Young's equation as $\gamma\cos(\theta)$. Also the solid–liquid interface area and the bubble volume will depend on θ. Consequently, also the critical radius and the free energy barrier E_aa should vary with θ. I assume that these effects have been discussed previously in the literature. Could the authors comment on this? In particular, I would be interested whether with increasing wetting angle a crossover between homogeneous and heterogeneous nucleation is expected.

Henry White replied: We agree with the above in principle. Whether the nucleation we observe is homogeneous or heterogeneous is unknown. As pointed out in the article, the radius of curvature of the critical nucleus is unaffected by the mechanism. The classical theory for the rate of heterogeneous nucleation of bubbles is $J = Z\exp\left(\frac{-16\pi\gamma^3\Phi(\theta)}{3kT(P_{\text{gas}} - P_{\text{ambient}})^2}\right)$. where Φ is a function of the contact angle, θ, given by $\Phi(\theta) = \frac{1}{4}(1 + \cos\theta)^2(2 \neg \cos\theta)$. For a contact angle of 0 degrees (measured through the liquid) $\Phi(\theta) = 1$ and the equation for homogeneous nucleation obtains.

Patrick Unwin commented: There is a growing literature and evidence of the importance of non-classical nucleation and growth processes, for example, in the growth of organic and inorganic materials[1] and metal nanoparticles (aggregative growth processes)[2–5], in which growth is by small clusters and aggregates coming together. Could such processes operate in nanobubble nucelation? Given the current magnitudes measured in these experiments, it should be possible to make dynamic measurements on quite a fast timescale (perhaps 100 kHz or better). Have you tried to make fast current–time measurements and, if so, do they provide any insights on the bubble nucleation and growth mechanism, for example, the dynamics of pre-nucleation cluster formation/dissolution?

1 P. R. Unwin, *Faraday Discuss.*, 2007, **136**, 409–416.
2 J. Ustarroz, X. Ke, A. Hubin, S. Bals and H. Terryn, *J. Phys. Chem. C*, 2012, **116**, 2322–2329.

3 J. Ustarroz, J. A. Hammons, T. Altantzis, A. Hubin, S. Bals and H. Terryn, *J. Am. Chem. Soc.*, 2013, **135**, 11550–11561.
4 S. C. S. Lai, R. A. Lazenby, P. M. Kirkman and P. R. Unwin, *Chem. Sci.*, 2015, **6**, 1126–1138.
5 Y.-R. Kim, S. C. S. Lai, K. McKelvey, G. Zhang, D. Perry, T. S. Miller and P. R. Unwin, *J. Phys. Chem. C*, 2015, **119**, 17389–17397.

Henry White responded: Bubble growth by nucleus aggregation is certainly possible. We are thankful for the educational list of references from the Unwin group in that regard although none of these articles noted by Professor Unwin actually describe gas bubble nucleation. We are currently studying rates of bubble nucleation at high recording frequencies that may shed additional light on these non-classical processes.

Andrew Ewing said: I think contact angle won't matter with huge pressure, and you say the gas covers the whole electrode. Is this correct? How do you measure pressure in the dynamic *vs* static case? Don't you have to take into account that hydrogen is flowing out of the bubble? Shouldn't there be a dynamic equilibrium between the hydrogen bubble and pressure of hydrogen in solution? How do you take this all in to account? Do you have to know the diffusional flux? Can you put this in context of where this would be something we could use for something else and what the interest is in this process?

Henry White answered: We believe the bubble covers nearly the entire electrode only after nucleation and growth when it reaches a steady-state size limited by the size of the electrode. Indeed, the bubble growth and dissolution are controlled by diffusional flux while the concentration at the bubble surface is determined by Henry's Law for equilibrium with the bubble's internal pressure. Further clarification on this dynamic equilibrium can be found in our previous work.[1–4] Nanobubbles have attracted attention for their applications in slip enhancement in fluidic channels, surface cleaning, froth flotation, templating nanomaterials, and as propulsion for nanodevices. Gaining an understanding of the nucleation processes and control over nanobubble localization and growth will be critical for these future developments. This knowledge will also aid the prevention of bubble formation where it can have adverse consequences as in cavitation, decompression sickness, and disruption of electrode processes in batteries and fuel cells.

1 L. Luo and H. S. White, *Langmuir*, 2013, **29**, 11169.
2 Q. Chen, L. Luo, H. Faraji, S. W. Feldberg and H. S. White, *J. Phys. Chem. Lett.*, 2014, **5**, 3539.
3 Q. Chen, H. S. Wiedenroth, S. R. German and H. S. White, *J. Am. Chem. Soc.*, 2015, **137**, 12064.
4 S. R. German, Q. Chen, M. A. Edwards and H. S. White, *J. Electrochem. Soc.*, 2016, **163**, 3160.

Patrick Unwin opened a general discussion of the paper by Jan Clausmeyer: In Fig. 3 of your paper you showed that many Ag nanoparticles and clusters were on the glass sheath around the carbon nanolectrode, but you were unsure about how they formed. I believe your observations are consistent with our recent measurements of silver nucleation and growth on highly oriented pyrolytic graphite (HOPG) electrodes,[1] where we showed that silver nanoparticles grew initially via the aggregation of silver ad-clusters, but that the nanoparticles detached from the surface after a few milliseconds (after reaching a size of about 25–30 nm in diameter). This process was characterised by a very regular periodicity for nucleation–growth–detachment (*ca.* 100 Hz). You see nanoparticles of about the same size around the nanoelectrode and I suspect they are formed on the electrode but detach and stick preferentially at the glass. So, your experiments lend further support to our mechanism. It would be interesting to run chronoamperometric measurements and see if you observe similar features.

1 S. C. S. Lai, R. A. Lazenby, P. M. Kirkman and P. R. Unwin, *Chem. Sci.*, 2015, **6**, 1126–1138.

Jan Clausmeyer replied: Yes, this observation matches our findings in a number of aspects. For instance, when we deposit silver at low overpotential, low silver ion concentration and for short times, the predominant nanoparticle size we obtain is consistent with the results in your paper. This is very interesting with regard to the electrosynthesis of nanoparticles with controlled size and shape. What factors influence the growth and detachment and thus the particle size? How could one use this to control the particle geometry?

However, in one respect our findings are different from what you observed: For long deposition times (many seconds) we do not see a periodic behavior of the cathodic current but rather a steady increase. As a result, the Ag clusters on the nanoelectrode get a cauliflower-like shape. This might be due to different mass transport rates for the two different experimental configurations, namely almost spherical diffusion for Ag growth on a pointy nanoelectrode *vs.* rather linear diffusion for Ag growth in the scanning capillary configuration in your work.

Julie MacPherson asked: First, have you looked at how the cathodic deposition current correlates with the anodic stripping current for Ag deposition and stripping? Assuming nothing else is going on cathodically then a larger cathodic current would support Pat Unwin's comment about Ag particle movement away from the electrode.

Second, given the previous discussions throughout the meeting on the use of TEM, this would be a much better technique for you to use than SEM as a means of assessing Ag particle size, have you considered using it?

Third, it was not clear what size C electrodes you are using. You just talk about a 10 pA current in the paper, can you comment further?

Jan Clausmeyer replied: Answering your first question: we have tried to correlate the transmitted charge for cathodic metal deposition with the charge of Ag stripping peaks and found that the deposition charge is consistently larger than one would expect when looking at the size of the deposited particles. In a different context we have made the same observation for the deposition of Au particles on nanoelectrodes. It seems to be a general phenomenon that any cathodically reduced material is very likely to escape from the electrode, bearing in mind that mass transport is very fast at such small electrode dimensions and that the interaction between the carbon material and the metal particles might be weak. Patrick Unwin's comment and our findings are in very good agreement with this picture.

Answering your second question: yes, there are a small number of papers proposing TEM as a tool for characterization of nanoelectrodes. We expect that

TEM analysis will improve the estimation of size and geometry of the electrodeposited metal clusters and reduce the uncertainty concerning the exact geometry of the dual barrel nanoelectrodes. We will try to establish TEM as a routine characterization tool for studies related to nanoelectrodes in future work.

Answering your third question: estimating the size of nanoelectrodes based on electrochemical data and electron microscopy data (see above) is not trivial. The common practice is to estimate the size from steady-state voltammograms, complemented by some electron micrographs. We believe that one should be very careful with making bold statements about the electrode size based on electrochemical data because, first, one needs to assume that electrodes are disk-shaped and co-planar with the insulating shield and, second, that classical continuum description of mass transport is valid at such dimensions. The latter point is addressed by Shengli Chen's paper presented in this Faraday discussion and earlier work.[1] For the nanoelectrodes used in our paper, which are made from Θ-shaped capillaries, it is very unlikely that they have a disk-shaped geometry. Hence, we only report the steady-state voltammetric currents to refer to a roughly estimated size. The 10 pA reported in the manuscript correspond to an electrode diameter of about 10 nm, if they were disk-shaped.

1 S. Chen and Y. Liu, *Phys. Chem. Chem. Phys.*, 2013, **16**, 635–652.

Christine Kranz asked: Could you please comment on the error bars in Fig. 6b of your manuscript? The measurements indicate that there is no significant difference for the different NaOH concentration given the huge standard deviations. Also this might be even more true for the lower concentration if more measurements would have been conducted, $n = 2$ seems very limited. Error bars should only reported for the same number of measurements. The general definition of "error bar" should be considered, otherwise it is confusing, if the standard deviation is not reported. Were repeat measurements conducted at the same electrode? What is the stability of individual Ag-nanocluster modified nanoelectrodes under the described conditions?

Jan Clausmeyer replied: Fig. 6b compares results obtained with different electrodes. The large error bars are a result of both the small number of experiments as well as some experimental uncertainty in the preparation and modification of the nanoelectrodes. Variations and some extent of uncertainty in the electrode geometry are inherent to any kind of nanofabrication protocol and can never be avoided, not even with high technical effort. The Ag nanoclusters are electrochemically stable for the typical time frame of the experiment (*i.e.* the voltammetric curves remained stable). However, due to mechanical instability of the single clusters attached to the nanoelectrodes, the probability of detachment increases with each time the electrode is transferred between different solutions. A possible solution would be to bury the Ag cluster in slight nanocavities.[1] Before validating a quantitative model that explains ORR currents at different electrolyte compositions, certainly more experiments would have to be performed. The aim of the present paper was to establish a new analytical technique to study mass and electron transfer at the three-phase boundary and to investigate electrochemical reactions at non-mass transport-limited conditions – until now, qualitatively.

1 J. Clausmeyer, P. Actis, A. López Córdoba, Y. Korchev and W. Schuhmann, *Electrochem. Commun.*, 2014, **40**, 28–30.

Thom Hersbach commented: In Fig. 6b, you display the oxygen reduction activity enhancement that occurs when increasing the oxygen concentration. This enhancement is largest at the intermediate electrolyte concentration of 5 M. In the article, you mention that this cannot just be explained by slower oxygen diffusivity at higher alkalinity; you tentatively suggest other factors that might influence the enhanced activity, such as a changing meniscus shape at higher NaOH concentrations. Can you elaborate on this and explain why you expect an optimum activity enhancement around concentrations of 5 M NaOH? Does an optimum occur because the changing meniscus shape counteracts the slower oxygen diffusivity?

Jan Clausmeyer responded: Yes, the height of a meniscus in a capillary depends on the surface tension, the contact angle and the density of the liquid – all quantities which are themselves functions of the NaOH concentration. We have evaluated the meniscus heights by just immersing quartz glass capillaries into the different solutions and found that the meniscus rose least in the case of 5 M NaOH. Hence, at this concentration the distance between the gas–liquid interface and the silver nanoparticle is shortest, which may explain the high ORR currents. The message is that one also needs to take into account hydrodynamics when trying to achieve an optimal current density. This is true for the present experimental configuration based on nanoelectrodes but in particular also for industrial porous gas diffusion electrodes.

Zhongqun Tian commented: The electrical conductivity of the electrolyte is not linearly related to the concentration. The highest concentration does not mean the highest conductivity. The maximum conductivity of concentration of NaOH, KOH and H_2SO_4 is around 5, 6 and 7 mol l^{-1} respectively in aqueous solution. For the even higher concentrations, the conductivity decreases significantly, because of the formation of an ion pair consisting of a cation and anion, which is neutral and does not contribute to the conductivity. There are no free water molecules in the solution and even the solvated water molecules are not enough. So cations and anions can contact with each other to form the neutral ion pair; the physicochemical properties of this kind of electrolyte solution are very unique and complex. Therefore the electrochemical behavior of 10 mol l^{-1} may not be better than the 5 mol l^{-1} system.

Jan Clausmeyer replied: Considering that the absolute currents for the ORR at the nanoelectrodes are very small (pA range), we do not expect the electric potential drop in solution to have a large impact on the electrocatalytic turnover. This is probably true for all methods involving micro- or nanoelectrodes for electrocatalytic studies. However, the point raised is very valid for a different reason: water is consumed as a reactant in the ORR. Taking into account that all water molecules are tied to the ions in an incomplete solvation shell (only ~2 water molecules are nominally available per ion) and thus there is very little or no free water, it is even more surprising how the ORR can run at high current densities. The formation of ion pairs may free some water molecules and is probably part of the answer to that question.

Wolfgang Schmickler opened a general discussion of the paper by Shengli Chen: During the 1980s Blum and Henderson modelled the electrolyte as an ensemble of hard spheres with charges (ions) or with dipole moments (solvent). Two important conclusions can be drawn from this model: first, the size of the solvent molecules is more important than that of the ions. Second, there is no special Stern layer. The latter is corroborated by computer simulations. Could you explain why you focus on the size of the ions, and why your model contains a Stern layer?

Shengli Chen answered: This is a very good point. Indeed, the sizes of the solvent molecules are very important. Due to the solvation effect, the sizes of different ions could be averaged. In our model, we actually include the size of the solvent molecules. When we consider the size of ions, we actually use the values for solvated ions. In this study, we focus on how the volume exclusion of electrolyte ions and their solvation shells as a whole affects the electric-double-layer structure and voltammetric responses of nanometer-sized electrodes. We don't emphasize any specific ions. As mentioned in our paper, the two sizes we have considered represent the typical sizes of the electrolyte ions in aqueous and organic media.

As for the Stern layer, we still believe there is a Stern layer (compact layer) adjacent to the electrode surface due to the interactions between electrode and solvent dipoles and the strong interaction between electrode charges and solvated counter ions.

Patrick Unwin asked: You have outlined the assumptions and approximations in the model. Referring to Fig. 1 in your manuscript, could you comment further on the implications of extending the double layer from the electrode over the (glass) sheath surrounding it, given that the glass may be (heavily) charged,[1] and – in reality – the double layer at that location will usually be different to that over the electrode.

1 See for example: D. Perry, R. Al Botros, D. Momotenko, S. L. Kinnear and P. R. Unwin, *ACS Nano*, 2015, **9**, 7266–7276.

Shengli Chen replied: This is a very expert comment and I agree with you on this point. In our model, it has been assumed that the glass sheath bears an identical compact double layer on its surface to that on the electrode surface, that is, both electrode and sheath surfaces bear a closely packed solvent dipole layer, adjacent which there is a closely packed solvated ion layer. We have to admit that this is a rather rough approximation. In a given electrolyte solution, the glass sheath surface would be fixedly charged; while the nature and density of charge on the electrode surface vary with potential. Therefore, the electric double layer above the glass sheath could be different to that on electrode. This would affect the ion profile in the region of the electrode edges. To this end, the results in this study may only qualitatively describe the effect of ion sizes on the double layer structures and voltammetric responses of nanoelectrodes. More detailed and accurate interfacial models should be used to get more quantitative information.

Henry White commented: Professor Shengli Chen's talk focused on the effect of interfacial electrical fields on ion transport and the observable current–voltage behavior of nanoscale electrochemical systems. This is a very different problem from previous theoretical work in 80s and 90s that focused more on the equilibrium interfacial structure using more atomistic modeling. Both are very challenging and important. There are a number of recent experimental systems of broad interest involving electron-transfer and ion transport within electric double layers in which we do not have a complete understanding. For instance, ion current rectification is observed in nanopores filled with aqueous electrolyte. Current theory suggests no rectification should be observed for systems in which the Debye thickness is much smaller than the radius of the nanopore, but experimentally, rectification is frequently reported under these conditions.[1] The importance of the electric double layer on transport is also amplified in the "Lemay thin-layer electrochemical cells", where the double layer can extend an appreciable distance across the electrolyte separating the two electrodes. These fields can have a dramatic influence on transport of redox molecules, even at moderate ionic strengths, as demonstrated in some of our recent work.[2–4] I believe that Professor Chen's work using continuum models is vital to understanding types of phenomena.

1 W.-J. Lan, M. A. Edwards, L. Luo, R. T. Perera, X. Wu, C. R. Martin and H. S. White, *Acc. Chem. Res.*, 2016, **49**, 2605–2613.
2 Q. Chen, K. McKelvey, M. A. Edwards and H. S. White, *J. Phys. Chem. C*, 2016, **120**, 17251–17260.
3 J. Xiong, Q. Chen, M. A. Edwards and H. S. White, *ACS Nano*, 2015, **9**, 8520–8529.
4 L. Fan, Y. Liu, J. Xiong, H. S. White and S. Chen, *ACS Nano*, 2014, **8**, 10426–10436.

Shengli Chen answered: It's pleasing that this theoretical work can receive such a positive comment. I agree with Professor White that the EDL would have much more pronounced effect on charge transfer and transport under nonequilibrium (dynamic) conditions than that expected from the equilibrium EDL models, especially in nanoscale interfacial electrochemical systems. There have been numerous theoretical and simulation works dealing with these problems during the last two decades. However, most of them have employed continuum equations and neglected the finite volumes of ions which could be very important at nanoscale electrochemical interfaces. This Faraday Discussion paper presents our preliminary results on how the finite sizes of ions impact the EDL structure and reactivity at the nanoscale electrochemical interface.

Philip Bartlett opened a general discussion of the papers by Jan Clausmeyer, Henry White and Shengli Chen, with a question for Henry White: The contact angle picture presented by Prof. Magnussen is reasonable but in practice experimentally contact angles can be quite irreproducible, with significant hysteresis between advancing and receding contact angle. Do you know how the bubble dissolves? In the literature, as I'm sure you are aware, there is much discussion of long lived nanobubbles and the presence of pancake shape nanobubbles on hydrophobic surfaces (see for example the review by Lhose and Zhang[1]). How quickly do your bubbles dissolve? Do they remain pinned at the glass/Pt interface as the hydrogen dissolves? Can you say anything about the capacitance current as the bubble dissolves?

In our experiments using regular structured surfaces on the micron scale we find that the gas bubbles nucleate on the adjacent insulating, hydrophobic, surface rather than on the metal; this is also seen in the literature (e.g. see Bussieux *et al.*[2]). Do you see any evidence for the bubble forming on the surrounding insulation rather than on the platinum electrode?

1 D. Lohse and X. Zhang, *Rev. Mod. Phys.*, 2015, **87**, 981.
2 C. Brussieux, P. Viers, H. Roustan and M. Rakib, *Electrochim. Acta*, 2011, **56**, 7194.

Henry White answered: The measured lifetimes (tens of milliseconds) of bubbles in our experimental system are 2–3 orders of magnitude longer than diffusion controlled theory predicts assuming either a pinned contact radius or a constant contact angle. We attribute the long lifetimes to a kinetic limitation of gas transfer and/or a mass transport resistance due to a recessed electrode geometry.

In our experiments, the surrounding insulator is glass and we do not anticipate, due to its hydrophilicity, that the bubbles would nucleate there. We do not know if the bubble remains pinned at the electrode insulator boundary as it dissolves. Experiments are being carried out to image the bubbles with atomic force microscopy that could help answer these questions. We have not measured a capacitive current as the bubble dissolves.

Sanli Faez remarked: Orrit and coworkers have measured dynamics of the generation and collapse of a nano-bubble initiated at a single gold nanoparticle with photo-thermal microscopy.[1] Knowing that the generation and collapse process can be so fast, a few nanoseconds, is it possible that the formation of small bubbles is left unnoticed in electrical current measurements?

1 L. Hou, M. Yorulmaz, N. R. Verhart and M. Orrit, *New J. Phys.*, 2015, **17**, 013050.

Henry White responded: Compared with thermally generated vapor bubbles, bubbles nucleated from dissolved gas exhibit slower growth and dissolution rates because they are governed by diffusion of the gas in solution. Nevertheless, these bubbles are still expected to have short lifetimes on the order of micro- to milliseconds. Observation of bubble signatures in electrochemistry depends first on the bandwidth of the electronic instrumentation and secondly, on the effect of bubbles on the electrochemical current. When recording at a ~50kHz and 10kHz filter frequency we observe a smoothly increasing current with minimal noise followed by a single drop indicative of the bubble covering a significant portion of the electrode. We cannot rule out possibility of the formation of bubbles too small/short-lived to observe electrochemically.

Kristina Tschulik asked: Regarding the reduced shrinkage rate you observed for the hydrogen nanobubbles, how sure are your that the Pt electrode does not serve as a hydrogen reservoir, *i.e.* is the solubility of hydrogen in Pt negligibly small, despite the high internal H_2 pressure and the applied potential? Recently we observed that the relaxation of the open circuit potential (OCP) with time after hydrogen bubble formation may reveal if the bubble remains attached to the electrode or not.[1] In this case, however, the micrometric electrode and bubble size allowed us to additionally follow the bubble shrinkage by optical means. Did you observe any similar features of the OCP time response that could reveal if a nano bubble is attached to your nanoelectrode? If so, could this serve as a measure at which minimum radius the bubble shrinkage makes it collapses?

1 F. Karnbach, X. Yang, G. Mutschke, J. Fröhlich, J. Eckert, A. Gebert, K. Tschulik, M. Uhlemann and K. Eckert, *J. Phys. Chem. C*, 2016 **120**, 15137–15146.

Henry White responded: Recently, we reported bubble lifetimes 2–3 orders of magnitude longer than diffusion-controlled theory predicts.[1] We do not attribute the longevity of the bubbles to a reservoir of dissolved hydrogen within the Pt electrode because, in contrast with Pd, hydrogen solubility is extremely low in Pt. We believe the long lifetimes to be affected more by kinetic limitations of gas transfer and/or mass transport resistance due to a recessed electrode geometry. The transient OCP response after bubble formation has not been studied in our experimental system, but we expect it would require an extremely high input impedance to record in nanoelectrode systems.

1 S. R. German, Q. Chen, M. A. Edwards and H. S. White, *J. Electrochem. Soc.*, 2016, **163**, 3160.

Sushanta Mitra returned to the paper by Shengli Chen: If one writes the free-energy for the problem, it will become evident that the ion size effect, entropic effect of ions, and solvent polarization are key contributions. Please refer to series of publications from our group.[1–2]

1 S. Das, S. Chakraborty and S. K. Mitra, *Phys. Rev. E*, 2012, **85**, 051508.
2 R. Mishra, S. Das and S. K. Mitra, *J. Chem. Phys.*, 2013, **138**, 114703.

Shengli Chen replied: You are totally right. In our model, we don't include the solvent polarization in the free energy formulation. What we are mainly concerned with is how the ion volume exclusion may affect the electric double layer structure and voltammetric responses. We have not attempted to give a comprehensive and accurate description of the electric double layer structure. So, we have ignored other possible contributions to the free energy at the interface. But the solvent polarization should be an important issue and it should be related to the ion sizes. Thank you for pointing out this and the references, we will look at them to improve our models in future works.

Andrew Mount addressed Henry White: The evidence in this work is that nanobubble formation is only weakly dependent on the nature of the gas. One driver to gas nanobubble formation and growth could therefore be the drive to minimise gas volume so as to maximise solvent–solvent intermolecular bonding (*i.e.* with the process being driven by maximising favourable solvent–solvent interactions). Have the authors carried out work *e.g.* on D_2 formation in D_2O systems and how does/would this give insight into the relative contributions of kinetics and thermodynamics to nanobubble nucleation and growth?

Henry White responded: Interestingly, the classical theory does not predict any dependence upon gas type. Instead, as the question suggests, the solution surface tension and the gas supersaturation are the dominant variables. We have not worked with the D_2/D_2O system; however, we are currently studying bubble nucleation in other solvents to further understand the process.

Michael Eikerling addresed Jan Clausmeyer: The triple-phase boundary is an ideal construction (a singularity). In your experimental system, what is the actual distribution of dissolved oxygen in the electrolyte surrounding the electrode? This distribution would determine the active surface area fraction of the electrode (and an effectiveness factor of surface area utilization). For a simple estimation, have you evaluated the diffusion length (or the reaction penetration depth) of oxygen in the electrolyte under the given experimental conditions? An estimate for this property could be easily obtained if the diffusion coefficient of oxygen in the electrolyte were known. Could you provide a value of this diffusion coefficient?

Jan Clausmeyer answered: Yes, oxygen is probably transported over a thin electrolyte film separating the gas phase and the solid catalyst. Unfortunately, we do not know the exact position of the gas–liquid interface with respect to the orifice of the nanocapillary. The diffusion coefficient in highly concentrated NaOH solution is in the range of some 10^7 cm^2 s^{-1} (for instance 8×10^7 cm^2 s^{-1} in 10 M NaOH.[1])[1] Based on that, the oxygen diffusion layer may still reach many micrometers into the solution. In ongoing experiments we are addressing this question using Scanning Electrochemical Microscopy (SECM). By approaching nanoelectrodes to gas diffusion electrodes one can determine both the diffusion length of oxygen around gas channels as well as the thickness of the depletion layer around electroactive catalyst domains.

1 C. Zhang, F.-R. F. Fan and A. J. Bard, *J. Am. Chem. Soc.*, 2009, **131**, 177–181.

Patrick Unwin asked: Have you thought about using a different configuration, for example, a theta pipette with an electrolyte-filled channel as an SICM distance sensor and an oxygen-delivery channel that you could bring up to a single nanoparticle on a surface or group of nanoparticles? This would seem to offer a well-defined arrangement for quantitative analysis.

Jan Clausmeyer replied: Yes, that would be a very good experimental arrangement which would allow the distance between the gas-liquid interface and the solid catalyst to be controlled with high precision and thus allow the oxygen diffusion layers to be imaged under these conditions. In fact, those experiments are currently ongoing in our laboratory. Several configurations are possible: first, approaching a double-barrel pipette with one oxygen-filled barrel under SICM control to Ag nanoparticles, second, approaching a Ag-modified nanoelectrode to an oxygen-permeable membrane and third, approaching an oxygen nanosensor to a gas diffusion electrode comprising both oxygen channels and catalyst spots. The presented work was only the first step in a series of different experimental approaches. Its advantage is that it does not require sophisticated positioning devices and software but only a laser puller and some tubing.

Wojciech Nogala asked Henry White: The number of molecules in the critical nucleus is calculated using the ideal gas law, assuming ambient temperature. Formation of gas bubbles on a gas-evolving electrode resembles cavitation.

The temperature of cavitating vapor bubbles differs significantly from the ambient temperature.[1,2] Is it possible that the formation of critical nuclei from an electrogenerated (supersaturated) gas solution, their growth and possible shrinking and collapse are adiabatic processes rather than isothermal? Then the temperature and the number of molecules would be different.

1 F. Lugli and F. Zerbetto, *Phys. Chem. Chem. Phys.*, 2007, **9**, 2447–2456
2 E. B. Flint and K. S. Suslick, *Science*, 1991, **253**(5026), 1397–1399.

Henry White responded: Rates of cavitation/evaporation and collapse/condensation of vapor bubbles are quite different from the growth and dissolution of gas-filled bubbles controlled by diffusion. The heats of vaporization and rate at which evaporation/condensation occur are extreme for vapor bubbles such that thermal diffusivity in water cannot maintain isothermal conditions. Because the bubbles we study are controlled by diffusive rates of gas in solution which is much lower than thermal diffusivity, we assume that nearly isothermal conditions are maintained. However, we do not have an experimental measure of temperature in our system and cannot be certain.

Frederic Kanoufi communicated: The pressure inside the nanobubbles is so high that I wonder if for certain systems you cannot evidence the formation of supercritical fluid droplets, for example with CO_2?

Henry White communicated in reply: We are aware that the formation of supercritical fluid droplets is an intriguing possibility. To date our studies of CO_2 bubbles have been hampered by the gases' high solubility and the concurrent generation of other gases. We are working to find the right conditions under which purely CO_2 bubbles can be nucleated.

Faraday Discussions

PAPER

Mechanistic aspects of hydrazine-induced Pt colloid instability and monitoring aggregation kinetics with nanoparticle impact electroanalysis

D. A. Robinson,[a] J. Duay,[a] A. M. Kondajji[a] and K. J. Stevenson*[ab]

Received 7th May 2016, Accepted 20th May 2016
DOI: 10.1039/c6fd00121a

Here we investigate the mechanistic aspects of Pt nanoparticle (NP) aggregation in solutions typically used for detecting NP/electrode impacts by electrocatalytic amplification (ECA). We previously proposed a general mechanism for Pt colloid destabilization that involved the participation of both the hydrazine redox probe and the pH buffer species as coagulants. Herein the Pt NP coagulation and aggregation mechanisms were further investigated with microscopic kinetic NP concentration monitoring and zeta potential measurements using nanoparticle tracking analysis (NTA), as well as open circuit potential experiments with a citrate-treated polycrystalline Pt surface to assess electrical double layer potential. After considering the combined results of these experiments we propose that the colloidal stability of citrate-capped platinum nanoparticles involves much more than the typical physicochemical interactions predicted by DLVO theory. A structure based on intermolecular H-bonding in the citrate capping layer is the most plausible explanation for the exceptional stability of large Pt NPs in high ionic strength buffers. Thus, the mechanism of Pt NP aggregation includes specific reactive contributions from hydrazine. The catalytic decomposition of hydrazine, in particular, is thought to occur to some extent at the citrate-coated Pt surface while the citrate remains adsorbed. Evolved gases such as ammonia and possible surface bound intermediates from Pt-catalyzed decomposition of hydrazine may disrupt the stability of the citrate layer, causing colloidal instability and thus promoting Pt NP coagulation. In the closing section, we demonstrate nanoparticle impact electroanalysis by ECA detection as a method to quantify Pt NP concentration with adequate time resolution for monitoring the kinetics of Pt NP coagulation.

[a]*Department of Chemistry, Center for Nano- and Molecular Science and Technology, The University of Texas at Austin, Austin, TX 78712, USA*

[b]*Skolkovo Institute of Science and Technology, Center for Electrochemical Energy Storage, 3 Nobel Street, Moscow 143026, Russia. E-mail: K.Stevenson@skoltech.ru*

Introduction

In a previous study, the colloidal stability of single particles during nanoparticle impact electroanalysis was examined.[1] In this report we expand upon these findings and extend the investigation into the colloidal instability of citrate-capped Pt nanoparticles (NPs) and establish the conditions necessary for detection of single NP impacts *via* electrocatalytic amplification (ECA). Almost a decade ago, Bard and coworkers pioneered the ECA strategy[2] for detecting single NP electrocatalysts in solutions containing reactive molecules such as hydrazine (N_2H_4) as ECA detection probes.[3] In amperometric ECA studies, an ultramicroelectrode (UME) is employed, and the best results are obtained when the UME substrate is electrochemically inert to the ECA probe molecules at the applied potential. A sharp burst in current is observed when the catalytic NP contacts the UME due to the resulting electrocatalytic oxidation or reduction of the reactive species in solution. For N_2H_4 oxidation, the characteristic shape of the ECA impact signal depends primarily on the composition and physical properties of the electrode material.

Among the various ECA detection strategies developed over the course of the past nine years, the most commonly employed scheme involves citrate-stabilized Pt NPs as the catalysts and hydrazine as the reactant species.[3–16] Koper and coworkers reported the first evidence of Pt NP aggregation while conducting ECA experiments with hydrazine and phosphate buffer at pH 8 using 3–5 nm Pt NPs and Au UMEs.[8] The authors showed that the Pt NPs were distributed as large branched aggregates on Au and pointed out large distributions of the measured current step ECA events which increase in polydispersity with increasing hydrazine concentration. Koper attributed the aggregation to specific interactions of hydrazine at the citrate/Pt interface and noted that the fractal dimension of the aggregates appeared to fit the growth model of diffusion-limited aggregation (DLA).[8]

The DLA model was developed based on the experimental observations of Meakin and coworkers when studying polystyrene, SiO_2, and citrate-capped Au colloids.[17] The authors experimentally established universal rate regimes for all three aggregating systems and correlated the kinetics to the observed fractal dimension of the resulting aggregates, which was common to all aggregated species regardless of chemical composition. The theory of universality for understanding colloid aggregation kinetics was then developed based on the fractal dimension of the aggregate structures as independent from the chemical nature of the colloid.[18,19] Aggregation rates in these systems are typically measured with minute-scale time resolution by traditional light scattering methods, which do not include direct NP quantification in the measurements.

Shumaker-Parry and coworkers recently elucidated the surface structure of citrate adsorbed to single crystal Au surfaces and Au NPs by combined infrared and X-ray photoelectron spectroscopic methods. They proposed that citrate adopts a unique bilayer structure with "dangling" dihydrogen citrate ions (monovalent species) coordinated by hydrogen-bonds, which sits atop a neutral hydrogen-bonded network at the citrate/Au interface.[20] The authors went on to suggest that the stability of this citrate/Au structure was crucial for Au NP growth during synthesis as well as the total stability of the colloid. In another recent study, Johnston and coworkers reported conditions in which a citrate-stabilized Au colloid preferentially assembles into small chain-like clusters.[21] The authors modeled their system based on Derjaguin–Landau–Verwey–Overbeek (DLVO) principles and proposed that attractive bridging interactions at the terminal ends of a small NP-chain directed the linear coagulation mechanism.

Nanoparticle tracking analysis (NTA) has recently been reported as a useful tool to characterize isosymmetric Au NPs[22,23] and asymmetric nanorods[22,24] in solution, with potential application to developing new immunoassay nanotechnologies that employ protein-conjugated Au NPs as detection probes.[22,23] NTA has also been evaluated for the analysis of NPs in complex matrices such as food and environmental samples.[25,26] Because NTA is a relatively new form of analytical instrumentation, many research groups have included it in multi-instrumental cross-correlative comparisons involving NP quantification and characterization.[26–28] NTA simultaneously provides quantification of particle concentration and hydrodynamic diameters in a single acquisition, in contrast to more commonly employed ensemble-based methods such as dynamic light scattering.

Our group recently addressed the importance of using lower ionic strength buffers to prevent hydrazine-induced Pt NP aggregation and obtain reliable impact results for individual 50 nm citrate-capped Pt NPs, using Hg/Pt as the UME.[1] In that study, we used kinetic NTA measurements to aid in the optimization of Pt colloids for single-NP detection by ECA nanoparticle impact analysis. We observed from the NP impact frequency that aggregating Pt NPs were found to collide with the Hg UME at a much slower rate than the optimized stable samples. The stable colloids, on the other hand, demonstrated the expected Langevin-derived diffusion-limited mass transport to the Hg UME, as correlated to NP hydrodynamic sizing by NTA and SEM measurements.[1]

The general mechanism for coagulation was found to involve the participation of both the hydrazine (N_2H_4) redox probe and the pH buffer species as coagulants. Chemically-induced colloid instability was attributed to chemisorption or partial decomposition of N_2H_4 on the catalytically active Pt NP surface, in line with earlier findings by Koper and coworkers.[8] However, we were able to stabilize the citrate-capped Pt NPs in the presence of N_2H_4 by simply decreasing the net ionic strength of the buffer. This NTA-guided systematic approach for optimizing colloid stability led to the discovery of solution conditions that were compatible for our single-NP sensing strategy based on the aforementioned ECA.[2,3] The resulting NP impact rate analysis was found to correlate well with sizing by NTA and scanning electron microscopy (SEM), as evidenced by Pt NP diffusion coefficient (D_{NP}) approximations based on the Stokes–Einstein–Sutherland relation.[29,30] The experimentally determined Pt NP impact rate of the most stable sample, while giving an expected result according to Einstein's kinetic relations,[1] contradicts previous theoretical models of Pt NP impact rates in the N_2H_4 solutions. The earlier suggested model, reported by Bard and Zhang, was based on impact rate analysis of citrate-capped Pt NP samples that were detected by ECA in high ionic strength conditions.[5] These experiments gave apparent diffusion coefficients that were two orders of magnitude slower than predicted based on diffusive transport to the UME. We suggested that the observed slow collision rates in earlier studies were attributed to hydrazine-induced particle coagulation, which is facilitated in high salt concentrations.[1]

In this report, we examine the mechanistic aspects of Pt NP aggregation in ECA solutions based on microscopic kinetic Pt NP concentration studies and zeta potential measurements using nanoparticle tracking analysis (NTA).[31,32] Zeta potential measurements were used to model the interparticle interaction energy based on DLVO theory and to study the effect of the citrate capping layer and Pt nanostructure on the colloidal stability. We evaluated citrate desorption from Pt with open-circuit potential (OCP) measurements and discovered evidence of preserved citrate–Pt stability in the presence of hydrazine undergoing Pt-catalyzed decomposition. Finally, we monitored the kinetics of Pt NP aggregation with NTA over longer time scales than previously reported. These NTA results revealed that the Pt NPs used in these experiments favor the aggregative assembly into linear chain structures as observed by scanning electron microscopy. This finding adds a new consideration to the overall discussion of "universality" in colloid aggregation,[17] specifically the effect of Pt NP surface structure on aggregative assembly. We demonstrate in the final section the utility of NP impact electroanalysis (NIE) for quantifying NP concentration and monitoring Pt NP aggregation kinetics.

Experimental

Materials

Mercury(I) nitrate dihydrate (ACS reagent), chloroplatinic acid hydrate (H_2PtCl_6, ≥99.9% trace metal basis), 64–65% hydrazine hydrate solution (reagent grade), and L-ascorbic acid (ACS reagent) were purchased from Sigma-Aldrich. ACS grade sodium phosphate buffer (SPB) salts, trisodium citrate, citric acid, concentrated sulfuric acid, and sodium borohydride were acquired from Fisher Scientific. All chemicals were used as received. Sulfuric acid was used to adjust the pH to 7.8 for each hydrazine-containing buffer. All water was purified to 18 MΩ cm resistivity with a Barnstead Epure System.

Synthesis of Pt NPs

The 50 nm Pt NPs used in this investigation were part of the same batch that was synthesized for a previous investigation.[1]

Nanoparticle tracking analysis

Nanoparticle tracking analysis (NTA) was performed using the ns500 instrument from Nanosight, Ltd. The NTA method of optical single-NP detection is based on dark-field scattering microscopy; the light source is a blue 488 nm laser. The time resolved size measurements were performed following the experimental procedure as reported earlier.[1] Zeta potential measurements were performed at an applied electrokinetic driving voltage of 24 V for samples in water, 8 V for 25 mM Tris sulfate and 10 mM sodium phosphate buffers,† and 12 V for all other samples.

† The zeta potential measurements performed in these high ion concentrations (>10 mM) are expected to be less reliable than those obtained in lower ionic strength because the electrokinetic driving voltage had to be reduced from 24 V (used for water and 10 mM ionic strength solutions) to 8 V in order to prevent a short circuit. The 8 V setting is the minimum recommended setting from the NTA manufacturer for obtaining reliable zeta measurements.

Open circuit potential measurements

All electrochemical experiments were performed with a two-electrode setup controlled by an electrochemical workstation (CH Instruments, Austin, TX, model 700). A saturated calomel electrode was used as reference. The OCP experiment is depicted schematically in Fig. 5.

Electron microscopy

Scanning transmission electron microscopy (STEM) was performed using a Hitachi-S5500 microscope operated at 30 kV equipped with a Bruker Quantax 4010 EDX detector. Samples were prepared for electron microscopy by dropping nanoparticle suspensions onto formvar-coated TEM grids and drying at 60 °C over the course of 3 days to fully evaporate water.

Scanning electron microscopy (SEM) was performed using a Hitachi-S5500 microscope operated at 5 kV. Samples were prepared by lowering a piece of stainless steel foil into two different ECA solutions (10 mM N_2H_4 and 5 mM SPB, pH 7.8 and 10 mM N_2H_4 and 50 mM SPB, pH 7.8) for two hours and allowing them to adsorb the Pt NP aggregates.

Nanoparticle impact electroanalysis

The NP impact data analyzed in this study along with the experimental protocol is reported in a previous publication.[1] The NP frequency was calculated by applying a custom moving sum-average with an Excel spreadsheet. Specifically, the frequency for the first ten impacts was calculated by dividing the sum total by the elapsed time after Pt NP introduction. All other events after the 10th impact were counted as a total of 10 events divided by the time in between the first event of the set and first event of the next ten impacts. This data processing was also applied in the backward direction starting at the end time and the two forward/backward results were averaged together to construct the plot in Fig. 10a. The scatter data in Fig. 10a was fit to the Smoluchowski coagulation equation as a custom function in the Origin 2016 data analysis package to produce the plot in Fig. 10b.

Results and discussion

Time-resolved NTA: Pt NP concentration measurements

Pt NP concentrations were monitored over time by NTA for three different buffer species: sodium phosphate buffer (SPB), Tris sulfate buffer (TSB), and hydrazine sulfate buffer (HSB), as shown in Fig. 1. TSB was chosen as a relevant system of interest to hydrazine because the Tris amino base has the same pK_a as hydrazine (8.1). It also works out that the ionic strength of both buffer species is equivalent to their respective formal concentrations at the pH maintained for these experiments. Therefore, the average valency of ions in solution is also equivalent for both buffers prepared at the same total ion concentration, so any difference in aggregation phenomena between the two conditions cannot be attributed to the relative degree of electrostatic charge-screening at the nanoparticle's electrical double layer. The left panel shows that for 10 mM ionic strength, the Pt colloid maintains its initial concentration in the presence of all three buffering species. We also performed these experiments with 25 mM ionic strength (not shown) to obtain a similar result as that for the 10 mM condition. As shown in the right

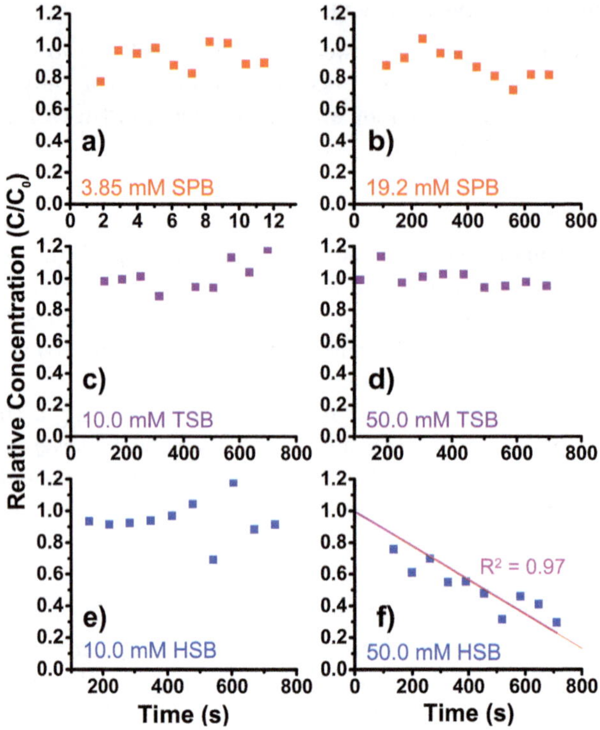

Fig. 1 Pt NP concentration measured over time by NTA. Ionic strength is 10 mM for panels (a, c and e); and 50 mM for (b, d and f) with respective buffer concentrations as labeled. The linear fit in part (f) was forced to intercept at C_0 (1.1 pM Pt NPs). Buffers were freshly prepared at pH 7.8 prior to acquisition.

panel of Fig. 1, the colloid also remains at a consistent concentration at 50 mM ionic strength for chemically inert SPB and TSB (Fig. 1b and d). In HSB, however, the Pt NP concentration does in fact decrease significantly over time (Fig. 1f), reducing to half of the initial concentration after 7–8 minutes. The aggregation rate is apparently linear at this time scale, and with an observed aggregation rate of -4.3×10^7 NPs mL^{-1} s^{-1}. The Pt colloid in 50 mM HSB therefore represents a unique experimental condition whereby aggregation is specifically induced by hydrazine, even when the ionic species for comparison (TSB) is a primary amine with similar basicity and assumedly comparable amine-Pt binding affinity. This result suggests that the reactivity of hydrazine at the Pt NP surface does indeed play a significant role in promoting aggregation.

NTA zeta potential and size measurements

To further understand the specific role of hydrazine in Pt NP aggregation, we employed the electrokinetic tracking features of NTA to measure zeta potentials of the individual particles in solution. The zeta potential is the potential at the slipping plane of the electrical double layer (EDL) on an individual NP surface in accordance with the Gouy–Chapman–Stern model.[33] The experimentally determined zeta potential values therefore serve as estimations for the activation energy of the pair interparticle interactions that are responsible for citrate-promoted charge stabilization of the Pt colloid. Fig. 2 shows the measured zeta potentials for a variety of solution conditions relevant to the ECA detection strategy, each plotted with respect to the approximated ionic strength. As expected, the Pt colloid in deionized water is the most stable sample, with a high average zeta potential of −100 mV. The Pt NP zeta potential decreases to less than half of this value for all other conditions tested in the range of ionic strength from 10 to 30 mM. The three buffer systems that were analyzed by NP concentration in the left panel of Fig. 1 (10 mM ionic strength) were also investigated. The average and standard deviation (measure of NP polydispersity and standard error) of the individual Pt NP zeta potential distributions for all three of these solution conditions are virtually identical (−46 ± 14 mV). This zeta potential for citrate-capped Pt NPs is sufficient for colloid stabilization, as verified earlier in the concentration measurements from Fig. 1. At higher buffer ionic strength (~25 mM; corresponding to 25 mM TSB and 10 mM SPB), the average zeta potential for the Pt NPs is decreased by ~20 mV with respect to the 10 mM ionic strength conditions and the zeta polydispersity increases as indicated by the vertical bars in Fig. 2.

The zeta result for the mixed hydrazine/buffer species condition of 5 mM SPB + 10 mM N_2H_4 is statistically similar to that of the three stable NP samples in 10 mM ionic strength (Fig. 2), which further supports our previously reported conclusions that the colloid is stable at this condition of important relevance to ECA detection of single Pt NPs.[1] The measured mean NP diameters are quite similar for all conditions tested. However, a slightly higher degree of aggregation is indicated for the SPB supported colloids according to the 4 nm difference between 3.85 mM SPB and the monoprotic weak base buffers, HSB and TSB. The hydrazine-containing sample has the largest NP size overall despite the fact that its ionic strength is much lower than the 10 mM SPB sample. This finding further supports our hypothesis that hydrazine is promoting aggregation by chemically interacting with the Pt NPs in solution, but that these interactions can be regulated by tuning the pH buffer concentration.

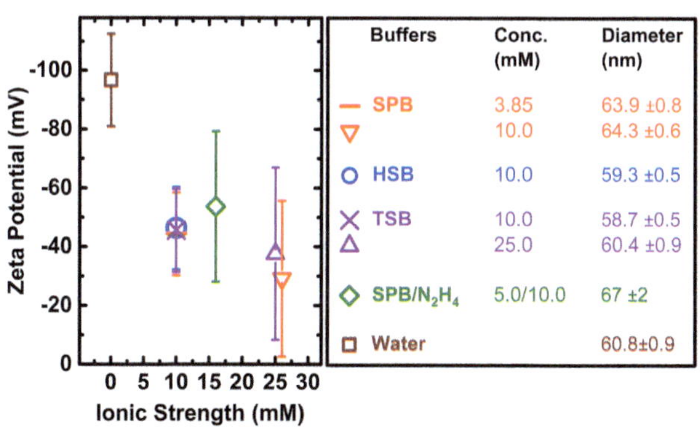

Fig. 2 NTA-measured Pt NP zeta potentials plotted as a function of ionic strength. Various solution conditions were investigated as labeled in the legend. The corresponding mean Pt NP hydrodynamic diameters and standard deviations (standard error) are listed in the legend. Buffers were freshly prepared at pH 7.8 prior to acquisition.

DLVO predictions and effect of Pt NP structure

The DLVO model of interparticle interactions is one of the most heavily applied theories to assess the stability of colloidal dispersions, and has been shown to correlate well with experimental measurements in various colloid systems. The DLVO approximation for the interaction energy between two approaching particles is based on the summation of the attractive van der Waals (V_{VDW}) and electrical double layer (V_{EDL}) contributions to obtain the pair interparticle potential (V_{T}) according to eqn (1).

$$V_{\mathrm{VDW}} + V_{\mathrm{EDL}} = V_{\mathrm{T}} \tag{1}$$

The Derjaguin approximation[34,35] serves as a simple model to estimate the van der Waals interaction energy as a function of interparticle separation distance (H) for hard spheres of equivalent radius (r) according to eqn (2),

$$V_{\mathrm{VDW}} = -\frac{A_{12}}{12h_1}\left[\frac{1}{1+h_2} + \frac{h_1}{1+h_1+h_3} + 2h_1 \ln\left\{h_1\left(\frac{1+h_2}{1+h_1+h_3}\right)\right\}\right] \tag{2}$$

where $h_1 = \frac{H}{r}$; $h_2 = \frac{H}{4r}$; $h_3 = \frac{H^2}{4r^2}$ and A_{12} is the Hamaker constant for the metal–water (1–2) interface.[36] One can also approximate the interaction energy of two overlapping double layers between two spherical NPs using eqn (3),

$$V_{\mathrm{EDL}} = 2\pi\varepsilon_{\mathrm{rel}}\varepsilon_0\zeta^2 r \ln[1 + \exp(-\kappa H)] \tag{3}$$

where $\varepsilon_{\mathrm{rel}}$ is the relative permittivity of the solvent, ε_0 is the permittivity of free space, ζ is the experimentally measured zeta potential from NTA, and κ is the inverse Debye length, calculated by eqn (4).

$$\kappa = \left[\frac{1000e^2N(2I)}{\varepsilon_{\mathrm{rel}}\varepsilon_0 k_{\mathrm{B}}T}\right]^{\frac{1}{2}} \tag{4}$$

Here e is the elementary charge, N is Avogadro's number, and I is the ionic strength. Eqn (3) is only valid when $\kappa r > 5$, such is the case for low potentials at high ionic strengths. At high zeta potentials, such as that observed for the 50 nm citrate-capped Pt NPs when dispersed in deionized water (Fig. 2), $\kappa r < 5$ and the expression for V_{EDL} then becomes

$$V_{\mathrm{EDL}} = \frac{4\pi\varepsilon_{\mathrm{rel}}\varepsilon_0}{H+2r}\left(\frac{k_{\mathrm{B}}TrY}{e}\right)^2 \exp(-\kappa H) \tag{5}$$

with

$$Y = \frac{8\tanh\left(\frac{e\zeta}{4k_{\mathrm{B}}T}\right)}{1 + \left[1 - \frac{(2\kappa r+1)}{(\kappa r-1)}\tanh^2\left(\frac{e\zeta}{4k_{\mathrm{B}}T}\right)\right]} \tag{6}$$

Fig. 3 shows the calculated prediction for V_{T} as a function of separation distance between two 50 nm Pt NPs at two different ionic strengths. The V_{EDL} at 10^{-3} mM ionic strength (blue solid curve) was calculated using eqn (5) and (6) because the calculated κr value is 3. The inverse of κ gives a Debye screening

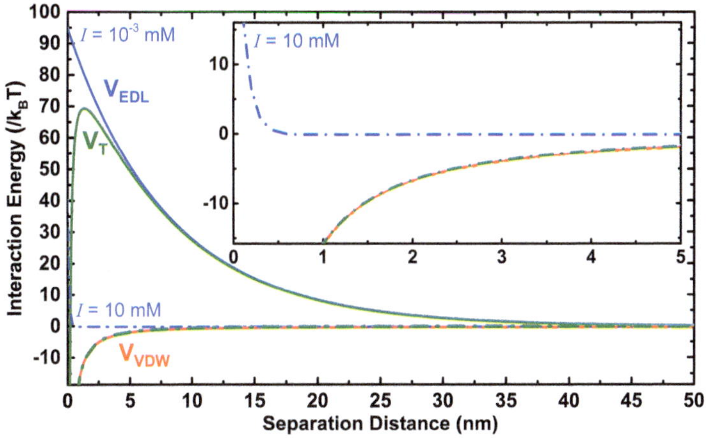

Fig. 3 Pair interparticle potentials calculated for Pt NPs based on eqn (1)–(6): van der Waals attraction, V_{VDW} (red), electrical double layer repulsion, V_{EDL} (blue), and total interaction energy, V_T (green) for 10^{-3} mM ionic strength (solid lines) and 10 mM ionic strength (dashed lines). Inset panel is expanded view for the 10 mM ionic strength condition. V_{EDL} was calculated using the corresponding measured zeta potential values from Fig. 2, with the zeta from deionized water applied for 10^{-3} mM ionic strength. V_{VDW} was approximated using an A_{12} value of 4×10^{-20} J based on the "crossed polarized wires" experimental determinations by Derjaguin and coworkers.[49]

length of 10 nm for this low electrolyte solution. V_{EDL} for 10 mM ionic strength (blue dashed curve) was calculated using the expression for $\kappa r > 5$ (eqn (3)). As shown in Fig. 3, V_T is predicted by DLVO theory to be a repulsive barrier at 10^{-3} mM ionic strength. For 10 mM ionic strength, however, the calculation predicts significant screening of the Pt surface charge to result in a net energy that is equivalent to V_{VDW}, essentially indicating that the 10 mM ion concentration is higher than the critical concentration of coagulation.[37–39] Our experimental results do not fit the traditional DLVO prediction of rapid coagulation for these colloid species. Experimental disagreements with DLVO theory are well documented in the colloid literature, and the various factors that the approximation does not take into account continue to be a heated topic of discussion.[40]

A more appropriate theoretical model must include another repulsive contribution to the V_T equation to explain our observations. We reason that the key to stability enhancement most likely involves the structure of the absorbed citrate on the Pt surface. We suspect this structure involves a substantial degree of H-bonding, such as that determined by Shumaker-Parry for Au,[20] with the addition of enhanced stability coming from Pt NP's hydrophilicity and more favorable binding energetics with the OH group of citrate due to the presence of Pt surface oxides/hydroxides. The Pt NPs used in this study have a nanostructured surface roughness of ~5 nm due to the aggregative growth mechanism during synthesis,[41] which we expect to promote a higher relative surface coverage of Pt oxides/hydroxides with respect to bulk polycrystalline Pt due to stronger Pt–OH interactions at NPs $\leq$ 5 nm in diameter.[42,43] Interestingly, Eychmüller and coworkers, who originally reported the synthesis of these Pt NPs, found it necessary to include a dielectric shell of 5 nm thickness in their computations of Pt NP extinction spectra in order to fit their experimental results.[41] If Pt nanostructure promotes H-bonded networks with citrate, then perhaps the thickness of this steric/charge stabilizing shell really is ∼5 nm. The −100 mV zeta value measured in pure water for our NPs is relatively high in comparison to citrate-capped Pt NPs of smaller size (∼5 nm, $\zeta = -40$–50 mV).[44] The zeta potential is also larger than that observed for Au[45] and Ag NPs[46] of comparable size. The zeta potential of citrate-stabilized Ag NPs, for example, has been reported to range from −27 to −53 mV for monodispersed colloids with an average diameter between 5 and 100 nm.[47] We suspect that Pt NP surface structure promotion of the H-bonding model with extended multi-citrate layer thickness accounts for this zeta potential discrepancy in relation to size, structure, and NP composition. The top panel of Fig. 4 shows an STEM image of two citrate NPs. The area of low contrast surrounding the NPs was observed to develop over time during image acquisition and was acquired after the layer had completed the thermally driven phase expansion caused by energetic losses to heat from the applied 30 kV accelerating voltage of the electron beam. We speculate that this shell is primarily composed of citrate and water that was trapped with the porous structure of the Pt NP during

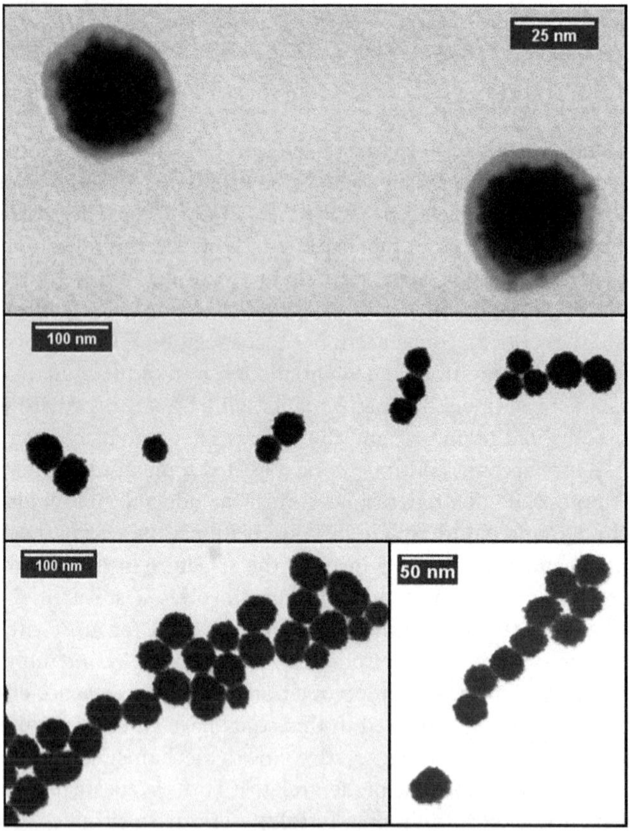

Fig. 4 Scanning transmission electron micrographs of citrate-stabilized Pt spherical NPs. The top panel shows two Pt NPs at high magnification and the film that develops around the NPs over time under exposure of the electron beam. The bottom three panels show representative images of chain-like Pt NP aggregates.

the synthesis and later expelled upon swelling under the electron beam. The coating material may also include oxidation products from the reducing agents, such as polymeric borate species[48] from sodium borohydride.

OCP experiments

OCP experiments were performed on a Pt wire to assess the stability of the citrate surface structure upon exposure to hydrazine solutions. Fig. 5a depicts the scheme for surface pretreatment experiments and Table 1 lists the corresponding OCP measurements. First, the OCP of a clean, flame-annealed Pt wire was measured in 50 mM SPB under ambient conditions to obtain a value of 260 mV *vs.* SCE, which is estimated to be ~0.95 V *vs.* RHE in pH 7.8 based on the Nernst equation (RHE = 0 V − 0.0591 × pH *vs.* NHE). This potential is the expected value based on the onset of the oxygen reduction reaction on Pt.[8] After cleaning the electrode again, it was then immersed in 100 mM HSB for 1000 s (step 1, Fig. 5a), giving an OCP of −510 mV which is poised between the electrocatalytic onsets for O_2 reduction and N_2H_4 oxidation. Importantly, macroscopic bubbles were observed to form on the surface of the electrode during this time period, indicating Pt-catalyzed hydrazine decomposition. The N_2H_4 decomposition reaction can occur by two alternative pathways, the direct pathway to form hydrogen and nitrogen gas as shown in eqn (7), and the alternative pathway leading to the formation of nitrogen gas and ammonia (eqn (8)).

$$N_2H_4 \rightarrow N_2 + 2H_2 \tag{7}$$

$$3N_2H_4 \rightarrow 4NH_3 + N_2 \tag{8}$$

Hydrazine decomposition on polycrystalline Pt and Pt NPs has been shown to significantly favor the ammonia production pathway in both acidic[50] and alkali[51]

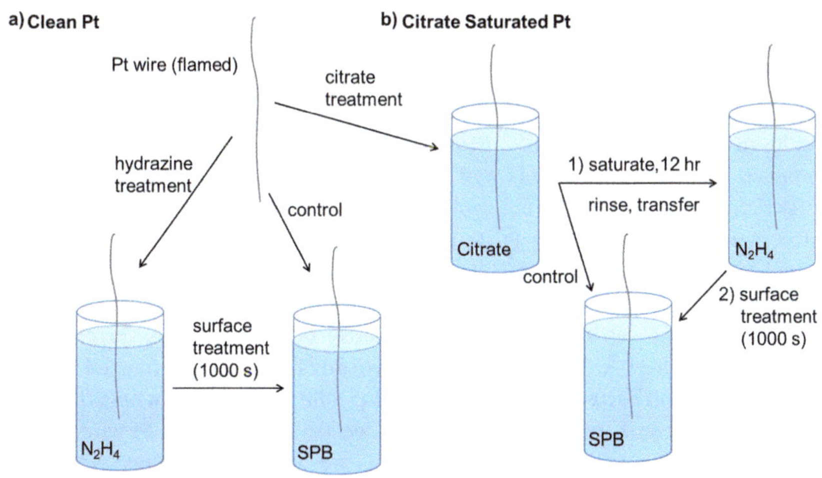

Fig. 5 Schematic representation of Pt surface pretreatment OCP experiments. Pt OCP was referenced to SCE.

Table 1 List of solution conditions outlined in Fig. 5 and measured OCP values. Experiments with listed error were performed in triplicate measurements or more

	Conditions	OCP (mV)
Clean Pt	Initial	255 ± 7
	100 mM HSB	-512 ± 6
	Post-treatment	140 ± 40^a
Citrate saturated Pt	Initial	87 ± 2
	100 mM HSB	-416 ± 5
	Post-treatment	-80 ± 50^a
	10 mM HSB	−490
	Post-treatment	87
	10 mM N_2H_4 + 50 mM SPB	−388
	Post-treatment	92

[a] For the post-treatment experiments in 100 mM HSB, the standard deviation in the OCP over a 1000 s time period was propagated with the standard error.

conditions. So while Pt is an excellent catalytic surface for efficient N_2H_4 decomposition, the reaction follows the less desirable NH_3 path making Pt a poor choice of catalyst for the efficient production of H_2 from N_2H_4.[51] We therefore assume that the observed bubbles are primarily composed of gaseous N_2 and NH_3. The catalytically evolved bubbles formed at a gradual rate and the OCP was not observed to fluctuate significantly over the course of a single measurement. In step 2 of Fig. 5a, the Pt wire from the N_2H_4 solution was quickly transferred to SPB to obtain a lower OCP than seen for the clean Pt surface. In this experiment, the potential *vs.* time was observed to fluctuate during the acquisition time and vary between trials. The cause of the observed fluctuation is unclear and could be due to several possible processes that are suggested to occur on deactivating Pt surfaces.[52]

Shown in Fig. 5b is the description of the citrate treatment method, the first step involving the saturation of the Pt wire surface with a Na citrate solution. After a gentle rinse, the electrode was transferred to 50 mM SPB to measure the effect of the citrate coating on the electrode potential (Table 1). We attribute the −178 mV difference between the OCPs measured for the clean and the citrate-treated Pt wires to the influence of the charged citrate layer adsorbed on the electrode. The potential shift is higher than the measured zeta potential for the Pt NPs (−100 mV) because the OCP measurement represents the potential at the compact double layer, which is well known to be greater in value than the zeta potential. We then tested the effect of 100 mM HSB on the adsorbed citrate by the same method used for clean Pt. Bubbles also gradually formed for the citrate-coated Pt wire in the 100 mM HSB condition, so the wire was still catalytically active for N_2H_4 decomposition despite the citrate-saturated surface. The equilibrium potential was shifted ~100 mV positive of that observed for clean Pt, indicating that N_2H_4 oxidation on Pt is kinetically hindered by attached citrate. For the OCP result corresponding to step 2 of Fig. 5b, we observed a negative shift that is roughly comparable in value to what we saw for the clean Pt wire experiment (~100–200 mV difference), so there is no apparent evidence from this experiment of citrate removal from the electrode after exposure to N_2H_4.

These experiments were also performed for the 10 mM N_2H_4 solutions that were studied previously with NTA and NP impact experiments.[1] Similar results are observed for both 10 mM N_2H_4 conditions with and without 50 mM SPB added, suggesting that ionic strength during the N_2H_4 pretreatment does not have an effect on the activity of citrate-coated Pt. There is no difference between the initial citrate-coated and post-treated Pt OCP values for the 10 mM N_2H_4 conditions, indicating that the citrate remains adsorbed to the electrode.

The observed bubble formation and OCP shift for 100 mM N_2H_4 warrants a discussion on the possibility of nanobubbles[53–55] forming from the products of catalytic N_2H_4 decomposition on these structured Pt NP surfaces. The properties of both the Pt surface and the encompassing citrate layer are expected to contribute to nanobubble stabilization, regardless of the decomposition product. We propose that a H-bonded citrate film allows the diffusion of most N_2H_4 to the NP surface and influences mass transport to some extent by electrostatic attraction of $N_2H_5^+$. This electrostatic effect could either enhance or hinder hydrazine transport to the Pt active sites by electromigration (enhanced transport) and electrostatic trapping (hindered transport) of incoming $N_2H_5^+$ at the charged carboxylate groups of citrate. The transport of hydrazine is therefore dependent on the porosity and thickness of the H-bonded polycitrate layer. We expect that most hydrazine molecules penetrate through the citrate "membrane" deep into the interstitial sites of the nanostructured Pt NP. The gaseous products that form, however, may become trapped in between the Pt surface and the H-bonded polymeric citrate coating layer. The nanobubble stabilizing mechanism is a working hypothesis at this juncture and beyond the scope of this study. However, we speculate that if nanobubbles are stabilized to some extent on these Pt NPs, they could very well influence the coagulation and direct it to occur at the most active Pt nanograins. Plausibly, the Pt catalyzed decomposition mechanism (eqn (8)) is the most likely explanation for the observed specific hydrazine-induced aggregation result from NTA in Fig. 1f.

Aggregation kinetics

Fig. 6a shows Pt NP size distributions over time as monitored by NTA for the aggregation inducing 50 mM SPB + 10 mM N_2H_4 condition. Fig. 6b shows the corresponding mean NP diameters plotted over a time of one hour. After 30 minutes of aggregate growth, a linear growth region is observed corresponding to an average aggregation rate of 3.8 nm min^{-1}. Fig. 6c shows the natural log of the aggregate diameter divided by the initial 50 nm diameter, $\ln(d/d_p)$, as a function of $\ln(t/\tau)$, where t is time and τ is the time constant for the rate of coagulation, determined to be roughly 240 s, which is the time when the NP concentration decreases to half its initial value (Fig. 6a). The ln–ln plot in Fig. 6c clearly shows two different rate regimes; the earlier one is slow followed by an abrupt transition to a faster rate regime after 30 minutes. The slow growth region indicates reversible coagulation. For the fast rate a fractal dimension (δ) of 1.0 from the inverse slope[56,57] of the plot in Fig. 6c is obtained, indicating a linear structure of the Pt NP aggregates. The inset electron micrograph in Fig. 6b shows a small linear chain of four Pt NPs that aggregated naturally during solvent evaporation on the hydrophobic carbon film coated TEM grid. Other Pt NP chains are shown in Fig. 4. The SEM images in Fig. 7 and 8 show the aggregate structures

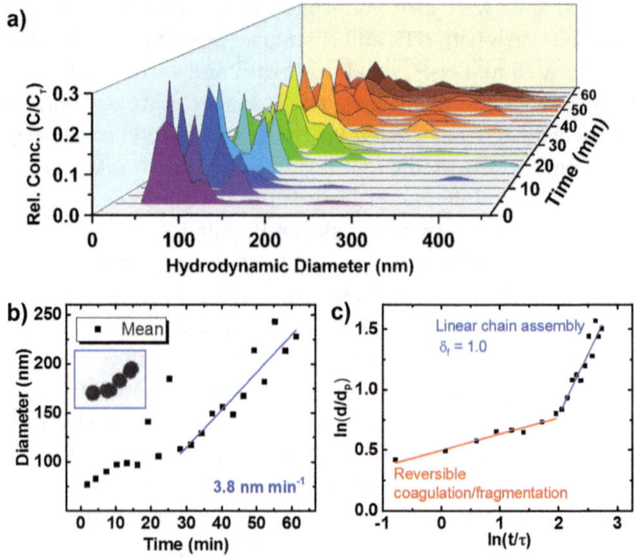

Fig. 6 NTA monitored Pt NP aggregation kinetics in 10 mM N_2H_4 + 50 mM SPB, pH 7.8 solution over a one-hour time period: (a) three-dimensional plot of hydrodynamic particle size distributions over time, (b) mean and mode diameters as a function of time, and (c) corresponding double logarithmic plot relating ratio of aggregate/primary Pt NP diameters (d/d_p) to time expressed as t/τ. The inset STEM image in (b) shows a chain of 4 primary 50 nm Pt NPs. Aggregation rate regimes are indicated by the figure labels in panel (c) for reversible coagulation/fragmentation (red) and chain-like assembly (blue) [diameters at 20 and 25 minutes were excluded as outliers by residuals analysis for fits in (b and c)].

formed in the presence of two ECA buffers that we've investigated previously with NTA and NP impact rate analysis. For these experiments, a piece of stainless steel foil was lowered into the ECA solution and left to adsorb the Pt NP aggregates that formed over the course of 2 hours. Fig. 7 shows a mixture of primary 50 nm Pt NPs and small aggregates containing no more than 5 primary NP units that adsorbed on the metal surface in 10 mM N_2H_4 + 5 mM SPB solution. In Fig. 8, we see larger aggregates that formed in the higher ionic strength 10 mM N_2H_4 + 50 mM SPB solution, as expected. Chain-like NP aggregates appeared in both figures along with some that adopted more closely packed arrangements.

We suggest that our NPs are directed to assemble in a linear fashion by stronger bridging interactions and large electrostatic repulsive contributions from the densely adsorbed citrate on the surface, in line with Johnston's proposed mechanism.[21] However, one must also consider the heterogeneity of the Pt nanograins on the NP surface. Some of these crystallites may be much more active than others for N_2H_4 decomposition. We are not sure exactly how this relates to aggregation. It could create pressure-driven repulsion if the rate of decomposition is sufficiently fast and bubbles are released at a continuous frequency. In this sense we are proposing that the Pt NPs may be behaving like nanomotors propelled by gas production.[58] If the gas is building up from within the Pt NP, then this could eventually rupture the stabilizing citrate layer at a specific site and further promote linear assembly. An alternate explanation for the nanobubble hypothesis is that hydrazine is becoming a part of the H-bonded citrate network,

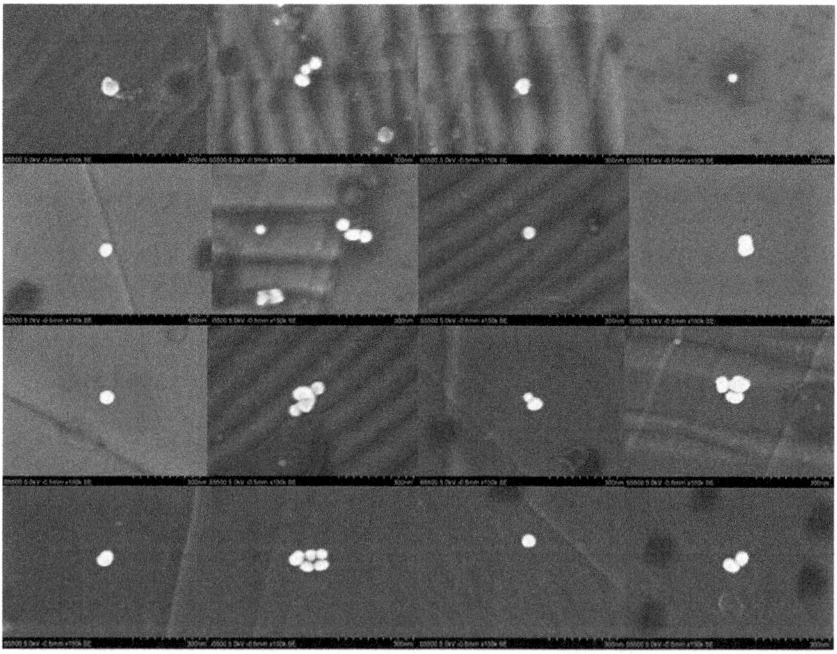

Fig. 7 SEM of Pt NPs adsorbed on stainless steel foil from a solution of 10 mM N_2H_4 + 5 mM SPB, pH 7.8 over the course of one hour at 1.1 pM Pt NP concentration.

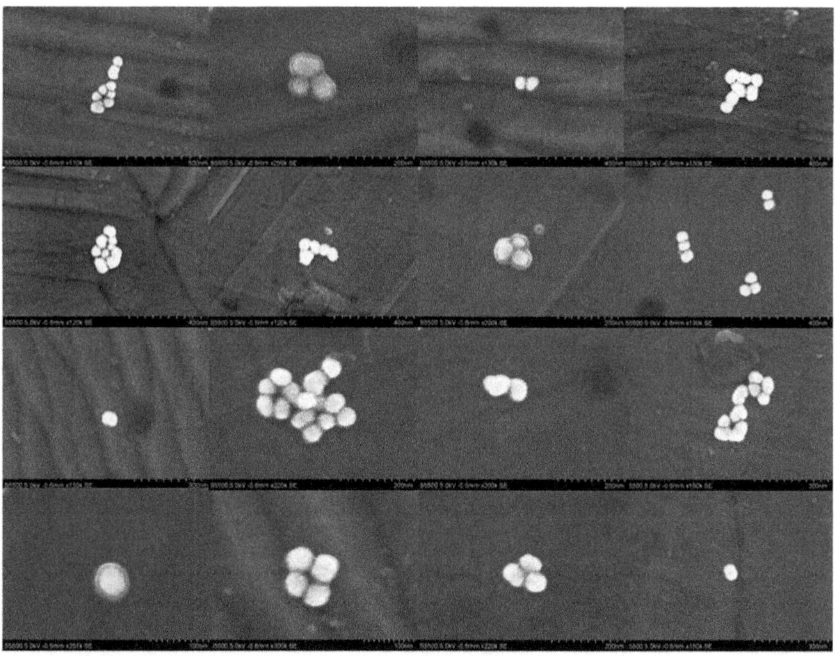

Fig. 8 SEM of Pt NPs that adsorbed on stainless steel foil from a solution of 10 mM N_2H_4 + 50 mM SPB, pH 7.8 over the course of one hour at 1.1 pM Pt NP concentration.

as proposed by Li and coworkers.[59] Indeed, N_2H_4 could very well function as both H-bond donor and acceptor. The N_2H_4 may be donating H to the carboxyl groups of citrate and create a strong bridging interaction,[21] thus cross-linking the Pt NPs by interparticle H-bonding to induce aggregation.

Monitoring aggregation kinetics with nanoparticle impact electroanalysis

Finally, utilizing our theory of Pt NP aggregation, we show that NP impact events detected by ECA can be calibrated to NP concentration, even for nonlinear fast coagulating conditions. The chronoamperometric trace in Fig. 9 shows the ECA impact events for a single measurement with 1.1 pM Pt NPs in 10 mM N_2H_4 + 50 mM SPB. As we pointed out previously, the NP impact frequency decreases significantly over time. This is shown graphically in Fig. 10a. For the spherical cap UME used in this study, the impact frequency is proportional to NP concentration according to eqn (9).

$$R_{\rm NP} = f_{\rm NP}C^{-1} = 2\pi\beta N_{\rm A}Da_{\rm UME} \tag{9}$$

$R_{\rm NP}$ is the impact rate, $f_{\rm NP}$ is the frequency of NP impacts at a given NP concentration (C), $N_{\rm A}$ is Avogadro's number, D is the NP diffusion coefficient, and a is the basal radius of the Hg cap (5 μm for Pt disk UME substrate). The term β is a size dependent constant based on the calculations by Alfred and Oldham.[60] For our Hg UME, $\beta = 1.35$.[1]

Fig. 10a shows the NP impact frequency as a function of time for five replicate measurements. We estimated $C_{\rm NP}$ by rearranging eqn (9) using the measured impact frequency and an average diffusion coefficient of $\sim 1 \times 10^{-7}$ cm^2 s^{-1} (50 nm diameter)[29,30] to calculate the scatter plot in Fig. 10a (blue), which is shown as relative concentration (C/C_0) with $C_0 = 1.1$ pM, the experimentally determined value in water by NTA. The measured C/C_0 values from NTA are included for comparison. The scatter plots were fitted using the general Smoluchowski coagulation function[61] (eqn (10)).

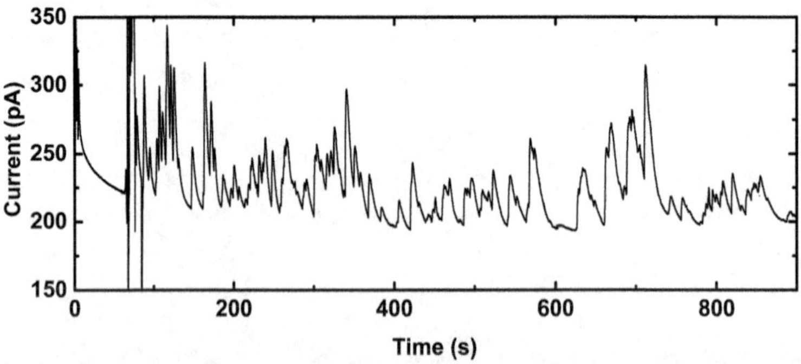

Fig. 9 Chronoamperometry of Pt NP impacts in 10 mM N_2H_4 and 50 mM SPB, pH 7.8. Pt NP concentration is 1.1 pM. The sharp current fluctuations at *ca.* $t \sim 60$ s mark the time when Pt NPs were introduced into the hydrazine/buffer solution (4 mL) as a 20 μL injection caused by opening the door on the Faraday cage.

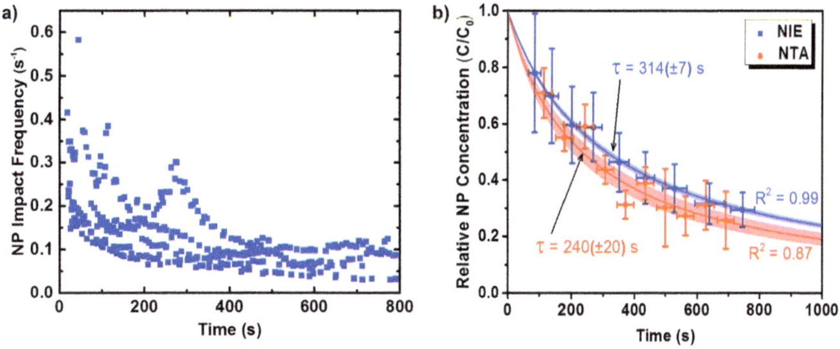

Fig. 10 Nanoparticle impact electroanalysis (NIE) of aggregating Pt NPs. Shown in blue are the NP impact frequency, (a), and relative concentration (averaged into discrete time intervals for every 35 NP impacts), (b), plotted as a function of time from five replicate trials with 1.1 pM Pt NPs (C_0), 10 mM N_2H_4 and 50 mM SPB, pH 7.8. The vertical error bars in (b) are the standard error and the horizontal bars are the time range extending from the midpoint time. The data shown in red comes from NTA and was constructed similarly from three replicate NTA trials. The curve fits for both experiments were derived from eqn (10). Calculated τ values are indicated in the figure labels. Adjusted R^2 values are also shown.

$$\frac{C}{C_0} = \frac{1}{1 + \frac{t}{\tau}} \tag{10}$$

Here $1/\tau$ represents the coagulation rate and τ is the time when the total particle concentration, C, is decreased to half of its initial value. As shown in Fig. 10b, both datasets match well to show the rate of coagulation, with NP impact electroanalysis giving a slightly slower coagulation rate and better fit with eqn (10) after averaging the raw data in Fig. 10a, as described in the Fig. 10 caption and Experimental section. These results demonstrate a proof-of-concept that by analyzing NP impact events, one can calculate the dynamic NP concentration during aggregation, even for nonlinear fast coagulating conditions.

Conclusions

The combined results of these experiments support that the stability of citrate on Pt involves much more than just electrostatic interactions. A structure based on intermolecular H-bonding in the citrate capping layer appears to be the most likely explanation for the exceptional stability of these large Pt NPs in 50 mM ionic strength in the absence of hydrazine and the stability in hydrazine-containing solutions of lower ionic strength (<20 mM). The stability was evidenced by the range of reversible metastability (agglomeration) before linear directed assembly of Pt NP chains. The mechanism of Pt NP aggregation includes specific reactive contributions from hydrazine, as indicated by the observed hydrazine decomposition gaseous products on a citrate-coated Pt wire. According to the OCP measurements, the citrate remains adsorbed on the Pt surface even after treatment with 10 mM and 100 mM N_2H_4. This is most likely true for the Pt NPs as well. Although citrate stays stuck to the Pt and does not undergo substantial ligand-exchange with hydrazine, we expect that the membrane-like structure is disrupted by evolved gaseous decomposition products or adsorbed intermediates and then ruptures, causing instability and promoting coagulation. The heterogeneity of the nanostructured Pt NP surface is expected to influence the aggregation mechanism and induce polarity by bridging interactions, which we expect to involve both a higher degree of interparticle H-bonding at the terminal primary NPs of the chain in the presence of hydrazine and the physicochemical contributions from Pt-catalyzed hydrazine decomposition products. In the last section, we demonstrated a proof-of-concept using nanoparticle impact electroanalysis by electrocatalytic amplification as a method for quantifying Pt NPs with suitable time resolution for studying kinetics of Pt colloid aggregation.

Acknowledgements

We thank the Defense Threat Reduction Agency (DTRA) (grant HDTRA1-11-1-0005) for financial support.

Notes and references

1 D. A. Robinson, A. M. Kondajji, A. D. Castañeda, R. Dasari, R. M. Crooks and K. J. Stevenson, *J. Phys. Chem. Lett.*, 2016, **7**, 2512–2517.
2 X. Xiao and A. J. Bard, *J. Am. Chem. Soc.*, 2007, **129**, 9610–9612.
3 X. Xiao, F.-R. F. Fan, J. Zhou and A. J. Bard, *J. Am. Chem. Soc.*, 2008, **130**, 16669–16677.
4 X. Xiao, S. Pan, J. S. Jang, F.-R. F. Fan and A. J. Bard, *J. Phys. Chem. C*, 2009, **113**, 14978–14982.
5 S. J. Kwon, H. Zhou, F.-R. F. Fan, V. Vorobyev, B. Zhang and A. J. Bard, *Phys. Chem. Chem. Phys.*, 2011, **13**, 5394–5402.
6 H. Zhou, J. H. Park, F.-R. F. Fan and A. J. Bard, *J. Am. Chem. Soc.*, 2012, **134**, 13212–13215.
7 S. J. Kwon and A. J. Bard, *J. Am. Chem. Soc.*, 2012, **134**, 10777–10779.
8 S. E. F. Kleijn, B. Serrano-Bou, A. I. Yanson and M. T. M. Koper, *Langmuir*, 2013, **29**, 2054–2064.
9 T. M. Alligrant, E. G. Nettleton and R. M. Crooks, *Lab Chip*, 2013, **13**, 349–354.
10 R. Dasari, D. A. Robinson and K. J. Stevenson, *J. Am. Chem. Soc.*, 2013, **135**, 570–573.
11 R. Dasari, B. Walther, D. A. Robinson and K. J. Stevenson, *Langmuir*, 2013, **29**, 15100–15106.
12 R. Dasari, K. Tai, D. A. Robinson and K. J. Stevenson, *ACS Nano*, 2014, **8**, 4539–4546.
13 T. M. Alligrant, M. J. Anderson, R. Dasari, K. J. Stevenson and R. M. Crooks, *Langmuir*, 2014, **30**, 13462–13469.
14 T. M. Alligrant, R. Dasari, K. J. Stevenson and R. M. Crooks, *Langmuir*, 2015, **31**, 11724–11733.
15 A. D. Castañeda, T. M. Alligrant, J. A. Loussaert and R. M. Crooks, *Langmuir*, 2015, **31**, 876–885.
16 D. A. Robinson, J. J. Yoo, A. D. Castañeda, B. Gu, R. Dasari, R. M. Crooks and K. J. Stevenson, *ACS Nano*, 2015, **9**, 7583–7595.
17 M. Y. Lin, H. M. Lindsay, D. A. Weitz, R. C. Ball, R. Klein and P. Meakin, *Nature*, 1989, **339**, 360–362.

18 M. Y. Lin, H. M. Lindsay, D. A. Weitz, R. Klein, R. C. Ball and P. Meakin, *J. Phys.: Condens. Matter*, 1990, **2**, 3093.

19 M. Y. Lin, H. M. Lindsay, D. A. Weitz, R. C. Ball, R. Klein and P. Meakin, *Phys. Rev. A*, 1990, **41**, 2005–2020.

20 J.-W. Park and J. S. Shumaker-Parry, *J. Am. Chem. Soc.*, 2014, **136**, 1907–1921.

21 R. J. Stover, E. Moaseri, S. P. Gourisankar, M. Iqbal, N. K. Rahbar, B. Changalvaie, T. M. Truskett and K. P. Johnston, *Langmuir*, 2016, **32**, 1127–1138.

22 B. Zhang, T. Lan, X. Huang, C. Dong and J. Ren, *Anal. Chem.*, 2013, **85**, 9433–9438.

23 A. E. James and J. D. Driskell, *Analyst*, 2013, **138**, 1212–1218.

24 J. G. Mehtala and A. Wei, *Langmuir*, 2014, **30**, 13737–13743.

25 J. A. Gallego-Urrea, J. Tuoriniemi and M. Hassellöv, *TrAC, Trends Anal. Chem.*, 2011, **30**, 473–483.

26 R. F. Domingos, M. A. Baalousha, Y. Ju-Nam, M. M. Reid, N. Tufenkji, J. R. Lead, G. G. Leppard and K. J. Wilkinson, *Environ. Sci. Technol.*, 2009, **43**, 7277–7284.

27 R. D. Boyd, S. K. Pichaimuthu and A. Cuenat, *Colloids Surf., A*, 2011, **387**, 35–42.

28 A. Dudkiewicz, S. Wagner, A. Lehner, Q. Chaudhry, S. Pietravalle, K. Tiede, A. B. A. Boxall, G. Allmaier, D. Tiede, R. Grombe, F. von der Kammer, T. Hofmann and K. Molhave, *Analyst*, 2015, **140**, 5257–5267.

29 W. Sutherland, *Philosophical Magazine Series 6*, 1905, **9**, 781–785.

30 A. Einstein, *Investigations on the Theory of the Brownian Movement*, Courier Corporation, 1956.

31 B. Carr and M. Wright, *Innovations Pharm. Technol.*, 2008, **26**, 38–40.

32 V. Filipe, A. Hawe and W. Jiskoot, *Pharm. Res.*, 2010, **27**, 796–810.

33 R. J. Hunter, R. H. Ottewill and R. L. Rowell, *Zeta Potential in Colloid Science: Principles and Applications*, Elsevier Science, 2013.

34 B. V. Derjaguin, L. Landau, *et al.*, *Acta Physicochim. URSS*, 1941, **14**, 633–662.

35 R. J. Hunter and L. R. White, *Foundations of Colloid Science*, Clarendon Press, 1987.

36 H. C. Hamaker, *Physica*, 1937, **4**, 1058–1072.

37 H. Schulze, *J. Prakt. Chem.*, 1882, **25**, 431–452.

38 W. B. Hardy, *Proc. R. Soc. London*, 1899, **66**, 110–125.

39 T. Cao, I. Szilagyi, T. Oncsik, M. Borkovec and G. Trefalt, *Langmuir*, 2015, **31**, 6610–6614.

40 D. Grasso, K. Subramaniam, M. Butkus, K. Strevett and J. Bergendahl, *Rev. Environ. Sci. Bio/Technol.*, 2002, **1**, 17–38.

41 N. C. Bigall, T. Härtling, M. Klose, P. Simon, L. M. Eng and A. Eychmüller, *Nano Lett.*, 2008, **8**, 4588–4592.

42 S. Mukerjee and J. McBreen, *J. Electroanal. Chem.*, 1998, **448**, 163–171.

43 E. Roduner, *Chem. Soc. Rev.*, 2006, **35**, 583–592.

44 G. Marzun, C. Streich, S. Jendrzej, S. Barcikowski and P. Wagener, *Langmuir*, 2014, **30**, 11928–11936.

45 S. Huo, H. Ma, K. Huang, J. Liu, T. Wei, S. Jin, J. Zhang, S. He and X.-J. Liang, *Cancer Res.*, 2013, **73**, 319–330.

46 G. Gicheva and G. Yordanov, *Colloids Surf., A*, 2013, **431**, 51–59.

47 S. Agnihotri, S. Mukherji and S. Mukherji, *RSC Adv.*, 2014, **4**, 3974–3983.

48 K. J. Stevenson, D. W. Hatchett and H. S. White, *Langmuir*, 1997, **13**, 6824–6828.

49 B. V. Derjaguin and T. N. Voropayeva, *J. Colloid Sci.*, 1964, **19**, 113–135.

50 A. V. Ananiev, J.-C. Broudic and P. Brossard, *Appl. Catal., A*, 2003, **242**, 1–10.

51 S. K. Singh and Q. Xu, *Inorg. Chem.*, 2010, **49**, 6148–6152.

52 A. J. Bard, *Anal. Chem.*, 1963, **35**, 1602–1607.

53 Q. Chen, L. Luo, H. Faraji, S. W. Feldberg and H. S. White, *J. Phys. Chem. Lett.*, 2014, **5**, 3539–3544.

54 Q. Chen, H. S. Wiedenroth, S. R. German and H. S. White, *J. Am. Chem. Soc.*, 2015, **137**, 12064–12069.

55 D. Seo, S. R. German, T. L. Mega and W. A. Ducker, *J. Phys. Chem. C*, 2015, **119**, 14262–14266.

56 T. A. Witten and L. M. Sander, *Phys. Rev. Lett.*, 1981, **47**, 1400–1403.

57 D. A. Weitz, J. S. Huang, M. Y. Lin and J. Sung, *Phys. Rev. Lett.*, 1984, **53**, 1657–1660.

58 Y. Wang, R. M. Hernandez, D. J. Bartlett, J. M. Bingham, T. R. Kline, A. Sen and T. E. Mallouk, *Langmuir*, 2006, **22**, 10451–10456.

59 Z. Zhao, G. Zhang, Y. Gao, X. Yang and Y. Li, *Chem. Commun.*, 2011, **47**, 12816–12818.

60 L. C. R. Alfred and K. B. Oldham, *J. Phys. Chem.*, 1996, **100**, 2170–2177.

61 J. R. Norris, *Smoluchowski's Coagulation Equation: Uniqueness, Non-uniqueness and a Hydrodynamic Limit for the Stochastic Coalescent*, Mathematical Sciences Research Inst., 1998.

Faraday Discussions

PAPER

Stochastic electrochemistry and photoelectrochemistry of colloidal dye-sensitized anatase nanoparticles at a Pt ultramicroelectrode†

Krishna K. Barakoti, Suman Parajuli, Pushpa Chhetri, Ganesh R. Rana, Rezvan Kazemi, Ryan Malkiewich and Mario A. Alpuche-Aviles*

Received 28th April 2016, Accepted 19th May 2016
DOI: 10.1039/c6fd00100a

We report the stochastic interactions between dye sensitized anatase nanoparticles, suspended in a colloid, and a Pt ultramicroelectrode (UME) that result in step-wise behavior in the current *vs.* time response. The stochastic currents are observed in the dark and under illumination. In the dark, the currents are anodic, consistent with the oxidation of the dye N719 at the Pt surface. The electrochemical behavior of the dye was investigated in MeOH and MeCN with a quasireversible cyclic voltammogram (CV) observed at 1 V s^{-1}. The anodic currents observed in the dark due to nanoparticles (NPs) at the Pt surface are consistent with the CVs in MeOH and MeCN. Under illumination cathodic steps are observed and assigned to the reduction of the oxidized form of the dye generated after electrons are injected into the TiO_2 NPs. The colloidal behavior is a strong function of the history of the colloid with illumination time increasing the size of the agglomerates and with larger agglomerates being less photoelectrochemically active. Agglomerates of *ca.* 100 nm in diameter are proposed to be photoactive entities with a higher probability of detection that contribute to the staircase photocurrent response.

Introduction

We present a study of the stochastic interactions between a Pt ultramicroelectrode and colloidal dye-sensitized nanoparticles (DSNPs). Recent efforts to perform electrochemical measurements on single entities have included the study of colloidal NPs with electrochemical methods, and our group has focused on the study of semiconductor nanoparticles (NPs) with the goal of measuring the photoelectrochemical rates of individual semiconductor NPs. We take our inspiration from the seminal work of stochastic interactions of colloidal NPs pioneered by Lemay[1] with electrocatalytic amplification[2–5] and nano-impact

Department of Chemistry, University of Nevada, Reno, 89557, USA. E-mail: malpuche@unr.edu

† Electronic supplementary information (ESI) available. See DOI: 10.1039/c6fd00100a

methods.[6–11] In electrocatalytic amplification, when the NPs collide with the electrode surface they enable an electrochemical reaction that is sluggish at the substrate electrode, typically an ultramicroelectrode (UME) with diameter of less than 30 μm. The reactions of choice include proton reduction,[2,12] oxygen evolution,[13] and hydrazine oxidation.[12] This approach has been modified to allow higher sensitivity in NP detection[14–16] and correlated with optical events.[17] Soft NP collisions by the Bard group[18,19] allowed electrochemical studies of emulsions. The technique has been coupled with magnetic NPs to increase the frequency of collision[20] and with electrochemiluminescence.[21] Studies of CeO_x nanoparticle collisions by Adreescu *et al.*[22] showed the effect of surface composition towards H_2O_2 disproportionation. Here, we present a study of suspended TiO_2 nanoparticles sensitized with di-tetrabutylammonium *cis*-bis(isothiocyanato)bis(2,2′-bipyridyl-4,4′-dicarboxylato)ruthenium(II), "N719", that display complex stochastic responses due to several possible electron transfer processes including oxidation/reduction of the dye, the mediator and electron transfer to/from the semiconductor NPs. We will discuss how these processes determine whether anodic or cathodic stochastic currents are observed under certain experimental conditions.

Recently, we reported on the complex behavior of colloidal DSNPs that we assigned to aggregates following a report where Compton and co-workers found that the nanoimpact method is sensitive to the colloidal behavior of NPs and aggregation or agglomeration.[9] Koper has also addressed the effect of aggregation when using hydrazine to detect Pt NPs.[23] With DSNPs, electrochemical oscillations were observed on a F-doped SnO_2 electrode[24] using catalytic amplification, distinct from the well-resolved step-wise[4,5,25] and blip[13] responses previously reported. In general a fast chronoamperometry technique is used to study the nature of the interactions between the NPs and the UME with the step-wise behavior assigned to the long-term, irreversible interactions with the electrode. On the other hand, the blip response is assigned to the reversible, short term interactions between NPs and the UME.[13,26] However, the ultimate shape of the i *vs.* t trace can also be influenced by the instrumentation and the electron transfer kinetics at the NP surface.[27,28]

Dye sensitization has been proposed to produce low cost photovoltaic devices since the seminal paper by O'Regan and Gratzel that reported the use of a sensitized film of TiO_2 NPs.[29] To study electron transfer on films with electrochemical techniques one has to take into account the electron transport through the complex NP network and the mass transport of reactants and products to and within the porous structure. Impedance techniques have been used for these studies that yield information of the electron lifetime, flat band potential and other properties of the NP/dye/electrolyte interface.[30–37] However, our goal is to isolate the problem of electron transfer from the complication of mass and electron transport through the film by studying the particle by particle contributions. Previous reports of slurries and colloids of TiO_2 and DSNPs have studied band alignment and other properties.[38] An additional motivation of our study is the possibility to study the dye/semiconductor/electrolyte because the identity of the electrolyte cation affects the rate of electron injection and we propose that this technique could be used to study the effect of electrolyte composition on the coupled electron transfer to regenerate the dye.[39] The dye N719 on TiO_2 is a model system given the high efficiency achieved in dye sensitized solar cells (>10%).[40]

Studies of the electrochemistry of the dye and similar Ru-based compounds (*e.g.*, N3) are complemented with spectroscopic techniques to determine the HOMO and LUMO.[40–42] To the best of our knowledge, the CV of N719 is not reversible in the usual electrochemical conditions. The use of N719/N3 as a model study for dye injection is common due to the fast electron injection into TiO_2 anatase,[43] even within femtoseconds of photon absorption.[44–47] Recent theoretical and experimental studies[44,48] *e.g.*, with ultrafast photoelectron spectroscopy study[49] demonstrate strong coupling between the dye orbitals and the conduction band of TiO_2.

In this paper, we report the stochastic behavior of colloidal DSNPs with a Pt ultramicroelectrode. The stochastic electrochemistry is strongly dependent on the conditions of the colloid and its history, *i.e.*, before and after illumination, with these differences assigned to changes in the distribution of agglomerates and/or aggregates. The stochastic steps are consistent with contact interactions with the Pt electrode[4,5,25] distinct from the oscillatory behavior observed with a F-doped SnO_2 microelectrode.[24] The individual steps are assigned to single entities that irreversibly attach to the electrode surface because of the step-wise discrete changes[4,5,25] that are mostly due to agglomerates' contributions rather than dispersed NPs. Because the colloidal behavior is a function of electrolyte composition and the illumination conditions, it is the main limitation factor rather than instrumental limitations.[27,28]

Experimental methods

Reagents and materials

All the reagents used were of analytical grades: tetrabutylammonium perchlorate (TBAP, electrochemical grade) was purchased from Sigma-Aldrich (Milwaukee, USA). Methanol (CH_3OH, HPLC grade) was purchased from Pharmco-AAPER. Polyethylene glycol (PEG, 20 000) and tetra-butyl ammonium perchlorate (TBAP, electrochemical grade) were purchased from Alfa Aesar. TBAP was recrystallized from 95% ethanol prior to use. All aqueous solutions were prepared in 18 MΩ cm water (Barnstead Nanopure, Thermo Scientific). Non-aqueous solutions for electrochemical measurements were prepared in MeOH or acetonitrile (CH_3CN, HPLC grade, EMD Millipore) with the solvents purified by incubation in activated alumina (Al_2O_3, Alumina N-Super 1, MP Biomedical) 18 mg Al_2O_3 per 1 ml of solvent, for at least 1 week in an Ar glove box. Titanium tetraisopropoxide (TTIP) was purchased from Sigma-Aldrich (97%) and used without further purification. N719 was purchased from Solaronix (Switzerland) and used without further purification.

Preparation of TiO_2 NPs

Anatase NPs were prepared by a modified procedure given by Zaban *et al.*[50] Briefly, NPs were prepared from a mixture of 2-propanol (6.01 g) and TTIP (7.23 g) in glacial acetic acid (33.6 g) and de-ionized water (100 ml). The mixture was pre-concentrated to a volume of 80 ml (*ca.* 6 hours) on a heating plate kept at 80 °C and transferred to an autoclave for hydrothermal growth at 230 °C for 12 hours. Details of the NP preparation are given elsewhere.[51] The as-prepared TiO_2 NPs were extensively purified, first with deionized water and finally with spectrophotometric grade methanol (as received). XRD and TEM measurements of the TiO_2 NPs showed a particle size of 18 nm.

Material and colloidal characterization

Every batch of synthesized TiO_2 NPs was characterized with powder X-ray diffraction (PXRD) to confirm the presence of anatase phase and that no evidence of other forms of TiO_2 was observed. The batch was also characterized by transmission electron microscopy (TEM) by transferring aliquots of the NPs to a TEM grid to determine the NP diameter and their distribution. Colloidal suspensions of the materials were characterized using dynamic light scattering (DLS) with a Particle Sizer 380 ZLS (Particle Sizing System, Santa Barbara, CA). The suspensions were prepared in spectrophotometric grade methanol filtered using a 20 nm syringe filter prior to dispersing the NPs. The suspensions were sonicated for 10 min prior to running the DLS. This procedure was applied to anatase NPs as prepared and sensitized.

Sensitization of TiO_2 NPs

The ruthenium based dye "N719" (mol. formula: $C_{58}H_{86}N_8O_8RuS_2$ and formula mass: 1188.55 g mol^{-1}) was used for sensitization of the anatase TiO_2 NPs colloidal suspension. As previously reported[52] a single TiO_2 NP of 18 nm size can hold ~600 N719 dye molecules; for sensitization the initial concentration of dye was 1000 times higher than that of TiO_2 NPs.[24] Typically, 40 ml of 200 μM solution of N719 dye was prepared in ethanol after adding a 2.6 ml aliquot of purified 3.07 μM TiO_2 suspension in ethanol in a brown vial to make a final volume of 40 ml with a final concentration of 200 nM of TiO_2 NPs and 200 μM in the dye.

The mixture was kept in the dark for one week and the excess dye was removed by centrifugation of the NPs at 10 000 rpm for 15 minutes and was washed by re-dispersing with neat methanol. This cycle of centrifugation and washing was repeated 8 times and yielded a clear supernatant. After removal of the unbound dye from the TiO_2 NPs, they were re-dispersed in MeOH to make a final volume of 40 ml. This stock suspension of dye sensitized TiO_2 NPs was used for the photoelectrochemical experiments. Typically, a final concentration of 2.5 nM of DSNPs was used for the study of stochastic interactions. The exposure of this stock suspension to light was minimized during manipulation and the colloid was stored in the dark.

Electrochemical and photoelectrochemical measurements

For the measurement with colloids the working electrode was a 25 μm Pt microelectrode prepared as described before.[53] The three electrodes were set up in a specially designed electrochemical cell made of polytetrafluoroethylene (PTFE) described before[24,51] with a platinum coil for the counter electrode (CE). A home-made reference electrode (RE) containing 10 mM I^-/I_3^- and 0.2 M TBAP in methanol as previously described[51] and a silver quasi reference electrode (Ag QRE) were used for electrochemical and photoelectrochemical measurements. The Ag QRE was calibrated against the I^-/I_3^- (10 mM in MeOH) reference. Measurements of the colloids were performed in MeOH with and without 0.1 M tetrabutylammoniumperchlorate (TBAP) as supporting electrolyte. Amperometric *i–t* experiments were performed for 2.5 nM DS TiO_2 solutions at an applied electrode potential, *e.g.*, $E_{app} = 0.6$ V *vs.* Ag QRE = 0.26 V *vs.* I^-/I_3^- (10 mM in MeOH) = 0.64 V *vs.* NHE.[51] Spectrophotometric grade neat methanol and methanol containing 0.1 M TBAP, and dye solutions were used as blanks unless otherwise stated. All solutions were purged with argon for 30 minutes and ultra-sonicated for at least 15 minutes before the experiments. For illumination, an ozone free Xe arc lamp, 150 W, along with an IR water filter was used. The potentiostat used for all electrochemical measurements was a CHI 760D Electrochemical Workstation (CH Instruments, Texas, USA) used without modification at the sampling rate of 16 ms per data point.

Results and discussion

We studied the stochastic behavior of suspensions of dye-sensitized nanoparticles (DSNPs) with a Pt ultramicroelectrode with stochastic steps that follow an interesting trend: the steps tend to be anodic in the dark and cathodic under illumination. This trend is observed when the colloids are prepared in the absence of electrolyte. The DSNPs are suspended in MeOH with the stochastic steps consistent with contact interactions with the electrode.[4,5,25] Therefore, the individual steps are assigned to single entities that irreversibly attach to the electrode surface. However, the differences between Fig. 1a and b are, to the best of our knowledge, unprecedented and are assigned to the availability of electrochemical and photoelectrochemical processes that occur at different rates in the dark and under illumination. Fig. 1a shows that in the dark the discrete current steps are mostly anodic, consistent with oxidations occurring at the DSNPs. Under illumination, the steps are in the cathodic direction, as shown in Fig. 1b. For these experiments, a 2.5 nM concentration of dye sensitized TiO_2 anatase nanoparticles was used to obtain *i vs. t* curves. The processes assigned to the steps are shown in Fig. 1c and d. In the dark, the electrode oxidizes the dye (N719), which in turn

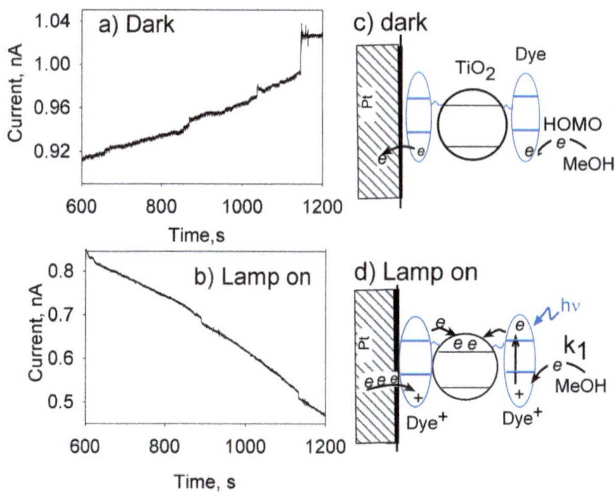

Fig. 1 (a) and (b) are *i vs. t* data showing anodic and cathodic stochastic steps for a 2.5 nM suspension of DSNPs in MeOH (a) in the dark, anodic steps and (b) under illumination of Xe arc lamp, cathodic steps. (c) and (d) are the schematics of the processes that are used to explain the different responses. $E_{app} = 0.64$ V *vs.* NHE.

oxidizes methanol. Several controls were run and are shown in the ESI, Fig. S1.† The steps were not observed when the blanks of CH_3OH and N719 dye were used in the same conditions. The concentration of the dye was chosen to approximate the expected dye content in the DSNPs. Therefore, we assign the current transients in Fig. 1 to stochastic interactions between the colloidal NPs and the Pt ultramicroelectrode.

Although Fig. 1 shows the archetypal response of the observed trends, we note that the results were dependent on the history of the colloid and the electrode. Fig. 2 shows the response for 2.5 nM DSNPs in neat methanol: Fig. 2a is the response in the dark showing anodic steps. For the same colloid, Fig. 2b corresponds to the full illumination conditions that result in cathodic and anodic steps. The inset shows a detail of the 0 to 50 s trend in Fig. 2b that displays mostly cathodic steps (negative direction in the graph) that are sharp transitions within the sampling rate of 16 ms per data point for our setup. For comparison, we recently reported the transient interactions between colloidal DSNPs with an FTO electrode, but the transients observed in those cases were very broad, even broader than expected based on the sampling rate and rising time of the potentiostat.[24] This is further evidence that the interactions in this case are long-term, irreversible interactions between the single elements in the colloidal suspension and the DSNPs. It is intriguing that the interactions under illumination appear to reverse signs. One complication is in the definition of a "step", which in this work is the sharp step-wise change in current; because the noise in our background is <1 pA, our steps are defined as a sharp discrete change of 1 pA or more. It has been demonstrated that these steps can be due to a single NP irreversibly attaching to a UME.[54] Here, we rely on the current *vs.* time trace to determine the nature of the interactions.

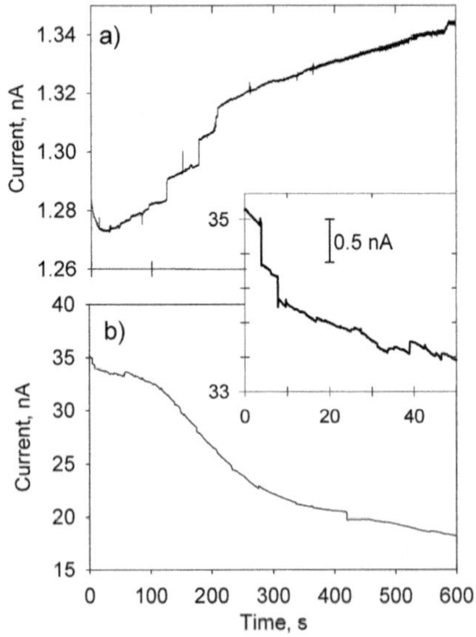

Fig. 2 Current *vs.* time curve for 2.5 nM DSNPs in neat methanol (a) in the dark showing anodic steps. (b) Under illumination showing cathodic and anodic steps. The inset shows the details of 0 to 50 s for trace (b). All other conditions as in Fig. 1.

To summarize, in the absence of electrolyte, the transients are anodic in the dark, but under illumination they are both in the anodic and cathodic direction. Fig. 3 shows the steps in the cathodic (Fig. 3a) and anodic direction (Fig. 3b). Note that the events are almost equally probable and most of the observed currents are <50 pA. In fact, within the 50 pA range, most of the currents are <10 pA (graph not shown). There are two possible explanations for the current steps observed in the reverse direction: (1) that the current change is caused by a NP that leaves the surface or that becomes de-activated, *e.g.*, by some by-product or impurity in solution, and (2) the arrival of a NP performing a different process that gives rise to cathodic currents. In this case, the anodic process could be the photooxidation of the dye which would give rise to a similar catalytic amplification like the one described before for bare TiO_2.[51] However, the assignment of these currents is not trivial given the electrochemical behavior of the dye in CH_3OH. Another interesting feature of these experiments is that when excess electrolyte is used there is no significant difference between the observed transient steps in the dark and under illumination as discussed below.

Under illumination we propose that the cathodic steps (*cf.* Fig. 1d) are explained due to the kinetics of electron injection from the excited state (Dye*) that are expected to yield the oxidized state of the dye (Dye^+). Dye^+ must react with CH_3OH to regenerate the reduced form of the dye. The fastest expected step is the injection of electrons into the TiO_2 NPs from Dye* which is expected to occur within <1 ps given the widespread agreement on the fast electron injection process.[43,44,48,55] This means that Dye^+ will be relatively stable at the DSNP surface and will become available for reduction when the illuminated NP approaches the electrode surface. Therefore, the reduction of Dye^+ results in cathodic steps under

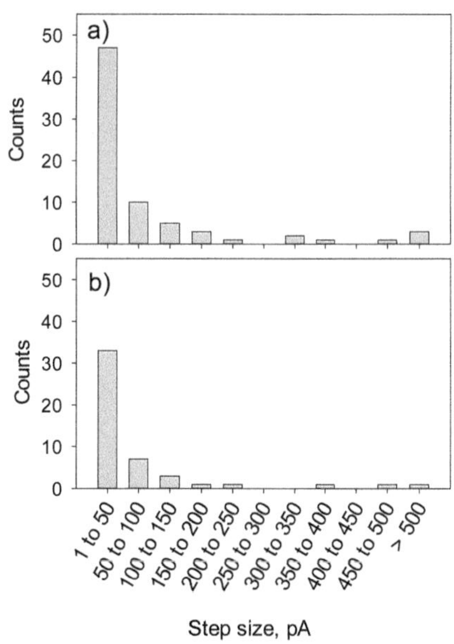

Fig. 3 Histogram of photoelectrochemical current steps observed under illumination for 2.5 nM DS TiO_2 suspended in neat CH_3OH. Total experiment time: 5400 s. (a) Cathodic step distribution (b) anodic step distribution. All other conditions as in Fig. 1.

illumination. This is consistent with the fact that the cathodic steps are not observed in the dark and decrease in frequency when the illumination power is reduced with neutral density filters (data not shown).

To explain the anodic steps in the dark we discuss the electrochemical behavior of the dye in CH_3CN and in MeOH. Fig. 4 shows the cyclic voltammograms of the oxidation of the dye in CH_3OH, curve (i), and CH_3CN, curve (ii), and the voltammogram in CH_3OH (curve i) is considerably shifted to cathodic potentials and the current of the same concentration, 1 mM, is considerably higher, consistent with the dye oxidizing the CH_3OH and the regeneration of the dye at a relatively high rate. Note that as expected the CV in CH_3CN is irreversible at the conditions of Fig. 4, $\nu = 100$ mV s^{-1}, (same data as in Fig. 5a) however, increasing the scan rate to 1 V s^{-1} yields a quasireversible CV in CH_3CN, shown in Fig. 5b. The CV in Fig. 5b shows an anodic peak around 0.6 V in the first segment (labeled 1, with the arrow indicating the initial anodic direction) of the two CVs shown in the figure. This peak is assigned to adsorbed species because the second CV does not show this peak. Note that segments 3 and 4 correspond to a quasireversible CV that agrees with a model after background subtraction (Fig. 2S, ESI†). The behavior in Fig. 5 is consistent with a coupled chemical reaction and a relatively sluggish electron transfer process ($k^0 = 9 \times 10^{-4}$ cm s^{-1}, see ESI† for fitting details with DigiElch, Elchsoft). Based on our results at the higher scan rates, we estimate the oxidation potential of the dye at 0.93 V *vs.* NHE, consistent with the values of the HOMO of the dye around 1 V *vs.* NHE.

The anodic steps observed in the dark were studied as a function of potential and are shown in Fig. 6. This behavior is consistent with the HOMO value above of 0.93 V *vs.* NHE, and with the fact that CH_3OH shifts the onset of oxidation currents as described above. The frequency of anodic steps in the dark increase in a manner consistent with the rate of dye oxidation seen in MeOH (Fig. 4 line i), with the probability of detecting anodic steps dropping sharply at potentials more

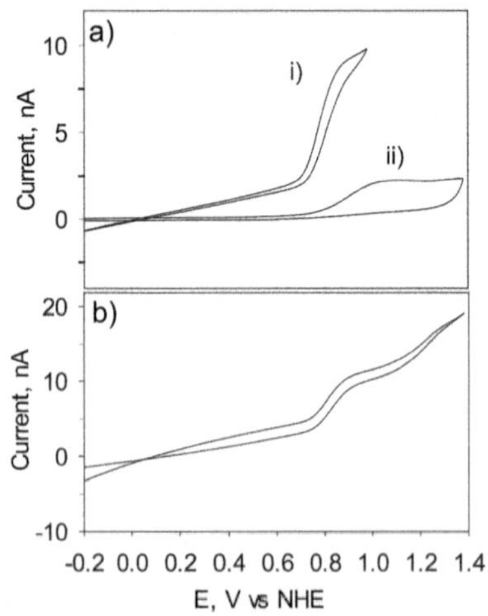

Fig. 4 Cyclic voltammograms of 1 mM N719 in CH_3OH (a, curve i) and CH_3CN (a, curve ii) at 100 mV s^{-1}. (b) corresponds to the same conditions as in (i) but over a larger potential window and shows an additional faradaic process at $E > 1$ V.

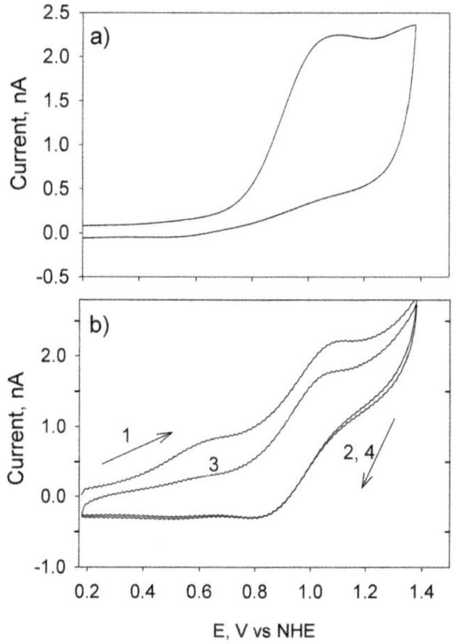

Fig. 5 Cyclic voltammograms of N719 in CH_3CN at the scan rates of (a) 100 mV s^{-1} and (b) 1 V s^{-1}. In (b) the first segment in the anodic direction shows a peak around 0.6 V assigned to an adsorption process that is not observed in segment 3. The arrows show the direction for the outgoing anodic segments (1 and 3) with the cathodic segments (2 and 4) closely overlapping.

negative than 0.6 V *vs.* NHE. At potentials more positive than 0.64 V, the frequency of detection does not continue to increase, likely due to competing faradaic processes seen both in CH_3OH and CH_3CN (Fig. 4 curves i and ii).

Effect of colloidal behavior

The trend of preferential anodic steps in the dark is not observed in colloids prepared with 0.1 M TBAP concentrations due to the difference in the colloidal

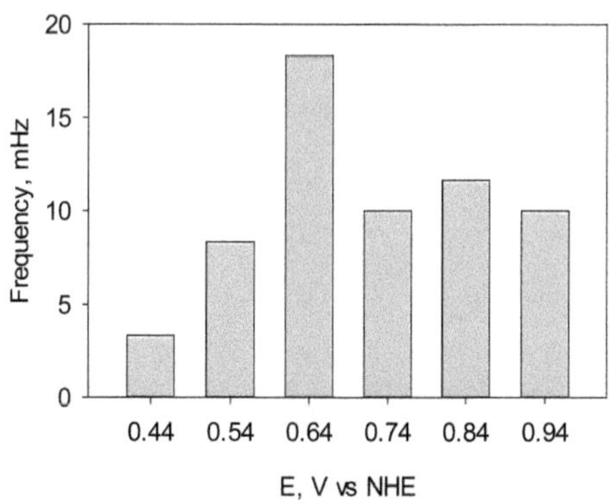

Fig. 6 Step frequency for anodic steps as a function of applied electrode potential observed in CH_3OH in the dark for 2.5 nM DSNPs. Total experimental time was 600 s.

behavior caused by the high concentration of supporting electrolyte. Fig. 7 shows the DLS behavior of the colloids without (Fig. 7a) and with 0.1 M TBAP (Fig. 7b). Note that the concentration used is large to force aggregation/agglomeration of the NPs. When the colloid is prepared in the absence of electrolyte, the light scattering distribution of the suspended particles is much smaller than the distribution of the colloid prepared in 0.1 M TBAP. Interestingly, both with and without excess TBAP the colloidal behavior is a strong function of the history of the colloid, with the distribution of size increasing with illumination time (red lines), and this is consistent with the tendency to lose our analytical signal during experiments. Given the size distribution seen in Fig. 7a we propose that the agglomerates of *ca.* 100 nm are more likely to be responsible for the steps that we observed. Further, addition of TBAP makes larger agglomerates that appear to have much larger recombination losses and do not show significant photocurrents under illumination. The size of the observed photocurrents is also consistent with agglomerates. For our illumination conditions, based on our measurements and on the lamp manufacturer's specifications, for $\lambda \leq 800$ nm, the power density of the beam is estimated to be 66 mW cm^{-2}. Assuming that all the photons interacting with one 20 nm anatase NP will be converted into electrons, the current for one NP will correspond to the order of 0.4 to 4×10^6 electrons per s depending on the energy cutoff of DSNP absorption. However, the currents of >1 pA observed correspond to $>6 \times 10^6$ electrons per s, which makes detection of single NPs possible but less likely given that every photon will have to be converted to an electron without recombination losses. Furthermore, photocurrents >100 pA were routinely observed (*cf.* Fig. 3).

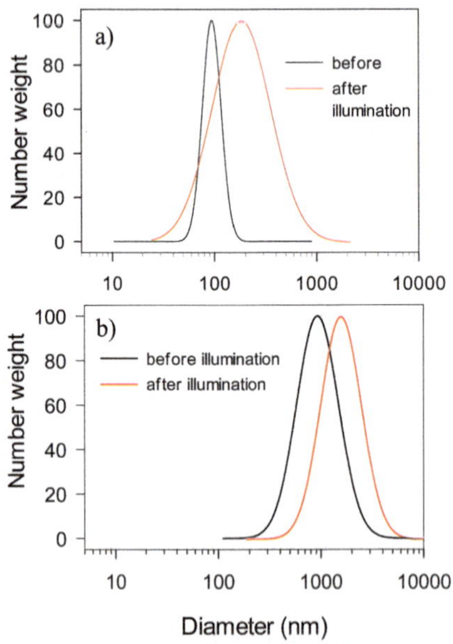

Fig. 7 Dynamic light scattering distribution of the colloidal 2.5 nM DSNPs in methanol. (a) In the absence of TBAP, the size distribution before illumination is 98 ± 20 nm (—) and after illumination is 229 ± 151 nm (—, red); (b) in presence of TBAP, the size distribution before illumination is 1037 ± 484 nm (—) and after illumination is 1741 ± 794 nm (—, red).

Conclusions

We report the stochastic interactions between DSNPs and a Pt ultramicroelectrode. The stochastic currents are observed in the dark and under illumination. In the dark, the currents are anodic, consistent with the oxidation of the dye at the Pt surface that in turn, drives the oxidation of MeOH. Under illumination we propose that the cathodic steps (*cf.* Fig. 1d) are explained as follows: dye photooxidation injects electrons into the TiO_2 NPs (a fast step)[43,44,48,55] which yields oxidized dye molecules (Dye^+) at the DSNP surface. The oxidized dye is reduced at the electrode surface resulting in cathodic steps that are not observed in the dark. The colloidal behavior is a strong function of the history of the colloid with illumination time increasing the size of the agglomerates and with larger agglomerates being less photoelectrochemically active. The steps are assigned to agglomerates of *ca.* 100 nm in diameter.

Acknowledgements

This project was supported by NSF CHE-1255387 Career award and by UNR startup for MAA.

References

1 B. M. Quinn, P. G. van't Ho and S. G. Lemay, *J. Am. Chem. Soc.*, 2004, **126**, 8360–8361.
2 X. Xiao and A. J. Bard, *J. Am. Chem. Soc.*, 2007, **129**, 9610–9612.
3 H. S. Ahn and A. J. Bard, *Angew. Chem., Int. Ed.*, 2015, **127**, 13957–13961.
4 A. J. Bard, H. Zhou and S. J. Kwon, *Isr. J. Chem.*, 2010, **50**, 267–276.
5 X. Xiao, F.-R. F. Fan, J. Zhou and A. J. Bard, *J. Am. Chem. Soc.*, 2008, **130**, 16669–16677.
6 Y. G. Zhou, N. V. Rees and R. G. Compton, *Angew. Chem., Int. Ed.*, 2011, **50**, 4219–4221.
7 Y. G. Zhou, N. V. Rees and R. G. Compton, *Chem. Phys. Lett.*, 2011, **511**, 183–186.
8 W. Cheng and R. G. Compton, *TrAC, Trends Anal. Chem.*, 2014, **58**, 79–89.
9 S. V. Sokolov, K. Tschulik, C. Batchelor-McAuley, K. Jurkschat and R. G. Compton, *Anal. Chem.*, 2015, **87**, 10033–10039.
10 Y. G. Zhou, N. V. Rees and R. G. Compton, *ChemPhysChem*, 2011, **12**, 2085–2087.
11 Y.-G. Zhou, N. V. Rees and R. G. Compton, *Angew. Chem., Int. Ed.*, 2011, **50**, 4219–4221.
12 X. Y. Xiao, F. R. F. Fan, J. P. Zhou and A. J. Bard, *J. Am. Chem. Soc.*, 2008, **130**, 16669–16677.
13 S. J. Kwon, F.-R. F. Fan and A. J. Bard, *J. Am. Chem. Soc.*, 2010, **132**, 13165–13167.
14 R. Dasari, D. A. Robinson and K. J. Stevenson, *J. Am. Chem. Soc.*, 2012, **135**, 570.
15 R. Dasari, K. Tai, D. A. Robinson and K. J. Stevenson, *ACS Nano*, 2014, **8**, 4539–4546.
16 R. Dasari, B. Walther, D. A. Robinson and K. J. Stevenson, *Langmuir*, 2013, **29**, 15100–15106.

17 S. E. Fosdick, M. J. Anderson, E. G. Nettleton and R. M. Crooks, *J. Am. Chem. Soc.*, 2013, **135**, 5994–5997.

18 B.-K. Kim, A. Boika, J. Kim, J. E. Dick and A. J. Bard, *J. Am. Chem. Soc.*, 2014, **136**, 4849–4852.

19 B.-K. Kim, J. Kim and A. J. Bard, *J. Am. Chem. Soc.*, 2015, **137**, 2343–2349.

20 D. A. Robinson, J. J. Yoo, A. D. Castañeda, B. Gu, R. Dasari, R. M. Crooks and K. J. Stevenson, *ACS Nano*, 2015, **9**, 7583–7595.

21 F.-R. F. Fan and A. J. Bard, *Nano Lett.*, 2008, **8**, 1746–1749.

22 N. P. Sardesai, D. Andreescu and S. Andreescu, *J. Am. Chem. Soc.*, 2013, **135**, 16770–16773.

23 S. E. F. Kleijn, B. Serrano-Bou, A. I. Yanson and M. T. M. Koper, *Langmuir*, 2013, **29**, 2054–2064.

24 A. Fernando, P. Chhetri, K. K. Barakoti, S. Parajuli, R. Kazemi and M. A. Alpuche-Aviles, *J. Electrochem. Soc.*, 2016, **163**, H3025–H3031.

25 H. Zhou, F.-R. F. Fan and A. J. Bard, *J. Phys. Chem. Lett.*, 2010, **1**, 2671–2674.

26 S. J. Kwon and A. J. Bard, *J. Am. Chem. Soc.*, 2012, **134**, 7102–7108.

27 C.-H. Chen, E. R. Ravenhill, D. Momotenko, Y.-R. Kim, S. C. S. Lai and P. R. Unwin, *Langmuir*, 2015, **31**, 11932–11942.

28 M. Kang, D. Perry, Y.-R. Kim, A. W. Colburn, R. A. Lazenby and P. R. Unwin, *J. Am. Chem. Soc.*, 2015, **137**, 10902–10905.

29 B. O'Regan and M. Gratzel, *Nature*, 1991, **353**, 737–740.

30 J. Bisquert, *J. Electroanal. Chem.*, 2010, **646**, 43–51.

31 J. Bisquert, F. Fabregat-Santiago, I. Mora-Seró, G. Garcia-Belmonte, E. M. Barea and E. Palomares, *Inorg. Chim. Acta*, 2008, **361**, 684–698.

32 J. Bisquert and V. S. Vikhrenko, *J. Phys. Chem. B*, 2004, **108**, 2313–2322.

33 J. Bisquert, A. Zaban, M. Greenshtein and I. Mora-Sero, *J. Am. Chem. Soc.*, 2004, **126**, 13550–13559.

34 F. Fabregat-Santiago, G. Garcia-Belmonte, J. Bisquert, A. Zaban and P. Salvador, *J. Phys. Chem. B*, 2002, **106**, 334–339.

35 F. Fabregat-Santiago, I. Mora-Seró, G. Garcia-Belmonte and J. Bisquert, *J. Phys. Chem. B*, 2003, **107**, 758–768.

36 J. García-Cañadas, F. Fabregat-Santiago, H. J. Bolink, E. Palomares, G. Garcia-Belmonte and J. Bisquert, *Synth. Met.*, 2006, **156**, 944–948.

37 E. Guillén, E. Azaceta, A. Vega-Poot, J. Idígoras, J. Echeberría, J. A. Anta and R. Tena-Zaera, *J. Phys. Chem. C*, 2013, **117**, 13365–13373.

38 W. W. Dunn, Y. Aikawa and A. J. Bard, *J. Electrochem. Soc.*, 1981, **128**, 222–224.

39 C. A. Kelly, F. Farzad, D. W. Thompson, J. M. Stipkala and G. J. Meyer, *Langmuir*, 1999, **15**, 7047–7054.

40 M. K. Nazeeruddin, A. Kay, I. Rodicio, R. Humphry-Baker, E. Mueller, P. Liska, N. Vlachopoulos and M. Graetzel, *J. Am. Chem. Soc.*, 1993, **115**, 6382–6390.

41 J. H. Snook, L. A. Samuelson, J. Kumar, Y.-G. Kim and J. E. Whitten, *Org. Electron.*, 2005, **6**, 55–64.

42 G. Liu, A. Klein, A. Thissen and W. Jaegermann, *Surf. Sci.*, 2003, **539**, 37–48.

43 S. Ardo and G. J. Meyer, *Chem. Soc. Rev.*, 2009, **38**, 115–164.

44 F. De Angelis, S. Fantacci, E. Mosconi, M. K. Nazeeruddin and M. Grätzel, *J. Phys. Chem. C*, 2011, **115**, 8825–8831.

45 T. Hannappel, B. Burfeindt, W. Storck and F. Willig, *J. Phys. Chem. B*, 1997, **101**, 6799–6802.

46 G. Benkö, J. Kallioinen, J. E. I. Korppi-Tommola, A. P. Yartsev and V. Sundström, *J. Am. Chem. Soc.*, 2002, **124**, 489–493.
47 B. Wenger, M. Grätzel and J.-E. Moser, *J. Am. Chem. Soc.*, 2005, **127**, 12150–12151.
48 A. Filippo De, F. Simona and S. Annabella, *Nanotechnology*, 2008, **19**, 424002.
49 M. Borgwardt, M. Wilke, T. Kampen, S. Mähl, W. Xiang, L. Spiccia, K. M. Lange, I. Y. Kiyan and E. F. Aziz, *J. Phys. Chem. C*, 2015, **119**, 9099–9107.
50 A. Zaban, S. Ferrere, J. Sprague and B. A. Gregg, *J. Phys. Chem. B*, 1997, **101**, 55–57.
51 A. Fernando, S. Parajuli and M. A. Alpuche-Aviles, *J. Am. Chem. Soc.*, 2013, **135**, 10894–10897.
52 B. C. O'Regan and J. R. Durrant, *Acc. Chem. Res.*, 2009, **42**, 1799–1808.
53 R. M. Wightman and D. O. Wipf, in *Electroanal. Chem.*, ed. A. J. Bard, 1989, vol. 15, pp. 267–353.
54 J. Kim, B.-K. Kim, S. K. Cho and A. J. Bard, *J. Am. Chem. Soc.*, 2014, **136**, 8173–8176.
55 T. A. Heimer, E. J. Heilweil, C. A. Bignozzi and G. J. Meyer, *J. Phys. Chem. A*, 2000, **104**, 4256–4262.

Faraday Discussions

PAPER

Electrochemistry at single bimetallic nanoparticles – using nano impacts for sizing and compositional analysis of individual AgAu alloy nanoparticles

En Ning Saw,[a] Viktoria Grasmik,[b] Christian Rurainsky,[a] Matthias Epple[b] and Kristina Tschulik*[a]

Received 2nd May 2016, Accepted 16th May 2016
DOI: 10.1039/c6fd00112b

The increasing interest in producing bimetallic nanoparticles and utilizing them in modern technologies sets the demand for fast and affordable characterization of these materials. To date Scanning Transmission Electron Microscopy (STEM) coupled to energy dispersive X-ray spectroscopy is usually used to determine the size and composition of alloy nanoparticles, which is time-consuming and expensive. Here electrochemical single nanoparticle analysis is presented as an alternative approach to infer the particle size and composition of alloy nanoparticles, directly in a dispersion of these particles. As a proof of concept, 14 nm sized $Ag_{0.73}Au_{0.27}$ alloy nanoparticles are analyzed using a combination of chronoamperometric single nanoparticle analysis and cyclic voltammetry ensemble studies. It is demonstrated that the size, the alloying and the composition can all be inferred using this approach. Thus, the electrochemical characterization of single bimetallic alloy nanoparticles is suggested here as a powerful and convenient complement or alternative to TEM characterization of alloy nanoparticles.

1. Introduction

Thanks to their many useful properties, nanoparticles are widely employed in a large number of consumer products, ranging from anti-microbial socks and medical implants to additives in sunscreen lotion, food or advanced optical devices.[1,2] Key to this success is the ability to obtain high surface to volume ratio and to control the electronic structure of these nanomaterials in contrast to conventional bulk materials, allowing the size of devices to be reduced, their efficiency to be increased, or their optical properties to be changed. While nanotechnology is often considered a recent discovery/development, the ancient

[a]Micro- & Nano-Electrochemistry and Center for Electrochemical Sciences (CES), Faculty of Chemistry and Biochemistry, Ruhr-University Bochum, D-44780 Bochum, Germany. E-mail: Kristina.Tschulik@rub.de

[b]Inorganic Chemistry and Center for Nanointegration Duisburg-Essen (CENIDE), University of Duisburg-Essen, Universitaetsstr. 5-7, D-45117, Essen, Germany

Romans had already used a combination of nanometer-sized gold and silver particles to produce dichroic glass.[3] Also, the red-coloured 'activated gold'[4] that Michael Faraday produced in the 1850s that led to the discovery of the Faraday–Tyndall effect was a suspension of gold nanoparticles, as we know today. Despite these early studies, only in the past few decades has the use of nanoparticles emerged greatly, thanks to the ability to image, and hence, finally systematically study them by transmission (TEM) or scanning electron microscopy (SEM).[5]

Thus, the research field of nanoscience emerged quickly in the past two decades, and nowadays the combination of different components and functionalities in a single nanoparticle, so called bimetallic and bifunctional nanomaterials, is state-of-the-art. Prominent examples are magnetic-core biocompatible-shell nanoparticles for medical application[6,7] or bimetallic alloy nanoparticles with tuneable catalytic activity[8,9] and optical properties.[10,11]

In many cases the synthesis of such bi- or multicomponent nanoparticles is not easy, and their characterization is equally challenging. Their physical properties and chemical reactivity depend not only on their size and shape, but also their spatial composition in terms of alloy *vs.* core–shell type arrangement of the constituting atoms may play an important role.[10,12] The most prominent examples for bimetallic nanoparticles are AgAu nanoparticles due to their many promising applications in medicine, optics and catalysis.[9,13,14] Characterization of these particles and their dispersions beyond sizing, however, is difficult. Standard solution-phase nanoparticle characterization techniques like dynamic light scattering (DLS), Nanoparticle Tracking Analysis (NTA) or Centrifugal Disc Sedimentation (DCS) are inherently insensitive to the particle composition. Investigations on dried nanoparticle samples, like nano-diffraction analysis in TEM investigations, are difficult due to the close similarity in crystal structures and lattice parameters of the two elements. Thus, electron dispersive spectroscopy (EDX) coupled to TEM analysis is the state-of-the-art analysis technique to characterize bimetallic nanoparticles.[15,16] As this requires expensive equipment, complex sample preparation and the removal of the solvent prior to the analysis, an alternative route to characterize bi- and multi-component nanoparticles is desirable.

It has been demonstrated recently that electrochemical analysis can be used to characterize ensembles of core–shell type nanoparticles in terms of shell quality and shell thickness.[17] Furthermore it was demonstrated that electrochemical nano impact experiments enable the sizing of the shell and the core of individual core–shell-type bimetallic nanoparticles.[16] In these nano impact experiments either the shell only, or the core plus shell, of individual Au core Ag shell particles were oxidized during their Brownian motion driven collision with an electrode, depending on the potential this electrode was held at.

Here we will demonstrate that the application of single nanoparticle impact experiments can be usefully extended to determine the size and the composition of individual bimetallic alloy nanoparticles. The observed relation between applied potential and measured charge per impact spike for individual alloy nanoparticles will be combined with alloy ensemble studies and compared to the behaviour of monometallic nanoparticles. Thus, electrochemical single nanoparticle analysis will be introduced as a new tool for alloy nanoparticle characterization in terms of the size and composition, using $Ag_{0.73}Au_{0.27}$ nanoparticles as a proof-of-concept example.

2. Theoretical background

Electrochemical differentiation between alloy- or core–shell-type bimetallic nanoparticles $A_{1-x}B_x$ (A having the lower standard potential, that is being less noble than B) can be illustrated based on the Nernst equation, which relates the equilibrium potential for the oxidation reaction to the activity of the component in the particle and the concentration of the ions in the solution

$$A \rightleftharpoons A^+ + e^- \tag{1}$$

$$E_{A/A^+} = E_{A/A^+}{}^{\emptyset} + \frac{RT}{F}\ln\frac{a(A^+)}{a(A)} \tag{2}$$

The activity of a pure metal in its standard state is 1, while in an alloy, by assuming the mixture of A and B to be ideal, the activity of A is given by its molar fraction. Thus, for a set activity of A^+ in solution, the oxidation potential of component A in an alloy will be shifted to a more positive potential than the pure metal. Fixing the activity of A^+ may usefully be achieved by choosing an electrolyte that contains an excess amount of an anion X^- that forms a poorly soluble compound AX and thus allows fixing of $a(A^+)$ by the solubility product (K_L) and the large excess amount of X^- in the solution.

$$AX(s) \rightleftharpoons A^+(aq) + X^-(aq) \tag{3}$$

$$E_{A/A^+} = E_{A/A^+}{}^{\emptyset} + \frac{RT}{F}\ln\frac{K_L}{a(A)\times a(X^-)} = \text{const} + \frac{RT}{F}\ln\frac{1}{a(A)} \tag{4}$$

Considering a bimetallic B-core A-shell nanoparticle, the outer shell of the particle exclusively contains A atoms, thus the mole fraction of A at the reaction front will be unity and so oxidation of the shell will occur at the same potential as observed for single monometallic A nanoparticles.[17] Once the A shell is oxidized, no faradaic current is seen until a second anodic peak is observed at potentials sufficiently high to facilitate oxidation of the remaining B core. For A-core B-shell nanoparticles, only one anodic peak is observed at the oxidation potential of B, comprising the oxidation of both B and A.

In the case of oxidation of an alloy $A_{1-x}B_x$ nanoparticle, leaching of the A component occurs at potentials above the oxidation potential of A and below the B oxidation potential. This so called 'de-alloying' increases the B content of the alloy, resulting in a continuous positive shift of the oxidation potentials of the remaining particle until the B oxidation potential is reached and the particle is fully oxidized. Thus, in a cyclic voltammogram a broad oxidation peak above the A oxidation potential and a subsequent B oxidation peak are observed. Depending on the experimental conditions (scan rate, temperature, electrolyte composition, *etc.*), surface reorganization of the remaining B on the partially leached nanoparticle might additionally occur, causing a further broadening of the oxidation peak.[18,19]

While electrochemical studies at ensembles of bimetallic nanoparticles enable the characterization of several billions of nanoparticles instantaneously, their individual size cannot be directly determined this way. Hence, electrochemical single entity studies of alloy particles are desirable to derive insights into both the size and the composition of individual bimetallic alloy nanoparticles. Herein, it will be demonstrated that nanoparticle impact experiments may provide a convenient route to gain these additional insights.

In a 'nano impact' experiment, the Brownian motion driven sporadic collision of dispersed nanoparticles with a potentiostated electrode is used either to mediate a reaction at the particle surface[20–28] or to reduce[29–31]/oxidize[16,32–35] the particle itself and thus, to derive information on a particle-by-particle basis (see Fig. 1). The oxidation of a nanoparticle impacting a suitably polarized electrode results in a burst of faradaic current, observed as a 'spike' in a current–time graph. The minimum potential required to drive the faradaic reaction is determined by the oxidation potential, and hence, the composition, of the impacting particle.

The charge Q exchanged during the impact event is determined by the number of atoms contained in the impacting nanoparticle N_{np} and the number z of electrons exchanged in the redox reaction:

$$Q = \int I\mathrm{d}t = ezN_{np} \tag{5}$$

For a spherical particle shape and granted that the material parameters (molar mass and density) are known, one can calculate the size of the impacting particle,

$$r_{np} = \sqrt[3]{\frac{3A_rQ}{4\pi eN_Az\rho}} \tag{6}$$

where r_{np} is the radius of the NP, A_r is the atomic mass, ρ is the density of the oxidized compound, and N_A is the Avogadro constant.[29]

For a AgAu alloy particle, due to the above-mentioned shift of the oxidation potential, a potential significantly higher than the Ag oxidation potential should be required to oxidize the Ag in the alloy particle. The associated Ag leaching further decreases the Ag content in the alloy at the reaction front. Thus, two predictions can be made:

(i) Anodic spikes of AgAu alloy particles should emerge at potentials more positive than the oxidation of pure Ag NPs under otherwise identical conditions (capping agent *etc.*) and (ii) increasing the potential applied, the charge exchanged per impact should steadily increase until the entire particle is oxidized at potentials sufficient to facilitate Au oxidation. These predictions will be experimentally tested in the following sections.

Fig. 1 Schematic drawing of nano impact experiments of bimetallic alloy nanoparticles.

3. Experimental

3.1. Nanoparticle synthesis

The bimetallic AgAu alloy nanoparticles used in this study were prepared as described in detail elsewhere.[15] In brief, the nanoparticles were synthesized by reduction with citrate and tannic acid in aqueous media. All preparations were carried out with double-distilled water. 300 mL of degassed water was heated to 100 °C. Then 0.9 mL of 10 mM $HAuCl_4$ and 2.1 mL of 10 mM $AgNO_3$ were added. A mixture of 3 mL of a solution of 102.6 mg trisodium citrate (116 mM) and 3 mL of a solution of 50.4 mg tannic acid (9.9 mM) was added using a syringe under vigorous stirring. The reaction mixture was then stirred for another 10 min at 100 °C before cooling to room temperature. Further functionalization was carried out by adding 2 mL of a solution of poly(*N*-vinylpyrrolidone) (PVP; 30 mg mL^{-1}) to the non-purified dispersion. After stirring for at least two more hours, the nanoparticles were separated from the unreacted material by ultracentrifugation (20 000 rpm, 30 min) and redispersed in ultrapure water by vortexing. The purification step was repeated twice to remove all synthesis by-products. Monometallic Ag and Au nanoparticles were produced using the same synthesis method to allow a direct comparison of the electrochemical behaviour of the mono- and the bimetallic particles, but 3.0 mL of either 10 mM $HAuCl_4$ or $AgNO_3$ solution were added.

3.2. NP characterization

Sizing of the nanoparticles was done using two solution-phase characterization techniques: differential centrifugal sedimentation (DCS) and dynamic light scattering (DLS). A narrow size-distribution and an average nanoparticle diameter of 14 nm of the bimetallic alloy nanoparticles were determined using DCS, as shown in Fig. 2a. DCS measurements were run at 24000 rpm (CPS Instruments Disc Centrifuge DC 24000), and 24 wt% sucrose solution to establish the density gradient.

In agreement with this, DLS characterization of the nanoparticles also showed a narrow size distribution with a modal particle size of 16 nm. DLS analysis was performed using a Malvern Instruments Ltd. Zetasizer.

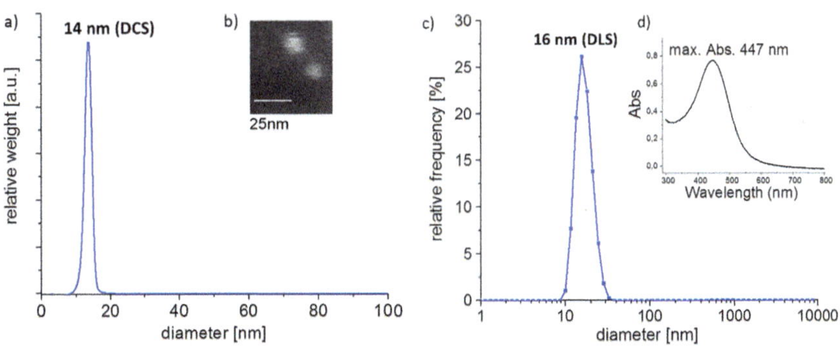

Fig. 2 (a) DCS analysis, (b) SEM image, (c) DLS analysis and (d) UV-vis spectrum of the $Ag_{0.73}Au_{0.27}$ NPs used in this work.

Atomic absorption spectroscopy (AAS) of the alloy nanoparticles dissolved in aqua regia revealed a composition of Ag : Au = 73 : 27 (Thermo Electron M-Series spectrometer, graphite tube furnace), which is in excellent agreement with the intended composition of $\approx Ag_{0.7}Au_{0.3}$ previously reported for the synthesis method used.[15] The size and compositional information were confirmed using SEM imaging (see Fig. 2b) and SEM-EDX analysis (FEI eSEM Quantax 3D FEG with EDAX Genesis XM2i detector), respectively.

For an extended X-ray diffraction and TEM-EDX analysis of the alloy nanoparticles we refer to Ristig *et al.*,[15,36] who demonstrated that the used synthesis method produces AgAu alloy nanoparticles of adjustable composition, with an increased Au content in the center of the particle. In agreement to their finding, the UV-vis absorption spectrum (Cary 60 UV-vis spectrophotometer by Agilent Technologies Inc.) of the nanoparticle suspension used herein showed a plasmon resonance absorption peak at 447 nm (see Fig. 2d).[15]

3.3. Electrochemical setup

Electrochemical experiments were run using a three-electrode setup comprising a Ag/AgCl, 3 M KCl reference electrode (potential = 0.197 V *vs.* SHE) and a Pt counter electrode at constant temperature (24 ± 0.5 °C). The entire cell was placed in a Faraday cage to reduce electronic noise. As a working electrode for nanoparticle ensemble studies a glassy carbon electrode (GCE) was used (Ø = 4 mm) and for nanoparticle impact experiments a carbon fibre (Ø = 7 μm) or Pt micro electrode (Ø = 10 μm) was employed.

3.4. Ensemble studies

Cyclic voltammetry (CV) ensemble studies were performed in 0.1 M HCl(aq) in the potential range from −0.2 V to 1.2 V *vs.* Ag/AgCl at a scan rate of 0.025 V s^{-1} using a potentiostat (Autolab PG Stat 12). CVs were recorded at blank and nanoparticle-modified GCEs. The modification with bimetallic nanoparticles was achieved by drop casting 2 μL of AgAu nanoparticle stock suspension on the GCE. For comparison pure Au, pure Ag and mixtures of Ag plus Au nanoparticles, respectively, were drop cast on an electrode to study the CV response of the monometallic particles and their mixture. To better resemble the capping agent free Au-rich remainder left by the oxidation of Ag from the bimetallic nanoparticles, PVP-capped Au monometallic nanoparticles were first immobilized on a GCE and cycled from −0.2 V to 1.2 V *vs.* Ag/AgCl in 0.1 M KNO_3 to de-cap the Au nanoparticle surface by oxidation and subsequent reduction of the particle surface.[37] Then these particles were analysed in HCl as described above. All CV measurements were repeated three times.

3.5. Single nanoparticle experiments

For chronoamperometric nano impact experiments a potentiostat (VA10, npi-electronics, linked to a computer using high speed DA/AD data acquisition cards, plug-in electronics) was used to apply different fixed potentials, ranging from 0 V to 1.2 V. The resulting current response was measured over a time of 50 s, recording 1200 data points per second. To filter the electronic noise without altering the spike charge, a 50 Hz 8 pole Bessel filter was applied. Chronoamperograms were recorded in 0.01 M HCl in the absence or the presence of

AgAu nanoparticles at particle concentrations of 10 pM. At each potential at least 5 measurements were performed, scattering the applied potential to omit any possible aging effect. At potentials at which spikes were seen, additional measurements were performed to record a sufficient number of impacts to obtain statistically relevant data. For comparison, Ag nanoparticle impacts were recorded in the same electrolyte.

Individual spikes were identified, setting the minimum required peak current to 3 times the noise level, and spike charges were determined using either SignalCounter software (provided by Dr D. Omanovic, Ruder Boscovic Institute Zagreb, Croatia) and Origin 2016 Pro. The mean charge and the error of the mean were determined for each potential applied.

4. Results and discussion

Cyclic voltammograms recorded at a scan rate of 0.025 V s^{-1} in 0.1 M HCl for the blank GCE and the GCE modified with monometallic Ag and Au nanoparticles are shown in Fig. 3a. An oxidation peak at 0.15 V and reduction peak at 0.05 V are associated with the formation of AgCl on the electrode and the reduction of this sparingly soluble salt.

$$Ag + Cl^- \rightleftharpoons AgCl + e^- \tag{7}$$

The second cycle (not shown) is identical to the first cycle in shape and position, but has a smaller peak area due to detachment of the AgCl from the electrode or chemical loss of AgCl from the electrode, forming soluble $[AgCl_{1+y}]^{y-}$ species.[38]

The oxidation of Au under the same experimental conditions occurs under release of a mixture of soluble Au^+ and Au^{3+} species, releasing 1.9 electrons per Au atom on average;[16,39] accordingly no peaks are seen in the second cycle of the CV.

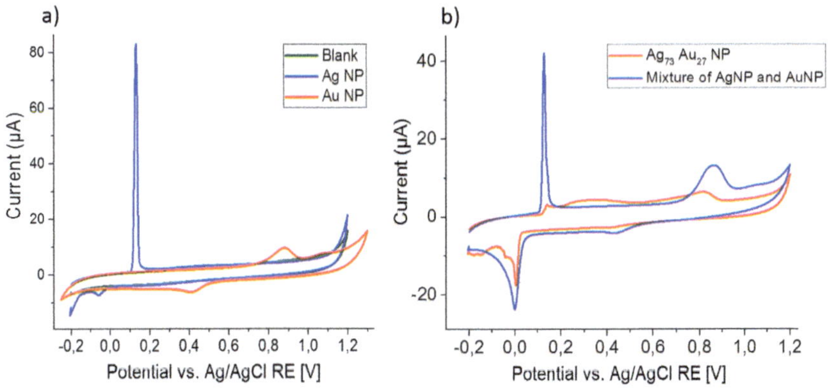

Fig. 3 Cyclic voltammogram of (a) a GCE modified with either Ag (blue) or Au (red) nanoparticles and a bare GCE (green); (b) a mixture of these monometallic nanoparticles (blue) and of the bimetallic AgAu alloy nanoparticles (red), recorded in 0.1 M HCl at a scan rate of 0.025 V s^{-1}.

$$Au + 2Cl^- \rightleftharpoons [AuCl_2]^- + e^- \quad (8)$$

$$Au + 4Cl^- \rightleftharpoons [AuCl_4]^- + 3e^- \quad (9)$$

CVs recorded using electrodes modified with a mixture of Ag and Au nanoparticles appear as a combination of these two CVs, giving peaks of the same shape and at the same positions as the monometallic nanoparticles (Fig. 3b).

Using the same experimental parameters as above, the bimetallic $Ag_{0.73}Au_{0.27}$ nanoparticles were analyzed and the recorded CV is plotted in Fig. 3b. While only a small signal is seen at the peak position for Ag oxidation (0.15 V), a very broad peak emerges at potentials slightly higher than that, eventually merging into the Au oxidation peak seen above 0.7 V. The second cycle comprises greatly-increased, sharp oxidation and reduction peaks at the typical position seen for monometallic Ag particles, revealing that the broad peak seen before must be associated to the oxidation of silver from the bimetallic nanoparticles. Similar to the monometallic Ag nanoparticles, this Ag oxidation peak in the second cycle is not suitable for quantitative analysis of the silver released in the first cycle, as it shows a large variation over repeating experiments and is about ten times smaller than expected for $Ag_{0.73}Au_{0.27}$ nanoparticles. As for the monometallic nanoparticles, no Au oxidation peak is seen in the second scan.

To exclude mass transport limitations as the cause of this peak broadening, repetitive scans were performed starting in smaller potential windows and successively increasing to the full potential range. The obtained set of CVs is shown in Fig. 4. Scanning from −0.2 V to 0.4 V, in the first scan a small Ag oxidation peak at 0.15 V is followed by a broad oxidation peak at more positive potentials, and the increased reduction signal shows that this broad peak is associated with the formation of AgCl. In the second cycle, sharp and increased peaks are seen at 0.15 V and 0 V. Increasing the potential window to 0.7 V in the subsequent third scan, the sharp Ag oxidation peak seen before is followed by a broad oxidation peak starting at the end of the previous potential window (0.4 V) and an increased reduction peak of AgCl, showing that AgCl was formed in this broad oxidation. Continued cycling in the same potential window again confirms

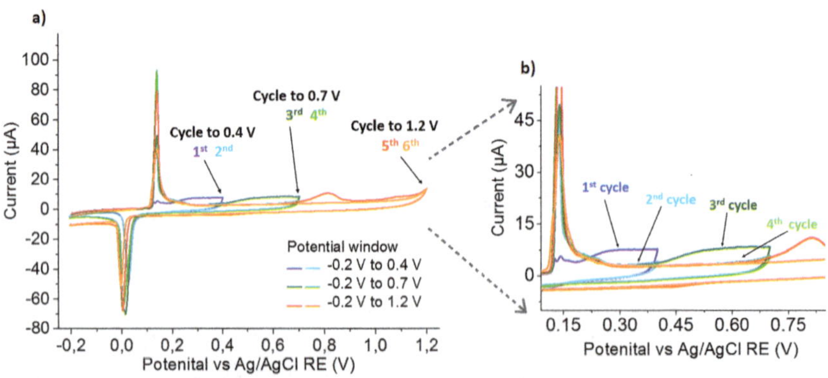

Fig. 4 (a) Series of successive CVs of the $Au_{0.73}Ag_{0.27}$ NPs immobilized on a GCE in 0.1 M HCl, scanning to increasing upper potential limits; (b) an enlarged plot of the region attributed to alloy oxidation.

the disappearing of the broad silver oxidation peak, and the constant (or slightly decreased) size of the peaks at 0.15 V and 0 V shows that no additional silver is released from the nanoparticle. Further increasing the potential window to 1.2 V, gold oxidation occurs. The charge transferred in the following AgCl reduction was found to be the same as in the previous scan, indicating that almost no silver was left in the leached particle at that stage. Repeated scanning in this potential window shows a slight decrease in the Ag oxidation and AgCl reduction peaks (again due to anticipated loss of material from the electrode by chemical formation of dissolved species), while the absence of a Au oxidation peak confirms the quantitative stripping in the previous (fifth) cycle.

This clearly proves that the broad Ag oxidation peak at potentials between 0.15 V and 0.7 V is not caused by mass transport limitations, but a consequence of alloying between Au and Ag in the bimetallic particle. Thus, the shift can be attributed to the thermodynamic shift of Ag oxidation to more positive values according to eqn (4), due to the decreased activity of Ag in an alloy $Ag_{1-x}Au_x$ ($a = 1 - x$; $0 < x < 1$) as compared to its activity in pure Ag ($a = 1$).[18]

The oxidative leaching of the Ag from the alloy continuously reduces the Ag content in the alloy, and hence successively increases x and the oxidation potential of Ag to form AgCl. The reduction of AgCl forms monometallic Ag entities on top of the bimetallic sample nanoparticles, as is visible from the increased oxidation peak at 0.15 V in the following cycle.

While it is clear from this ensemble analysis that the bimetallic nanoparticles are alloy particles, the quantitative determination of the composition is not possible due to the broad oxidation peak of the Ag in the first cycle making a quantitative peak integration very difficult and the loss of Ag from the electrode prior to the sharp Ag oxidation in the second cycle. Naturally for the ensemble studies, also no information on the size of the individual nanoparticles can be gained. Thus, the nano impact method will be exploited in the following paragraph to study single bimetallic AgAu particles on a particle-by-particle scale.

Chronoamperometric nano impact experiments were performed using a dispersion of the bimetallic AgAu nanoparticles in 0.01 M HCl(aq) and a particle concentration of *ca.* 10 pM. The current transients recorded at potentials ≥0.6 V showed distinct current spikes of several millisecond duration, while at lower potentials no such features were observed, as shown in Fig. 5a and b. Control experiments in the absence of nanoparticles did not contain any of these features, irrespective of the applied potential, verifying that these spikes originate from charge transfer during nanoparticle impact events (Fig. 5a, 'blank').

Analysis of the average charge per impact as a function of the potential applied yielded the data summarized in Fig. 5c. Taking eqn (6), the charge expected for quantitative oxidation of a 14 nm sized $Ag_{0.73}Au_{0.27}$ nanoparticle (7 nm radius), is 1.67×10^{-14} C, while the exclusive oxidation of Ag should result in a charge of 0.98×10^{-14} C (as indicated by the red dashed lines). Also indicated is the detection limit as the grey dashed line.

From this it is evident that spikes recorded at potentials smaller than the Au oxidation potential (≤0.7 V) gave a charge that is in excellent agreement with the charge expected for $Ag_{0.73}Au_{0.27}$ nanoparticles of 14 nm diameter.

Notably the minimum potentials at which spikes were recorded was significantly above that reported for monometallic citrate-capped Ag nanoparticles in chloride containing solutions (≈0.05 V), and also higher than that seen in the

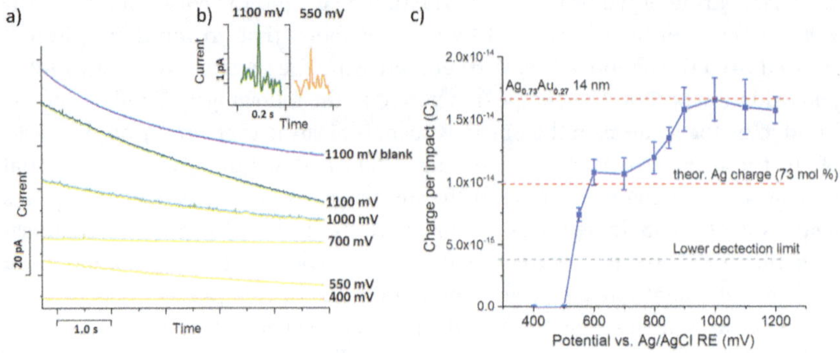

Fig. 5 (a) Chronoamperograms recorded in 0.01 M HCl at different potentials in the presence and absence of $Au_{0.73}Ag_{0.27}$ NPs, shifted along the *y*-axis for better clarity; (b) typical spike shapes and sizes seen at two different potentials and (c) detected average spike charge as a function of applied potential.

ensemble studies shown above (≈0.1 V). In a previous work employing cetyl trimethylammonium bromide CTAB-capped nanoparticles, a dominating role of the capping agent in the particle oxidation potential was reported.[40] Thus, to quantify the stabilizing effect of the PVP capping agent potentially causing this shift, nano impact experiments using monometallic Ag nanoparticles synthesized using the same procedure as for the AgAu particle synthesis were tested in the same electrolyte. Oxidative Ag impacts were already seen at potentials ≥0.2 V (not shown), proving that this increase in the Ag oxidation potential can not be attributed to a mere PVP capping agent-induced effect. Conversely, this demonstrates that alloying of silver in the bimetallic AgAu nanoparticles can be detected on an individual particle scale, due to the associated thermodynamic increase of the Ag oxidation potential. These findings are in excellent agreement with the nanoparticle ensemble studies described above.

The further increase of the spike charge at potentials exceeding 0.7 V reveals the onset of the Au oxidation in the impacting particles. At potentials ≥0.9 V a constant average charge per impact is detected, marking the oxidation of the entire AgAu nanoparticle. As indicated in Fig. 5c, this is in excellent agreement with the charge expected for the quantitative oxidation of the 14 nm sized bimetallic AgAu particles used. This clearly demonstrates that nano impact experiments can provide both size and compositional analysis of bimetallic alloy nanoparticles on an individual particle scale.

Notably the observed relatively shallow increase of the spike charge above this potential is different from the charge-potential dependence reported in previous works for the oxidation of Ag, Au and Au-core Ag-shell nanoparticles. In these cases a steep, step-like increase of the spike charge above a threshold potential was seen in Cl^- containing electrolytes.[16] It is speculated that this reflects either an increase in the Au oxidation potential in an alloy, due to the fast time scale of the nano impact oxidation as compared to the slower process of de-alloying or a diffusional limitation of Au dissolution due to the quickly-formed porous AgCl network, which hinders the diffusion of Cl^- and $[AuCl_{1-y}]^-$. Further investigation using bimetallic alloy particles of different Ag : Au compositions and in different electrolytes are planned to gain further insight into this aspect.

5. Conclusions

In this work the single entity methodology of electrochemical nano impact experiments was used to study 14 nm sized bimetallic AgAu alloy nanoparticles dispersed in an aqueous electrolyte. It was found that both the sizing and compositional analysis of individual nanoparticles is possible by virtue of chronoamperometry at different potentials. The presence of an alloying effect stabilising the silver in the bimetallic particles and thus shifting the onset of silver oxidation to more positive potentials was also demonstrated on this individual particle basis. Cyclic voltammetry ensemble studies confirm the results of these single entity experiments.

Thus, the work highlights electrochemistry as a powerful tool to characterise bimetallic alloy nanoparticles and suggests nano impact analysis as a convenient complement or alternative to vacuum characterization techniques used today. It is expected that this methodology can readily be applied to study a variety of different bimetallic and multi-component nanoparticles directly and conveniently in a dispersed state.

Acknowledgements

The authors are grateful to Dr Dario Omanovic (Ruder Boscovic Institute Zagreb, Croatia) for providing the peak analysis software SignalCounter and Sandra Schmidt (RUB) for assistance in the SEM imaging. This work was financially supported by the Ministry of Innovation, Science and Research of the State of North Rhine-Westphalia (NRW Rückkehrerprogramm, K. T.) and the Cluster of Excellence RESOLV (EXC 1069) funded by the German Research Foundation (DFG, K. T. and E. N. S.). M. E. thanks the Deutsche Forschungsgemeinschaft (DFG) for funding of the project Ep 22/44-1.

References

1 C. N. R. Rao, G. U. Kulkarni, P. J. Thomas and P. P. Edwards, *Chem. Soc. Rev.*, 2000, **29**, 27–35.
2 R. Kessler, *Environ. Health Perspect.*, 2011, **119**, A120–A125.
3 I. Freestone, N. Meeks, M. Sax and C. Higgitt, *Gold Bull.*, 2007, **40**, 270–277.
4 M. Faraday, *Phil. Trans. R. Soc.*, 1857, **147**, 145–181.
5 A. Howie, L. D. Marks and S. J. Pennycook, *Ultramicroscopy*, 1982, **8**, 163–174.
6 H. Amiri, L. Bordonali, A. Lascialfari, S. Wan, M. P. Monopoli, I. Lynch, S. Laurent and M. Mahmoudi, *Nanoscale*, 2013, **5**, 8656.
7 P. B. Santhosh and N. P. Ulrih, *Cancer Lett.*, 2013, **336**, 8–17.
8 J. R. Croy, S. Mostafa, L. Hickman, H. Heinrich and B. R. Cuenya, *Appl. Catal., A*, 2008, **350**, 207–216.
9 D. M. Alonso, S. G. Wettstein and J. A. Dumesic, *Chem. Soc. Rev.*, 2012, **41**, 8075–8098.
10 P. Mulvaney, M. Giersig and A. Henglein, *J. Phys. Chem.*, 1993, **97**, 7061–7064.
11 X. López Lozano, C. Mottet and H.-C. Weissker, *J. Phys. Chem. C*, 2013, **117**, 3062–3068.
12 G. Guisbiers, R. Mendoza-Cruz, L. Bazán-Díaz, J. J. Velázquez-Salazar, R. Mendoza-Perez, J. A. Robledo-Torres, J.-L. Rodriguez-Lopez, J. M. Montejano-Carrizales, R. L. Whetten and M. José-Yacamán, *ACS Nano*, 2016, **10**, 188–198.

13 A. J. Logsdail and R. L. Johnston, *J. Phys. Chem. C*, 2012, **116**, 23616–23628.
14 H. Zhang and N. Toshima, *Appl. Catal., A*, 2012, **447**, 81–88.
15 S. Ristig, O. Prymak, K. Loza, M. Gocyla, W. Meyer-Zaika, M. Heggen, D. Raabe and M. Epple, *J. Mater. Chem. B*, 2015, **3**, 4654–4662.
16 L. R. Holt, B. J. Plowman, N. P. Young, K. Tschulik and R. G. Compton, *Angew. Chem. Int. Ed.*, 2016, **55**, 397–400.
17 K. Tschulik, K. Ngamchuea, C. Ziegler, M. G. Beier, C. Damm, A. Eychmueller and R. G. Compton, *Adv. Funct. Mater.*, 2015, **25**, 5149–5158.
18 A. L. Pawlowych, D. L. Pile, H. W. Pickering and K. G. Weil, *Z. Phys. Chem.*, 1998, **207**, 113–126.
19 J. Erlebacher, M. J. Aziz, A. Karma, N. Dimitrov and K. Sieradzki, *Nature*, 2001, **410**, 450–453.
20 A. J. Bard, H. Zhou and S. J. Kwon, *Isr. J. Chem.*, 2010, **50**, 267–276.
21 S. J. Kwon, H. Zhou, F.-R. F. Fan, V. Vorobyev, B. Zhang and A. J. Bard, *Phys. Chem. Chem. Phys.*, 2011, **13**, 5394–5402.
22 S. E. F. Kleijn, B. Serrano-Bou, A. I. Yanson and M. T. M. Koper, *Langmuir*, 2013, **29**, 2054–2064.
23 R. Dasari, K. Tai, D. A. Robinson and K. J. Stevenson, *ACS Nano*, 2014, **8**, 4539–4546.
24 D. A. Robinson, J. J. Yoo, A. D. Castañeda, B. Gu, R. Dasari, R. M. Crooks and K. J. Stevenson, *ACS Nano*, 2015, **9**, 7583–7595.
25 X. Li, C. Batchelor-McAuley, S. A. I. Whitby, K. Tschulik, L. Shao and R. G. Compton, *Angew. Chem. Int. Ed.*, 2016, **55**, 4296–4299.
26 J. J. Yoo, J. Kim and R. M. Crooks, *Chem. Sci.*, 2015, **6**, 6665–6671.
27 M. Kang, D. Perry, Y.-R. Kim, A. W. Colburn, R. A. Lazenby and P. R. Unwin, *J. Am. Chem. Soc.*, 2015, **137**, 10902–10905.
28 C.-H. Chen, E. R. Ravenhill, D. Momotenko, Y.-R. Kim, S. C. S. Lai and P. R. Unwin, *Langmuir*, 2015, **31**, 11932–11942.
29 K. Tschulik and R. G. Compton, *Phys. Chem. Chem. Phys.*, 2014, **16**, 13909.
30 K. Shimizu, K. Tschulik and R. G. Compton, *Chem. Sci.*, 2016, **7**, 1408–1414.
31 W. Cheng and R. G. Compton, *Angew. Chem. Int. Ed.*, 2014, **53**, 13928–13930.
32 Y.-G. Zhou, N. V. Rees and R. G. Compton, *Angew. Chem. Int. Ed.*, 2011, **50**, 4219–4221.
33 C. S. Lim, S. M. Tan, Z. Sofer and M. Pumera, *ACS Nano*, 2015, **9**, 8474–8483.
34 H. S. Toh, K. Jurkschat and R. G. Compton, *Chem.–Eur. J.*, 2015, **21**, 2998–3004.
35 V. Brasiliense, A. N. Patel, A. Martinez-Marrades, J. Shi, Y. Chen, C. Combellas, G. Tessier and F. Kanoufi, *J. Am. Chem. Soc.*, 2016, **138**, 3478–3483.
36 S. Ristig, S. Chernousova, W. Meyer-Zaika and M. Epple, *Beilstein J. Nanotechnol.*, 2015, **6**, 1212–1220.
37 S. G. D. Shackleford, C. Boxall, S. N. Port and R. J. Taylor, *J. Electroanal. Chem.*, 2002, **538–539**, 109–119.
38 C. Batchelor-McAuley, A. Martinez-Marrades, K. Tschulik, A. N. Patel, C. Combellas, F. Kanoufi, G. Tessier and R. G. Compton, *Chem. Phys. Lett.*, 2014, **597**, 20–25.
39 Y.-G. Zhou, N. V. Rees, J. Pillay, R. Tshikhudo, S. Vilakazi and R. G. Compton, *Chem. Commun.*, 2011, **48**, 224–226.
40 B. J. Plowman, K. Tschulik, N. P. Young and R. G. Compton, *Phys. Chem. Chem. Phys.*, 2015, **17**, 26054–26058.

Faraday Discussions

PAPER

Electrochemical transformation of individual nanoparticles revealed by coupling microscopy and spectroscopy†

Vitor Brasiliense,[a] Pascal Berto,[b] Catherine Combellas,[a] Robert Kuszelewicz,[b] Gilles Tessier[b] and Frédéric Kanoufi*[a]

Received 27th April 2016, Accepted 9th May 2016
DOI: 10.1039/c6fd00098c

Although extremely sensitive, electrical measurements are essentially unable to discriminate complex chemical events involving individual nanoparticles. The coupling of electrochemistry to dark field imaging and spectroscopy allows the triggering of the electrodissolution of an ensemble of Ag nanoparticles (by electrochemistry) and the inference of both oxidation and dissolution processes (by spectroscopy) at the level of a single nanoparticle. Besides the inspection of the dissolution process from optical scattering intensity, adding optical spectroscopy reveals chemical changes through drastic spectral changes. The behaviours of single NPs and NP agglomerates are differentiated: in the presence of thiocyanate ions, the transformation of Ag single nanoparticles to AgSCN is investigated in the context of plasmonic coupling with the electrode; tentative interpretations for optically unresolved groups of nanoparticles are proposed.

Introduction

Optical microscopies, which allow a rich mode of interrogation of chemistry or electrochemistry at the nanoscale, are generating promising perspectives for the *in situ* and *in operando* characterization of nanomaterials. In the field of single nanoparticle (NP) electrochemical activity, the few emerging studies have already demonstrated the potential of optical monitoring to provide valuable complementary information. Most of these coupled opto-electrochemical studies are, however, at the proof of concept stage since they have targeted model NP systems, mostly plasmonic nanomaterials such as Au or Ag. In this respect, the optical strategy usually adopted exploits the ability of the NP to generate localized surface plasmon resonance (LSPR) and the ability of electrochemistry to modulate or

[a]*Sorbonne Paris Cité, Paris Diderot University, ITODYS, CNRS-UMR 7086, 15 rue Jean de Baif, F-75013 Paris, France. E-mail: frederic.kanoufi@univ-paris-diderot.fr*

[b]*Sorbonne Paris Cité, Université Paris Descartes, Neurophotonics Laboratory, CNRS-UMR 8250, 45 rue des Saints-Pères, F-75006 Paris, France*

† Electronic supplementary information (ESI) available: Movie 1. See DOI: 10.1039/c6fd00098c

interrogate this LSPR.[1–7] Plasmonic NPs are then used as nanoantennae or nanoreporters of their local chemical or electrochemical environment. In this respect, in an electrochemical context, the presence of the electrode, as with any kind of surface,[8] plays a significant role in the sensitivity of the optical detection. This can be exploited to enhance the optical signal sensitivity, for example, by using a few nanometer thin Au electrode, which can propagate surface plasmon resonance. The coupling between plasmons of the NP and of the electrode leads to local electromagnetic field enhancements (hot-spots) which are highly sensitive reporters with promising applications in sensing platforms, such as in surface enhanced Raman spectroscopy (SERS). The optical strategies then rely on the visualization by a CCD camera of individual NPs through direct local measurement of the LSPR and of their modification upon electrochemically driven charge transfer processes.

The simplest optical setup for visualizing electrochemistry at single plasmonic NPs uses simple dark field (DF) illumination and direct optical detection by a standard microscope of the light scattered by the plasmonic objects. This was recently proposed by Hill and Pan[9] for the monitoring of Ag NP electrodeposition and electrodissolution processes. Richer information can be obtained through the spectroscopic analysis associated to the electrochemical event.[3,4] Then the light scattered by individual NPs is spectroscopically analysed, giving more insight, but at the expense of longer acquisition times (>several seconds) and therefore restricted to the case of slow and reversible transformations. The use of a propagating surface plasmon electrode is an asset for the enhancement of the optical signal, through plasmon coupling between, for example, an Au NP and an Au film electrode, which allows interrogation of the electroadsorption of ions at single Au NPs.[4] This field enhancement effect is exploited through the direct monitoring of the surface plasmon resonance of a thin Au layer during NP electrochemistry, as developed by Tao's group.[7,10] It is likely the most sensitive system. However, because it relies on the plasmonic coupling between the NP and the electrode or the perturbation of the electrode plasmon by electrochemical events, it is spatially limited since regions <200 nm from the electrode are effectively probed. In all the related approaches, so far, the individual optical signals are used to reconstruct an electrochemical signature of individual nanoobjects.[1–4,7,9,10] The electrochemistry is that of a large ensemble and no direct coupled optical and electrochemical individual signature was obtained.

Our groups have developed a 3D dark field approach using holographic microscopy and applied it to visualize electrochemically-triggered processes occurring at single Ag NPs.[11] Like most works mentioned above, it uses changes in the optical signal at single NPs to infer single electrochemical transformation. In the context of Ag NPs, it characterizes the oxidative electrodissolution process. The 3D optical monitoring is particularly appealing as it allows a full inspection of the volume of the solution surrounding the electrode (typically $100 \times 100 \times 40$ μm^3) and therefore NP transport/sizing. Then, by implementing a microfabricated ultramicroelectrode, UME,[11c] with the optical monitoring, we could visualize in 3D the whole process of Ag NP electrochemical nanoimpacts. The synchronization of the electrochemical signal associated to Ag NP electrodissolution strongly supports the concept of single electrochemical impacts on UMEs for the characterization of NP electrochemistry.[12] In addition, we have shown that the use of precipitating ligands in the solution may retard the dissolution process through the possible separation of a phase transformation (a crystal formation, associated to the charge injection) and of a slower dissolution process. In this complex situation, electrochemistry allowed identification of the charge transfer (an electrochemical process) associated to the formation of an AgSCN NP while optical microscopy allowed monitoring of the dissolution (a chemical process) of this same AgSCN NP.

Even if the same phenomenon was also indirectly detected during the electrodissolution of AgCl nanoclusters, it is striking that the transformation of metallic Ag NPs into non-metallic AgCl or AgSCN nanodomains could not be monitored optically. Indeed, as a first guess from the large difference in the refractive indices of the metallic Ag and AgCl or AgSCN materials, one would expect that the electroinduced chemical phase transformation is associated with detectable changes in the intensity of the scattered light. Since scattering intensity variations do not reveal these changes, at least with a $\lambda = 532$ nm probe wavelength, complementary optical information must be sought.

Herein, we propose a spectroscopic overview of the processes occurring during the oxidation of individual Ag NPs in a solution of SCN^-. The Ag NP oxidation is performed at an UME in the electrochemical nanoimpact experiment context. The UME surface is optically monitored under dark field (DF) illumination, and associated to spectrometric imaging in order to get further insights into the transformation (electrodissolution) mechanism of the NPs (single NPs or NP agglomerates).

Results and discussion

A. Experimental principle and setup

Fig. 1a presents the experimental configuration used to spectro-electro-chemically monitor the oxidative electrodissolution of individual Ag NPs in a 50 mM KSCN solution. A 50×50 μm^2 semi-transparent gold UME was built by lithography on a glass microscope slide. It was assembled into a *ca.* 100 μm thick fluidic cell using a low-temperature melting spacer (parafilm layer). The inlet and the outlet

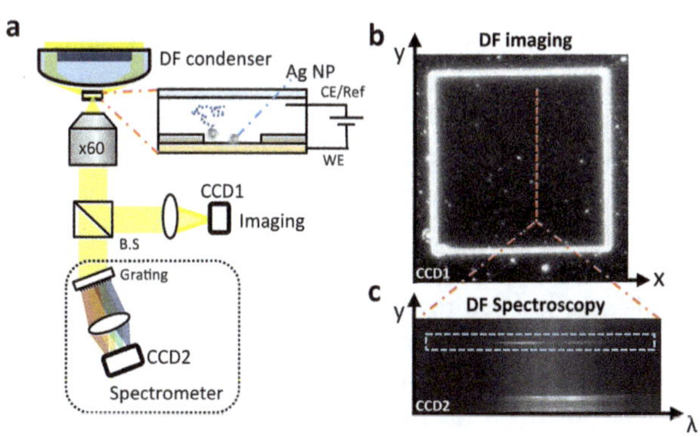

Fig. 1 Schematic representation of dual dark field microscopy and spectroscopic monitoring of individual Ag NP oxidation in a 50 mM KSCN aqueous solution. (a) Optical setup. B.S: Beam Splitter. (b) DF image (recorded on CCD1) – the square ultramicroelectrode (UME) is 50×50 μm^2. (c) DF spectroscopic image (on CCD2).

of the cell were made of micropipettes in which an Ag wire was inserted to function as both quasi-reference (Ag QRE) and counter electrode in the electrochemical system. The fluidic cell was illuminated from the top by white light using an air DF condenser. Only the light scattered by the different objects present on the UME is collected by a low numerical aperture microscope objective (60×, NA = 0.7) and directed towards two optical pathways. One is collected directly by a first camera (CCD1), giving a DF image of the UME region (Fig. 1b). The second branch is directed through a line-spectrometer operating in the visible range and sent to a second camera, CCD2. The image obtained with CCD2 (Fig. 1c) directly presents the spectra of the different scattering objects present on the different pixels of a given line (the red line schematically drawn in the DF image) of the plane of the UME.

In this configuration, CCD1 monitors the NPs landing and the reactivity on the whole UME surface, while CCD2 monitors the spectroscopic changes associated to the reactivity of selected NPs present in a defined region (line) of the UME. To ascertain that the spectroscopic characterization of the NP oxidation can be monitored, we needed to allow a significant number of NPs to adsorb on the UME, in the absence of electrical polarization. The position of the line can then be chosen to encompass a sufficient number of adsorbed NPs whose landings were monitored before the UME polarization.

A typical example of Ag NP oxidation in a 50 mM SCN^- solution at the UME is presented in Movie 1, provided as ESI.† Movie 1a† recorded with CCD1 presents a DF visualization of the NP dissolution: the NP disappearance is associated to the decay of the scattered light intensity, I_{sc}. Simultaneously, Movie 1b† recorded with CCD2 presents the change in the scattered spectra of 3 different NPs present in the selected line during the oxidation process. The first images of both movies are given in Fig. 1b and c and clearly show that a large number of NPs (*ca.* 44) are present at the UME surface before their oxidation and 3 NPs are individually analysed by spectrometry, one of which is shown in Movie 1c.†

B. Dark field monitoring of Ag NP electrodissolution

B.1. Optical nanoparticle sizing. Owing to the rather long time of adsorption prior to the oxidation process, it is likely that NP agglomeration has taken place in solution. The NPs adsorbed on the UME then comprise single NPs (*ca.* 50 nm radius) as well as their agglomerates. However, contrary to the 3D holographic microscopy described previously,[11] DF optical monitoring allows the inspection of a relatively thin depth of field (<3 μm compared to 40 μm in holography) above the UME with a much lower acquisition frequency (1 Hz) since our white-light delivers much less optical power in the region of interest than the laser used in holography. Under such conditions, we were not able to track the Brownian motion of NPs from the bulk solution before they landed on the UME. Without the dynamic motion of individual NPs, we do not access their individual hydrodynamic sizing before they collide with the UME. However, an approximate estimate of the size of the individual NPs adsorbed on the UME can be obtained indirectly from I_{sc} measured from the DF images. Indeed, Mie's theory relates the scattering cross section (directly proportional to I_{sc}) of objects to their size, and Hill and Pan recently verified this correlation in the context of the DF monitoring of the electrochemistry of immobilized Ag nanodomains.[9a] Fig. 2a presents the distribution

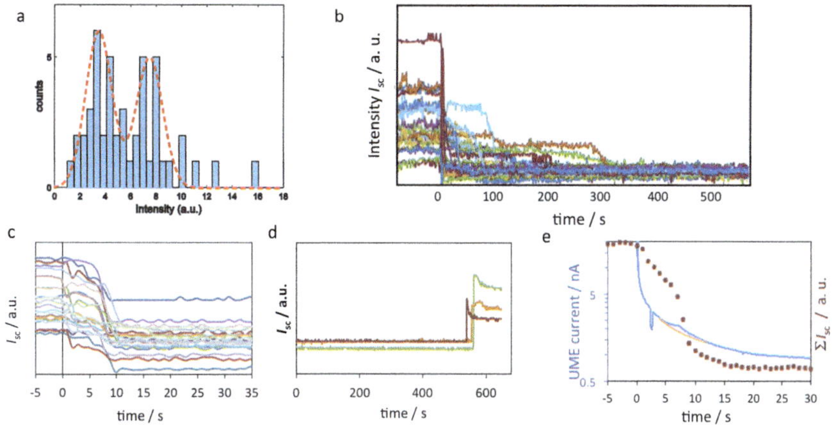

Fig. 2 Dark field study of Ag NP electrodissolution in 50 mM KSCN. (a) NP population based on their scattered intensities. (b) Examples of intensity profiles showing successive sets of dissolutions. (c) Examples of dissolving NPs extracted from Movie 1† at the early stage of the electrode polarization at 0.4 V *vs.* Ag QRE (started at $t = 0$). (d) Example of NP agglomerates landing on the UME. (e) Comparison of the electrochemical current flowing at the UME (blue line; extrapolated background, orange trace) with the sum of I_{sc} of all dissolving NPs (red symbols).

of I_{sc} measured for different individual NPs on the UME surface. It can be clearly decomposed into two distributions of I_{sc}, suggesting that 2 size populations of NPs are detected. In agreement with the reported[9a] size–I_{sc} correlation, the most abundant population is single 50 nm radius NPs detected with the lowest intensities, and less abundantly, dimers with intensities *ca.* twice as high. It is noteworthy that the I_{sc} signal of the dimers is actually close to the CCD1 detector saturation; henceforth, trimers may also be present, but are difficult to differentiate from dimers under our operating conditions.

B.2. DF monitoring of NP dissolution. Next, the oxidation of the ensemble of Ag NPs (single and agglomerates) present on the UME was triggered and the electrodissolution of individual NPs was detected by CCD1. As previously reported, the dissolution of individual NPs is associated to a decrease in the scattered light intensity, I_{sc}. Fig. 2b (and Fig. 2c for a shorter timescale) gives an example of the evolution of I_{sc} for different NPs. It is described here for a large number of adsorbed NPs ($N = 65$ in total for 2 electrochemical experiments). During chronoamperometry (CA) experiments, the electrode potential was suddenly changed at $t = 0$, from 0 to 0.4 V *vs.* Ag QRE. As seen in Fig. 2c, this is associated, for some NPs, to a noticeable (up to 15%) increase or decrease of I_{sc}, in agreement with other reports.[4] It is then followed, 1 to 5 s after the electrode potential application, by a more significant decrease in I_{sc}, attributed to the dissolution process. The dissolution is relatively slow, lasting between 2 and 10 s. For NP agglomerates, the dissolution process can also be described by two other scenarios: (i) the dissolution is not completed as some light scattering material is left on the UME for a long period of time and, (ii) the dissolution is incremental, suggesting that different individual NPs in the agglomerate are sequentially dissolved.

Dynamics of dissolution of ensembles of NPs. Even though a large proportion of NPs (60%, $N = 27$) simultaneously dissolve within the first 10 s of the potential application, a second set of NPs dissolves several tens of seconds after the first burst. This is attributed to the time required for sufficient dilution of the soluble Ag^+ ions after the first burst of dissolution. Indeed, assuming the dissolution of a single 50 nm radius NP releases *ca.* 50 attomoles of Ag^+ ions (equivalent charge of *ca.* 5 pC), the quasi-simultaneous dissolution of 27 NPs within about 5 s will release about 1 femtomole of Ag^+ ions in the vicinity of the UME. The transport of these ions over the UME region can be compared to the transport of redox active species electrogenerated at a micrometric array of nanoelectrodes. The modelling[13] of this situation showed that different diffusion regimes are explored depending on the time and separation distance between the nanosources: after radial diffusion from each individual nanosource, the individual hemispherical diffusion layers quickly overlap into a planar diffusion. Over longer times, and for micrometre-sized arrays, which is the size range of the UME used here, the diffusion again becomes radial at a larger scale, following a hemispherical diffusion over the UME geometry. In this context, the dissolution of 27 single Ag NPs at the UME can potentially generate above the UME, in a hemispherical layer of solution, Ag^+ cations with >μM levels (10^5 times the Ag^+ solubility in a SCN^- solution). This suggests that AgSCN will precipitate from the supersaturated Ag^+ solution electrogenerated by the UME, as discussed in B.3., and the soluble Ag^+ ions (mainly $Ag(SCN)_2^-$) will follow a similar release pattern, explaining the intrinsically slow dissolution processes. Moreover, for other NPs to dissolve, the solution adjacent to the UME should contain $Ag(SCN)_2^-$ species far below their solubility level. The evacuation of the dissolved electrogenerated ions from the UME region will proceed by diffusion. In a confined cell, a dilution of the solution adjacent to the UME by a factor of 1000 requires >20 s '(according to Comsol simulation, not shown), in rather good agreement with the duration of the intervals between different successive sets of dissolution at adsorbed Ag NPs (Fig. 2b).

Electrodissolution of landing NPs. Due to the restricted cell height, *ca.* 100 μm, the NP pre-adsorption step has almost completely depleted the solution of NPs and the electrodissolution of a NP upon landing on the UME becomes a rare event ($N = 9$ over 1400 s). Some examples of I_{sc} profiles corresponding to landing NPs are given in Fig. 2d. Here again, the I_{sc} profiles suggest that the NP may not simultaneously dissolve upon landing on the UME, or may even not be dissolved.

This study has provided new insights, particularly for the dissolution of an ensemble of Ag NPs, from individual optical monitoring. In addition, all the characteristics of the dissolution process measured optically here – slowness, onset time, sequential character, and incompleteness – confirm previous 3D holography inspection performed at a lower UME coverage of NPs.[11]

Electrochemistry of Ag NP electrodissolution. From an electrochemical viewpoint, the long pre-adsorption time and therefore the significant depletion of NPs within the cell volume also affects the nature of the electrochemical events that can be encountered. Indeed, the massive dissolution of NPs in the early stages of the potential application precludes the recording of reliable individual EC signals, but the EC response of an ensemble of NPs is rather expected. The time evolution of the EC current recorded during the first seconds after the electrode potential application is presented in Fig. 2e. It is compared to the time evolution of the scattered light intensity variation of the ensemble of the dissolving NPs. The initial intense EC current transient blurs the measurement of the ensemble EC response associated to the triggering of *ca.* 27 NPs dissolution. It would indeed correspond to the flow of about 140 pC over a few second period, making it difficult to discern from the noise level. Since the occurrence of EC impacts is also too rare and spans a large time-range, it could not be assigned with certainty. This also agrees with our previous conclusion on the slowness of the oxidation process, making a broader and much less intense EC signature of individual Ag NPs oxidation in a SCN^- solution.

Experimentally, however, a large oxidation peak is observed (Fig. 2e) between 5 and 12 s after the electrode actuation. Its integration (Fig. 3c) corresponds to 1.1 nC, exceeding by about 5 to 10 times the charge expected from the optically detected dissolution of 27 single NPs. Moreover, the variation of the charge precedes the decrease of I_{sc} for the ensemble of NPs by *ca.* 1 s. If both phenomena seem related, the exchanged charge is neither the direct signature of the NPs dissolution nor of their oxidation.

B.3. Growth and dissolution of micrometric deposits. A new process is also monitored in Movies 1a and b:† the electrodissolution of Ag NPs is accompanied with the appearance, growth and further disappearance of 3 bright disk-shaped micrometer-sized patterns on the UME surface. The overall process lasts for *ca.* 30–70 s. The onset time of the appearance of these patterns follows the onset of the optical dissolution of most NPs by 1–2 s. The dynamics of the formation-disappearance of such patterns is presented in Fig. 3. The growth is rather fast with an approximate initial radial expansion of 0.7 $\mu m\ s^{-1}$ until the patterns have reached a steady radius of the order of 2–3 μm before they finally decompose much more slowly. Interestingly, such patterns could only be detected when

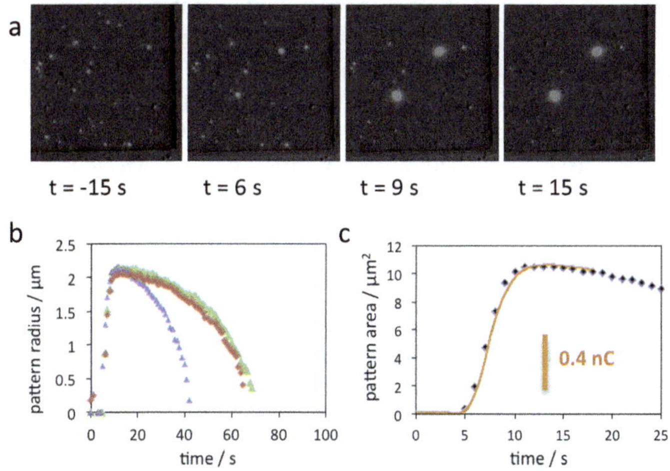

Fig. 3 Growth and dissolution of micrometric deposits at the UME surface during Ag NP electrodissolution. (a) 35 × 35 μm^2 images showing the simultaneity of dissolution of *ca.* 20 NPs with the pattern formation. (b) Growth and dissolution of the 3 patterns formed in Movie 1.† (c) Comparison of the total area of patterns (symbols) in (b) and of the charge (orange line) of the oxidation peak from Fig. 2e.

a large number of NPs were dissolved. The effect was also not detected on the same UME in the absence of NPs, showing that it is not due to parasitic processes at electrode defect sites. The simultaneity of the massive disappearance of individual NPs with the growth and further disappearance of these patterns suggests that this phenomenon is related to the formation of precipitate domains. We propose that they are the manifestation of the local precipitation of AgSCN domains from the super-saturated solution liberated upon massive Ag NP dissolution. Similar dissolution and redeposition processes have indeed been postulated[14] and observed by AFM during the formation of silver halide nanocrystals from the oxidation of Ag NPs in a halide anion solution.[14a] Interestingly, even if the location of the seeding Ag NPs could be tracked by AFM before the oxidation step, after oxidation, a silver halide nanocrystal was formed on a different location of the electrode, suggesting a similar dissolution, transport and recrystallization from a locally super-saturated solution. We attribute the observations obtained *in situ* and in real time in Movie 1† and Fig. 3 to this phenomenon.

It is indeed well known from late silver-based photography that light exposure can convert silver salts (AgX where X is a precipitating anion of Ag^+) into Ag metallic atoms or clusters. This photoreduction process was recently transposed to various AgX NPs,[15] including AgSCN ones.[16] It yields the formation of a core–shell like Ag@AgX structure, where metallic Ag atoms cover the external surface of the AgX NP. The latter Ag@AgX core–shell NPs exhibit efficient light absorption at redshifted wavelengths. Typically, the exposure of AgSCN NPs to the intense light from a halogen lamp successfully produced Ag@AgSCN NPs, which presented a strong new absorption in a broad 400–600 nm range.[16]

Interestingly, the growth of these micrometric patterns also coincides with the oxidation peak depicted in the previous section. Fig. 3c shows that the variation of the overall pattern surface area fits the variation of the charge exchanged during this oxidation peak. The charge flow could also be explained by the formation of the Ag@AgSCN material. It was indeed demonstrated that Ag@AgX nanomaterials efficiently harvest electric (and electrochemical) energy from light. Under an appropriate illumination, corresponding to the band gap of the AgX semiconducting material, an ITO electrode coated with $Ag@Ag_3PO_4$ NPs allows charge separation processes, which could be detected by the appearance of an anodic photocurrent.[15] We believe the oxidation current detected in the present work to have the same photoelectrochemical origin.

C. Spectroscopic monitoring of Ag NP electrodissolution

C.1. Monitoring single Ag NPs. We now turn to the inspection of the spectroscopic changes during the oxidation of Ag nanoparticulates in a SCN^- solution. Upon dissolution of Ag NPs, which can bear LSPR, Mie's theory predicts that in a homogeneous aqueous medium, the NP shrinking is accompanied by a blue-shift and a decrease in intensity of the overall scattering. However, in an electrochemical context, during oxidation, the plasmonic Ag NP is in contact with a thin Au layer, which can bear surface plasmon-polaritons. Coupling between the LSPR of the Ag NP and the surface plasmon-polaritons of the Au metallic film is therefore expected. A way of understanding this coupling consists of considering that the Au layer behaves as a partial mirror creating a mirror image of the Ag NP facing it, classically described as an image dipole. Like real NP pairs, which have spacing-dependent spectral properties used in "plasmonic ruler"[17] experiments, this virtual pair of NPs allows the evaluation of the distance between the plasmonic structures (the NP–film or the NP–NP distance) within the nm range. From a spectroscopic point of view, by coupling NP–film or NP–NP systems, new peaks corresponding to new hybridized modes of plasmon resonance appear.[8] The wavelengths of these new modes characterize the strength of the coupling, which results from the chemical nature and size of the gap between the NP and the film or between the NPs. The spectroscopy of metallic NPs under electrochemical operation is then a complementary mechanistic tool to decipher single NP electrochemistry.

The signal recovered in CCD2 provides a spectroscopic image of the plasmon couplings between single Ag NPs or Ag NP agglomerates and the thin Au UME during the electrodissolution process. Fig. 4a and c present the spectral evolution (f = 1 Hz) of an oxidizing single NP during a CA experiment. The spectral evolution is also presented in Movie 1b and c.† Prior to the electrode polarization, the $I_{sc}(\lambda)$ spectrum is constant and presents two scattering modes with peaks at 500 and 700 nm respectively. As in Au NP–Au film,[2] or Ag NP–Ag film[18] systems, the higher energy peak essentially corresponds to the LSPR of the Ag NP.

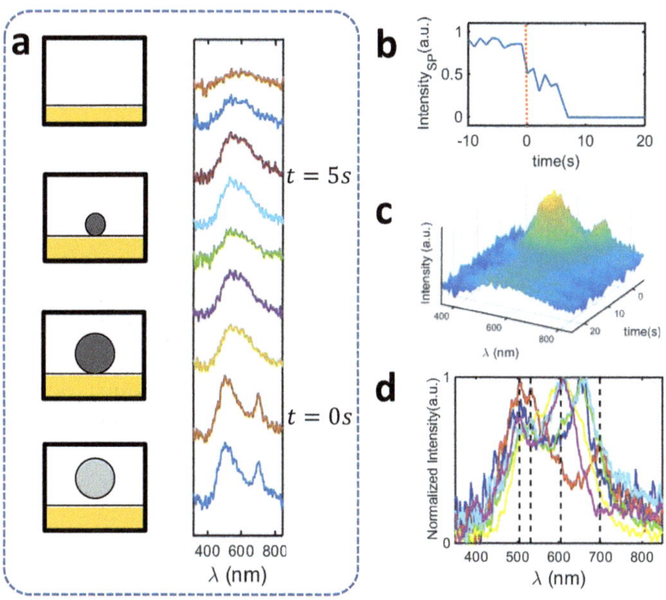

Fig. 4 Spectroscopic monitoring of the electrodissolution of a single Ag NP. (a) Time evolution (f = 1 Hz) of the spectrum of the scattered light upon oxidation and schematic representation of the Ag → Ag@AgSCN process describing subsequent NP dissolution. Time t = 0 is related to the beginning of the electrode polarization. The oxidative transformation lasts for *ca.* 1 s, the dissolution starts after 4 s and lasts for 3 s. The broad spectrum corresponds to the Ag@AgSCN nanocrystal formed upon oxidation and photoreduction of the surface of AgSCN. (b) Time evolution of the scattered intensity and (c) spectrum of scattered light in 3D. (d) Comparison of the spectra of all ascribed single Ag NPs in the vicinity of the Au UME, showing the distance-modulation of their plasmonic coupling from the peak-splitting.

Conversely, the spectral distance between these two peaks would reflect the NP–film coupling and therefore the nature and size of the gap. Upon electrode polarization ($t = 0$), the double-peak spectrum is transformed into a unique broadband peak centered at 550 nm, suggesting that the Ag NP–Au film coupling is now lost. The shape and position of the band then remain unchanged but its intensity gradually fades out, suggesting some nanomaterial dissolution.

We ascribe the spectrum change upon electrode polarization to the oxidation of the Ag NP into an AgSCN nanocrystal, subsequently slowly dissolved (*ca.* 10 s). The observed time scales match our previous studies of the Ag → AgSCN transformation and dissolution.[11c] Interestingly, the spectrum of this AgSCN material is too redshifted compared to the broad absorption band at 300 nm of a solution of AgSCN NPs. Instead, it might correspond to the formation under illumination of the core–shell Ag@AgSCN nanocrystal postulated in B.3. It is noteworthy that this study suggests that if the Ag@AgSCN nanocrystal dissolution can be monitored optically, the electrochemical transformation of the Ag NPs into the Ag@AgSCN nanocrystal – *i.e.* the oxidation process – is more difficult to decipher without a spectroscopic analysis. Indeed, if the integration of the spectra over all wavelengths is monitored ($\int I_{SC}(\lambda)d\lambda$, shown in Fig. 4b), it does not vary significantly upon oxidation under white light DF illumination. Similarly, based on the given spectra, $I_{sc}(\lambda)$, it is anticipated that the monochromatic monitoring at 532 nm of the same oxidation process would not significantly change (by less than 20% in Fig. 4) during the Ag → Ag@AgSCN transformation, in agreement with our previous results.[11c] A full electromagnetic modelling of the optical response of these systems will be necessary in order to confirm these scenarios. However, broad behavior schemes for the NP–surface plasmonic interaction can be used to propose tentative analyses of more complex systems, as shown below.

C.2. Towards complex Ag NP agglomerate oxidation analysis. Based on the scattered intensity, I_{sc}, measured on CCD1, the other NPs monitored by spectroscopy are NP agglomerates. Fig. 5 presents an example of the evolution of their scattering spectrum upon oxidation. As for the former case of Ag NP–Au film coupling, the coupling between two metallic NPs (*e.g.* Ag nanocubes,[19] or other NP assemblies[20]) typically manifests as peak splitting. Here, the plasmonic coupling is more intricate than for a single Ag NP–Au film, or for a dimer of Ag NPs, since a three (or more)-body interaction is under consideration, which can be governed by the different inter-body separation distances, explaining qualitatively the observed broadening of the peaks. Particularly in the case of a NP dimer near the electrical contact region, the higher energy resonance (characteristic of the NP) is expected to blueshift upon coupling with a metallic film.[4b] In Fig. 5a, we have monitored the spectra of a NP agglomerate that was sitting on the UME surface before the beginning of the experiment. Along with the total scattered intensity, the complicated shape of the spectrum strongly suggests that the agglomerate is made of several individual NPs. Before the oxidation, the spectrum of the Ag NP agglomerate in contact with the Au UME can be roughly related to the broadened double-peak spectrum (peaks around 520 and 650 nm). Upon application of the anodic potential, the spectrum changes dramatically, within 1–2 s, which is again consistent with the oxidation time previously observed in single electrochemical NP impacts for the Ag → AgSCN transformation.[11c] The new spectrum shows a peak at 550 nm tailing in the red region. This intense single peak, redshifted compared to the single Ag NP–Au film, is consistent with the presence of

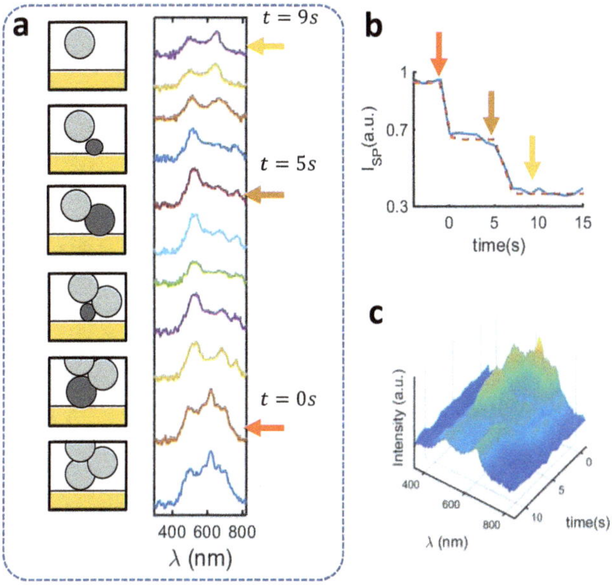

Fig. 5 Spectroscopic monitoring of NP agglomerate. (a) Scheme illustrating a possible scenario explaining the spectra taken at a 1 Hz sampling rate. The red arrow indicates electrode polarization. (b) Intensity profile obtained by integrating the spectra and showing the stepwise oxidation of the agglomerate. (c) 3D profile illustrating the dynamics of the evolution of the spectra over time.

a NP-dimer.[4b] About 3–4 s later, the agglomerate undergoes another radical transformation. The spectral change is again observed in 1–2 s, with the progressive building of a new peak at 650 nm (which blueshifts by 50 nm during the transformation) and the progressive blueshift by 50 nm and broadening of the low energy peak.

Moreover, if the total intensity of the spectrum (obtained by integration of the spectrum over all wavelengths) is monitored, we notice that changes of the spectra are accompanied by a steady decrease in intensity, forming a step-like pattern (Fig. 5b). This is also consistent with the hypothesis that the particle is indeed an agglomerate and that the NPs within the agglomerate are oxidized sequentially.

All these observations suggest that the NPs in the agglomerate are not in electrical contact with the Au UME or with one another. This is also supported by our previous works, in which the formation of a dimer in solution and its later impact with a polarized electrode were monitored, yielding a stepwise oxidation, *i.e.* oxidation of one NP at a time.[11c] A good agreement is observed comparing the persistent spectrum of that (and other) oxidized agglomerates to the spectrum of a single NP, shown in Fig. 4d, which supports that the NP agglomerates are incompletely oxidized into a single particle. Qualitatively, it confirms that the incompletely oxidized aggregates optically behave as Ag NPs separated from the Au electrode. In principle, the particle–plane distance could be obtained from the coupling strength estimated from their spectra. This will require the modelling of the optical response of the situation, keeping in mind that the role of the AgSCN material, which likely controls the gap between the two plasmonic resonators, may be difficult to determine. It is then difficult to go beyond these qualitative arguments at the present stage.

Experimental

D. Chemicals

Colloidal solutions of 50 nm radius, citrate stabilized Ag NPs were obtained from Sigma-Aldrich. Their size distribution was characterized by dynamic light scattering (DLS, Zetasizer 3000HS, Malvern Instruments). Ultrapure water of resistivity not less than 18.2 MΩ cm was used to prepare all solutions. The colloidal electrolytic solutions were prepared just before use. All reagents were of analytical grade from Sigma-Aldrich.

E. Electrode fabrication

The transparent gold UMEs were prepared by lithography in clean room facilities. First, the glass slides were carefully cleaned with acetone and isopropanol, then covered with a photosensitive resin (TI-35 ES MicroChemicals) by spin coating (500 rpm for 3 s, then 4000 rpm for 30 s). Next, the electrode pattern was fixed on the resin (using a MJB4 mask aligner in the soft contact mode). The resin + mask ensemble was exposed to UV light for 20 s (12 mW cm^{-2} at 365 nm), followed by heating (40 °C for 10 min and 130 °C for 2 min) and flood exposure for 60 s (12 mW cm^{-2} at 365 nm). The pattern was developed by immersion in a developer solution (MIF326 Microchemicals). A 5 nm Ti adhesion layer and a 45 nm Au layer were then deposited using a Plassys MEB 550S evaporator. The resist was removed using acetone, leaving the gold layer only on the areas not exposed. A 400 nm silicon nitride layer was then deposited by plasma-enhanced chemical vapor deposition (Corial D250), over which a thin layer of positive resin S1818 (MicroChemicals) was spin-coated and later exposed to UV light for 20 s (12 mW cm^{-2} at 365 nm). Using a MJB4 mask aligner in the soft contact mode and a custom made chrome mask the exposure was blocked on 60 × 60 μm^2 squares, which later become the UME regions. The silicon nitride layer was then plasma-etched (RIE Corial 200R) and the remaining resist was removed using acetone and oxygen plasma cleaning.

F. Microfluidic cell assembly

The microfluidic cell assembly is described in detail elsewhere.[11a] In short, parafilm was placed with a cylindrical hole of *ca.* 2 cm diameter to define the limits of the cell. Capillary pipettes, which are used to inject and evacuate solution from the cell, were introduced and the cell was closed with a thin glass slide and sealed by melting the parafilm upon heating up to 50 °C.

G. Optical setup

All spectra were acquired linking a grating-based spectrometer (ImSpector V8E Enhanced Model) with an EMCCD camera (Andor Ixon 897). Electrochemical measurements were performed with a PalmSens 3 portable potentiostat.

Conclusions

Spectroscopic detection was coupled to dark field imaging to study electro-oxidation of Ag NPs in a thiocyanate solution. It confirms that the oxidation process is accompanied by a slow phase transformation, evidenced through the *in situ* and real time monitoring of a radical change in the visible spectrum of individual nanodomains. This transformation is followed by a disappearance of the nanodomain, attributed to its dissolution. Owing to the spatially limited detection, adsorption of a relatively large number of NPs (*ca.* 0.02 NP μm^{-2}) was needed. Most of the NP oxidation occurs during the first seconds of the electrode polarization (chronoamperometry experiments), and the ensemble electrochemical response is then blurred in the electrical noise. The individual signature can then be obtained from the individual light scattering intensity. Moreover, during the course of the concomitant oxidation of a large number of NPs, the growth of micrometer sized bright domains was observed. It is attributed to the crystallization of some AgSCN material from the local supersaturated solution and it provides an indirect optical visualization of the role of the overlapping of diffusion layers generated by random arrays of closely separated nanosources of Ag^+ ions.

Finally, the spectrometric interrogation of individual Ag NPs (single NPs or NP agglomerates) during their oxidation allows qualitative discrimination between the behaviour of single NPs and that of NP agglomerates. Particularly, the incomplete oxidation of agglomerates results from the formation of loose contacts between the NPs and the UME. This study shows the interest in implementing complementary information for the understanding of complex chemical processes occurring during electrochemical reactions. Besides the qualitative demonstration of the potential of the method, more quantitative insight will require the modelling of the spectroscopic response of plasmonic hybridization in multiple-body systems.

Acknowledgements

The authors acknowledge funding by the Agence Nationale pour la Recherche (ANR) through the NEOCASTIP ANR-15-CE09-0015-02 project. J. Médard is thanked for clean room fabrication.

Notes and references

1 (*a*) D. Y. Wu, J. F. Li, B. Ren and Z. Q. Tian, *Chem. Soc. Rev.*, 2008, **37**, 1025–1041; (*b*) W. Wang and N. Tao, *Anal. Chem.*, 2014, **86**, 2–14; (*c*) C. M. Hill, D. A. Clayton and S. Pan, *Phys. Chem. Chem. Phys.*, 2013, **15**, 20797–20807.

2 T. Sannomiya, H. Dermutz, C. Hafner, J. Voros and A. B. Dahlin, *Langmuir*, 2010, **26**, 7619–7626.

3 (*a*) M. Chirea, S. S. E. Collins, X. Wei and P. Mulvaney, *J. Phys. Chem. Lett.*, 2014, **5**, 4331–4335; (*b*) C. Novo, A. M. Funston, A. K. Gooding and P. Mulvaney, *J. Am. Chem. Soc.*, 2009, **131**, 14664–14666.

4 (*a*) C. P. Byers, B. S. Hoener, W.-S. Chang, M. Yorulmas, S. Link and C. F. Landes, *J. Phys. Chem. B*, 2014, **118**, 14047–14055; (*b*) C. P. Byers,

B. S. Hoener, W.-S. Chang, S. Link and C. F. Landes, *Nano Lett.*, 2016, **16**, 2314–2321.

5 Y. Li, J. T. Cox and B. Zhang, *J. Am. Chem. Soc.*, 2010, **132**, 3047–3054.

6 S. K. Dondapati, M. Ludemann, R. Müller, S. Schwieger, A. Shwemer, B. Händel, D. Kwiatkowski, M. Djiango, E. Runge and T. A. Klar, *Nano Lett.*, 2012, **12**, 1247–1252.

7 (*a*) Y. Fang, W. Wang, X. Wo, Y. Luo, S. Yin, Y. Wang, X. Shan and N. Tao, *J. Am. Chem. Soc.*, 2014, **136**, 12584–12587; (*b*) X. Wo, Z. Li, Y. Jiang, Y. Su, W. Wang and N. Tao, *Anal. Chem.*, 2016, **88**, 2380–2385.

8 F. Le, N. Z. Lwin, J. M. Steele, M. Käll, N. J. Halas and P. Norlander, *Nano Lett.*, 2005, **5**, 2009–2013.

9 (*a*) C. M. Hill and S. Pan, *J. Am. Chem. Soc.*, 2013, **135**, 17250–17253; (*b*) C. M. Hill and S. Pan, *J. Phys. Chem. C*, 2015, **119**, 6760–6768.

10 X. Shan, I. Diez-Perez, L. Wang, P. Wiktor, Y. Gu, L. Zhang, W. Wang, J. Lu, S. Wang, Q. Gong, J. Li and N. Tao, *Nat. Nanotechnol.*, 2012, **7**, 668–672.

11 (*a*) C. Batchelor-McAuley, A. Martinez-Marrades, K. Tschulik, A. N. Patel, C. Combellas, F. Kanoufi, G. Tessier and R. G. Compton, *Chem. Phys. Lett.*, 2014, **597**, 20–25; (*b*) A. N. Patel, A. Martinez-Marrades, V. Brasiliense, D. Koshelev, M. Besbes, R. Kuszelewicz, C. Combellas, G. Tessier and F. Kanoufi, *Nano Lett.*, 2015, **15**, 6454–6463; (*c*) V. Brasiliense, A. N. Patel, A. Martinez-Marrades, J. Shi, Y. Chen, C. Combellas, G. Tessier and F. Kanoufi, *J. Am. Chem. Soc.*, 2016, **138**, 3478–3483.

12 (*a*) Y. Zhou, N. Reels and R. G. Compton, *Angew. Chem., Int. Ed.*, 2011, **50**, 4219–4221; (*b*) X. Xiao and A. J. Bard, *J. Am. Chem. Soc.*, 2007, **129**, 9610–9612; (*c*) S. E. F. Kleijn, S. C. S. Lai, M. T. M. Koper and P. R. Unwin, *Angew. Chem., Int. Ed.*, 2014, **53**, 3558–3586.

13 (*a*) N. Godino, X. Borrisé, F. X. Muñoz, F. J. del Campo and R. G. Compton, *J. Phys. Chem. C*, 2009, **113**, 11119; (*b*) C. Adam, F. Kanoufi, N. Sojic and M. Etienne, *Electrochim. Acta*, 2015, **179**, 45–56.

14 (*a*) U. Hasse, S. Fletcher and F. Scholz, *J. Solid State Electrochem.*, 2006, **10**, 833–840; (*b*) P. Singh, K. L. Parent and D. A. Buttry, *J. Am. Chem. Soc.*, 2012, **134**, 5610–5617.

15 Z. Yi, J. Ye, N. Kikugawa, T. Kako, S. Ouyang, H. Stuart-Williams, H. Yang, J. Cao, W. Luo, Z. Li, Y. Liu and R. L. Withers, *Nat. Mater.*, 2010, **9**, 559–564.

16 J.-W. Hu and X. Zhang, *Powder Technol.*, 2014, **267**, 240–244.

17 C. Sönnischen, B. M. Reinhard, J. Liphardt and A. P. Alivisatos, *Nat. Biotechnol.*, 2005, **23**, 741–745.

18 J. Lee, S. Yoo, M. Shin, A. Choe, S. Pak and H. Ko, *J. Mater. Chem. A*, 2015, **3**, 11230–11235.

19 T. A. F. König, P. A. Ledin, M. Russell, J. A. Geldmeier, M. A. Mahmoud, M. El-Sayed and V. V. Tsukruk, *Nanoscale*, 2015, **7**, 5230–5239.

20 (*a*) A. Lombardi, M. P. Grzelczak, A. Crut, P. Maioli, I. Pastoriza-Santos, L. M. Liz-Marzan, N. Del Fatti and F. Vallée, *ACS Nano*, 2013, **7**, 2522–2531; (*b*) P. Nordlander, C. Oubre, E. Prodan, K. Li and M. I. Stockman, *Nano Lett.*, 2004, **4**, 899–903; (*c*) L. Slaughter, W. Chang and S. Link, *J. Phys. Chem. Lett.*, 2011, **2**, 2015–2023.

Faraday Discussions

PAPER

Challenges in nanoelectrochemical and nanomechanical studies of individual anisotropic gold nanoparticles

P. Knittel, O. Bibikova and C. Kranz*

Received 10th May 2016, Accepted 6th June 2016
DOI: 10.1039/c6fd00128a

The characterization of nanoparticles and the correlation of physical properties such as size and shape to their (electro)chemical properties is an emerging field, which may facilitate future optimization and tuning of devices involving nanoparticles. This requires the investigation of individual particles rather than obtaining averaged information on large ensembles. Here, we present atomic force – scanning electrochemical microscopy (AFM-SECM) measurements of soft conductive PDMS substrates modified with gold nanostars (*i.e.*, multibranched Au nanoparticles) in peak force tapping mode, which next to the electrochemical characterization provides information on the adhesion, deformation properties, and Young's modulus of the sample. AFM-SECM probes with integrated nanodisc electrodes (radii < 50 nm) have been used for these measurements. Most studies attempting to map individual nanoparticles have to date been performed at spherical nanoparticles, rather than highly active asymmetric gold nanoparticles. Consequently, this study discusses challenges during the nanocharacterization of individual anisotropic gold nanostars.

Introduction

Among metal nanoparticles, gold nanoparticles (AuNPs) are the most commonly applied nanomaterials in (bio)analytical sensing, electrocatalysis, and health-related diagnostics and therapies due to their unique catalytic, electric, and optical properties.[1–4] In comparison to other noble metal nanoparticles, they are readily modified *via* versatile thiol-based chemistry and a wide variety of shapes is available including spheres, cubes, triangles, octahedras, stars, rods, and cages.[5] The most commonly used AuNPs in (electro)analytical applications are spherically shaped (*i.e.*, isotropic NPs), which are conventionally obtained by simply reducing gold chloride in stabilizing solutions.[6,7] Recently anisotropic AuNPs such as nanorods, as well as nanobranched and nanostar-shaped AuNPs have gained considerable attention.[8–10] Asymmetric AuNPs (*i.e.*, anisotropic NPs) are

Institute of Analytical and Bioanalytical Chemistry, Ulm University, Albert-Einstein Allee 11, 89081 Ulm, Germany. E-mail: christine.kranz@uni-ulm.de

characterized by a high surface area, an advantageous mass-to-surface ratio, and multiple resonances due to their complex structures resulting in different resonance modes of electrons. Hence, Au nanostars (AuNSts) with their multibranched structure provide unique optical properties, which render them highly suitable as multifunctional biolabels,[11] for optical sensing applications,[12] and in surface enhanced Raman scattering (SERS) applications.[13,14] The multibranched structures also influence their magnetic, and (electro)catalytic properties. More importantly, the shape and particle size is also suspected to influence their cytotoxicity,[3] and their electrochemical properties,[15] which are not yet well characterized. Of particular interest to the present study is the correlation of the nanoelectrochemical and nanomechanical properties of such anisotropic AuNPs for modification of flexible electrodes like conductive composites of polydimethylsiloxane (PDMS).[16,17] For example, macroscopic electrochemical experiments are highly suitable to analyze the exposed surface area and the dominant surface structure by underpotential deposition (UPD) and subsequent stripping. It has been shown for various nanoparticles that depositions preferentially occur at certain facets.[18–21] For AuNPs, typically UPD of lead (Pb) is used, as Pb is predominantly deposited at the ⟨110⟩ and ⟨100⟩ facet.

Microscopic techniques are used to characterize the size and shape of nanoparticles ranging from vacuum techniques such as high-resolution scanning electron microscopy (HR-SEM), and transmission electron microscopy (TEM) to scanning probe techniques, such as atomic force microscopy (AFM), scanning tunneling microscopy (STM), and scanning electrochemical microscopy (SECM).[22] Electrochemical imaging techniques have gained significant interest, as they allow the correlation of structural properties and particle size to the (electro) chemical functionality/activity of the nanoparticles beyond mere morphological information.[23,24] For example, Meier *et al.* investigated the catalytic activity towards the hydrogen evolution of single supported palladium nanoparticles in dependence of their size using electrochemical STM.[25] SECM studies involve a wide range of experiments regarding nanoparticles, which range from localized deposition of *e.g.*, AuNPs with various shapes[26,27] to collision and landing experiments at ultramicroelectrodes (UMEs) providing transport processes and collision dynamics of NPs with solid electrode surfaces. Bard and co-workers investigated single PtNP collisions with a carbon microelectrode by recording the current transients due to the hydrogen evolution reaction, which was only observable at the surface of the PtNPs, once they were in contact with the electrode surface.[28] SECM studies were also targeted towards imaging the catalytic activity of various noble metal NPs with different shapes.[24,29–32] In these studies, typically localized assemblies or aggregates of nanoparticles were investigated. For example, AuNP spot arrays have been investigated towards the electrocatalytic activity for the oxygen reduction reaction in dependence of the shape. Sánchez-Sánchez *et al.*[30] showed that cubic AuNPs are the most active particles toward oxygen reduction in basic solution followed by spherical AuNPs, and Au nanorods, which the authors related to a higher ratio of Au ⟨100⟩. Wain[32] also investigated fairly large arrayed spots of AuNPs with varying sizes (50–5 nm) towards the effectiveness of oxygen reduction using the redox competition mode[33] and the tip collection mode. An increase in reactivity with decreasing particle diameter was observed, which again was related to an increased Au ⟨110⟩ to Au ⟨111⟩ ratio for smaller particles. In these SECM studies, conventional micrometer-sized

SECM electrodes were employed, which limited the achievable spatial resolution and required additional characterization techniques such as TEM and UPD of lead to determine the size distribution and facets. High-resolution SECM investigations of single AuNPs with electrode radii as small as 3 nm for static experiments, and radii of 14 nm for dynamic imaging experiments were recently presented by Mirkin and co-workers.[34] Single nanoparticles were imaged by scanning the sample at constant height in the *x*, *y*-plane. Besides imaging in feedback mode, the catalytic activity towards the hydrogen evolution of AuNPs in a size range of 10–20 nm has also been investigated.

As the size and shape crucially influences the reactivity, and may even change during targeted applications such as electrocatalysis, studying the relationship between functionality and morphology requires that samples are investigated sequentially by high-resolution techniques such as HR-SEM, AFM or STM, *etc.*, which is challenging in single nanoparticle analysis. More suitably, high-resolution morphological information should in fact be simultaneously obtained along with (electro)chemical information. Significant progress has recently been achieved on electrochemical hybrid techniques based on dual microcapillaries, which were initially introduced for patterning experiments in scanning ion conductance microscopy.[35] Based on such dual capillaries, Unwin and coworkers introduced scanning electrochemical cell microscopy (SECCM),[36] which provides the topography of the sample based on the measured ion current. For the small orifice of the capillary, high-resolution topographical images may be obtained. The catalytic properties towards the oxygen reduction reaction of individual platinum nanoparticles (size 100 nm $\pm$ 14 nm) deposited onto carbon nanotubes was investigated with SECCM.[37] It could be shown that at the single particle level, similar particles revealed different reactivity based on small differences of their morphology. O'Connell *et al.*[29] studied the hydrogen peroxide generation during the oxygen reduction at individual AuNPs within NP ensembles (approx. 1 μm in size). A dual-barrel capillary with one barrel filled with carbonaceous material was used as an imaging probe[38] that was subsequently modified with platinum (Pt) by electrodeposition.[39] As an alternative to hybrid SICM-SECM, the combination of AFM-SECM[40] has been used to study nanoparticles. Demaille *et al.* used conical AFM-SECM probes produced from etched and bent gold nanowires.[41] AuNPs deposited at an amino-thiolated gold substrate were modified with redox mediator molecules *via* flexible polyethylene glycol (PEG) chains, which prevents the diffusional dispersion of the mediator.[42] AFM measurements in intermittent mode provide topographical information whereas biasing the conical electrode at the tip apex resulted in imaging the modified gold nanoparticles in feedback mode SECM. The achieved spatial resolution in these mediator-confined measurement scenarios was on the order of several tens of nanometers.

A particular advantage of AFM hybridized with electrochemical/electroanalytical techniques is that besides topographical and electrochemical information, nanomechanical properties such as adhesion, deformation, and Young's modulus of the sample can also be simultaneously mapped based on the recorded force information.[43,44] In peak force tapping™ (PFT) mode, the AFM tip is usually oscillating with amplitudes in the range from less than 100 up to 3000 nm at frequencies of 0.25 to 2.0 kHz.[45] For example, using the Derjaguin–Muller–Toporov (DMT) model the Young's modulus of the scanned sample can be derived with spatial resolution.[46] Moreover, this mode is highly suitable for imaging soft samples such as cells and polymers as the force is controlled at each pixel, which avoids drifts in imaging and impedes high lateral forces. Another important aspect is that changes of the sample or substrate due to aging or due to *e.g.*, applying mechanical or chemical forces to the sample, can be mapped. Recently, nanoparticle-modified soft materials have gained importance in various application fields such as flexible SERS substrates, in microfluidic devices or nanoparticle modified scaffolds for cell proliferation. Hence, for such materials not only changes in topography but also in the mechanical properties are of utmost importance. Finally, the interaction of nanoparticles with biological samples such as cells has been investigated in peak force tapping mode.

Consequently, in the present study we report on simultaneous topographical, nanomechanical, and nanoelectrochemical investigations (QNM-AFM-SECM) of anisotropic gold nanoparticles (*i.e.*, gold nanostars) deposited at the surface of a soft conductive PDMS substrate and the experimental challenges of such measurements. While the catalytic properties towards the oxygen reduction reaction of various shaped AuNPs have been investigated in detail,[1] gold nanostars have only been investigated in depth for their optical properties, yet little has been reported on their electrochemical characterization and their performance, *e.g.*, in catalysis.[15,47] In order to obtain spatially resolved electrochemical information suitable for individual nanoparticle mapping, nanodisc electrodes have been integrated into AFM-SECM probes, as previously reported by Pust *et al.*[48] and Lee *et al.*[49] In contrast to these approaches, the AFM-SECM probes used for the presented experiments have AFM tips with well-defined curvature radii suitable for nanomechanical mapping and high-resolution electrochemical imaging. Also, to date such bifunctional probes have not been applied for investigating individual nanoparticles. Imaging at a soft substrate providing nanomechanical information such as Young's modulus and adhesion forces along with spatially resolved electrochemical properties of such asymmetric gold nanoparticles is reported herein for the first time.

Experimental

Fabrication of the Au nanostars

Nanostars (NSts) were fabricated by a seed-mediated growth method with slight modifications following the procedure of Yuan *et al.*[50] Nanospheres (NSps) with diameters of 15 nm were used as seeds. 10 μL of 1 M HCl and 100 μL of the NSps solution were added to 10 mL of 0.25 mM $HAuCl_4$ aqueous solution at room temperature under vigorous stirring. Then, 100 μL of 2 mM $AgNO_3$ and 50 μL of 0.1 M ascorbic acid were simultaneously added to the solution, which then was stirred for 30 s while its color turned from light red to dark gray. Immediately afterwards, NSts were centrifuged and washed (300 g, 20 min) to concentrate the nanoparticles. After this step, the AuNSts were re-dispersed in water. The so-produced AuNSts had an outer diameter of approximately 80 nm with 8–10 thorns with lengths of 20–30 nm.

Fabrication of the carbon–PDMS substrate

To prepare soft conductive substrates, a 10 : 1 mixture of PDMS (base: curing agent, Sylgard 184, Dow Corning, US) was thoroughly mixed with carbon exhibiting particle sizes of approx. 50 nm (Carbon black, Vulcan XC72-R, Cabot Inc., US). Then, the mixture of carbon and PDMS 25% w/w (CPDMS) was degassed for 30 min and stored at −15 °C until usage. For the CPDMS mixture used in the presented experiments, a conductivity of approx. 10 S·m^{-1} was obtained.[51] The CPDMS was coated onto a macroscopic gold disc electrode (CH Instruments, Texas, US) and cured at 80 °C for 12 h prior to usage. Subsequently, the CPDMS surface was cleaned *via* ozone treatment and modified by immersion in 1 mM (3-mercaptopropyl)trimethoxysilane (MPTMS, Sigma-Aldrich, DE) in ethanolic solution to increase the adhesion of the gold nanoparticles. AuNSts were then drop-casted onto the surface of MPTMS-modified or unmodified CPDMS from a 30 mg L^{-1} solution. For the spherical Au nanoparticles (BBI Solutions, US), the same procedure was applied using a 57 mg L^{-1} solution.

AFM measurements

All AFM measurements were performed with a Bioscope Catalyst AFM (Bruker, DE) in PFT mode. For the characterization of the CPDMS substrate and the AuNSts in air, a sharp silicon nitride tip with a nominal tip curvature radius of 2 nm (MSNL, Bruker, DE) was used. The CPDMS-modified gold electrodes were mounted in the AFM electrochemical liquid cell using a customized holder, which also allowed electrically connecting the tip-integrated electrode and the macro-electrode (as shown in Fig. 1). Prior to the AFM measurements in PFT mode, the AFM probes were calibrated on a hard silicon substrate and a soft PDMS model sample (Young's modulus, $E = 2.5$ MPa, Bruker, DE) to determine the actual spring constant (spring constants provided by the manufacturer are nominal values and may significantly vary from the actual value) and the contact radius of the tip, required to determine the Young's modulus. As the tip penetrates and deforms the soft sample, the contact radius differs from the tip curvature radius

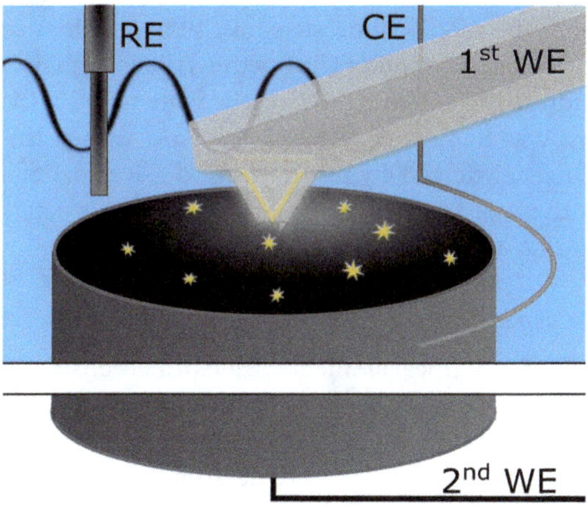

Fig. 1 Scheme of the QNM-AFM-SECM setup. The modified gold electrode is mounted perpendicularly in a custom-made holder, which is completely sealed to allow measurements with an AFM-SECM probe (1st WE) in solution. From the bottom, the electrode can be connected as a (second) working electrode (2nd WE). An Ag/AgCl reference (RE) and Pt wire as a counter electrode (CE) are integrated into the electrochemical cell.

especially for sharp probes. The images were obtained with a peak force of 800 pN and an amplitude of 200 nm at a frequency of 1 kHz.

Fabrication of AFM-SECM probes with integrated nanoscopic gold electrodes

For the fabrication of the AFM-SECM probes, triangular silicon nitride cantilevers with silicon nitride tips (Olympus TR800PSA, JP) were modified with a Ti/Au coating after removal of the initial metal coating and insulated with mixed silicon nitride/silicon dioxide layers according to a procedure described elsewhere.[52,53] The tip-integrated gold nanoelectrode was exposed in a two-step process *via* focused ion beam milling (FEI Quanta 3D FEG dual-beam FIB/SEM, NL). First, the gold layer is exposed by removing the insulation in a rectangular region at an angle of 42° to the electron column (ion-column angle is 52° to the electron column) leaving an L-shaped tip with a tilt of 10° (see Fig. 4a) compensating for the mounting angle of the AFM. The sharp thorn (*i.e.*, the insulating AFM tip), which is required for high-resolution topographical imaging was achieved by milling the residual insulation into a triangular structure. In the second step, the tip is milled from the side to form a pyramidal tip. An optional third step to sharpen the tip and clean the electrode surface can be performed. Thus produced AFM-SECM probes have a spring constant of 1.5–2.5 N m^{-1} (approx. 10-times the nominal spring constant of 0.15 N m^{-1} as specified by the supplier), which is correlated to the additional deposition of layers, a tip curvature radius of less than 50 nm and an integrated gold electrode with a radius below 50 nm. For the measurements, the probes were mounted onto a custom holder and thoroughly insulated using UV curable glue.

Electrochemical characterization

All electrochemical measurements were performed with an Ivium CompactStat (Ivium, NL) using an Ag/AgCl (sat.) reference electrode and a Pt wire as auxiliary electrode. All potentials specified in this paper are reported *versus* Ag/AgCl (sat.). Characterization of the modification steps of CPDMS such as ozone treatment, modification with (3-mercaptopropyl)trimethoxysilane and modification with AuNSts was performed in 0.1 M KCl *via* cyclic voltammetry (CV) in a potential range of −2 to +2 V and at a scan rate of 100 mV s^{-1} with at least 5 consecutive cycles. The electrocatalytic activity of the electrodes was determined in 10 mM H_2O_2 solution with 0.1 M KCl as supporting electrolyte in a potential range of −1 to +1 V at a scan rate of 100 mV s^{-1}.

QNM-AFM-SECM experiments

Prior to QNM-AFM-SECM measurements, the spring constant and the contact radius of the AFM-SECM probes were determined and the integrity of the insulation, in particular the electrical contact pad, was evaluated in 5 mM $Fc(MeOH)_2$/0.1 M KCl. Contact radii of the tips were determined in air on a model PDMS sample ($E = 2.5$ MPa). Approach curves were recorded on a SiO_2 substrate at a ramp rate of 0.1 Hz and a force set point force of 30 nN. For feedback measurement of the AuNSt-modified MPTMS–CPDMS substrate, the tip-integrated electrode was biased at +0.5 V. The QNM-AFM-SECM measurements were carried out with a peak force of 8 nN, and an amplitude of 20 nm at a frequency of 1 kHz. All electrochemical measurements are Fourier-filtered to remove 50 Hz noise. The obtained electrochemical image was further smoothed with respect to the electrode size (all pixels within 100 nm are averaged).

Results and discussion

Characterization of AuNSts-modified CPDMS substrates

Soft AuNSt-modified substrates were characterized with respect to their nanomechanical properties, which is a relevant aspect in terms of the aging of such materials or potential changes occurring during electrochemical experiments. Hence, SEM and AFM studies were performed to characterize the surface of the AuNSts and of the CPDMS, which is the crucial interface during electrochemical experiments. The characterization of nanostructures *via* AFM – or more general, scanning probe microscopies – can be rather challenging, as the resolution of the obtained images is dependent on the tip geometry of the applied AFM probe. Consequently, overestimated sample dimensions may be derived. This effect is known as 'tip dilation' or 'tip convolution', and is of particular importance when characterizing small objects such as nanoparticles, which may approach in their dimensions the tip of the probe.[54] For macroscopic structures, such convolution effects may be negligible (*i.e.*, of course dependent on the feature size that should be resolved), as conventional microfabricated AFM probes have tip curvature radii usually on the order of magnitude smaller than the investigated sample. However, once the structure or feature size approaches the dimensions of the tip, artifacts will be unavoidable, thereby leading to a distorted topographic image.[54]

AuNSts deposited onto CPDMS substrates, as investigated herein, are even more prone to tip artifacts given that high aspect ratio structures extending from the AuNSts surface can be as small as 1–10 nm in diameter with a length up to 30 nm (see SEM image in Fig. 2a). For obtaining accurate topographical images, the probe tip would actually have to penetrate these structures, which is impossible using conventional AFM probes with a typical conical or pyramidal shape at cone/pyramid half angles of 15–35° along with tip radii of up to 60 nm.

For the measurements shown in Fig. 2, a sharp silicon nitride probe with a nominal tip curvature radius of 2 nm was used. As nanomechanical information is recorded, the tip contact radius was calibrated prior to the studies facilitating subsequent fitting according to the DMT model. For mapping elasticity data, such sharp tips are less suited, as the substrate needs to be deformed potentially leading to tip wear during extended measurements. Also, the deformation of the sample may lead to changes in penetration depth, and hence, a different contact radius affecting the DMT model and the associated accuracy of the elasticity data.[55] Fig. 2b shows the topography image of AuNSts physisorbed onto CPDMS. As the tip radius is on the order of magnitude of the diameter of the AuNSts features, the high aspect ratio structures may not be imaged, yet the overall particle shape is clearly recognizable. The adhesion force determined for the gold structure was determined at 1–2 nN, which is significantly lower compared to the value of 6–7 nN obtained for the CPDMS substrate (Fig. 2c). As the AFM studies were performed in air, the dominating forces are capillary forces. However, the obtained difference in adhesion force is most probably related to mechanical adhesion, as the tip penetrates deeper into the soft conductive composite (Fig. 2d). Likewise, the obtained Young's modulus is higher for AuNSts at E_{AuNSts} = 4.5, 3.8, and 3.5 MPa, respectively, which is expected, as gold is significantly

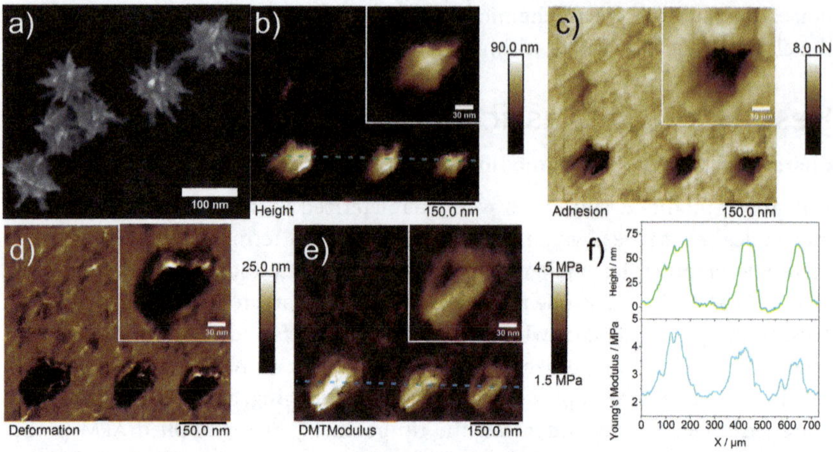

Fig. 2 (a) SEM image of AuNSts revealing the strongly branched high-aspect ratio structures at dimensions of a few up to a few tens of nanometers extending from the core AuNP. (b) AFM topography image of three individual AuNSts physisorbed onto a CPDMS substrate with simultaneously obtained (c) adhesion, (d) deformation, and (e) Young's modulus images, as obtained by DMT fitting. The insets show a zoomed view of the smallest nanostar located at the right corner of the respective images. In addition, (f) shows exemplary profiles that have been extracted from the topography image (green dashed line in (b)) and the Young's modulus image (blue-dashed line in (e)). AFM images were recorded in air at a frequency of 1 kHz, at a peak force of 800 pN, and at an amplitude of 200 nm with a sharp silicon nitride AFM probe with a nominal tip curvature radius of 2 nm.

stiffer than the CPDMS substrate with $E_{CPDMS} = 2.3$ MPa (Fig. 2e).[56] The DMT modulus image also suffers from tip artifacts, which are observable in Fig. 2e. From the extracted profiles shown in Fig. 2f, it is evident that the width of the AuNSt affects their elasticity, as all AuNSts have the same average height (*i.e.*, approx. 70 nm). The Young's modulus for CPDMS of approx. 2.3 MPa is in excellent agreement with literature values reported for a 1 : 10 mixture of PDMS.[57] The carbon filling material of the CPDMS appears to be of minor influence on the overall elastic modulus of the surface, although more variance in the elasticity compared to pure PDMS substrate surfaces was observed (data not shown).

Electrochemistry of CPDMS and AuNSts-modified CPDMS

Carbon-based materials are commonly used in electroanalysis and energy-related applications (*e.g.*, fuel cell electrodes). Recently, soft conductive materials such as CPDMS have gained importance in soft electronic devices and lab-on-a-chip applications.[58] Insulating PDMS films can be converted into conductive polymers by adding, *e.g.*, carbon black, at a concentration exceeding 10 wt%.[16,17] Thus prepared CPDMS composites with 25 wt% have a rather low conductivity of $10\ S\cdot m^{-1}$.[58] AFM images of the material have revealed that the carbon particles with dimensions of 50 nm and below are readily incorporated and distributed within the PDMS matrix, which may limit electron transfer to/from these conductive centers. Hence, the electrochemical performance of such composite materials is expected to be rather poor at low concentration.[51] Consequently, it was anticipated that appropriate surface modifications, *e.g.*, *via* decoration with nanoparticles such as AuNSts not only improves the electrocatalytic properties of such materials, but increases the surface electroactivity and surface area. PDMS is a well-established material applied in soft lithography, *e.g.*, for microfluidic devices,[16] and a wide variety of surface modifications have been reported.[59] One straightforward strategy is the attachment of thiol-terminated silanes onto a plasma-treated substrate surface[60] followed by the tethering of gold nanoparticles *via* the thiol moiety to the immobilized silane layer.

Fig. 3a illustrates the electrochemical characterization of a CPDMS substrate in 0.1 M KCl before and after various surface modification steps. The unmodified CPDMS layer revealed small capacitive currents with an onset of hydrogen evolution at −1.5 V, and of oxygen evolution at approx. +2 V, thus providing a potential window of −1.0 to +1.5 V, which compares to other carbon-based electrodes at neutral pH values. After ozone treatment, a small increase in capacitive current was observed, and the onset of oxygen evolution was shifted to lower potentials. It is hypothesized that due to the ozone treatment an amorphous hydrophilic surface layer is formed similar to oxygen plasma treatment,[61] which leads to an increase in capacitive current. After modification with MPTMS, the capacitive current further increases, which is mainly attributed to the formation of a thin silica layer at the CPDMS surface.

After incubation with AuNSts, a decrease of the capacitive current is evident. In addition, a peak at +1.4 V is observed, which is associated with the oxidation of gold. To determine whether the intermolecular forces between CPDMS and AuNSTs are sufficient for ensuring a robust and lasting modification of such electrode surfaces in solution, CVs in 10 mM H_2O_2 were recorded. It was shown that the conductive composite substrate does not reveal any oxidation current for hydrogen peroxide within the scanned potential range. After immobilization of the AuNSts, two distinct peaks are evident at +0.3 V and +0.1 V. Furthermore, the onset of the oxygen reduction is shifted by +0.5 V to a less negative potential.

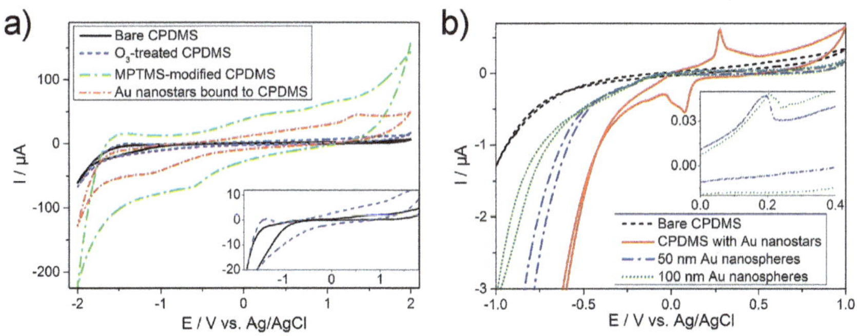

Fig. 3 Cyclic voltammetric characterization of the CPDMS substrate after several surface modification steps. The voltammograms in (a) show the increasing capacitive currents recorded in 0.1 M KCl at CPDMS (black solid curve), after ozone treatment (blue doted curve), after modification with MPTMS (green dashed curve), and after physisorption of AuNSts (red dashed curve). (b) CVs obtained in 10 mM H_2O_2/0.1 M KCl before and after modification with spherical AuNPs with 50 nm diam. (blue dotted curve), spherical AuNPs with 100 nm diam. (green dotted curve), and AuNSts with approx. 80 nm in diam. (red solid curve) physisorbed onto the CPDMS layer. The CVs evidence enhanced electrocatalytic activity obtained for the AuNSts immobilized at the soft CPDMS substrate (inset shows the oxidation peaks obtained at the spherical AuNPs with a magnified current scale).

Hence, the obtained data clearly evidences the presence of AuNSts at the CPDMS surface. In a complementary set of experiments, the soft substrate was modified with spherical AuNPs. Although the concentration of the spherical AuNPs was almost double that of the AuNSt solution, the observed oxidation current for such isotropic particles was several orders of magnitude lower and shifted by 0.1 V, as shown in the inset of Fig. 3b. This observation may be attributed to (i) a lower coverage of spherical AuNPs compared to AuNSts, as the AuNSts adhere more effectively to the substrate due to the thin gold spikes that may penetrate the soft CPDMS layer, and due to their inherently higher surface area. (ii) It is expected that the reduced current also results from a lower electrochemical activity of the spherical AuNPs compared to the AuNSts, which also reveal enhanced optical properties in comparison to spherical AuNPs. In both cases (*i.e.*, with isotropic AuNPs and anisotropic AuNSts), subsequent cycling of the modified substrates between either −1 and +1 V did not reveal any changes in peak height, thereby indicating stable electrode surfaces.

Fabrication of advanced AFM-SECM probes for nanoelectrochemical imaging

In recent years, electrochemical imaging at the nanoscale has emerged as a particularly relevant yet challenging topic requiring substantial efforts towards reliable and reproducible measurements.[24] In order to obtain electrochemical surface information, *e.g.*, feedback currents at single nanoscopic entities, the electrode has to be precisely positioned in close proximity to the sample surface at a distance of a few electrode radii. Focused ion beam (FIB)-based fabrication methods for AFM-SECM probes with electrodes recessed from the tip apex facilitate the tailoring of such bifunctional probes for specific electrode-to-sample distances. This is achieved by deliberately selecting the length of the insulating AFM tip used for topographical imaging, which simultaneously acts as a spacer for the recessed electrode. Additionally, due to the high positioning accuracy inherent to AFM, and its high-resolution topographical imaging capability, the distance of the substrate to the electrode may be accurately determined. However, it has to be noted that particularly for imaging nanoscale objects, high aspect ratio probes are advantageous for limiting potential tip artifacts of the obtained AFM images. The shape of such tips is readily controlled during FIB fabrication, albeit the radius of the tip curvature remains limited to few nanometers by the nature of the FIB milling process and the tip material. By exposing the gold electrode (see Fig. 4a), a plateau with a side length of approx. 2 μm is created at the apex of the AFM-SECM probe entailing a 900 nm thick insulation layer. It should be noted that obtained RG values for such integrated nanoscopic disc electrodes are comparatively high with values in the range of approx. 20. As the AFM cantilever is mounted at an angle of 9° within the probe holder of the AFM, at decreasing length of the tip, it is increasingly probable that the front edge of the plateau may touch the sample surface *in lieu* of the re-shaped AFM tip. For an ideally flat sample surface, the critical length of the AFM tip was calculated at approx. 250 nm. To avoid this issue, the mounting angle was taken into account during the FIB fabrication process by milling the plateau at a tilt compensation angle of 10° (Fig. 4a). A re-shaped AFM tip with a height of approx. 200 nm is exemplarily shown in Fig. 4b. Fig. 4c illustrates a top view of the final AFM-SECM probe providing a distance between the nanoscopic disc electrode and the

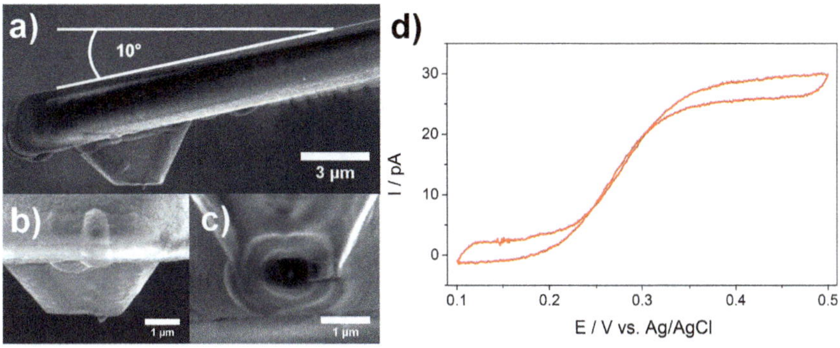

Fig. 4 (a) SEM images of an exemplary AFM-SECM probe with integrated nanoscopic gold electrode. (a) For compensating the mounting angle within the AFM nose cone (9°), the plateau exposing the electrode was tilt-compensated at an angle of 10°. The frontal view clearly shows the finally obtained insulating AFM tip (b) located next to the recessed disc nanoelectrode, which was used for topographical imaging (length approx. 200 nm). The recessed gold electrode has a radius of 50 nm (c). An exemplary CV obtained during the electrochemical characterization in 5 mM $Fc(MeOH)_2$/0.1 M KCl is shown in (d) (scan rate: 100 mV s^{-1}).

reshaped AFM tip apex of 200 nm. It was calculated that the re-shaped AFM tip next to the disc electrode at the specified distance will result in a decrease of the faradaic current by <5% due to the diffusion limitation imposed by the probe, in comparison to the total diffusion towards a planar disc electrode (unpublished results). The obtained tip-integrated electrode and the integrity of the insulated electrical contact were verified by recording CVs with $Fc(MeOH)_2$/0.1 M KCl serving as electrolyte solution (Fig. 4d).

The CV shown in Fig. 4d was obtained with the AFM-SECM probe mounted in the AFM instrument immersed into a bulk solution of 5 mM $Fc(MeOH)_2$/0.1 M KCl above a SiO_2 substrate. From the obtained steady-state current, an effective disc electrode diameter of 60 nm was calculated. In order to obtain simultaneously quantitative force data *via* bifunctional AFM-SECM probes, the force constant of the cantilever was determined using the thermal noise method.[62] During the calibration routine, the faradaic current response was simultaneously obtained at the integrated disc nanoelectrode while recording the corresponding force curves. At an insulating SiO_2 substrate, a negative current feedback of 1 pA was obtained (Fig. 5). The on-set of the negative feedback current was determined already at a distance of 500 nm between nanodisc electrode and sample surface, which results from the relatively large RG value of 20. The small change in current correlates well with the length of the microfabricated AFM tip. From the retract part of the force–distance curve, the adhesion of the tip to the substrate can be derived. The insulating SiO_2, shows only minute adhesion to the AFM tip, which was expected as the radius of the tip curvature is small and the tip is not penetrating into such a solid substrate. In addition, low adhesion forces are also evidence for successful tilt compensation. If the edge of the silicon nitride insulation would be the point of contact with the sample surface, a significantly higher adhesion force would be observed, as reported by Kopycinska-Müller *et al.*[63]

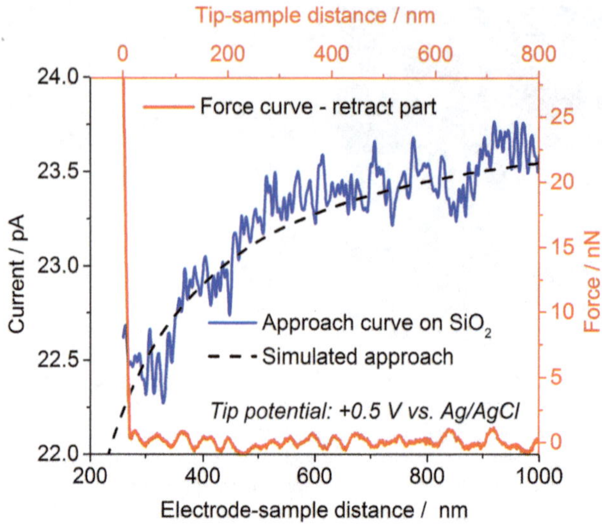

Fig. 5 Approach curve recorded *via* AFM-SECM probes towards an insulating SiO_2 surface in 5 mM $Fc(MeOH)_2$/0.1 M KCl (blue curve) and simulated approach curve (black dashed, RG = 20). The corresponding force distance curve (retract part) is shown in red.

QNM-AFM-SECM measurements of AuNSts in SECM feedback mode

A first feasibility study on the utility of AFM-SECM probes for simultaneous nanoelectrochemical and nanomechanical mapping was performed at MPTMS-modified CPDMS substrates decorated with AuNSts. The electrochemical measurements were executed in feedback mode SECM using 5 mM $Fc(MeOH)_2$. After calibrating the AFM-SECM probe for nanomechanical mapping, the topography, feedback current, elasticity, and adhesion force of AuNSts at the surface of MPTMS-modified CPDMS substrates were simultaneously imaged. The obtained topography image revealed an agglomeration of AuNSts with an isolated individual nanostar evident in Fig. 6a and d. The simultaneously recorded feedback current shows an increase of approx. 500 fA above the particles *vs.* the CPDMS substrate (Fig. 6e). As the AFM tip is located next to the integrated nanodisc electrode, a minute shift of the imaged particle between the topographical and the electrochemical image of a single nanostar is observed. The feature in the current image is shifted slightly to the left in comparison to the topography image, which readily correlates with the probe orientation and scan direction (Fig. 6a). As already observed in air, also in solution AuNSts reveal less adhesion to the silicon nitride AFM tip in comparison to the CPDMS substrate (Fig. 6b and f).

The Young's modulus image revealed that the AuNSts have a modulus of approx. 200 MPa. In comparison to unmodified CPDMS, the MPTMS-modified CPDMS has an average modulus of 20 MPa, but also contains regions with a higher modulus of up to 400 MPa. As previously discussed, the surface modification step with ozone treatment and MPTMS modification resulted in an amorphous PDMS layer comprising a thin silica layer. Such layers increase the elastic modulus, as previously observed.[64] The zoomed views (Fig. 6d–g) reveal the topographical features of an individual AuNSt, although the high aspect ratio features remain unresolved along with some tip artifacts. As the tip serving for

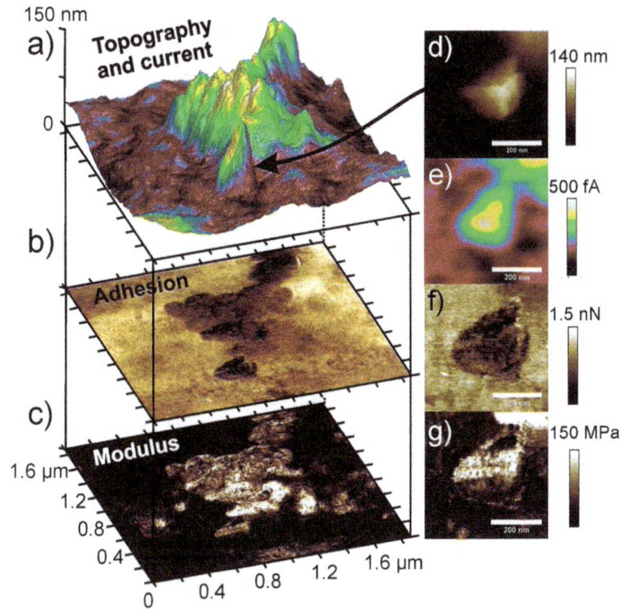

Fig. 6 QNM-AFM-SECM measurements at a MPTMS-modified CPDMS substrate after incubation with AuNSts in 5 mM $Fc(MeOH)_2$/0.1 M KCl with a peak force of 8 nN, and an amplitude of 20 nm at a frequency of 1 kHz. The AFM tip integrated nanodisc electrode was biased at 0.5 V. The 3D topography image (a) shows the topographical feature (relief shown in red) overlaid with the color-coded feedback current response (increased currents are denoted in white and yellow). It reveals a minute shift between topography and current signal, as expected due to the gap between the integrated nanodisc electrode and the silicon nitride AFM tip. The adhesion force (b) observed at the AuNSts is lower than the adhesion force at the CPDMS substrate surface; likewise, the obtained Young's modulus (c) is significantly higher for AuNSts, as anticipated. The insets show a zoomed view of an individual AuNSt: (d) topography, (e) current, (f) adhesion, and (g) Young's modulus.

topographical imaging has a rather steep incline towards the nanodisc electrode, and a less inclined facet at the opposite side (Fig. 4b), the imaged structure appears slightly stretched to the left. It should be noted though that due to the thorough characterization of AFM-SECM probes *via* high-resolution SEM images – which is difficult to achieve for capillary-based probes – such artifacts are readily recognized, and post-imaging data treatment algorithms may be applied. The increased insulating nature of the MPTMS-modified CPDMS substrate allows a clear distinction of the AuNSts. Using an unmodified substrate, the conductivity was evidently sufficient such that no change in feedback current between AuNSts and the conductive substrate were observed (data not shown).

Conclusions

Nanomechanical and nanoelectrochemical properties of gold nanoparticles may be simultaneously investigated using combined AFM-SECM. As the functionality of nanoparticles is correlated to the size and shape, such investigations are of great relevance to many applications in electrocatalysis and sensing as the physical and (electro)chemical properties may change over time. Soft substrates such as carbon-filled PDMS modified with nanoparticles require several modification steps for improving the adhesion of the nanoparticles, which in turn may alter their physical properties. Hence mapping such nanomechanical and electrochemical properties is important for designing appropriate substrates for various targeted applications. Mapping the properties of individual asymmetric particles, such as the gold nanostars shown here, in terms of their nanomechanical properties is rather challenging given their surface morphology. This would require high-aspect ultrasharp probes, yet such probes may induce severe artifacts in the modulus and deformation information. Hence, the AFM probe design has to be adapted to find a compromise optimizing the topographical resolution and minimizing the artifacts in the nanomechanical information. Here we use FIB technology to fabricate such specialized AFM-SECM probes with nanoscopic disc electrodes. The individual modification steps of the substrate were investigated in terms of adhesion properties, Young's modulus and deformation properties along with the bulk electrochemical investigation with respect to the electrocatalytic properties towards hydrogen peroxide oxidation. A comparison between spherical and multibranched nanoparticles indicates a higher electrocatalytic activity of the nanostars. In addition, the electrochemical properties in combination with the nanomechanical properties of an individual Au-nanostar were mapped with QNM-AFM-SECM using feedback mode SECM. Although the topographical image shows some distortion, the shape of the nanostar is clearly observed. The electrochemical response clearly shows a contrast in the feedback signal. As expected, there is a small shift in location of the electrochemical response in respect to the topography. This can be explained by the shape of the AFM-SECM probe with the AFM tip located next to the nanodisc electrode. A clear advantage of the AFM-SECM probe is that high-resolution SEM imaging allows the determination of the geometry of the AFM-SECM probe, which is rather challenging for combined SICM-SECM probes or nanoscopic SECM electrodes. Hence, the obtained images can be in future corrected for such artifacts, which is a standard procedure in AFM imaging.

Current research is focused on mapping the electrocatalytic properties of individual asymmetric nanoparticles in comparison to *e.g.*, spherical nanoparticles of similar shape. In addition, changes in physical and (electro)chemical properties over time may provide important information on long-term applications of such nanomaterials.

Acknowledgements

The authors acknowledge J. Haas for the SEM images of the AuNSts. P. K. acknowledges the German Federal Ministry of Education and Research (BMBF – NanoMatFutur: 13N12545) for financial support. Also, the FIB Center UUlm at the Institute of Analytical and Bioanalytical Chemistry, and the Cleanroom & Workshop at Ulm University are acknowledged.

References

1 P. Rodriguez and M. T. M. Koper, *Phys. Chem. Chem. Phys.*, 2014, **16**, 13583–13594.

2 *Gold Nanoparticles in Analytical Chemistry*, ed. M. Valcarcel and A. I. Lopez-Lorente, Elsevier, Amsterdam, 2014.

3 N. Khlebtsov and L. Dykman, *Chem. Soc. Rev.*, 2011, **40**, 1647–1671.

4 C. Fenzl, T. Hirsch and A. J. Baeumner, *TrAC, Trends Anal. Chem.*, 2016, **79**, 306–316.

5 M. R. Langille, M. L. Personick, J. Zhang and C. A. Mirkin, *J. Am. Chem. Soc.*, 2012, **134**, 14542–14554.

6 G. Frens, *Nature (London), Phys. Sci.*, 1973, **241**, 20–22.

7 J. Turkevich, P. C. Stevenson and J. Hillier, *Discuss. Faraday Soc.*, 1951, **11**, 55–75.

8 N. Li, P. Zhao and D. Astruc, *Angew. Chem., Int. Ed.*, 2014, **53**, 1756–1789.

9 O. Bibikova, A. Popov, A. Bykov, A. Fales, H. Yuan, I. Skovorodkin, M. Kinnunen, S. Vainio, T. Vo-Dinh, V. Tuchin and I. Meglinski, *IEEE J. Sel. Top. Quantum Electron.*, 2016, **22**, 4600808.

10 S. E. Lohse and C. J. Murphy, *Chem. Mater.*, 2013, **25**, 1250–1261.

11 M. Schütz, D. Steinigeweg, M. Salehi, K. Kömpe and S. Schlücker, *Chem. Commun.*, 2011, **47**, 4216–4218.

12 L. Polavarapu, J. Pérez-Juste, Q.-H. Xu and L. M. Liz-Marzán, *J. Mater. Chem. C*, 2014, **2**, 7460–7476.

13 C. Hrelescu, T. K. Sau, A. L. Rogach, F. Jäckel and J. Feldmann, *Appl. Phys. Lett.*, 2009, **94**, 153113.

14 C. Khoury and T. Vo-Dinh, *J. Phys. Chem. C*, 2008, **112**, 18849–18859.

15 H. Liao, Y. Jiang, Z. Zhou, S. Chen and S. Sun, *Angew. Chem., Int. Ed.*, 2008, **47**, 9100–9103.

16 M. A. Unger, H. P. Chou, T. Thorsen, A. Scherer and S. R. Quake, *Science*, 2000, **288**, 113–116.

17 X. Niu, S. Peng, L. Liu, W. Wen and P. Sheng, *Adv. Mater.*, 2007, **19**, 2682–2686.

18 F. J. Vidal-Iglesias, R. M. Arán-Ais, J. Solla-Gullón, E. Herrero and J. M. Feliu, *ACS Catal.*, 2012, **2**, 901–910.

19 A. Hamelin, *J. Electroanal. Chem.*, 1984, **165**, 167–180.

20 J. Hernández, J. Solla-Gullón and E. Herrero, *J. Electroanal. Chem.*, 2004, **574**, 185–196.

21 J. Solla-Gullón, P. Rodríguez, E. Herrero, A. Aldaz and J. M. Feliu, *Phys. Chem. Chem. Phys.*, 2008, **10**, 1359–1373.

22 C. Kranz and B. Mizaikoff, in *Gold Nanoparticles in Analytical Chemistry*, ed. M. Valcarcel and A. I. Lopez-Lorente, Elsveier, Amsterdam, 2014, pp. 257–299.

23 J. B. Sambur and P. Chen, *Annu. Rev. Phys. Chem.*, 2014, **65**, 395–422.

24 S. E. F. Kleijn, S. C. S. Lai, M. T. M. Koper and P. R. Unwin, *Angew. Chem., Int. Ed.*, 2014, **53**, 3558–3586.

25 J. Meier, K. A. Friedrich and U. Stimming, *Faraday Discuss.*, 2002, 365–372; discussion 441–462.

26 R. G. Fedorov and D. Mandler, *Phys. Chem. Chem. Phys.*, 2013, **15**, 2725–2732.

27 T. Danieli and D. Mandler, *J. Solid State Electrochem.*, 2013, **17**, 2989–2997.

28 X. Xiao and A. J. Bard, *J. Am. Chem. Soc.*, 2007, **129**, 9610–9612.

29 M. A. O'Connell, J. R. Lewis and A. J. Wain, *Chem. Commun.*, 2015, **51**, 10314–10317.

30 C. M. Sánchez-Sánchez, F. J. Vidal-Iglesias, J. Solla-Gullón, V. Montiel, A. Aldaz, J. M. Feliu and E. Herrero, *Electrochim. Acta*, 2010, **55**, 8252–8257.

31 C. M. Sánchez-Sánchez, J. Solla-Gullón, F. J. Vidal-Iglesias, A. Aldaz, V. Montiel and E. Herrero, *J. Am. Chem. Soc.*, 2010, **132**, 5622–5624.
32 A. J. Wain, *Electrochim. Acta*, 2013, **92**, 383–391.
33 K. Eckhard, X. Chen, F. Turcu and W. Schuhmann, *Phys. Chem. Chem. Phys.*, 2006, **8**, 5359–5365.
34 T. Sun, Y. Yu, B. J. Zacher and M. V. Mirkin, *Angew. Chem., Int. Ed. Engl.*, 2014, **53**, 14120–14123.
35 K. T. Rodolfa, A. Bruckbauer, D. Zhou, Y. E. Korchev and D. Klenerman, *Angew. Chem., Int. Ed.*, 2005, **44**, 6854–6859.
36 N. Ebejer, M. Schnippering, A. W. Colburn, M. A. Edwards and P. R. Unwin, *Anal. Chem.*, 2010, **82**, 9141–9145.
37 S. C. S. Lai, P. V. Dudin, J. V. Macpherson and P. R. Unwin, *J. Am. Chem. Soc.*, 2011, **133**, 10744–10747.
38 Y. Takahashi, A. I. Shevchuk, P. Novak, Y. Zhang, N. Ebejer, J. V MacPherson, P. R. Unwin, A. J. Pollard, D. Roy, C. A. Clifford, H. Shiku, T. Matsue, D. Klenerman and Y. E. Korchev, *Angew. Chem., Int. Ed.*, 2011, **50**, 9638–9642, S9638/1–S9638/5.
39 K. McKelvey, B. P. Nadappuram, P. Actis, Y. Takahashi, Y. E. Korchev, T. Matsue, C. Robinson and P. R. Unwin, *Anal. Chem.*, 2013, **85**, 7519–7526.
40 C. Kranz, *Analyst*, 2014, **139**, 336–352.
41 J. Abbou, C. Demaille, M. Druet and J. Moiroux, *Anal. Chem.*, 2002, **74**, 1627–1634.
42 J. Abbou, A. Anne and C. Demaille, *J. Am. Chem. Soc.*, 2004, **126**, 10095–10108.
43 F. Gaboriaud, B. S. Parcha, M. L. Gee, J. A. Holden and R. A. Strugnell, *Colloids Surf., B*, 2008, **62**, 206–213.
44 A. Rosa-Zeiser, E. Weilandt, S. Hild and O. Marti, *Meas. Sci. Technol.*, 1999, **8**, 1333–1338.
45 B. Pittenger, N. Erina, and C. Su, *Bruker Appl. Note #128*, 2011.
46 B. V. Derjaguin, V. M. Muller and Y. P. Toporov, *J. Colloid Interface Sci.*, 1975, **53**, 314–326.
47 M. Chirea, *Catalysts*, 2013, **3**, 288–309.
48 S. E. Pust, M. Salomo, E. Oesterschulze and G. Wittstock, *Nanotechnology*, 2010, **21**, 105709/1–105709/12.
49 E. Lee, M. Kim, J. Seong, H. Shin and G. Lim, *Phys. Status Solidi RRL*, 2013, **7**, 406–409.
50 H. Yuan, C. G. Khoury, H. Hwang, C. M. Wilson, G. A. Grant and T. Vo-Dinh, *Nanotechnology*, 2012, **23**, 075102.
51 M. Brun, J. F. Chateaux, A. L. Deman, P. Pittet and R. Ferrigno, *Electroanalysis*, 2011, **23**, 321–324.
52 C. Kranz, G. Friedbacher, B. Mizaikoff, A. Lugstein, J. Smoliner and E. Bertagnolli, *Anal. Chem.*, 2001, **73**, 2491–2500.
53 J. Moon, H. Shin, B. Mizaikoff and C. Kranz, *J. Korean Phys. Soc.*, 2007, **51**, 920–924.
54 C. Wong, P. E. West, K. S. Olson, M. L. Mecartney and N. Starostina, *JOM*, 2007, **59**, 12–16.
55 B. V Derjaguin, V. M. Muller and Y. P. Toporov, *J. Colloid Interface Sci.*, 1975, **53**, 314–326.
56 D. R. Lide, *CRC Handbook of Chemistry and Physics*, Taylor and Francis, London, 2008.

57 Z. Wang, A. A. Volinsky and N. D. Gallant, *J. Appl. Polym. Sci.*, 2014, **131**, 41050–41054.
58 X. Niu, W. Wen and Y.-K. Lee, *Appl. Phys. Lett.*, 2005, **87**, 243501.
59 J. Zhou, D. A. Khodakov, A. V. Ellis and N. H. Voelcker, *Electrophoresis*, 2012, **33**, 89–104.
60 D. Liu, R. K. Perdue, L. Sun and R. M. Crooks, *Langmuir*, 2004, **20**, 5905–5910.
61 N. Bowden, W. T. S. Huck, K. E. Paul and G. M. Whitesides, *Appl. Phys. Lett.*, 1999, **75**, 2557.
62 J. L. Hutter and J. Bechhoefer, *Rev. Sci. Instrum.*, 1993, **64**, 1868–1873.
63 M. Kopycinska-Müller, R. H. Geiss and D. C. Hurley, *Ultramicroscopy*, 2006, **106**, 466–474.
64 P. Knittel, B. Mizaikoff and C. Kranz, *Anal. Chem.*, 2016, **88**, 6174–6178.

Faraday Discussions

PAPER

Imaging electrocatalytic processes on single gold nanorods†

Chao Jing,[ab] Zhen Gu[a] and Yi-Tao Long*[a]

Received 10th April 2016, Accepted 5th May 2016
DOI: 10.1039/c6fd00069j

Imaging electrochemical processes has attracted increasing attention in past decades. Particularly, monitoring electrochemical reactions rapidly at the nano-scale is still a challenge due to the ultra-low current detection and long scanning time required. The development of optical techniques provide a new way to demonstrate electrochemical processes through optical signals which enhance sensitivity and spatial resolution. Herein, we developed a novel method to image electrocatalytic processes on single gold nanorods (GNRs) during Cyclic Voltammetry (CV) scanning based on plasmon resonance scattering information by using dark-field microscopy. The electrocatalytic oxidation of hydrogen peroxide was selected as a typical reaction and the catalytic mechanism was revealed using the obtained spectra. Notably, observation on single GNRs avoided the averaging effects in bulk systems and confirmed that the individual nanoparticles had variable catalytic properties with different spectral change during the reaction process. Furthermore, a color-amplified system was introduced to convert light intensity into imaging information *via* the Matlab program which was able to image thousands of nanoparticles simultaneously. This approach offered the statistical intensity distribution of all of the nanoparticles in a dark-field image which dramatically enhanced the detection accuracy and avoided random events.

Introduction

Noble metal nanoparticles including gold, silver and copper have attracted considerable attention due to their chemical, physical, optical and catalytic properties.[1–4] Particularly, their localized surface plasmon resonance (LSPR) bands afford them with unique absorption and scattering spectroscopic properties which have been applied in many fields.[5–7] When the diameter of plasmonic nanoparticles are far smaller than the wavelength of incident light, their surface free electrons will oscillate with the light wave. When the oscillation frequency of

[a]*Key, Laboratory for Advanced Materials and Department of Chemistry East China University of Science and Technology, 130 Meilong Road, Shanghai 200237, P. R. China. E-mail: ytlong@ecust.edu.cn*

[b]*Physik-Department E20 Technische Universität München, James-Franck-Str. 1, D-85748 Garching, Germany*

† Electronic supplementary information (ESI) available: Contains details of dark-field microscopy setup and the modification procedure. See DOI: 10.1039/c6fd00069j

nanoparticles is equal to the frequency of the incident light, plasmon resonance occurs resulting in the enhanced movement of electrons. The plasmon resonance bands of nanoparticles are dependent on their morphology, composition, surface electron density and surrounding medium.[8–10] Therefore, numerous kinds of plasmonic nanoparticles have been synthesized such as nanorods, nanotriangles, core–shell structures and satellites with various resonance bands to meet the requirements of sensors.[11–14] Due to the development of fabrication techniques, especially nanotechnology, nanoparticles with controllable size and shape can be designed and prepared which significantly expands the applications of plasmonics. For instance, by changing the aspect ratio of nanorods, their plasmon resonance bands could be modulated in a wide range from the visible light region to the near-infrared and infrared regions.[15–17]

Aside from their optical properties, plasmonics have lots of free electrons on their surface that can provide abundant active sites for reactions and have acted as excellent catalysts for organic chemistry and contaminant degradation, *etc.*[18,19] Their catalytic ability plays important roles in the fields of energy sources, life sciences and industry.[20,21] Notably, since the catalysis of gold nanoparticles for the oxidation of CO was discovered at low temperature, gold-catalyzed reactions have become a hot topic and undergone rapid development.[22] It has been reported that gold nanoparticles exhibit glucose oxidase (GOx)-like activity that can catalyze the oxidation of glucose into gluconic acid and H_2O_2 in the presence of O_2.[23] The catalytic ability is surface modulated so when some molecules are adsorbed on the nanoparticle surface, GOx-like activity would be confined. Thus ultra-sensitive sensors have been constructed for the detection of biomolecules such as DNA and ATP using surface controlled catalysis.[24] Notably, plasmonics, especially gold nanoparticles, have been widely applied in electrocatalysis including as sensitive biosensors and enzyme electrodes.[25–27] Gold nanoparticles with the benefits of high conductivity, facile modification and active surfaces could enhance the reaction efficiency and act as conductive wires to promote the electron transfer process.[28] It is worth noting that every nanoparticle can function as a nano-reactor which offers active areas for reactions. Because of the inhomogeneity of nanoparticles, it is important to monitor the reaction on a single nanoparticle to eliminate the average effects in the bulk system.[29] The averaged results of all the nanoparticles may not reveal accurate reaction information. The signal from a single nanoparticle is beneficial to analyze the kinetics and dynamics of mechanisms to demonstrate the quantum effects.[30] It will also provide good guidance to modulate the parameters of nanoparticles to enhance their catalytic ability. Many advanced techniques have been developed to monitor reactions at the single nanoparticle/molecule level.[31–33] However, for single nanoparticle analysis, there are still disadvantages including the random events of a single particle and the possibility of obtaining results with large errors. Hence, it is necessary to get statistical data of a large amount of single nanoparticles to avoid random events. Therefore, monitoring electrochemical reactions on single nanoparticles with high throughput has become a hot topic in recent years. Researchers have paid increasing attention to image redox reactions at the nanoscale.[34,35] Conventional electrochemical imaging methods such as scanning electrochemical microscopy (SECM) have been developed for many years and achieve determination at micro/nano sizes.[36,37] However, traditional electrochemical imaging methods suffer from the disadvantages of slow scan rate. It is time-consuming to obtain an image for a large area and it is difficult to detect fast reactions and fast potential scanning processes such as Cyclic Voltammetry (CV). Therefore, developing fast and simultaneous imaging techniques with nano-scale resolution has promising potential in the understanding of the mechanism of electrochemical reactions and enhancing detection sensitivity.

Recently, several methods have been proposed to image electrochemical processes in real-time on a single nanoparticle utilizing surface plasmon resonance properties.[38] Tao monitored electrocatalytic reactions on single Pt nanoparticles *via* a surface plasmon resonance (SPR)-based imaging technique which was able to obtain the current on the particle surface according to changes in the local medium.[39,40] Also, this method has been applied to observe the oxidation process of a single silver nanoparticle during a collision on a gold-chip electrode.[41] Pan proposed an approach to image the electrochemical deposition of hundreds of silver nanoparticles simultaneously through scattering spectra under dark-field microscopy.[42] Based on the relationship between resonance intensity and particle diameter, it is capable of mapping the size of nanoparticles during the deposition process with high time-spatial resolution which is in good agreement with the generated faradic current. As the resonance intensity of plasmonic nanoparticles is easily affected by their morphology, surface electron density and surrounding dielectric medium, it is possible to exploit the scattering intensity of nanoparticles to investigate surface reactions and electron transfer processes during electrochemical voltammetry scanning.[43–45] Compared with the scattering wavelength, the signal of intensity is easier to capture without requiring a spectrometer, and also the time resolution will be enhanced dramatically. Moreover, it allows us to detect the intensity of numerous individual nanoparticles with high throughput rather than capturing the scattering spectra of several nanoparticles at one time through a conventional spectrometer. Herein, we developed a novel method to image an electrocatalytic reaction in real-time on single gold

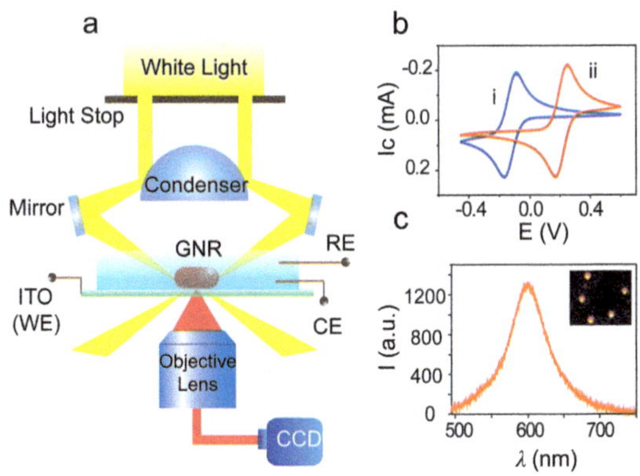

Fig. 1 (a) Setup of dark-field microscopy combined with an electrochemistry workstation. The GNR modified ITO slide acts as the working electrode (WE) and two Pt wires function as the counter electrode (CE) and quasi-reference electrode (RE), respectively. (b) Cyclic voltammograms (CVs) of 1 mM $K_3[Fe(CN)_6]$ solution on the bare ITO electrode. Electrolyte: 0.1 M KNO_3 solution. Scan rate: 100 mV s^{-1}, *vs.* Pt quasi-reference electrode (i) and saturated calomel electrode (SCE) (ii). (c) Scattering spectrum of a single gold nanorod (GNR), insert is a typical dark-field image of mono-dispersed GNRs.

nanoparticles based on the resonance scattering intensity by using dark-field microscopy (Fig. 1a). The potential of the Pt reference electrode was calibrated using potassium ferricyanide solution as depicted in the Experimental section. In this work, the oxidation process of gold nanoparticles and hydrogen peroxide were investigated.[46] The intensity changes of single nanoparticles were converted into easily-recognized colors to achieve the imaging of the reaction process. Furthermore, the statistic intensity distributions of hundreds to thousands of nanoparticles in a dark-field image were calculated which offered a semi-quantitative analysis of nanoparticles. This imaging method enabled the detection of redox reactions on single nanoparticles with high time-spatial resolution, high throughput and accuracy which has great potentials in the investigation of electrochemistry at the nanoscale.

Results and discussion

Redox reaction on a single gold nanorod (GNR)

In this work, gold nanorods were selected as nanoprobes given their strong plasmon resonance scattering spectroscopy properties. The scattering band of nanorods could be modulated in a wide range to satisfy the requirements of different applications by adjusting their aspect ratios.[47] The scattering cross section C_{sca} of a nanorod is depicted below:[48–50]

$$C_{csa} = \frac{k^4}{6\pi}\left|\alpha^4\right| \tag{1}$$

$$\alpha = 4\pi abc\sum_i \frac{\varepsilon_{Au} - \varepsilon_m}{\varepsilon_m + L_i(\varepsilon_{Au} - \varepsilon_m)} \tag{2}$$

where k is the wave vector, α is the polarizability of the nanoparticle, ε_m and ε_{Au} are the dielectric functions of the surrounding medium and the metal material, a, b, and c are the three major axes of the nanorod. L_i is the shape-dependent depolarization factor which is calculated using the following equations:

$$L_1 = \frac{1-e^2}{e^2}\left(\frac{1}{2e}\ln\left(\frac{1+e}{1-e}\right) - 1\right) \tag{3}$$

$$L_2 = L_3 = \frac{1-L_1}{2} \tag{4}$$

$$e = (1 - 1/\varphi^2) \tag{5}$$

here $\varphi = a/b$ is the aspect ratio of the nanorod. The scattering spectra of nanorods have more sensitivity to changes in the surrounding medium and surface electron density than nanospheres due to their lower depolarization factor.[51] From eqn (1)–(5), it could be concluded that the plasmon resonance bands of GNRs are dependent on the particle size, aspect ratio and surrounding medium. It was also reported that the dielectric function of gold ε_{Au} is correlated to the potential and surface electron density as depicted below:[52]

$$\varepsilon_{Au} = \varepsilon_{JC}(\omega) + \frac{{\omega_p}^2}{\omega^2 - i\omega\Gamma_0} - \frac{\left(\omega_{p,mod}(E)\right)^2}{\omega^2 - i\omega(\Gamma_0 + \Gamma_S + \Gamma_C(E))} \tag{6}$$

$$\omega_{\text{p,mod}}(E) = \sqrt{\frac{n(E)e^2}{\varepsilon_m}} \tag{7}$$

where $\varepsilon_{\text{JC}}(\omega)$ is an empirical dielectric constant, ω_{p} is the plasma frequency, Γ_0 is the damping in bulk gold. Γ_{S} reveals the interaction of the plasmon with the GNR surface and $\Gamma_{\text{C}}(E)$ reflects the adsorbate damping with the solvent and ions which is dependent on the applied potential E. In eqn (7), $\omega_{\text{p,mod}}(E)$ is a modified plasma frequency related to the applied potential, e is the elementary charge and m is the effective mass of the sp-band electrons, and $n(E)$ is the surface electron density which is dependent on the applied potential E. Therefore, using the plasmon resonance signals enables us to demonstrate electrochemical processes including electron transition, ion adsorption and chemical reactions. Under dark-field microscopy, GNRs (*ca.* 40 × 84 nm) showed orange or red color with scattering spectral peak positions around 600 to 700 nm as shown in Fig. 1c. Nanoparticles showed different scattering properties because of the heterogeneity of their size and shape. The SEM images of mono-dispersed nanorods are shown in Fig. S1 in the ESI.†

In order to investigate the effect of electric field on the scattering spectra of GNRs, a constant positive potential of 1.00 V was applied. Fig. 2a shows the scattering peak intensity (I_{max}) and wavelength (λ_{max}) changes of a single GNR under the potential. In the first 180 s, the scattering peak position of the GNR gradually red shifted *ca.* 12 nm along with an intensity decrease. According to previous reports, gold atoms could be oxidized into gold hydroxide and oxide at potentials of more than 0.5 V.[46] Under these conditions, both the lowering of surface electron density and formation of gold hydroxide/oxide could induce the energy loss of plasmon resonance. The oxidation reaction is the main factor influencing the scattering spectra that the produced gold hydroxide/oxide dramatically increased the refractive index of the medium surrounding the gold surface.[52] It is worth noting that, when the applied potential stopped (after 180 s), the spectra gradually changed back to the initial state. This result indicated that the generated gold hydroxide/oxide species are not stable at open circuit voltage (*ca.* −0.13 V) and are easily reduced into gold atoms which confirmed the reversibility of the redox reaction of gold nanoparticles. As a control, the scattering peak intensity and wavelength change under open circuit were performed to confirm the influence of environment as shown in Fig. S2 (ESI†), suggesting the validity of the experiment. Then, the scattering peak intensity I_{max} alternation of a single GNR under CV scanning from −0.10 V to 1.00 V was observed. From Fig. 2b, the I_{max} of GNRs showed reversible change upon applied potentials. From −0.10 V to 0.50 V, I_{max} decreased slowly for the discharging of surface electrons and ions adsorption of double layer.[53] At this potential range, the intensity decrease was approximately in proportion to the applied potential with a decreasing rate ($\Delta I/I_0$) of *ca.* 20% per volt. This result indicated the quasi-linear relationship between I_{max} and potential-dependent surface electron density and double layer charging. From 0.50 V to 0.75 V, I_{max} showed a fast non-linear decrease during the oxidation process and then came to a maximum at higher potential. The averaged intensity change rate was *ca.* 40% per volt which was much higher than the discharging process. The oxidation of gold reduced the plasmon resonance volume of the GNR as a portion of gold atoms lost electrons.

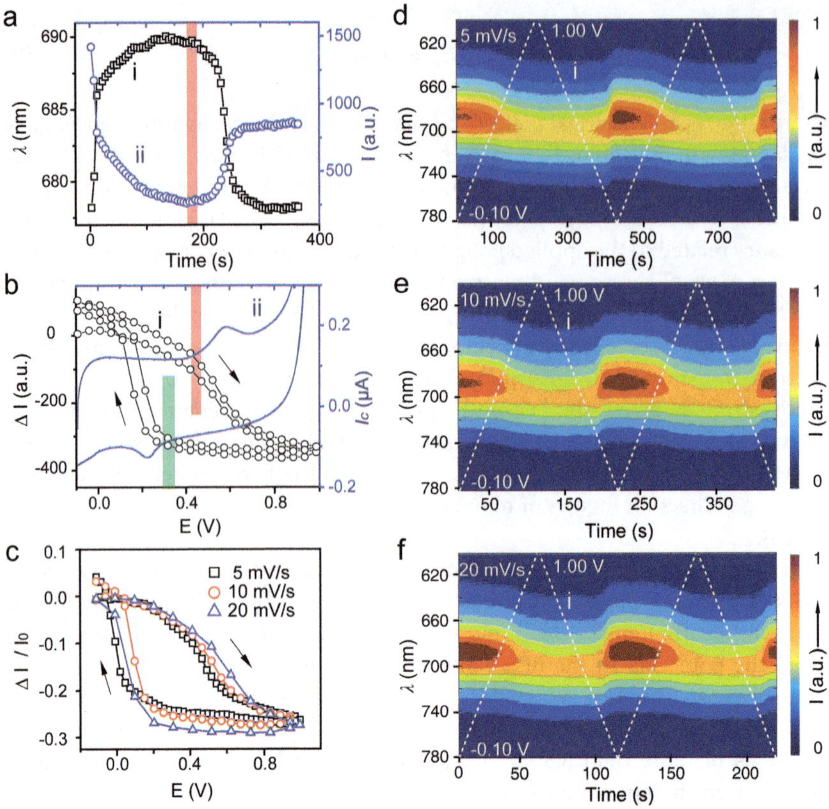

Fig. 2 (a) Changes of plasmon resonance scattering peak wavelength (i) and peak intensity (ii) of a single GNR under a constant potential of 1.00 V in 0.1 M KNO_3 solution *vs.* a Pt quasi-reference electrode. (b) Scattering peak intensity change of a single GNR (i, two circles) under CV scanning (ii) in 0.1 M KNO_3 solution from −0.10 V to 1.00 V, at a scan rate of 10 mV s^{-1} *vs.* Pt quasi-reference electrode. (c) Scattering peak intensity change of a single GNR under CV scanning in 0.1 M KNO_3 solution from −0.10 V to 1.00 V, at scan rates of 5 mV s^{-1} (black), 10 mV s^{-1} (red) and 20 mV s^{-1} (blue) *vs.* a Pt quasi-reference electrode. (d–f) Scattering spectra change of a single GNR under CV scanning in 0.1 M KNO_3 solution from −0.10 V to 1.00 V, at scan rates of 5 mV s^{-1} (d), 10 mV s^{-1} (e) and 20 mV s^{-1} (f) *vs.* a Pt quasi-reference electrode.

Also, the formed gold hydroxide/oxide increased the refractive index of gold surface.[54] The scattering intensity remained constant until the reduction potential of 0.20 V in the reverse segment for the saturated oxidation of gold. After 0.20 V, I_{max} increased quickly to the beginning value with a rate of 80% per volt. These results show good agreement with previous reports.[52] The reversible intensity change confirmed that the scattering intensity could be used as a reporting signal to characterize electrochemical processes in real-time due to the sensitivity to the applied potential and redox reactions. The scattering spectra of a single GNR under CV scanning at different scan rates were also studied as shown in Fig. 2c–f. Both the scattering intensity and peak wavelength displayed cycling change, however, they did not show clear differences at different scan rates of 5, 10 and 20 mV s^{-1}. As the redox reaction has a primary effect on the plasmon resonance, the scattering spectra is strongly dependent on the redox peak potential E_P. Therefore, this phenomenon may be because that for most reversible reactions, E_P is independent of the scan rates which showed no correlation with the scattering spectra of GNRs during the reactions.

To image the electrochemical process according to the dark-field images of the GNRs, the Matlab program was introduced to rapidly map the scattering intensity change of each nanoparticle. Fig. 3 displayed the mapping images and statistical

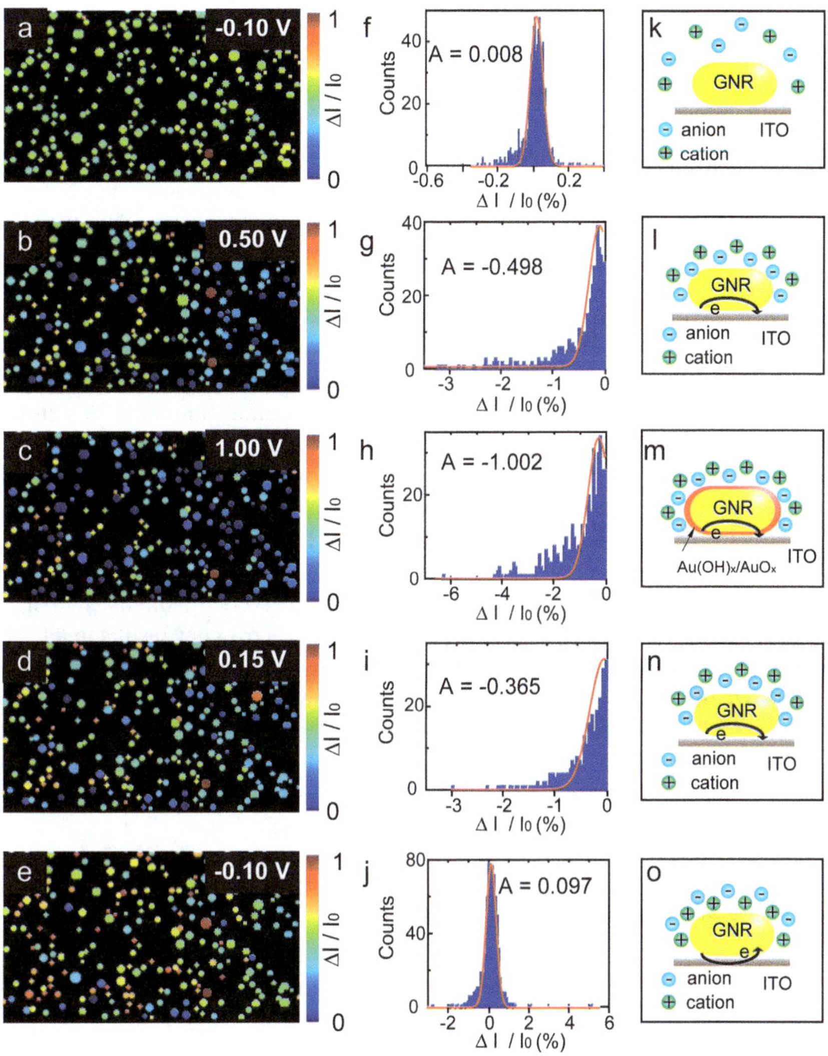

Fig. 3 (a–e) Calculated intensity change ($\Delta I/I_0$) of different nanoparticles in a dark-field image during CV scanning from −0.10 V to 1.00 V in 0.1 M KNO_3 solution, at a scan rate of 10 mV s^{-1} *vs.* a Pt quasi-reference electrode. (f–j) Statistical intensity change distributions and averaged results of every single nanoparticle in the dark-field image during CV scanning corresponding to the potential in (a–e). Results with increasing intensity in (g–i) were not taken into consideration. (k–o) Proposed redox reaction process of gold nanoparticles during CV scanning corresponding to the potential in (a–e), anions and cations are from the electrolyte.

intensity distributions of numerous nanoparticles in a dark-field image under CV scanning from −0.10 V to 1.00 V. The color of nanoparticles in Fig. 3a–e represent the averaged $\Delta I/I_0$ of every pixel in a spot. At the beginning of −0.10 V (Fig. 3a and f), GNRs did not show distinct intensity change with an averaged $\Delta I/I_0$ less than 0.01%. As the potential increased to 0.50 V, the color of GNRs changed from yellow to green and blue with decreased intensity due to double layer formation and electrons discharging (Fig. 3b and g). The averaged intensity change is *ca.* −0.5%. Here it should be noted that the GNRs with increased intensity were not taken into consideration when calculating the averaged results and statistical distribution. As the experiments were performed in an open system, the intensity of GNRs may be influenced by the surrounding environment leading to the occurrence of some random events.

Hence, to obtain accurate results, GNRs with abnormally increased changes were removed by the calculation program. At the potential of 1.00 V, the intensity of GNRs decreased obviously with blue color as Fig. 3c shows. The averaged intensity change is *ca.* 1% for the oxidation of nanoparticles (Fig. 3h). The calculated intensity change from the image was much lower than the results from the peak intensity obtained through spectroscopy. This is because the particle-recognition process included background pixels which have no change under the applied potential leading to fewer differences. Also, the intensity change from the image included all the wavelengths of the scattering light, while the result from spectroscopy is the intensity of peak. When the potential came to 0.15 V in the cathodic back scan, the intensity of the GNRs increased after the reduction of gold hydroxide/oxide (Fig. 3d and i). Then, the intensity increased continuously until the potential of −0.10 V was reached as shown in Fig. 3e and j. The mechanism of potential based procedures are illustrated in Fig. 3k–o. The averaged $\Delta I/I_0$ changes of all the nanoparticles in the dark-field image were displayed in Fig. S3 in the ESI.† In this process, most of GNRs showed reversible intensity changes which were in good agreement with the scattering spectroscopy results in Fig. 2b. However, for different nanoparticles, the intensity change ratio and response to applied potentials are greatly different confirming the inhomogeneity of individual particles. Therefore, this method not only provides an approach to image the reaction process of every single nanoparticle, but also offers a statistical distribution of numerous particles which greatly enhances the calculation accuracy.

Electrocatalytic oxidation of H_2O_2 on the surface of single GNRs

After the addition of 1 mM H_2O_2 in solution, CV scanning was performed from −0.10 V to 1.00 V to investigate the electrocatalytic oxidation process. The CV curves showed an obvious oxidation peak current of H_2O_2 catalyzed by gold nanoparticles (Fig. 4a). Fig. 4b shows the scattering spectra of a single GNR under CV scanning. However, the scattering spectra did not exhibit distinct red shift compared with the conditions in the absence of H_2O_2. The detailed scattering intensity and peak wavelength change of the single GNR after Lorenz fitting are displayed in Fig. 4c. The intensity decreased nearly linearly as the potential increased from −0.10 V to 1.00 V and increased gradually during 1.00 V to −0.10 V in the reverse segment. The scattering spectra only showed 3–4 nm shifts of peak wavelength and *ca.* 20% intensity decrease. The change of Full Width at Half

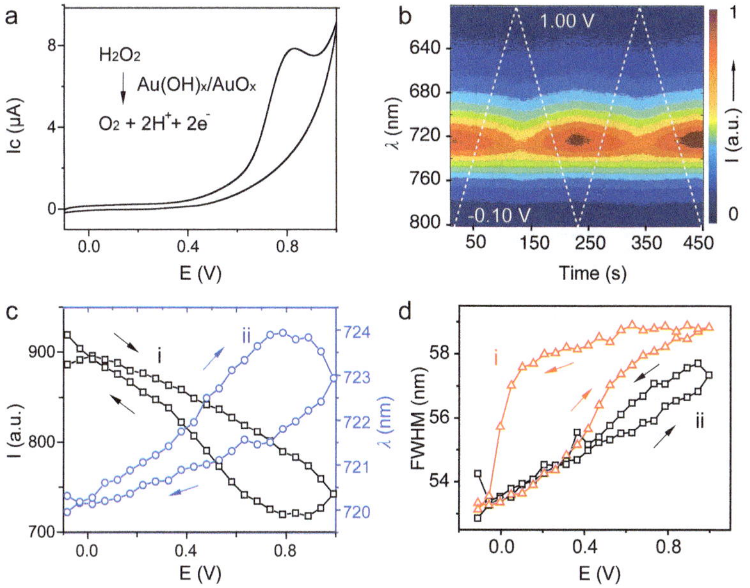

Fig. 4 (a) CV curve of electrocatalytic oxidation of 1 mM H_2O_2 in 0.1 M KNO_3 solution on the GNR-modified ITO electrode scanning from −0.10 V to 1.00 V *vs.* a Pt quasi-reference electrode, scan rate: 10 mV s^{-1}. (b) Scattering spectra change of a single GNR under CV scanning of (a). (c) Changes of plasmon resonance scattering peak intensity (i) and peak wavelength (ii) of a single GNR in (a). (d) Changes of Full Width at Half Maximum (FWHM) of plasmon resonance scattering spectra of a single GNR under CV scanning in 0.1 M KNO_3 solution without (i) and with (ii) 1 mM H_2O_2.

Maximum (FWHM) under these two conditions also indicated that the redox reactions of gold nanoparticles by electrochemistry were not addressed in the presence of H_2O_2 which suggested that produced gold hydroxide/oxide was reduced into gold atoms during the reaction.[55] The electron discharging and ion adsorption/desorption were the main factor that affected the spectra of gold nanoparticles. From these results, it could be deduced that the catalytic activity of GNRs was attributed to their oxidized production. In this procedure, gold hydroxide/oxide was first generated by the applied oxidation potential. Then, gold hydroxide/oxide reacted with H_2O_2 and reduced into gold atoms, meanwhile, H_2O_2 was oxidized into oxygen irreversibly. This reaction is a dynamic process with a fast reaction rate, therefore the scattering spectra did not show obvious changes illustrating the oxidation of gold. Fig. S4 in the ESI† displays the scattering spectra change under different scan rates which also show good reversibility. In this experiment, the oxidation potential of H_2O_2 and spectral change showed a weak correlation to the scan rates.[46]

The mapping images of the electrocatalytic oxidation process of H_2O_2 on single GNRs are given in Fig. 5a–e. It is obvious that nearly all the nanoparticles in the dark-field image showed cycling intensity alternation during the CV scanning which is similar to the results of the single nanoparticle change obtained from the spectroscopy in Fig. 4c. At the beginning, nanoparticles also showed slight intensity change with green color. When the potential came to 0.5 V, their intensity reduced and the spots of particles exhibited a light blue color with an

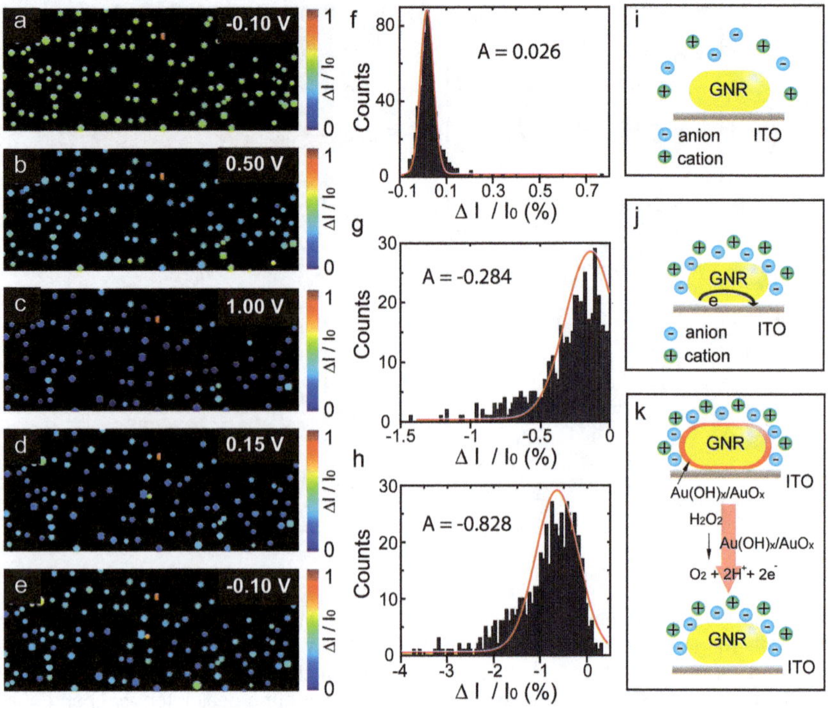

Fig. 5 (a–e) Intensity change of different nanoparticles in a dark-field image during CV scanning from −0.10 V to 1.00 V in 0.1 M KNO_3 solution containing 1 mM H_2O_2, at a scan rate of 10 mV s^{-1} *vs.* a Pt quasi-reference electrode. (f–h) Statistical intensity change and averaged results of every single nanoparticle in the dark-field image during CV scanning corresponding to the potentials in (a–c). Results with increased intensity in (g) were not taken into consideration. (i–k) Proposed interaction process between gold nanoparticles and H_2O_2 during CV scanning corresponding to the potentials in (a–c).

averaged intensity change of *ca.* −0.3%. Under the potential of 1.00 V, nanoparticles showed a dark blue color and the averaged intensity change of hundreds of nanoparticles is *ca.* −0.8%. As the potential decreased to −0.10 V in the reverse segment, the intensity of scattering light increased continuously as shown in Fig. 5d–e. The reaction mechanism at different potentials is illustrated in Fig. 5i–k. When the potential is less than 0.5 V, the intensity of the nanoparticles were mainly affected by the discharge of surface electron density and double layer adsorption which is similar to the conditions without H_2O_2. Under the oxidation potential of H_2O_2, the electrochemical reaction generating gold hydroxide/oxide was reduced by H_2O_2, so the main effect on the resonance scattering property was still the electron discharging through application of potentials. Therefore, the nanoparticles showed less intensity decrease compared with Fig. 3a–e. The averaged intensity changes of all the nanoparticles are displayed in Fig. S5 (ESI†) which are in good agreement with Fig. 4c. Notably, during the entire electrochemical process, different nanoparticles displayed extremely various intensity changes indicating the reaction heterogeneity. The diverse color change of nanoparticle spots could be attributed to several reasons including the electron transfer efficiency in bulk solution and the catalytic ability of nanoparticles. It is also possible to observe the inhomogeneous chemical reactions at different areas due to the low conductivity of ITO electrode. This method not only allows us to detect the catalytic reaction on single nanoparticles, but also provides a novel method to analyze the electrochemical process concerning multiple parameters to reduce system errors and improve the understanding of electrochemistry at the nanoscale.

Accurate redox reaction of single GNRs

It is important to note that the electrochemical catalytic reaction provides a smart method to investigate the accurate oxidation process of a single gold nanoparticle under the application of potentials. During conventional CV scanning (Fig. 2b-ii), the electrochemical current could not reveal the reactions on a single nanoparticle and showing an average result of the bulk system. With respect to the scattering spectra changes, the signals from a single GNR under the potentials include the procedures of electron density charging, double layer adsorption and redox reactions. However, in this catalytic reaction, the generated gold hydroxide/oxide was reduced by H_2O_2 in a very short period of time. The scattering spectra of GNRs in the presence of H_2O_2 could be considered as a background only containing the electrons charging and double layer adsorption. Therefore, the differences between these two processes are the main oxidation process of gold

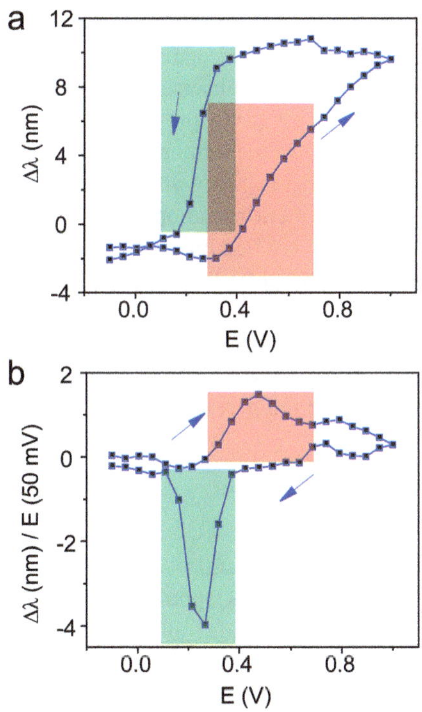

Fig. 6 (a) Subtraction difference of spectral peak wavelength change ($\Delta\lambda_{max,gold} - \Delta\lambda_{max,H_2O_2}$) of a single GNR during the CV scanning from −0.10 V to 1.00 V in the absence ($\Delta\lambda_{max,gold}$) and presence ($\Delta\lambda_{max,H_2O_2}$) of H_2O_2. (b) Differential results of spectral peak wavelength change in (a).

atoms. Fig. 6a shows the subtraction difference of the scattering spectral peak wavelength of a single GNR indicating the production (red area) and reduction (green area) of gold hydroxide/oxide on the nanoparticle surface. As expected, after differential, the scattering peak wavelength showed similar reaction results as CV curves as shown in Fig. 6b, which presented the "real" reaction process of a single nanoparticle. Therefore, this method makes it possible to address the redox reactions of single nanoparticles with high precision by avoiding mass transition and electron charging processes. Furthermore, the scattering spectra changes during CV scanning of different nanoparticles under the conditions with and without H_2O_2 are shown in Fig. S6 in the ESI† which demonstrated the specification of individual nanoparticles. It is distinct that different nanoparticles displayed various plasmon resonance changes due to the heterogeneity of nanoparticles and reaction surroundings. This work has enabled us to investigate electrochemical reactions at the single nanoparticle level, which is significant in their applications of catalysis.

Conclusions

In conclusion, we have developed a novel method to image the electrochemical redox reactions and catalytic processes on the surface of a single nanoparticle. Compared with traditional electrochemical imaging methods, this plasmon resonance scattering light-based approach has opened up a new way to achieve the fast imaging of hundreds/thousands of nanoparticles with high time and spatial resolution. In this work, after converting the scattering intensity of single nanoparticles into visible and easily-recognized color information, the electrochemical processes could be readily observed. The heterogeneity of individual nanoparticles was revealed which could provide an efficient way to screen the catalytic ability of nanoparticles. Furthermore, the high-throughput calculation procedure achieved the statistics of spectra differences to offer the distribution of intensity and averaged values of entire nanoparticles in a dark-field image. The obtained multiple parameters including imaging process, statistical and averaged results eliminated errors generated by the random events of nanoparticles and the averaged effect of a bulk system which enhanced the analysis accuracy dramatically. We believe that this colorful imaging method has promising applications in the investigation of electrochemistry at the nanoscale and provides a new insight into the redox reactions of plasmonic nanoparticles.

Experimental section

Materials

All reagents were of analytical grade and used as received without any further purification. Ethanol (AR), acetone (AR), potassium nitrate (AR) were purchased from Sigma (USA). Hydrogen peroxide (30%) was purchased from Aladdin (Shanghai, China). Gold nanorods (*ca.* 40 nm × 84 nm; 10 O.D.) were purchased from Nanoseedz (Hong Kong, China). Ultrapure water with a resistivity of 18.2 MΩ cm was produced using a Milli-Q apparatus (Millipore, USA) and used in the preparation of all the solutions. The indium tin oxide (ITO)-coated glass sheets (sheet resistances 20–30 Ω sq^{-1}; 1 mm thickness) were purchased from Shenzhen Laibao hi-tech (Shenzhen China).

Electrochemistry

All electrochemical measurements were performed using a CHI 660 (Chenhua, Shanghai, China) electrochemistry workstation with a three-electrode system. The ITO slide acted as the working electrode, and two Pt wires functioned as the counter and quasi-reference electrodes, respectively. All measurements were performed at 25 °C. The potential of the Pt reference electrode was calibrated *via* potassium ferricyanide as shown in Fig. 1b. The $E_0^{1/2}$ of 1.00 mM $K_3[Fe(CN)_6]$ solution using a Pt quasi-reference electrode and saturated calomel electrode (SCE) was −0.13 V and 0.21 V, respectively. In this work, all of the electrochemistry scans were performed with a Pt quasi-reference electrode unless otherwise stated.

Calculation procedures

The calculation process is illustrated in Fig. S8 (ESI†). At first, a contrast image (I_0) was captured at open circuit as a background. During the electrochemical process, dark-field images (I_n) were captured with a rate of 2*f*/*s*. All of the captured dark-field images were converted into grey scale format. The nanoparticles in dark-field images were recognized automatically in red rectangle *via* a threshold segmentation calculation method. Then the intensity of every pixel in the obtained images (I_n) was subtracted from the values in the contrast image (I_0) to reveal the intensity change. As the differences between I_n and I_0 may be minus values, an amplified method was introduced to the convert the results into positive values. The values of $((I_n - I_0) + 1)/2$ were used in the imaging process. Furthermore, the statistical data for all the nanoparticles in a dark-field image were obtained to give a distribution of intensity change and an averaged result to enhance the analysis accuracy.

Acknowledgements

This research was supported by the National Natural Science Foundation of China (21327807, 21421004), the Program for Professor of Special Appointment (Eastern Scholar) at Shanghai Institutions of Higher Learning (YJ0130504), and the Postdoctoral Council of China (2015 International Postdoctoral Exchange Fellowship Program).

References

1 W. J. Qi, C. Z. Huang and L. Q. Chen, *Talanta*, 2010, **80**, 1400.
2 J. Wang, L. Wang, X. Liu, Z. Liang, S. Song, W. Li, G. Li and C. Fan, *Adv. Mater.*, 2007, **19**, 3943.
3 L.-X. Qin, Y. Li, D.-W. Li, C. Jing, B.-Q. Chen, W. Ma, A. Heyman, O. Shoseyov, I. Willner, H. Tian and Y.-T. Long, *Angew. Chem., Int. Ed.*, 2012, **51**, 140.
4 X. Gan, T. Liu, J. Zhong, X. Liu and G. Li, *ChemBioChem*, 2004, **5**, 1686.
5 K. Li, W. Qin, F. Li, X. Zhao, B. Jiang, K. Wang, S. Deng, C. Fan and D. Li, *Angew. Chem., Int. Ed.*, 2013, **52**, 11542.
6 L. Shi, C. Jing, W. Ma, D. W. Li, J. E. Halls, F. Marken and Y. T. Long, *Angew. Chem., Int. Ed.*, 2013, **52**, 6011.
7 K. A. Willets and R. P. Van Duyne, *Annu. Rev. Phys. Chem.*, 2007, **58**, 267.
8 A. M. Brown, M. T. Sheldon and H. A. Atwater, *ACS Photonics*, 2015, **2**, 459.

9 C. Jing, Z. Gu, Y.-L. Ying, D.-W. Li, L. Zhang and Y.-T. Long, *Anal. Chem.*, 2012, **84**, 4284.
10 L. Zhang, Y. Li, D.-W. Li, C. Jing, X. Chen, M. Lv, Q. Huang, Y.-T. Long and I. Willner, *Angew. Chem., Int. Ed.*, 2011, **50**, 6789.
11 C.-J. Heo, S.-H. Kim, S. G. Jang, S. Y. Lee and S.-M. Yang, *Adv. Mater.*, 2009, **21**, 1726.
12 L. Zhang, S. J. Zhen, Y. Sang, J. Y. Li, Y. Wang, L. Zhan, L. Peng, J. Wang, Y. F. Li and C. Z. Huang, *Chem. Commun.*, 2010, **46**, 4303.
13 M. Grzelczak, J. Pérez-Juste, P. Mulvaney and L. M. Liz-Marzán, *Chem. Soc. Rev.*, 2008, **37**, 1783.
14 M. G. Blaber, A.-I. Henry, J. M. Bingham, G. C. Schatz and R. P. Van Duyne, *J. Phys. Chem. C*, 2012, **116**, 393.
15 M. L. Pedano, S. Li, G. C. Schatz and C. A. Mirkin, *Angew. Chem., Int. Ed.*, 2010, **49**, 78.
16 C. Jing, L. Shi, X. Liu and Y. T. Long, *Analyst*, 2014, **139**, 6435.
17 H. Chen, L. Shao, Q. Li and J. Wang, *Chem. Soc. Rev.*, 2013, **42**, 2679.
18 P. Pyykko, *Angew. Chem., Int. Ed.*, 2004, **43**, 4412.
19 A. Leyva-Perez and A. Corma, *Angew. Chem., Int. Ed.*, 2012, **51**, 614.
20 A. Ghosh, S. Basak, B. H. Wunsch, R. Kumar and F. Stellacci, *Angew. Chem., Int. Ed.*, 2011, **50**, 7900.
21 A. S. K. Hashmi and G. J. Hutchings, *Angew. Chem., Int. Ed.*, 2006, **45**, 7896.
22 M. Haruta, T. Kobayashi, H. Sano and N. Yamada, *Chem. Lett.*, 1987, **16**, 405.
23 X. Zheng, Q. Liu, C. Jing, Y. Li, D. Li, W. Luo, Y. Wen, Y. He, Q. Huang, Y. T. Long and C. Fan, *Angew. Chem., Int. Ed.*, 2011, **50**, 11994.
24 Q. Liu, C. Jing, X. Zheng, Z. Gu, D. Li, D.-W. Li, Q. Huang, Y.-T. Long and C. Fan, *Chem. Commun.*, 2012, **48**, 9574.
25 P. Rodriguez and M. T. M. Koper, *Phys. Chem. Chem. Phys.*, 2014, **16**, 13583.
26 F. Raimondi, G. G. Scherer, R. Kotz and A. Wokaun, *Angew. Chem., Int. Ed.*, 2005, **44**, 2190.
27 M. Brust and G. J. Gordillo, *J. Am. Chem. Soc.*, 2012, **134**, 3318.
28 X. Luo, A. Morrin, A. J. Killard and M. R. Smyth, *Electroanalysis*, 2006, **18**, 319.
29 A. Kumar, F. Ciucci, A. Morozovska, S. Kalinin and S. Jesse, *Nat. Chem.*, 2011, **3**, 707.
30 M. A. O'Connell, J. R. Lewis and A. J. Wain, *Chem. Commun.*, 2015, **51**, 10314.
31 S. M. Oja, Y. Fan, C. M. Armstrong, P. Defnet and B. Zhang, *Anal. Chem.*, 2016, **88**, 414.
32 S. C. S. Lai, P. V. Dudin, J. V. Macpherson and P. R. Unwin, *J. Am. Chem. Soc.*, 2011, **133**, 10744.
33 I. L. Buurmans and B. M. Weckhuysen, *Nat. Chem.*, 2012, **4**, 873.
34 S. Liu, Q. Li and Y. Shao, *Chem. Soc. Rev.*, 2011, **40**, 2236.
35 D. A. Walsh, K. R. J. Lovelock and P. Licence, *Chem. Soc. Rev.*, 2010, **39**, 4185.
36 F. P. Filice, M. S. M. Li, J. D. Henderson and Z. Ding, *J. Phys. Chem. C*, 2015, **119**, 21473.
37 P.-Y. Blanchard, T. Sun, Y. Yu, Z. Wei, H. Matsui and M. V. Mirkin, *Langmuir*, 2016, **32**, 2500.
38 W. Wang and N. Tao, *Anal. Chem.*, 2014, **86**, 2.
39 X. Shan, I. Díez-Pérez, L. Wang, P. Wiktor, Y. Gu, L. Zhang, W. Wang, J. Lu, S. Wang, Q. Gong, J. Li and N. Tao, *Nat. Nanotechnol.*, 2012, **7**, 668.
40 X. Shan, U. Patel, S. Wang, R. Iglesias and N. Tao, *Science*, 2010, **327**, 1363.

41 Y. M. Fang, W. Wang, X. Wo, Y. S. Luo, S. W. Yin, Y. X. Wang, X. N. Shan and N. J. Tao, *J. Am. Chem. Soc.*, 2014, **136**, 12584.
42 C. M. Hill and S. L. Pan, *J. Am. Chem. Soc.*, 2013, **135**, 17250.
43 Y. Leroux, J. C. Lacroix, C. Fave, G. Trippe, N. Félidj, J. Aubard, A. Hohenau and J. R. Krenn, *ACS Nano*, 2008, **2**, 728.
44 C. Novo and P. Mulvaney, *Nano Lett.*, 2007, **7**, 520.
45 J. N. Anker, W. P. Hall, O. Lyandres, N. C. Shah, J. Zhao and R. P. Van Duyne, *Nat. Mater.*, 2008, **7**, 442.
46 C. Jing, F. J. Rawson, H. Zhou, X. Shi, W. H. Li, D. W. Li and Y. T. Long, *Anal. Chem.*, 2014, **86**, 5513.
47 P. K. Jain, K. S. Lee, I. H. El-Sayed and M. A. El-Sayed, *J. Phys. Chem. B*, 2006, **110**, 7238.
48 K. S. Lee and M. A. El-Sayed, *J. Phys. Chem. B*, 2006, **110**, 19220.
49 S. Eustis and M. El-Sayed, *J. Phys. Chem. B*, 2005, **109**, 16350.
50 Y. Peng, B. Xiong, L. Peng, H. Li, Y. He and E. S. Yeung, *Anal. Chem.*, 2015, **87**, 200.
51 C. Novo, A. M. Funston and P. Mulvaney, *Nat. Nanotechnol.*, 2008, **3**, 598.
52 S. K. Dondapati, M. Ludemann, R. Müller, S. Schwieger, A. Schwemer, B. Händel, D. Kwiatkowski, M. Djiango, E. Runge and T. A. Klar, *Nano Lett.*, 2012, **12**, 1247.
53 C. Novo, A. M. Funston, A. K. Gooding and P. Mulvaney, *J. Am. Chem. Soc.*, 2009, **131**, 14664.
54 S. J. Xia and V. I. Birss, *J. Electroanal. Chem.*, 2001, **500**, 562.
55 M. Hu, C. Novo, A. Funston, H. Wang, H. Staleva, S. Zou, P. Mulvaney, Y. Xia and G. V. Hartland, *J. Mater. Chem.*, 2008, **18**, 1949.

Faraday Discussions

DISCUSSIONS

Electrochemistry of single nanoparticles: general discussion

Tim Albrecht, Julie MacPherson, Olaf Magnussen, David Fermin, Richard Crooks, Justin Gooding, Thom Hersbach, Frederic Kanoufi, Wolfgang Schuhmann, Cameron Bentley, Nongjian Tao, Sushanta Mitra, Katharina Krischer, Kristina Tschulik, Sanli Faez, Wojciech Nogala, Patrick Unwin, Yitao Long, Marc Koper, Zhongqun Tian, Mario A. Alpuche-Aviles, Henry White, Vitor Brasiliense, Christine Kranz, Wolfgang Schmickler, Keith Stevenson, Chao Jing and Martin Edwards

DOI: 10.1039/c6fd90068b

Sushanta Mitra opened a general discussion of the paper by Keith Stevenson: For the nanoparticle impact study, if there are surface charges present on the nanoparticle, will there be any screening effect? The reason for such an issue is that often they are modeled as particle-collector (Happel Cell model) where there could be valence shell electron-pair repulson.

The nanoparticle interactions are beyond DLVO theory. Please see the recent discussions by C. A. Silvera Batista *et al.*[1] Hence, one needs to account for such potentials in order to obtain an adequate picture for the interactions. This is very different to colloidal interaction (which is typically in micron size particles) where DLVO is valid.

More importantly, the emphasis was for sensor development using nanoparticles, where the surface enhanced resonance plasmonics could be an important issue. Under such cases, the particle size are of the order of a few nanometers, which again brings into the aspects of re-visiting the existing DLVO theory.

1 C. A. Batista, R. G. Larson and N. A. Kotov, *Science*, 2015, **350**, 1242477.

Keith Stevenson answered: Thank you for directing us to this Happel Cell model. We will consider in our future work.

Julie MacPherson remarked: It was interesting to see that hydrazine could be decomposed by Pt under OCP conditions using a citrate coated Pt wire. Following on from this how does citrate surface coverage effect this process (and direct electron transfer)? Will you have similar coverage (per unit area) when you compare the wire with a NP? Finally what potential do you hold at when recording the Pt NP impact data in **Fig. 9**?

Keith Stevenson answered: We are not sure how the citrate coverage influences hydrazine catalysis nor the electron transfer process. As mentioned in the paper there is evidence that the citrate layer is thick but permeable. With other Pt nanoparticles with stronger capping ligands (PPV, thiols, DNA) we do not see hydrazine catalysis. The potential for recording the Pt NP is $E_{appl} = -50$ mV *vs.* Ag/AgCl.

Sushanta Mitra said: For the nanoparticle impact study, if there are surface charges present on the nanoparticle, will there be any screening effect?

Keith Stevenson responded: Yes, there can be screening effects which cause aggregation of nanoparticles in solution. The main point addressed in a separate paper[1] was to establish concentrations of hydrazine and buffer (ionic strength) to unambiguosly observe single nanoparticle collisions and not of aggregates. For the *Faraday Discussions* paper we then used conditions to follow aggregation kinetics using nanoparticle impacts.

1 D. A. Robinson, A. M. Kondajji, A. D. Castañeda, R. Dasari, R. M. Crooks and K. J. Stevenson, *J. Phys. Chem. Lett.*, 2016, 7, 2512–2517.

Patrick Unwin remarked: A very nice feature of your measurements, compared to solid electrodes, is that the poisoning of the impinging Pt nanoparticles means that you avoid complications of multiple impacts, making this a very powerful analytical platform. Nanoparticle impacts with solid electrodes can be very revealing of the interactions of nanoparticles with surfaces, but need very careful measurement (high bandwidth–low noise) and analysis.

Olaf Magnussen opened a general discussion of the paper by Mario A. Alpulche Aviles: I do not understand the charge transfer process proposed for the electrode under illumination (**Fig. 1d**). This would imply that in the cathodic steps, massive charging of the TiO_2 nanoparticles occurs. How could the buildup of such a charge be rationalized? In particular, the charge is supposed to be located in the TiO2 conduction band. What should hinder the transfer back into the conduction band of the metal electrode to which the particles are attached?

Mario A. Alpuche-Aviles answered: I agree that charging of the NPs will have to be involved to explain the large currents that we see here. In the paper we discuss that the currents are too large to be assigned to single NPs. The buildup of charge in colloidal semiconducting NPs is known to occur[1] at the order of 200 electrons/TiO_2 NP. I believe some details of the charge accumulation are less understood. For example, the charge is many times thought to reside in the conduction band (CB) but it is possible for it to reside in states below the CB. I do not believe there is a directionality for the charge built into the Dye/NP/Pt electrode interface. What we propose is that the scheme in **Fig. 1d** is the more probable and therefore, the most frequently observed direction for the current. In the paper we discuss trends in the current, because the opposite direction (anodic in this case) is also observed. We propose that the longer lifetime of the oxidized form of the dye is what makes the process described in **Fig. 1d** statistically more favorable.

1 V. Subramanian, E. E. Wolf and P. V. Kamat, *J. Am. Chem. Soc.*, 2004, **126**, 4943–4950.

David Fermin remarked: In your experiments, illumination is carried out with an Arc-Xe lamp which is capable of direct photoexcitation of TiO_2 nanoparticles. Holes in the TiO_2 particles can readily oxidise methanol under these conditions. Can you elaborate on the difference in the collision signals in the experiments with and without dye species?

Mario A. Alpuche-Aviles replied: The effect of the dye sensitizing the nanoparticle (NP) is that the photocurrent is larger. In this paper we discussed experiments on Pt ultramicroelectrode and in this case the photocurrent for bare anatase is typically <10 pA[1] while with the dye much larger photocurrents are larger, with photocurrents > 100 pA routinely observed (**Fig. 3** in the *Faraday Discussions* paper). Furthermore, in experiments with microelectrodes prepared with F-doped tin oxide (FTO), the photocurrents were > 5× to 10× larger for the dye sensitized NPs with respect to the bare NPs.[2] Note that in the case where the electrode substrate is FTO the NPs response is in the form of oscillations and not discrete steps.

1 A. Fernando, S. Parajuli and M. A. Alpuche-Aviles, *J. Am. Chem. Soc.*, 2013, **135**, 10894–10897
2 A. Fernando, P. Chhetri, K. K. Barakoti, S. Parajuli, R. Kazemi and M. A. Alpuche-Aviles, *J. Electrochem. Soc.*, 2016, **163**, H3025–H3031.

Justin Gooding remarked: Towards the end of the paper you suggest you are only seeing collisions from some of the larger aggregates. Do you think you will be able to measure single particle collisions? What are the limitations? Do you think the reaction or the laser power is promoting aggregation?

Mario A. Alpuche-Aviles answered: Yes, I think it will be possible to measure single nanoparticle collision. Right now we are limited by illumination power and I believe that by increasing illumination power we will be able to resolve the differences between single particles and aggregates. The limitations, we think based on some unpublished results, are colloidal behavior. As in the case of the metal NPs discussed by Stevenson in this session, the NPs do not have to exist as single NPs or agglomerates/aggregates. They could be in an equilibrium between the two states and we should be able to adjust the conditions to observe the single nanoparticles. Regarding the second point of your question, we illuminate the sample using an arc lamp, not a laser, but it is still possible that the illumination power has an effect on aggregation. We are currently investigating this issue.

David Fermin asked: Is there oxygen present in the electrolyte solution? This would open a pathway for removing electrons from the conduction band of TiO_2 under illumination. Otherwise, it is difficult to rationalise how electrons accumulate in the particles under these conditions.

Mario A. Alpuche-Aviles responded: We need to remove oxygen from the suspension to observe these photocurrents. We purge the suspensions with Ar and in some cases, we setup the cell in an Ar globe box and we perform the experiments on the bench. I agree that in the presence of oxygen one should not expect the accumulation of charge under illumination.

Justin Gooding commented: You made a comment in the paper that, if using high electrolyte concentrations, the responses were the same under light/dark conditions. Could you expand on what you think is happening here? Secondly, you mentioned that the particles stick to the electrode surface. Does the frequency of collisions therefore go down as a function of time? I imagine as the particle start covering the electrode, this would be the case.

Mario A. Alpuche-Aviles replied: With high electrolyte concentrations we drive the formation of larger agglomerates/aggregates. These larger entities probably have higher recombination losses and therefore photoelectrochemistry in these larger aggregates becomes less likely.

On your second point, yes, we have seen that at larger times (>1 h) we tend to lose our collisions. As you suggested, we assigned this to the particles eventually blocking access to the electrode.

Julie MacPherson opened a general discussion of the paper by Kristina Tschulik: You state in your paper that the process described is an alternative to time-consuming STEM-type techniques. However, I would argue that this process could also end up being time consuming. For example, you would need to find yourself a 'friendly electrochemist' who was very familiar with nanoparticle impact experiments, and then presumably you would need to already know what metals constituted the bimetallic particle so the electrochemist could run the baseline individual metals response in order to elucidate how the metals were arranged in the particle (and the size). The process described was for AgAu alloys, but will this approach also work for all metals?

Kristina Tschulik replied: I see the strength of the electrochemical approach in the comparably small investment cost of a few thousand Euros in contrast to the significant investment and running cost of an STEM. Indeed, knowledge of the constituting metals and an initial identification of a suitable electrolyte and experimental conditions is required. However, once suitable oxidation potentials and electrolytes are identified for a given bimetallic system, then the performance and analysis of the experiments is fast and easily done by a non-electrochemist. Thus, I see a suitable application as a standard characterization method and for quality control in nanoparticle synthesis laboratories.

The method can readily be applied to study a variety of bimetallic particles. The only requirement is that both constituents undergo oxidation at different electrode potentials, which can be controlled by the used electrolyte solution.

Justin Gooding asked: Part of the argument for your approach is that you do not need to do TEM. However, we have heard lots about particle aggregation. So if one does not do TEM, one would worry that the nanoimpacts would not tell you about aggregation. Can nanoimpacts tell us about the aggregation if we don't do TEM? What questions can nanoimpacts answer that TEM cannot?

Kristina Tschulik responded: We have indeed used nanoimpact experiments to study the agglomeration of nanoparticles in different electrolytes.[1] When adding the nanoparticles to the solution considering their small concentration (typically in the order of picomolar) their agglomeration (2nd order reaction) will occur at speeds in the order of seconds and minutes. Consequently, impact charges at short times will be smaller and impacts will occur more frequently than those occurring at later times if agglomeration takes place. Thus, chronoamperograms recorded at different times after adding the nanoparticles to a solution will allow one to differentiate single nanoparticles from the agglomerates they form.[2] Among other examples, the agglomeration of commercial silver nanoparticles in authentic seawater was studied[3] and the magnetic field induced agglomeration of Fe_3O_4 superparamagnetic nanoparticles in phosphate buffer has been studied.[4] Both types of study would have been impossible using conventional TEM where the samples have to be dried prior imaging. Also TEM imaging of particles dried from an electrolyte containing very few particles (pM) and high salt loads (>10 mM) is extremely difficult.

Other examples where nano impact studies provided information not available by TEM are the analysis of potentials of zero charge in solution[5] or measuring the electrostatic interaction of nanoparticles and electrodes in solutions of different ionic strengths.[6]

1 J. Ellison, K. Tschulik, E. J. E. Stuart, K. Jurkschat, D. Omanović, M. Uhlemann, A. Crossley and R. G. Compton, *ChemistryOpen*, 2013, **2**, 69–75.
2 E. J. E. Stuart, Y.-G. Zhou, N. V. Rees and R. G. Compton, *RSC Adv.*, 2012, **2**, 6879–6884.
3 E. J. E. Stuart, K. Tschulik, D. Omanović, J. T. Cullen, K. Jurkschat, A. Crossley and R. G. Compton, *Nanotechnology*, 2013, **24**, 444002.
4 K. Tschulik and R. G. Compton, *Phys. Chem. Chem. Phys.*, 2014, **16**, 13909–13913.
5 J. Poon, C. Batchelor-McAuley, K. Tschulik and R. G. Compton, *Chem. Sci.*, 2015, **6**, 2869–2876.
6 K. Tschulik, W. Cheng, C. Batchelor-McAuley, S. Murphy, D. Omanović and R. G. Compton, *ChemElectroChem*, 2015, **2**, 112–118.

Patrick Unwin asked: I am a bit puzzled by the data in **Fig. 5**. For 10 fC charge (the range of part c of the figure), a current of 10 pA would flow for 1 ms (of the order of the data acquisition rate). Yet, the noise level on the inset (part b) looks like it is a few hundred pA and the signal itself on part a and part b is about 1 nA or more and lasts for perhaps tens of ms. The corresponding charge would be much too large for nanoparticles of the size studied. Could you please comment?

Separately, a charge of 10 fC, or so, seems about right for the oxidation of nanoparticles of this size, but to detect it would require extraordinarily low background currents, low noise and the detection of a fast oxidation process. Given the morphology and magnitude of the background current–time traces in **Fig. 5a**, how is this achieved?

Kristina Tschulik responded: Thank you very much for pointing out this mistake. Indeed there was an error in the labelling of the the figure. The *y*-scale values are given in pA (not nA as stated in the the plot). This error will be corrected in the revised version of the paper. As explained in more detail in the experimental part of the paper, we have a peak-to-peak noise of about 0.5 pA. This was achieved by using an analog potentiostat featuring a 3-electrode head amplifier

(NPI Electronic) using an 8 pole Bessel filter (to conserve the charge of the spike) at a 50 Hz cut-off frequency.

Our system allows us to sample at up to 50 kHz and with cut-off frequencies between 20 Hz and 5 kHz. If we aim at analysing the peak shape then we perform high bandwidth studies and if we aim at low noise to detect small charges (as in this work), then we apply low filter frequencies.

Patrick Unwin communicated in connection to the previous comment: We (Prof. Patrick Unwin, Dr. Jon Ustarroz and Minkyung Kang) would like to report some of the data that is related to the poster presented at the *Faraday Discussion* by Minkyung Kang ("Stochastic phenomena in the electrodissolution of individual Ag nanoparticles").

It is evident from these data that destructive NP impacts on solid electrodes are complex processes leading to a wide range of very different and distinctive current–time (*i*–*t*) transients during the same potentiostatic experiment with NPs of the same size. Whereas it has been claimed that the charge consumed in single *i*–*t* transients can be used to determine NP size with extraordinary precision[1–6] *via* electro-oxidation impact coulometry, we have found that this not the case, with significant implications for the Single Nanoparticle Electrochemical Impacts (SNEI) field. Although some NPs may be consumed in a single impact, many are consumed in a series of 'bites' that represent the electrochemical dissolution of a very small fraction (sometimes less than 1%) of the NP. The effect is more significant the bigger the NP is. Hence, NP electrodissolution impacts cannot necessarily be used as a general method for sizing colloidal NPs. Fig. 1a shows a representative overview of an *i*–*t* transient of a solution containing 25 mM $NaNO_3$ and 28 pM AgNPs solution (d = 60 nm). A potential of E = +0.6 V *vs.* Ag/AgCl was applied to a glassy carbon electrode which acts as the working electrode in an SECCM configuration (further details are given elsewhere).[7,8] Fig. 1b, 1c and 1d are magnified *i*–*t* traces in the time range indicated by the blue arrows in Fig. 1a. Each of these three examples show that each "*i*–*t* event" seen in

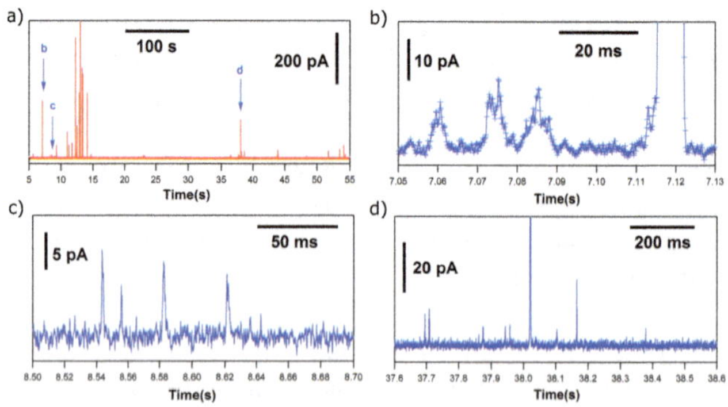

Fig. 1 Representative current transients obtained by applying E = + 0.6 V (*vs.* Ag/AgCl wire quasi-reference electrode, QRE) to a glassy carbon electrode with a solution containing silver nanoparticles with nominal diameter, *d*, of 60 nm. The blue transients (b, c, d) are magnified sections of the red transient (a) in the time range shown by the blue arrows. The data were recorded with a scanning electrochemical cell microscopy (SECCM) format (meniscus contact cell).

the overview (Fig. 1a), is actually a sequence of consecutive partial stripping events in which the charge consumed may be orders of magnitude lower ($Q = 1 \times 10^{-14}$ C) than the charge needed for the stripping of a Ag NP of $d = 60$ nm, $Q \approx 1 \times 10^{-12}$ C. A comprehensive report based on these results will be given elsewhere.[8]

Prof. Yi-Tao Long, Dr. Yilun Ying and Hui Ma would like to show some of the data that is related to the poster presented at the *Faraday Discussion* by Hui Ma ("Recognition of Electro/Chemical Reactions at Ultramicroelectrode Surface by Single Entity Collision"). Using a high bandwidth electrochemical measurement platform, the complex electrochemical oxidation of individual Ag NPs on a gold ultraelectrode (UME) surface was monitored, resulting in significantly distinct current traces. Both the experimental results and Monte Carlo simulations demonstrate that the majority individual NPs *e.g.* 80% at NP diameter of 80 nm undergo a sequential multistep dissolution rather than a one-step simple oxidation. Especially, the charge consumed from each electrochemical measurement of NPs is less than that calculated by the size analysis, particularly for large NPs ($d > 40$ nm). Therefore, the complex motion trajectories of individual nanoparticles should be seriously considered for further applications in SNEI field.

Fig. 2 shows an *i–t* transient of individual AgNPs with diameter of 80 nm suspended in 25 mM phosphate buffer (pH = 7.4) on an Au UME (diameter of UME: 12.5 μm). A low-noise measurement platform with bandwidth of 3 kHz and sampling frequency of 250 kHz was used to acquire the individual transient of NPs (further details are given elsewhere).[9,10] Fig. 2b, 2c and 2d are the enlarged characteristic signals of AgNPs corresponding to the crosses with different colours in Fig. 2a. These current traces showed a spike with undulating terrain or multi-spikes revealing multiple distinct motion trajectories of single AgNPs. All of these four examples show that the duration of the first spike of spike cluster is consistent with the timescale of a single encounter, but the entire duration extends to tens of milliseconds. This result is attributed to a series of partially oxidized stages of a AgNP. A detailed report based on these results can be found elsewhere.[10]

1 Y.-G. Zhou, N. V. Rees and R. G. Compton, *Angew. Chem. Int. Ed. Engl.*, 2011, **50**, 4219–4221.
2 J. C. Lees, J. Ellison, C. Batchelor-Mcauley, K. Tschulik, C. Damm, D. Omanović and R. G. Compton, *ChemPhysChem*, 2013, **14**, 3895–3897.
3 T. R. Bartlett, S. V. Sokolov and R. G. Compton, *ChemistryOpen*, 2015, **4**, 600–605.
4 H. S. Toh, K. Jurkschat and R. G. Compton, *Chem. Eur. J.*, 2015, **21**, 2998–3004.

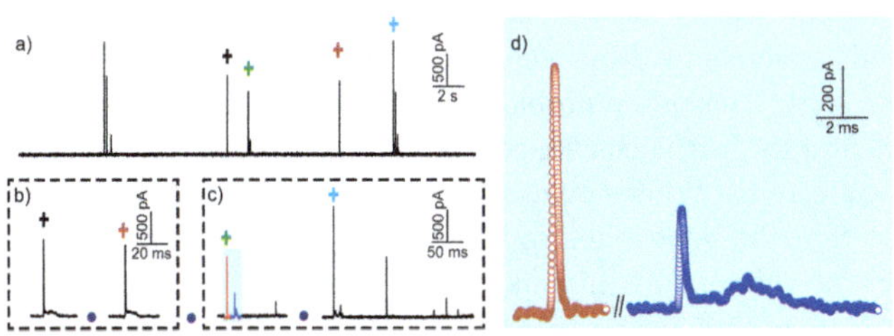

Fig. 2 Representative current transients of individual AgNPs with nominal diameter of 80 nm. (a) *i–t* transients showing oxidation for individual AgNPs collisions on the Au UME (diameter of UME: 12.5 μm). (b, c) Close-ups of the representative spike clusters shown by the crosses with different colors. (d) Enlarged sections of the red and blue transient in (c). Data acquired in 25 mM phosphate buffer (pH = 7.4) at an applied potential of +0.6 V *vs.* Ag/AgCl QRE in the presence of AgNPs.

5 C. Batchelor-McAuley, J. Ellison, K. Tschulik, P. L. Hurst, R. Boldt and R. G. Compton, *Analyst*, 2015, **140**, 5048–5054.
6 S. V. Sokolov, K. Tschulik, C. Batchelor-McAuley, K. Jurkschat and R. G. Compton, *Anal. Chem.*, 2015, **87**, 10033–10039.
7 M. Kang, D. Perry, Y. R. Kim, A. W. Colburn, R. A. Lazenby and P. R. Unwin, *J. Am. Chem. Soc.*, 2015, **137**, 10902–10905.
8 J. Ustarroz, M. Kang, E. Bullions and P. R. Unwin, 2016, submitted.
9 L-J. Zhao, R-C. Qian, W. Ma, H. Tian and Y-T. Long, 2016, *Anal. Chem.*, 2016, **88**, 8375–8379.
10 W. Ma, H. Ma, J.-F. Chen, Y.-Y. Peng, Z.-Y. Yang, H.-F. Wang, Y.-L. Ying, H. Tian and Y.-T. Long, 2016, submitted.

Olaf Magnussen remarked: Is it possible in these experiments to go beyond measurements of the average particle size and composition? Specifically, could you measure the size and composition distribution?

Kristina Tschulik responded: Yes, thanks to the fact that we analyse several hundreds of individual particles we usually plot this data in histograms to obtain the size distribution.

Marc Koper communicated: Regarding the voltammogram: why is the peak for silver so sharp, but for gold so broad?

Kristina Tschulik answered: The sharp oxidation of silver is typically observed when chloride is present in the solution and the oxidation occurs by formation of AgCl(s). In contrast, Au nanoparticles are oxidized to diffusional species $[AuCl_2]^-$(aq) and $[AuCl_4]^-$(aq) in HCl, causing a broader peak.

David Fermin asked: Have you modelled the plasmonic responses of the Au–Ag core-shell nanoparticles in order to extract structural information that can be cross-correlated with collision experiments?

Kristina Tschulik replied: We have not yet done this. Since the UV–Vis spectra of different nanoparticle compositions show a clear absorption band in the visible range, cross-correlation of this data with a modelled response could be easily done. Thank you for this suggestion. We will look for a group experienced in modelling of plasmonic responses.

Richard Crooks opened a discussion of the three papers in the session: What does one actually learn from collision experiments? One common finding is that the materials analyzed by these techniques are heterogeneous, but you know that before you start. The other problem is that the techniques you describe are destructive and the particles change their properties throughout their residence time on the electrode. What would be nice is to sort the 'good' nanoparticles from the bad so that the good ones could undergo additional characterization and testing. Do you think something like that is possible?

Keith Stevenson replied: This is a very good question. 'Heterogeneous' can apply not only in terms of nanoparticle size, shape, and surface chemistry but also to their catalytic, optical and electrical properties which influence the electroanalytical response. At this stage we are just at the beginning to understand the the nanoparticle impact electronalysis parameters for a well controlled system using a chosen set of benchmarks (citrate capped Pt nanoparticles, hydrazine, buffer type, ionic strength). In fact we did these experiments so that we could understand collision experiments with more monodisperse NP properties. In the paper presented here we demonstrate that we can use this method to follow non-linear aggregation behavior for systems that you can't typically follow with other methods (*e.g.* dynamic light scattering or nanoparticle tracking analysis).

Mario A. Alpuche-Aviles replied: I agree that the collision experiments should be developed to the point where we can learn more about kinetics. Now, we can see that there is a distribution of rates. We determine that certain particles are for example 10 times more active than the average. From an ensemble measurement you would be able to extract the average of the rate and not the distribution. We would like to, but we cannot yet, assign the rates to specific NPs. For example, certain surface planes should be more active but we are not there yet. Trapping NPs at an electrode surface does not have to be destructive: a particle can be stable on a certain microelectrode material. It is possible that NPs change with time, but our goal will be to match the conditions of our measurements to the support of the NP in a certain application, *e.g.*, TiO_2/FTO rather than TiO_2/Pt. The challenge there is the fabrication of microelectrodes of different materials. I think it is also possible to sort out the good from bad nanoparticles. The first approach will be to control the synthesis and study for example TiO_2 NPs of different exposed surface planes. Later, I hope we can watch for a highly active NP to collide with the electrode surface and then come back and study the NP with another technique like scanning electrochemical microscopy (SECM).

Kristina Tschulik responded: Information can be gained on the reactivity of individual particles, their agglomeration or dissolution in a suspension (including mixtures). This is not readily accessible by other techniques. Indeed, purification of particle mixtures, by sorting them according to different chemical properties, would be attractive for lab on chip applications *etc.* This will be possible in micro- or nanofluidic devices, once attractive particle-wall interactions are minimized to overcome the problem of unwanted particle loss.

Frederic Kanoufi opened a general discussion of the paper by Kristina Tschulik: Could the approach you propose be used in a synthetic aspect to prepare nanostructured nanoparticles? In your study you address the potential-selective oxidation of Ag in alloy or core–shell Ag/Au NPs. In KNO_3 solution, in the absence of Cl^- anions you could produce nanostructured Au nanomaterials by dissolution of Ag (rather than formation of AgCl). Could it be used as a synthetic strategy?

Moreover, have you performed the same experiment in the absence of Cl^-, and would you detect the total electrodissolution of Ag, and would you be able to detect the formation of Au oxide at the remaining Au nanomaterial (this is actually detected by SECCM as suggested by a poster presented by Unwin *et al.*)?

Kristina Tschulik answered: The nanoimpact method could be used as a synthetic method in principle. However, for a synthetic approach one would likely aim at having a large total number of particles reacting. Instead of using a microelectrode and following individual particles (leading to only a few thousands of particles reacting per minute), one would most likely prefer using a macroscopic electrode to convert billions of particles in the same time, yet at the cost of not being able to resolve individual particle impacts. Alternatively arrays of microelectrodes could be used to perform single particle analysis at high throughput.

Indeed we studied selective oxidation of the Ag from these particles in nitrate solution using both single particle and ensemble studies. We saw similar quantitative leaching of the silver in that case. Due to the small size of the used particles and the relatively small Au content, we were not able to resolve the formation of an Au surface oxide in our single particle studies. Our collaborators currently work on the size-selective synthesis of bigger AgAu particles to facilitate studies in this direction.

Cameron Bentley remarked: Could you explain Fig. **5c** (Charge per impact *vs.* Potential) in more detail? Specifically, the gradual increase between the two plateau regions?

Kristina Tschulik responded: The figure shows the average charge transferred during oxidation of an individual $Ag_{0.73}Au_{0.27}$ particle as a function of the applied electrode potential. The gradual increase of the charge from zero to the first plateau is due to the increasing amount of silver leaching out of the particle during the impact. This can be explained from the Nernst equation (see **eqn 4**). Taking $Ag_{-x}Au_x$ as an ideal mixture, then x determines the activity of the Au and Ag oxidation, which be shifted to higher values than expected for pure Ag. Hence, the higher the applied electrode potential the more Ag can be leached out of the particle, until almost all silver is leached out and a charge plateau is reached. Only once Au oxidation sets in at more positive potentials (*ca.* 800 mV *vs.* Ag/AgCl), then the oxidation charge per particle increases further until the full charge expected for quantitative oxidation of the $Ag_{0.73}Au_{0.27}$ particle is reached. Thus, a second plateau of in the charge–potential graph is seen.

Sanli Faez opened the discussion of the papers by all three speakers: To look at the aggregation of very small particles, optical methods can be very helpful. To combine this with these electrochemical measurements, one would most probably need transparent electrodes. What are the prospects of making transparent (semiconductor) micro-electrodes for nano-impact measurements?

Keith Stevenson answered: In fact, we have recently published several papers on the development of optically transparent carbon based electrodes and optically transparent carbon disk array electrodes. For more information see the following.[1–3]

1 J. Duay, J. Elliot, J. B. Shear and K. J. Stevenson, *Anal. Chem.*, 2015, **87**, 10109–10116.
2 J. Duay, J. M. Goran, and K. J. Stevenson, *Anal. Chem.*, 2014, **86**, 11528–11532.
3 E. K. Walker, D. A. Vanden Bout and K. J. Stevenson, *Langmuir*, 2012, **28**, 1604–1610.

Mario A. Alpuche-Aviles responded: We would like to do some of these experiments. However, making a microelectrode of a transparent material is not straightforward. We have made some progress[1] in making microelectrodes of F-doped SnO_2 and we continue to improve the method of preparation.

1 A. Fernando, P. Chhetri, K. K. Barakoti, S. Parajuli, R. Kazemi and M. A. Alpuche-Aviles, *J. Electrochem. Soc.*, 2016, **163**, H3025–H3031.

Kristina Tschulik replied: There is no major obstacle in the fabrication and use of optically transparent microelectrodes. We have conducted Ag nanoparticle impact experiments using ITO micro electrodes in the past. Since these experiments are very sensitive to electrode contamination, we preferentially use C or noble metal electrodes that can be cleaned easily, as long as we do not need optical transparency.

Justin Gooding remarked: Just a comment on terminology.We have heard the words agglomeration, aggregation and reversible agglomeration. What are the correct terms? I would have thought reversible agglomeration should be called flocculation according to traditional colloidal chemistry.

Keith Stevenson answered: While we have tried to be consistent in the use of our terminology in the presentation of our paper, there are several conflicting definitions used by national and international standards and also by certain disciplines (nanoscience, powder technologists, pharmaceuticals, geology). Here we intend to use a term aggregate as collection smaller sized particles that are still suspended in solution and they can be dynamic and reversible. An agglomerate is an irreversible collection of a mass of smaller particles that and can either remain suspended but can also settle out of solution over time. Floculation is the process of irreversible agglomeration where a flocculant is added to solution to force the suspended agglomerates to settle out of solution usually due to gravitational forces.

David Fermin opened a general discussion of the paper by Keith Stevenson: It is hard to rationalise how a citrate ligand can induced a spontaneous quasi-linear assembly of nanoparticles. I would argue that free electrons in the metallic nanostructures can induce collective oscillations as two particles approach, generating a dipolar effect which will promote the approach of a third particle in a linear configuration. Such spontaneous assembly will be dynamic and dampened at long ranges due to entropic contributions.

Keith Stevenson answered: This is an interesting point. We direct you to related work describing linear assembly of citrate-capped Au nanoparticles.[1] Here they propose additional bridging interactions due to diffuse counter ions.

R. J. Stover, E. Moaseri, S. P. Gourisankar, M. Iqbal, N. K. Rahbar, B. Changalvaie, T. M. Truskett and K. P. Johnston, *Langmuir*, 2016, **32**, 1127–1138.

Patrick Unwin made a general comment: Another way to do these impact experiments is to adopt a scanning electrochemical cell microscopy (SECCM) format,[1–3] which allows an electrochemical cell to be brought into meniscus contact with a very wide range of substrates, which one does not need to encapsulate as an ultramicroelectrode. One can thus select collector electrode materials that have very small background currents/noise and can be quite exotic, like suspended graphene[4] or a transmission electron microscopy grid.[1] This even allows subsequent TEM imaging of NPs that are studied electrochemically.[1]

1 S. E. F. Kleijn, S. C. S. Lai, T. S. Miller, A. Yanson, M. T. M. Koper and P. R. Unwin, *J. Am. Chem. Soc.*, 2012, **134**, 18558–18561.
2 C.-H. Chen, E. R. Ravenhill, D. Momotenko, Y.-R. Kim, S. C. S. Lai and P. R. Unwin, *Langmuir*, 2015, **31**, 11932–11942.
3 M. Kang, D. Perry, Y.-R. Kim, A. W. Colburn, R. A. Lazenby and P. R. Unwin, *J. Am. Chem. Soc.*, 2015, **137**, 10902–10905.
4 G. Zhang, A. G. Güell, P. M. Kirkman, R. A. Lazenby, T. S. Miller and P. R. Unwin, *ACS Appl. Mater. Interfaces*, 2016, **8**, 8008–8016.

Henry White opened a general discussion of the paper by Keith Stevenson: One challenge of metal nanoparticle collision experiments is understanding how the particle electrochemical activity correlates with particle size and geometry. The vast majority of nanoparticle collision experiments employ solutions of polydisperse particles. We have been developing methods for delivering individual particles of known size to the electrode surface, such that the electrochemical response can be correlated with particle size. These one-particle-at-a-time experiments also allow the dynamics of particle interactions with the surface to be controlled.[1] The multipass resistive pulse method we have reported also allows particle size to be measured with sub-namometer resolution.[2–4] These methods may be of interest to the nanoparticle community.

1 K. McKelvey, M. A. Edwards and H. S. White, *J. Phys. Chem. Lett.*, 2016, **7**, 3920–3924.
2 M. A Edwards, S. R. German, J. E. Dick, A. J. Bard and H. S. White, *ACS Nano*, 2015, **9**, 12274–12282.
3 S. R. German, T. S. Hurd, H. S. White and T. L. Mega, *ACS Nano*, 2015, **9**, 7186–7194
4 Y. Zhang, M. A. Edwards, S. R. German and H. S White, *J. Phys. Chem. C.*, 2016, **120**, 20781–20788

Yitao Long opened a general discussion of the paper by Mario A. Alpulche Aviles by communicating: Question 1: How do you understand the cathodic and anodic steps you obtained instead of spikes when dye-sensitized anatase nanoparticles at a Pt ultramicroelectrode?

Question 2: Is it possible to detect the intrinsic photoinduced electron transfer from the dye to an ultramicroelectrode? And how can you achieve it?

Mario A. Alpuche-Aviles responded: In response to question 1: we assign the step-wise current observed to long-term adsorption of the nanoparticles at a Pt ultramicroelectrode. In our previous work we showed that on F-doped SnO_2 the particles show spikes, which we assigned to short NP/electrode interactions.

In response to question 2: If I understand the question correctly, this should be addressed in the context of the relevant material and photoelectrochemical cycle (*e.g.*, TiO_2/dye/iodide as in the dye sensitized solar cell). I believe it would be possible to measure the intrinsic rate of injection from the dye to the electrode material of interest. The electrode will probably be a semiconductor. As discussed in some of the previous questions, the problem is preparing the microelectrode of a semiconductor.

Nongjian Tao opened a general discussion of the paper by Vitor Brasiliense: What is the smallest particle that you can detect with your dark field imaging setup? What determines the detection limit? Would it help if you increase the light illumination intensity?

Vitor Brasiliense replied: With our current setup and keeping the same dynamics, it should be possible to detect a particle down to 30 nm (radius) at most, based on our current signal-to-noise ratio. Of course this depends on how well the spectrum should be resolved, and how small the spectral details are that we want to observe. To enhance the intensity it is a possibility to improve the resolution, but of limited practical utility. While the scattered cross-section of the particle is proportional to the incident light power, it scales as r^6, therefore a 10 fold improvement on the incident power will lead to improvement of a factor $10^{\frac{1}{6}}$ (= 1.46) on our size resolution. The best available sources offer an output power of at most a few watts, which represents a factor of 100 with respect to the one we used, therefore with modern technology, our setup could analyze particles of 20 nm. Moreover, when enhancing the intensity of the source one needs to be careful not to alter its activity by photochemical effects. Actually, it also opens an interesting route for the electrochemical and optical characterization of the photochemical activity of NPs. For our specific chemical system here, silver and (pseudo) silver halogenates, is known to be sensitive to light effects (principle of argentic photography), as evidenced in our work. There is a trade-off between being able to observe particles and not damaging them. Another way to improve the intensity of the detected spectra is at the expense of the time resolution, by increasing the integration time for the image acquisition. This is a common dilemma in making measurements. We can see better or faster, but rarely both. Finally I would like to point out that there are a few improvements that could have been implemented to significantly improve our sensitivity. For example, matching optical indexes between the dark field condenser and the sample (using an oil instead of an air condenser).

Sanli Faez addressed Vitor Brasiliense and Nongjian Tao: The main limitation for detection of small non-fluorescent nanoparticles in a typical dark-field microscope is background scattering by the substrate. Using a fiber based method, dielectric objects as small as 19 nm and single CCMV viruses of 1.5 MDa have been measured.[1]

By using interferometric techniques (iSCAT,[2] absorption[3]) or a lock-in method of photo-thermal detection,[4] much smaller objects such as single molecules can also be detected without fluorescence detection.

1 S. Faez, Y. Lahini, S. Weidlich, R. F. Garmann, K. Wondraczek, M. Zeisberger, M. A. Schmidt, M. Orrit and V. N. Manoharan, *ACS Nano*, 2015, **9**, 12349–12357.
2 P. Kukura, H. Ewers, C. Müller, A. Renn, A. Helenius and V. Sandoghdar, *Nat. Methods*, 2009, **6**, 923–927.
3 M. Celebrano, P. Kukura, A. Renn and V. Sandoghdar, *Nat. Photonics*, 2011, **5**, 95–98.
4 A. Gaiduk, M. Yorulmaz, P. V. Ruijgrok and M. Orrit, *Science*, 2010, **330**, 353–356.

Vitor Brasiliense answered: Thank you for your comment and insights into promising optical detection methods that might be combined with electrochemical processes.

However, some of these techniques are hardly compatible with full-field 3D imaging, especially when studying moving particles. Modulation-based imaging methods (*e.g.* -photothermal or extinction-based) require multiple images for demodulation. Correct localization is only possible if the particle stays in the same position during the acquisition of these multiple images, which is incompatible with small, fast-moving particles. For example, using a full-field holographic version of a configuration similar to Gaiduk *et al.*, our group was able to detect 10 nm diameter particles,[1] but this was achieved on immobilized particles, with long integration times, and attempts to extend this concept to moving particles were not convincing.

The background is indeed a strong limitation. Nevertheless, when working in liquids, a weak surface/solution refractive index contrast can strongly reduce surface scattering, and one can choose to work away from any surface. In such cases, we believe that the ultimate limit to the sensitivity of most optical techniques is the amount of light which the particle can withstand before modification or destruction by thermal effects through light absorption in the particle or the liquid.

1 E. Absil, G. Tessier, M. Gross, M. Atlan, N. Warnasooriya, S. Suck, M. Coppey-Moisan and D. Fournier, *Opt. Express*, 2010, **18**, 780–786.

Kristina Tschulik asked: Do you think that it is possible to determine the kinetics of the chemical dissolution of the electrochemically formed AgSCN, by reducing it back to silver and comparing the intensity of your optical signal before and after this oxidation/ reduction cycle? As the intensity scales with r^6 this should provide a very precise size information if the drift of your system is small. By variation of the scan rate, that is the time between AgSCN formation and back-reduction to Ag, the rate for the chemical dissolution of AgSCN could be extracted.

Vitor Brasiliense responded: I think this is a great idea, it would be interesting to try. The only drawback I can eventually foresee is the question of the reversibility of the Ag/AgSCN transformation. I think this experiment would highlight well the utility of having the spectroscopic information, as it would allow us to be sure we have reduced AgSCN back to Ag. Just based on scattered light at a fixed wavelength, it is sometimes impossible to distinguish Ag from AgSCN, as shown previously.[1] You are also familiar with similar Ag/AgCl electrochemical transformation that we probed optically.[2] In this work we have performed such an oxidation/reduction cycle and did not record an appreciable change in the scattered light by individual NPs during the Ag oxidation (as for Ag oxidation into AgSCN) or during the reverse AgCl reduction. Our present contribution suggests that spectroscopy might detect, during cyclic voltammetry, the changes associated to Ag oxidation or AgX (X = SCN or Cl) reduction and why not answer to what extent such transformation is reversible.

1 V. Brasiliense, A. Patel, A. Martinez-Marrades, J. Shi, Y. Chen, C. Combellas, G. Tessier and F. Kanoufi, *J. Am. Chem. Soc.*, 2016, **138**, 3478–3483.
2 C. Batchelor-McAuley, A. Martinez-Marrades, K. Tschulik, A. Patel, C. Combellas, F. Kanoufi, G. Tessier and R. Compton, *Chem. Phys. Lett.*, 2014, **597**, 20–25.

Mario A. Alpuche-Aviles asked: In your experiments you correlate the observation of NP landing on the electrode and the electrochemistry signal. Does every particle that lands get dissolved? Do some bounce off? How fast do the particles travel? Are they moving by diffusion or is there a migration contribution and how do you control the migration effects?

Vitor Brasiliense replied: No, not all of them. It is actually quite common to see particles that arrive at the electrode and remain unreactive, while others appear to sit on the surface for a while. In a previous paper where 3D microscopy was used,[1] we noticed that particles that appear to be on the electrode surface are sometimes just in its vicinity. Superlocalisation of these particles reveal movement on a 100 nm lenght scale, they appear to be randomly bouncing on the electrode surface.[2] So far we have interpreted this movement as a target search strategy to find an active site on the particle (or electrode) surface. This leads to a time delay between arrival and reaction of a few seconds. In either case, however, we have not yet seen a particle that bounces off to the point it goes back to solution.

In our experiments, particle transport is purely diffusive in the usual case, as evidenced through Mean Square displacement analysis. Electrophoretic forces are negligeable due to relatively high electrolyte concentration (50 mM), screening off the electric field influence. In previous work[3] we did however use holography to put forth diffusiophoretic transport towards an electrode. A gradient of chemical concentration was created through the electrochemical oxidation of the gold electrode, releasing protons and creating a pH gradient. This situation creates a phoretic force analogous to osmotic pressure, capable of compelling nanoparticles towards the electrode. In this case we notice that particles arrive at the surface much faster, almost ballistically.

1 V. Brasiliense, A. Patel, A. Martinez-Marrades, J. Shi, Y. Chen, C. Combellas, G. Tessier and F. Kanoufi, *J. Am. Chem. Soc.*, 2016, **138**, 3478–3483.
2 V. Brasiliense, P. Berto, C. Combellas, G. Tessier and F. Kanoufi, *Acc. Chem. Res.*, 2016, **49**, 2049–2057.
3 A. Patel, A. Martinez-Marrades, V. Brasiliense, D. Koshelev, M. Besbes, R. Kuszelewicz, C. Combellas, G. Tessier and F. Kanoufi, *Nano Lett.*, 2015, **15**, 6454–6463.

J Gooding said: Do you ever see the particles hit the surface and then diffuse across the surface as distinct from collisions with the surface and remain in the same position?

Vitor Brasiliense responded: Yes, using 3D microscopy to monitor the arrival of the particle[1] we were able to monitor several kinds of events that differed from the expected arrival-reaction idealization. For instance, some particles arrive at the polarized electrode but do not react. We attribute this kind of behavior to the presence of a native layer of oxide coating the NP which may have been formed photochemically at the laser power used, showing that care must be taken when designing the experiment to minimize photochemical activity. Among the particles that do react, we observed a delay between the arrival time and reaction. By superlocalizing the particle, we were able to put forward that sub diffraction limit movement close to the surface, as if the particle is bouncing on the surface, and looking for the right position to interact with the electrode surface. The NP actually moves a few ten nanometers in the z direction and in the electrode plane, the xy-directions.[2] Such bouncing behaviours have been recently confirmed and presented during the conference in the poster section (Stochastic phenomena in the electrodissolution of individual Ag nanoparticles, K. Minkyoung, J. Ustarroz, E. Bullions and P. R. Unwin), where, as an alternative explanation, a single AgNP can oxidize in multiple small partial oxidation steps. Those partial oxidation steps would have been difficult to detect electrochemically using our present electrochemical current.

1 V. Brasiliense, A. Patel, A. Martinez-Marrades, J. Shi, Y. Chen, C. Combellas, G. Tessier and F. Kanoufi, *J. Am. Chem. Soc.*, 2016, **138**, 3478–3483.
2 V. Brasiliense, P. Berto, C. Combellas, G. Tessier and F. Kanoufi, *Acc. Chem. Res.*, 2016, **49**, 2049–2057.

Patrick Unwin said: We have also seen the transient interaction of silver nanoparticles with electrode surfaces, during anodic dissolution by nanoparticle impacts (see my earlier comment). Electro-oxidation occurs in a series of bites, and that process becomes more pronounced the larger the particle.

Your measurements provide considerable detail on the complexity of the electrodissolution of silver nanoparticles, and I was wondering about the *z*-resolution and whether that could be improved further?

Vitor Brasiliense replied: Using 3D holography, our resolution in *z*-displacement is currently around 50 nm, and depends on signal-to-noise ratio and on the exposition time.[1] We can in theory improve the resolution by enhancing laser power but it could damage the particle. Therefore, using only the super-localization strategy, there is little room for improvement.

However, when the silver particle is near a layer of gold (the electrode), we show in this work that there is plasmonic coupling between them, which leads to a peak splitting. A new peak appears in the spectrum that is highly dependant on the distance between the particle and the electrode. Unpublished modelling shows that, in principle, this would improve our *z*-position resolution by one to two orders of magnitude. We are currently investigating this aspect.

1 V. Brasiliense, P. Berto, C. Combellas, G. Tessier and F. Kanoufi, *Acc. Chem. Res.*, 2016, **49**, 2049–2057.

Thom Hersbach opened a general discussion of the paper by Christine Kranz: The electrolyte used in the measurement in **Fig. 3b** is 10 mM H_2O_2. Why did you choose to use H_2O_2? Additionally, you do not seem to use a supporting electrolyte in this measurement. Why is a supporting electrolyte not needed to ensure good enough conductivity in this experiment?

Christine Kranz answered: We have chosen hydrogen peroxide as an analyte to investigate the electroactive properties of Au nanostars, as hydrogen peroxide is an important analyte not only in biomedical applications but also in the food industry. These preliminary results are targeted towards investigating if a higher electrocatalytic activity is achieved for Au nanostars in comparison to conventional spherical Au nanoparticles. Interesting electrocatalytic properties of Au nanostars towards hydrogen peroxide reduction has been recently reported.[1]

Unfortunately, the information on the electrolyte was missing in the initial manuscript, which will be corrected during the galley proof process. We have carefully checked the experimental conditions again, and in fact the experiments were conducted in 10 mM H_2O_2 with 0.1 M KCl as supporting electrolyte.

Y. Li. J. Ma and Z. Ma, *Electrochim. Acta*, 2013, **108**, 435–440.

Julie MacPherson asked: You state in the answer to the previous question that **Fig. 3b** does not contain supporting electrolyte, only hydrogen peroxide. For the data of CPDMS with Au nanostars, which shows the enhanced current activity, the key features in the CV seem a little strange in shape (sharp peak oxidatively and almost a double peak reductively)? I believe hydrogen peroxide can be both oxidised and reduced on gold. Are these features thus visible on the CV? Which potential does the CV start scanning from? If you believe Ag stripping is playing a role why is this only visible in this CV and not the others which are also Au functionalised?

Christine Kranz answered: Please compare this answer to the already raised question regarding **Fig. 3b.** All bulk experiments (*i.e.*, also the measurements in H_2O_2 solution) were performed with 0.1 M KCl as supporting electrolyte. This missing information was added during the galley proof process. We completely agree that the shape of the oxidation and reduction peak appears distorted, and we are aware that H_2O_2 can be oxidized and reduced at AuNPs (see response to question regarding the same figure and the recent cited paper on H_2O_2 reduction in dependence of the electrolyte anion). We assume that we observe a convoluted signal during the present experiments in respect to silver ion impurities, which may be present due to the synthesis of the Au nanostars. The CVs were started at 0 V in the positive scan direction and the potential range was from −1 V to +1 V *vs.* Ag/AgCl as illustrated in **Fig. 3b**. The strange shape of the CV recorded at the Au nanostar-modified substrate needs to be re-evaluated, and we are currently investigating an electrochemical route for producing Au nanostars without using $AgNO_3$; thereafter, will re-evaluate the electrocatalytic properties towards H_2O_2 conversion. To the best of our knowledge, the synthesis of spherical AuNPs does not involve $AgNO_3$. Hence, no oxidation and reduction peak for this couple was evident in the bulk CVs recorded at the Au NP-modified CPDMS. Only a small oxidation peak was observable.

Frederic Kanoufi said: Your work is a very nice demonstration of the current potentiality of topography/activity nanoscale imaging by SECM/AFM nanotips. I wonder to what extent you can deconvolute topography and activity in your feedback image? For example was it performed at constant distance or constant height?

Moreover for comparison you provide the approach curve of an insulating SiO_2 surface. An approach curve of the conductive PDMS could have been more appropriate to really seize the Au nanostar activity.

Christine Kranz answered: The insulating silicon nitride tip has a length of 200 nm, and provides the topographical information in constant force mode with the tip following the topography. Hence, imaging of the electrochemical information based on the tip-recessed electrode is obtained in constant distance SECM mode. A line scan reflecting the topographical information along with the electrochemical feedback current is shown in Fig. 3.

This figure clearly shows that the asymmetric re-shaped AFM tip results in an imaging artifact in the topographical information, which is not reflected in the electrochemical response revealing a symmetric signal reflecting an enhanced current signal for the Au nanoparticle. For quantitative evaluations, the

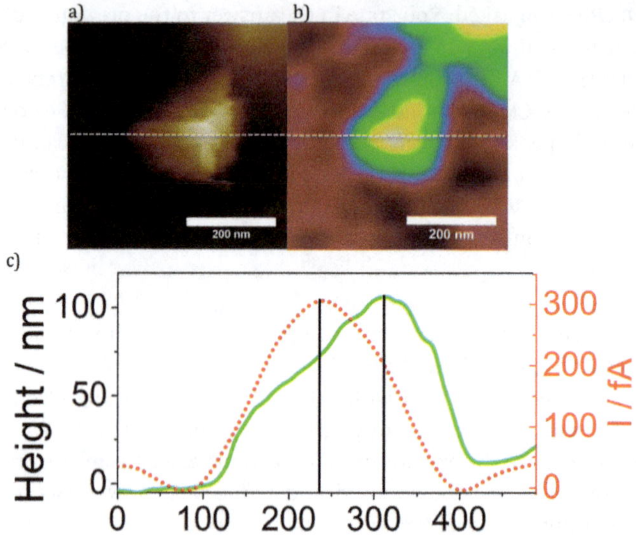

Fig. 3 AFM-SECM image of an individual Au nanostar recorded in 5 mM $Fc(MeOH)_2$/0.1 M KCl in PFT mode AFM and feedback mode SECM. a) Topography, b) electrochemical image and c) extracted line scans from topography and current (as marked with the dotted line) reflecting the slight asymmetry of the re-shaped AFM tip as well as the lateral distance between the AFM tip and integrated disc nanoelectrode.

experiments have to be accompanied by simulations. Since the geometry of the AFM-SECM probes can be accurately derived from SEM images, this appears rather straightforward. Regarding the deconvolution: the topographical data are not convoluted once the tip and electrode are either above the substrate or above the nanoparticle. An area where the tip is already above a particle but the electrode remains above the substrate entails a change in height, which effect is dependent on the size and the distance of the electrode to the sample, and the nature of the sample (*i.e.*, steep or shallow flanks, *etc.*) For the given example, this can be hardly resolved given the diffusion-based process.

We agree that an approach curve should be recorded above an individual Au nanostar; we are currently performing such experiments. At the time of submission, we were not able to determine a difference between the non-modified CPDMS and the gold nanostar in $Fc(MeOH)_2$/0.1 M KCl solution. Right now, this is evaluated for MPTS-modified CPDMS.

Patrick Unwin remarked: At the start of your presentation you gave a brief overview of electrochemical imaging techniques that have independent distance control. Could you briefly compare these techniques and their relative strengths and weaknesses? AFM-SECM has obviously now been around for a long time. How robust is the technique these days? Additionally, an interesting feature of batch-fabricated tips is the possibility of integrating measurement electronics onto the AFM chip. For example, one could envisage integrating a current preamplifier. Has this been done or have you investigated this?

Christine Kranz answered: The most commonly used approaches in SECM using independent distance control are shear-force based positioning, hybrid techniques like AFM-SECM, and pipette-based techniques including SECCM and SICM-SECM. At the moment, given the body of published work I would state that shear-force based positioning is the most frequently used technique given the fact that it can be fairly easily implemented within existing instrumentation; commercial systems also offer this. The topographical information is obtained in a non-contact mode approach, however, given the dimensions of the probes, the achievable topographical resolution is not comparable to techniques such as SECCM, AFM-SECM, and SICM-SECM. Nonetheless, nanoelectrodes have been used therein, and real-world applications have been demonstrated. In recent years, pipette-based techniques have gained tremendous popularity, as the probes can be easily fabricated in a laboratory environment at rather low cost, if a laser pipette puller is available. More sophisticated approaches using multiple pipettes are conducted with custom-built instrumentation, and hence, are not yet available to a broader scientific community. In terms of applications, nanopipette-based approaches have distinct advantages such as non-contact operation, which renders them highly suitable for biological/biomedical applications, in particular using multichannel probes with associated modifications (see the contribution of Lane Baker, DOI: 10.1039/c6fd00133e). In addition, the topographical resolution is determined by the orifice of the pipette, which may be in the range of a few tens of nanometers (around 50 nm appears to be fairly reproducibly achievable), and the probes can be well characterized *via* TEM. With respect to materials science and surface characterization, for applications involving 'hard' samples, hybrid AFM-SECM has significant advantages not only in the unsurpassed, highly resolved topographical information it can yield, but also by the fact that force spectroscopic measurements, and hence, material properties such as the Young's modulus, adhesion, elasticity *etc.* can be obtained simultaneously with electrochemical information (see contribution of Christine Kranz, DOI: 10.1039/c6fd00128a). Meanwhile, two AFM companies commercially offer such dual-function probes, and it will be interesting to see which direction the technique evolves from there. Regarding the timeline, 10–15 years from the first scientific reports until commercialization seems normal compared to other developments. For sure, a disadvantage is the relatively high cost for such probes, as the fabrication process is labor-intensive and involves more sophisticated processes compared to pipettes. However, AFM instrumentation is mature and readily available in many laboratories. Integrating the electronics on chip is a very interesting approach, which is not yet available for AFM-SECM probes. We had some initial discussions with the Institute of Electron Devices and Circuits here at Ulm University, however, we have not yet realized such probes.

Julie MacPherson asked: There are many advantages to be had from coupling electrochemistry with AFM, one of which, as you describe nicely in your paper, is the ability to elucidate the mechanical properties of the substrate from the AFM component of the probe. We have already discussed in this meeting the issues surrounding electrode fouling, but what about AFM tip fouling when making indentation measurements, especially, as is the case here, you are working with a soft and potentially sticky substrate (carbon PDMS).

Christine Kranz answered: This is a very good point, which always has to be considered since any modification of the AFM tip either by adhering contaminations or tip wear may contribute to imaging artifacts. Our approach for AFM-SECM using a re-shaped silicon nitride AFM tip as well as commercial AFM probes surely faces this problem. In fact, this holds true for almost all scanning probe techniques operated outside UHV conditions. In case of the sample investigated herein, due to the softness of the sample, tip wear is not expected to be a significant issue. However, adhesion of material to the AFM tip may occur. A facile way to detect such tip modifications is by imaging a hard sample with known surface features after the measurements at the soft sample. Thereby, tip artifacts generated by adhering material are usually recognizable by repeating regular features in the AFM image, which are not correlated to the known topography of the substrate.

Nongjian Tao opened a general discussion of the paper by Chao Jing: Your images show many particles (rods) on the surface. Do you expect that the reaction at one particle affects the one next to it by correlation analysis? What is the closest distance between two adjacent particles in your image? What is the radius of the diffusion sphere for the time scale of your measurement?

Chao Jing replied: Actually I did not consider the effect of adjacent nanoparticles before, thank you very much for your questions. The closest distance between two adjacent particles is around 3 μm. In this work, two reactions were involved. For the oxidation of gold nanoparticles, the products are gold oxide/hydroxide which were adsorbed on the nanoparticle surface. This surface-controlled reaction should be independent for every single nanoparticle. For the oxidation of hydrogen peroxide, the diffusion radius of hydrogen peroxide was calculated using the equation $r_d^{\,2} = r_0(\pi Dt)^{\frac{1}{2}} + r_0^{\,2}$ through the combination of linear diffusion and radial diffusion on a single nanoparticle surface.[1] The capture time of the image t is 2 s, D is the diffusion coefficient of hydrogen peroxide, which is *ca.* 1.4×10^{-5} cm^2 s^{-1}, r_0 is the radius of nanoparticle (*ca.* 80 nm), the calculated radius of diffusion sphere is 2.7 μm which is larger than the half of the closest distance of adjacent nanoparticles. Therefore, for concentrated nanoparticles, the electrochemical process will have influence on other particles. The distance between nanoparticles is better to be larger than 10 μm with a low disperse density to avoid the inter-particle effect. We will pay attention to this phenomenon in further experiments.

1 B. R. Scharifker, *J. Electroanal. Chem.*, 1988, **240**, 61–76.

Katharina Krischer commented: In the last video of your representation, one could discern a wave propagating from right to left. Could you comment on the origin of this wave? Could it be a cooperative phenomeon resulting from the coupling of the gold particles under reaction conditions through diffusion or the electrostatic potential?

Chao Jing responded: Thank you very much for your questions. The intensity change wave may be generated from the low conductivity of the ITO electrode and the position of the counter electrode which could induce the inhomogeneous current density and potential propagation on the ITO surface. As the concentration of hydrogen peroxide is very high, in my opinion, the diffusion process is not the main reason for the wave-like intensity change.

Justin Gooding remarked: You made a statement in the paper that the difference in nanoparticles intensity change ratio is due to inhomogeneity of individual particles. Is it possible that the difference between how nanoparticles respond relates to differences in potential across the electrode surface?

Chao Jing answered: It is quite possible that the potential across the electrode surface will have an effect on the response of nanoparticles under potential scanning. In this work, there are several factors that could affect the scattering intensity including the properties of nanoparticles and surrounding environment. For instance, the shape, size and surface modifications of nanoparticles (nanorods were synthesized in CTAB solution) will have influence on the reaction area, catalytic ability and also sensitivity to the medium change. On the other hand, the environmental conditions including the electrolyte concentration, defects on the ITO surface, the contact between nanoparticle and electrode and the potential across the electrode surface will extensively impact on the reaction process and spectral response of the nanoparticles.

Wojciech Nogala made a comment: A wave of changing light intensity from polarized gold nanorods was seen in the movie presented during the meeting. It propagates from one side of the image to another. This phenomenon is also visible in **Fig. 3(a–c)** in the paper, as the light intensity change, presented in blue color, propagates from right to left. Besides this, a possible non-uniform current density in the electrolyte caused by ohmic drop and a non-symmetrical position of the auxiliary electrode and substantial sheet resistance of the ITO electrode (20–30 Ohm^2) may also cause this phenomenon. If the ITO-coated glass sheath is connected to the working electrode (WE) electric clamp from potentiostat cable in one point, then a noticeable ohmic drop will occur in the ITO layer. This causes a non-uniform potential on the ITO surface when the current flows. This could be verified by connecting the WE cable to ITO at the opposite side of the cell. A wave propagating in the opposite direction would confirm this hypothesis.

Chao Jing answered: Thank you very much for your valuable comments which explain the phenomenon in my experiments well. I will pay more attention to the effect of all three electrodes in electrochemical detection.

Marc Koper posed a question to all three speakers: One of the goals of single entity electrochemistry is not just to see individual events but correlate the structure and activity of individual particles. Can you comment on the prospects of your techniques for obtaining individual structure–activity relationships?

Christine Kranz responded: AFM has been used as a tool to characterize the morphology of single nanoparticles. Hence, adding electrochemical functionality to the AFM probe *via* an integrated nanoelectrode could provide the correlation between the topography of individual particles and their activity. We have just started this endeavor, and little is published on the correlation of electrocatalytic activity and morphology using scanning probe techniques. However, I am convinced that AFM-SECM is ideally suited for such investigations given a well-characterized bifunctional probe with defined tip curvature radius, electrode geometry, *etc.* Further correlation with HR-TEM and spectroscopic data may aid in obtaining a fundamental understanding on single particle activity/morphology correlation.

Vitor Brasiliense answered: If single-particle spectroscopic measurements are available, a lot of information about structure and activity can be addressed for many particles in parallel, as stressed in the talks of Prof. Tao and Chao Jing. In our work, we show how a NP spectroscopic signature changes during an electrochemical reaction and further dissolution. We could deliver information about size (from the intensity of scattered light) and chemical composition. Developing this further in the case of plasmonic materials, extensive literature exists relating particle geometry and spectrum, including core–shell, asymmetric shapes and agglomerates.[1,2] It was also shown that spectroscopic data may also bear information about the chemical environment of the particle and the surface state of the particle.[3] Information about electronic structure can be derived as well. In fact, the localized surface plasmon resonance of nanoparticles is so sensitive that electron density variations corresponding to $\approx$ 5000 elementary charges per second ($\approx$ 1fA) have been monitored,[4] a performance comparable to electrochemical measurements. Another way of accessing structural information consists of tracking the particle movement in solution (single particle tracking, or nanoparticle tracking analysis), as shown in the previous section by Prof. Stevenson (DOI: 10.1039/c6fd00121a). Indeed, in previous work we have employed 3D microscopy techniques to track particles moving in solution, thus being able to access its hydrodynamic size and aggregation state.[5] In principle, it is possible to address its shape as well.

Finally, in my poster (V. Brasiliense, J. Clausmeyer, C. Combellas, G. Tessier, W. Schuhmann and F. Kanoufi, "In operando optical monitoring of single particles during Oxygen Evolution Reaction"), I describe a slightly different strategy, where single nanoparticles are grown from a nanoelectrode, while their optical and electrochemical properties are addressed, revealing how changes in particle size and morphology might impact catalytic activity.

1 H. Zang, D. W. Brandl, P. Nordlander and N. J. Halas, *Acc. Chem. Res.*, 2007, **40**, 53–62.
2 A. Crut, P. Maioli, N. Fatti and F.Vallée, *Chem. Soc. Rev.*, 2014, **43**, 3921–3956.
3 C. P. Byers, B. S. Hoener, W. Chang, S. Link and C. F. Landes, *Nano Lett.*, 2016, **16**, 2314–2321.
4 C. Novo, A. M. Funston and P. Mulvaney, *Nat. Nanotechnol.*, 2008, **3**, 598–602.
5 V. Brasiliense, A. Patel, A. Martinez-Marrades, J. Shi, Y. Chen, C. Combellas, G. Tessier and F. Kanoufi, *J. Am. Chem. Soc.*, 2016, **138**, 3478–3483.

Chao Jing answered: Thank you very much for your question. As dark-field microscopy can only monitor the scattering light of nanoparticles, we can estimate the structures of nanoparticles in a rough way through the relationship between the morphology and scattering spectroscopy of nanoparticles. For instance, it is possible to calculate the size of nanospheres through the scattering spectra of nanoparticles.[1] In addition, nanoparticles with different shapes such as triangles, rods, wires, cubes, show various scattering bands which enable the recognition of the structures of particles.[2–4] On the other hand, we can also use other characterization methods such as SEM to obtain the corresponding structures of particles. By comparing the optical signals and SEM images, it is not difficult to achieve the analysis of relationships between structures and activities.

1 C. Jing, Z. Gu, Y.-L. Ying, D.-W. Li, L. Zhang and Y.-T. Long, *Anal. Chem.*, 2012, **84**, 4284–4291.
2 J. J. Mock, M. Barbic, D. R. Smith, D. A. Schultz and S. Schultz, *J. Chem. Phys.*, 2002, **116**, 6755.
3 K. A. Willets and R. P. Van Duyne, *Annu. Rev. Phys. Chem.*, 2001, **202**, 60–65.
4 C. Jing and Y.-T. Long in *Localized Surface Plasmon Resonance Based Nanobiosensors*, 2014, Springer.

Zhongqun Tian opened a general discussion of the paper by Chao Jing: In surface plasmon reasonance (SPR), especially localised SPR (LSPR), the electromagnetic near-field around plasmonic materials is not uniformly distributed, but highly localized in the spatially narrow regions (so called 'hotspots'), such as, nanotips, inter-particle nanogaps or particle–substrate nanogaps. The size and shape of plasmonic materials (*e.g.*, Au, Ag and Cu) are essential in order to create hotspots. For example, nanoparticles will mainly absorb light and generate heat when the size is smaller than 20 nm while they will mainly scatter light when the size is from 30 nm to 200 nm. As different shapes can create different hotspots, it's necessary to take the advantage of fast progress of nanoscience in order to use suitable plasmonic nanostructures not only for spectroscopic imaging but also for chemical/photochemical reactions. SPR can simultaneously cause electromagnetic near-field enhancement, heat generation and excitation of a hot-electron in a sharp narrow region. Recently, SPR mediated (enhanced) chemical reactions at the hotspot have gained attention. With the unique feature of high spatial resolution (typically 5–10 nm), LSPR could be useful to study various nano-reactors in the field of single entity electrochemistry.

Chao Jing answered: Thank you very much for your valuable comments. The near-field enhancement of plasmonic materials plays important roles in the fields of optics, chemistry, physics and biology, especially after the advent of SERS and TERS techniques. Utilizing functional nanostructures such as core–shells and asymmetric particles has significantly improved the detection sensitivity and expanded the application of plasmonics. The hotspot based on plasmon resonance enhancement can provide an active reaction area due to the strong electron storm and intensive electromagnetic field on the surface/gap of nanoparticles. This active area makes plasmonics a perfect catalyst to promote chemical/electrochemical/photochemical reactions. SPR- and LSPR-dependent optical techniques such as SPR imaging, dark-field scattering spectroscopy and SERS are promising methods in the application of ultra-sensitive detection of molecules and monitoring of chemical/electrochemical reactions at the single nanoparticle level. In particular, the TERS technique overcame the optical diffraction limit and enabled observation at the single molecule level. Improving the time–spatial resolution of SPR/LSPR is greatly meaningful in the development of single entity detection.

Tim Albrecht made a general comment: Following on from the previous comment by Prof. Tian, I would like to emphasize that, in my view, simulations of the dynamics of nanoobjects, including nanoparticles, nanorods or other species, close to the electrode surface would be useful, in order to shed light into the expected timescales, the adsorption behaviour, the orientation during the collision (facet dependence?), the influence of the local electric field and other factors, potentially important in the impact process. Focus needs to be not only on average properties, but also distributions and event statistics, including multiple collisions by a single particle, inter-event time distributions *etc.* This will help to develop a more comprehensive, perhaps 'molecular' level understanding.

Nongjian Tao responded: Statistical analysis will certainly add new information. It would be nicer to correlate the measured properties with the size, shape, structure, and collision location of the individual entities. Optical imaging can provide size, collision location and possibly shape information, and Prof. White's nanopipette particle counter can provide precise size information, but structural (atomic scale) information of the individual entities seems difficult to obtain *in situ*.

Patrick Unwin opened a general discussion of the paper by Christine Kranz: An important issue in electrocatalysis at arrays, or ensembles, of nanoparticles is the diffusion (and subsequent reaction) of soluble intermediates between neighbouring nanoparticles. How far away are we from being able to track inter-nanoparticle fluxes of intermediates with electrochemical imaging tools?

Vitor Brasiliense replied: With a complementary visualization of the electrochemical process at the nanoscale, one should be able to track such inter-nanoparticle fluxes. It depends on the strategy to track the intermediates released by the electrochemical activation of individual NPs. For example, Prof. Tao showed in his talk (DOI: 10.1039/c6fd00180g) how SPR microscopy (or other refractive index imaging) could image diffusion layers. In our work, the simultaneous dissolution of a large number of Ag NPs into $Ag(SCN)^{2-}$ intermediates of low solubility is accompanied by the formation of a precipitate pattern that scatters light. It is definitely indirect evidence of the overlapping of diffusion profiles generated by closely-spaced nanometric sources. Even though the chemical system here is likely complex, similar optical imaging should be able to track inter-particle diffusional cross-talk. Of course, supports from models of both the diffusional process and the optical response of vicinal nanoparticle are needed.

Christine Kranz responded: This is a very interesting and important question, as it points toward the complexity of electrocatalysis involving nanoparticles. As far as I know, little to none has been published in respect to electrochemical imaging of the diffusion of soluble intermediates between individual nanoparticles. I believe that the field is moving in the right direction with improved probes, their reliable characterization, and improved hardware, thereby allowing data acquisition with high spatial and temporal resolution. Are we there yet? My informed guessed is, not yet; however, the community is heading there.

Patrick Unwin remarked: As we have heard quite a lot of talks on nanoparticle impacts at this Discussion, I wondered whether you had every tried modifying a conventional AFM tip so that there would be a nanoparticle at the end that one could place (land or impact) in different environments with very good control? As there is quite a lot of interest in support effects in electrocatalysis, this could provide an opportunity to place the same nanoparticle in different environments and measure the electrochemistry.

Christine Kranz responded: We have not pursued any experiments like that, but this would be a really interesting approach. In fact, AFM provides excellent control on targeted positions. In a different context, experiments where a single nanoparticle was picked up and delivered have already been shown.[1] Hence, such experiments appear entirely feasible.

Y. Wang, Y. Zhang, B. Li, J. Lü and J. Hu, *Appl. Phys. Lett.*, 2007, **90**, 133102.

Wolfgang Schuhmann made a general comment: By using electrochemistry to measure single entities or aiming to measuring single molecules we have the disadvantage compared to optical methods that we still need to transfer a few thousand electrons to get a measurable current. However, the instrumental development in the past few years allows us to think today about possible experiments which were unthinkable 5 years ago. There has been huge progress in nanoscale electrochemical measurements.

Nongjian Tao answered: Combining the strengths of electrochemical, optical and plasmonic methods into one measurement seems to be a promising approach to study single entity electrochemistry. One example shown by Prof. Crooks is to measure the electrochemical current of single nanoparticles during collision under an optical microscope.

Tim Albrecht answered: Perhaps one could use tunnelling currents across individual particles, too, as we have done in the past.[1] I have always been intrigued by the idea of performing the same experiment on a catalytically active nanoparticle, to see whether catalytic activity could be detected in the tunnelling current noise. The tunnelling current can easily be in the nA range in such an experiment and thus presumably higher than the Faradaic current. Hence, the bandwidth could be accordingly higher.

1 T. Albrecht, S. F. L. Mertens and J. Ulstrup, *J. Am. Chem. Soc.*, 2007, **129**, 9162–9167.

Wojciech Nogala replied: Low current measurement problems determine a bottleneck of nanoelectrochemistry. In the case of studies on single molecules of catalyst (*e.g.* enzyme molecules), until the availability of apparatus that allows current measurements with noise level low enough to conclude that the recorded current corresponds to a single catalyst molecule, we could focus on the detection of electroactive products of a catalytic reaction instead of wiring catalyst molecules to a nanoelectrode. Then the catalytic reaction does not have to be a redox process. One can increase the sensitivity of the detection of the electroactive product *e.g.* by using stripping voltammetry, or entrapping it in a small volume and following exhaustive electrolysis. An example based on redox cycling of quinone generated by single tyrosinase molecule has been presented by Lemay's group.[1] I would be grateful for any suggestions of a specific catalytic reaction

(S → P) catalyzed by an enzyme or other molecular catalyst, where its electroactive product (P) or its electroreduced/electrooxidized form can be adsorbed on the electrode and be electrochemically stripped. Nevertheless, I would like to notice in the next few years that further progress in high-bandwidth measurement of low currents open even more possibilities in single entity electrochemistry. The ultimate dream is recording of single faradaic electrons.

1 L. Rassaei, K. Mathwig, S. Kang, H. A. Heering and S. G. Lemay, *ACS Nano*, 2014, **8**, 8278–8284.

Martin Edwards replied: While it might seem unrealistic to be sensing single molecules due to the 1 or 2 electrons transferred during a typical redox process there is in fact a significant current amplification step occurring in the ion channel detection scheme. While the conductivity of the channel typically results in a current of several picoamps – the current difference between the channel in its two conductivity states is also typically of the pA range. This therefore relates to millions of ions per second and is measureable in a relatively straightforward manner. In this way, single molecule detection is routinely performed in a number of laboratories.

Frederic Kanoufi opened a general discussion of the paper by Chao Jing by communicating: In your paper, for example **Fig. 3f–j** you are presenting an intensity distribution in the 0–3% range. However in **Fig. 2** the changes in optical intensity were more in the 20–40% range. I wonder how statistically significant are the reported distributions, or if there is any error in these plots? If not, could you explain what is your instrumentation strategy to reach such low level of optical noise?

Chao Jing communicated in reply: Thank you very much for your questions. Actually, the reaction condition of nanoparticles in **Fig. 2** and **3** are the same. In theory, the signal from the image should be similar to the data from the spectra. However, in **Fig. 2**, the intensity changes are from the peak intensity in the scattering spectra while the signals in **Fig. 3f–j** are from the averaged values of intensity at all wavelengths in the image without light splitting. In addition, by the auto-recognition process, the calculated averaged intensity also includes some pixels in the dark background area whose intensity is very low. Therefore, the intensity change from the image is much smaller than the data from the spectral peak intensity due to the calculation method. From both of the results of a single nanoparticle and the statistic values, the calculated intensity changes from the image are very low but show similar trends compared with the results from the peak intensity change which confirm the feasibility of this method. The statistical intensity could reveal the activity distributions of nanoparticles that give the occurrence probability of electrochemical reactions on a nanoparticle surface. Compared with results from a single nanoparticle, the statistical distribution is more valid by avoiding random and abnormal events.

Yitao Long opened a general discussion of the papers by Vitor Brasiliense by communicating: It is significant and essential for us to achieve an optical and high-resolution electrical detection synchronously of the oxidation of single nanoparticles. Do you have a strategy to meet this goal?

Vitor Brasiliense communicated in reply: That depends on what is meant by high resolution. Every chemical process is a multiscale process, with many timescales involved. In a previous work with silver nanoparticles[1] the particle dissolution and charge transfer responses fell on the ms–s timescales. This could be resolved simultaneously with an EM-CCD camera and standard electrochemical potentiostat (PalmSens). If more detailed information is needed, such as the full visible spectrum for instance, the optical signal is in general weaker and therefore the time resolution shrinks. In our present work, for example, where we acquire one spectrum per second, we are able to resolve the dissolution process (~10s) but not the oxidation process.

1 V. Brasiliense, A. Patel, A. Martinez-Marrades, J. Shi, Y. Chen, C. Combellas, G. Tessier and F. Kanoufi, *J. Am. Chem. Soc.*, 2016, **138**, 3478–3483.

Faraday Discussions

PAPER

Interactions of anions and cations in carbon nanotubes

L. Mohammadzadeh,[ab] P. Quaino[b] and W. Schmickler[a]

Received 12th April 2016, Accepted 19th April 2016
DOI: 10.1039/c6fd00076b

We consider the insertion of alkali-halide ion pairs into a narrow (5,5) carbon nanotube. In all cases considered, the insertion of a dimer is only slightly exothermic. While the image charge induced on the surface of the tube favors insertion, it simultaneously weakens the Coulomb attraction between the two ions. In addition, the anion experiences a sizable Pauli repulsion. For a one dimensional chain of NaCl embedded in the tube the most favorable position for the anion is at the center, and for the cation near the wall. The phonon spectrum of such chains shows both an acoustic and an optical branch.

1 Introduction

Carbon nanotubes hold great promise as electrode materials for batteries and supercapacitors: they have high surface area, low weight, exist with semiconducting and conducting modifications, and can be doped by defects or impurities. They also pose a fundamental problem for double-layer theory: the dimensions of nanotubes, with diameters in the order of 1 nm, are smaller than the characteristic lengths for double layers at planar surfaces. Hence the familiar concepts of double layer theory no longer apply, and new concepts are being called for. Given the small spatial extensions of nanotubes, any new model must be based on an atomistic description. Therefore, our group has embarked on a project to develop models for electrolytes in narrow tubes using a bottom-up approach, starting with single ions.[1–3] We use density functional theory (DFT) as a tool to explore the electronic properties of the system, and on this basis we construct physical models for the interaction of ions with nanotubes.

Our previous work in this area has been devoted to single ions; here we go one step further and first consider ion pairs in nanotubes, and later a one-dimensional salt, *i.e.* a chain of anions and cations in alternating positions embedded in a tube. The central question is: how does the nanotube affect the interaction between the ions?

Throughout this work we focus on nanotubes with diameters of less than 1 nm. This range is of particular interest, because such tubes have an

[a]*Institute of Theoretical Chemistry, Ulm University, Germany. E-mail: wolfgang.schmickler@uni-ulm.de*
[b]*PRELINE, FIQ, Universidad Nacional del Litoral, Santa Fe, Argentina*

unexpectedly high capacity.[4,5] There is presently a debate of whether the capacity per unit area of these tubes is actually higher than that of wider tubes – for a recent contribution regarding this question see ref. 6. A central difficulty is the measurement of the surface area of narrow tubes. However, the main point is that these narrow tubes have a high capacity, even though the ions have to lose a large part of their solvation shell when they enter the tube. The physical reason for this effect was pointed out by Kondrat and Kornyshev:[7] ions in narrow tubes create an image charge on the adjacent walls, which lowers their energy and compensates for the loss of solvation energy. Furthermore, the image charge effectively screens the Coulomb potential along the axis of the tube; this weakens the interaction between ions of the same sign, and allows a denser packing of charge. The same group has also developed two- and three-state Ising models for ionic liquids stored in carbon nanotubes (CNT) and calculated their charge-capacity characteristics.[8–10]

The rest of this article is organized as follows: in order to make this article self-contained we shall first briefly review a few results for the insertion of single ions. Then we shall present DFT results for a few ion pairs, which we shall interpret in terms of a simple model. Next we shall examine a chain of NaCl as an example of a one-dimensional salt embedded in a tube, and finally calculate the phonon spectrum of such chains.

2 A few results for the insertion of single ions

In our previous work[1–3] we have investigated the insertion of simple alkali and halide atoms in a variety of CNTs, all with diameters less than 1 nm. In all cases investigated the inserted atoms were completely ionized, and did not form chemical bonds with the CNT. The charge that had been transferred to the CNT forms an image charge that surrounds the ion – see Fig. 1.

The energetics of ion insertion are governed by competition between the image interaction and Pauli repulsion. The image interaction favors insertion and forms a large part of the total insertion energy; it is more favorable near the wall. In contrast, Pauli repulsion is weakest at the center. As a consequence, for larger ions, especially the halide ions, the optimum position in these narrow tubes is at the center, while the smaller alkali ions Li^+ and Na^+ are usually incorporated off-center.

Our calculations show that the image charge screens the Coulomb potential of the inserted ions and thus reduces the electrostatic interaction. This confirms the

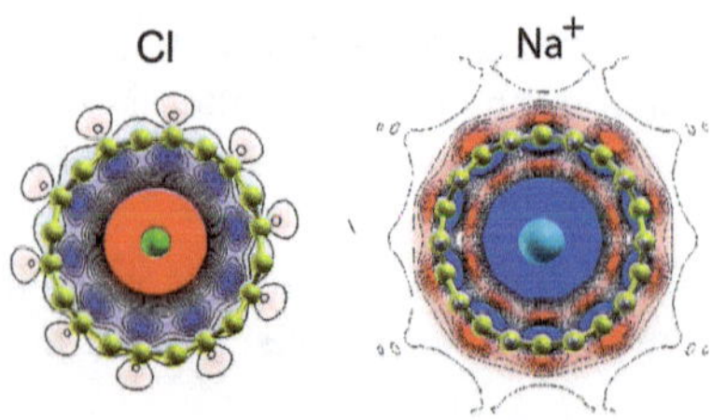

Fig. 1 Charge difference plots for a chlorine and a sodium atom in a (10,0)CNT. Red (blue) indicates an excess of negative (positive) charge.

notion of a superionic state postulated by Kondrat and Kornyshev on the basis of classical electrostatics.[7] Fig. 2 shows the screened Coulomb potential along the axis generated by a Li^+ ion for two CNTs with different properties: the (8,0)CNT is semiconducting and has a radius of 3.18 Å, while (6,3)CNT is conducting and has a very similar radius of 3.10 Å. In spite of their different conductivity, the two tubes show practically the same screening. In fact, the insertion of the Li atom has made the previously semiconducting (8,0)CNT conductive, because it has taken up an extra electron.[3]

We have calculated the screened Coulomb potential for a series of alkali and halide ions; outside the ionic core, the potential is independent of the nature of the ion.[1,2] This shows that the screening is a purely electrostatic effect. In order to characterize the screening properties of a tube, we have introduced the notion of the *effective image radius of the tube*, which is the radius of a classical metal tube which produces the same screening as the real tube under consideration. For the two tubes shown in Fig. 2 the values are: $R_{im} = 2.14$ Å for the (8,0)CNT, and $R_{im} =$ 2.01 Å for the (6,3)CNT. Thus, the effective image radius is smaller than the physical radius of the tubes as defined by the position of the carbon atoms. This confirms the visual impression of Fig. 1 that the image charge sits inside the tube. We shall return to the concept of screening when we discuss the interaction between ion pairs.

3 Insertion of ion pairs

As a model tube we chose the (5,5)CNT, which is conductive and has a diameter of 6.78 Å. Note that in this paragraph we consider insertion into the center of the tube, because in this case the symmetry helps in understanding the energetics. Positions of the ions off center will be considered later. Fig. 3 shows the pair NaCl inserted into the tube.

We inserted the ion pairs NaCl, LiCl, and LiF into this tube and calculated the equilibrium distances between the ions and the energetics of insertion using DFT. Table 1 gives the distances for the inserted pairs and compares them with the results of a model, described in the next section, with the bond distance of the

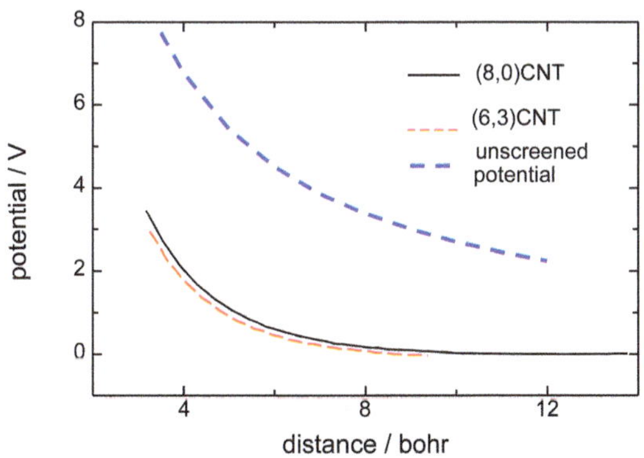

Fig. 2 Electrostatic potential generated by a Li^+ ion along the axis of a (8,0)CNT and (6,3) CNT; the ion sits at the center of the tube. For comparison, we also show the unscreened Coulomb potential.

Fig. 3 NaCl inserted into a (5,5)CNT; the small blue sphere indicates the position of Na^+, the greenish sphere is Cl^-.

diatomic molecule and the bond distance of the corresponding salts, all of which have a NaCl structure. In all cases the bond distance for the pair in the tube is somewhat larger than for the molecule. We ascribe this to the screening of the ionic charge, and will discuss this effect in greater detail below. In the salt the distance is even larger; this well known effect is caused by the repulsion from the six nearest neighbors.

As usual we define the insertion energies of the ion pairs with respect to the atoms:

$$E_{\text{ins}}^{\text{atom}} = \text{energy(tube + ion pair)} - \text{energy(tube)} - \text{energy(alkali atom)} - \text{energy(halogen atom)} \quad (1)$$

Alternatively we can consider the insertion energy of the dimer:

$$E_{\text{ins}}^{\text{dimer}} = \text{energy(tube + ion pair)} - \text{energy(tube)} - \text{energy(dimer in vacuum)} \quad (2)$$

In both equations all subsystems considered are at equilibrium.

The insertion energies for the two atoms are quite favorable (see Table 2); they are almost equal for NaCl and LiCl, but much more favorable for LiF. Obviously, the Pauli repulsion, which is lower for F^- than for Cl^-, plays an

Table 1 Equilibrium distances (in Å) for the ion pairs in the (5,5)CNT, the molecule and the salt; we also give the values calculated from a model described in the next section

Pair	Tube (DFT)	Tube (model)	Molecule	Salt
NaCl	2.45	2.54	2.36	2.82
LiCl	2.07	2.15	2.02	2.57
LiF	1.60	1.63	1.56	2.01

Table 2 Various energies associated with the insertion of ion pairs: E_{ins}^{atom} is the insertion energy referring to the insertion of the two atoms as defined by eqn (1); E_{ins}^{dimer} is the insertion energy referring to the insertion of the dimer as defined by eqn (2); E_{mol} is the binding energy of the binary molecule in the vacuum; E_{bind}^{t} is the binding energy of the pair in the tube. E_a and E_c are the insertion energies for a single anion or cation. $E_t = E_a + E_c + E_{bind}^{t}$, and should roughly equal the insertion energy E_{ins}^{atom}

Pair	E_{ins}^{atom}	E_{ins}^{dimer}	E_{mol}	E_{bind}^{t}	$E_a + E_c$	E_t
NaCl	−4.43	−0.22	−4.211	−1.36	−2.65	−4.01
LiCl	−4.96	−0.14	−4.82	−2.07	−2.73	−4.80
LiF	−6.60	−0.65	−5.95	−2.71	−3.73	−6.44

important role. In contrast, the energy gained by inserting the dimer is quite low. The reason for this can be seen when we look at the binding energy E_{bind}^{t} for the pair inside the tube, *i.e.* the gain in energy when the two ions approach each other inside the tube from infinite separation to the equilibrium distance. This is much smaller than the formation energy in vacuum. As we shall explain in greater detail in the next section, this is caused by the screening of the charge by the image charge on the walls of the tube. In other words: what the ion pair gains in energy from the image charge, it loses in terms of binding energy.

To gain an impression of the screening we consider the charge redistribution caused by the insertion. Again there are two different ways to define them. The first is to refer them to the two inserted atoms:

$$\Delta\rho_1 = \rho(\text{tube} + \text{ion pair}) - \rho(\text{tube}) - \rho(\text{alkali atom}) - \rho(\text{halogen atom}) \quad (3)$$

Alternatively, it can be referred to the dimer in vacuum:

$$\Delta\rho_2 = \rho(\text{tube} + \text{ion pair}) - \rho(\text{tube}) - \rho(\text{dimer in vacuum}) \quad (4)$$

Both charge differences are shown in Fig. 4. The first quantity shows the ionization of the two atoms and the image charge; the latter shows the screening of the dimer more clearly.

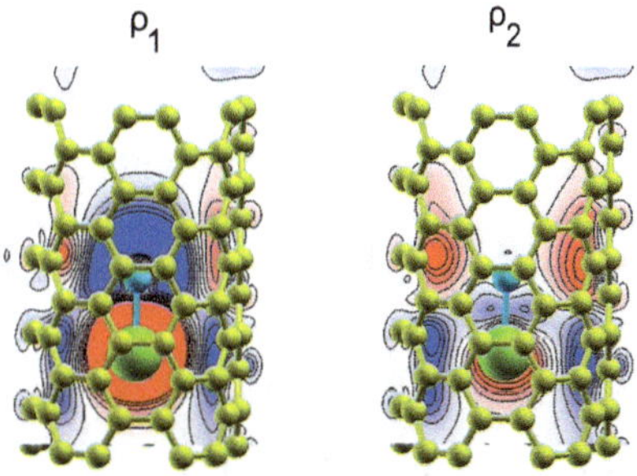

Fig. 4 Charge differences for NaCl inserted into a (5,5)CNT; the small blue sphere indicates the position of Na^+, the greenish sphere is Cl^-. The l.h.s. refers to the insertion of the two atoms, the r.h.s. to the insertion of the dimer – see eqn (3) and (4).

It is of interest to compare the energies of insertion of the ion pairs with those of the individual ions. As we have argued before, for an infinite tube we can write the energy of insertion for an alkali atom as:

$$I_c = I_1 - \Phi + E_c \tag{5}$$

where I_1 is the first energy of ionization, Φ is the work function of the tube, and E_c is the interaction energy of the ion with the tube, which contains both the image energy and the Pauli repulsion. Likewise, for a halide atom we obtain:

$$I_a = -A + \Phi + E_a \tag{6}$$

where A is the electron affinity of the atom, and E_a the interaction energy of the anion with the tube. Taking the sum we obtain:

$$I_a + I_c = I_1 - A + (E_a + E_c) \tag{7}$$

where the work function has cancelled. For a finite system with relatively few free electrons, the energy required to take away an electron is not exactly equal to the energy gained by adding an electron; so eqn (7) is an approximation, which is useful to obtain the order of magnitude for the sum of the interaction terms $(E_a + E_c)$, which is also given in Table 2. The values that we use here are for the center of the tube; this is the optimum position for the two anions considered, but not for the two cations. Therefore the repulsion energy should be determined by the anions, which is in line with the fact that $(E_a + E_c)$ are almost equal for NaCl and LiCl, but more favorable for LiF with the smaller anion.

For comparison we note that the image energy of a univalent ion at the center of the tube is:

$$E_{im} \approx \frac{-0.436}{R_{im}} \text{ in a.u.} \tag{8}$$

where R_{im} is the effective image radius of the tube, which in this case is $R_{im} = 3.8$ (a.u.) $\approx$ 2 Å, so that $E_{im} \approx -3.12$ eV. In all cases $|(E_a + E_c)|$ is much smaller than $|2E_{im}|$, so the effect of Pauli repulsion is considerable.

The sum $(E_a + E_c)$ refers to the insertion of two individual, non-interacting ions. As a first approximation, it should differ from the insertion energy of the ion pair by the binding energy E^t_{bind} for the pair in the tube:

$$E_t = E_a + E_c + E^t_{bind} \approx E^{atom}_{ins} \tag{9}$$

As can be seen from the last column in Table 2, this relation holds to first order. Deviations are caused by the fact that Pauli repulsion also deforms the electronic structure of the tube, which in turn effects the work function – in other words, the various interactions are not strictly additive. Therefore the deviation is greatest for NaCl, the largest pair of ions.

4 Extension of the Rittner model

Alkali halide molecules are good model systems for understanding ionic bonds. Therefore, they were a popular topic of research before the advent of computational chemistry, and several models have been constructed for their binding properties. The simplest model contains only the Coulomb attraction and a repulsive term,[11] but better versions include terms for the ion-induced dipole interactions and a van der Waals term. A good parametrization of the latter kind of model has been proposed by Rittner,[12] where the interaction potential is written in the form:

$$V(R) = -\frac{1}{R} - \frac{(\alpha_1 + \alpha_2)}{2R^4} - \frac{(2\alpha_1\alpha_2)}{R^7} + A\exp -\frac{R}{\rho} - \frac{C}{R^6} \tag{10}$$

where R is the separation between the two ions, α_1 and α_2 are their polarizabilities, and A and ρ are constants; atomic units have been used. The meaning of the individual terms is as follows: (1) Coulomb attraction; (2) and (3): ion-induced dipole interaction; (4) Pauli repulsion, and (5) van der Waals attraction. Within a nanotube the Coulomb attraction is strongly screened,[7] so that the first term must be adapted to the new surroundings. The charge on the ions gives rise to a compensating image charge on the walls of the tube. As we have discussed above, the resulting screened potential can be described by an effective image radius R_{im}, which has the following interpretation: the screened Coulomb potential is the same as that produced by a classical metal tube of radius R_{im}. Considering ions along the center of the tube, this potential can be written as:

$$\phi(z) = \frac{2}{R}\sum_{m=1}^{\infty}\frac{\exp(-k_m z/R)}{k_m|J_1(k_m)|^2} \tag{11}$$

where z is the distance from the center of the ion, k_{m} denotes the roots of the Bessel function $J_0(k_{\mathrm{m}}) = 0$, and J_1 is the Bessel function of first order. This formula is easily programmed, and we have used it here. It is, however, not very practical for computer simulations, and therefore it is sometimes approximated by the first and leading term – see *e.g.* ref. 13 – at larger distances. However, a better and even simpler approximation is to spread the image charge uniformly on an infinitely thin ring surrounding the ion:

$$\phi(z) \approx \frac{1}{z} - \sqrt{\frac{1}{z^2 + R^2}} \quad \text{for large } z \tag{12}$$

In order to test if the concept of screening can be applied to the ion pairs in CNTs, we have replaced the first term in eqn (10) by $-\Phi(R)$ as given by eqn (11). For explicit calculations we have taken the parameters for the various ions from ref. 14; for the effective image radius of the (5,5)CNT we have used the value of $R_{\mathrm{im}} = 2.0$ Å as determined in ref. 2. In this way we have calculated the potential energy curves for NaCl and LiCl in the nanotube; the results are shown in Fig. 5, where they are compared with the results from the DFT calculations. Near the minima the two curves agree quite well; in particular the bond distance is well represented – see also Table 1. At larger distances the agreement is not so good, and the binding energies are too low by a few tenths of an eV. We surmise that this discrepancy is caused by the ion-induced dipole terms in eqn (10), which are based on the assumption that the ion potential is unscreened. We have refrained from introducing corrections, because the original values from ref. 14 are semi-empirical. In any case the model calculations explain the great reduction in

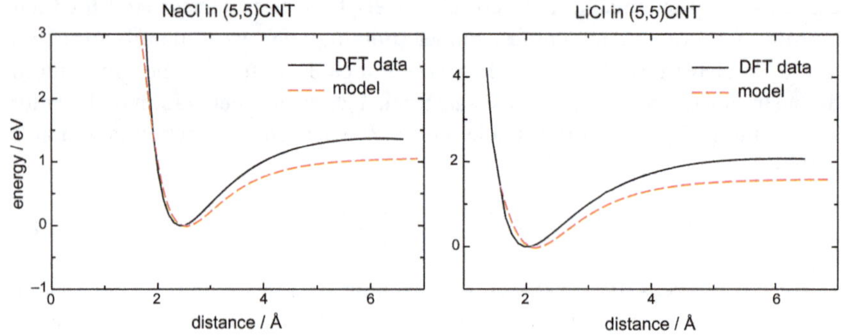

Fig. 5 Potential energy curves for the ion pairs in a (5,5)CNT; the energy at the minimum has been set to zero.

binding energy from the molecule in a vacuum (−4.211 eV for NaCl and −4.82 eV for LiCl) caused by the screening.

5 One-dimensional NaCl

Up to now we have considered only ions positioned in the center of the tube; this greatly helped in the interpretation of the results, and allowed the application of simple models. The center is indeed the optimum position for the anions considered,[3] but not for small cations like Li^+ and Na^+, which prefer to reside near the wall. So in this section we shall study a truly one-dimensional chain of NaCl in our tube, and investigate two configurations: (1) all ions lined up along the axis; (2) all ions at their optimum position; the latter gives rise to a zig-zag configuration, in which the anion is at the center and the cation is at the side. The corresponding arrangements are shown in Fig. 6.

In order to compare the insertion energies of the chains with those of the ion pairs we define:

$$\Delta E = [\text{energy(tube + chain)} - \text{energy(tube)} - m\,\text{energy(Na)} - m\,\text{energy(Cl)}]/m \quad (13)$$

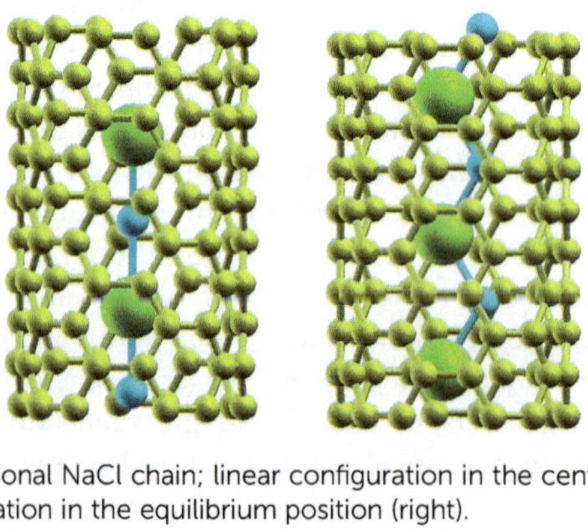

Fig. 6 One-dimensional NaCl chain; linear configuration in the center of the tube (left), and zigzag configuration in the equilibrium position (right).

where m is the number of ion pairs per unit cell. As expected, the zigzag configuration is more stable than the linear, but the difference is quite small. In both cases the insertion energy is more favorable than for the ion pair, where it is −4.01 eV. This is understandable, because here each ion has two neighbors. If the interaction of an ion with its two neighbors were additive, ΔE would be equal to the sum of the insertion energy plus the ion formation energy E^{t}_{tube} inside the tube, which is $-4.01 - 1.36$ eV $= -5.37$ eV. In fact, the binding is somewhat less favorable.

For the zigzag geometry the bond distance is slightly smaller than for the ion pair, where it is 3.45 Å. In contrast, in the linear geometry it is slightly larger, and it varies between the ion pairs, because the positions of the ions are not equivalent. In all cases the bond distance is larger than that for a molecule in vacuum and smaller than that for the three-dimensional salt (see Table 1).

Such one-dimensional salts embedded in a CNT have indeed been observed experimentally: Senga *et al.*[15] have formed a 1-D chain of CsI in a doublewalled nanotubed and imaged the system with a high-resolution transmission electron microscope. They accompanied their experimental work with DFT calculations. Their results are quite compatible with ours: they observed a binding distance of about 3.4 Å, a little larger than the value for the CsI molecule in a vacuum (3.3 Å), and smaller than for the CsI crystal. Their calculations indicate that the anion is fixed at the center, while the cation can move more freely in the axial direction. In this context the earlier work of Fan *et al.*[16] on a chain of iodine atoms embedded in a CNT is also of interest. They observed that the optimum position of iodine is near the wall, and the chains had a helical structure. These nanotubes had a diameter of 13.6 Å and were thus wider than those studied by Senga *et al.* and by us. So there is no contradiction. As we have pointed out before,[3] for sufficiently wide tubes the optimum position must always be near the wall. Finally, we mention that in earlier works by Green *et al.*[17,18] the chains embedded in CNTs were still a few atoms wide.

6 Phonon spectrum

In a recent communication one of us[19] has considered the capacity of a nanotube permeable to only one kind of ion. In this context, he also considered the phonon spectrum of the vibrations along the axis of the tube. Because of the screening of the Coulomb force it was sufficient to consider the interaction between nearest neighbors only. The problem then reduces to a textbook problem and the resulting spectrum consists of one acoustic branch. In our case the spectrum is more interesting, since we have two kinds of ions with different masses. Specifically we consider a one-dimensional salt consisting of a chain of two kinds of

Table 3 Bond distances and insertion energies for zigzag and linear geometry

Geometry	d/Å	ΔE/eV
Zigzag	2.40	−4.94
Linear	2.53–2.62	−4.87

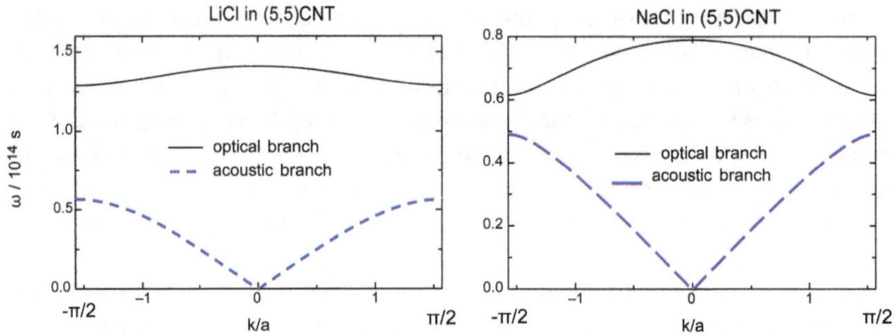

Fig. 7 Phonon spectra for LiCl and NaCl.

ions aligned along the axis of the tube. The equations of motions for the ions can be written in the form:

$$m_1 \frac{d^2 s_n^{(1)}}{dt^2} = -f s_n^{(1)} - s_n^{(2)} - s_{n-1}^{(2)} \tag{14}$$

$$m_2 \frac{d^2 s_n^{(2)}}{dt^2} = -f s_n^{(2)} - s_n^{(1)} - s_{n+1}^{(1)} \tag{15}$$

where the superscripts (1) and (2) refer to the two kinds of ions, s denotes a deviation from the equilibrium position, m a mass, and f is the force constant. Again, this is a textbook problem that is explored on a number of webpages – see *e.g.* ref. 20. The spectrum now has two branches given by:

$$\omega_+{}^2 = \frac{f}{\tilde{m}} \pm f \sqrt{\frac{1}{\tilde{m}} - \frac{4}{m_1 m_2} \sin^2 \frac{\boldsymbol{k}a}{2}} \tag{16}$$

where $\tilde{m}$ is the reduced mass, a the lattice constant and $\boldsymbol{k}$ the wave vector. Because of the different masses of the two ions the spectrum now has an acoustic branch, which corresponds to the minus sign in eqn (16), and an optical branch corresponding to the plus sign. Fig. 7 shows the spectra for LiCl and NaCl, aligned along the center of the tube, as examples. Since the lithium ion has a smaller mass, the frequencies are higher for LiCl than for NaCl. Such phonon spectra have been measured for thin films of salts,[21] but not yet for one-dimensional salts.

7 Conclusions

In this work we have considered the simultaneous insertion of an alkali and a halide atom into a narrow (5,5)CNT. The energies gained by inserting two individual atoms are quite high, of the order of several eV. However, the gain by inserting the alkali–halide dimer from the vacuum is quite small, less than one eV. One of the reasons for this is the Pauli repulsion, which the larger halide ions experience in particular. In addition, the image charge formed on the wall of the tube has two contrary effects: the interaction of the ions with the image charge favors insertion, but on the other hand it weakens the binding between the two ions.

The same principles govern the formation of a one-dimensional salt embedded in the tube. In particular, we have studied two configurations for the NaCl chain: aligned along the center, and the optimum configuration in which the cation is displaced towards the wall. The energy difference between these configurations is quite small, indicating that the cation has some freedom to move radially. The phonon spectrum of such 1D-salts shows both an acoustic and an optical branch.

8 Technical details of the DFT calculation

The DFT calculations have been performed in the same way as in ref. 3. We briefly summarize the main points. The correlation and exchange functionals were described within the generalized gradient approximation (GGA) in the Perdew, Burke and Ernzerhof (PBE) flavor.[22] The electron–ion interactions were represented through ultrasoft pseudopotentials,[23] and a plane wave basis set was used to describe the valence electrons. The basis set was expanded to a kinetic energy cutoff of 400 eV (450 eV for the density). Brillouin zone integration was performed using the gamma point. The carbon nanotubes were studied by using VASP[24] code. Periodic boundary conditions were used in order to correctly represent the infinite nanotubes. A separation of 15 Å between neighboring systems was imposed in the directions perpendicular to the tubes and in order to avoid interactions between neighbor images. All the nanotubes and alkali or halide atoms were fully relaxed until the total forces were less than 40 meV A^{-1}. All the systems used were neutral, but we confirmed the loss or gain of charge in the central atom by using the Bader analysis method.[25] We used a dipole-correction scheme[26] in the systems that are not completely symmetric, in order to have a well-defined vacuum potential. The electrostatic part (ionic and Hartree potentials) of the local potential was calculated, but the exchange–correlation was not added.

Acknowledgements

Financial support by the Deutsche Forschungsgemeinschaft under Schm344/48-1 is gratefully acknowledged. We also thank the DAAD-DAHZ for funding the exchange between Ulm University and the Universidad del Litoral, Santa Fe. P. Q. thanks PICT-2014-1084, CONICET, UNL and Santa Fe Science Technology and Innovation Agency (ASACTEI, grant 00010-18-2014) for continuous support. W. S. thanks CONICET for continuous support.

References

1 A. Goduljan, F. Juarez, L. Mohammadzadeh, P. Quaino, E. Santos and W. Schmickler, *Electrochem. Commun.*, 2014, **45**, 48.
2 L. Mohammadzadeh, A. Goduljan, F. Juarez, P. Quaino, E. Santos and W. Schmickler, *Electrochim. Acta*, 2015, **162**, 11.
3 L. Mohammadzadeh, A. Goduljan, F. Juarez, P. Quaino, E. Santos and W. Schmickler, *ChemPhysChem*, 2016, **17**, 78.

4 J. Chmiola, G. Yushin, Y. Gogotsi, C. Portet, P. Simon and P. L. Taberna, Anomalous increase in carbon capacitance at pore sizes less than one nanometer, *Science*, 2006, **313**, 1760.
5 C. Largeot, C. Portet, J. Chmiola, P.-L. Taberna, Y. Gogotsi and P. Simon, *J. Am. Chem. Soc.*, 2008, **130**, 2730.
6 N. Jäckel, M. Rodner, A. Schreiber, J. Jeongwook, M. Zeiger, M. Aslan, D. Weingarth and V. Presser, *J. Power Sources*, 2016, **326**, 660.
7 S. Kondrat and A. Kornyshev, *J. Phys.: Condens. Matter*, 2011, **23**, 022201.
8 A. Lee, S. Kondrat and A. Kornyshev, *Phys. Rev. Lett.*, 2014, **113**, 048701.
9 A. Kornyshev, *Faraday Discuss.*, 2013, **164**, 117.
10 C. Rochester, A. Lee, G. Pruessner and A. Kornyshev, *ChemPhysChem*, 2013, **14**, 4121.
11 N. W. Ashcroft and N. D. Mermin, *Solid State Physics*, Saunders College, Philadelphia, 1976.
12 E. S. Rittner, *J. Chem. Phys.*, 1951, **19**, 1030.
13 A. Kornyshev, *Faraday Discuss.*, 2013, **164**, 117.
14 T. Gowda and S. Benson, *J. Phys. Chem.*, 1982, **86**, 847.
15 R. Senga, H.-P. Komsa, Z. Liu, K. i. Hirose-Takai, A. V. Krasheninnikov and K. Suenaga, *Nat. Mater.*, 2014, **13**, 1050.
16 X. Fan, E. C. Dickey, P. C. Eklund, K. A. Williams, L. Grigorian, R. Buczko, S. T. Pantelides and S. J. Pennycook, *Phys. Rev. Lett.*, 2000, **84**, 4621.
17 R. R. Meyer, J. Sloan, R. E. Dunin-Borkowski, A. I. Kirkland, M. C. Novotny, S. R. Bailey, J. L. Hutchison and M. L. H. Green, *Science*, 2002, **289**, 1324.
18 J. Sloan, A. I. Kirkland, J. L. Hutchison and M. L. H. Green, *Chem. Commun.*, 2002, 1319.
19 W. Schmickler, *Electrochim. Acta*, 2015, **173**, 91.
20 www.itp.tu-berlin.de/fileadmin/a3233/upload/SS12/TheoFest2012/Kapitel/Chapter7.pdf.
21 T. G. Fiske and L. B. Coleman, *Phys. Rev. B: Condens. Matter Mater. Phys.*, 1992, **45**, 1414.
22 J. P. Perdew, K. Burke and M. Ernzerhof, *Phys. Rev. Lett.*, 1996, **77**, 3865.
23 D. Vanderbilt, *Phys. Rev. B: Condens. Matter Mater. Phys.*, 1990, **41**, 7892.
24 (*a*) G. Kresse and J. Hafner, *Phys. Rev. B: Condens. Matter Mater. Phys.*, 1993, **47**, 558; (*b*) G. Kresse and J. Hafner, *Phys. Rev. B: Condens.Matter Mater. Phys.*, 1994, **49**, 14251.
25 (*a*) R. F. W. Bader, P. E. Cade and P. M. Beddall, *J. Am. Chem. Soc.*, 1971, **93**, 3095; (*b*) G. Henkelman, A. Arnaldsson and H. Jonsson, *Comput. Mater. Sci.*, 2006, **36**, 354.
26 L. Bengtsson, *Phys. Rev. B: Condens. Matter Mater. Phys.*, 1999, **59**, 12301.

Faraday Discussions

PAPER

Theory of electrostatic phenomena in water-filled Pt nanopores

Jun Huang,[ab] Jianbo Zhang*[ac] and Michael Eikerling*[b]

Received 26th April 2016, Accepted 28th April 2016
DOI: 10.1039/c6fd00094k

The emerging field of nanoprotonics is concerned with controlling proton distribution and transport in nanoporous media. These phenomena, dictated by the surface charging properties of the host medium, are of vital importance in porous electrodes for fuel cells, electrolysers, supercapacitors and nanofluidic devices. In this theoretical study, we explore the interplay of the metal charging relation with the proton density and oxygen reduction activity in a water-filled nanopore with walls made of platinum. We exploit a non-monotonic charging behavior derived from a refined structural model of the Pt-solution interface. This charging relation replaces the oversimplified linear relation that has been widely used in practical applications. The water-filled pore, with one opening interfacing with a polymer electrolyte membrane as a proton source, always possesses negative surface charge in the potential range of 0–1.0 V (RHE). Therefore, its proton conductivity can be several orders higher than that of pure water. We obtain an analytical expression for the oxygen reduction activity of the nanopore and parameterize it using the polarization data of an ionomer-free thin-film Pt electrode. The structure *vs.* performance relation of the water-filled Pt nanopore is examined.

1 Introduction

Ion transport confined in nanoscale pores has been gaining wide attention in the emerging fields of energy harvesting and conversion,[1,2] nanofluidic devices[1,3] and biochemical sensing.[4,5] The surface charge repels ions of the same sign while it attracts ions of the opposite sign. Therefore, the surface charging properties of the nanoporous host material largely dictate its functionality, such as ion selectivity and conductivity. In the realm of polymer electrolyte fuel cells and water electrolysers, water-filled nanopores with walls made of Pt constitute the active media for the oxygen reduction and evolution reactions.[6–10] In these systems, protons are supplied or transported away *via* structural diffusion in

[a]*Department of Automotive Engineering, State Key Laboratory of Automotive Safety and Energy, Tsinghua University, Beijing 100084, China*
[b]*Department of Chemistry, Simon Fraser University, Burnaby, BC V5A 1S6, Canada. E-mail: meikerl@sfu.ca*
[c]*Beijing Co-innovation Center for Electric Vehicles, Beijing Institute of Technology, Beijing 100081, China*

bulk water. The fundamental questions are: what is the proton density in the nanopore and how does it vary with materials properties and electrode potential? In answering these questions, the charging relation of the host material, *viz.* a nanoporous medium with Pt deposited at the pore walls, is the main property of interest.

The potential of zero charge (pzc) has been widely employed to parameterize surface charging phenomena.[11] For the case of Pt in the potential range where surface oxide formation prevails, the pzc concept is highly oversimplified and insufficient. As it turns out, the effective surface charge density is strongly dependent on the extent of surface oxide formation and the alignment of interfacial water molecules.[12] For a planar Pt electrode, the modified model of the electrochemical interface presented in ref. 12 revealed a non-monotonic charging behavior with consecutive transitions from negative surface charge to positive surface charge and again to negative surface charge upon increasing the metal phase potential through the region of oxide formation. This finding invalidates the role of the pzc as a key descriptor of metal charging phenomena in electrochemical systems.

The present study investigates the interplay between the charging relation, proton conductivity and electrocatalytic performance of a cylindrical water-filled Pt nanopore. In our previous work, we considered an extended planar metal electrode that was immersed in an electrolytic solution with fixed ion concentration in the bulk phase.[12] The salient feature of a water-filled nanopore is that the ion density distribution in it could be entirely governed by double layer effects. In addition, the nanopore geometry brings into play the pore radius and length as key variables and it implies more complex boundary conditions at the nanopore openings towards the polymer electrolyte membrane (PEM) on one side and a gas transport medium (GTM) on the other side.

In the remainder of this article, we formulate the problem, and then simplify it based on rational assumptions corroborated by numerical calculations. Analytical solutions are then used to investigate the electrostatic properties, including the proton distribution and proton conductivity, of a water-filled Pt nanopore. Afterwards, an analytical expression is derived and parameterized for the ORR current of the nanopore. Finally, the structure *vs.* performance relation of the water-filled Pt nanopore is investigated.

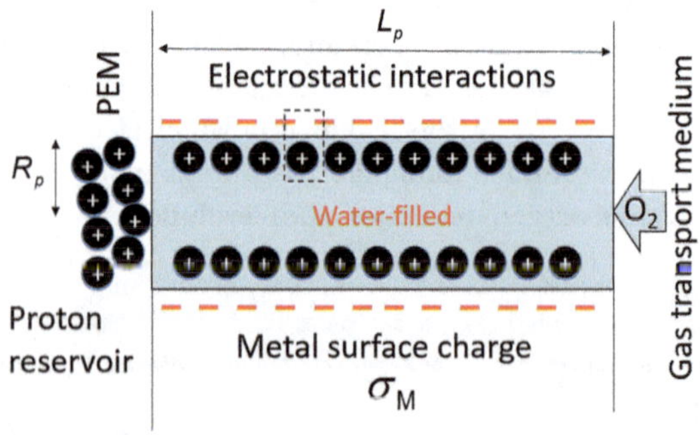

Fig. 1 Schematic of a water-filled Pt nanopore.

2 Problem formulation

The model system is illustrated in Fig. 1. It consists of a straight cylindrical nanopore of length L_p and radius R_p. The walls of the pore are made of Pt. One opening of the nanopore, at $z = 0$, forms an interface with a PEM, which acts as a proton reservoir. The other pore opening at $z = L_p$ interfaces with a GTM through which oxygen is supplied. As we will see, end effects due to proton spillover at the PEM interface are negligible and the axial distribution of the proton concentration is nearly uniform. The model is formulated and solved under steady state conditions.

2.1 Poisson–Nernst–Plank equations

Proton distribution and transport in the pore are described by the Poisson–Nernst–Planck (PNP) equations,

$$\nabla\cdot(\varepsilon_S\nabla\phi_S) = -F\sum_i z_i c_i, \tag{1.1}$$

$$\frac{\partial c_i}{\partial t} = -\nabla\cdot\mathbf{N}_i = 0, \tag{1.2}$$

$$\mathbf{N}_i = -D_i\left(\nabla c_i + \frac{z_i F c_i}{RT}\nabla\phi_S\right), \tag{1.3}$$

where R is the gas constant, T the temperature, F the Faraday constant, ε_S the dielectric constant of water (solution), ϕ_S the solution phase potential, and z_i, c_i, N_i, D_i refer to the charge number, concentration, flux and diffusion coefficient of species i (i = H^+, OH^-, O_2), respectively.

The only ionic species present in water in the nanopore are protons (H^+, in the form of hydronium ions) and hydroxide ions (OH^-). At steady state, concentrations of H^+ and OH^- are related by

$$c_{H^+}c_{OH^-} = K_w, \tag{2}$$

where K_w is the dissociation constant of water. The independent variables to be solved for are c_{H^+}, c_{O_2} and ϕ_S.

Boundary conditions to close the PNP equations are,

(1) Symmetrical conditions at the axis, $r = 0$,

$$\frac{\partial c_{H^+}}{\partial r} = 0, \quad \frac{\partial c_{O_2}}{\partial r} = 0, \quad \frac{\partial \phi_S}{\partial r} = 0. \tag{3.1}$$

(2) At the Pt wall, $r = R_p$, the fluxes of oxygen and proton are determined by the local ORR current, j_{ORR}, and the potential gradient is related to the free metal surface charge density, σ_M, at the Pt wall,

$$D_{H^+}\frac{\partial c_{H^+}}{\partial r} = -\frac{j_{ORR}}{F}, \quad D_{O_2}\frac{\partial c_{O_2}}{\partial r} = -\frac{j_{ORR}}{4F}, \quad \varepsilon_S\frac{\partial \phi_S}{\partial r} = \sigma_M. \tag{3.2}$$

(3) At the opening to the PEM, $z = 0$, c_{H^+} is assumed to be the volume-averaged proton concentration of the bulk PEM, $c^0_{H^+}$, ϕ_S is assigned as the reference potential, and, as usual, oxygen cannot penetrate the PEM,

$$c_{H^+} = c^0_{H^+}, \quad D_{O_2}\frac{\partial c_{O_2}}{\partial z} = 0, \quad \phi_S = 0. \tag{3.3}$$

(4) At the opening to the GTM, $z = L_p$, c_{O_2} is given as the constant oxygen concentration at the GTM/nanopore interface, $c^0_{O_2}$, and proton flux and electric field are zero,

$$D_{H^+}\frac{\partial c_{H^+}}{\partial z} = 0, \quad c_{O_2} = c^0_{O_2}, \quad \frac{\partial \phi_S}{\partial z} = 0. \tag{3.4}$$

2.2 Metal charging relation

As in eqn (3.2), the potential gradient is related to σ_M, ensuring that the total charge in the nanopore is commensurate to the metal surface charge. Another requirement is that the potential must be continuous from the interior of the metal phase to the bulk water phase. Fig. 2 depicts the schematic of the metal–water interface with the potential profile plotted on the right hand side.

The details of this structural model are provided in ref. 12. Here, we briefly discuss the essential features. On the solution side, interfacial water molecules with field-dependent orientation are located at the inner Helmholtz plane (IHP), followed by counter-ions lining up at the outer Helmholtz plane (OHP), and a diffuse layer that stretches towards the bulk water phase. On the metal side, an oxide layer forms at the metal surface with the oxide coverage being a function of the electrode potential and solution pH. The oxide layer is abstracted as a Helmholtz-type capacitor with positively charged metal atoms and negatively

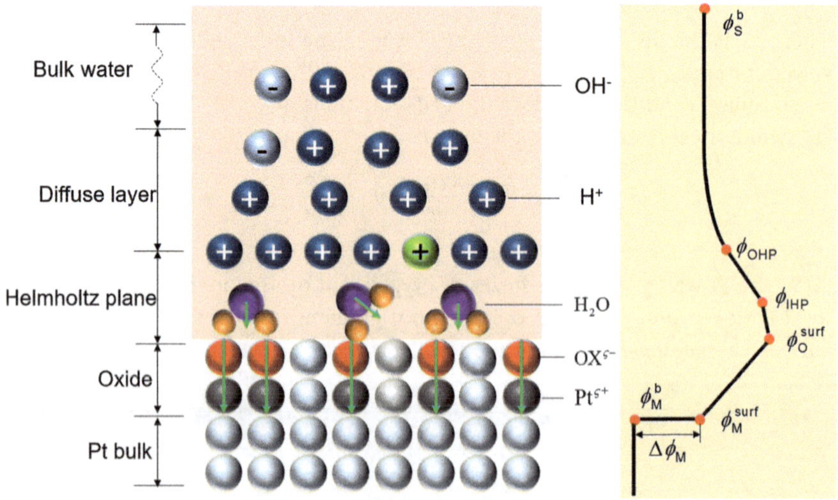

Fig. 2 Schematic of the Pt–solution interface.

charged oxygen atoms lining up rigidly at two planes with a distance corresponding to the thickness of the oxide layer. The formation of surface oxide commences at a certain potential. Below this potential, we set the thickness of the oxide layer to be zero.

The oxidation reaction is,

$$H_2O_{ad} \leftrightarrow O(\upsilon H)_{ad}(Pt) + (2-\upsilon)H^+ + (2-\upsilon)e^-, \tag{4}$$

where $\upsilon = 0$ or 1.[12] Equalizing the electrochemical potentials of involved species on the two sides of eqn (4) leads to,

$$(2-\upsilon)\frac{F\left(\phi_M^b - \phi_S^{surf} - E_{OX}^0\right)}{RT} + 2.3(2-\upsilon)pH^{surf} = \ln\left(\frac{\theta_{OX}}{1-\theta_{OX}}\right) + \frac{\xi_{OX}\theta_{OX}}{RT}, \tag{5}$$

where ϕ_S^{surf} is the interfacial potential in the water phase, ϕ_M^b the potential in the bulk Pt phase, θ_{OX} the normalized coverage of Pt oxide, E_{OX}^0 the equilibrium potential of reaction eqn (4) at standard state, pH^{surf} the surface pH, and ξ_{OX} a lateral interaction parameter. The Frumkin correction term $\xi_{OX}\theta_{OX}$ accounts for deviations from Langmuir-type adsorption behavior.

The oxide dipole moment, μ_{PtO}, is written as,

$$\mu_{PtO} = N_{tot}(\theta_{OX}\varsigma e\delta_{PtO}), \tag{6}$$

where N_{tot}, (m^{-2}), is the Pt atom density, approximated by $4\Big/\left(\sqrt{3}a_{Pt}^2\right)$ (a_{Pt} is the lattice constant), δ_{PtO} the thickness of the Pt oxide layer, ς the partial charge number of ions in the oxide layer and e is the elementary charge.

Considering a constant potential drop at the metal surface due to the electron spillover and the potential change in the oxide layer due to μ_{PtO} and σ_M, the potential at the oxide surface, ϕ_O^{surf}, is,

$$\phi_O^{surf} = \phi_M^b - \Delta\phi_M - \frac{\mu_{PtO}}{\varepsilon_{PtO}} - \frac{\sigma_M\delta_{PtO}}{\varepsilon_{PtO}}. \tag{7}$$

The OHP potential, ϕ_{OHP}, is given by,

$$\phi_{OHP} = \phi_M^b - \Delta\phi_M - \frac{\mu_{PtO}}{\varepsilon_{PtO}} + \frac{N_{tot}\mu_w}{\varepsilon_{IHP}}\tanh(X) - \sigma_M\left(\frac{\delta_{PtO}}{\varepsilon_{PtO}} + \frac{\delta_{OHP}}{\varepsilon_{OHP}} + \frac{\delta_{IHP}}{\varepsilon_{IHP}}\right), \tag{8}$$

where μ_w is the water dipole moment and the variable X is calculated using the two-state water dipole model,[13]

$$\left(\frac{0.6}{\pi\delta_{IHP}^2} + N_{tot}\right)\frac{N_A\mu_w^2}{\delta_{IHP}\varepsilon_{IHP}RT}\tanh(X) - X = \frac{N_A\mu_w\sigma_M}{\varepsilon_{IHP}RT}, \tag{9}$$

with Avogadro's number N_A.

On the solution side, given a certain σ_M, ϕ_{OHP} can be obtained by solving the PNP equations. Then, eqn (8) results in a new expression for σ_M. In short, σ_M can be determined by iterative solution to fulfill the potential continuity condition. Under certain assumptions, the PNP equations can be solved analytically, and an analytical expression for σ_M can be found.

2.3 Oxygen reduction reactions

Fig. 3 illustrates the complex interplay between the surface charging effects, oxide formation and surface potential and solution pH. In short, the surface conditions, *viz.* solution pH and potential, determine the oxide formation, then, the surface oxide dipole would impact σ_M, which in turn dictates the distribution of potential and proton concentration in the nanopore. The ORR current, j_{ORR}, as a term in the boundary conditions in eqn (3.2), is regulated by these properties according to the Butler–Volmer (BV) equation,

$$j_{ORR} = j^0_{ORR}\left(\frac{c^{surf}_{O_2}(z)}{c^0_{O_2}}\right)^{\gamma_{O_2}}\left(\frac{c^{surf}_{H^+}(z)}{c^0_{H^+}}\right)^{\gamma_{H+}}\exp\left(-\frac{\alpha_c F}{RT}\left(\phi^b_M - \phi^{surf}_S - E^0_{ORR}\right)\right), \quad (10)$$

where γ_{O_2}, γ_{H^+} are the reaction orders of oxygen and protons, respectively, $c^0_{O_2}$ and $c^0_{H^+}$ the concentrations of oxygen and protons under reference conditions that correspond to j^0_{ORR}, and E^0_{ORR} and α_c are the equilibrium potential and the effective transfer coefficient of the ORR, respectively.

2.4 Summary of assumptions

The assumptions are:

(1) Structure: the nanopore is straight without tortuosity; the metal wall is smooth and continuous; intrinsic properties along the nanopore are homogeneous.

(2) Solution: the dilute solution limit applies as the proton concentration remains well under 0.1 M, as we will see below; the dielectric constant of water is constant.

(3) Water dipole: water molecules with field-dependent orientations reside at the IHP; we employ the two-state model of ref. 13, with water molecules found in either H-up or H-down orientation.

(4) Oxide layer: all oxide anions line up rigidly at the outer side of the oxide layer, forming a Helmholtz-type double layer capacitor.

(5) Metal: the potential drop due to electron spillover at the metal surface is constant.

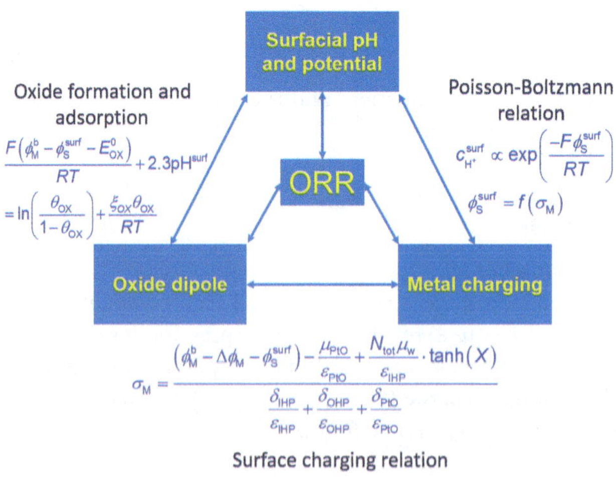

$$\frac{F(\phi^b_M - \phi^{surf}_S - E^0_{OX})}{RT} + 2.3\mathrm{pH}^{surf} = \ln\left(\frac{\theta_{OX}}{1-\theta_{OX}}\right) + \frac{\xi_{OX}\theta_{OX}}{RT}$$

$$c^{surf}_{H^+} \propto \exp\left(\frac{-F\phi^{surf}_S}{RT}\right) \qquad \phi^{surf}_S = f(\sigma_M)$$

$$\sigma_M = \frac{\left(\phi^b_M - \Delta\phi_M - \phi^{surf}_S\right) - \frac{\mu_{PtO}}{\varepsilon_{PtO}} + \frac{N_{tot}\mu_w}{\varepsilon_{IHP}}\cdot\tanh(X)}{\frac{\delta_{IHP}}{\varepsilon_{IHP}} + \frac{\delta_{OHP}}{\varepsilon_{OHP}} + \frac{\delta_{PtO}}{\varepsilon_{PtO}}}$$

Fig. 3 Schematic of the factors affecting the ORR and their interrelations. Note that the surface charging relation is restricted to an extended surface case as in ref. 12.

(6) Two interfaces at the nanopore openings: the proton concentration at the PEM interface is constant and the potential at this position is set to zero, as a reference; the concentration of oxygen at the interface with the GTM is constant; the proton flux at the interface with the GTM is zero; the oxygen flux at the PEM interface is zero.

3 Numerical simulation and analytical approximation

3.1 Model parameterization

Parameters used in this model are listed in Table 1. Interfacial parameters for the metal surface charging relation are adopted from ref. 12. In short, we determined δ_{PtO} and ς from DFT calculations and found E^0_{OX} and ξ_{OX} by fitting the curve of the total surface charge density *versus* $\phi^{\text{b}}_{\text{M}}$ using experimental data from ref. 14. Other parameters of the electrified interface have their typical values according to the literature, see ref. 12. The structural parameters and transport properties of the nanopore are adopted from ref. 8. The ORR kinetic parameters are discussed in Section 4. For the remaining parameters, we used typical values from the literature.

3.2 Numerical solution

The PNP equations are solved in COMSOL, coupled with the metal surface charging relation solved in MATLAB. More details about the numerical calculation are given in Appendix A.1.

3.3 Analytical solution

The full 2D problem can be reduced to a 1D + 1D problem by invoking two assumptions. The first assumption concerns the axial distributions of c_{H^+} and ϕ_{S}, which are considered to be uniform along the nanopore. This assumption is valid when the nanopore is completely in the electrochemical doubler layer region, in which case the distributions of c_{H^+} and ϕ_{S} are dictated by σ_{M} which is uniform at the Pt wall. In the radial direction, the Boltzmann relation is used to correlate c_{H^+} and ϕ_{S}. This correlation is usually applicable in nanoscale geometries since the magnitude of the diffusion flux and of electromigration are several orders larger than the magnitude of the net flux of ions at the interface.

The second assumption concerns the radial distribution of c_{O_2}, which is considered to be uniform. In other words, only the variation in c_{O_2} in axial direction is considered. This assumption is rationalized by calculating the difference in c_{O_2} between the nanopore center and the metal wall. Assuming a linear concentration distribution in radial direction, we have $\Delta c_{\text{O}_2} = (R_{\text{p}} j_{\text{ORR}})/(4FD_{\text{O}_2})$ with the oxygen diffusion coefficient in water, D_{O_2}. Considering a current density of $j_{\text{ORR}} = 1\ \text{A}\ \text{cm}_{\text{Pt}}^{-2}$ gives $\Delta c_{\text{O}_2} = 0.01\ \text{mol}\ \text{m}^{-3}$, which is negligibly small.

Using the parameters listed in Table 1, we solved the problem numerically at an electrode potential of 0.2 V, corresponding to high current density operation with large gradients of c_{H^+}, ϕ_{S} and c_{O_2}. This potential is chosen to examine the above two assumptions for an extreme scenario. Fig. 4(a) shows that c_{H^+} is almost uniform in the axial direction. In addition, c_{H^+} has larger values near the Pt wall

Table 1 A list of model parameters

Category	Item	Value	Note
General	Gas constant, R	8.314 J K^{-1} mol^{-1}	Constant
	Faraday constant, F	96 485 C mol^{-1}	Constant
	Temperature, T	353.15 K	Typical value
	Elementary charge, e	$1.60217662 \times 10^{-19}$ C	Constant
	Avogadro's number, N_A	6.02×10^{23} mol^{-1}	Constant
	Water dipole moment, μ_w	3.1 D	Typical value
	Dissociation constant of water, K_w	1×10^{-8} mol^2 m^{-6}	Constant
	Pt atom density, N_{tot}	1.5×10^{19} m^{-2}	Calculated using $\alpha_{Pt} = 3.92$ Å
	Vacuum permittivity, ε_0	8.85×10^{-11} F m^{-1}	Constant
Nanopore structure	Radius, R_p	10 nm	Typical value
	Length, L_p	100 nm	Typical value
Oxide formation	Equilibrium potential of Pt oxidation, E^0_{OX}	0.716 V	Ref. 12
	Lateral interaction parameter of oxide adsorption, ξ_{OX}	$64.2(1 - \theta_{OX})$ kJ mol^{-1}	
Parameters of the interface	Permittivity of the oxide layer, ε_{PtO}	$30.0\varepsilon_0$	Ref. 12
	Permittivity of the IHP, ε_{IHP}	$6.0\varepsilon_0$	
	Permittivity of the IHP, ε_{OHP}	$30.0\varepsilon_0$	
	Permittivity of bulk water, ε_S	$78.5\varepsilon_0$	
	Thickness of the oxide layer, δ_{PtO}	0.2 nm	
	Charge number of oxide dipole, ς	0.8	
	Thickness of the IHP, δ_{IHP}	0.275 nm	
	Thickness of the OHP, δ_{OHP}	0.515 nm	
	Potential drop at Pt surface, $\Delta\phi_M$	0.3 V	
ORR parameters	Equilibrium potential, E^0_{ORR}	1.203 V	Ref. 8
	Transfer coefficient, α_c	1.0	
	Oxygen reaction order, γ_{O_2}	1.0	
	Proton reaction order, γ_{H^+}	1.1	Fitted
	Exchange current density, j^0_{ORR}	2.3×10^{-3} A m_{Pt}^{-2}	
	Oxygen concentration at the GTM opening, $c^0_{O_2}$	0.32 mol m^{-3}	Ref. 8
	Proton concentration a the PEM opening, $c^0_{H^+}$	1250 mol m^{-3}	
Transport properties	Diffusion coefficient of H^+, D_{H^+}	1.8×10^{-8} m^2 s^{-1}	
	Diffusion coefficient of O_2, D_{O_2}	1.0×10^{-8} m^2 s^{-1}	

because σ_M is negative at this potential. More clearly, Fig. 4(b) indicates that c_{H^+} has a significant gradient at the PEM opening, followed by an almost uniform distribution with less than 1% variation outside of the PEM interfacial region that is confined to be within 5 nm.

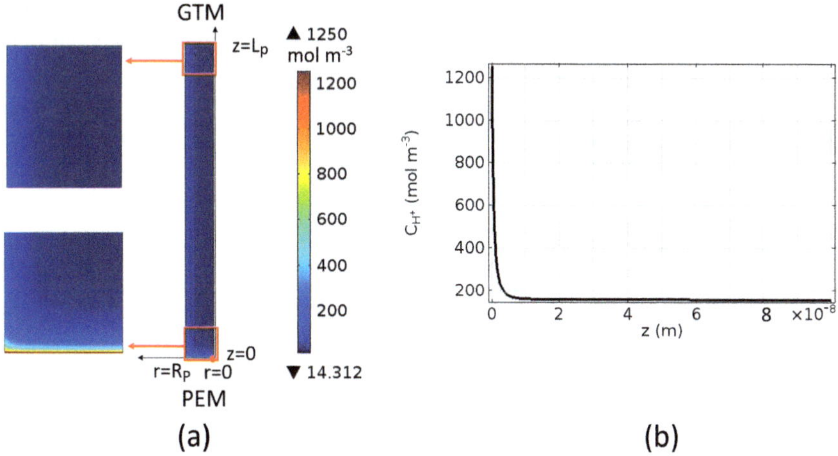

Fig. 4 Distribution of proton concentration in the water-filled Pt nanopore at 0.2 V: (a) 2D distribution; (b) distribution along the Pt wall.

Fig. 5(a) shows the distribution of c_{O_2}, revealing that c_{O_2} has a nearly uniform distribution in the radial direction and exhibits a strong variation in the axial direction. Fig. 5(b) further demonstrates that the difference in c_{O_2} between the nanopore centre and the Pt wall is negligible.

Based on the above two assumptions, the problem is simplified to a 1D + 1D problem, where the first part refers to the variation of proton density and electric potential in the radial direction and the second part refers to the variation of the oxygen concentration in the axial direction. In this scenario, the problem can be solved analytically.

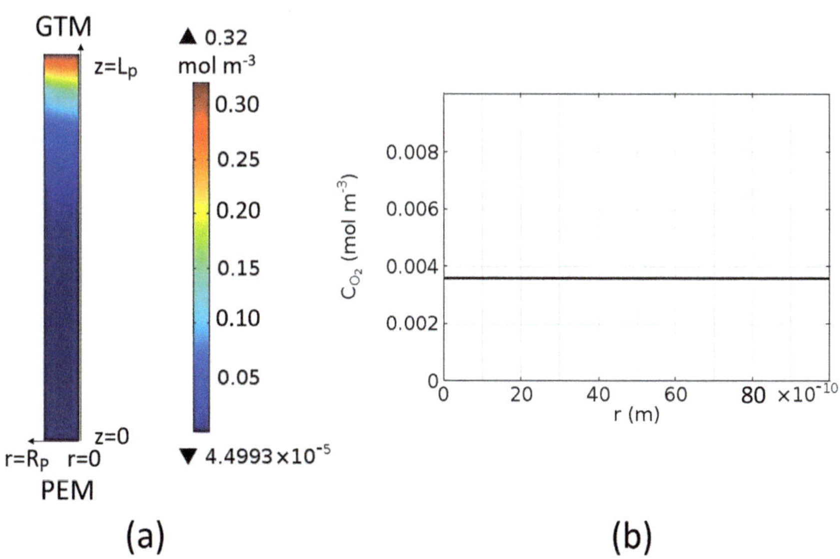

Fig. 5 Distribution of oxygen concentration in the water-filled Pt nanopore at 0.2 V: (a) 2D distribution; (b) radial oxygen distribution at half height of the nanopore.

Analytical expression for the surface charge density. Using the Boltzmann relation translates the PNP equations to the Poisson–Boltzmann (PB) equation:

$$-\frac{\partial}{\partial r}\left(\varepsilon_S r\frac{\partial \phi_S}{\partial r}\right) = Fc_{H^+}{}^{c}\exp\left(-\frac{F}{RT}\left(\phi_S - \phi_S{}^{c}\right)\right), \tag{11}$$

where the superscript '*c*' denotes the variables at the nanopore center.

The distribution of ϕ_S can be obtained by solving eqn (11),

$$\phi_S - \phi_S{}^{c} = \frac{2RT}{F}\ln\left(1 - \frac{r^2}{8\lambda_D{}^2}\right), \tag{12}$$

and the distribution of c_{H^+} is given by,

$$c_{H^+} = \left(1 - \frac{r^2}{8\lambda_D{}^2}\right)^{-2} c_{H^+}{}^{c}, \tag{13}$$

where λ_D is the Debye length,

$$\lambda_D = \frac{\varepsilon_S RT}{F^2 c_{H^+}{}^{c}}, \tag{14}$$

and the central proton concentration, $c_{H^+}{}^{c}$, is expressed as,

$$c_{H^+}{}^{c} = \frac{-8\sigma_M}{4FR_p - \dfrac{F^2 R_p{}^2}{RT\varepsilon_S}\sigma_M}. \tag{15}$$

Substituting eqn (12) into (8) leads to,

$$\begin{aligned} &-\frac{RT}{F}\ln\left(\frac{-2\sigma_M}{FR_p c^0_{H^+}}\left(1 - \frac{FR_p}{4\varepsilon_S RT}\sigma_M\right)\right) + \sigma_M\left(\frac{\delta_{PtO}}{\varepsilon_{PtO}} + \frac{\delta_{OHP}}{\varepsilon_{OHP}} + \frac{\delta_{IHP}}{\varepsilon_{IHP}}\right) \\ &= \left(E_{app} - \Delta\phi_M - \frac{\mu_{PtO}}{\varepsilon_{PtO}}\right) + \frac{N_{tot}\mu_w}{\varepsilon_{IHP}}\tanh(X), \end{aligned} \tag{16}$$

where E_{app} is the applied electrode potential *versus* the reference potential that is defined at the PEM opening. σ_M can be determined using eqn (16) concomitantly with eqn (5) and (9). Fig. 6 compares σ_M calculated using the numerical model of eqn (1), (5) and (8) and using the analytical expression in eqn (16) in the potential range from 0 to 1.0 V. Note that, in Fig. 6, water dipole moment and ORR current are not considered in either numerical or analytical calculations, *viz.* we set X in eqn (16) to zero to simplify the calculation. Quantitative agreement supports the simplifications based on the above assumptions.

Analytical expression for the ORR current. Based on the second assumption, the controlling equation of oxygen transport is simplified to

$$\frac{\partial^2 c_{O_2}}{\partial z^2} = \Lambda^2 c_{O_2}, \tag{17}$$

with

$$\Lambda^2 = \frac{j^0_{ORR}}{2FR_p D_{O_2} c^0_{O_2}}\left(\frac{c^{surf}_{H^+}(z)}{c^0_{H^+}}\right)^{\gamma_{H^+}}\exp\left(-\frac{\alpha_c F}{RT}\left(\phi^b_M - \phi^{surf}_S - E^0_{ORR}\right)\right). \tag{18}$$

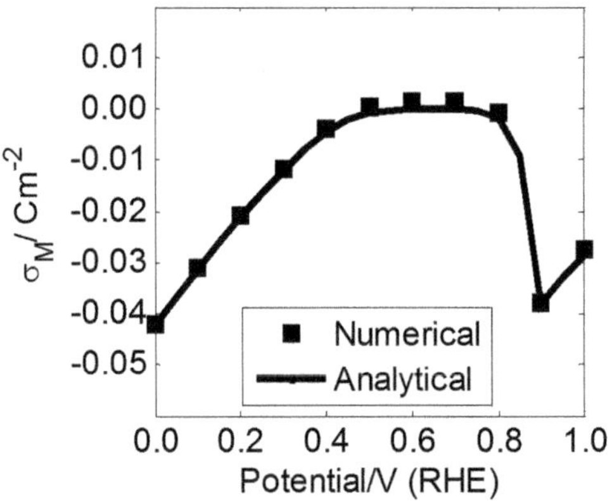

Fig. 6 Comparison between the free metal surface charge density calculated by the numerical model and that calculated by the analytical model.

Closed by the boundary conditions in eqn (3), eqn (17) results in,

$$c_{O_2}(z) = c_{O_2}^0 \frac{\cosh(\Lambda z)}{\cosh(\Lambda L_p)}. \tag{19}$$

Then, the ORR current generated by the nanopore is given by,

$$J_p = (2\pi R_p L_p) j_{ORR}^0 \frac{\tanh(\Lambda L_p)}{\Lambda L_p} \left(\frac{c_{H^+}^{surf}(z)}{c_{H^+}^0}\right)^{\gamma_{H+}} \exp\left(-\frac{\alpha_c F}{RT}\left(\phi_M^b - \phi_S^{surf} - E_{ORR}^0\right)\right). \tag{20}$$

An electrostatic effectiveness factor is defined as,[8]

$$\Gamma_{elec} = \frac{J_p}{J_p\left(c_{H^+}^{surf} = c_{H^+}^0\right)}. \tag{21}$$

where $J_p = (c_{H^+}^{surf} = c_{H^+}^0)$ refers to the ORR current when $c_{H^+}^{surf} = c_{H^+}^0$.

4 Results and discussion

4.1 Electrostatic properties of a water-filled Pt nanopore

Using the analytical model, we first examine the electrostatic properties of the nanopore with the basic parameter set as listed in Table 1. Starting from a refined structural model for an electrified interface, a non-monotonic charging relation is obtained for the nanopore model considered in this article, as shown in Fig. 7(a). Remarkably, σ_M always remains negative. In addition, σ_M decreases in the oxide region due to the oxide dipoles formed at the Pt surface. The increase in σ_M above 0.9 V is ascribed to the saturation of oxide species.

For the case of a planar Pt electrode considered in ref. 12, we found that σ_M changes from negative to positive and then to negative as the potential increases from the hydrogen adsorption region, to the intermediate double layer region, and further to the oxide formation region. However, for the case of a water-filled nanopore, the intermediate region with $\sigma_M > 0$ region disappears, and the whole charging relation is shifted to the negative domain. The key difference between

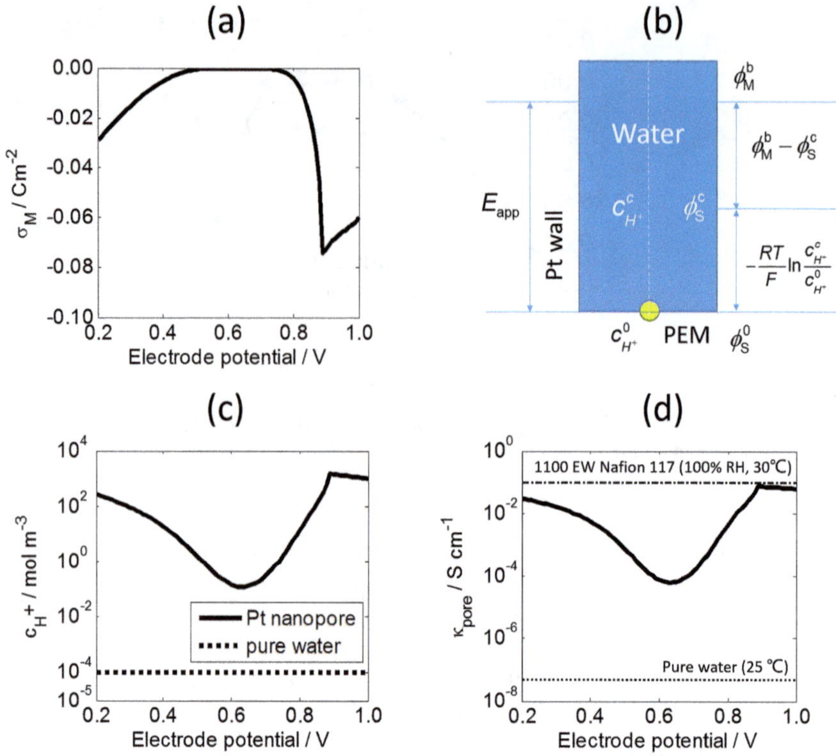

Fig. 7 (a) Free metal surface charge density of the water-filled Pt nanopore in the potential range of 0–1.0 V (other parameters are given in Table 1); (b) the potential relation in the water-filled nanopore; (c) the surface proton concentration as a function of electrode potential; (d) proton conductivity of the water-filled Pt nanopore as a function of electrode potential.

a planar electrode and this case is that the proton density and potential distributions here are entirely controlled by double layer effects. In this regard, the potential at the nanopore center, $\phi_S{}^c$, is different from the reference potential defined at the PEM opening (the bulk solution), as illustrated in Fig. 7(b). Therefore, the potential difference between the Pt bulk and the nanopore center, $\phi_M^b - \phi_S{}^c$, is distinct from the applied electrode potential *versus* the reference point, ϕ_M^b (*vs.* RHE). We take an intermediate potential of 0.6 V as an example. Fig. 7(c) shows that $c_{H^+}{}^c$ has a value of around 10^{-2} mol m^{-3} at 0.6 V. The Boltzmann gives $\phi_M^b - \phi_S{}^c = 0.298$ V, which is slightly smaller than $\Delta\phi_M = 0.3$ V. As a result, the Pt wall at 0.6 V has $\sigma_M < 0$, according to eqn (16). Intuitively speaking, the pore opening to the PEM, as a proton source, ensures that the nanopore contains an excess of protons, that is, $\sigma_M < 0$.

Fig. 7(c) depicts the interfacial proton concentration, $c_{H^+}^{surf}$, in the nanopore. It can be seen that $c_{H^+}^{surf}$, in the range between 10^{-2} to 10^2 mol m^{-3}, is highly potential-dependent and two to six orders of magnitude higher than the proton concentration in pure water, owing to the negatively charged Pt wall. In line with the behavior of σ_M, $c_{H^+}^{surf}$ increases due to the oxide dipoles with increasing the potential from the intermediate region to the oxide formation region. The decreasing trend with further increasing the potential is attributed to the saturation of oxide species.

The proton conductivity of the nanopore, κ_{pore}, is a key parameter.[1,3,9] We assume that the proton mobility remains close to the bulk water value. Consequently, κ_{pore} is proportional to the average proton concentration in the nanopore, $c_{H^+}^{avg}$, which is given by,

$$c_{H^+}^{avg} = \frac{1}{\pi R_p^2}\int_0^{R_p} 2\pi r c_{H^+}(r)\mathrm{d}r = -\frac{2\sigma_M}{FR_p}. \tag{22}$$

Based on $c_{H^+}^{avg}$ derived from eqn (22), κ_{pore} is calculated and compared to by referring to the conductivity of pure water, $\kappa_w = 5.4 \times 10^{-8}$ S cm^{-1} (25 °C), shown in Fig. 7(d). Two findings can be concluded: (1) κ_{pore} is two to five orders larger than κ_w; (2) κ_{pore} in the oxide region can be close to the proton conductivity of Nafion membrane (1100 EW PFSA Nafion 117, 30 °C).[16] Thompson and Baker measured the proton conductivity of a PTFE-bound Pt black electrode under different RH using electrochemical impedance spectroscopy.[16] The measurements were performed at 0.2 V. They reported that the proton conductivity of the Pt black electrode at 100% RH could approach a value of 0.06 S cm^{-1}, in good correspondence to the value of 0.04 S cm^{-1} in Fig. 7(d).

In summary, for a water-filled Pt nanopore interfacing with a PEM as its opening, σ_M is non-monotonic and exhibits negative values in the potential range of 0 to 1.0 V. In this configuration, surface charging is not governed by the pzc. In addition, the nanopore has a two to five orders larger proton conductivity compared to pure water.

4.2 Electrocatalytic performance of a water-filled Pt nanopore

In this subsection, the ORR performance of a water-filled Pt nanopore is investigated. The results are meaningful for both conventional and ultra-thin catalyst layers (CLs), as discussed below.

Conventional CLs comprise Pt catalyst supported on carbon. The porous structure is impregnated with Nafion ionomer. Thanks to advances in characterization techniques, hierarchical structures of conventional CLs have been uncovered and depicted as follows: (1) micropores (<2 nm) exist in primary carbon particles that have diameters of ~30 nm; (2) these primary particles assemble into agglomerates of ~100–300 nm within which mesopores of 2–20 nm exist; and (3) agglomerates coagulate into blocks of ~1–3 μm with a continuous network of macropores (>20 nm).[17,18] Nafion ionomer cannot penetrate into agglomerates and it resides predominantly within the macropores by covering the outer periphery of agglomerates.[19–22] On the contrary, many Pt particles of 2–4 nm deposit in the mesopores inside of the agglomerates and locate at the necks of the micropores.[17,18] Recently attention has been paid to the study of the activity of intra-agglomerate Pt particles. Soboleva *et al.* found that a CL with ultra-low (5 wt%) ionomer content can produce surprisingly high performance under high relative humidities (RHs), indicating that intra-agglomerate Pt particles are electrochemically active and contribute a significant portion of the current.[18] Furuya *et al.* reported that the ORR activity of conventional CLs decreases with increasing ionomer coverage.[23]

Fig. 8(a) shows a schematic of an agglomerate in a conventional CL. The intra-agglomerate Pt particles line up at the wall of the water-filled nanopore. Neglecting the discreteness of Pt, we assume that a continuous Pt film covers the wall of the water-filled nanopore. Modelling work considering the discreteness of Pt particles and charging effects of carbon is undergoing.

Ultra-thin CLs (UTCLs), *viz.*, CLs with a thickness less than 1 μm, are a promising system in view of achieving markedly improved mass activity of Pt in PEFCs.[6,15] Examples of UTCLs are plentiful, including nanoporous Pt layers directly sputtered on the gas diffusion layer[24,25] or the PEM,[26] and 3M's nanostructured thin film (NSTF) electrode fabricated by depositing a polycrystalline thin Pt (alloy) film onto an ordered nanostructured support.[15] These UTCLs can be fabricated ionomer-free.[15,24–26] In this scenario, water acts as the proton transport medium.[8–10] In this regard, a water-filled Pt nanopore can be regarded as an elementary unit in the UTCL, as shown in Fig. 8(b).

Fig. 9(a) shows the comparison between calculated and experimental polarization curves of a membrane electrode assembly (MEA) with an NSTF catalyst layer at the cathode.[15] Note that the current density is defined with respect to the electrochemical surface area of Pt. The voltage is IR-free, while it is not corrected for mass transport in the gas diffusion layer and the CL.[15] The model parameters are mainly taken from ref. 8: $R_p = 20$ nm, on the scale of the whisker size, and $L_p = 270$ nm, on the scale of the CL thickness; $c_{O_2} = 1$, $\gamma_{H^+} = 1.1$, and $\alpha_c = 1.0$, assuming that the relevant range of cell voltages, 0.7–0.9 V, lies in the high-potential regime for the ORR kinetics;[28] and parameters of the electrified interface and Pt oxide formation are inherited from ref. 12, as listed in Table 1. For the exchange current density, we used $j^0_{ORR} = 2.3 \times 10^{-2}$ A m^{-2}, in good correspondence to literature values.[27–29]

Fig. 9(b) shows Γ_{elec} as a function of cell voltage. Ref. 8 adopted a linear charging relation, $\sigma_M = C_{dl}(\phi^b_M - \phi^{pzc})$, parameterized using the pzc, $\phi^{pzc} = 0.7$ V. In that article, Γ_{elec} was found to monotonically decrease with increasing

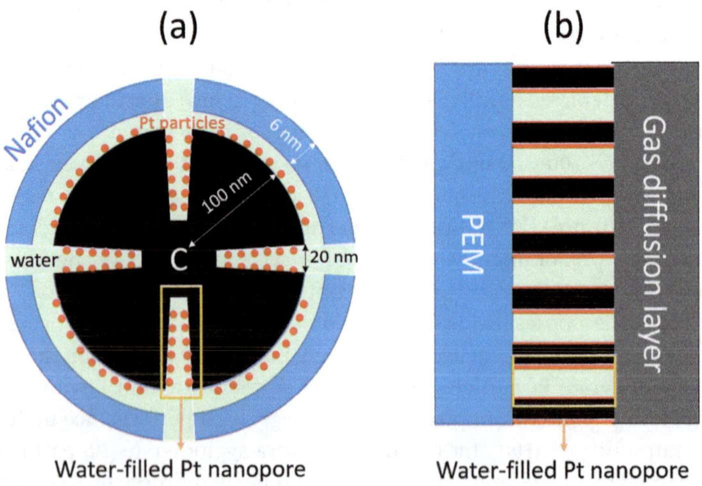

Fig. 8 (a) Schematic of an agglomerate in conventional catalyst layers, and (b) schematic of an ultra-thin catalyst layer.

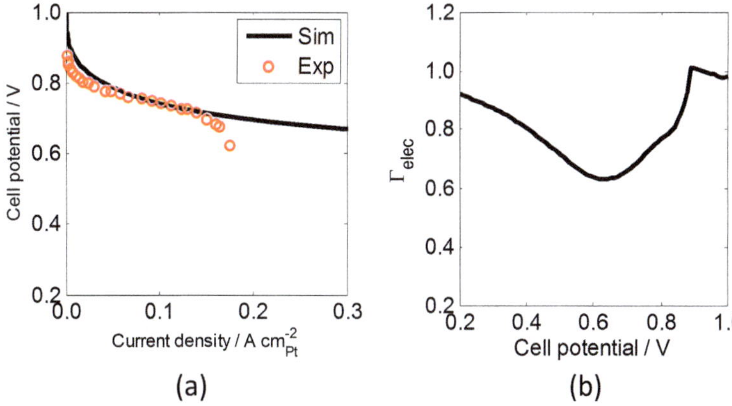

Fig. 9 (a) Model fitting of the polarization curve of a NSTF MEA, and (b) the electrostatic effectiveness factor as a function of the cell potential.

ϕ_M^b due to the monotonic increase in σ_M. When $\phi_M^b > 0.7$ V, Γ_{elec} decreased below 0.1 and it approached zero with further increasing ϕ_M^b. The revised metal charging relation employed in this article gives a non-monotonic relation between Γ_{elec} and ϕ_M^b, as shown in Fig. 9(b). Due to the reduced magnitude of σ_M in the potential range of 0.4–0.8 V, a trough of Γ_{elec} can be seen in this region.

Note that the relation between Γ_{elec} and ϕ_M^b depends on the magnitude of γ_{H^+} and α_c. Using the Boltzmann relation, one obtains

$$\Gamma_{elec} \propto \exp\left(\frac{F\phi_S^{surf}}{RT}(\alpha_c - \gamma_{H^+})\right) \propto \left(\frac{c_{H^+}^{surf}}{c_{H^+}^0}\right)^{(\gamma_{H^+} - \alpha_c)}. \tag{23}$$

According to eqn (23) and the relation, $\gamma_{H^+} > \alpha_c$, Γ_{elec} increases with $c_{H^+}^{surf}$. Since $c_{H^+}^{surf} < c_{H^+}^0$ in the potential range of 0.2–0.9 V, Γ_{elec} increases with decreasing γ_{H^+}. Based on eqn (10), the apparent transfer coefficient, α_c^{app}, can be expressed as,

$$\alpha_c^{app} = \alpha_c - (\gamma_{H^+} - \alpha_c)\frac{RT}{Fc_{H^+}^{surf}}\frac{\partial c_{H^+}^{surf}}{\partial \phi_M^b}. \tag{24}$$

Fig. 7(c) reveals that $\partial c_{H^+}^{surf}/\partial \phi_M^b > 0$ for ϕ_M^b in the range of 0.6–0.9 V. Therefore, it is expected that $\alpha_c^{app} < 1$ in this potential range, which is consistent with the experimentally observed increase in the Tafel slope at lower ϕ_M^b.

4.3 Pore size effect

The pore size parameters, namely R_p and L_p, can be tuned during electrode fabrication. Therefore, understanding how these parameters modulate the surface charging relation, the ORR activity and utilization of Pt, and the ORR performance of a nanopore and a CL is of practical interest.

Based on eqn (16), σ_M is dependent on R_p. $\partial\sigma_M/\partial R_p$ is given by,

$$\frac{\partial\sigma_\mathrm{M}}{\partial R_\mathrm{p}}=\frac{-\left(\frac{RT}{FR_\mathrm{p}}\right)^2\frac{1}{\frac{RT}{FR_\mathrm{p}}-\frac{\sigma_\mathrm{M}}{4\varepsilon_\mathrm{S}}}}{\left(\frac{\delta_\mathrm{PtO}}{\varepsilon_\mathrm{PtO}}+\frac{\delta_\mathrm{OHP}}{\varepsilon_\mathrm{OHP}}+\frac{\delta_\mathrm{IHP}}{\varepsilon_\mathrm{IHP}}\right)-\frac{RT}{F}\frac{1}{\sigma_\mathrm{M}}-\frac{N_\mathrm{tot}\mu_\mathrm{w}}{\varepsilon_\mathrm{IHP}}\frac{1}{\cosh(X)^2}\frac{\partial X}{\partial\sigma_\mathrm{M}}}. \tag{25}$$

In Appendix A.2, we prove that,

$$\frac{N_\mathrm{tot}\mu_\mathrm{w}}{\varepsilon_\mathrm{IHP}}\frac{1}{\cosh(X)^2}\frac{\partial X}{\partial\sigma_\mathrm{M}}<\frac{\delta_\mathrm{IHP}}{\varepsilon_\mathrm{IHP}}. \tag{26}$$

Considering, $\sigma_\mathrm{M} < 0$, we have, $\partial\sigma_\mathrm{M}/\partial R_\mathrm{p} < 0$. That is, the magnitude of σ_M will increase with increasing R_p, as shown in Fig. 10(a).

Fig. 10(b) shows the effect of R_p on the ORR activity of Pt. The ORR activity is higher for nanopores with a larger R_p, because of the increased surface charge density of the Pt surface.

As shown in Fig. 4(b), there are negligible distributions of $c_{\mathrm{H}^+}(z)$ and $\phi_\mathrm{S}(z)$ in the axial direction in most of the nanopore. Significant gradients are found only in the vicinity of the PEM opening within $L_\mathrm{end} = 5$ nm. In this regard, Γ_elec is independent of L_p if $L_\mathrm{p} \gg L_\mathrm{end}$.

On the other hand, L_p influences the oxygen transport. Ref. 8 introduced an effectiveness factor corresponding to oxygen transport, Γ_{O_2}. From eqn (20), we have,

$$\Gamma_{\mathrm{O}_2}=\frac{\tanh(\Lambda L_\mathrm{p})}{\Lambda L_\mathrm{p}}, \tag{27}$$

where Λ, defined in eqn (18), is the reciprocal of the reaction penetration length. Λ depends on D_{O_2}, $c^0_{\mathrm{O}_2}$, and the cell potential. A positive correlation between Λ^{-1} and the cell current density is expected. At 0.8 V, $\Lambda^{-1} \approx 100$ nm.

Asymptotic behaviors of Γ_{O_2} with respect to L_p are,

$$\Gamma_{\mathrm{O}_2}\cong 1, \tag{28.1}$$

when $\Lambda L_\mathrm{p} \ll 1$, and,

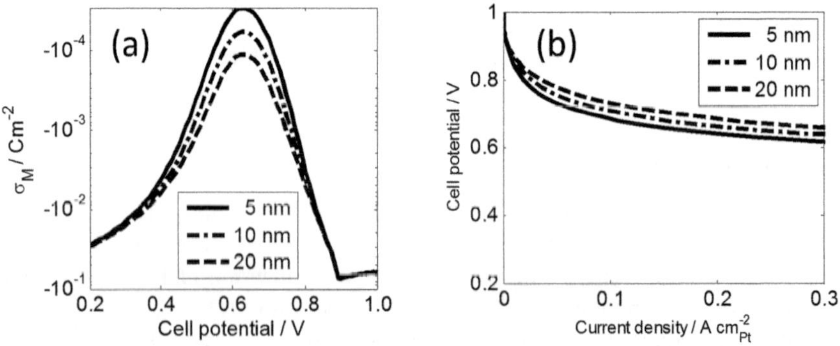

Fig. 10 Effect of pore radius on: (a) the free metal surface charge density, and (b) the ORR current with respect to Pt area.

$$\Gamma_{O_2} \cong (\Lambda L_p)^{-1}, \tag{28.2}$$

when $\Lambda L_p \gg 1$. Eqn (28) indicates that the utilization of Pt could be limited by oxygen transport when $L_p \gg \Lambda^{-1}$. The same conclusion can be found in ref. 8. Combining the electrostatic and oxygen transport effects, one will obtain a range of L_p,

$$L_{end} \ll L_p \ll \Lambda^{-1}, \tag{29}$$

in which the ORR current with respect to Pt area in the nanopore is nearly independent of L_p.

The ORR current of a catalyst layer composed of a parallel array of water-filled Pt nanopores, j_{cl} [A cm_{geo}^{-2}], is given by,

$$j_{cl} = \frac{\varepsilon_p J_p}{\pi R_p^{\ 2}} \propto \frac{2\varepsilon_p L_p}{R_p} \Gamma_{elec} \Gamma_{O_2}, \tag{30}$$

where ε_p is the porosity of the catalyst layer. Eqn (30) indicates that j_{cl} will increase with increasing L_p or decreasing R_p.

5 Conclusions

Based on PNP equations and a refined structural model for the Pt-solution interface, we have examined the surface charging phenomena, electrostatic properties, and ORR activity and performance of a water-filled Pt nanopore. The main findings are:

(1) The intrinsically two-dimensional problem can be simplified to a 1D + 1D problem by assuming that the axial distributions of c_{H^+} and ϕ_S are uniform, and the radial distribution of c_{O_2} is negligible. Analytical solutions can be obtained in the simplified case.

(2) The water-filled pore, with one opening interfacing with a PEM as a proton reservoir, always possesses negative surface charge in the potential range of 0–1.0 V (RHE). In the oxide region, σ_M is shifted to more negative values due to the formation of the surface dipole moment. The obtained non-monotonic charging relation helps to rectify earlier modelling studies of the ORR activity in water-field nanopores that employed a linear charging relation parameterized with the pzc of the metal.

(3) The proton conductivity of the nanopore was found to be 2 to 5 orders of magnitude higher than that of pure water. In the range of low potentials, *i.e.* below 0.3 V (RHE), corresponding to large negative surface charge, the proton conductivity is found to be of the order of 0.01 S cm^{-1}. A similar value of the proton conductivity is calculated for the potential range of 0.8–1.0 V. In the potential range of 0.4–0.8 V, the proton conductivity is smaller, with values in the range of 10^{-3} to 10^{-4} S cm^{-1} due to the reduced magnitude of σ_M. Conductivity measurement would be the most direct way of measuring the metal charging relation and validating this model.

(4) The high proton density is the underlying reason for the decent ORR performance of a water-filled Pt nanopore. An analytical expression for ORR current is parameterized using the polarization data of a NSTF MEA. A non-

monotonic relation between the electrostatic effectiveness factor, Γ_{elec}, and electrode potential is obtained.

(5) The magnitude of σ_{M} increases with increasing R_{p}, leading to better ORR activity of Pt in the nanopore. There exists a range of pore lengths, $L_{\text{end}} \ll L_{\text{p}} \ll \Lambda^{-1}$, in which Pt utilization is nearly independent of L_{p}. Otherwise, the Pt utilization decreases with increasing L_{p}. The ORR performance of a catalyst layer composed of water-filled nanopores can be improved by increasing L_{p} or decreasing R_{p}.

(6) The model developed in this study can be used as a mathematical tool to evaluate the structure–property–performance relation of a water-filled nanopore with charged metal walls, so as to guide the material selection and structural design of porous electrodes in polymer electrolyte fuel cells and water electrolysers.

6 Appendix

A.1 Numerical calculation

The PNP equations are solved in COMSOL, coupled with the metal surface charging model solved in MATLAB. In the COMSOL model, a 2D axisymmetric rectangle is built and the width and height are $(R_{\text{p}} - \delta_{\text{IHP}} - \delta_{\text{OHP}})$ and L_{p}, respectively. The 'electrostatic' module is used to describe the Poisson equation, and the 'transport of dilute species' module is employed to formulate the Nernst–Plank equations. The space charge density in the 'electrostatic' module is $c_{\text{H}^+} - K_{\text{w}}/c_{\text{H}^+}$ and the surface charge density is given by a Matlab function to solve eqn (5), (9) and (16). The inputs required from the COMSOL model to feed the MATLAB function are $\phi_{\text{S}}^{\text{surf}}$ and $c_{\text{H}^+}^{\text{surf}}$. As mentioned before, iteratively using the potential continuity condition will determine the three independent variables of the system, namely c_{H^+}, c_{O_2} and ϕ_{S}, at a certain electrode potential. The mesh properties are as follows: first, the rectangle is mapped with a distribution of 20 elements (element ratio: 10) in both the width and height directions; second, boundary layers (number: 10, stretching factor: 2, thickness of the first layer: 0.2 nm) are set at the Pt wall and the PEM opening to refine the mesh because remarkable gradients exist at these two boundaries. The whole model is programmed in the MATLAB interface.

A.2 Analysis of the metal charging relation

The derivative of eqn (9) with respect to σ_{M} is written as,

$$\left(\frac{0.6}{\pi\delta_{\text{IHP}}^2} + N_{\text{tot}}\right)\frac{N_{\text{A}}{\mu_{\text{w}}}^2}{\delta_{\text{IHP}}\varepsilon_{\text{IHP}}RT}\frac{1}{\cosh(X)^2}\frac{\partial X}{\partial \sigma_{\text{M}}} - \frac{\partial X}{\partial \sigma_{\text{M}}} = \frac{N_{\text{A}}\mu_{\text{w}}}{\varepsilon_{\text{IHP}}RT}, \tag{A-1}$$

then, $\dfrac{\partial X}{\partial \sigma_{\text{M}}}$ is given by,

$$\frac{\partial X}{\partial \sigma_{\text{M}}} = \frac{1}{\left(\dfrac{0.6}{\pi\delta_{\text{IHP}}^2} + N_{\text{tot}}\right)\dfrac{\mu_{\text{w}}}{\delta_{\text{IHP}}}\dfrac{1}{\cosh(X)^2} - \dfrac{\varepsilon_{\text{IHP}}RT}{N_{\text{A}}\mu_{\text{w}}}}. \tag{A-2}$$

Magnitude analysis gives,

$$\frac{N_{\rm tot}\mu_{\rm w}}{\varepsilon_{\rm IHP}}\frac{1}{\cosh(X)^2}\frac{\partial X}{\partial\sigma_{\rm M}} = \frac{1}{\left(\frac{0.6}{\pi\delta_{\rm IHP}^2}+N_{\rm tot}\right)\frac{\varepsilon_{\rm IHP}}{N_{\rm tot}\delta_{\rm IHP}} - \frac{\varepsilon_{\rm IHP}{}^2RT}{N_{\rm tot}\mu_{\rm w}{}^2N_{\rm A}}\cosh(X)^2} < \frac{\delta_{\rm IHP}}{\varepsilon_{\rm IHP}}. \tag{A-3}$$

Acknowledgements

J. Huang and J. Zhang gratefully acknowledge financial support from Toshiba-Tsinghua project (No. 20153000019), and State key laboratory of automotive energy and safety (No. ZZ2014-031). M. Eikerling gratefully acknowledges financial support toward this project from the NSERC APC network CaRPE-FC. J. Huang thanks the China Scholarship Council for providing financial support for his stay in Simon Fraser University as a visiting PhD student.

Notes and references

1 H. Daiguji, *Chem. Soc. Rev.*, 2010, **39**, 901–911.
2 L. Cao, W. Guo, W. Ma, L. Wang, F. Xia, S. Wang, Y. Wang, L. Jiang and D. Zhu, *Energy Environ. Sci.*, 2011, **4**, 2259–2266.
3 P. Gao and C. R. Martin, *ACS nano*, 2014, **8**, 8266–8272.
4 S. Nasir, M. Ali, P. Ramirez, V. Gómez, B. Oschmann, F. Muench, M. N. Tahir, R. Zentel, S. Mafe and W. Ensinger, *ACS Appl. Mater. Interfaces*, 2014, **6**, 12486–12494.
5 D. D. Ordinario, L. Phan, W. G. Walkup IV, J. M. Jocson, E. Karshalev, N. Hüsken and A. A. Gorodetsky, *Nat. Chem.*, 2014, **6**, 596–602.
6 M. Eikerling and A. Kulikovsky, *Polymer Electrolyte Fuel Cells: Physical Principles of Materials and Operation*, CRC Press, 2014.
7 S. Holdcroft, *Chem. Mater.*, 2013, **26**, 381–393.
8 K. Chan and M. Eikerling, *J. Electrochem. Soc.*, 2011, **158**, B18–B28.
9 Q. Wang, C. S. Cha, J. Lu and L. Zhuang, *ChemPhysChem*, 2012, **13**, 514–519.
10 Q. Wang, C. S. Cha, J. Lu and L. Zhuang, *Phys. Chem. Chem. Phys.*, 2009, **11**, 679–687.
11 O. A. Petrii, *Russ. J. Electrochem.*, 2013, **49**, 401–422.
12 J. Huang, A. Malek, J. Zhang and M. Eikerling, *J. Phys. Chem. C*, 2016, **120**, 13587–13595.
13 N. F. Mott and R. J. Watts-Tobin, *Electrochim. Acta*, 1961, **4**, 79–107.
14 R. Rizo, E. Sitta, E. Herrero, V. Climent and J. M. Feliu, *Electrochim. Acta*, 2015, **162**, 138–145.
15 M. K. Debe, *J. Electrochem. Soc.*, 2013, **160**, F522–F534.
16 E. L. Thompson and D. Baker, *ECS Trans.*, 2011, **41**, 709–720.
17 T. Soboleva, X. Zhao, K. Malek, Z. Xie, T. Navessin and S. Holdcroft, *ACS Appl. Mater. Interfaces*, 2010, **2**, 375–384.
18 T. Soboleva, K. Malek, Z. Xie, T. Navessin and S. Holdcroft, *ACS Appl. Mater. Interfaces*, 2011, **3**, 1827–1837.
19 K. Malek, T. Mashio and M. Eikerling, *Electrocatalysis*, 2011, **2**, 141–157.
20 M. Eikerling, *J. Electrochem. Soc.*, 2006, **153**, E58–E70.
21 L. Guetaz, M. Lopez-Haro, S. Escribano, A. Morin, G. Gebel, D. A. Cullen, K. L. More and R. L. Borup, *ECS Trans.*, 2015, **69**, 455–464.

22 M. Lopez-Haro, L. Guétaz, T. Printemps, A. Morin, S. Escribano, P. H. Jouneau, P. B. Guillemaud, F. Chandezon and G. Gebel, *Nat. Commun.*, 2014, 5, DOI: 10.1038/ncomms6229.
23 Y. Furuya, H. Iden, T. Mashio, A. Ohma and K. Shinohara, In Meeting Abstracts, *Electrochem. Soc.*, 2012, **40**, 1522.
24 W. Mei, T. Fukazawa, Y. Nakano, Y. Akasaka and K. Naito, *ECS Trans.*, 2013, **50**, 1377–1384.
25 W. Mei, T. Fukazawa, T. Yang, N. Yoshinaga and Y. Kanai, *ECS Trans.*, 2015, **69**, 755–759.
26 R. O'Hayre, S. J. Lee, S. W. Cha and F. B. Prinz, *J. Power Sources*, 2002, **109**, 483–493.
27 A. Parthasarathy, S. Srinivasan, A. J. Appleby and C. R. Martin, *J. Electroanal. Chem.*, 1992, **139**, 2856–2862.
28 D. B. Sepa, M. V. Vojnovic and A. Damjanovic, *Electrochim. Acta*, 1981, **26**, 781–793.
29 R. K. Ahluwalia, X. Wang, A. Lajunen, A. J. Steinbach, S. M. Hendricks, M. J. Kurkowski and M. K. Debe, *J. Power Sources*, 2012, **215**, 77–88.

Faraday Discussions

PAPER

Nanocapillary electrokinetic tracking for monitoring charge fluctuations on a single nanoparticle

Sanli Faez,*[a] Sela Samin,[b] Dashdeleg Baasanjav,[a] Stefan Weidlich,[cd] Markus Schmidt[d] and Allard P. Mosk[a]

Received 27th April 2016, Accepted 24th May 2016
DOI: 10.1039/c6fd00097e

We introduce *nanoCapillary Electrokinetic Tracking* (nanoCET), an optofluidic platform for continuously measuring the electrophoretic mobility of a *single* colloidal nanoparticle or macromolecule *in vitro* with millisecond time resolution and high charge sensitivity. This platform is based on using a nanocapillary optical fiber in which liquids may flow inside a channel embedded inside the light-guiding core and nanoparticles are tracked using elastic light scattering. Using this platform we have experimentally measured the electrophoretic mobility of 60 nm gold nanoparticles in an aqueous environment. Further, using numerical simulations, we demonstrate the underlying electrokinetic dynamics inside the nanocapillary and the necessary steps for extending this method to probing single biomolecules, which can be achieved with existing technologies. This achievement will immensely facilitate the daunting challenge of monitoring biochemical or catalytic reactions on a single entity over a wide range of timescales. The unique measurement capabilities of this platform pave the way for a wide range of discoveries in colloid science, analytical biochemistry, and medical diagnostics.

Introduction

More than a century after Robert Millikan had reported on measuring the elementary charge, his method is still being used for experiments such as the high precision detection of fractional charges at ambient conditions[1] or single charging events on an optically-trapped colloid particle in apolar solvent.[2] In this method, by observing the drift velocity of a particle forced by an external electric field, one obtains its electrical mobility, which has usually a well-defined relation with the net charge of the particle. In parallel, measuring electrophoretic mobility

[a]*Debye Institute for Nanomaterials Research, Center for Extreme Matter and Emergent Phenomena, Utrecht University, Princetonplein 5, 3584CC Utrecht, The Netherlands. E-mail: s.faez@uu.nl*
[b]*Institute for Theoretical Physics, Center for Extreme Matter and Emergent Phenomena, Utrecht University, Princetonplein 5, 3584CC Utrecht, The Netherlands*
[c]*Heraeus Quarzglas GmbH and Co. KG, 63450 Hanau, Germany*
[d]*Leibnitz Institute of Photonic Technology, Albert-Einstein-Str. 9, 07754, Jena, Germany*

has become one of the most common methods in analytical (bio)chemistry and colloid science. The electrophoretic mobility (drift velocity in a viscous medium under application of external electric field) of proteins and other macromolecules is a very sensitive indicator of their internal charge state, their interaction with the surrounding environment, and their hydrodynamic radius. As such, it has been used for sorting a mixture of macromolecules[3] or studying their conformational changes due to chemical reactions or physical adsorption to other molecules. However, common electrophoresis measurement techniques are based on measuring ensembles. In an ensemble measurement, the observation of intermediate states is often impossible because of averaging over unsynchronized and inhomogeneous processes.

Single particle measurements, on the contrary, are not prone to disturbance by impurities, which can be singled out in data analysis. Furthermore, such measurements can be repeated as many times as necessary for collecting reliable and informative statistics. In other words, when the physicists' granular view of a reaction on a single particle is reconstructed by a direct measurement, the characteristic kinetics and thermodynamics of the reaction is no more obscured by material inhomogeneity or stochastic dynamics over the metastable intermediate states. Important chemical reactions, such as photo-catalysis or enzyme-catalysis, involve multiple steps of largely differing timescales. Furthermore, their reaction rates can be heterogeneous within an ensemble of catalyst particles and can even vary over time for a single catalyst particle. In principle, there is no fundamental barrier for studying a full chain of chemical reactions on a single molecule or nanoparticle through monitoring its electrophoretic mobility, if (and that is indeed a big if) one could track the single molecule motion both on the shortest necessary timescales and as long as the reaction is complete. In other words, if one could perform Millikan's experiment on objects as small single molecules, she or he could visualize the reaction kinetics of that single object. In such a measurement, one can directly "watch" the intermediate steps of a reaction as they happen.

The single most challenging step for mobility measurement at this small scale on a single nanoparticle is tracking its motion with sufficient spatial and temporal resolution. On these criteria, optical methods generally score much higher than all-electronic techniques. The optical signal (for example scattering intensity) from sub-wavelength particles, however, decreases steeply as the particle size gets smaller. Very recently, we have demonstrated a new particle tracking method that provides the necessary sensitivity and speed for tracking sub-20 nm particles.[4] In this article we report on building a single particle optical capillary electrophoresis device based on that achievement. We dub this technique nanoCapillary Electrophoretic Tracking (nanoCET).

We present the details of our experimental setup and report the first measurements of electrophoretic mobility on a single particle in this platform with ms time resolution. As a typical colloidal particle we perform these measurements on a single gold nanoparticle of 60 nm in diameter in an aqueous environment, using the nanoCET technique. We explain some of our observations by comparing the measurements with numerical simulations using the finite element method. Based on these simulations, we present the necessary follow-up steps for extending this method to probing single biomolecules.

Experimental

We have used nanofluidic optical fibers, which were previously introduced in ref. 4, for visualizing the motion of nanoparticles inside the liquid. In these single-mode step-index fibers a hollow channel of 600 nm in diameter is embedded concentric to the light guiding doped-silica core. The channel can be filled with liquid containing the nanoparticles. Laser light is coupled into the fiber and the elastically scattered light is detected from the side of the fiber. The particle positions are imaged using the elastically scattered light from the fiber-core through a home-built microscope with an overall magnification of 100× connected to a high-speed sCMOS camera (Orca Flash4.0). Recorded sequences of images are processed for tracking the particles similar to the method presented in ref. 4. For the measurements reported here, we have additionally placed two micro-electrodes in-plane and orthogonal to the optical fiber as sketched in Fig. 1. The width of each electrode is 100 μm and the tip-to-tip distance between the electrodes is roughly 600 μm. The electrodes were connected to an alternating high-voltage amplifier ($|V - G| < 200$ volts) which was fed by an alternating square waveform of frequency ν.

Gold nanoparticles (BBI-solutions) of 60 nm in diameter suspended in water were inserted into the nanocapillary of 50 mm length. The suspension was taken directly from the bottle provided by the manufacturer without any extra purification step. For each measurement, between 1 and 10 particles were visible inside the field of view of 200 μm from the center of the fiber with a signal to noise ratio of better than 100. Light at a vacuum wavelength of 670 nm from a diode laser (ventus, Laser Quantum) is coupled into the core of the fiber. The coupled intensity inside the fiber is estimated to be around 5 mW resulting in a net scattering signal of higher than 50 million electronic counts per second from each nanoparticle on the camera. This signal provides a localization accuracy of 60 nm for the center of the particle along the fiber axis, at the highest measurement rate and considering the sub-optimal point spread function of the imaging optics. In addition to the Brownian motion, the

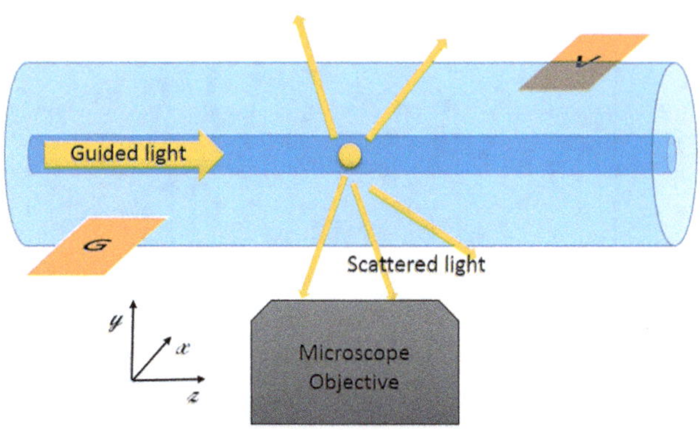

Fig. 1 Schematic representation of the experimental conditions. Electrophoretically actuated thermally diffusing particles are tracked inside the liquid-filled core of a nano-capillary optical fiber. The electric field for actuation is generated by metal electrodes placed outside the fiber as depicted.

particles are driven back and forth under the influence of the alternating potential applied to the electrodes with a square waveform using a function generator and a low-noise amplifier. The particle tracks are recorded at a frame rate of 2 kHz for driving frequencies between 1 and 200 Hz.

To present a typical measurement, the velocity of a single particle as a function of time is plotted in Fig. 2 under actuation at 50 Hz. The particles oscillates synchronously with the alternating electric field. The average speed of the particle in the first millisecond of each period is roughly 30 μm s^{-1}. Deviation from the average is mainly due to Brownian motion. We observed no long term change in the particle response even after tens of minutes, indicating the stability of the experimental conditions.

The particle speed reduces to half its peak value 10 ms after the start of each period. This behavior is attributed to the restructuring of the double layer inside the capillary, which screens the electric field.

The electrophoretic dynamics of the particle and the surrounding ions are best visible in the tracks recorded at low driving frequency (Fig. 3). At the start of each cycle, particles move at an average drift velocity of 30 μm s^{-1}. The drift velocity decreases as the solvent ions reorganize to shield the electrostatic field inside the channel. After $t = 100$ ms through one cycle the particle displacement due to external field is undistinguishable from Brownian movements, but averaging over many cycles demonstrates persistence of nonzero average drift up to at least 500 ms. This curve is plotted in Fig. 3 for different combinations of the axis scaling. This decrease in drift velocity is clearly not behaving like a single exponential.

To determine the single particle mobility, and hence the charge of the particle from the drift velocity, we need to know the electrostatic field and the fluid velocity inside the channel. For this analysis, we need to know the influence of the electro-osmotic flow of the surrounding liquid and also separate the time dependence of the electrostatic field at the position of the particle due to ion screening dynamics. To this end, we have used finite

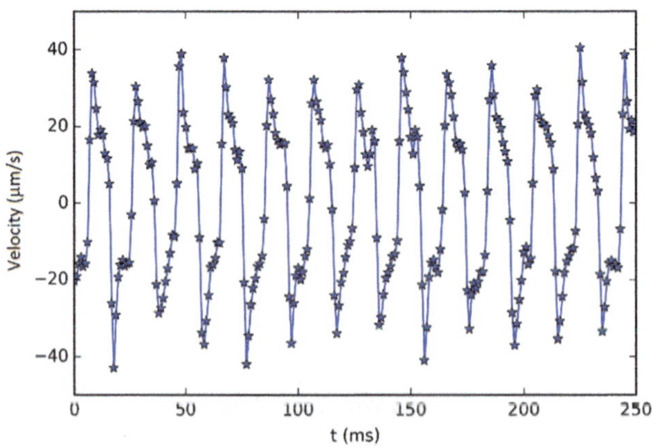

Fig. 2 Drift velocity of a 60 nm gold nanoparticle inside the capillary under actuation by external electric field with a square waveform alternating at 50 Hz. The peak to peak variations are due to Brownian motion.

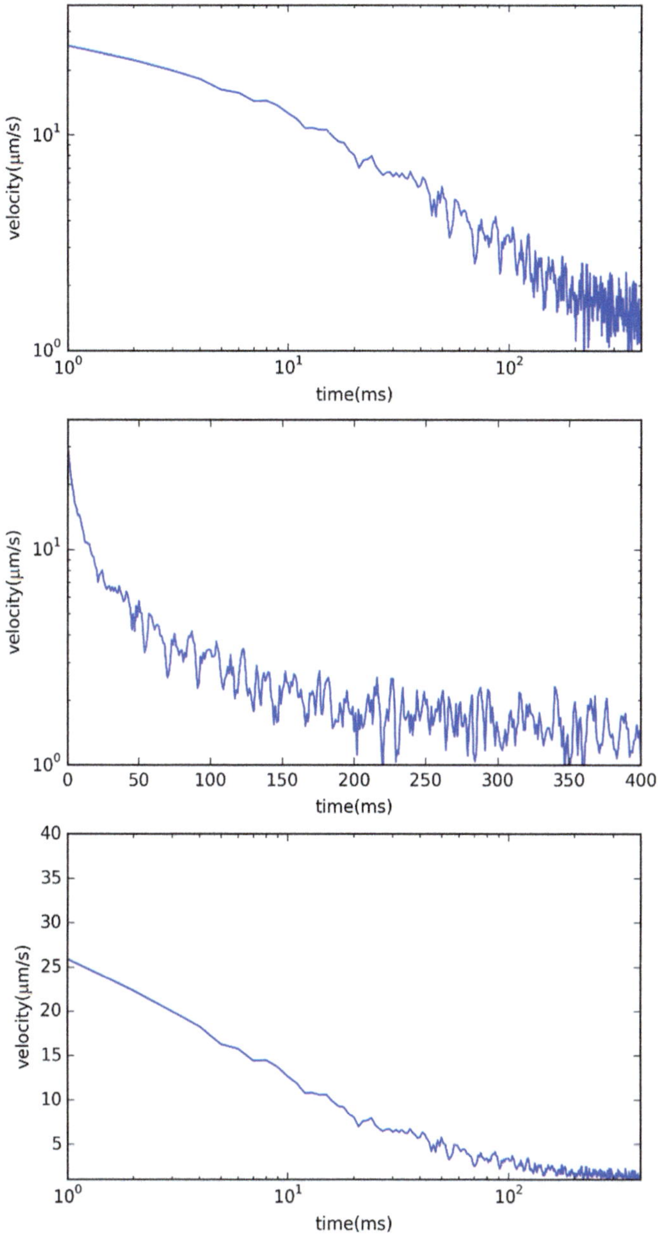

Fig. 3 The average drift velocity of a single particle under constant applied potential decaying as function of time. The values are averaged over 10 cycles. Each panel show the same data with a different scaling on the axes. The decay behaviour is clearly not of a single exponential type.

element modelling to simulate the ion movement in a geometry that resembles our experimental conditions, but is smaller in size to make the computation tractable.

Modelling

The electrokinetic flow inside the capillary is described within the well-established framework of the Poisson–Nernst–Planck (PNP) equations[5] for the dynamics of point-like ions in combination with the Stokes equations for the fluid dynamics, assuming that $\mathrm{Re} \ll 1$, where Re is the Reynolds number. The equations were solved using the finite-elements software COMSOL Multiphysics (v5.2). A schematic illustration of the computational domain is shown in Fig. 4. In order to make calculations in long capillaries computationally feasible, we take advantage of the system symmetry and set-up a 2D cylindrical coordinate system (r, z). At the origin, we place a long cylindrical capillary with length $L = 10$ μm and radius $R = 0.2$ μm, connected at each end to a cylindrical material reservoir of length $2L$ and the same radius R. At the reservoir edges we set the electrostatic potential to zero, the salt concentration to c_0 and the total fluid stress to zero, in order to mimic a bulk fluid at rest. At $r = 0$ we impose symmetry boundary conditions while for the solid walls at $r = R$ we impose zero flux of the ionic species and the no-slip boundary condition for the fluid. The reservoir walls are insulating (zero displacement field) while the capillary wall carries a constant charge density $\sigma = 6.1 \times 10^{-4}$ C m^{-2}, typical of a glass water interface.[5] The capillary is also surrounded by a silica matrix in which two oppositely-charged metal ring electrodes with a radius R are embedded. The electrodes are positioned at $(10R, \pm 0.4L)$. These electrodes model the spatially non-uniform applied external field in the experimental set-up and their potential $\pm V$ is set such that the generated electric fields are similar in magnitude to those in the experiment.

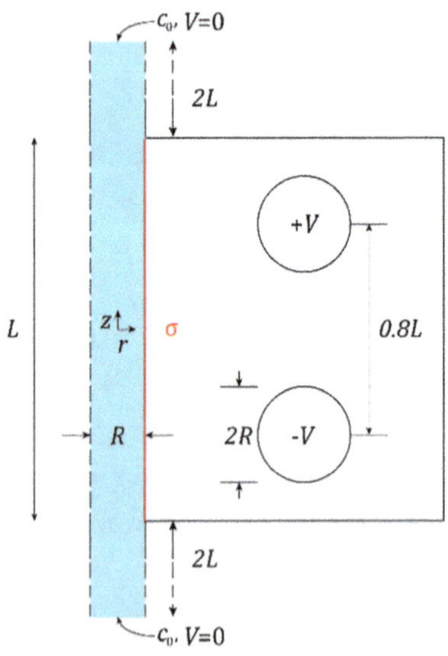

Fig. 4 Schematic illustration of the computational domain in the numerical calculations. A charged capillary of radius R and length L is embedded in a silica matrix and connected to two large electrolyte reservoirs. Two ring electrodes with voltages $\pm V$ mimic the experimentally applied external field. See text for values of dimensions and other parameters. The illustration in not drawn to scale due to the large aspect ratio of the capillary.

We start modelling from an equilibrated system under an external field where the electrodes' potential is $V = \pm 3$ V. In this state, a relaxed diffuse double-layer of ions with a characteristic thickness λ_D, the Debye length, is formed at the capillary wall, and the fluid is quiescent. We then abruptly reverse the electrodes' potential, $V \rightarrow -V$, following the square-wave function used in the experiments. The ions must then rearrange according to the new external field leading to an electro-osmotic flow in the capillary as they drag the fluid with them. Eventually, a new diffuse layer of ions will reform and the electric field will be screened. During this dynamic process, however, the field outside the double-layer is significant.

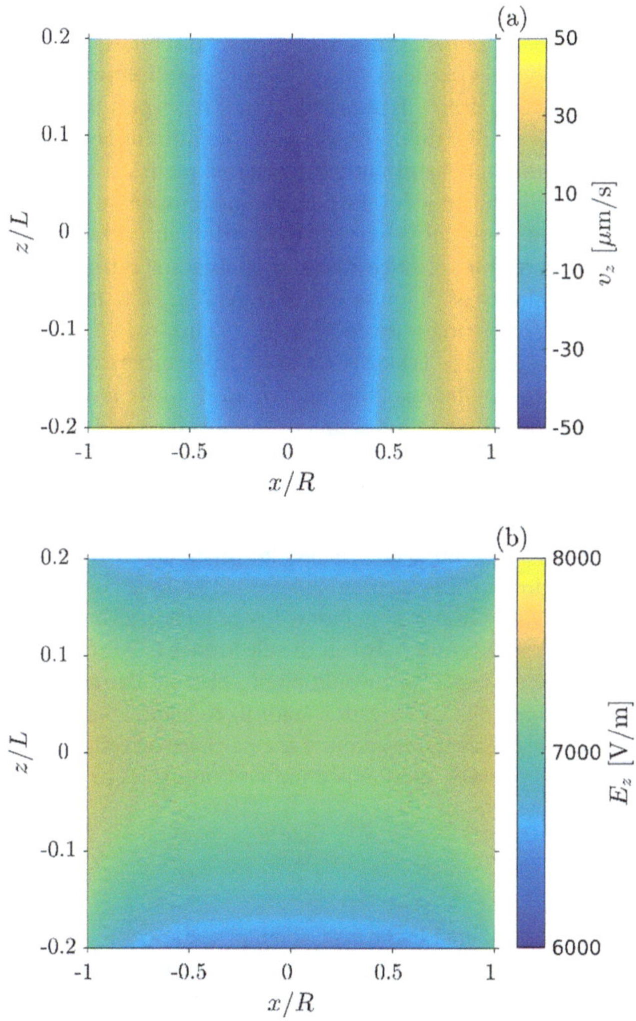

Fig. 5 (a) The axial component of the fluid velocity and (b) the axial component of the electric field in the x–z plane at time $t = 0.03$ ms after the reversal of a voltage on the electrodes. The bulk salt concentration is $c_0 = 0.1$ mM. Here and in other figures we used the properties of water at 298 K for the solution inside the capillary, those of silica for the surrounding medium, and both cations and anions have a diffusion constant $D = 10^{-9}$ m^2 s^{-1}.

A snapshot of the resulting flow and electric field in the axial direction, shortly after the application of the voltage step, is presented in Fig. 5. Clearly, a complex electro-osmotic flow develops inside the capillary, with the axial component of the velocity v_z presented in Fig. 5(a) showing a change of sign at an intermediate distance from the capillary wall. Nevertheless, v_z is rather uniform in the axial direction and since $\gg R > a$, the diffusing colloidal particle only experiences the velocity gradients in the radial direction. We calculated the average velocity (averaged over the capillary cross-section) $\langle v_z \rangle$ for $z = 0$ in Fig. 5(a) and found that it does not vanish but is small, about 0.15 times the maximal velocity at $r = 0$. In contrast, Fig. 5(b) shows that the axial electric field E_z is uniform both axially and radially, especially at the capillary center. Assuming that the small particle does not disturb the flow substantially, we conclude that the particle experiences a uniform external electric field, parallel to the capillary axis, but also a variable local flow. On the measurement timescale of 1 ms, however, the particle traverses randomly the whole cross section of the channel, due to Brownian motion, and the net influence of this drag will be averaged out. We expect this net flow in the capillary to be much smaller for the experimental conditions because the channel is 3 orders of magnitude longer than the simulated model.

To characterize the dynamics of the particle environment, we plot in Fig. 6 the time dependence of the axial velocity and electric field at the pore center for various salt concentrations. Notice that the simulation can access short time scales where the velocity at the pore center is very large. Both the velocity and electric field decay exponentially except for $c_0 = 0.02$ mM, where there is a distinct long tail. The relaxation time is slower for lower salt concentrations, but is short compared to experiments, being of the order of 0.1–1 ms for low values of c_0. This discrepancy is attributed to the limitation of short capillaries in the simulations. Analysis of diffuse charge dynamics by many authors[6–8] has shown that the primary time scale that controls bulk field screening by electrodes is $\tau = \lambda_D l/D$, where D is the diffusion constant of the ions and l is the separation between the electrodes. In a similar set-up to ours, a detailed analysis which demonstrates the exponential relaxation outside the double layer with a time constant τ was given by Bazant *et al.*[9] Clearly in our simulation $l \sim L$. Hence, if one could simulate a system in which the distance between the electrodes is 3 orders of magnitude bigger, as in the experiments, one would obtain also similar relaxation times of the order of 10–100 ms. This observation leads us to believe that our calculations capture qualitatively the dynamics around the particle correctly, at least in times that directed electrophoretic motion is dominant over the Brownian motion ($t <$ 20 ms in Fig. 3).

Discussion

The measurements presented in the experimental section were mainly performed to demonstrate the technical feasibility of continuously monitoring electrophoretic mobility on a single nanoparticle using our nanoCET platform. We are not claiming to be the first group who has measured electrophoretic mobility on a single nanoparticle. To our knowledge, however, no qualitatively similar measurement have been performed on similarly small single nanoparticles for such a long period of time and with such a high recording speed. Such mobility measurements have been done frequently in the past, in different geometries

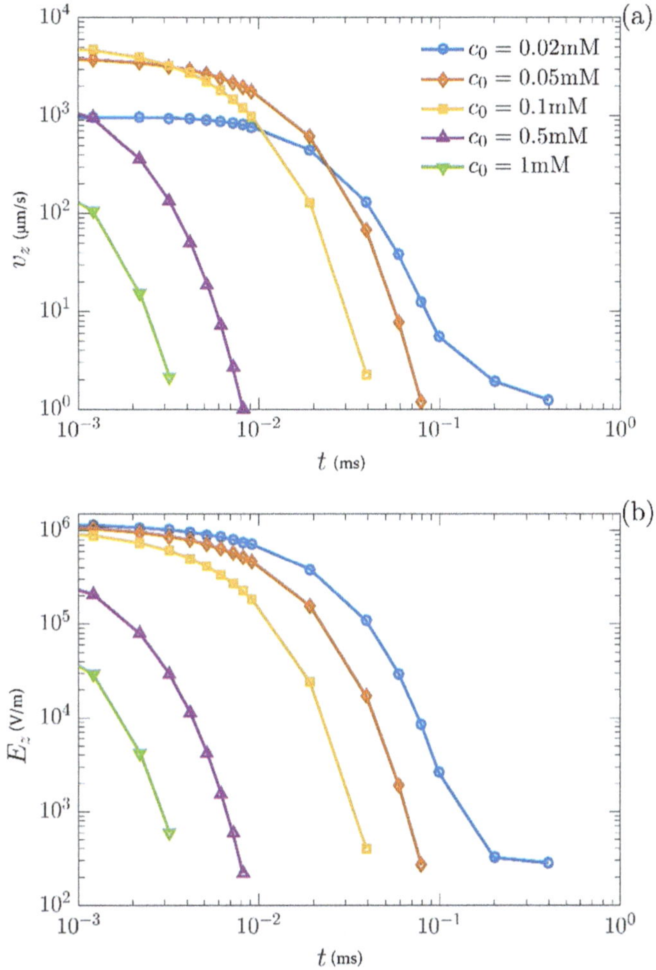

Fig. 6 (a) The axial component of the fluid velocity and (b) the axial component of the electric field at the pore centre as a function of time for various bulk salt concentrations c_0. Both quantities decay exponentially except for $c_0 = 0.02$ mM. The relaxation time increases with decreasing salt concentration.

using fluorescence microscopy. The most remarkable experimental platform in our view is the anti-Brownian electrokinetic (ABEL) trap developed by Cohen and Moerner,[10] which is able to measure the transient mobility of single proteins down to 1 nm in size and even a single fluorescent dye molecule. Inherent reliance of the ABEL trap on fluorescence signal, however, limits the time-resolution and maximum duration of tracking a single particle in that platform. In 2014, Wang and Moerner[11] have used a much optimized ABEL trap to extract both mobility and diffusivity of a single protein with a time resolution of 200 ms using time-stamped single photon counting. Despite these outstanding achievements, the time resolution of such measurements in the ABEL trap is hard to improve further because the photon flux is strictly limited by the fluorescence emission rate and the imperfect collection efficiency.[12] Furthermore, because of photo-bleaching of the fluorescent labels, the maximum duration of those measurements are usually less than a few seconds.

Using the scattering signal on a remarkably low background in the nanoCET platform is the key improvement over previous methods that will allow enhancement of time resolution and extension of measurement duration to a level that one can go beyond transient mobility measurements and monitor the charge kinetics on a single nanoparticle. To reach this goal however, certain improvements have to be made on our current setup, which we discuss here.

We consider a spherical particle as a model system for our discussion. As a typical example, we take a negatively charged colloidal particle with a radius of $a = 5$ nm and a typical surface potential of $\psi_s = -25$ meV in 1 mM (1 : 1) aqueous electrolyte solution. For this system, the dimensionless size parameter $a/\lambda_D \approx 0.5$. Using the well-known relation $q_s = \varepsilon\psi_s(1/\lambda_D + 1/a)$ for surface charge density in this geometry, the total surface charge of the particle is determined to be $Q \approx -10e$ with e being the fundamental charge. Using the Smoluchowski relation, the electrophoretic mobility of this particle $\mu \approx 1.8 \times 10^{-8}$ in SI units. Applying the Henry correction factor is not necessary for the current discussion and we consider the zeta potential has the same value as the above mentioned surface potential. The particle drift velocity under a realistic (unscreened) electric field of 100 kV m^{-1} is 1.8 μm ms^{-1}. This drift velocity can be measured in our current nanoCET setup with an accuracy better than 1 part in 20. This sensitivity translates to sensing surface charge variations as small as $e/2$ or surface potential change of $k_BT/20$, with a time resolution of 1 ms.

Because electrophoretic mobility in this regime depends only on the surface potential and is independent of the particle size, measuring smaller particles is just a quest of enhancing the optical signal. In future experiments, we expect to achieve more than a 100 fold increase in optical sensitivity by using holographic background reduction. This will allow us to monitor changes in electrophoretic mobility of some large proteins such as fibrinogen, catalase or hemoglobin, and short DNA strands. The scattering signal from smaller molecules can be enhanced by using fluorescent labels for enhancing the particle polarisability due to their large near-resonant dipole moments. As a benchmark, it is worth noting that the polarisability of a single Rhodamine 6G molecule, excited on the red side of its absorption line, is comparable with that of a latex particle of 10 nm in diameter or a gold nanocluster of 5 nm in size that is frequently used as a contrast agent for electron microscopy.

Our numerical modelling demonstrates that the electric field inside the channel is spatially uniform. The exact temporal behavior can be simulated in an up-to-scale model, which requires larger computational resources, or be measured on calibrated tracer particles. The result can then be used to determine the unknown mobility of the particles under investigation. The radial variation of the background fluid motion poses a greater challenge for determining the mobility since it adds extra noise to particle velocity measurements. This noise is more prominent for faster readouts because the particle does not get enough time to sample the whole cross-section of the channel. To correct for this noise, we can implement a three dimensional spatial localization protocol to perform a point-by-point background flow correction.

Conclusions

In summary, we have investigated the feasibility of performing electrophoretic mobility measurements on individual nanoparticle in a closed aqueous environment with numerical simulations and proof-of-concept experiments. We have proposed a new tracking experiment, nanoCET, based on a nanocapillary optical fiber platform combined with high speed actuation and readout. This measurements can be readily performed on high refractive index metallic particles as small as 10 nanometers. For measuring smaller particles such as biomolecules, directly, one needs to enhance the optical readout sensitivity, which can be achieved by interferometry.

The electrophoretic mobility of particles can be extracted at high measurement rates of several kHz. For such high rates three dimensional localization and careful consideration of the background fluid flow is necessary. Using numerical simulations of the PNP–Stokes equation on a smaller model system, we have presented the qualitative behavior of the fluid flow, charge screening, and electric field inside the capillary as a function of time. Our numerical results qualitatively agree with our measurements, but also prove that a true scale calculation with larger computational resources is necessary for quantitatively describing the experiments.

Acknowledgements

S. F. thanks Michel Orrit for hosting the experiment during an early stage of this project and Emiel Wiegers for technical support. This work is financially supported by the European Research Council, Project No. 279248. S. S. acknowledges funding from the European Union's Horizon 2020 under the Marie Skłodowska-Curie grant agreement No. 656327.

Notes and references

1 N. M. Mar, E. R. Lee, G. R. Fleming, B. C. K. Casey, M. L. Perl, E. L. Garwin, C. D. Hendricks, K. S. Lackner and G. L. Shaw, *Phys. Rev. D: Part. Fields*, 1996, **53**, 6017–6032.

2 F. Beunis, F. Strubbe, K. Neyts and D. Petrov, *Phys. Rev. Lett.*, 2012, **108**, 016101.

3 I. Gitlin, J. D. Carbeck and G. M. Whitesides, *Angew. Chem., Int. Ed.*, 2006, **45**, 3022–3060.

4 S. Faez, Y. Lahini, S. Weidlich, R. F. Garmann, K. Wondraczek, M. Zeisberger, M. A. Schmidt, M. Orrit and V. N. Manohoran, *ACS Nano*, 2015, **9**(12), 12349.

5 J. H. Masliyah and S. Bhattacharjee, *Electrokinetic and Colloid Transport Phenomena*, John Wiley & Sons, 2006, DOI: 10.1002/0471799742.

6 J. Gunning, D. Y. C. Chan and L. R. White, *J. Colloid Interface Sci.*, 1995, **170**, 522.

7 M. Scott, R. Paul and K. V. I. S. Kaler, *J. Colloid Interface Sci.*, 2000, **320**, 377.

8 A. D. Hollingsworth and D. A. Saville, *J. Colloid Interface Sci.*, 2003, **257**, 65.

9 M. Z. Bazant, K. Thornton and A. Ajdari, *Phys. Rev. E: Stat., Nonlinear, Soft Matter Phys.*, 2004, **70**, 021506.

10 A. E. Cohen and W. E. Moerner, *Proc. Natl. Acad. Sci. U. S. A.*, 2006, **103**, 4362–4365.

11 Q. Wang and W. E. Moerner, *Nat. Methods*, 2014, **11**, 555–558.

12 M. Kumbakhar, D. Hähnel, I. Gregor and J. Enderlein, *Pramana*, 2014, **82**, 121–134.

Faraday Discussions

PAPER

High-bandwidth detection of short DNA in nanopipettes†

Raquel L. Fraccari,[a] Marco Carminati,[b] Giacomo Piantanida,[b] Tina Leontidou,[a] Giorgio Ferrari[b] and Tim Albrecht*[a]

Received 2nd May 2016, Accepted 10th May 2016

DOI: 10.1039/c6fd00109b

Glass or quartz nanopipettes have found increasing use as tools for studying the biophysical properties of DNA and proteins, and as sensor devices. The ease of fabrication, favourable wetting properties and low capacitance are some of the inherent advantages, for example compared to more conventional, silicon-based nanopore chips. Recently, we have demonstrated high-bandwidth detection of double-stranded (ds) DNA with microsecond time resolution in nanopipettes, using custom-designed electronics. The electronics design has now been refined to include more sophisticated control features, such as integrated bias reversal and other features. Here, we exploit these capabilities and probe the translocation of short dsDNA in the 100 bp range, in different electrolytes. Single-stranded (ss) DNA of similar length are in use as capture probes, so label-free detection of their ds counterparts could therefore be of relevance in disease diagnostics.

Introduction

The transport of biopolymers such as DNA or RNA through nanometre-sized pores and narrow channels is a fundamental process in biology, but is of increasing technological interest, too. In nanopore sensing, for example, the process is recreated in a device consisting of an electrolyte-filled cell that is separated into two compartments by a highly insulating membrane. The latter typically features a single, or in some cases small number of, well defined pore(s) that allow for ion, liquid or analyte transport such as DNA or proteins between the two compartments. Transport across the pore may be induced by concentration, potential or pressure differences between the compartments, in accordance with the electrochemical potential gradient. Detection is typically either optical (fluorescence, SER(R)S) or electric (ion or tunnelling current),[1–5] where the duration, magnitude or temporal evolution of the signal is then used to extract information

[a]*Imperial College London, Department of Chemistry, Exhibition, Road, London SW7 2AZ, UK. E-mail: t.albrecht@imperial.ac.uk*

[b]*Politecnico di Milano, Dipartimento di Elettronica, Informazione e Bioingegneria, P.za, Leonardo da Vinci 32, Milano, Italy*

† Electronic supplementary information (ESI) available. See DOI: 10.1039/c6fd00109b

about the analyte, such as length, conformation, structure or composition.[6,7] In the latter context, prospects of fast and label-free sequencing of DNA, RNA and even proteins have inspired the field since its beginnings, with remarkable success in recent years (*vide infra*).[8,9]

Broadly speaking, one may distinguish three classes of nanopore sensors that have been used for single-molecule detection, namely biological nanopores (>1990s), solid-state 'chip-based' nanopore sensors (early 2000s) and nanopipettes (since ~2010). Hybrid devices aiming to combine the respective strengths have also been reported.[10–12]

In biological nanopore devices, the pore is made up of a pore forming protein, such as α-hemolysin (α-HL), which is typically embedded in a lipid-bilayer membrane.[3] In the case of α-HL, the channel is rather narrow, so that single-stranded (ss)DNA can pass through, but dsDNA or larger biomolecular complexes cannot. The pores are highly reproducible, in terms of size and sensor performance.

In chip-based devices, the membrane is usually made from highly insulating dielectric materials, such as SiO_2 or Si_3N_4, or two-dimensional materials such as graphene or MoS_2, and the pore is milled using a focused ion or electron beam.[4,13–15] Compared to biological pores, the pore size is less reproducible, but at the same time the design is more flexible, in that larger pores may be used to probe larger analytes, such as protein/DNA complexes.[16,32] In conventional chip-based devices, Si has been used as the base material, which however exhibits rather unfavourable electric characteristics, including high capacitance and therefore high electric noise. As a consequence, recent efforts have been directed to replacing Si with glass or quartz as the base material, with remarkable reduction in device capacitance to below 10 pF, and in some cases even below 1 pF.[17–19]

Finally, small micro- or nanopipettes have been used in patch clamping to probe ion transport across cell membranes since the 1970s,[20,21] and for surface imaging in Scanning Ion Conductance Microscopy (SICM) since the late 1980s.[22,23] Only recently have they been employed for the single-molecule detection and characterisation of DNA or proteins.[7,24,25,42] Compared to conventional chip-based devices, advantages include ease of fabrication outside of the clean room, favourable wetting properties (in aqueous electrolytes) and low device capacitance (facilitating high-bandwidth electric recording). While nanopipettes with pore diameters of around 20 nm can be fabricated directly using a pipette puller, making single-nm pores is difficult and typically requires post-modification, *e.g.* by Atomic Layer Deposition (ALD),[26] or electron beam-induced reshaping.[24] This can be desirable, for example to enhance the signal-to-noise ratio for the detection of small analytes, such as proteins.

It is difficult to characterize the shape of the pore channel on the inside of the pipette directly, and in particular close to the sensing region. However, it may be reconstructed from the current-time trace by translocating geometrically well-defined analytes, such as spherical nanoparticles.[27] In practice, it is often assumed that the channel is conical until the very end of the tip, with good agreement between theory and experiment.

In any case, nanopipettes now constitute a well-established platform for the study of fundamental ion- and liquid transport processes, and as a biophysical tool. In the present paper, we build on our recent work on the development of high-bandwidth low-current detection electronics and translocation of "long" DNA in the kilo basepair (kbp) range. We present a revised electronic circuitry with additional features, including built-in large-amplitude bias reversal ('zapping') and fully integrated software control. We focus on the translocation of short, 100 and 200 bp dsDNA (geometric length $L_{DNA} \sim 34$ and 68 nm, respectively) in nanopipettes with pore diameters between 17 and 26 nm. The DNA lengths are thus comparable to the persistence length of dsDNA, which is on the order of 35 nm and higher (depending on the electrolyte conditions),[28,29] and the translocating analyte more akin to a rigid rod than a flexible, long polymer. Accordingly, as we show below, we no longer find different folding states in the translocation data, which were reported previously for dsDNA in the kbp range.[6,7] ssDNA of similar length are also used as capture probes in disease diagnostics,[30] so the label-free detection of their double-stranded counterparts may find applications in biomedical sensing.

Materials and methods

DNA samples

Linear 100 bp dsDNA was obtained by hybridization of two complementary ssDNA strands. For the hybridization, 100 picomoles of each ssDNA strand was added to 5 μL of PerfectHyb™ Plus hybridization buffer (Sigma Aldrich, Dorset, UK) and 43 μL of UltraPure™ DNase/RNase-Free distilled water (Life Technologies, CA, USA). The mixture was first heated to 95 °C for 5 minutes then incubated at 37 °C for five hours then finally cooled to 4 °C for 20 minutes before storing a −20 °C until use. Linear 200 bp dsDNA was purchased from Fisher Scientific (Leicestershire, UK) and stored at −20 °C until use. The DNA concentrations were measured by UV-vis spectroscopy (NanoDrop 2000c spectrophotometer, Thermo Scientific, MA, U.S.A) and the purity of the DNA samples confirmed on a 2% agarose gel, see ESI Fig. S1.†

Nanopipette fabrication

Quartz capillaries (O.D. 1 mm, I.D. 0.5 mm, length 7.5 cm) were first plasma cleaned and then pulled on a laser pipette puller (P200, Sutter Instruments, Novato, USA) using one of two programs. The programme parameters were either: HEAT: 575, FILAMENT: 3, VELOCITY: 35, DELAY: 145 and PULL: 75 for the first step, then HEAT: 900, FILAMENT: 0, VELOCITY: 15, DELAY: 128 and PULL: 200 for the second step ("Programme 57") or alternatively HEAT: 700, FILAMENT: 5, VELOCITY: 35, DELAY: 150 and PULL: 75, then HEAT: 700, FILAMENT: 0, VELOCITY: 15, DELAY: 128 and PULL: 200 ("Programme 99"). Current–voltage (I–V) curves between −1 V and +1 V were used to determine the pore conductance. Using this measured pore conductance the diameter of the nanopores was estimated using the following equation:[31]

$$d_{\text{pore}} = \frac{4Gl + \frac{\pi}{2}GD_{\text{i}}}{D_{\text{i}}\pi g(c) - \frac{\pi}{2}G} \tag{1}$$

where D_{i} is the inner diameter at the capillary base (0.5 mm), G the conductance of the nanocapillary, $g(c)$ the conductivity of the electrolyte (measured for 1 M KCl:

101 mS cm^{-1}; 2 M LiCl: 100.1 mS cm^{-1}; 4 M LiCl: 167 mS cm^{-1}) and l the taper length (which was either 2 or 3 mm, depending on the pulling programme used, see above).

Translocation experiments

Glass vials ($V = 2$ mL) were rinsed with 70% ethanol (EtOH) and ultrapure water then filled with 1 mL of the respective electrolyte solution (*e.g.* 4 M LiCl, 10 mM Tris–HCl pH 8), as indicated. The nanopipettes were backfilled with the same electrolyte as the vial and fixed to the cap. For each experiment two freshly prepared 0.25 mm Ag/AgCl electrodes were used; the counter electrode (CE) was inserted inside the nanopipette and the working electrode (WE) placed in the electrolyte solution surrounding the nanopipette. The DNA concentrations ranged from 3–6 nM and a negative bias voltage applied to the working electrode (*i.e.* electrophoretic transport of DNA is from the outside to the inside of the pipette).

Ionic current measurements

As discussed in more detail below, a low-noise, wide-bandwidth CMOS current amplifier was employed for the ionic current measurements.[32,42] Briefly, the custom-built amplifier splits the measured ionic current into a slowly responding ('DC channel') that contains the open pore current of the nanopore, and a fast responding ('AC channel') where short current transients such as translocation events are displayed. For translocation experiments, the AC channel was filtered at up to 100 kHz and sampled at 1 MS s^{-1}. The low-noise amplifier is placed inside a Faraday cage as close as possible to the nanopipette so that the parasitic capacitance of connecting wires and the coupling of interferences would be minimized.

Data analysis

The ion current transients were analyzed with a custom written MATLAB script and Origin 9.0. Translocation events were defined as current transients with a depth (or peak amplitude) I_p greater than 4 or 5σ, as indicated (σ is the standard deviation of noise in the AC channel). The dwell time, τ was defined as the time difference between the data points where the current crossed a 1σ cutoff for each event. Such a threshold was chosen to preserve the overall shape of each translocation event as much as possible, while reducing the effect of (small) local baseline fluctuations.

Results

We start with a discussion of the electronics design. A key role in achieving the ultimate performance of a nanopore sensing system is played by the current amplifier employed to measure the ionic current. The two most important features of such an amplifier are its current resolution (*i.e.* its noise) and its speed (*i.e.* its bandwidth BW). High resolution is required in order to retrieve information from the details of the pulse produced by the molecule translocation, while speed is clearly useful to detect short molecules and fast events. Unfortunately, these properties are strictly related and cannot be chosen independently:

the higher the bandwidth, the higher the noise. Advanced circuit topologies can be adopted in order to design novel amplifiers that partially relax this fundamental trade-off, which is extremely stringent in classical transimpedance amplifiers.[33] In any case, the minimization of the amplifier noise should represent a primary goal in the development of advanced nanopore platforms.[34] An important guideline to achieve this target is the minimization of the amplifier input capacitance. This capacitance is the sum of several contributions, including the capacitance of the shielded coaxial cable connecting the working electrode to the amplifier (proportional to the cable length with a specific capacitance of 50–100 pF per meter), and the stray capacitance of the device itself. Thus, an accurate design of both the sensing setup (minimization of cable length) and of the device (choice of proper materials and geometries) is crucial. In the literature, few examples demonstrate enhanced noise performance achievable by accurate device design, including the choice of glass substrates[35–37] and thick isolation from the ionic solution.[38]

One effective solution that provides significant advantages is the implementation of the current amplifier into a single microelectronic chip.[39] Thanks to its miniaturization (area of a few mm^2), it can be placed in very close proximity of the electrodes. Furthermore, the ultimate level of connection shrinking is achievable when a metal pad (properly treated) of the chip serves as working electrode.[38] Finally, the straightforward integration of several amplifiers in parallel in a single chip enables multi-channel operation.

Recently, we have presented an integrated CMOS current amplifier, specifically designed for low-capacitance nanopore applications offering unrivalled performance: DC to 3.5 MHz bandwidth, 6 pA_{rms} resolution with BW = 100 kHz (*i.e.* the max. BW of Axopatch 200B, having twice the noise), and 50 pA_{rms} resolution with BW = 1 MHz (the limit of the Chimera CMOS amplifier,[40] having twice the noise) in open circuit condition (input capacitance ≈ 3.5 pF). The architecture of the amplifier leverages a self-compensating integrator–differentiator topology based on matched-MOSFETs,[41] endowed with an additional feedback network for handling large DC signals (±100 nA) and thus zooming with high gain (470 MV A^{-1}) on the details of the translocation signal. This amplifier allows achieving state-of-the-art speed with low-capacitance devices (ideally <1–10 pF) such as glass nanopipettes.[42]

Here we introduce an original and complete electronic platform built around the integrated amplifier for versatile nanopore sensing. In fact, once the amplifier noise has been minimized, it becomes necessary to minimize also the noise of the voltage generator used to bias the counter electrode, since it is also injected in the signal path, contributing to the total noise. Thus, the main novelty of the current platform is to include an on-board low-noise voltage generator along with auxiliary circuitry for versatile operation. As shown in Fig. 1, a microcontroller (Arduino DUE) drives a 16-bit DAC (AD5570) whose output is abundantly low-pass filtered (at 1 Hz). The microcontroller allows application of both a constant bias and triangular waveforms for cyclic voltammetry. A digitally-controlled precision analog switch (ADG5204, MUX in Fig. 1A) then allows the application to the counter electrode one of the four signals: (1) unfiltered DAC output for applying fast signals (such as a "zapping" transient voltage pulses to unclog the pore), (2) the filtered DAC output, (3) an inverted version of the DAC output for fast bias reversing,[43,44] (4) an external signal such as a small sinusoidal voltage, useful for

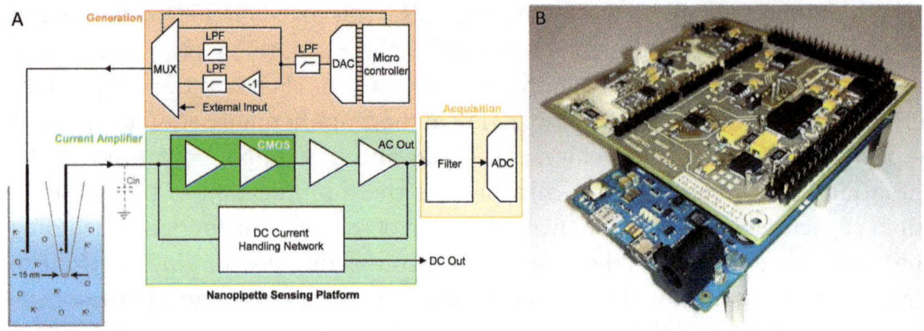

Fig. 1 (A) Block scheme of the dedicated nanopore characterization platform combining ultra-low noise current reading and voltage biasing along with additional features such as zapping and automatic reverse biasing. (B) Photograph of the compact prototype platform (8 cm by 8 cm), dissipating 125 mA at ±15 V plugged onto the Arduino DUE micro-controller that is managing all the operating modes.

impedance measurement during the device characterization. The switch commutation time is very fast (<300 ns), thus allowing abrupt bias reversal, while preserving the low-noise quality of both low-pass-filtered and locally generated biasing potentials. A fast control logic based on selectable thresholds allows automatic reverse biasing when translocation is detected. The prototype platform of this single-board and versatile platform is extremely compact (8 cm by 8 cm, Fig. 1B). The data presented in this manuscript were detected either with this new setup or the one reported previously (*cf.* ESI†).[42]

Based on previous data, we first estimated the minimum dsDNA length $L_{DNA,m}$ that we would be most likely to be able to detect under the conditions used here (1 M KCl electrolyte solution, similar pore diameter, filter frequency ~100 kHz), assuming that the scaling factor of the translocation time with DNA does not change significantly.‡ Further details of this assessment can be found in the ESI.† Briefly, we first extracted the I_p values for four DNA lengths (4, 5.31, 10, and 48.5 kbp) at four different V_{bias} (−200, −300, −400, and −800 mV), as well as the AC root-mean-square (rms) noise I_{rms}^{AC} of the AC channel under the respective filter conditions (Table S1, in the ESI†). We found the signal-to-noise ratio (S/N) I_p/I_{rms}^{AC} to be 7.4 on average, and approximately independent of V_{bias} and L_{DNA} (note the different filter frequencies, however). In a second step, we took the most probable translocation times τ_{mp} for each L_{DNA} and V_{bias}, and calculated the corresponding scaling law, $\log(\tau_{mp})$ *vs.* $p \log(L_{DNA})$. p is the scaling factor and was found to be between 1.16 and 1.24. With S/N approximately length independent, we then calculated the minimum $L_{DNA,m}$, for which $\tau = 10$ μs, for each V_{bias}. This yielded $L_{DNA,m}$ values of 326, 480, 684 and 1152 bp for $V_{bias} = -200, -300, -400$ and −800 mV. The approximately linear dependence between $L_{DNA,m}$ and V_{bias} (Fig. S4†) further allowed for extrapolation to $V_{bias} = -50$ and −100 mV, yielding $L_{DNA,m} = 152$ and 220 bp. Thus, we would expect to be able to detect translocation of dsDNA in the few 100 bp range, in 1 M KCl and at low V_{bias}. Events shorter than

‡ This is arguably a somewhat crude approximation, and different scaling factors between DNA length and translocation time have indeed been reported previously.[19] The estimated values nevertheless provide a useful guideline, as we show in the main text.

10 μs might still be detected, but not fully resolved due to filter convolution. Smaller pore diameters, alternative electrolytes (such as LiCl, *vide infra*) or surface modification would potentially enable detection of even shorter DNA segments with nanopipettes.[45–47]

Indeed, we were subsequently able to confirm this rather crude estimate experimentally, based on translocation experiments with 200 bp dsDNA in 1 M KCl ($V_{bias} = -300$ mV), as shown in Fig. 2 (we were unsuccessful in detecting translocation of 100 bp dsDNA, data not shown). In order to differentiate noise features from actual translocation events, we analyzed the current–time trace with regards to positive and negative event polarity. While DNA translocation events result in a temporal decrease in the current (negative spikes in the AC channel), random noise is expected to be equally distributed around the mean. An excess of negative spikes thus points to translocation events, even if event duration τ and magnitude I_p are close to the detection limit of the setup under given conditions. For the data set shown in Fig. 2, we found 8008 events with positive polarity and 15 069 events with negative polarity (4σ cutoff), corresponding to 7061 events formally assigned to DNA translocation (580 s recording time, $c_{DNA} = 3$ nM), which corresponds to a translocation frequency of 4 events per s per nM, which is comparable to those reported in ref. 48.

This is also reflected in the τ- and I_p-histograms shown on the right in Fig. 2. The prominent, sharp peak feature at $\tau_{mp,1} = 9$ μs and $I_{mp,1} = 63$ pA, respectively, appears in both positive and negative polarity traces. These events represent random noise, presumably convoluted with the filter response. For negative polarity, however, the DNA translocation events appear as a second distribution at longer τ/higher I_p. Formally, we obtained $\tau_{mp,2} = 16$ μs and $I_{mp} \approx 80$–90 pA. The distribution of the latter is surprisingly wide, which we speculate could be related to the orientation of the (short) DNA during translocation.[49]

For data recorded at $V_{bias} = -100$ mV and -200 mV (not shown), we obtained for the DNA translocation events $\tau_{mp,2} = 22$ μs and 20 μs, respectively. While there

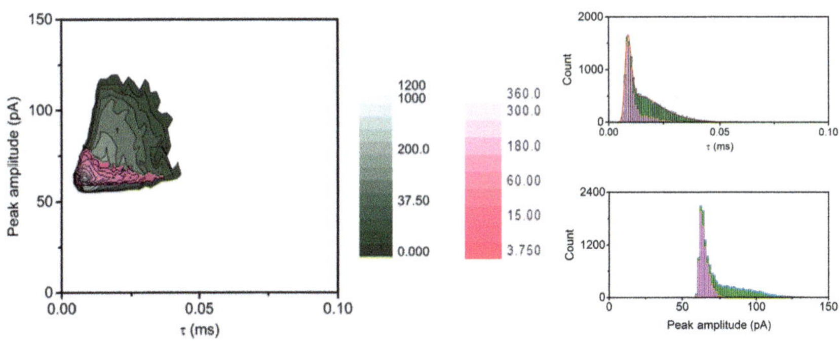

Fig. 2 Translocation of 200 bp dsDNA in 1 M KCl at $V_{bias} = -300$ mV, filter frequency: 70 kHz. Green (magenta): events with negative (positive) polarity. Left: contour plots for each data set, I_p *vs.* τ. Right top/bottom: τ- and I_p-histograms. A double log-normal distribution fit (red line) is included for the τ-histogram at negative event polarity (distribution 1: $\tau_{mean,1} = 9$ μs, $\tau_{mode,1} = \tau_{mp,1} = 9$ μs, std. dev. = 0.159; distribution 2: $\tau_{mean,2} = 19$ μs, $\tau_{mode,2} = \tau_{mp,2} = 16$ μs, std. dev. = 0.453). Pore conductance: 31 nS, estimated pore diameter, acc. to eqn (1): 17 nm. See further discussion in the main text.

seems to be a small increase in translocation time with decreasing V_{bias}, the small change suggests some filter convolution (filter frequencies were 50 and 60 kHz, respectively).

In an effort to slow down the translocation process, we changed the electrolyte from 1 M KCl to 2 and 4 M LiCl, respectively, which has previously been shown to increase the translocation time by up a factor of 10.[45,46] This has been rationalized in part by differences in ion pairing between K^+ and Li^+ and the DNA backbone. Interestingly, Bell *et al.* have recently shown that the change in electrolyte can also have more subtle consequences, such as a transition from flux- to barrier-limited translocation.[48]

In any case, we were able to reproduce the qualitative trend of slowed translocation in highly concentrated LiCl electrolytes also for short dsDNA in nanopipettes. Translocation data for 100 bp DNA in 2 M LiCl electrolyte are shown in Fig. 3, at $V_{bias} = -100$ mV (left, blue; 40 kHz filter frequency, 5σ cutoff) and -200 mV (right, purple; 50 kHz filter frequency, 5σ cutoff). Both contour plots (top) show a single population of translocation events and no significant additional noise features. Accordingly, we did not observe any indication of different folding states of the DNA, which is unsurprising given their short length, in

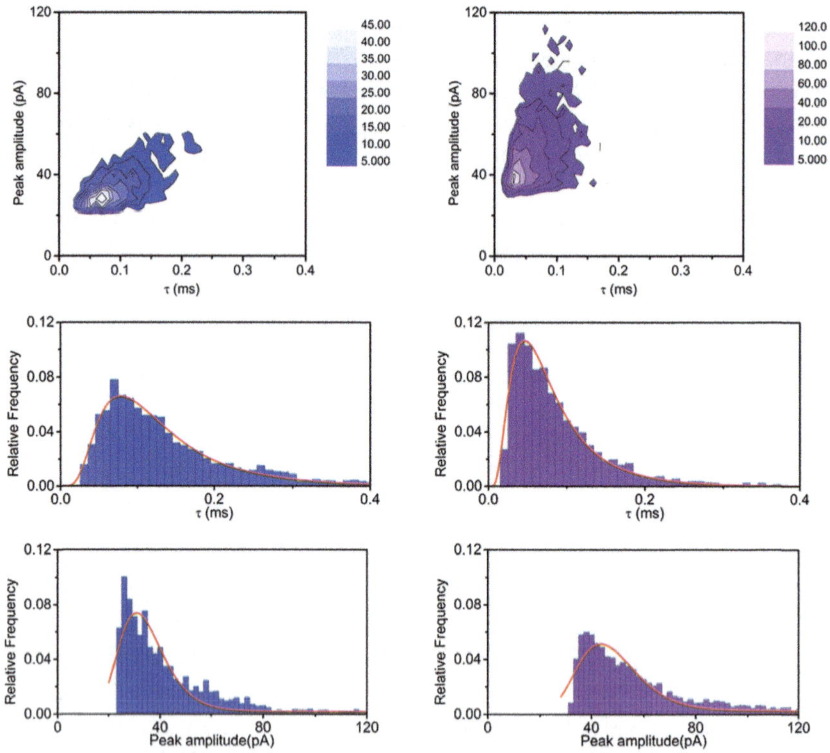

Fig. 3 Translocation experiments with 100 bp dsDNA in 2 M LiCl electrolyte, at $V_{bias} = -100$ mV (left, blue; 1650 events) and -200 mV (right, purple; 1682 events). Contour plot of I_p *vs.* τ (top); τ-(middle) and I_p-histogram (bottom), including log-normal distribution fits (red, solid lines). Filter frequency: 40 kHz ($V_{bias} = -100$ mV) and 50 kHz ($V_{bias} = -200$ mV), respectively. Pore conductance: 27 nS, estimated pore diameter: 22 nm.

comparison with the persistence length. The τ-histogram is represented very well by a single log-normal distribution with maxima at $\tau_{mp} = 79$ µs for $V_{bias} = -100$ mV and $\tau_{mp} = 46$ µs for $V_{bias} = -200$ mV, respectively. While the amount of data is clearly too limited for a systematic analysis of the bias dependence, we note that these values are roughly in-line with the V_{bias}^{-1}-dependence of τ_{mp} reported previously for translocation of long DNA in nanopipettes (see ref. 42 and references therein). As above, the I_p-histogram is relatively broad and does not feature a distinct maximum (even though the contour plot does show a distinct event cluster, *vide supra*). The most probable I_p values are between 30 and 40 pA and appear to be higher for larger V_{bias}, as expected.

Kowalczyk *et al.* have previously quantified the increase in translocation time for long DNA upon changing the electrolyte from 1 M KCl to 1 M LiCl and 2 M LiCl with a chip-based nanopore.[45] They found a combined effect of $4.8 \times 1.5 = 7.2$, relative to the translocation time in 1 M KCl. Assuming a similar increase here, we estimate the τ_{mp} values for 100 bp dsDNA in 1 M KCl to be approximately 11 and 6 µs at $V_{bias} = -100$ mV and -200 mV, which would explain why this short fragment remained undetected. Correspondingly, we would expect $\tau_{mp} \approx 22$ and 12 µs for 200 bp DNA at these bias voltages, which is similar to the experimental values obtained for this sample in 1 M KCl, see above.

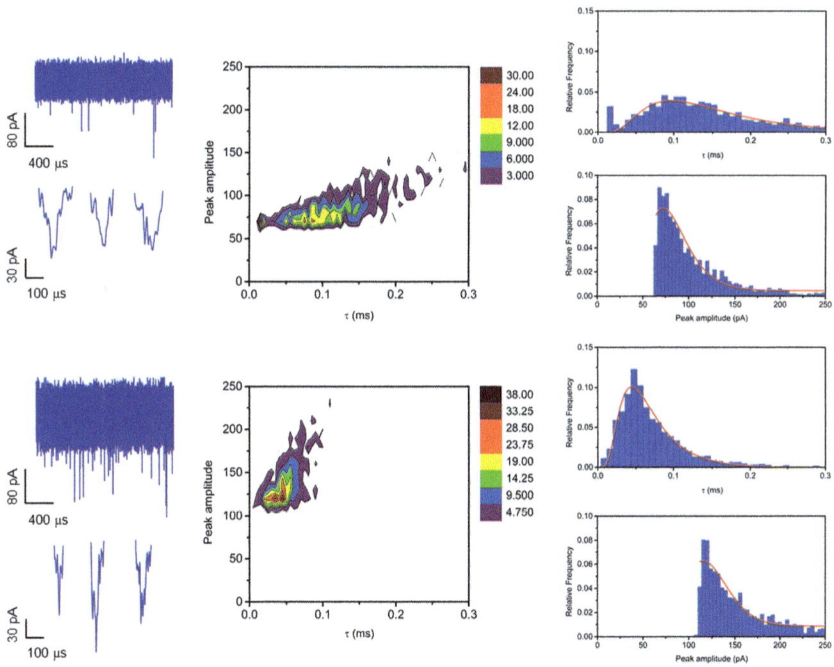

Fig. 4 Translocation experiments with 100 bp dsDNA in 4 M LiCl electrolyte, at $V_{bias} = -100$ mV (top, 2150 events) and -200 mV (bottom, 4012 events). Left: sections of $I(t)$ traces in the presence of DNA in solution; translocation events are downward spikes. Middle: contour plots of I_p *vs.* τ. Right: τ- and I_p-histograms, including log-normal distribution fits (red, solid lines). Filter frequency: 50 kHz ($V_{bias} = -100$ mV) and 100 kHz ($V_{bias} = -200$ mV), respectively. Pore conductance: 43 nS, estimated pore diameter: 21 nm.

A further increase in the translocation time of approximately 33% was observed upon increasing the LiCl concentration to 4 M.[45] While a direct comparison with the above data is complicated by device-to-device variation (*i.e.* the data was measured with different pipettes), we observed similar or slightly increased τ_{mp} values under these conditions (Fig. 4). For $V_{bias} = -100$ mV, we found $\tau_{mp} = 85 \pm 9$ µs and for $V_{bias} = -200$ mV, $\tau_{mp} = 42 \pm 2$ µs (averages of 3). The bias dependence of τ_{mp} again appears to be in-line with the ${V_{bias}}^{-1}$ dependence observed previously (*vide supra*), with τ_{mp} halved as V_{bias} is doubled.

In conclusion, we have presented a new electronics design for the high-bandwidth/low-current detection of translocation events in quartz nanopipettes with pore diameters in the region of 20 nm. In comparison to our previous work, this design features on-board voltage generation, triggered bias reversal and 'zapping' capabilities and software with fully integrated data acquisition and instrument control. Building on this work, we first theoretically assess and then demonstrate detection of 200 bp dsDNA in 1 M KCl. The translocation time is on the order of 20 µs or shorter and thus at the limit of detection. Upon changing the electrolyte from KCl to LiCl and increasing its concentration from 1 to 2 and 4 M, respectively, we slow down the translocation process (τ_{mp} between 50 and 100 µs, depending on V_{bias}) and show that the detection of 100 bp dsDNA, and potentially even shorter fragments, is feasible with pores in this size range. A further slowing down of the process may be achieved using surface functionalization, *e.g.* silanization,[47] which we will explore in future work. Preliminary results for long DNA indeed show the expected effect (*cf.* ESI†), which also highlights the important role of DNA/surface interactions before or during the translocation process.[42] In addition, an increase in the S/N ratio may be achieved with a reduction of the pore diameter.

While resolving features at the level of individual DNA bases (or base pairs) is beyond reach, and not the immediate aim of our work, the label-free detection of DNA fragments in 100 bp range may offer interesting prospects for disease detection and biomedical sensing, especially when coupled with a suitable workflow.

References

1 C. A. T. Chansin, R. Mulero, J. Hong, M. J. Kim, A. J. deMello and J. B. Edel, *Nano Lett.*, 2007, 7, 2901–2906.

2 M. P. Cecchini, A. Wiener, V. A. Turek, H. Chon, S. Lee, A. P. Ivanov, D. W. McComb, J. Choo, T. Albrecht, S. A. Maier and J. B. Edel, *Nano Lett.*, 2013, **13**, 4602–4609.

3 J. J. Kasianowicz, E. Brandin, D. Branton and D. W. Deamer, *Proc. Natl. Acad. Sci. U. S. A.*, 1996, **93**, 13770–13773.

4 J. Li, D. Stein, C. McMullan, D. Branton, M. J. Aziz and J. A. Golovchenko, *Nature*, 2001, **412**, 166–169.

5 A. P. Ivanov, E. Instuli, C. M. McGilvery, G. Baldwin, D. W. McComb, T. Albrecht and J. B. Edel, *Nano Lett.*, 2011, **11**, 279–285.

6 D. Fologea, M. Gershow, B. Ledden, D. S. McNabb, J. A. Golovchenko and J. Li, *Nano Lett.*, 2005, **5**, 1905–1909.

7 L. J. Steinbock, O. Otto, C. Chimerel, J. Gornall and U. F. Keyser, *Nano Lett.*, 2010, **10**, 2493–2497.

8 D. Branton, *et al.*, *Nat. Biotechnol.*, 2008, **26**, 1146–1153.
9 J. Quick, *et al.*, *Nature*, 2016, **530**, 228–232.
10 A. R. Hall, A. Scott, D. Rotem, K. K. Mehta, H. Bayley and C. Dekker, *Nat. Nanotechnol.*, 2010, **5**, 874–877.
11 E. C. Yusko, J. M. Johnson, S. Majd, P. Prangkio, R. C. Rollings, J. Li, J. Yang and M. Mayer, *Nat. Nanotechnol.*, 2011, **6**, 253–260.
12 N. A. W. Bell, C. R. Engst, M. Ablay, G. Divitini, C. Ducati, T. Liedl and U. F. Keyser, *Nano Lett.*, 2012, **12**, 512–517.
13 A. J. Storm, J. H. Chen, X. S. Ling, H. W. Zandbergen and C. Dekker, *Nat. Mater.*, 2003, **2**, 537–541.
14 G. F. Schneider, S. W. Kowalczyk, V. E. Calado, G. Pandraud, H. W. Zandbergen, L. M. K. Vandersypen and C. Dekker, *Nano Lett.*, 2010, **10**, 3163–3167.
15 K. Liu, J. Feng, A. Kis and A. Radenovic, *ACS Nano*, 2014, **8**, 2504–2511.
16 D. Japrung, A. Bahrami, A. Nadzeyka, L. Peto, S. Bauerdick, J. B. Edel and T. Albrecht, *J. Phys. Chem. B*, 2014, **118**, 11605–11612.
17 A. Balan, B. Machielse, D. Niedzwiecki, J. Lin, P. Ong, R. Engelke, K. L. Shepard and M. Drndić, *Nano Lett.*, 2014, **14**, 7215–7220.
18 A. Balan, C.-C. Chien, R. Engelke and M. Drndić, *Sci. Rep.*, 2015, **5**, 17775.
19 M.-H. Lee, A. Kumar, K.-B. Park, S.-Y. Cho, H.-M. Kim, M.-C. Lim, Y.-R. Kim and K.-B. Kim, *Sci. Rep.*, 2014, **4**, 7448.
20 E. Neher and B. Sakmann, *Nature*, 1976, **260**, 799–802.
21 B. Sakmann and E. Neher, *Annu. Rev. Physiol.*, 1984, **46**, 455–472.
22 P. K. Hansma, B. Drake, O. Marti, S. A. C. Gould and C. B. Prater, *Science*, 1989, **243**, 641–643.
23 C.-C. Chen, Y. Zhou and L. A. Baker, *Annu. Rev. Anal. Chem.*, 2012, **5**, 207–228.
24 L. J. Steinbock, S. Krishnan, R. D. Bulushev, S. Borgeaud, M. Blokesch, L. Feletti and A. Radenovic, *Nanoscale*, 2014, **6**, 14380–14387.
25 T. R. Gibb, A. P. Ivanov, J. B. Edel and T. Albrecht, *Anal. Chem.*, 2014, **86**, 1864–1871.
26 J. Y. Y. Sze, S. Kumar, A. P. Ivanov, S. H. Oh and J. B. Edel, *Analyst*, 2015, **140**, 4828–4834.
27 W.-J. Lan, D. A. Holden, B. Zhang and H. S. White, *Anal. Chem.*, 2011, **83**, 3840–3847.
28 P. J. Hagerman, *Annu. Rev. Biophys. Biophys. Chem.*, 1988, **17**, 265–286.
29 S. Brinkers, H. R. C. Dietrich, F. H. de Groote, I. T. Young and B. Rieger, *J. Chem. Phys.*, 2009, **130**, 215105.
30 E. M. LeProust, B. J. Peck, K. Spirin, H. Brummel McCuen, B. Moore, E. Namsaraev and M. H. Caruthers, *Nucleic Acids Res.*, 2010, **38**, 2522–2540.
31 L. J. Steinbock, A. Lucas, O. Otto and U. F. Keyser, *Electrophoresis*, 2012, **33**, 3480–3487.
32 P. Nuttall, K. Lee, P. Ciccarella, M. Carminati, G. Ferrari, K.-B. Kim and T. Albrecht, *J. Phys. Chem. B*, 2016, **120**, 2106–2114.
33 T. Albrecht, M. Carminati, G. Ferrari, P. Nuttall, W. Pitchford and A. J. Rutkowska, Electrochemical applications of nanopore systems in *Electrochemistry: Nanosystems Electrochemistry*, Royal Chemical Society 2014, vol. 12, pp. 155–186, ISBN 9781849735810, 2014.
34 M. Carminati, G. Ferrari, A. P. Ivanov, T. Albrecht and M. Sampietro, *Analog Integr. Circuits Signal Process*, 2013, **77**(3), 333–343.

35 M. Carminati, M. Vergani, G. Ferrari, L. Caranzi, M. Caironi and M. Sampietro, *Sens. Actuators, B*, 2012, **174**, 168–175.

36 M. H. Lee, A. Kumar, K. B. Park, S. Y. Cho, H. M. Kim, M. C. Lim, Y. R. Kim and K. B. Kim, *Sci. Rep.*, 2014, **4**, 7448.

37 A. Balan, C.-C. Chien, R. Engelke and M. Drndić, *Sci. Rep.*, 2015, **5**, 17775.

38 J. K. Rosenstein, M. Wanunu, C. A. Merchant, M. Drndic and K. L. Shepard, *Nat. Methods*, 2012, **9**, 487–489.

39 M. Crescentini, M. Bennati, M. Carminati and M. Tartagni, *IEEE Trans. Biomed. Circuits Syst.*, 2014, **8**(2), 278–292.

40 J. Larkin, R. Y. Henley, M. Muthukumar, J. K. Rosenstein and M. Wanunu, *Biophys. J.*, 2014, **106**, 696–704.

41 P. Ciccarella, M. Carminati, G. Ferrari, R. Fraccari and A. Bahrami, *Proc. IEEE PRIME*, 2014, DOI: 10.1109/PRIME.2014.6872701.

42 R. L. Fraccari, P. Ciccarella, A. Bahrami, M. Carminati, G. Ferrari and T. Albrecht, *Nanoscale*, 2016, **8**, 7604–7611.

43 M. Gershow and J. Golovchenko, *Nat. Nanotechnol.*, 2007, **2**, 775–779.

44 C. Plesa, L. Cornelissen, M. W. Tuijtel and C. Dekker, *Nanotechnology*, 2013, **24**, 475101.

45 S. W. Kowalczyk, D. B. Wells, A. Aksimentiev and C. Dekker, *Nano Lett.*, 2012, **12**, 1038–1044.

46 J. Uplinger, B. Thomas, R. Rollings, D. Fologea, D. McNabb and J. Li, *Electrophoresis*, 2012, **33**, 3448–3457.

47 Y.-R. Kim, J. Min, I.-H. Lee, S. Kim, A.-G. Kim, K. Kim, K. Namkoong and C. Ko, *Biosens. Bioelectron.*, 2007, **22**, 2926–2931.

48 N. A. W. Bell, M. Muthukumar and U. F. Keyser, *Phys. Rev. E*, 2016, **93**, 022401.

49 Y. Qiu, *et al.*, *ACS Nano*, 2015, **9**, 4390–4397.

Faraday Discussions

PAPER

Energetics of base flipping at a DNA mismatch site confined at the latch constriction of α-hemolysin†

Robert P. Johnson, Rukshan T. Perera, Aaron M. Fleming, Cynthia J. Burrows* and Henry S. White*

Received 21st March 2016, Accepted 29th March 2016
DOI: 10.1039/c6fd00058d

Unique, two-state modulating current signatures are observed when a cytosine–cytosine mismatch pair is confined at the 2.4 nm latch constriction of the α-hemolysin (αHL) nanopore. We have previously speculated that the modulation is due to base flipping at the mismatch site. Base flipping is a biologically significant mechanism in which a single base is rotated out of the DNA helical stack by 180°. It is the mechanism by which enzymes are able to access bases for repair operations without disturbing the global structure of the helix. Here, temperature dependent ion channel recordings of individual double-stranded DNA duplexes inside αHL are used to derive thermodynamic (ΔH, ΔS) and kinetic (E_A) parameters for base flipping of a cytosine at an unstable cytosine–cytosine mismatch site. The measured activation energy for flipping a cytosine located at the latch of αHL out of the helix (18 ± 1 kcal mol^{-1}) is comparable to that previously reported for base flipping at mismatch sites from NMR measurements and potential mean force calculations. We propose that the αHL nanopore is a useful tool for measuring conformational changes in dsDNA at the single molecule level.

Introduction

The DNA helix is a flexible molecule that continually undergoes a series of conformational changes over a range of timescales.[1–4] The dynamics of DNA are extremely important in regulating overall function and recognition of individual base pairs.[5,6] On the nanosecond timescale, DNA undergoes a series of bending and twisting,[7–9] breathing[6,10,11] and fraying[12,13] motions that affect DNA protein interactions and consequently have an important role in gene expression.[5]

On the millisecond timescale, more significant conformational changes in DNA are observed. For example, individual bases are capable of rotating out of

Department of Chemistry, University of Utah, 315 South 1400 East, Salt Lake City, UT, 84112-0850, USA. E-mail: burrows@chem.utah.edu; white@chem.utah.edu

† Electronic supplementary information (ESI) available: Extended current-time traces and additional residence time histograms as an accompaniment to Fig. 2 and 3, respectively. See DOI: 10.1039/c6fd00058d

a DNA helical stack, in a process known as 'base flipping'.[14,15] A growing body of evidence has shown that base flipping is a biologically important mechanism that plays a fundamental role in DNA maintenance and repair,[16–18] primarily because the rotation of a single base out of a helical stack exposes the base to the solvent and permits access to proteins. In the context of a flipped-out base, these proteins are able to carry out repair (*e.g.*, base recognition and excision)[18,19] or maintenance (*e.g.*, methylation)[16] without disrupting the global helical structure.

In the standard B-form conformation of double-stranded DNA (dsDNA), the DNA bases are held together in pairs through hydrogen bonding and π–π stacking interactions.[20] Evidence from fluorescence,[21] NMR[22–24] and potential mean force calculations[25–27] suggests that the energy required to disrupt the interactions that stabilize bases within the helix, and thus rotate an individual base out into the surrounding solvent, is large, between 10 and 20 kcal mol^{-1}. The precise energy required depends on the nature of the base-pair[22,28] (mismatched base-pairs are less stable) and the sequence context,[24] (which controls the strength of the stacking interactions).

Among the most common methods used to study spontaneous base flipping and base-pair opening are NMR measurements.[22–24,29–32] In these measurements, rotation of a DNA base out of the helical stack and exposure to solution permit the exchange of the imino protons on the bases guanine and thymine with protons in solution, resulting in measurable changes that indicate intrahelical lifetimes on the order of milliseconds[30,31] and extrahelical lifetimes on the order of microseconds.[19] In a biological context, a number of proteins critical for DNA repair and function are believed to make direct use of base flipping to access individual bases. Because the lifetime of the extrahelical state in spontaneous base flipping is short, many proteins are believed to be actively involved in stabilizing the extrahelical state[33,34] and/or lowering the kinetic energy barrier for base flipping. However, experimentally measuring such protein–DNA interactions is extremely challenging.

In our recent work, we have discovered that ion channel recordings with the protein nanopore αHL can be used to quantitatively measure conformational changes in dsDNA.[35] These ion channel recordings can be used to identify changes to DNA structure such as damage,[36–38] methylation[39] and mismatch[35] sites. In a typical experiment, a DNA helix, modified with a threading poly-T tail at the end of one strand, is driven into the αHL nanopore (embedded in a lipid bilayer) under an applied potential (Fig. 1A). As the diameter of dsDNA (2 nm)[40] is greater than the diameter of the central constriction (1.4 nm),[41] the DNA is held within the nanopore for tens of milliseconds up to seconds before 'unzipping' into its constituent single-stranded components.[42]

The residual current during DNA residence within αHL is highly dependent on the composition of the bases situated at the latch constriction of αHL (Fig. 1A).[36,37,39] A DNA duplex containing a single mismatched base-pair positioned at the latch constriction results in a current signature that modulates between two distinct states (I_1 and I_2). In contrast, for a fully-complementary reference duplex, no modulation is observed (Fig. 1B). We have speculated that the modulating signatures are a result of a cytosine base flipping in and out of the helical structure.[35] Base flipping is expected to be significantly more prominent at mismatch sites because of their dynamic structure relative to a fully complementary duplex.[21,22,30]

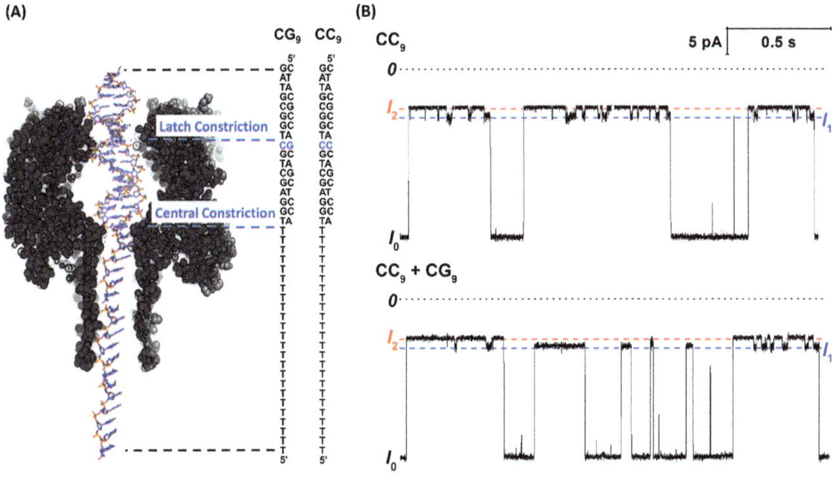

Fig. 1 A duplex within the αHL nanopore gives unique modulating current signatures when a CC mismatch is located at the latch constriction. (A) A dsDNA duplex with a 5′-poly-T tail is driven into αHL up to the central constriction under an applied potential, with the mismatch site located in proximity to the 2.4 nm latch constriction. (B) (top) *i*–*t* trace for the capture and analysis of dsDNA containing a single CC mismatch (3 capture and unzipping events are shown in this trace). When a CC mismatch pair is present at the latch, the residual blocking current modulates between two distinct current states, I_1 and I_2. (bottom) *i*–*t* trace recorded in a solution containing a mixture of dsDNA molecules with either the CC mismatch or the CG Watson–Crick base pair. Current modulation is not observed for a fully-complementary duplex (only the I_1 state is observed, *e.g.*, the second capture event in the trace). Data were recorded in a 10 mM phosphate buffer (pH 7.5) with 0.25 M KCl at 25 °C. A potential of 120 mV was applied across the αHL channel.

Herein, we present measurements of the temperature dependence of base flipping at a mismatch site in a DNA duplex held within the protein nanopore αHL. We extract kinetic (activation energy, E_A) and thermodynamic (enthalpy, ΔH, and entropy, ΔS) parameters for the observed DNA conformational changes inside αHL. We demonstrate that these measured base pair lifetimes and activation energies are in good agreement with previously published base flipping dynamics data from NMR and fluorescence measurements. We discuss the evidence that supports our speculative theory of base flipping, and the potential for using αHL to measure conformational changes in DNA at the single molecule level.

Results

Measuring base flipping kinetics for a DNA mismatch site at the latch constriction of αHL

When a single DNA helix is captured inside the α-hemolysin (αHL) nanopore, there is significant attenuation of the ion flux and the measured electrical current decreases. Furthermore, when a DNA helix containing a cytosine–cytosine (CC) mismatch is positioned at the αHL latch constriction during capture, a distinct modulation of the electrical signal is observed that is described by two well-defined residual current states (I_1 and I_2), as shown in Fig. 1 and 2. These states

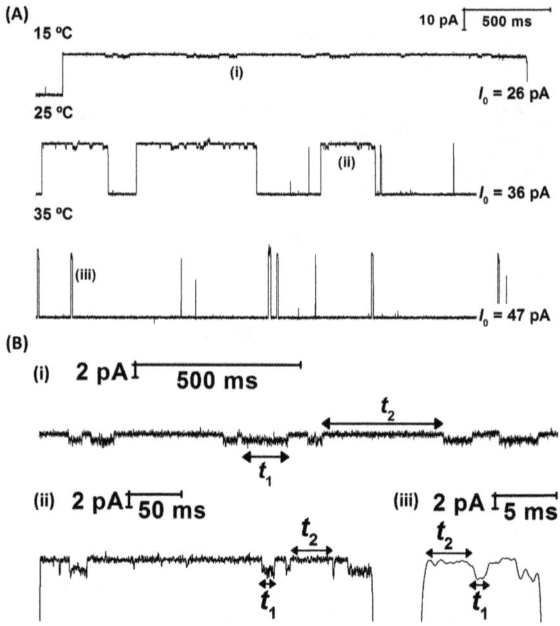

Fig. 2 Effect of temperature on measuring base flipping kinetics inside the αHL nanopore. (A) Representative current–time traces showing the effect of temperature on measuring base flipping of DNA molecules captured with αHL. (B) Expanded view of one event each at (i) 15 °C (ii) 25 °C and (iii) 35 °C highlighting the significant change in lifetimes of states I_1 and I_2 as a function of temperature. Further examples are shown in Fig. S1–S5.† Data were recorded in a 10 mM phosphate buffer (pH 7.5) with 0.25 M KCl. A potential of 120 mV was applied across the αHL channel.

represent two distinct configurations of the dsDNA inside the αHL channel that result from base flipping. The focus of our discussion is on the energetics and dynamics of transitioning between these two chemical states *via* single molecule measurements.

The sequences of the two dsDNA molecules used in these studies, shown in Fig. 1A, differ by a single-base change that results in a CC mismatch at position 9, as measured from the end of the duplex captured within the ion channel vestibule. A poly-T tail, 24 nucleotides long, is attached to the 5′ end of one ssDNA strand in order to thread the DNA by electrophoresis into the αHL channel under an applied voltage of 120 mV using Ag/AgCl electrodes placed on opposite sides of the channel. All measurements were performed in aqueous solutions containing 0.25 M KCl, with temperature control of ±1 °C.

The latch of αHL, which is just 0.4 nm larger in diameter than DNA, constitutes the tightest constriction relative to the duplex and is a major source of resistance in the protein channel.[38,41] Assuming that the measured residual current is primarily determined by the DNA conformation at the latch constriction, we assign I_1 in Fig. 1 to a DNA conformation in which all of the base-pairs are intrahelical and contained within the duplex. This assignment is based on the identical residual currents observed for state I_1 for the duplex CC_9 and the sole state observed for a fully complementary reference duplex, as shown in Fig. 1B. The residual current for the duplexes CC_9 and CG_9 are expected to be very similar when the conformation/structure of the base-pairs at the latch constriction is similar.

We observe that the lifetimes of the I_1 and I_2 states are highly dependent on temperature (Fig. 2), with a 100-fold decrease in the lifetimes of each state as the temperature is increased from 15 °C to 35 °C. The overall dwell times of DNA inside the αHL nanopore prior to unzipping also increase at lower temperatures. At lower temperatures, the dsDNA within αHL is stabilized and less likely to unzip into its constituent single-stranded components.[43] The temperature also affects the measured current of the two blocked current states (I_1 and I_2) and the open channel current (I_0). In each case, the current rises approximately linearly with temperature (Fig. S6†), because of an increase in ion mobility.[36]

The lifetimes of the two residual current states I_1 and I_2 are well described by first-order rate kinetics at all temperatures. We extracted lifetime constants (τ) for each state from histograms of the state lifetimes, examples of which are shown in Fig. 3. For a given experiment, the fraction (n/n_T) of the measured states I_x ($x = 1$ or 2) with lifetime τ is given by:

$$\frac{n(I_x)}{n\left(I_{x(\mathrm{T})}\right)} = \mathrm{e}^{-k_x t} \tag{1}$$

where $n(I_x)$ is the number of times the residual current is in state x, and $n(I_{x(\mathrm{T})})$ is the total number of measured states. The rate k (s^{-1}) describes the kinetics of transitions between states, and is inversely related to τ.

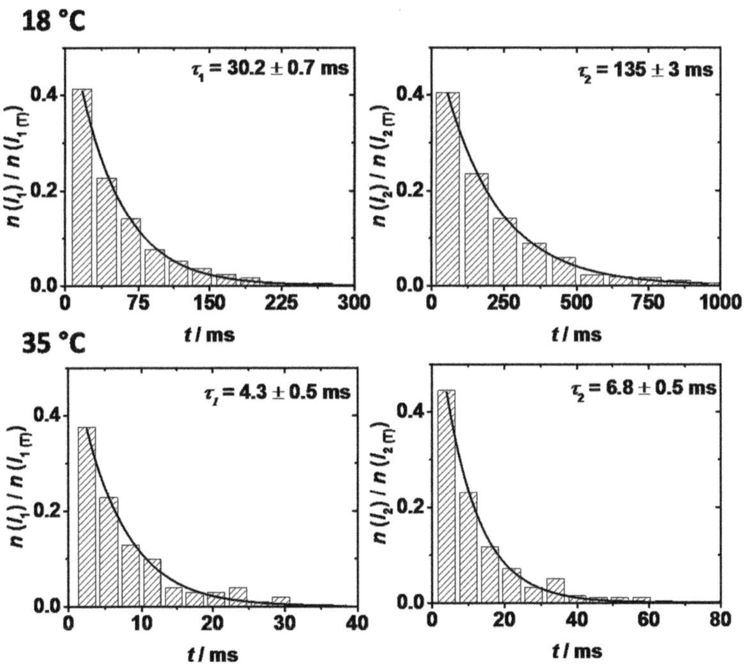

Fig. 3 Effect of temperature on the lifetimes of states I_1 and I_2. Representative lifetime histograms (examples shown above for 18 °C and 35 °C) were used to extract first-order lifetimes (τ) and thus, rate constants (k) as a function of temperature. Lifetime histograms for 15, 20, 25 and 30 °C temperatures are shown in Fig. S7.†

$$\tau_x = 1/k_x \tag{2}$$

While states I_1 and I_2 equate to possible conformations of DNA inside the protein pore, transitions between the states represent transformations from one conformation to another (Fig. 4A). For state I_1, two transitions are possible, either base flipping into state I_2, which we have assigned to a DNA conformation in which one of the cytosine bases is extrahelical, or I_0, which corresponds to unzipping of the dsDNA into the constituent single-stranded components, leading to the open channel without DNA inside (Fig. 4B). The transition from I_1 to I_0 thus describes the well-known unzipping process for dsDNA inside αHL.[35] The dwell time of the DNA inside αHL prior to unzipping is dependent on the duplex stability, and for 17 base-pair duplexes used in this study, unzipping times of 10–100 ms are expected.[44–46]

As discussed previously, we speculate that the transition from I_1 to I_2 represents a base flipping out of the helix. In our model, the rotation of a single cytosine base out of the helical stack constitutes a conformational change in the DNA structure that gives rise to a measurable change in the residual current. The extrahelical base has a lifetime constant τ_2, with the I_2 to I_1 transition representing the base flipping back into the helical structure. Curiously, we never

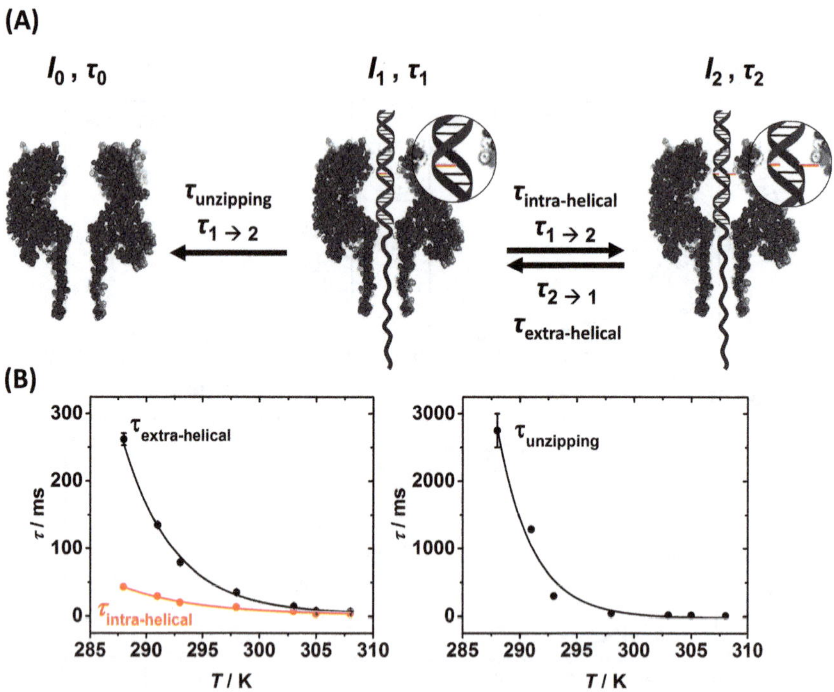

Fig. 4 The effect of temperature on the kinetics of base flipping and unzipping. (A) Model for the base flipping and unzipping processes. Transitions between states I_1 and I_0 are unzipping of the DNA, while transitions between states 1 and 2 represent base flipping. Unzipping from state I_1 is not observed. (B) Effect of temperature on extracted time constants for base flipping and unzipping.

observe duplex unzipping from state I_1 (to I_0), implying that the extrahelical base is interacting with the latch constriction in such a way as to raise the energy barrier for unzipping the helix. We discuss this topic in more detail later.

The intrahelical lifetime (the time before the base flips out of the helix) can be separated from the unzipping time (the dwell time in the pore prior to unzipping) using simple probability analysis. The time constant for state I_1 (τ_1) comprises both the unzipping time and the intrahelical lifetime:

$$\frac{1}{\tau_1} = \frac{1}{\tau_{\text{unzipping}}} + \frac{1}{\tau_{\text{intra-helical}}} \tag{3}$$

The unzipping and intrahelical lifetime constants can be separated from one another by calculating the fraction of transitions along a particular pathway as a fraction of all the transitions. The intrahelical lifetime constant is given by:

$$\frac{1}{\tau_{\text{intra-helical}}} = \frac{n_{1\to 2}}{n_{1\to 0} + n_{1\to 2}} \frac{1}{\tau_1} \tag{4}$$

where $n_{1\to 2}$ represents transitions between states I_1 and I_2 (the base flipping out of the helix) and $n_{1\to 0}$ represents transitions between states I_1 and I_0 (unzipping of the helix). Because the only transition from state I_2 is to I_1, the lifetime constant τ_1 is precisely the extrahelical lifetime:

$$\tau_{\text{extra-helical}} = \tau_2 \tag{5}$$

The intrahelical, extrahelical and unzipping lifetime constants are shown for the CC-mismatch containing duplex as a function of temperature in Fig. 4B. In all three cases, the change in lifetime constants with temperature is described by an exponential function. The lifetime constants that describe the base flipping process, $\tau_{\text{extrahelical}}$ and $\tau_{\text{intrahelical}}$, converge at a value of approximately 6 ms at 35 °C. Although αHL is stable over a wide temperature range,[47] we were unable to resolve base flipping at higher temperatures because the dwell time of DNA inside the pore prior to unzipping becomes comparable to the intrahelical lifetime ($\tau_{\text{unzipping}} \approx \tau_{\text{intrahelical}}$), and in many cases the DNA unzips before the base flips out. In addition, the increased noise associated with the current measurement at higher temperatures[38] increases the difficulty of distinguishing between the two residual current states I_1 and I_2.

The measured unzipping times as a function of temperature are comparable to those published in other unzipping experiments with duplexes of a similar length and stability.[48] The unzipping time decreases as a function of increasing temperature because of changes to the global stability of the helix, where higher temperatures provide more energy required to break the stacking and hydrogen-bond interactions that hold the helical structure together.[20] Overcoming these interactions is a pre-requisite to unzipping the helical structure into its constituent components.

The lifetime constant I_1, which we speculate represents the intrahelical lifetime, ranges from 50 ms at 15 °C to 6 ms at 35 °C. Values for the intrahelical lifetimes reported from NMR measurements, which are generally performed at low temperatures to improve resolution, range from 3 ms up to 300 ms at 15 °C, consistent with our measurements within the αHL nanopore. Such a wide range of intrahelical lifetimes are reported from NMR measurements because the stability of individual base-pairs is highly dependent on both the surrounding sequence context[24] and the nature of the base-pair itself,[22,30] with, for example, a GC pair being more stable than AT because of the additional hydrogen bond.

It should be noted that recently published values from fluorescence measurements have suggested that base-pair lifetimes may be considerably longer, with values in the tens of seconds reported at 25 °C.[21] The discrepancy is attributed to differences in each measurement, with shorter lifetimes reported from NMR measurements because imino-proton exchange, which occurs when a base is exposed to the surrounding solvent, can occur before a base is flipped fully out of the helix.[49,50] However, whether an individual base must rotate out by 180° to fulfil the biological function of 'base flipping', that is, to permit access to maintenance and repair proteins, is debatable.

The extrahelical lifetimes measured from the αHL nanopore measurements range from 250 ms at 15 °C to 6 ms at 35 °C. These times are significantly longer than those reported in bulk solution from NMR measurements,[22] but comparable to those reported by Zhao and co-workers who measured extrahelical lifetimes between 6 and 30 ms at 25 °C using fluorescence, comparable to the values we report here.

Extracting kinetic and thermodynamic parameters for base flipping

The derived time constants for the base flipping process were used to extract activation energies for flipping a base into or out of the DNA helix (Fig. 5). The activation energy, E_A, is related to the rate constant, k, at a given temperature, T, by the Arrhenius equation:

$$\ln\left(\frac{1}{\tau}\right) = \ln(k) = \ln C + \frac{E_A}{RT} \tag{6}$$

where k is the rate constant for flipping the base in or out of the helix, R is the gas constant and C is the constant 'pre-exponential' factor.

The activation energy required to flip a cytosine base out of the helix at the latch constriction of αHL was found to be $E_{A(\text{flip-out})} = 18 \pm 1$ kcal mol^{-1}, which is

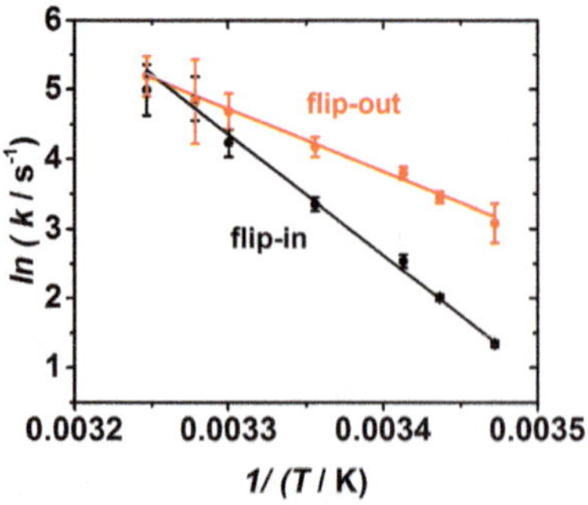

Fig. 5 Arrhenius plots were used to derive the activation energy required flip a base out of the helix (red line), and the activation energy required to overcome interactions between the latch constriction of αHL and the extrahelical cytosine base in order flip the base back into the helical stack (black line). Values of the activation energy were $E_{A(\text{flip-out})} = 18 \pm 1$ kcal mol^{-1} and $E_{A(\text{flip-in})} = 34 \pm 1$ kcal mol^{-1}.

within the range of base flipping activation energies experimentally measured by NMR and fluorescence techniques (6 to 21 kcal mol^{-1}).[21,22,24] Activation energies between 18 and 20 kcal mol^{-1} have also been predicted from potential mean force calculations.[25–27] The high activation energy barriers for the base flipping process are indicative of the significant molecular forces that must be disrupted to remove a base from the helical stack, including breaking of hydrogen bond(s) and overcoming the base-stacking interactions.[18]

Within the αHL nanopore, we determined the flip-in activation energy to be $E_{A(\text{flip-in})} = 34 \pm 1$ kcal mol^{-1}. The higher activation energy for flipping the base into the helix, relative to flipping the base out, indicates that the extrahelical state is the more stable conformation, which is not the case in the absence of a protein.

Base flipping in and out within the αHL nanopore occurs readily on the experimental timescale of our measurement, and as such, the derived rate constants can be used to compute an equilibrium constant, K_{open}.

$$K_{\text{open}} = \frac{k_{\text{out}}}{k_{\text{in}}} = \frac{\tau_{\text{extra-helical}}}{\tau_{\text{intra-helical}}} \tag{7}$$

In eqn (7), k_{in} and k_{out} are the rate constants for flipping the base in and out of the helix, respectively, and K is the equilibrium constant in the direction of flipping out the base.

From the temperature-dependent equilibrium constants, we determined the enthalpy and entropy of the base flipping process using the linearized version of the van't Hoff equation (Fig. 6). The van't Hoff equation relates the equilibrium constant to the changes in the enthalpy, ΔH, and entropy, ΔS, between the flipped-in and flipped-out states:

$$\ln(K_{\text{open}}) = -\frac{\Delta H}{RT} + \frac{\Delta S}{R} \tag{8}$$

Values for the enthalpy and entropy changes that occur when a base flips out of the helix at the αHL latch were determined as $\Delta H = -13.6 \pm 0.3$ kcal mol^{-1} and $\Delta S = -44 \pm 1$ kcal mol^{-1} K^{-1}, respectively, from the slope and intercept of Fig. 6. The positive slope of the Van't Hoff plot (Fig. 6) indicates that flipping out the base inside the αHL is exothermic, while the negative intercept indicates that the process is entropically disfavored. The overall change in Gibbs free energy that

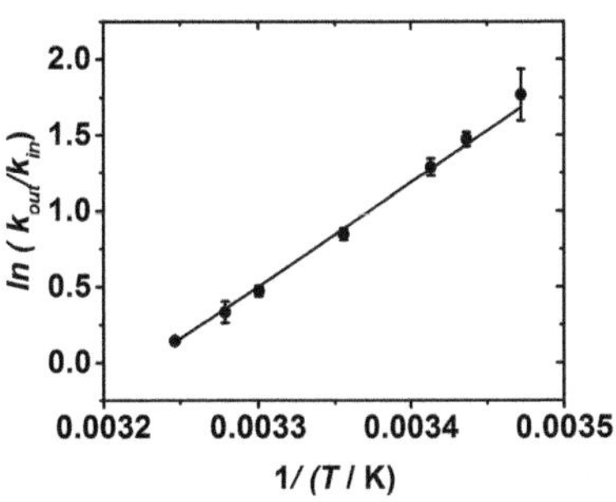

Fig. 6 Van't Hoff plot used to determine values for the enthalpy and entropy of base flipping, which were calculated as $\Delta H = -13.6 \pm 0.3$ kcal mol^{-1} and -44 ± 1 kcal mol^{-1} K^{-1}.

accompanies the base flipping out of the helix at 25 °C is -0.50 ± 0.02 kcal mol^{-1}, indicating that the process is marginally spontaneous. The difference in stability of the DNA when all of the bases are intrahelical, or one cytosine base is flipped out of the helical stack is small, equivalent to the Gibbs free energy that accompanies the formation of a single hydrogen bond between water molecules (*ca.* 0.5 kcal mol^{-1} at 25 °C),[51] or the free energy of stacking between a CC mismatch and a neighbouring base pair (0.6–1.2 mol^{-1} at 25 °C).[52]

Discussion

Before attributing the modulating current states observed in our experiments to base flipping, we considered a number of other conformational changes that might give rise to a measurable change in ion flux through αHL while DNA is inside the pore. Based on our observations, several criteria must be met for any proposed mechanism:

1. The conformational change must be large because of the significant energy barrier (as measured from the activation energy) that separates the two states.

2. Each of the two conformations must be relatively stable to account for the millisecond lifetimes.

3. The conformational change must be specific to the identity of the base-pair of the latch, and localized to the latch constriction itself.

While we cannot completely discount that conformational changes other than base flipping give rise to the observed modulating current signatures, there are no other conformational changes in DNA that meet all three of the above criteria. The majority of DNA conformational changes, such as bending, breathing and twisting, are on very short timescales.[1,2] While interactions between the inner ion channel surface and the DNA backbone are expected,[53] it seems unlikely that they would change the DNA conformation significantly enough to generate a measurable change in the ion flux. In any case, interactions with the phosphate backbone should be the same regardless of the base-pair identity. As another potential mechanism, we did also consider DNA sliding vertically within the pore, but found the kinetics to be independent of voltage;[35] the driving force against diffusion that holds the DNA within the pore. Thus, our discussion herein assumes base flipping as the process observed from our measurements of DNA inside the αHL nanopore.

The lifetime of a base outside the helical stack for spontaneous base flipping (that is, in the absence of an interaction that stabilizes the extrahelical state) is shorter than the lifetime of the conformation in which all of the bases are held within the helical stack.[33,34] This is not surprising given that stacking and hydrogen-bonding interactions will stabilize intrahelical base pairs.[20,52] However, we observe long extrahelical lifetimes for the cytosine base in the confined environment of the latch constriction of αHL. Taken together with the large activation energy barrier required to flip the base back into the helix, our data strongly suggest a stabilizing interaction between the internal surface of the αHL latch and an extrahelical cytosine base.

Analysis of the crystal structure of αHL indicates that the latch constriction consists of a ring of lysine residues that will be positively charged at pH 7.5.[41] Interactions between the nucleobase cytosine and lysine amino acid residue have been previously reported.[54–56] We speculate here that the latch constriction is capable of stabilizing the extrahelical cytosine base, possibly through hydrogen bonding interactions. The precedent for such an interaction is found in the proteins that are known to utilize base flipping to repair or alter DNA base-pairs as a part of cell functions.[33,34] Examples of such proteins include UDG (uracil-DNA glycosylase) and DNMT (DNA methyltransferase). NMR measurements have suggested that extrahelical lifetimes in bulk solution last less than a microsecond at 25 °C, and so, for proteins such as UDG to be able to carry out their function, they must be capable of interacting and stabilizing the extrahelical base.[33,34]

Base flipping in the presence of repair proteins has been found in some cases to be active, and in others, passive. In active cases, the kinetic barrier for base flipping is lowered, while in the passive case, the protein stabilizes the extrahelical state without altering the activation barrier. UDG falls into the latter category, with a body of evidence demonstrating that the enzyme can alter the equilibrium constant for opening base-pairs by stabilizing the extrahelical state by up to 23-fold.[19,33,34] The activation energies we measure for outward base flipping inside αHL are comparable to those in bulk solution. We therefore speculate that the lysine residues at the latch constriction are not actively flipping the base out of the helix, but interactions between cytosine and lysine stabilize the extrahelical base once it is already flipped out. In this scenario, the resulting interactions between the flipped-out base and the αHL latch replace the base-stacking and hydrogen bonding that stabilizes the intrahelical state, resulting in a shift in the equilibrium constant that stabilizes the extrahelical base relative to the intrahelical conformation.

Conclusions

We have measured the temperature-dependent kinetics of base flipping at a DNA cytosine–cytosine mismatch site located at the latch constriction of α-hemolysin. Values for the intrahelical lifetime of the base-pairs, as well as the activation energy required to flip a base out of the helix, are in agreement with values previously reported from NMR measurements of DNA in bulk solution. We have speculated that an extrahelical cytosine is stabilized by interactions with the positively charged lysine residues that comprise the latch constriction in αHL. Thus, the latch constriction of α-hemolysin constitutes a useful tool for the measurement of biologically relevant conformational changes in double-stranded DNA at the single molecule level.

Experimental

DNA preparation & purification

DNA was prepared from commercially available phosphoramidites (Glen Research, Sterling, VA) by the DNA Core Facility at the University of Utah. Purification was *via* an anion-exchange HPLC column running a linear gradient of B from 25% to 100% over 30 min while monitoring UV absorbance at 260 nm ($A =$ 20 mM NaPi, 1 M NaCl, pH 7 in 10% CH_3CN/90% ddH_2O, $B =$ 10% CH_3CN/90% ddH_2O, flow rate = 3 mL min^{-1}).

Chemicals & materials for nanopore measurements

Unless otherwise stated, all chemicals were purchased from Sigma-Aldrich and used without further purification. The buffer solutions were prepared as 10 mM phosphate, 0.25 M KCl at pH 7.5. Wild-type αHL was purchased from List Biological Laboratories in the monomer form of lyophilized powder and dissolved in water at 1 mg mL^{-1}. 1,2-Diphytanoyl-*sn-glycero*-3-phospho-choline (DPhPC) was purchased from Avanti and dissolved in decane at 10 mg mL^{-1} and used to form the bilayer. The bilayer was supported by a glass nanopore membrane of diameter 600 nm, the fabrication of which has been described previously.[57,58] Glass nanopore membranes were modified with a 2% (v/v) (3-cyanopropyl) dimethylchlorosilane in acetonitrile to create a moderately hydrophobic surface. The DNA duplexes were annealed by mixing the 41-mer and 17-mer at a 1 : 1 mole ratio, followed by heating in a 90 °C water bath for 5 min and then cooling to room temperature over 3 hours.

Current–time recordings

Current–time (*i*–*t*) recordings were performed using the low noise Nanopatch system (Electronic BioSciences, Inc., San Diego, CA). A voltage was applied across the GNM between two Ag/AgCl electrodes placed inside and outside of the capillary. A lipid bilayer was deposited across the GNM orifice as previously described. A pressure of 60 to 80 mmHg was applied to the inside of the GNM capillary using a syringe, allowing the lipid bilayer to be functional for the protein channel reconstitution.[58] Next, 0.2 μL of αHL monomer solution at 1 mg mL^{-1} was added to the *cis* (external) side of the GNM (a volume of 350 μL). Protein reconstitution into the lipid bilayer was indicated by a single jump in the current of approximately 0.3 pA mV^{-1} at 25 °C in 0.25 M KCl (pH 7.5) electrolyte. The duplex DNA (15 μM) was added to the *cis* (external) side of the GNM prior to measurement. A voltage of 120 mV was applied *trans vs. cis*, (Ag/AgCl electrode inside the capillary *vs.* Ag/AgCl electrode placed in the external solution (*cis* negative)).

Temperature control was achieved by a Peltier heater/cooler (custom PID control of a small-scale thermoelectric cooler (CUI Inc., CP20151)) situated directly below the *cis* (external reservoir). The temperature is measured by a thermocouple (0.1 °C precision) residing in the same solution as the pore (volume 350 μL). The distance between the pore orifice and the thermocouple is <2 mm. Thermal equilibrium at each desired temperature was reached in approximately 2 min, and each measurement reported was taken after leaving a minimum of 5 min to equilibrate, after which the solution (internal and external) and thermocouple are expected to be at the same temperature.

Data collection and analysis

I–*t* blockades that lasted longer than 1 ms in total were identified as DNA unzipping events, and those less than 1 ms were identified as single-stranded translocations. Events were extracted using QuB version 2.0.023 (freeware, http://www.qub.buffalo.edu). QUB was used to assign one of three states (0, 1 or 2) to continuous current–time data using the software's inbuilt 'idealization' SKM (segmental K-means) method, which uses mathematical models to predict changes in state based on the current amplitude. State 0 corresponds to the open channel current. States 1 and 2 are the two modulating states observed when the duplex is inside αHL. The lifetimes and average current of each state were used to plot the histograms shown in this article. The number of transits between each state (*e.g.* 1 to 2 or 1 to 0) was extracted using QUB's 'transit' function. Origin Pro 2015 (OriginLab) was used to plot histograms of the event time and determine lifetime constants.

Acknowledgements

RPJ acknowledges funding from a Marie Curie International Outgoing Fellowship under the EU FP7 program (Project No. 625984). This work was funded in part by a grant from the NIH (R01 GM093099). The authors thank Electronic BioSciences Inc. (San Diego, CA) for donating the ion-channel recording instrument and software, and Dr M. A. Edwards (University of Utah) for assistance with the statistical and probability analysis.

References

1 R. Galindo-Murillo, D. R. Roe and T. E. Cheatham III, *Nat. Commun.*, 2014, **5**, 5152.
2 R. Galindo-Murillo, D. R. Roe and T. E. Cheatham III, *Biochim. Biophys. Acta*, 2015, **1850**, 1041–1058.
3 M. D. Frank-Kamenetskii and S. Prakash, *Phys. Life Rev.*, 2014, **11**, 153–170.
4 A. Vologodskii, *Phys. Life Rev.*, 2006, **3**, 119–132.
5 C. Lavelle, *Curr. Opin. Genet. Dev.*, 2014, **25**, 74–84.
6 P. H. von Hippel, N. P. Johnson and A. H. Marcus, *Biopolymers*, 2013, **99**, 923–954.
7 T. M. Okonogi, A. W. Reese, S. C. Alley, P. B. Hopkins and B. H. Robinson, *Biophys. J.*, 1999, **77**, 3256–3276.
8 S. Rüdisser, A. Hallbrucker and E. Mayer, *J. Am. Chem. Soc.*, 1997, **119**, 12251–12256.
9 Ł. Szyc, M. Yang, E. T. J. Nibbering and T. Elsaesser, *Angew. Chem., Int. Ed.*, 2010, **49**, 3598–3610.
10 A. Krueger, E. Protozanova and M. D. Frank-Kamenetskii, *Biophys. J.*, 2006, **90**, 3091–3099.
11 J. Fei and T. Ha, *Proc. Natl. Acad. Sci. U. S. A.*, 2013, **110**, 17173–17174.
12 D. Andreatta, S. Sen, J. L. Pérez Lustres, S. A. Kovalenko, N. P. Ernsting, C. J. Murphy, R. S. Coleman and M. A. Berg, *J. Am. Chem. Soc.*, 2006, **128**, 6885–6892.
13 S. Nonin, J.-L. Leroy and M. Gueron, *Biochemistry*, 1995, **34**, 10652–10659.
14 R. J. Roberts and X. Cheng, *Annu. Rev. Biochem.*, 1998, **67**, 181–198.
15 J. T. Stivers and Y. L. Jiang, *Chem. Rev.*, 2003, **103**, 2729–2760.
16 N. Huang, N. K. Banavali and A. D. MacKerell, *Proc. Natl. Acad. Sci. U. S. A.*, 2003, **100**, 68–73.
17 L. Larivière and S. Moréra, *J. Biol. Chem.*, 2004, **279**, 34715–34720.
18 J. T. Stivers, *Prog. Nucleic Acid Res. Mol. Biol.*, 2004, **77**, 37–65.
19 J. T. Stivers, *Chem.–Eur. J.*, 2008, **14**, 786–793.
20 J. SantaLucia, *Proc. Natl. Acad. Sci. U. S. A.*, 1998, **95**, 1460–1465.
21 Y. Yin, L. Yang, G. Zheng, C. Gu, C. Yi, C. He, Y. Q. Gao and X. S. Zhao, *Proc. Natl. Acad. Sci. U. S. A.*, 2014, **111**, 8043–8048.
22 J. G. Moe and I. M. Russu, *Biochemistry*, 1992, **31**, 8421–8428.
23 D. Coman and I. M. Russu, *Biophys. J.*, 2005, **89**, 3285–3292.

24 E. Folta-Stogniew and I. M. Russu, *Biochemistry*, 1994, **33**, 11016–11024.
25 B. Bouvier and H. Grubmüller, *Biophys. J.*, 2007, **93**, 770–786.
26 M. Fuxreiter, N. Luo, P. Jedlovszky, I. Simon and R. Osman, *J. Mol. Biol.*, 2002, **323**, 823–834.
27 N. K. Banavali and A. D. MacKerell Jr, *J. Mol. Biol.*, 2002, **319**, 141–160.
28 D. J. Krosky, F. P. Schwarz and J. T. Stivers, *Biochemistry*, 2004, **43**, 4188–4195.
29 A. E. Every and I. M. Russu, *J. Mol. Recognit.*, 2013, **26**, 175–180.
30 P. K. Bhattacharya, J. Cha and J. K. Barton, *Nucleic Acids Res.*, 2002, **30**, 4740–4750.
31 M. Gueron, M. Kochoyan and J.-L. Leroy, *Nature*, 1987, **328**, 89–92.
32 M. Guéron and J.-L. Leroy, *Methods Enzymol.*, 1995, **261**, 383–413.
33 C. Cao, Y. L. Jiang, J. T. Stivers and F. Song, *Nat. Struct. Mol. Biol.*, 2004, **11**, 1230–1236.
34 C. Cao, Y. L. Jiang, D. J. Krosky and J. T. Stivers, *J. Am. Chem. Soc.*, 2006, **128**, 13034–13035.
35 R. P. Johnson, A. M. Fleming, L. R. Beuth, C. J. Burrows and H. S. White, *J. Am. Chem. Soc.*, 2016, **138**, 594–603.
36 R. P. Johnson, A. M. Fleming, Q. Jin, C. J. Burrows and H. S. White, *Biophys. J.*, 2014, **107**, 924–931.
37 Q. Jin, A. M. Fleming, R. P. Johnson, Y. Ding, C. J. Burrows and H. S. White, *J. Am. Chem. Soc.*, 2013, **135**, 19347–19353.
38 R. P. Johnson, A. M. Fleming, C. J. Burrows and H. S. White, *J. Phys. Chem. Lett.*, 2014, **5**, 3781–3786.
39 Y. Ding, A. M. Fleming, H. S. White and C. J. Burrows, *ACS Nano*, 2015, **9**, 11325–11332.
40 H. R. Drew, R. M. Wing, T. Takano, C. Broka, S. Tanaka, K. Itakura and R. E. Dickerson, *Proc. Natl. Acad. Sci. U. S. A.*, 1981, **78**, 2179–2183.
41 L. Song, M. R. Hobaugh, C. Shustak, S. Cheley, H. Bayley and J. E. Gouaux, *Science*, 1996, **274**, 1859–1866.
42 A. F. Sauer-Budge, J. A. Nyamwanda, D. K. Lubensky and D. Branton, *Phys. Rev. Lett.*, 2003, **90**, 238101.
43 C. E. Angevine, S. J. Seashols-Williams and J. E. Reiner, *Anal. Chem.*, 2016, **88**, 2645–2651.
44 Q. Jin, A. M. Fleming, Y. Ding, C. J. Burrows and H. S. White, *Biochemistry*, 2013, **52**, 7870–7877.
45 T. C. Sutherland, M. J. Dinsmore, H.-B. Kraatz and J. S. Lee, *Biochem. Cell Biol.*, 2004, **82**, 407–412.
46 Y. Wang, K. Tian, L. L. Hunter, B. Ritzo and L.-Q. Gu, *Nanoscale*, 2014, **6**, 11372–11379.
47 X. F. Kang, L.-Q. Gu, S. Cheley and H. Bayley, *Angew. Chem., Int. Ed.*, 2005, **44**, 1495–1499.
48 Q. Jin, A. M. Fleming, C. J. Burrows and H. S. White, *J. Am. Chem. Soc.*, 2012, **134**, 11006–11011.
49 E. Giudice, P. Várnai and R. Lavery, *Nucleic Acids Res.*, 2003, **31**, 1434–1443.
50 U. D. Priyakumar and A. D. MacKerell, *Chem. Rev.*, 2006, **106**, 489–505.
51 K. A. T. Silverstein, A. D. J. Haymet and K. A. Dill, *J. Am. Chem. Soc.*, 2000, **122**, 8037–8041.
52 N. Peyret, P. A. Seneviratne, H. T. Allawi and J. SantaLucia, *Biochemistry*, 1999, **38**, 3468–3477.

53 W. Vercoutere, S. Winters-Hilt, H. Olsen, D. Deamer, D. Haussler and M. Akeson, *Nat. Biotechnol.*, 2001, **19**, 248–252.
54 N. M. Luscombe, R. A. Laskowski and J. M. Thornton, *Nucleic Acids Res.*, 2001, **29**, 2860–2874.
55 V. Bruskov, *Stud. Biophys.*, 1978, **67S**, 43–44.
56 M. Akeson, D. Branton, J. J. Kasianowicz, E. Brandin and D. W. Deamer, *Biophys. J.*, 1999, **77**, 3227–3233.
57 B. Zhang, J. Galusha, P. G. Shiozawa, G. Wang, A. J. Bergren, R. M. Jones, R. J. White, E. N. Ervin, C. C. Cauley and H. S. White, *Anal. Chem.*, 2007, **79**, 4778–4787.
58 R. J. White, E. N. Ervin, T. Yang, X. Chen, S. Daniel, P. S. Cremer and H. S. White, *J. Am. Chem. Soc.*, 2007, **129**, 11766–11775.

Faraday Discussions

PAPER

Protein detection using tunable pores: resistive pulses and current rectification†

Emma L. C. J. Blundell,‡ Laura J. Mayne,‡ Michael Lickorish, Steven D. R. Christie and Mark Platt*

Received 11th April 2016, Accepted 20th April 2016
DOI: 10.1039/c6fd00072j

We present the first comparison between assays that use resistive pulses or rectification ratios on a tunable pore platform. We compare their ability to quantify the cancer biomarker Vascular Endothelial Growth Factor (VEGF). The first assay measures the electrophoretic mobility of aptamer modified nanoparticles as they traverse the pore. By controlling the aptamer loading on the particle surface, and measuring the speed of each translocation event we are able to observe a change in velocity as low as 18 pM. A second non-particle assay exploits the current rectification properties of conical pores. We report the first use of Layer-by-Layer (LbL) assembly of polyelectrolytes onto the surface of the polyurethane pore. The current rectification ratios demonstrate the presence of the polymers, producing pH and ionic strength-dependent currents. The LbL assembly allows the facile immobilisation of DNA aptamers onto the pore allowing a specific dose response to VEGF. Monitoring changes to the current rectification allows for a rapid detection of 5 pM VEGF. Each assay format offers advantages in their setup and ease of preparation but comparable sensitivities.

Introduction

Interest in nanoscale channels within synthetic materials have grown over the last two decades.[1–3] These channels have applications in biosensing,[4–6] material characterisation,[7,8] quantification of ligand–target interactions,[9–11] drug delivery,[12] and mimicking biological systems enabling the study of ionic transport within confined geometries.[13–16] These nanopores have been created in a range of materials including graphene,[17–20] polymers,[21] silicon nitride[22] and glass.[13,14,23,24] The transport of the ions or analyte through the channel can by controlled by tuning the applied potential, pore wall charge, pore size, supporting electrolyte concentration and composition, with a further degree of selectivity by modifying the pore walls with selective ligands.[25–27]

Department of Chemistry, Loughborough University, Loughborough, LE11 3TU, United Kingdom. E-mail: m.platt@lboro.ac.uk; Tel: +44 1509 222 573

† Electronic supplementary information (ESI) available. See DOI: 10.1039/c6fd00072j

‡ These authors contributed equally to the work.

In addition to the pore properties, the translocation speed and frequency of materials such as small molecules, proteins or nanosized particles are also governed by the analyte size and charge,[5,28–31] Fig. 1a. Nanopore sensors can be categorised into two general areas: Resistive Pulse Sensors, RPS, where the translocation of the analyte creates a characteristic change in resistivity within the pore, Fig. 1a and b, or current flux/rectification studies that monitor current–voltage, I–V, and can be dominated by the charge on the pore wall to measure, Fig. 1c. The flux of material through the pore is determined by the small and large pore geometries, d_s and d_l, pore length, L, and analyte charge. By controlling the aspect ratio of the pore, RPS has been used to measure analytes that range from single molecules, DNA, proteins and cellular vesicles to cells, bacteria and viruses; detailed reviews on the types of analytes and applications can be found elsewhere.[4,5,28,30]

We present a comparison between RPS and rectification ratios on a tunable pore platform. The pore is made of polyurethane, PU, allowing manipulation (stretching) in real time to suit the sample.[32] The pores are conical in shape and here typically $d_s > 700$ nm. In the first example we utilise an aptamer-modified nanoparticle to detect Vascular Endothelial Growth Factor, VEGF. By measuring the translocation velocities of the aptamer-modified particles the VEGF protein was detected down to 18 pM, equating to *circa* 10 proteins per particle. In comparison a second strategy was tested by modifying the pores directly with the anti-VEGF aptamer and monitoring the current rectification ratio (measured at ±1.6 V) in the presence of the VEGF protein. The surfaces of the pores were easily modified using a layer-by-layer (LbL) assembly of polymers, such as polyethylene amine, PEI, and polyacrylic acid–maleic acid and PAAMA. The use of PEI and PAAMA allowed for easy modification, and reversible surface charge of the pores giving a pH- and ionic strength-controlled current flow, with current rectification ratios as high as 3. The LbL assembly was shown to be stable for days, allowing the modification of the pore wall with DNA *via* standard carbodiimide chemistry. The

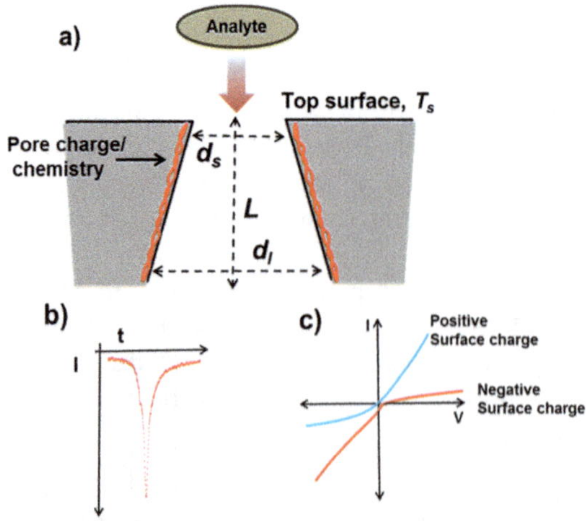

Fig. 1 (a) Schematic of a conical PU pore with small, large pore diameter and pore length, d_s, d_l and L respectively. The top surface of the pore is labelled T_S. (b) Example of a resistive pulse. (c) Schematic of I–V curves for conical pores with different surface charges.

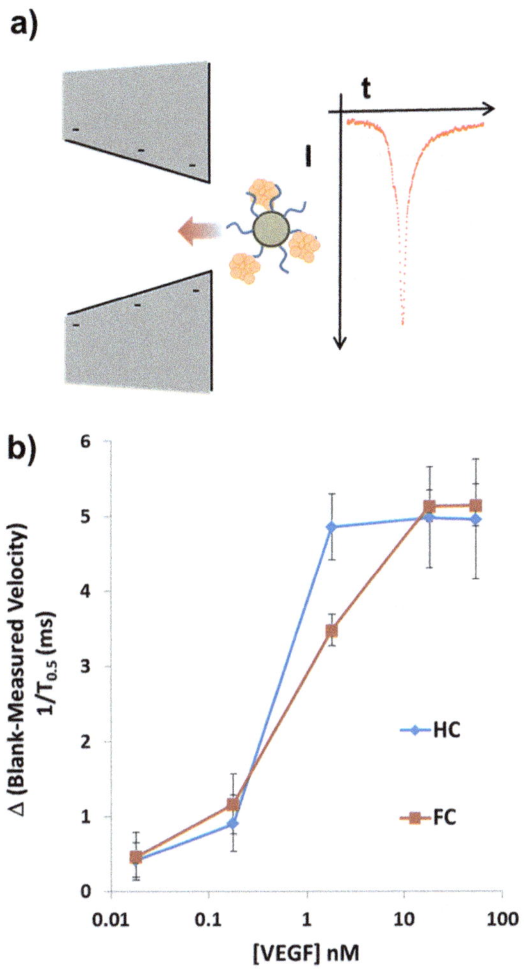

Fig. 2 (a) Schematic of the protein bound to the aptamer-modified particle and typical resistive pulse. (b) Plot of change in translocation time (relative to a blank = 0) *versus* VEGF concentration for aptamer-modified beads, FC aptamer coverage – red, HC – blue circles. Each assay was run in triplicate with over 500 particles counted in each run.

current rectification assay allows for the detection of VEGF down to 5 pM. Despite this comparable sensitivity, RPS offer a larger dynamic range. The scope and ease of each assay format allows for a versatile technology that can be tailored to suite the target analyte.

Materials and methods

Poly(ethyleneimine), PEI, low molecular weight, LMW (LMW PEI $M_w \sim 2000$ g mol^{-1}, 50% wt, 408 700) and high molecular weight, HMW, (HMW PEI M_w 750 000 g mol^{-1}, analytical standard, 50% wt, P3143), poly(acrylic acid-*co*-maleic acid) (PAAMA, $M_w \sim 3000$ g mol^{-1} 50% wt, 416 053), phosphate buffered saline solution (P4417 (0.01 M phosphate buffer, 0.0027 M potassium chloride, 0.137 M

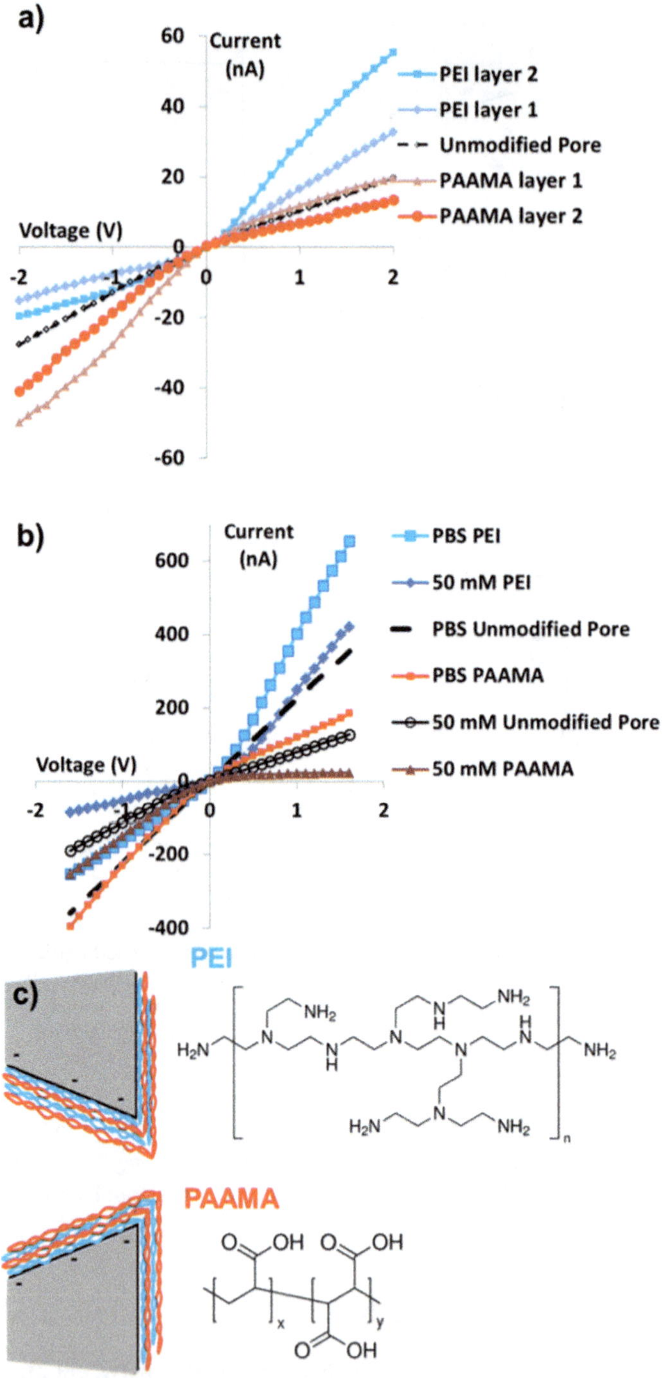

Fig. 3 (a) Current–voltage curves for HMW PEI and PAAMA layers alongside a blank unmodified pore in 5 mM KCl. (b) Current–voltage curves for layer 2 bilayers alongside a blank unmodified pore. (c) Schematic of the layer-by-layer assembly of PEI and PAAMA.

Table 1 Rectification ratios for pores measured in KCl solution at *circa* pH 7. Also listed is the rectification value of the same surface in 1× PBS

	Rectification ratio				
[KCl] mM	Unmodified pore	PEI – layer 1	PAAMA – layer 1	PEI – layer 2	PAAMA – layer 2
5	1.38	0.46	2.41	0.37	2.86
10	1.09	0.67	1.69	0.29	2.44
50	1.09	0.90	1.10	0.73	2.09
PBS (pH 7.4)	1.15	0.81	1.10	0.93	1.29

sodium chloride, pH 7.4)), bovine serum albumin (BSA, lyophilized powder, ≥96%, A2153) and 2-(*N*-morpholino)ethanesulfonic acid hydrate (MES hydrate, ≥99.5%, M2933) fibrinogen from human plasma (F3879), albumin from human serum (A9511) and γ-globulin from human blood (G4386) were purchased from Sigma Aldrich, UK. Tunable conical pores (NP200) were purchased from Izon Science (Christchurch, NZ). Carboxylated polystyrene particles with a mean nominal diameter of 220 nm were purchased from Bangs Laboratories, US and are denoted as CPC200. Potassium chloride (KCl, >99%, P/4240/60) and potassium hydroxide (KOH, 0.1 M, >85%, P/5600/60) were purchased from Fisher Scientific, UK. Hydrochloric acid (HCl, 0.5 M, 37%) was purchased from VWR, UK. 1-Ethyl-3-(3-dimethylaminopropyl)carbodiimide hydrochloride (EDC, 22 980) and Recombinant Human Vascular Endothelial Cell Growth Factor (VEGF, lyophilised, >95%, PHC9394) were purchased from Thermo Scientific, UK. Streptavidin-coated superparamagnetic particles (120 nm, 4352 pmol mg^{-1} binding capacity, 03 121) were purchased from Ademtech, France. The custom DNA oligonucleotide 5′TGTGGGGGTGGACGGGCCGGGTAGATTTTT (V7t1 amine),[33] was purchased as a lyophilised powder (100 pmol μL^{-1}) from Sigma Aldrich, UK. The sequence was synthesised, with a biotin or amine functional group at the 3′ end.

All reagents were used without further purification and all solutions were prepared in purified water with a resistance of 18.2 MΩ cm (TKA, Smart2Pure). pH of solutions were altered using HCl and KOH and the solutions were measured using a Mettler Toledo easy five pH meter with a Mettler Toledo InLab microelectrode.

Particle assay

120 nm diameter streptavidin-coated particles were diluted to a concentration of approximately 5 × 10^9 particles per mL. The diluted particle solutions were then vortexed for 30 s, and sonicated for 2 min, to ensure they were well dispersed. The biotinylated aptamer was added to the streptavidin coated particles (4352 pmol mg^{-1} binding capacity – determined by the supplier) at 113 and 226 nM for 50 and 100% DNA coverage per particle, respectively. The samples were then placed on a rotary wheel for 30 min. Any unbound DNA remaining in solution was then removed *via* magnetic separation by placing the samples onto a Magrack (GE Healthcare, UK) for 30 min. The supernatant was then removed and replaced with new buffer (PBS). The VEGF was added at the required concentration and then placed on a rotary wheel for 30 min before being analysed.

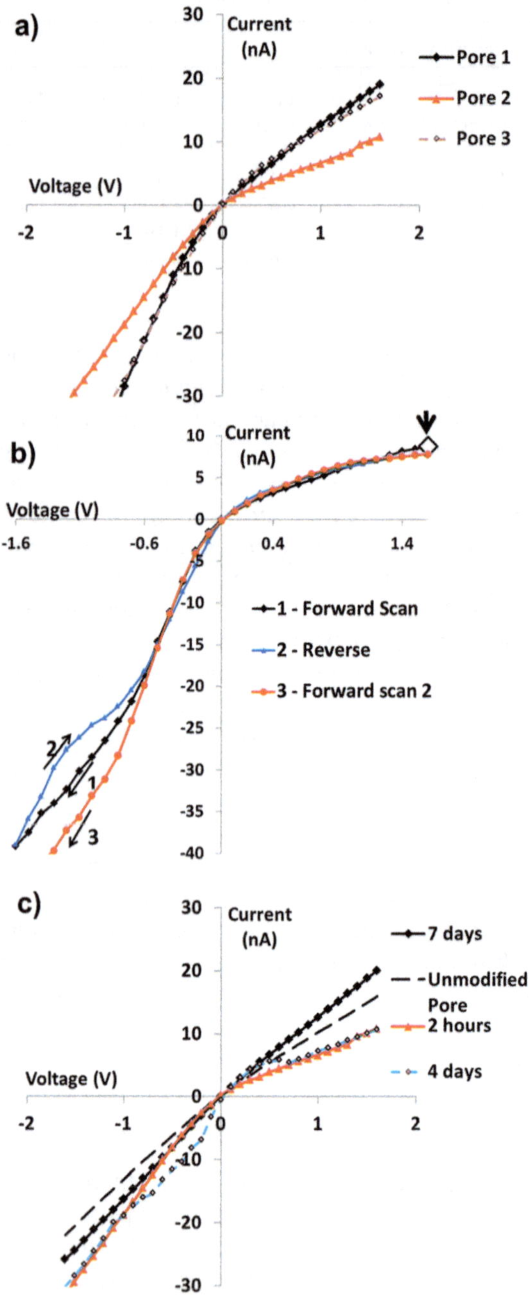

Fig. 4 (a) I–V curves for three different NP200 pores coated with two bilayers. (b) Consecutive scans across the voltage range; starting point is indicated with an arrow. A forward scan refers to moving from +1.6 V to −1.6 V. Each current was recorded after a 5 second wait to the new voltage. (c) I–V curves for an individual pore coated with two bilayers; each day the pore was set up and I–V curves recorded before being washed with DI water and air dried. All data obtained at 45 mm stretch, pH 6.8 and 5 mM KCl.

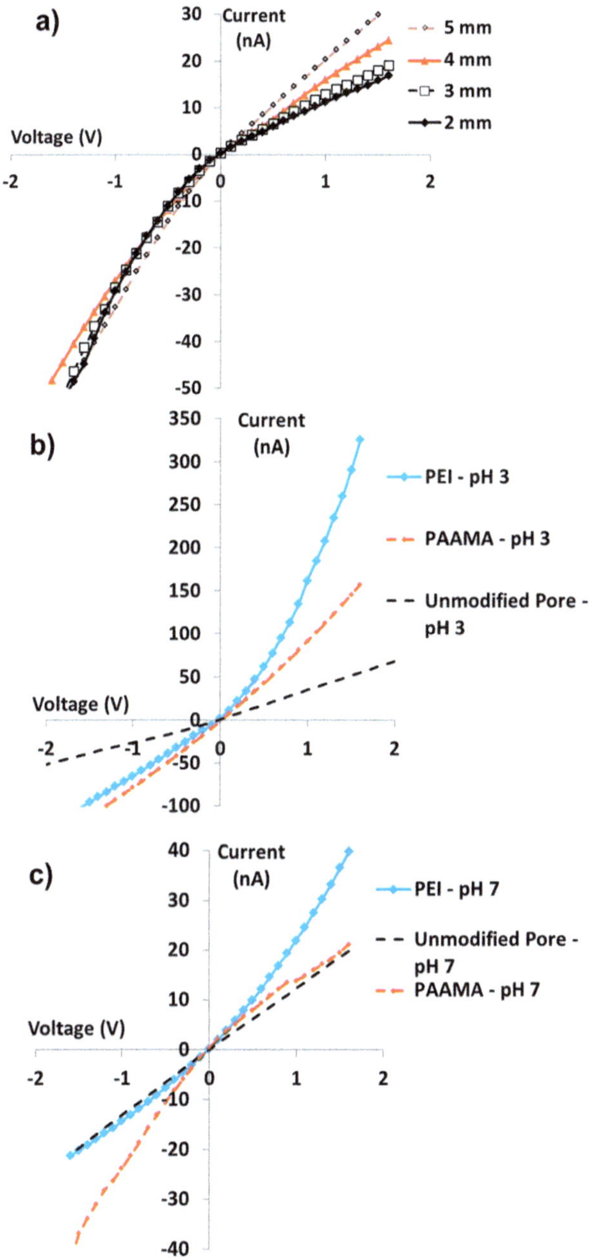

Fig. 5 (a) I–V curves for a NP200 pore coated with two bilayers. Measured at pH 6.8 and 5 mM KCl. (b) I–V curves for an unmodified pore, PAAMA (two complete bilayers) and PEI (1.5 bilayers). Obtained at 45 mm stretch, pH 3 and 5 mM KCl. (c) I–V curves for an unmodified pore, PAAMA (two complete bilayers) and PEI (1.5 bilayers). Obtained at 45 mm stretch, pH 7 and 5 mM KCl.

Table 2 Rectification ratios for pores with different surface chemistry as a function of pH. The concentration of KCl was 5 mM for all experiments. HMW-PEI was used for the bilayers

	Rectification ratio		
pH	Unmodified pore	PEI – layer 2	PAAMA – layer 2
3	0.83	0.31	0.73
7	1.40	0.52	1.98

TRPS set-up

All measurements were conducted using a qNano (Izon Sciences Ltd, NZ) instrument combining tunable nanopores with data capture and analysis software, Izon Control Suite v.3.2. The lower fluid cell contains the electrolyte (75 μL).

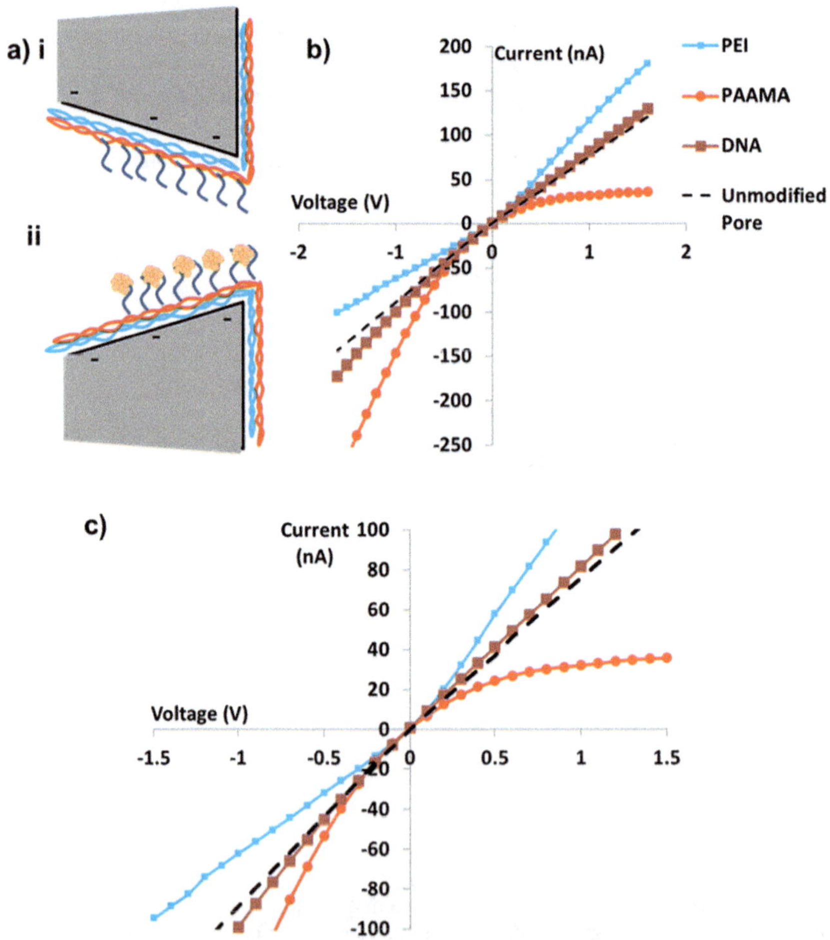

Fig. 6 (a – i) Schematic of the DNA-modified pore and (ii) aptamer-VEGF interaction. (b) I–V curves for each layer in the DNA immobilisation. (c) Enlarged section of (b). All data are recorded at 50 mM KCl, pH 6.8, 45 mm stretch.

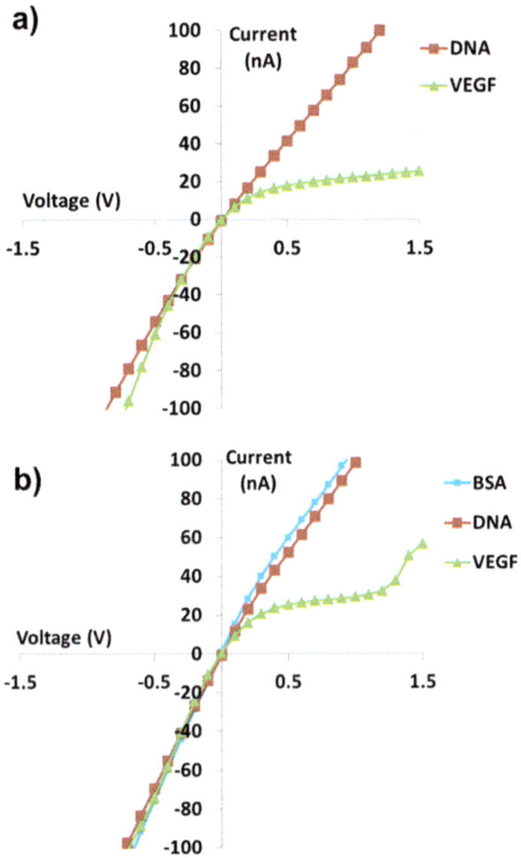

Fig. 7 (a) *I*–*V* curves for aptamer (DNA) and VEGF incubated pores. VEGF was at 50 nM. (b) Equivalent data set for conditions given in a, for a second pore. The pore was first incubated with BSA at 50 nM. KCl is 50 mM, stretch – 45 mm.

The upper fluid cell contains 40 μL of sample (which was suspended in the electrolyte) and an inherent pressure on the system (47 Pa) was present when making a measurement. After each sample run, the system was washed by placing 40 μL of the run electrolyte into the upper fluid cell several times with various pressures applied to ensure there were no residual particles remaining and therefore no cross contamination between samples. The membranes are placed into jaws on the qNano instrument and are capable of being stretched.[34] The applied stretch is quantified by the distance between the jaws, with an upstretched distance being 42 mm, any values quoted below are the additional distances in mm applied to the membranes. Most experiments were run at 45 mm. The pore diameters measured in the SEM and in the text were calculated from the current at 45 mm stretch.

Translocation speed

The velocity of the particle was calculated by extracting the pulse width. Blockade duration events are recorded from the peak of the blockade back to the baseline

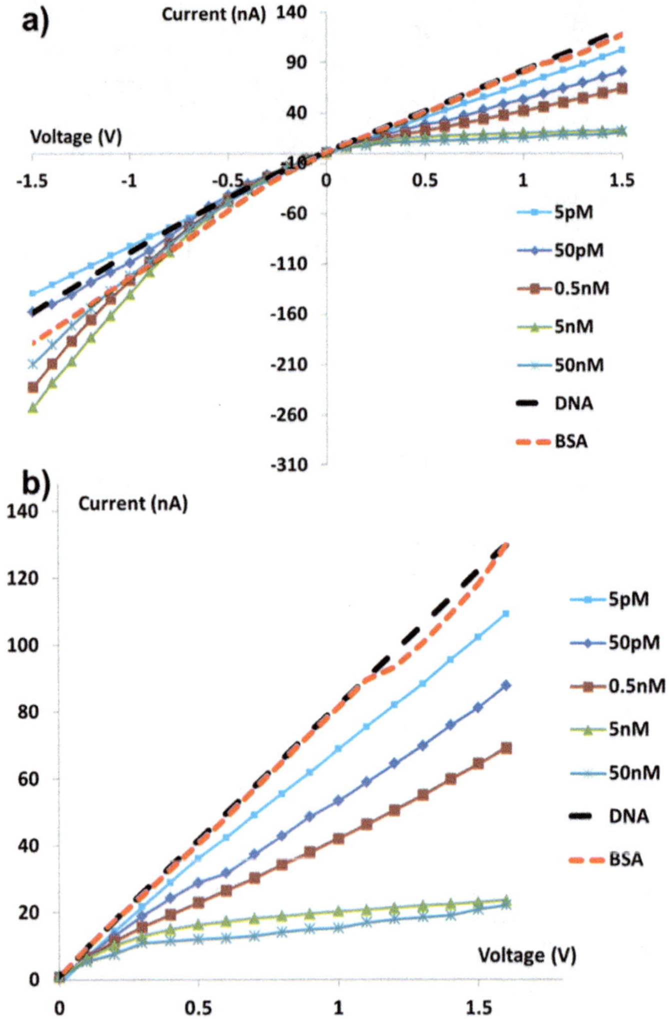

Fig. 8 (a) Plot of rectification ratio *versus* VEGF concentration for a third pore. (b) Magnified section of the current between 0 and 1.6 V. KCl 50 mM, pH 6.8. Stretch 45 mm. Each measurement was taken in series on the same pore.

Table 3 Rectification ratios measured for the data shown in Fig. 8

Concentration of VEGF	Rectification ratio
5 pM	1.36
50 pM	1.93
0.5 nM	3.64
5 nM	11.60
50 nM	10.12
DNA	1.33
BSA	1.53

current; the total time gives the blockade duration. Nine time points are recorded along the peak,[29] relative to different positions within the pore and are denoted $T_{0.90}$, $T_{0.80}$, $T_{0.70}$ *etc.*, here we use one measurement $1/T_{0.5}$ to represent the particle velocity.

Pore modification

Conical pores were modified by incubating the pore in polymer solution (5% wt in H_2O) at a stretch of 45 mm for two hours, followed by rinsing the pores with deionised water. The pores were then incubated with a second polymer layer for 2 hours, and again washed with deionised water. This process was repeated until the required number of layers was achieved.

Modification of PAAMA modified pores with DNA

The aptamer was dissolved in 100 mM MES buffer (pH 5.9) containing 1 mg mL^{-1} EDC. The final concentration of the DNA was 220 nM. The pores were incubated with the DNA and EDC solution for 2 hours.

VEGF *I–V* assay

VEGF was suspended in PBS buffer to give the desired concentration. DNA-modified pores were incubated in VEGF solution, and in each experiment the VEGF solution was only placed on the side of the pore with the small pore opening D_s. When multiple solutions of different concentrations of VEGF were used, the lowest concentration was measured first. The VEGF solution was in contact with the pore for 30 minutes with the pore being rinsed with water ×3, and PBS ×3 after each protein concentration. The current rectification property of the pore was then measured, in a range of KCl solutions starting with 5 nM first and working up to 50 mM. When a BSA control was used, 50 nM BSA was incubated first for 30 minutes, with the rectification properties being measured in KCl solutions before adding VEGF to the pore.

I–V measurements

The pores were mounted between two fluid cells which contain an electrolyte solution. Current–voltage (*I–V*) curves were recorded using Izon control suite v3.2, the potential was stepped in 100 mV increments from +1.6 to −1.6 V and the resulting current measured.

Results and discussion

Assay 1 – measuring translocation times from resistive pulses

To facilitate sample handling and fast assays some RPS strategies have included nanomaterials by either immobilising the target analytes onto their surface facilitating an analyte-induced aggregation,[35,36] or measuring the particle translocation speed/frequency upon the binding of the analyte.[11,30,37–40] With the charge of the particle being a contributing factor in pore translocation speed and frequency, the use of DNA-modified materials and pores has become increasingly popular.[11,30,41–43] The polyanionic backbone of the DNA can enhance or determine the translocation speed/direction and frequency of the current.[28]

In the first assay format 120 nm particles were modified with a VEGF aptamer.[33] In two variations on this experiment different surface coverages of VEGF aptamer were used. Under condition one, the streptavidin binding sites on the particles are saturated with biotinylated aptamer (termed FC); in the second, *circa* half of binding sites (defined by the suppliers specifications) were filled with the aptamer (termed HC). The change in translocation time relative to a particle that was not incubated with the protein, *i.e.* a blank, is shown in Fig. 2. In previous work we have shown how the translocation velocity can be measured from the pulse width and then converted to zeta potentials using Henry's law.[29] The technique was able to distinguish between DNA coverage, in addition to length and structure *i.e.* ssDNA or dsDNA, on the particle surface, with the detection of dsDNA down to 10 nM LOD.[29]

In the strategy reported here rather than converting to zeta potentials, we simply measure changes to the relative velocities and infer the presence of the DNA and proteins. The hypothesis was that the binding of the aptamer to the protein target, Fig. 2a, should result in a change in translocation time. Fig. 2b shows the change in translocation times, relative to the blank, *versus* VEGF concentration for the two coverages of DNA.

There is a decrease in velocity for both sets of aptamer-modified beads as the VEGF concentration increases. The change in aptamer shape from ssDNA to one of a folded tertiary structure upon binding the protein, bringing the DNA closer to the particle surface, could lead to a higher surface change density on the particle. This may result in an increase in particle speed. Conversely, as shown in Fig. 2b, a decrease in particle velocity upon the addition of the protein is observed. The velocity of the particles with HC remains constant at VEGF concentrations above 2 nM. This is attributed to the binding capacity of the beads being reached, and that any additional protein to the solution does not change the numbers on the particle surface, or if it does increase they cannot be measured *via* this technique. The velocity of the FC beads continues to decrease until a concentration of 18 nM is reached. By tuning the aptamer concentration on the particle surface, the dynamic range of the assay can be extended.

The decrease in velocity is thought to be due several parameters that shield the negative charge on the phosphate backbone, increasing the counter ion condensation onto the DNA.[43] Firstly the conformational change to the DNA structure as it binds the protein requires an increased number of counter ions to stabilise the tertiary structure, and secondly the protein pI is 8.5, and therefore is positively charged at the pH used in the experiment. Both particle assays see a decrease in speed, respective to their blanks, at a concentration of 18 pM.

Previous work using aptamer-modified nanomaterials on the TRPS platform measured a change in particle translocation frequency as the transduction signal.[11,39] The benefit of the method presented here is in the ability to use data from each individual particle as they translocate the pore. Thus it does not rely upon averaging data across hundreds of particles per min, reducing run times and bias in samples with different particle concentrations. A potential limitation for future assays within complex biological solutions is the tendency of proteins to foul the pore wall or particle surface. The particle translocation speed is a contribution of the electrophoretic speed of the particle and electroosmotic contribution of the pore wall. If the pore wall charge changes throughout the experiment it may impact on the assay's reproducibility. A solution to this would be to develop a non-fouling coating on the particle and pore wall, preventing the nonspecific adsorption of proteins. Alternatively, the modification of the pore directly could facilitate the detection of the analyte of choice. Other nanopore systems have either utilised the natural surface functionality of the material or modified the pore walls with gold to allow the facile attachment of materials.[44] To date there has been no reported modification of tunable polyurethane (PU) pores for RPS studies, although strategies for the modification of PU are available.[45–47]

Assay 2 – modification of the pore walls and current rectification

An alternative to a particle-based assay is to use the change in ionic current through the pore, which can be controlled through electrostatic interactions *via* the pore wall. This effect is seen to a greater extent within conical pores, and typically recorded for pore openings where the diameter, D_S, is equivalent to the electrical double layer (DL) thickness,[24,48] although the DL does not need to fully extend across the pore opening,[24] and larger pores exhibit current rectification effects.[23,49] Rectifying properties of conical pores are described in detail elsewhere,[48] with the degree of rectification being defined as the ratio of absolute currents recorded at a given negative potential and the identical absolute positive potential. Typically conical pores with charged surfaces do not exhibit ohmic behaviour at lower electrolyte concentrations, and the magnitude of the current through the nanopore at negative potentials is greater or less than the current at positive potentials. The ratios can be tuned by changing the supporting electrolyte, pH, ionic strength and applied voltage, with the current–voltage, I–V, curves recording a preferred direction of current flow[24,48] (Fig. 1c).

Tunable pores fabricated from PU have a small negative surface charge[29,49] at pH > 5, and are conical in shape. The unmodified pores used here had an approx. pore diameter, d_S, of 800 nm, at a stretch of 45 mm, calculated from the measured current in 50 mM KCl. This is likely to be an averaged dimension, as in some instances the pore will not be spherical in shape, Fig. S1.†

In Fig. 3a, the black dashed line shows the I–V curves of an unmodified pore in 5 mM KCl. A weak current rectification ratio of 1.38 is recorded. The I–V curve for an unmodified pore in 50 mM KCl is shown in Fig. 3b, and here the rectification value is 1, illustrating the return of the ohmic response at higher ionic strengths.

In an attempt to introduce a facile method for modifying the surface chemistry of the pore, a LbL assembly using PEI/PAAMA was investigated. This system is well studied having been previously used to modify a range of materials.[50–55] We favoured this technique over other PU treatments such as plasma or the incorporation of grafting polymers into the matrix *via* swelling,[47,56,57] as it allows for a simple and rapid dip-coating strategy. In addition, the LbL assembly would allow the thickness and even the porosity of the PEI/PAAMA bilayer to be controlled in the future.[53] Here we have opted to use a similar system as Yang *et al.*[54] and Fu *et al.*[53] that have shown how the thickness of the bilayers can be controlled ensuring that they do not extend across the pore opening and that the thickness of the bilayers remains in the order of a few nanometres.

Fig. 3c shows a schematic of the bilayer construction with consecutive PEI/PAAMA layers. The resulting I–V curves in 5 mM KCl are shown in Fig. 3a alongside the unmodified pore. The addition of PEI onto the pore surface resulted in a change in the preferred direction of current flow, with a reduced and enhanced current flow through the pore under a negative and positive applied potential, respectively. This is indicative of a positive surface charge.[58] Upon coating with PAAMA, the surface charge switches to being negative, resulting in the preferred direction of current flow being inverted (Fig. 3a). All rectification ratios for the HMW PEI bilayers are listed in Table 1, and are measured at ±1.6 V. The addition of each layer of the LbL assembly caused the preferred current direction, or "on state", to be switched. The magnitude of the rectification can be used to access the presence and quality of the pore coating and as the number of bilayers increased the rectification ratios improved. It is interesting to note that even with two bilayers added to the pore walls, the pore opening remains unobstructed, as the addition of 210 nm particles to the upper fluid cell results in a standard resistive pulse response as they traverse the pore (data not shown).

Fig. 3b shows the effect of increasing the ionic strength of the solution on the rectification ratio, again values are listed in Table 1. Increasing the ionic strength reduced the rectification value, however it is interesting to note that even at >100 mM KCl with a pore diameter *circa* 800 nm some rectification was observed. The current rectification may have been enhanced by the nature of the modification of the pores here. The PU pores have a large area (diameter of *circa* 2.5 mm) on the top surface, T_s, as shown in Fig. 1. This surface is also modified along with the inner pore walls *via* the LbL route. A combination of top surface and pore wall modification has been shown to have a larger effect on the rectification ratio.[24]

A similar set of *I–V* curves were obtained using LMW PEI, shown in Fig. S2.† However the rectification ratios for each subsequent PEI and PAAMA layer were smaller. This is likely due to the nature of the PEI layer placed onto the surface. The mechanism for LbL assembly goes through island formation before forming a full layer and it is hypothesised that when using the LMW PEI it takes longer to form a fully coated surface layer.

The LbL approach was reproducible on multiple pores (Fig. 4a) and shows little hysteresis through multiple cycles, as shown in Fig. 4b. The LbL coating was also stable over short periods of time even after multiple uses, washes, and being dried and stored overnight, as shown in Fig. 4c. Although multiple uses and storage for over a week resulted in a deterioration of the surface chemistry (Fig. 4c), the same pore could be reused and recoated with comparable results (data shown in Fig. S3).†

Given the flexible nature of the PU pore it should be possible to change the rectification ratio by subjecting the pore to a smaller stretch producing a higher rectification ratio. Fig. 5a shows the effect on the *I–V* curves as the stretch is decreased from 5 to 2 mm. As the pore opening, d_s, is reduced, the observed current at positive potentials decreases. In addition to the pore size, both the PEI and PAAMA have functional groups that should allow a pH controlled current flow. Fig. 5b and c show the effect of pH on the *I–V* relationship for both the PEI and PAAMA surfaces, for comparison the data for an unmodified pore is given alongside. Lowering the pH of the solution to 3 increases the charge density on the PEI surface as more amines become protonated; this results in an increased current flow at positive potentials and a rectification ratio as low as 0.31. Conversely, increasing the pH to 7 reduced the positive charge on the PEI and the current rectification ratio increases to 0.52. Table 2 lists the rectification ratios for PEI and PAAMA pores at pH 3 and 7. For the PAAMA surface at pH 3, the charge density across the carboxyl groups is reduced and the response which is similar to the unmodified pore is recorded (Table 2). Increasing the pH to 7 for the PAAMA pores increases the negative charge density of the PAAMA pore wall, with an observed decrease in current flow at positive potentials and a current rectification of 1.98. The unmodified pore shows a current rectification of 0.8 at a pH of 3, indicating the surface groups have become protonated.

For all pores, a particle with 210 nm diameter suspended in PBS was passed through the opening as we built up the layers. The particle's ability to traverse the opening was interpreted to mean that the pore remained unblocked with a thick bilayer mesh, or that the bilayers have not restricted the pore orifice to a large extent.

The current flow through the pore has two contributing factors; the electro-osmotic flow, EOF, across the pore surface and migration of ions through the centre. Whilst others have shown the EOF to be a small contributing factor to current rectification in smaller pores,[24] and that a combined effect of the charge on the pore wall and top surface has the largest contribution to the current rectification.[24] In the work reported here, a large area of T_s is also modified with the bilayers and future work may be needed to ascertain if this has a larger contribution in the PU pores. The observation of rectification behaviour does illustrate that the enriched ion zone at the pore mouth used to describe the rectification properties of smaller pores[24,42] sufficiently exerts an influence across the larger opening, even when the electrical DL is much shorter than the opening D_s. This effect has also been attributed to biphasic pulse behaviour in TRPS.[49]

The reported setup allows for the easy modification of the pore wall, and by using different polymers of different thickness it may be possible to further tune these larger pores to be ion selective.

Modification of the pore walls with ssDNA

The main motivation for modifying the pore walls with LbL polymers was to introduce a functionality to which biomolecules could be easily coupled to the pore wall. Similar aspirations have led to the modification of pores with Au allowing thiol-terminated ligands to be placed along the pore wall or the utilisation of the carboxyl groups.[43,59] Our method of immobilisation uses EDC chemistry placing the DNA across the T_s and pore surface, shown schematically in Fig. 6a. Fig. 6b shows the I–V plots for a pore having gone through the LbL and DNA assembly, a magnified section through the origin is given in Fig. 6c. In the case where DNA was immobilised onto the pore wall, only one bilayer was used. The reason being that one bilayer was sufficient to introduce the carboxyl groups on the surface, and stopping at one bilayer reduced the number of preparation stages. The data in Fig. 6 was recorded at 50 mM KCl. At higher ionic strengths (>50 mM KCl), the DNA-modified pores do not exhibit current rectification; equivalent 5 mM KCl curves are given in Fig. S4.† The reduction in rectification for the DNA surface over the PAAMA layer may be due to the high number of DNA strands on the pore surface and the counter ion condensation shielding the high charge densities; in addition the EDC chemistry may have crosslinked the PAAMA to the underlying PEI causing a reduction in surface charge density.

Incubation of the pore with the target VEGF results in a strong rectification; shown in Fig. 7a (equivalent data for 5 mM KCl are given in Fig. S5†). The decrease in current at positive potentials was shown to be specific to the aptamer–VEGF interaction in the presence of BSA (Fig. S6†), Fig. 7b, and a mixture of other proteins whose pI values range from 5–7, Fig. S7.†

A similar trend in rectification properties was observed for the lysozyme system in cylindrical pores,[59] an explanation for the behaviour was attributed to the high pI of the proteins, and the binding of the protein to the DNA inverting the surface charge. This resulted in a positive surface on one side of the pore and the asymmetric surface charge resulted in current rectification. Our experiment may have produced a similar effect, *i.e.* the introduction of the protein on one side of the pore produced a different surface charge density from top to bottom. However, if in our study this resulted in a positively charged narrow pore opening, studies into diode-like conical pores would suggest the rectification properties would produce the opposite of the observed effects *i.e.* an increase in current at positive potentials,[60] which is not the observed result.

Our hypothesis for the observed rectification behaviour also relies upon the high pI of VEGF (8.2), and that the protein is positively charged. The inclusion of the protein to the DNA at the pore mouth and T_s increases the cation cloud density at the pore mouth. In addition to this, the folding of the aptamer into a tertiary structure to bind to the VEGF must require an increase in counter ions to stabilise the structure, and these two factors contribute to a decrease in conductivity at the pore mouth for positive potentials, as described elsewhere.[24]

Fig. 8 shows the concentration dose response of the DNA-modified pore at 50 mM KCl, equivalent 5 mM curves are show in Fig. S8.† As the concentration of VEGF is increased the current measured at 1.6 V decreases, and appears to saturate or remain unchanged beyond 5 nM. The rectification ratios for the 50 mM KCl experiment are given in Table 3. The detection of the VEGF at such low levels, when compared to RPS, may be aided by the inherent nature of the membrane to pre-concentrate cations in the pore mouth.

Conclusions

Presented is a comparison between a resistive pulse and current rectification aptamer assays using the same technology platform. To enable this we present the first LbL modification of PU pores that allows a pH- and ionic strength-controlled current flow. We feel that the tunable pore is unique in this ability to be easily adapted to both, and that the LbL assay reported here offers a simple and reusable method for modifying the surface chemistry of the pores. The RPS assay offers a format that can be easily prepared and multiplexed by pairing each target with a unique particle size/shape. Whilst it offers a wider dynamic range, its sensitivity is limited by the number of proteins required to change the particle surface charge sufficiently to detect a measurable change in translocation velocity. Further improvement to the sensitivity may require the specific design of the particle size and surface to allow a limited and controlled number of aptamers, although this offers a complication in assay time as the flux of material through the pore decreases with smaller particles and pores, or the use of Janus particles to localise the biorecognition.[35]

The LbL assembly of the pore offers a pH- and ionic strength-dependent rectification behaviour and is the first reported method on such a system. It allows the first observation of strong rectification on pores of this dimension. The strong rectifying properties illustrate that the electric double layer does not need to significantly extend across the pore opening, although it is acknowledged that further studies are required to assess what the effects of the modification of the top surface of the pore have on the rectification mechanism, and the total contrition of electroosmotic flow in these systems.

Acknowledgements

This work was supported by the European Commission for Research (PCIG11-GA-2012-321836 Nano4Bio), Loughborough University Chemistry Department (Start-up fund). The authors thank Izon Scientific Ltd for their support and the Centre for Analytical Science at Loughborough University. M. P. is a member of the National Institute for Health Research (NIHR) Diet, Lifestyle & Physical Activity Biomedical Research Unit based at University Hospitals of Leicester and Loughborough University. The views expressed are those of the authors and not necessarily those of the NHS, the NIHR or the Department of Health. E. L. C. J. B. is supported by Izon Science Ltd. L. M. is a member of the DISTINCTIVE consortium.

Notes and references

1 C. Dekker, *Nat. Nanotechnol.*, 2007, **2**, 209–215.
2 D. Branton, D. W. Deamer, A. Marziali, H. Bayley, S. A. Benner, T. Butler, M. Di Ventra, S. Garaj, A. Hibbs, X. Huang, S. B. Jovanovich, P. S. Krstic, S. Lindsay, X. S. Ling, C. H. Mastrangelo, A. Meller, J. S. Oliver, Y. V. Pershin, J. M. Ramsey, R. Riehn, G. V. Soni, V. Tabard-Cossa, M. Wanunu, M. Wiggin and J. A. Schloss, *Nat. Biotechnol.*, 2008, **26**, 1146–1153.
3 L. T. Sexton, L. P. Horne and C. R. Martin, *Mol. BioSyst.*, 2007, **3**, 667–685.
4 L. Sexton, L. Horne and C. Martin, in *Molecular- and Nano-Tubes SE – 6*, ed. O. Hayden and K. Nielsch, Springer US, 2011, pp. 165–207.
5 H. Bayley and C. R. Martin, *Chem. Rev.*, 2000, **100**, 2575–2594.
6 H. Bayley and P. S. Cremer, *Nature*, 2001, **413**, 226–230.
7 D. Kozak, W. Anderson, R. Vogel and M. Trau, *Nano Today*, 2011, **6**, 531–545.
8 G. S. Roberts, S. Yu, Q. Zeng, L. C. L. Chan, W. Anderson, A. H. Colby, M. W. Grinstaff, S. Reid and R. Vogel, *Biosens. Bioelectron.*, 2012, **31**, 17–25.
9 P. Kohli, C. C. Harrell, Z. Cao, R. Gasparac, W. Tan and C. R. Martin, *Science*, 2004, **305**, 984–986.
10 K. B. Jirage, J. C. Hulteen and C. R. Martin, *Science*, 1997, **278**, 655–658.
11 E. R. Billinge, M. Broom and M. Platt, *Anal. Chem.*, 2014, **86**, 1030–1037.
12 P. M. Sinhal, G. Valco, S. Sharma, X. Liu and M. Ferrari, *Nanotechnology*, 2004, **15**, S585.
13 W.-J. Lan, D. A. Holden, J. Liu and H. S. White, *J. Phys. Chem. C*, 2011, **115**, 18445–18452.
14 W.-J. Lan, D. A. Holden, B. Zhang and H. S. White, *Anal. Chem.*, 2011, **83**, 3840–3847.
15 K. A. Whitehead, R. Langer and D. G. Anderson, *Nat. Rev. Drug Discovery*, 2009, **8**, 129–138.
16 S. Lee, Y. Zhang, H. S. White, C. C. Harrell and C. R. Martin, *Anal. Chem.*, 2004, **76**, 6108–6115.

17 S. Nam, I. Choi, C. Fu, K. Kim, S. Hong, Y. Choi, A. Zettl and L. P. Lee, *Nano Lett.*, 2014, **14**, 5584–5589.

18 R. R. Henriquez, T. Ito, L. Sun and R. M. Crooks, *Analyst*, 2004, **129**, 478–482.

19 T. Ito, L. Sun, R. R. Henriquez and R. M. Crooks, *Acc. Chem. Res.*, 2004, **37**, 937–945.

20 L. Sun and R. M. Crooks, *J. Am. Chem. Soc.*, 2000, **122**, 12340–12345.

21 G. S. Roberts, D. Kozak, W. Anderson, M. F. Broom, R. Vogel and M. Trau, *Small*, 2010, **6**, 2653–2658.

22 J. B. Heng, A. Aksimentiev, C. Ho, P. Marks, Y. V. Grinkova, S. Sligar, K. Schulten and G. Timp, *Biophys. J.*, 2006, **90**, 1098–1106.

23 W.-J. Lan, C. Kubeil, J.-W. Xiong, A. Bund and H. S. White, *J. Phys. Chem. C*, 2014, **118**, 2726–2734.

24 H. S. White and A. Bund, *Langmuir*, 2008, **24**, 2212–2218.

25 E. A. Heins, Z. S. Siwy, L. A. Baker and C. R. Martin, *Nano Lett.*, 2005, **5**, 1824–1829.

26 Z. Siwy, L. Trofin, P. Kohli, L. A. Baker, C. Trautmann and C. R. Martin, *J. Am. Chem. Soc.*, 2005, **127**, 5000–5001.

27 M. Ali, P. Ramirez, M. N. Tahir, S. Mafe, Z. Siwy, R. Neumann, W. Tremel and W. Ensinger, *Nanoscale*, 2011, **3**, 1894–1903.

28 E. L. C. J. Blundell, L. J. Mayne, E. R. Billinge and M. Platt, *Anal. Methods*, 2015, **7**, 7055–7066.

29 E. L. C. J. Blundell, R. Vogel and M. Platt, *Langmuir*, 2016, **32**, 1082–1090.

30 E. Weatherall and G. R. Willmott, *Analyst*, 2015, **140**, 3318–3334.

31 Y. Qiu, C. Yang, P. Hinkle, I. V. Vlassiouk and Z. S. Siwy, *Anal. Chem.*, 2015, **87**, 8517–8523.

32 G. R. Willmott, R. Vogel, S. S. C. Yu, L. G. Groenewegen, G. S. Roberts, D. Kozak, W. Anderson and M. Trau, *J. Phys.: Condens. Matter*, 2010, **22**, 454116.

33 Y. Nonaka, K. Sode and K. Ikebukuro, *Molecules*, 2010, **15**, 215–225.

34 R. Vogel, W. Anderson, J. Eldridge, B. Glossop and G. Willmott, *Anal. Chem.*, 2012, **84**, 3125–3131.

35 M. Platt, G. R. Willmott and G. U. Lee, *Small*, 2012, **8**, 2436–2444.

36 Y. S. Ang and L. Y. L. Yung, *ACS Nano*, 2012, **6**, 8815–8823.

37 E. R. Billinge, J. Muzard and M. Platt, *Nanomaterials and Nanosciences*, 2013, **1**, 1.

38 E. R. Billinge and M. Platt, *Anal. Methods*, 2015, **7**, 8534–8538.

39 E. R. Billinge and M. Platt, *Biosens. Bioelectron.*, 2015, **68**, 741–748.

40 O. A. Alsager, S. Kumar, G. R. Willmott, K. P. McNatty and J. M. Hodgkiss, *Biosens. Bioelectron.*, 2014, **57**, 262–268.

41 C. C. Harrell, P. Kohli, Z. Siwy and C. R. Martin, *J. Am. Chem. Soc.*, 2004, **126**, 15646–15647.

42 S. F. Buchsbaum, G. Nguyen, S. Howorka and Z. S. Siwy, *J. Am. Chem. Soc.*, 2014, **136**, 9902–9905.

43 G. Nguyen, S. Howorka and Z. S. Siwy, *J. Membr. Biol.*, 2011, **239**, 105–113.

44 M. Ali, S. Nasir, Q. H. Nguyen, J. K. Sahoo, M. N. Tahir, W. Tremel and W. Ensinger, *J. Am. Chem. Soc.*, 2011, **133**, 17307–17314.

45 L. Maisonneuve, O. Lamarzelle, E. Rix, E. Grau and H. Cramail, *Chem. Rev.*, 2015, **115**, 12407–12439.

46 F. Hileman, R. Sievers, G. Hess and W. Ross, *Anal. Chem.*, 1973, **45**, 1126–1130.

47 A. Kulkarni, W. Diehl-Jones, S. Ghanbar and S. Liu, *J. Biomater. Appl.*, 2014, **29**, 278–290.

48 Z. S. Siwy, *Adv. Funct. Mater.*, 2006, **16**, 735–746.

49 E. Weatherall and G. R. Willmott, *J. Phys. Chem. B*, 2015, **119**, 5328–5335.

50 B. Steitz, H. Hofmann, S. W. Kamau, P. O. Hassa, M. O. Hottiger, B. von Rechenberg, M. Hofmann-Amtenbrink and A. Petri-Fink, *J. Magn. Magn. Mater.*, 2007, **311**, 300–305.

51 J. J. O'Mahony, M. Platt, D. Kilinc and G. Lee, *Langmuir*, 2013, **29**, 2546–2553.

52 Y.-H. Yang, M. Haile, Y. T. Park, F. A. Malek and J. C. Grunlan, *Macromolecules*, 2011, **44**, 1450–1459.

53 J. Fu, J. Ji, L. Shen, A. Küller, A. Rosenhahn, J. Shen and M. Grunze, *Langmuir*, 2009, **25**, 672–675.

54 Y.-H. Yang, L. Bolling, M. Haile and J. C. Grunlan, *RSC Adv.*, 2012, **2**, 12355–12363.

55 P. Podsiadlo, M. Michel, J. Lee, E. Verploegen, N. Wong Shi Kam, V. Ball, J. Lee, Y. Qi, A. J. Hart, P. T. Hammond and N. A. Kotov, *Nano Lett.*, 2008, **8**, 1762–1770.

56 A. B. Jozwiak, C. M. Kielty and R. A. Black, *J. Mater. Chem.*, 2008, **18**, 2240–2248.

57 M. Gultekinoglu, Y. Tunc Sarisozen, C. Erdogdu, M. Sagiroglu, E. A. Aksoy, Y. J. Oh, P. Hinterdorfer and K. Ulubayram, *Acta Biomater.*, 2015, **21**, 44–54.

58 Z. Siwy, E. Heins, C. C. Harrell, P. Kohli and C. R. Martin, *J. Am. Chem. Soc.*, 2004, **126**, 10850–10851.

59 M. Ali, S. Nasir and W. Ensinger, *Chem. Commun.*, 2015, **51**, 3454–3457.

60 I. Vlassiouk and Z. S. Siwy, *Nano Lett.*, 2007, **7**, 552–556.

Faraday Discussions

ROYAL SOCIETY OF CHEMISTRY

DISCUSSIONS

Nanopores: general discussion

Andrew Mount, Minkyung Kang, David Fermin, Tim Albrecht, J. Gooding, Richard Crooks, Nongjian Tao, Wolfgang Schmickler, Julie MacPherson, Shengli Chen, Paolo Actis, Olaf Magnussen, Lane Baker, Philip Bartlett, Sanli Faez, Jan Clausmeyer, Bradley Thomas, Philip A. Ash, Frederic Kanoufi, Yitao Long, Patrick Unwin, Marc Koper, Serge Lemay, Andrew Ewing, Zhongqun Tian, Robert Johnson, Michael Eikerling and Mark Platt

DOI: 10.1039/C6FD90069K

Serge Lemay opened a general discussion of the paper by Wolfgang Schmickler: Since a nanotube effectively forms a Faraday cage, it is not possible to generate a longitudinal DC electric field inside the nanotube. Any externally applied potential difference between the two reservoirs at the ends of the tube would thus cause the electric field to be localized near the two ends only. Can this behavior be accounted for in the present description?

Wolfgang Schmickler answered: Yes, in our calculations the surface of the tube has a constant potential throughout.

Changing the electrode potential changes the electrochemical potential of the ions inside the tube with respect to a bulk electrolyte. Transport inside the tube is by diffusion. Migration can play a role at the entrance to the tube.

Bradley Thomas asked: First, could you possibly shed more light on your comments concerning the lack of difference between conducting and semi-conducting nanotubes? Within the literature, there are experimental examples that show a large difference between the two types. A prime example of this would be work recently published by colleagues in my own group.[1] Here, a new coulombic procedure is developed to experimentally determine the electronic structure of guest@nanotube materials, with fundamentally different behaviour seen between the two nanotube types. Your reasoning seems to be based upon previous calculations and the observed phenomenon that lithium atoms undergo direct electron transfer with the host semiconducting nanotubes on encapsulation. But surely the fact that this occurs shows how different conductive and semi-conductive nanotubes are? Rather than suggesting they are the same, as the produced material is electronically comparable to that produced from encapsulating lithium cations into a conducting nanotube, they are fundamentally different. If lithium cations are encapsulated into a semiconducting nanotube, this electron transfer surely cannot occur as the ions no longer have the required electron to donate. Therefore, for the nanotube to then become conducting you would require an external source of electrons (presumably from the electrical circuit) – something that would not be required in the conducting example.

Secondly, you discussed, both in the manuscript and in your discussion with the delegates, the solvation energy of ions in solution, yet the references you cite in the manuscript for the formation of one-dimensional salts (*e.g.* J. Sloan, ref. 18 of your manuscript) use molten salts primarily due to problems faced in achieving such uniform ionic arrays using solution filling. Have you taken these issues, such as the preferential filling of one ion, into account beyond simply determining the loss of solvation energy upon encapsulation when performing your calculations?

1 R. L. McSweeney, T. W. Chamberlain, M. Baldoni, M. A. Lebedeva, E. S. Davies, E. Besley and A. N. Khlobystov, *Chem. – Eur. J.*, 2016, **22**, 13540.

Wolfgang Schmickler responded: Of course, semiconducting and conducting CNTs behave differently.

However, upon inserting a cation or an anion into the tube, the corresponding image charge flows onto the tube. This shifts the Fermi level near the ion, and the tubes become conductive locally.[1]

There are various ways of filling the tubes. Loss of solvation energy occurs when the ions are inserted from an electrolyte solution. An alternative is molten salts, where the two kinds of ions often have very different sizes. This may entail that only one kind of ion can be inserted. I have discussed this in a previous publication.[2]

1 L. Mohammadzadeh, A. Goduljan, F. Juarez, P. Quaino, E. Santos and W. Schmickler, *ChemPhysChem*, 2016, **17**, 78.
2 W. Schmickler, *Electrochim. Acta*, 2015, **173**, 91.

Philip Bartlett commented: In applications, for example in a supercapacitor, the nanotubes can be in bundles or with solvent and ions in contact with the outside of the carbon nanotube as well as with ions inside the tube. Can you say something about how the environment on the outside of the tube would affect the behavior?

Wolfgang Schmickler answered: This is a good question. At the moment, we are investigating if an ion adsorbed outside of the tube interacts with an ion inside. Please ask me again in a few weeks.

Patrick Unwin addressed Wolfgang Schmickler: In studying electrochemistry at single-walled carbon nanotubes, the main interest is in the double layer external to the nanotube and what effect this might have on the electrochemistry. Especially in the context of Chen's paper, could you comment on whether you would expect such double layer effects to be significant? Rate constants and the voltammetric characteristics for several redox couples at approximately 1 nm diameter single-walled carbon nanotubes[1,2] appear to be broadly similar to those on larger radius (although still nanoscale) platinum disc electrodes.[3]

1 I. Heller, J. Kong, H. A. Heering, K. A. Williams, S. G. Lemay and C. Dekker, *Nano Lett.*, 2005, **5**(1), 137–142.

2 A. G. Güell, K. E. Meadows, P. V. Dudin, N. Ebejer, J. V. Macpherson and P. R. Unwin, *Nano Lett.*, 2014, **14**(1), 220–224.
3 P. Sun and M. V. Mirkin, *Anal. Chem.*, 2006, **78**(18), 6526–6534.

Wolfgang Schmickler answered: Our main interest is in nanoconfinement, *i.e.* in ions inside the tube. Outside the tube the electric field will be strong and affect the double layer properties.[1] However, the effect on electron transfer reactions should be minor.

1 E. Leiva, P. Velez, C. Sanchez and W. Schmickler, *Phys. Rev. B*, 2006, **74**, 035422.

David Fermin remarked: Have you performed calculations of protons or acid incorporation into nanotubes and does this follow the trends observed for NaCl and LiCl?

Wolfgang Schmickler answered: The insertion of a proton is on our list of things to do. We shall model it as a $H_7O_2^+$. However, the energy of ionization of the hydrogen atom is quite high, so we are not sure that DFT will give us a real proton. This is the reason why we started with alkali ions, whose ionization energies are much lower. A previous attempt to model a proton in front of a planar surface was not successful.[1]

1 P. Quaino, N. B. Luque, G. Soldano, R. R. Nazmutdinov, E. Santos, T. Roman, A. Lundin, A. Groß and W. Schmickler, *Electrochim. Acta*, 2013, **105**, 341.

Philip Bartlett said: Carbon nanotubes often have defects of different types. Can you say something about the effects of these defects on the ions inside the nanotube?

Wolfgang Schmickler responded: We have first results for nitrogen-doped nanotubes, but it is too early to comment on them.

Marc Koper asked: How should we think about the interaction of water with the inside of these pores? How would that affect your calculations?

Wolfgang Schmickler responded: We are investigating this question at the moment. We have data, but we still have to evaluate them. Obviously, the ions will lose a part of their solvation shell. It is a competition between solvation and image interaction.

Philip Bartlett opened a general discussion of the paper by Michael Eikerling: In your paper you refer to results for planar platinum (ref. 12 of the manuscript, Huang *et al.*) showing a non-monotonic charging behavior. Can you briefly summarize this work for us?

Michael Eikerling responded: The article referred to is centred on the charging relation of platinum, the catalyst of choice for PEM-based electrochemical technologies such as fuel cells and water electrolysers. For details of the physical-mathematical model as well as the solution, I would like to kindly ask you to look up the original article. The main features are:

Firstly, by revealing the non-monotonic charging behavior of Pt from first principles, the presented theoretical work solves a long-standing puzzle in electrochemistry. It takes the theory of electrified interfaces a decisive step beyond the concept of potential of zero charge (pzc), famously coined by A. N. Frumkin in 1928 and called "one of the most fundamental ideas in electrochemistry".[1]

Secondly, the presented model for the electrified interface at the Pt surface incorporates oxide and ordered water layers. The model can be adapted for other oxide-forming metal surfaces with a plethora of applications in electrocatalysis, semiconductor electrochemistry, and nanofluidics.

Thirdly, the presented work represents a framework for deeper forays into the field of first principles electrochemistry that must account self-consistently for the coupling of metal charging phenomena and vital descriptors of electrocatalyst activity and stability.

1 J. M. Bockris, M. A. Devanathan and K. Muller, *Proc. R. Soc. London, Ser. A*, 1963, **274**, 55–79.

Tim Albrecht asked: Could your model be adapted to include hydrodynamic and bipolar effects in small Pt pores that connect two liquid compartments? If so, can you speculate about the effect those processes might have on the charge distribution inside the pore channel, and potentially rectification effects in the current? Several papers might add some experimental context to the question,[1,2] and in regards to bipolar effects.[3]

1 M. Ayub, A. Ivanov, J. Hong, P. Kuhn, E. Instuli, J. B. Edel and T. Albrecht, *J. Phys.: Condens. Matter*, 2010, **22**(45), 454128.
2 A. Rutkowska, J. B. Edel and T. Albrecht, *ACS Nano*, 2012, **7**(1), 547–555.
3 A. Rutkowska, K. Freedman, J. Skalkowska, M. J. Kim, J. B. Edel and T. Albrecht, *Anal. Chem.*, **87**(4), 2337–2344.

Michael Eikerling replied: Solution of the model, as presented in the Faraday Discussions article, gives the equilibrium distributions of the solution phase potential and proton density in the water-filled nanochannel. These distributions are controlled by the free surface charge density of the metal, which is a non-monotonic function of the metal-phase potential. We use these equilibrium distributions to calculate the activity of the nanochannel for the oxygen reduction reaction (ORR). This approximative treatment is based on the assumption that the finite proton flux and the consumption of protons, required to maintain the ORR current, cause negligible perturbation of the equilibrium proton density and potential. We have verified the validity of this assumption in a previous article,[1] where we compared the analytical solution of the simplified set of equations (assuming equilibrium distributions) with the numerical solution of the fully coupled dynamic problem (based on Poisson–Nernst–Planck equation). The reason for the weakness of hydrodynamic coupling in our case is that rates of proton transport and proton consumption, required to maintain the ORR current, are very small. We only consider one mobile charged species in the channel (protons); thus the model system considered is not suitable to study current rectification effects. However, the non-linear and non-monotonic charging relation of the metal surface, shown in Fig. 6 of the manuscript, should give rise to interesting ionic current rectification effects in systems containing mobile cations and anions.

1 K. Chan and M. Eikerling, *J. Electrochem. Soc.*, 2011, **158**, B18–B28.

Wolfgang Schmickler said: Your charge–potential curves look quite interesting. Could you explain them in greater detail? Is there a potential of zero charge, or a change of local pH?

Michael Eikerling replied: Please consider the curves shown in Fig. 5b and Fig. 6 of our related work in *J. Phys. Chem. C*[1] as a reference for the charge–potential curves shown in our *Faraday Discussions* article. As seen in that work, the charging relation of an extended metal surface forming an interface with a concentrated electrolyte exhibits consecutive transitions from negative to positive and again to negative free surface charge upon increasing potential (in agreement with experimental findings of Frumkin and Petrii,[2] and Garcia-Araez *et al.*[3]). The non-monotonic behaviour and transition to a negative charging region at high potential represents a paradigm change in comparison to the monotonic behaviour found in classical double layer models.[4] The peculiar aspect of the model considered in the Faraday Discussions article is that no explicit electrolyte is added to the solution filling the pore. The only source of excess ions, protons in this case, is the PEM that forms an interface with the pore at one of its openings. This situation forces the metal charging relation of the water-filled nanopore to remain in the negative domain at all potentials. Fig. 7c shows the excess proton concentration as a function of potential. The pore is highly protophilic at small potentials. With increasing potential it becomes less protophilic; however, the proton concentration remains high compared to that of free bulk water. The spatial distribution of potential and proton density in the axial direction of the nanopore is negligible (except for a steep decline in a double layer region at the pore opening to the PEM which is very thin compared to the length of the pore). The water-filled nanopore does not possess a potential of zero charge, as evident from Fig. 7a. In the case of an extended Pt surface interfacing with an explicit electrolyte, a potential of zero charge exists (in the double layer region of the metal), but the charging phenomena are more complex than what could be described by a pzc and a constant double layer capacitance alone.

1 J. Huang, A. Malek, J. Zhang and M. Eikerling, *J. Phys. Chem. C*, 2016, **120**(25), 13587–13595.
2 A. N. Frumkin and O. A. Petrii, *Electrochim. Acta*, 1975, **20**, 347–359.
3 N. Garcia-Araez, V. Climent and J. Feliu, *J. Phys. Chem. C*, 2009, **113**, 9290–9304.
4 J. O'M. Bockris, M. A. V. Devanathan and K. Muller, *Proc. R. Soc. Lond. A*, 1963, **274**, 55–79.

Philip Bartlett commented: Using electrodeposition from lyotropic liquid crystal templates, it is possible to make platinum and other metal films with uniform diameter pores in the range 1.8 to around 3.5 nm diameter with control over the pore diameter.[1,2] Birkin *et al.*[3] and Jiang and Kucernak[4,5] have studied oxygen reduction on this type of mesoporous platinum electrode.

Can you extend your model to smaller pores and to include the presence of electrolyte in order to compare to these experiments?

1 G. S. Attard, P. N. Bartlett, N. R. B. Coleman, J. M. Elliott, J. R. Owen and J. H. Wang, *Science*, 1997, **278**, 838.
2 J. M. Elliott, P. R. Birkin, P. N. Bartlett and G. S. Attard, *Langmuir*, 1999, **15**, 7411.
3 P. R. Birkin, J. M. Elliott and Y. E. Watson, *Chem. Commun.*, 2000, 1693.
4 J. Jiang and A. Kucernak, *Electrochem. Solid State Lett.*, 2000, **3**, 559.
5 J. Jiang and A. Kucernak, *J. Solid State Electrochem.*, 2012, **16**, 2571.

Michael Eikerling replied: Our continuum modeling approach can be expected to give meaningful results down to pore radii in the range of 2 nm. For smaller radii, specific molecular effects at interfaces will become dominant which must be treated by molecular simulations.

Serge Lemay opened a general discussion of the paper by Sanli Faez: The longitudinal electric field is applied by capacitive coupling using external electrodes. A rough estimate of the capacitance between these electrodes and the solution yields approximately 10^{-13} F. Combined with a channel resistance of approximately 2×10^{12} Ohms for a 0.1 mM solution, this yields an RC time of approximately 0.2 s, which is long enough for the electric field not to be screened on the time scale of the experiment. If the solution had physiological salt concentrations (~0.1 M) on the other hand, the electric field would be screened by ions on the sub-ms time scale. Will this limit the technique to low salt concentrations only?

Sanli Faez replied: Indeed, without a major improvement in the measurement speed, increasing the salt concentration is problematic. However, measuring with microsecond time resolution is quite feasible by using a faster camera or a linear array photo detector.

Olaf Magnussen remarked: You assumed in the comparison of your simulations with the experimental data that the particle rapidly samples all axial positions with equal probability and that the measured velocity thus is the average value over the tube diameter. Is this a good assumption? There is evidence that Brownian motion is not uniform, but that particles have an increased residence time near walls. Because of the large differences in the liquid flow near the walls and in the center of the tube, this might have large effects. Is there a way one can experimentally measure the residence times at different axial positions?

Sanli Faez answered: The smallest time step we have considered is 10 ms which is enough for the particle to diffuse the entire cross section, so averaging can be justified. The hydrodynamically hindered diffusion only occurs at distances from the wall comparable with the particle size, so considering that the channel diameter is roughly ten times larger than the particle diameter, for most parts, the diffusion is similar to bulk. Furthermore, the particle and the channel wall are both negatively charged and the Debye layer is relatively long due to low ionic concentration.

Because of the special shape of the optical mode profile, the radial position of the particle determines the instantaneous scattering intensity. By rapidly collecting the intensity statistics, one can obtain the residence time at each radial position.

Philip Bartlett commented: In the video you showed the particles appear as two bright spots rather than one. Can you explain this for us?

Sanli Faez replied: This is due to the optical aberration caused mainly by the cylindrical interfaces of the fiber and imperfect index matching to the immersion oil.

Nongjian Tao asked: How does the detected light scale with the size of the particle? Is it with the square of the volume? What is the smallest size of particle can you detect? Is the detection limit determined by some sort of background scattering? How does your approach compare with zeta potential measurement?

Sanli Faez replied: Yes, the scattering intensity is proportional to the volume squared. For such small particles, the Rayleigh scattering formula is a very good description. The smallest detectable size depends on the refractive index of the material as well. The smallest detectable scattering signal from mobile particles is limited by the shot-noise and technical noise of the background scattering from the inherent roughness of the channel surface. By implementing interferometric detection and active intensity stabilization, this limitation can also be overcome.

By tracking particles that are actuated in an external electric field, one can measure the electrophoretic mobility and from there determine the zeta potential considering a proper model for the charge double layer.

Philip A. Ash remarked: You mention that labelling with fluorescent tags will be necessary to extend the tracking technique to biomolecules. Would you be able to expand on this comment? Many proteins contain cofactors that already have relatively intense absorbances (and sometimes naturally fluoresce) in the UV-visible region; iron-sulphur centres, haem groups, flavin centres etc. What is the specific requirement of these absorbances that will allow direct application of your technique to biomolecule samples? If the requirement is simply a large scattering cross section is it possible, for example, that the intense Soret absorbance of cytochromes (many of which catalyse important reactions) or the fluorescent or light sensing properties of green fluorescent protein or blue light sensing BLUF domain proteins might be directly applicable to your technique with no labelling necessary?

Sanli Faez responded: The determining factor for tracking smaller entities with their elastic scattering signal is their polarizability at the excitation frequency. For optically active molecules the peak in absorbance is often close to the peak in the scattering cross section in the spectrum. The Lorentz oscillator is often a very good model for describing such a dipole resonance. For a large scattering signal, in the semi classical picture, decay from the excited state should be mostly radiative. For fluorescent molecules this translates to saying that the quantum efficiency should be high.

Zhongqun Tian asked: For your interesting approach, what is the smallest diameter you can reach in the channel of the optical fibre? Can you fix a nanoparticle at the end of the optical fibre robustly? What is the best way to increase the success rate? The reason I am asking is that my group is developing a SPR tip (a nanoparticle-based tip) method for local imaging and reaction.[1] Fig. 1 illustrates that the plasmonic nanoparticle, such as a gold nanoparticle, is fixed at the sharp end of fibre/pipette, and the gap between the nanoparticle and substrate surface can be precisely controlled by the piezo with the feedback signal of the scattering light from the nanogap (hotspot) formed by the nanoparticle and the surface. The SPR peak wavelength and light intensity depend critically on the gap width, *i.e.*, the strong coupling can result largely in red-shift resonance and strongly enhanced intensity.[2] This method could be utilized to investigate the heterogeneous surface and interfacial properties such as the electrochemical

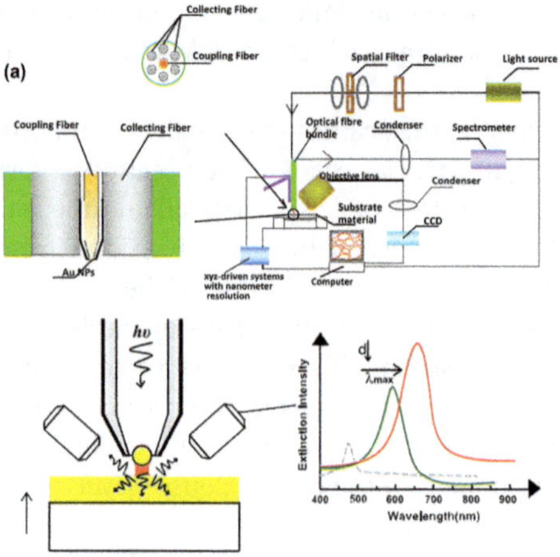

Fig. 1 The schematic illustration of working principle and setup of tip-enhanced dark field microscopy.

double-layer structure and SPR mediated reaction. It may also be applied to study the photoelectrochemistry of single dye sensitized nanoparticles as well as for nanoparticle impact electroanalysis because the particle can be mechanically controlled to hit and leave the electrode surface precisely with or without light illumination.

1 Z. Q. Tian, F. F. Wang, D. P. Zhan and J. Z. Zhou, 'Tip-enhanced dark field microscopy, electrochemical measurement and leveling device', Chinese Patent CN 102798735 B, March 2015.
2 F. F. Wang, J. Z. Zhou, S. Y. Ding, D. P. Zhan and Z.Q. Tian, in preparation.

Sanli Faez responded: The smallest channel diameter that can be controlled for an extended length of the fiber is roughly 100 nm. It is possible to locally shrink the inner diameter even further by post-processing.

Immobilizing a gold nanoparticle inside or at the end of the fiber is also possible.

Tim Albrecht remarked: As a comment, I would like to add that you might consider small heme-containing proteins as model systems for your work, for example cytochrome c, cytochrome c4 or cytochrome c3. These are small proteins in the low-nanometer size range, which feature a different number of heme (iron porphyrin) groups. Such heme groups would act 'intrinsic' staining agents, enhance the scattering cross sections of the protein, and may be relatively easy to study. In contrast to nanoparticle samples, proteins typically have a well-defined molecular weight, often the crystal structure is available (for the above small metalloproteins anyway), and are otherwise well-characterised (charge, surface properties *etc.*). They might also be modified using site-directed mutagenesis, if required.

Shengli Chen returned to the paper by Wolfgang Schmickler: In the charge difference diagram, can you see charge transfer between the tube walls and ions besides the induced image charge?

By the way, what does Pauli repulsion between the nanotube wall and ions mean here?

Wolfgang Schmickler replied: In the case of a single ion, a unit charge is transferred to the walls of the tube. The result is an ion inside the tube and an image charge on the walls of the tube. There is no other charge transfer. Pauli repulsion is a consequence of the Pauli principle, which states that no two electrons can be in the same state. Therefore electronic overlap is unfavourable, unless a chemical bond is formed. In the van der Waals potential, Pauli repulsion is represented by the repulsive branch.

Julie MacPherson said: You state in your paper that ion insertion results in the same charge screening effect be the tube metallic or semiconducting. However, is this a consequence of the way the model works *i.e.* you assume the ion enters as an atom and then transfers an electron to become charged which results in a change in the position of the Fermi level? What would happen if you modelled the process as an ion entering and not an atom?

Wolfgang Schmickler responded: In the calculation we place an atom inside the tube, and DFT converts this into an ion: the charge is transferred to the tube where it forms the image charge.

In an electrochemical experiment an ion enters the tube. Since the tube must stay neutral, charge flows through the external circuit onto the tube and forms the image charge.

The final state is the same in both cases.

In the slab configuration, which we use to model infinite tubes, DFT works for neutral systems only. If we place an ion inside the tube, DFT automatically compensates this by a uniform background charge of opposite sign. The result would be an artificial charge distribution.

Sanli Faez said: If we assume the ion on the axis is in the ground state, what would be the energy in the first excited state of the ion orbiting the tube axis? At what temperatures can one expect to observe quantum confinement effects?

Wolfgang Schmickler responded: I assume you are not talking about electronic excitations. The ion has three degrees of freedom, two of which correspond to vibrations perpendicular to the axis of the tube. These vibrations are quantized. The third corresponds to motion along the axis. If there is only a single ion, this motion has the nature of a translation, but it does experience the corrugation of the surrounding tube. If the tube is filled with ions, motion along the axis is also vibrational. This is in the classical regime, and the frequencies depend on the filling of the tube. Some of these effects have been discussed in a previous publication.[1]

1 W. Schmickler, *Electrochim. Acta*, 2015, **173**, 91.

Shengli Chen asked: In your calculation, the neutral metal atom inserted in a narrow carbon nanotube can be ionised, which should be because of the strong image interaction between metal ions and the tube wall. What do you think it happens if you put the metal atom outside of the tube? Will it be ionised? If you don't have water outside the tube, I don't think the metal atom will be ionised.

Wolfgang Schmickler responded: You are right, for a particle outside the tube the image interaction is weaker, and ionization, within DFT, is more difficult.

Corresponding calculations are running...

See also our recent article.[1]

1 P. Quaino, N. B. Luque, G. Soldano, R. R. Nazmutdinov, E. Santos, T. Roman, A. Lundin, A. Groß and W. Schmickler, 2013 *Electrochim. Acta*, **105**, 341.

Zhongqun Tian said: Your approach is very interesting and I wonder if your studies can be further extended to metallic atoms/clusters/nanoparticles inside carbon nanotubes (CNTs) or underneath graphene. Deng *et. al.* encapsulated Fe clusters/nanoparticles within pod-like CNTs for the oxygen reduction reaction.[1] The direct contact of nonprecious metal particles with harsh environments including acid media, oxygen, and sulfur contamination is avoided. It does not impede the high activation of O_2 but has long-term stability even in the presence of SO_2 poison. However, it's not easy to use CNTs to encapsulate various metals with a high density of active sites. It's much more feasible to introduce metal clusters/nanoparticles between graphene layers and electrode substrates.[2,3] More importantly, all these systems may provide a defined model for understanding the synergetic (nanoconfinement) effect of metal–CNTs and metal–graphene for tuning them to have a better electronic structure of the localized active sites for electrocatalysis. May I have your comments on the relevant theoretical modeling and calculation?

1 D. H. Deng, L. Yu, X. Q. Chen, G. X. Wang, L. Jin, X. L. Pan, J. Deng, G. G. Sun and X. H. Bao, *Angew. Chem. Int. Ed.*, 2013, **52**, 371–375.

2 R. T. Mu, Q. Fu, L. Jin, L. Yu, G. G. Fang, D. Tan and X. H. Bao, *Angew. Chem. Int. Ed.*, 2012, **51**, 4856–4859.

3 D. H. Deng, K. S. Novoselov, Q. Fu, N. F. Zheng, Z. Q. Tian and X. H. Bao, *Nat. Nanotech.*, 2016, **11**, 218–230.

Wolfgang Schmickler replied: Many thanks for your interesting comments. One of my coauthors has started to perform calculations for iron encapsulated in carbon nanotubes. I think she was inspired by the articles you mention. It is too early to comment on her results.

Bradley Thomas commented: This comment is in relation to a discussion you had with another delegate on the need for simulations to determine whether electron transfer, as seen with lithium metal encapsulation, can still occur with the metal electron source outside the nanotube. Within the literature this has been somewhat answered already.[1,2] To give just a couple of many examples, one of the most favoured functionalisation methods for carbon nanotubes involves adding nanotubes to a solution of sodium (or lithium) metal dissolved in liquid ammonia. The sodium is oxidised in this process, with the nanotubes acting as an electron sink. In this electron-rich state the nanotubes are able to act as nucleophiles in reactions with molecules containing groups such as organic halides. The fact that this functionalisation works regardless of whether the nanotubes have been "opened" suggests encapsulation is not required.

1 D. Wunderlich, F. Hauke and A. Hirsch, *J. Mater. Chem.*, 2008, **18**, 1493–1497.
2 J. J. Stephenson, A. K. Sadana, A. L. Higginbotham and J. M. Tour, *Chem. Mater.*, 2006, **18**(19), 4658–4661.

Wolfgang Schmickler responded: The energies of ionization of Li and Na are so low that they are expected to be ionized when they get into contact with a CNT, no matter whether they enter the tube or stay outside. The previous discussion was about DFT, which does not always give the correct charge. When the Li or Na ion is inside the tube, the image energy is larger, and hence DFT gives the correct charge for small tubes. When they are outside, the image energy is lower, and DFT may have difficulties in obtaining the correct charge.

Philip A. Ash asked: I am interested in proton coupled electron transfer reactions in biomolecules, which often involve transfer of both H^+ and H_2O *via* a series of hydrophobic or hydrophilic interactions. The two 'nanopore' systems you have studied seem to represent models for both hydrophobic and hydrophilic channel environments within proteins. A recent paper by Voth and coworkers[1] suggests a new mechanism for proton transport through hydrophobic channels *via* the creation of temporary water wires by hydrated excess protons. Hydrated protons were found to induce wetting within a hydrophobic nanotube, vastly reducing the free energy barrier to proton tunneling (*via* a Grotthuss mechanism) through the nanotube. Conversely, other monatomic cations were found to have the opposite effect and wetting of the nanotube was disfavoured. Would you be able to comment on proton mobility through such confined hydrophobic or hydrophilic channels and how it might be affected by the presence of other ions?

1 Y. Peng, J. M. J. Swanson, S.-g. Kang, R. Zhou and G. A. Voth, *J. Phys. Chem. B*, 2015, **119**, 9212.

Wolfgang Schmickler answered: I think that the wetting is induced by the strong interaction of the proton with its image charge. This induces the creation of favorable proton transfer channels along the wall. This favorable arrangement can be disturbed by monatomic cations, which form strong solvation shells, and thus inhibit the formation of favorable water channels.

Andrew Mount addressed Wolfgang Schmickler and Michael Eikerling: The calculations of the stabilities of the systems of inserted ions (and the progress to calculating different systems with and without added water in the pores) in nanotubes are extremely interesting. These give insight into the most thermodynamically stable system *e.g.* for each cation after ion insertion but I suspect that for the applications envisaged (*e.g.* supercapacitor charge storage with associated ion insertion), the practical charging rate determines which structure is in fact formed, as this is likely to be kinetically controlled (likely determined by the size of the activation barrier *e.g.* of paying the energy penalty of at least partial ion desolvation prior to ion insertion at the nanotube end). Given the high energy price to pay for full ion desolvation, is this not likely to mean that partially desolvated ions are likely to be inserted initially, followed by a subsequent relaxation (solvent equilibration and potential ejection from the pore) as this metastable state relaxes towards the thermodynamically most stable state? How can your ion and/or solvent calculations be extended to give insight into such relaxation processes?

Wolfgang Schmickler replied: The ion loses at least a part of its solvation sphere, which costs energy. It gains the image energy, and it experiences the potential drop between the solution and the nanotube, so a suitable potential can drive the ions into the tube. I agree that probably at first the ions will enter in a partially hydrated form – this was discussed in one of our papers.[1]

We are performing calculations for solvated ions in nanotubes, but the results are not yet ready for publication.

I agree that the entry of an ion into the tube may require an activation energy; at the same time as the ion starts to lose a part of its solvation shell it begins to feel the image force, and the activation energy will be determined by the balance as the ions enter. We have an idea on how to model this, but, again, it is too early to talk about it. Our results for ions can be applied to ionic liquids, where solvation is not an issue.[2]

1 A. Goduljan, F. Juarez, L. Mohammadzadeh, P. Quaino, E. Santos and W. Schmickler, *Electrochem. Comm.*, 2014, **45**, 48.
2 W. Schmickler, *Electrochim. Acta*, 2015, **173**, 91.

Frederic Kanoufi communicated: In Fig. 7 of your manuscript you are suggesting the intercalation of ions in a nanotube will be accompanied by an optical phonon. Could you comment a bit more on that? Do you mean a way to evidence the extent of such intercalation could be probed spectroscopically, typically the IR based on the frequency you are reporting?

Wolfgang Schmickler communicated in reply: In principle these optical phonon branches could be detected by IR spectroscopy. I do not know if this is feasable experimentally. If it is, it will give information about the filling of the tube.

Frederic Kanoufi communicated a question regarding the paper by Sanli Faez: This is an interesting platform to monitor a single nanoparticle in an electric field. To complement on the comment of Professor Tian regarding the potential of your system for electrochemistry at a single nanoparticle, I was wondering to what extent one could use it for performing bipolar electrochemistry at a single nanoparticle level. This would actually require increasing the applied electric field by 1 to 2 orders of magnitude, is it experimentally feasible?

Richard Crooks communicated in reply: I think it is possible to perform bipolar electrochemistry using nanopores, but as you say some redesign of the system would be required. In fact, I believe there are people working on this experiment.

Sanli Faez communicated in reply: Thank you for the interesting suggestion. One can apply as an high electric field as silica glass can withstand without breakdown.

Yitao Long communicated a question for Wolfgang Schmickler: Firstly, how do you ensure that a one-dimensional salt is embedded in a tube, rather than separated anions and cations in the tube?

Secondly, does the insertion of ion result from free diffusion or electric field stress?

Wolfgang Schmickler communicated in reply: A one-dimensional salt is the energetically stable form. Separated anions and cations would be highly unfavorable because of the Coulomb repulsion.

The insertion results from a gradient of the electrochemical potential.

Sanli Faez opened a general discussion of the paper by Tim Albrecht: Is the circuit design for your special amplifier available as open-source hardware? If so, under which license?

Tim Albrecht answered: We are more than happy to collaborate in this regard and make the designs available as much as possible.

Paolo Actis asked: How is this amplifier different than Chimera Instruments VC100?

Tim Albrecht replied: The noise performances of our device(s) and the Chimera amplifier are quite comparable – we believe our setup does a little better at high frequencies. It also allows for higher input currents (up to 100 nA, rather than 50 nA) and the AC/DC channel splitting means that translocation events in the AC channel can be analysed without complex baseline correction. A proper, point-by-point comparison would be necessary to address this question in full though. Some characterisation of our devices is shown in our contribution to this Faraday Discussion and in our previous paper in Nanoscale[1](http://pubs.rsc.org/en/content/articlehtml/2016/nr/c5nr08634e), cf. Supporting Information. There are also references to the electronics design, which was published by the Milan group separately.

1 R. L. Fraccari, P. Ciccarella, A. Bahrami, M. Carminati, G. Ferrari and T. Albrecht

J. Gooding inquired: With non-biological nanopores, can we use surface modification to slow down the translocation through the pores to alleviate issues related to the translocation speed? Controlling surface chemistry so it proceeds inside the pore only has been achieved in a few ways.[1,2] Can you illuminate us on the merits versus the fast amplifier? I agree that there are issues with regards to silanization and stability.

1 K. A. Kilian, T. Böcking, K. Gaus and J. J. Gooding, *Angew. Chem. Int. Ed.* 2008, **47** 2697–2699.
2 B. Guan, S. Ciampi, G. Le Saux, K. Gaus, P. J. Reece and J. J. Gooding, *Langmuir*, 2011, **27** 328–334.

Tim Albrecht responded: Tuning the surface properties of a solid-state nanopore sensor can be beneficial to the sensing performance. Careful control of the interaction strength is of paramount importance, however, since otherwise pore clogging or crowding of the analyte can occur. An interesting special case seems to be the translocation of protein/DNA complexes, where the protein appears to adsorb to the sensor surface much more strongly than the DNA itself (see examples from our own work involving p53/dsDNA, low-noise nanopore chips[1] and SSB/ssDNA, nanopore chips[2]). Hence, controlling both at the same time might add additional challenges. We have also recently shown theoretically that surface friction may affect the scaling law between DNA length and translocation time.[3] The work of Michael Mayer *et al.* around lipid-bilayer-coated solid-state nanopore sensors offers another elegant approach for controlling the translocation process (; protein translocation).

Generally, a well-controlled translocation process combined with high-performance detection electronics will enable new capabilities, depending on the specific application (better resolution, smaller analytes etc.).

1 P. Nuttall, K. Lee, P. Ciccarella, M. Carminati, G. Ferrari, K-B. Kim and T. Albrecht, *J. Phys. Chem. B*, 2016, **120**(9), 2106–2114.
2 D. Japrung, A. Bahrami, A. Nadzeyka, L. Peto, S. Bauerdick, J. B. Edel and T. Albrecht, *J. Phys. Chem. B*, 2014, **118**(40), 11605–11612.
3 R. L. Fraccari, P. Ciccarella, A. Bahrami, M. Carminati, G. Ferrari and T. Albrecht, *Nanoscale*, 2016, **8**, 7604–7611.
4 E. Yusko, J. M. Johnson, S. Majd, P. Prangkio, R. C. Rollings, J. L. Li, J. Yang and M. Mayer, *Nat. Nanotech.*, 2011, **6**, 253–260.

Yitao Long asked: You did show the detection of the short DNA in nanopipettes using the high-bandwidth CMOS current amplifier, what about applying this system in a biological nanopore? Can the bandwidth go higher, and what is the highest bandwidth it can get to?

Tim Albrecht answered: The best achievable performance depends on the input capacitance during the measurement and the magnitude of the current modulation, relative to the background current noise (in the AC channel in our case). These are specific to each experiment, which would have to be optimized accordingly. At present I do not see any reason why the setup cannot be used in conjunction with a biological pore, provided the DC current does not exceed 100 nA and the input capacitance is sufficiently low.

I should add that the internal bandwidth of the amplifier is 3.5 MHz, and our analog filter is currently limited to 2 MHz. However, in measurements we have used up to 400 kHz so far, at S/N ratios similar to the one reported in the paper (long dsDNA translocation in nanopipettes) – these results are unpublished.

Patrick Unwin commented: I'd like to comment on the reliability of eqn 1 in your paper to estimate the aperture size of nanopipettes. While this is widely used, we have tested this equation recently and carried out a through geometric and mass transport analysis of nanopipettes.[1] There are significant changes in the local taper angle with distance into pulled nanopipettes, and the internal and external angles can differ significantly. Such effects need to be taken into account to understand in detail the ion current response of these devices and this information can be obtained very nicely with TEM analysis.

1 D. Perry, D. Momotenko, R. A. Lazenby, M. Kang and P. R. Unwin, *Anal. Chem.*, 2016, **88**(10), 5523–5530.

Tim Albrecht responded: I think your study in *Analytical Chemistry* is very important in this context. It will be interesting to investigate the effect of these parameters on the translocation characteristics of DNA or proteins, for example.

We usually take the conductance as a parameter to assess the similarity of different pipettes (pulled using the same programme), but being a composite quantity this does not rule differences in the actual pore diameter, shape of the channel and so forth. Unfortunately, we are not in a position to perform TEM imaging on every pipette we use (ideally before and after!) Having said this, using the conductance criterion typically provides good reproducibility of the translocation characteristics in our studies. The errors given in our Discussions paper as well as in our previous publication in Nanoscale are normally the standard errors (of the most probable translocation time, for example), based on three independent measurements (hence three different pipettes). These are relatively small, perhaps a reflection of the fact that the very end of the pipette typically has the largest effect on the overall conductance (*i.e.* the sensing region).

Philip Bartlett said: In your paper you say that the translocation of the DNA through the pore can be slowed down by using LiCl electrolyte rather that KCl. Why is that? Has anyone looked at using double-stranded DNA–PNA rather than DNA–DNA duplexes, as this would also be a way to change the charge on the duplex?

Tim Albrecht replied: The reason for the slowing down of DNA translocation in LiCl (and some other electrolytes), compared to KCl, has been associated mainly with differences in cation interactions with the negatively charged DNA backbone (ion pairing). Ref. 45 and 46 in our manuscript constitute rather detailed studies in this regard. Our ref. 48 highlights differences in the transport mechanism, namely flux- *vs.* barrier-limited translocation.

Meller *et al.* used PNA hybridized to dsDNA in tagging/bar coding experiment[1], which is closest to what is suggested in the question, as far as I am aware.

1 A. Singer, M. Wanunu, W. Morrison, H. Kuhn, M. Frank-Kamenetskii and A. Meller, *Nano Lett.*, 2010, **10**(2), 738–742.

Minkyung Kang remarked: As others have said, we also often experienced the change of the glass pipette (diameter < 50 nm) by measuring the *I–V* curve within a day. This change could be significant to the experiment with nanopore devices as the geometry of the tip affects the current responses, as followed by our previous work.[1]

1 D. Perry, D. Momotenko, R. A. Lazenby, M. Kang and P. R. Unwin, *Anal. Chem.* 2016, **88**(10), 5523–5530.

Tim Albrecht replied: Yes, indeed, we also observe equilibration effects in some nanopipettes. For example, in some cases we detect rectification, which tends to disappear after a few bias cycles. For translocation experiments we normally use nanopipettes that show linear and stable *I–V* characteristics (over hours), as shown in the manuscript (ESI).

Richard Crooks opened a general discussion of the paper by Robert Johnson: Proteins like α-hemolysin are structurally dynamic, they misfold, and they are often mobile when in bilayers. How do these types of effects affect the noise and reproducibility of the types of measurements you make? What is the effect of the charges inside the barrel of the protein on translocation?

Robert Johnson responded: The α-hemolysin pore is structurally dynamic. In a typical experiment, we add monomer units that are commercially available and wait for these to diffuse into the lipid bilayer and form a protein channel. The α-hemolysin pore should comprise seven monomer units when correctly formed. However, we do often see mis-folding and incorrect formation during our experiments. Identification of the correctly formed protein channel is achieved through measuring the conductance, which is *ca.* 1 nS at 25 °C in 1 M KCl. With our system, we have the added advantage of pressure control over the protein insertion rate and if a mis-folded protein forms in the lipid bilayer then lowering the pressure can be used to remove it without destroying the bilayer. Measurements with the correctly formed protein channel are highly reproducible.

Charges inside the protein pore are important in governing the interactions between the internal protein walls and DNA held confined within the channel. For example, the latch constriction of α-hemolysin comprises a ring of positively charged lysine groups. Interactions between these groups and the DNA are likely involved in modulating the base-flipping effect that we observe for a DNA mismatch site situated at the latch constriction.

Tim Albrecht commented: The discussion of the advantages and disadvantage of biological *vs.* solid-state pores is a complex one and cannot be exhaustive here. Biological pores have probably no match in the solid-state world, when it comes to well-defined surface properties and the reproducibility of the pore dimensions. They are also very sensitive, in that even small chemical changes inside the pore channel can be detected and studied in detail. They can also be modified by site-directed mutagenesis. A disadvantage can be that the pore size cannot be changed dramatically. So while α-hemolysin and MspA pores are able to pass single-stranded DNA, translocation of larger analytes, such as double-stranded DNA, many proteins or protein/DNA complexes, is not possible without structural alterations (such as separation of the strands in dsDNA). Biological pores in their membrane environment can also stable enough for operation in the field, as demonstrated by Oxford Nanopore Technologies recently in the context of the Ebola epidemic in Africa.[1]

1 Quick *et al.*, *Nature*, 2016, **530**, 228–232.

Philip Bartlett said: I think that it is very impressive that you can carry out kinetic measurements on single DNA molecules and that you can carry out elegant temperature dependence studies to extract activation energies. From your results the base flipping is driven by interaction between the flipped out base and the wall of the latch region. How does this compare with what is happening in the solution? I noted from your paper that there is some disagreement in the literature between the results from NMR and from fluorescence.

Robert Johnson answered: While the activation energies and timescales for a base flipping out of the DNA helix are similar to those published for DNA in bulk solution (between 10 and 20 kcal mol^{-1}), the energies and time scales for flipping a base back into the helix are not. This strongly suggests that the extra-helical state is in someway stabilized relative to what would be observed in bulk solution. At the latch constriction of α-hemolysin, there is a ring of positively charged lysine groups and it is highly likely that these are able to interact with an extra-helical base through electrostatic and hydrogen-bonding interactions. Some recent ionic-strength dependent studies that we have done in our lab support this, and do show that ionic strength does effect the base-flipping kinetics within the pore. We plan to publish this work shortly.

In a biological context, base-flipping is understood to occur both passively and actively. In the active case, enzymes may stabilize the extra-helical conformation through binding and this type of interaction may be similar to what we observe when the base-flipping occurs at the latch constriction of alpha-hemolysin.

Andrew Ewing asked: When you look at summary data and get the variance from the data, if you do that in one channel *vs* several channels, is the variation the same? I think all proteins are not the same; crystal structures don't tell the whole story and proteins fold differently depending on how they are made. So a little variability can occur, so purification process must pick the same variation. How do the commercial sources make their protein that you use?

Robert Johnson said in reply: The α-hemolysin protein can be purchased commercially as monomer units that are secreted from *Staphylococcus aureus* and subsequently isolated. We place these monomer units into a buffer solution that is in contact with the lipid bilayer. Subsequent diffusion into the bilayer, followed by the combination of seven monomer units, form the heptameric protein used for ion channel recordings. At 25 °C, this pore has a conductivity of 1 ns $\pm$ 0.04 ns in 1 M KCl.[1] Occasionally, the monomers units do not form the correct protein channel and when the conductance of the open channel is out of this range we reject the channel.

1 J. J. Kasianowicz, E. Brandin, D. Branton and D. W. Deamer, *Proc. Natl. Acad. Sci.*, 1996, **93**, 13770.

Lane Baker asked: Can you comment on use of this approach to measure single base changes in unknown samples?

Robert Johnson replied: This is a really interesting point. The approach that we are presenting relies of knowing what single-base change you are looking for. In an application, one would design a probe DNA strand to bind to a target DNA sequence such that the base site of interest is aligned with the alpha-hemolysin latch constriction. Then, DNA molecules can be analyzed within the pore for the presence or absence of this base change one molecule at a time. Actually, the real power in our approach is in counting how many DNA molecules within a population have a known base change. This is important for determining heterogeneity within a gene sample. Nanopores have a real advantage here, because of their inherent ability to measure single entities of DNA one at a time rather than relying on averaging across hundreds or thousands of molecules.

Philip Bartlett opened a general discussion of the paper by Mark Platt: You mention in the Conclusions to your paper that an advantage of the restive pulse assay is that it can be multiplexed. How can this be achieved and what is a realistic value for the number of different analytes you could detect simultaneously in a multiplexed measurement?

Mark Platt answered: Multiplexed assays have been done with two analytes by the group previously.[1] You need particles of narrow size distributions.

1 E. R. Billinge and M. Platt, *Biosens. and Bioelectron.*, 2015, **68**, 741.

J. Gooding asked: You present to us two assay formats where the particles are the sensing element or where modification of the pore is used. Why would you do the latter? What's the advantage of doing this when with the particle assay you can sample a lot more solution space?

Mark Platt responded: There were two motivations for developing the second – *i.e.* the aptamer modified pore wall experiment. The first was to continue to explore the nature of the rectification behavior of these larger pores. Having seen some increase in current rectification behavior we were curious to how and if DNA and aptamer walls changed the current flow, in the presence of the analyte. This had a natural lead into some other research in the group looking at blood assays where we might want to simply drop some sample into the upper fluid cell and confirm the presence of an analyte by monitoring the current flow and not have to use particles. The second was in our work looking at environmental samples for example heavy metal ions in water. Again the sample could be dropped into the upper fluid cell for analysis, but long term we are exploring adding this into fluid chips with automated sample handing and the pores might serve as a means to provide a rapid read out.

Philip Bartlett questioned: One of the questions that has already come up in the Discussion is about the effects and extent of overlap of the double layers of adjacent surfaces. In this case in your work you have a very large pore (> 700 nm) and yet you see current rectification when you might not expect to. Have you some idea why this should be?

Mark Platt replied: I think the early literature suggested the pore opening had to be of similar size to the double layer, but there is a growing number of papers where larger pores are still producing rectification behavior, and even biphasic pulses in resistive pulse measurements are seen on the same system.[1] This all suggests that the double layer does not need to be comparable to the pore size. H. White *et. al* have suggested the electroosmosis plays a larger role then initially suspected and I think in our pores we will see a much larger contribution of this,

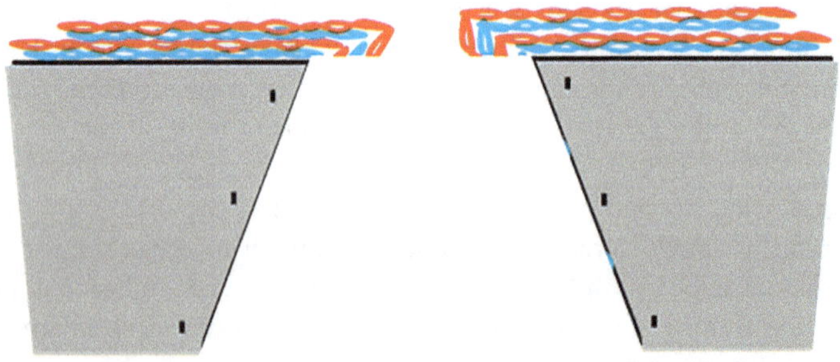

Fig. 2 Schematic of the layer-by-layer assembly on the top pore surface.

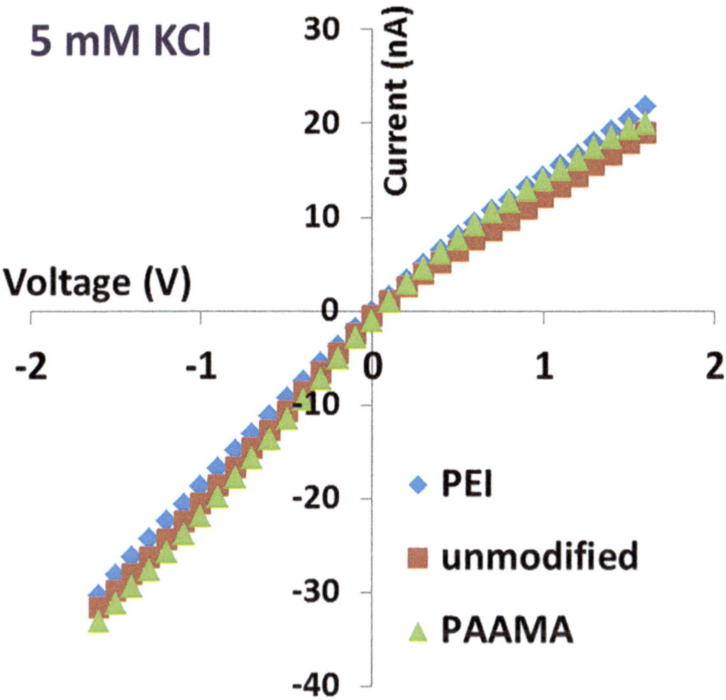

Fig. 3

but modelling is required for me to say that with certainty. We did hypothesize in the paper that the pore top surface might contribute to the behavior, but studies done by us since the submission of the paper (Fig. 2 and 3 and Table 1 below) shows this is not the case.

1 E. Weatherall and G. R. Willmott, *J. Phys. Chem. B*, 2015, **119**(16), 5328–5335.

Patrick Unwin commented: The membrane you use to make the pore is rather thick. I noticed that with unmodified pores, the current-voltage response was close to ohmic and that significant rectification was seen with modified pores. Is that correct? Could the rectification in modified pores be magnified by sub-surface constrictions from polymer strands or plugs? Further, do these polymeric pores change their dimensions over time?

Table 1 Table showing the current rectification ratios (measured at ± 1.6V) for each modification of the pore surface

[KCl] / mM	Unmod	PEI	PAAMA
5	1.64	1.38	1.65
10	1.12	1.00	1.10
50	1.05	0.93	1.04
137	0.97	0.94	1.06

Mark Platt replied: The standard unmodified pores gave a small rectification behavior. If a smaller pore size is used this is increased; however we opted to use a larger pore opening for the study as it's the same size we use for the resistive pulse experiments. The increase in rectification behavior with the polymers could be due to the polymer forming islands or protrusions in the pore, but with two bilayers Ithink this would still be only a few nm in size. However given that we are never sure of the exact features in each pore, more work has to be done to characterize the cause of behavior in the *I–V* curves. One of the areas of research we are just starting is the production of bespoke pores, with more controlled features, and this might help us understand the mechanism.

The pores do change dimensions over time by becoming more flexible, and as we continue to stretch the polymer over the course of weeks we often find that the baseline current increases as the pore openings become larger. We only used the pores in this study a few times, in some cases only once (each DNA experiment was performed on a new pore).

Tim Albrecht asked: In order to explain the occurrence and magnitude of current rectification effects in pore channels (including nanopipettes), it might be necessary, under some conditions, to consider hydrodynamic (electroosmotic) effects as well (see, *e.g.*Keyser *et al.*)[1]

1 N. Laohakunakorn, V. V. Thacker, M. Muthukumar and U. F. Keyser, *Nano Lett.* 2015, **15**, 695–702.

Mark Platt answered: Thanks for the comment, I agree the contribution of the EOF may well be unknown here. White and co-workers also looked at this and whilst we alluded to their work to indicate that EOF may be important, we have no idea yet as to the extent of its contribution. This looks like an excellent reference to add to the search and our future work hopes to look at pores of differing geometries via modelling and through experimental work.

Patrick Unwin addressed Tim Albrecht: Apologies for stating the obvious, but while the Debye length might be a reasonable rule of thumb for the size of the double layer, compared to, say, the diffusion layer at most of the electrodes used in electrochemistry, when dealing with nanoscale devices, such as nanopipettes, it is important to consider the entire potential–distance profile. Several people at this Discussion, including us, have looked at this and current–voltage curves are predicted that are closely in line with what is seen experimentally.[1–5]

1 H. S. White and A. Bund, *Langmuir*, 2008, **24**, 2212–2218.
2 W.-J. Lan, C. Kubeil, J.-W. Xiong, A. Bund and H. S. White, *J. Phys. Chem. C*, 2014, **118**, 2726–2734.
3 N. Sa, W.-J. Lan, W. Shi and L. A. Baker, *ACS Nano*, 2013, 7, 11272–11282.
4 K. McKelvey, S. L. Kinnear, D. Perry, D. Momotenko and P. R. Unwin, *J. Am. Chem. Soc.*, 2014, **136**, 13735–13744.
5 D. Perry, D. Momotenko, R. A. Lazenby, M. Kang and P. R. Unwin, *Anal. Chem.*, 2016, **88**(10), 5523–5530.

Tim Albrecht replied: I fully agree – considering the charge distribution will implicitly include the actual geometry of the channel, which is obviously important, too. It is also important, however, to differentiate between preferential ion transport *(i.e.* differences in the transport numbers) and current rectification (bias asymmetry of the current) – the former does not automatically lead to the latter. The point I was trying to make was that electroosmotic effects can lead to such a current asymmetry, which is perhaps not as widely appreciated as electrostatic effects.

Serge Lemay commented: While current rectification was originally often described in terms of double-layer overlap, more general modeling has shown that it is sufficient to have spatially inhomogeneous transference numbers for cations and anions along the length of the channel. This can be the result of double-layer overlap but also of, for example, surface conduction.

J. Gooding asked Mark Platt: You talked about the unexpected rectification for a large nanopore and I was wondering whether the open and porous nature of the layer by layer films, which swell in solution, may contribute to this effect considering there may be strands of polymer floating around in solution and possibly close to the pore entrance? I appreciate it is hard to modify a pore but perhaps something like plasma reacting where the chemistry is maintained very close to the surface might help answer such a question.

Mark Platt replied: An excellent point and to understand the true mechanism of the rectification behavior we will have to use a more specific chemistry to ensure we do not have any polymer extending out from the surface. In the supplementary material we do use a lower molecular weight PEI solution (although the same MW PAAMA) and observe similar behavior, although the rectification is smaller. This was initially attributed to the PEI forming islands before completely coating the pore wall, and we can not exclude the fact that the polymer may extend out and swell. I would comment that we have not optimized the layer-by-layer assembly, in fact other groups often place the polyelectrolytes down in specific pH, high ionic strength solutions to ensure they polymers go down in a dense coating and we might observe much stronger and varied behavior if we varied this. For resistive pulse measurements a swollen polymer would prohibit particle translocations, but such a mechanism might be very powerful for other types of assays. Only by knowing the actual mechanism can we exploit this and more work is needed.

Paolo Actis returned to discussion of the paper by Tim Albrecht: In the last slide of your presentation you proposed a detection mechanism very similar to the one published by Bell and Keyser.[1] How is your approach different?

1 N. A. W. Bell and U. F. Keyser, *Nat. Nanotech.*, 2016, **11**, 645.

Tim Albrecht replied: The idea is very similar, but perhaps more akin to Plesa *et al.*,[1] where dsDNA overhangs on a long dsDNA track are being detected. The purpose of the latter study was to investigate velocity fluctuations during translocation, whereas our current idea is to use ssDNA overhangs to 'fish' for particular target sequences in a sample. This is similar to a DNA assay, hence the term 'DNA assay on a string'. The idea of detecting features incorporated in or bound to DNA is obvious not new, see for example Singer *et al.*,[2] , for PNA/DNA complexes.

1 C. Plesa, N. van Loo, P. Ketterer, H. Dietz and C. Dekker, *Nano Lett.*, 2015, **15**(1), 732–737.
2 A. Singer, M. Wanunu, W. Morrison, H. Kuhn, M. Fank-Kamenetskii and A. Meller, *Nano Lett.*, 2010, **10**(2), 738–742.

Richard Crooks addressed Tim Albrecht, Robert Johnson and Mark Platt: I have two questions. First, how come the nanopores rarely seem to clog up? There must be a very high concentration of "junk" in the solution. Second, an expert in nanopore DNA sequencing told me there was no way that it would ever be possible to sequence DNA due to the noise level. Yet, it seems to be that this is what people are doing. Can you reconcile these two points of view?

Tim Albrecht responded: First, they do! The more contaminated the electrolyte solutions, the more likely they clog. Sometimes this clogging is reversible, *i.e.* it can be removed by a large reverse voltage pulse ('zapping'), sometimes it is not (and the device may have to be re-cleaned or discarded).

Regarding the second question, I think it depends on how one defines the question or rather the term 'nanopore DNA sequencing'. To this date, no-one has demonstrated DNA sequencing with a plain, unmodified solid-state device. The DNA bases pass too quickly, and the differences between the bases are indeed drowned in the noise. On the other hand, it has been shown that modified biological pores can successfully be used. It is necessary to control (slow down) the translocation process (*e.g. via* a processive motor enzyme, such as a helicase) and to enhance the specificity of the 'read head' inside the pore (for example, by adding a cyclodextrin using site-directed mutagenesis). Combining these with the appropriate sample processing, electronics development and other factors will result in a functional sequencing device, as demonstrated for ebola by Oxford Nanopore Technologies[1] or for that demonstrated for sequencing in space.[2] The latter is based on alpha-hemolysin. The Gundlach group at the University of Washington have been developing an alternative technology around the MspA pore (apparently with involvement from Illumina). The Kasianowicz group at NIST, in collaboration with a group at Cornell, have developed a very interesting, nanopore-based 'sequencing by synthesis' method, which may not be quite as far developed towards a commercial product, but does provide very good differentiation between bases (since it is actually a PEG tag that's being detected, not the base itself).

However, this does not mean that this technology is necessarily the best for every application. At present, it seems that nanopore-based technologies possess higher error rates than other technologies (partly because they read more than one base at a time consecutively and the actual composition must be reconstructed using appropriate statistics), but they are very compatible with field- or 'point of care' operation. The answer given here cannot be comprehensive in a sense that it captures every aspect of this diverse and rapidly developing field; it is to the best of the author's kowledge at this stage.

1 Quick *et al.*, *Nature*, 2016, **530**, 228–232.
2 https://www.nasa.gov/mission_pages/station/research/news/dna_sequencing

Robert Johnson replied: First, occasionally we see clogging with protein nanopores. We do take care to remove possible contaminants by filtering our buffer solutions and using HPLC to purify DNA samples prior to measurement.

Secondly, with the alpha-hemolysin pore, the noise level of our measurements tends to be down around 1 pA or less, and we can observe dynamic changes in a DNA molecule held within the pore that give rise to current changes of just 1–2 pA. Indeed, noise can actually be beneficial to us in our measurements. For example, we have found that the noise associated with the blocking current when dsDNA is inside the pore to be dependent on the stability of the DNA.[1] We have speculated that the noise associated with the blocking current may be a result of short timescale motions of the DNA duplex inside the pore, such as breathing and bubble formation.

1 R. P. Johnson, A. M. Fleming, Q. Jin, C. J. Burrows and H. S. White, *Biophys. J.*, 2014, **107**(4), 924–931.

Mark Platt answered: Dealing with blocked pores would be the a typical week for the group and for the larger PU pores the lifetime is correlated to how quickly we block them with particles. If we have a high concentration of nanomaterials then the pores block within seconds and can't be recovered. This can mean you need some experience/information on the sample you are analysing to ensure you don't block the pore immediately, and this of course is not always possible.

Jan Clausmeyer addressed Tim Albrecht: The presented electronic amplifier might be very useful as a tool to address questions related to nanoparticle electrochemistry. For experiments exploiting collisions of nanoparticles with electrode surfaces (nanoimpacts), the dilemma is that one only has a very short time to obtain information about individual particles. In most cases, some kind of transient current signal is detected upon collision of the particle but it is very unlikely to find that very particle again. Often, one must decide whether to acquire information regarding the particle size OR regarding electrocatalytic activity of particles. However, it would be very valuable to correlate the structure of individual particles with their activity.

The fast and automatic switching of the applied voltage would be very useful to get more information out of nanoparticle impact experiments. One may test several electrochemical reactions at different electrode potentials (for instance electrocatalysis followed by coulometry for particle sizing) during the very short residence time of the particle. However, electrode capacitance would certainly be a problem for that. Do you think it is possible to implement such an electronic amplifier for a typical experimental setup used for nanoimpact measurements? Which factors need to be taken into account?

Tim Albrecht responded: In fact, we have just begun to explore those possibilities, *i.e.* to try to use the nanopore setup for nanoimpact experiments. However, some previous nanoimpact studies in the literature have been performed with very powerful electronics, too, so we will have to see whether we can improve on this aspect further. It will be key to minimize the capacitance of the electrochemical cell as much as possible, ideally to significantly below 10 pF, since this is important for maximizing the time resolution also with our nanopore setup. If further performance enhancement can be achieved, it remains to be seen whether the improved time resolution provides significant new insight into the nanoimpact process. Nevertheless, this is currently of great interest to us.

Switching experiments, *e.g.* trying to recapture particles after bouncing off the electrode surface, are a more complicated matter, because the switching time will depend on the RC time constant. Hence, it is not clear yet whether fast enough switching can be achieved (even if C is very small, R may be significant, too).

Olaf Magnussen asked: One of the very fundamental questions in single entity electrochemistry is the ultimate limit of charge transfers that can be measured at electrochemical interfaces. Currently, the limit seems to be on the order of femtocoulombs. Is there a chance to go to significantly lower values, *e.g.* one or two orders of magnitude lower? Otherwise, every measurement will always have to rely on strong amplification effects, which strongly limits the scope of these measurements.

Tim Albrecht replied: Indeed, in our experiments on 200 bp dsDNA we obtain a most-probable translocation time of about 20 μs at event currents of 60 pA – this yields 1.2 fA. A single electron would be 0.16 aF, so that's still a significant way off.

Single-nanoparticle capacitances of approximately 1 aF have been measured at the single-particle level using electrochemical STM,[1] but in those experiments the tunnelling current acts in some way as an amplifier. Moreover, the configuration is rather specific.

1 T. Albrecht, S. F. L. Mertens and J. Ulstrup, *J. Am. Chem. Soc.*, 2007, **129**(29), 9162.

Patrick Unwin inquired: How are you generating the AC signal? I could not find many details in the paper. Could you explain this in a bit more detail?

Tim Albrecht answered: A full explanation of how the pipette current is handled through the different stages of the detection electronics would be difficult to provide here and a separate, electronics-focused publication is currently in preparation. However, we did describe the basic principle in our Nanoscale paper,[1] (see figure 1 in particular) and provide further references therein. An important point is that current amplification is achieved without (noisy) feedback resistors, but rather by matched transistor/capacitor circuits, combined with a transimpedance amplifier. This part of the setup is described in detail in reference 25 in our Nanoscale paper.[2] On the other hand, the large DC current is handled by a low-frequency DC feedback network, as described by P. Ciccarella *et. al.*[3]

In effect, a fast path is provided by the current amplifier/transimpedance amplifier, which amplifies the short pulses given by the pore blockade irrespective of the DC current (and low-frequency fluctuations therein). The DC current is determined by the voltage drop across a known resistor prior to the DC feedback network.

1 R. L. Fraccari, P. Ciccarella, A. Bahrami, M. Carminati, G. Ferrari and T. Albrecht, *Nanoscale*, 2016, **8**(14), 7604–7611.
2 G. Ferrari, M. Farina, F. Guagliardo, M. Carminati and M. Sampietro, *Electron. Lett.*, 2009, **45**, 1278–1280.
3 M. Carminati, G. Ferrari, R. Fraccari and A. Bahrami, An Integrated Low-Noise Current Amplifier for Glass-Based Nanopore Sensing, *10th Conference on Ph.D. Research in Microelectronics and Electronics (PRIME 2014)*, Grenoble, France, June 29–July 30, 2014, 1–4.

Paolo Actis asked: Nanopipettes can be integrated with nanomanipulators to comprise a scanning ion conductance microscope. Are you planning to use your protocols to deliver short nucleotides into living cells?

Tim Albrecht replied: I imagine they could be used in this context, but we do not have any immediate plan to do so.

Yitao Long communicated a question for Robert Johnson: First, when focusing on Fig. 1 of your manuscript, the control experiments show that the CC9 mismatch contributes to two well-defined residual current states (I1 and I2). However, how can you make sure that well-defined residual current states were not caused by dsDNA with CG9 base pair?

Secondly, according to Fig. 4 of your manuscript, why is $\tau_{\text{intra-helical}}$ less dependent on temperature?

Robert Johnson communicated in reply: Regarding the first question, three experiments were run with either just a fully complementary duplex, just a CC mismatch containing duplex and both the complementary and mismatch containing duplexes. Modulation between I1 and I2 states is observed only for the CC containing duplex, and all capture events for this duplex exhibit a modulating signature.

Regarding the second question, this is an indication that the activation energy of the two processes are different. As shown in Fig. 5 of the manuscript, the activation energy required to flip the base into the helix is higher, and so, the extra helical lifetime changes more with temperature.

Faraday Discussions

PAPER

Concluding remarks: single entity electrochemistry one step at a time

Richard M. Crooks

Received 4th October 2016, Accepted 10th October 2016
DOI: 10.1039/c6fd00203j

This article provides a summary of the Faraday Discussion on single entity electrochemistry held in York, U.K., in early September, 2016. The introduction provides some context for thinking about electrochemical studies of single entities. The next four sections follow the themes of the meeting as they relate to single-entity electrochemistry: (1) nanoparticles, nanotubes, and nanowires; (2) nanopores and nanofluidics; (3) complex surfaces and reactions at the nanoscale; and (4) molecular electroanalysis. Each paper presented at the Discussion is summarized, and some personal thoughts as to the significance of the findings, the technical advances that made the experiments possible, and common themes between articles are interspersed. Finally, at the end, I round-up my impressions and provide them in a succinct list.

Introduction

At the invitation of the meeting chairman, Dr Patrick Unwin, these remarks represent my personal reflections on the many outstanding papers, posters, and presentations, and the ensuing discussions, that comprised Faraday Discussion no. 194. This Discussion, which many participants, myself included, proclaimed as one of the most intellectually stimulating of their career, focused on "Single-Entity Electrochemistry". Many aspects of this forward-looking topic were discussed at the meeting, including (very broadly defined): nanomaterials, nanopores, and electrochemical detection. The unifying theme tying together the vast range of materials and methods that engaged the participants was to step out of the world of ensemble electrochemistry and into the realm of single particles, molecules, reactions, and, well, "entities"!

The meeting was held on the campus of the University of York in North Yorkshire, England, in September, 2016. York is an ancient city, founded by the Romans in 71 A.D. The central role of York in the affairs of the U.K. over the last 200 or so years, and its vibrant tourism industry, are largely due to George Hudson, a railway entrepreneur, who established York as a hub. Though Mr

Department of Chemistry, The Texas Materials Institute, The University of Texas at Austin, 105 E. 24th St., Stop A5300, Austin, TX 78712-1224, USA. E-mail: crooks@cm.utexas.edu; Tel: +1-512-475-8674

Hudson went bankrupt building his railroad empire, I can report that York is still one of the most accessible and diverse cities in the U.K. thanks to his efforts. York is truly a beautiful city too, and anyone would be well-advised to put it on their bucket list of places to visit.

Although the temptations of York were alluring, the Discussion itself was downright stimulating. I thought the most interesting question that was discussed was "why study single entities using electrochemical methods?", which really was the main theme of the meeting. It's premature to get into specifics now, but we will get there by the end of this article. At the outset, however, I will say that part of the reason is now "we can" and before "we couldn't", and the difference between then and now has a lot to do with advances in instrumentation and simply thinking about electrochemistry in a different way.

For electrochemists, I think a case can be made that the 1960s and 1970s were the decades of mechanistic organic electrochemistry, and this area of research had very little, if anything, to do with single molecules. There were three major developments in the 1980s, however, that began to change the world-view of electrochemists: the ultramicroelectrode (UME),[1] typically a conductive disk having a diameter of around 5–25 μm, the scanning tunneling microscope (STM),[2] and the atomic force microscope (AFM).[3] The UME got the community thinking about doing electrochemistry in small volumes, like rat brains,[1] while the STM and AFM gave electrochemists their first look at the atomic-level detail of electrode surfaces. The invention of the scanning electrochemical microscope (SECM) in the mid-1980s,[4,5] which has greatly influenced electrochemical imaging[6] and measurements in small volumes, was a direct consequence of the appearance of the STM and AFM.

In the 1990s, new electrochemical techniques that involved pores on the order of 2 nm, rather than electrodes, were reported, and this made it possible to analyze nanomaterials, and even molecules, one at a time as they passed through the pore.

Like the farm boy who sees the big city lights, there was no going back: the race was on to probe ever smaller volumes and numbers of molecules with ever smaller electrodes and pores. Nowadays electrodes having diameters of less than 10 nm are fairly commonplace, and even smaller ones have been reported by the experts.[7] These tiny electrodes open up amazing possibilities for analyzing the properties of single entities.

Given this brief historical context, I will do my best in these concluding remarks to summarize the content of the wide-ranging Faraday Discussion on single-entity electrochemistry. The stage was set for the meeting by N. J. Tao who provided an outstanding overview of single-entity electrochemistry, a field which he has had a major hand in inventing. The remainder of the Discussion was organized around four themes: (1) nanoparticles (NPs), nanotubes, and nanowires; (2) nanopores and nanofluidics; (3) complex surfaces and reactions at the nanoscale; and (4) molecular electroanalysis. I'm going to follow this same general flow, but there is quite a bit of overlap in these themes and, as with all good science, it is sometimes difficult to pigeon-hole a particular experiment. At the end of the article I will put into perspective some of the challenges electrochemists face moving forward.

Nanoparticles, nanotubes, and nanowires

I have a dream. I dream that it will be possible to characterize single, complex, electrocatalytically active NPs in the 1–2 nm size range, like the one shown in Fig. 1,[8] *in operando*. The reason for this is that my research group has been studying well-defined metallic and bimetallic NPs for 25 years[9,10] and there is still far more that we don't know about these materials than we do know. In fact, the more we learn about them, the more we realize how little we know! The problem is that these materials are really small: on the order of 30–200 atoms, and that makes them very difficult to characterize at all, much less one at a time in an electrocatalytic environment.

The bigger the particle, the easier it is to characterize, so why not just forget about 1–2 nm particles and focus on materials that are an order of magnitude larger? There are two reasons. First, the most interesting changes in materials properties occur in the <3 nm size range. Larger sizes are dominated by facets rather than edge and corner atoms, and in part for this reason they display more-or-less bulk properties. Second, my group's focus is on correlating first-principles calculations to experimental measurements, and for particles containing more than 200–300 atoms this is not feasible with the present level of computing power. Clearly, it would be wonderful to be able to calculate catalyst structures that yield just the right specificity and turnover rate, but until calculations are tested against experimental measurements on models that correlate directly to the theoretical constructs, that's not going to be possible.

Our approach, and that of most other groups, for correlating theory and experiment is to devise really good synthetic methods that make it possible to synthesize billions of identical NPs, characterize the resulting ensemble, and then extrapolate back to the structure of individual NPs. In many respects this is a fool's errand, because there is no way to make billions of identical particles like that shown in Fig. 1. I used to envy biochemists who study enzymatic catalysis, because after all every enzyme of a particular type is synthesized using the same DNA template. Therefore, it is possible to study ensembles and deduce the function of individual enzymes, right? Wrong. Even enzymes having exactly the same molecular structure have different properties (Zhan *et al.*, DOI: 10.1039/c6fd00061d). What this means is that we electrochemists have to break away

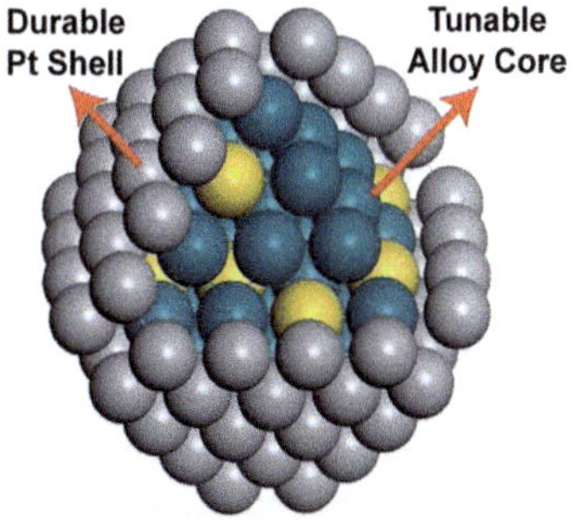

Fig. 1 An illustration of a complex NP consisting of an alloy core covered with a single monolayer shell.

from the ensemble model and begin to focus on individual entities. That's part of what this Discussion was about, and it was quite evident that amazing progress is being made characterizing single NPs, nanotubes, and nanowires.

One of the most interesting methods for analyzing single NPs was first reported by Lemay and coworkers in 2004,[11] and the impact of their approach was evident during the Discussion. In their "collision experiment", a microelectrode was held at a potential that maintained a steady-state current (arising from a reversible redox couple). When polymeric beads having diameters as small as 300 nm were added to the solution, quantized decreases in current were observed periodically. The authors correctly interpreted this result in terms of sticking collisions between the beads and the electrode surface. Specifically, individual NPs blocked part of the electrode surface, thereby reducing the surface area, and this is why the current decreased.

A few years after Lemay's groundbreaking publication, Xiao and Bard showed that a related type of experiment, which they called electrocatalytic amplification (ECA), could yield information about significantly smaller metallic NPs.[12] In their experiment, a catalytically inactive electrode was held at a potential at which little or no current flowed, even though a kinetically slow redox molecule was present in solution. When catalytic metal NPs were added to the solution, however, quantized increases in current were observed. These results were interpreted in terms single nanoparticles striking the electrode and acting as individual electrocatalytic sites that sped up the kinetics of the redox molecule. Each time another NP struck the electrode, another current pulse or step was observed.

These ECA experiments seem to have potential for realizing part of my dream for characterizing individual NPs during electrocatalytic reactions, but at this stage in their development many problems remain. The most significant of these, which has yet to be resolved, is that such experiments provide no structural information about the colliding nanoparticle. In fact, as Stevenson and coworkers showed in their Discussion paper (DOI: 10.1039/c6fd00121a), the situation is even worse than that. Because characterization is electrochemical, relatively high ionic strength solutions are required for implementation, and this usually leads to NP aggregation. In other words, even if NPs having a fairly well-defined size and composition are used for ECA experiments, by the time collisions occur it is likely they will have morphed into new configurations. One might think it possible to coat the NPs with a stabilizer, such as DNA, but it turns out that this kills their catalytic properties[13] so no signal is observed.

Another challenge for ECA and related experiments is that the nature of the collision between the NP and the electrode is not understood. For example, sometimes the resulting electrocatalytic reaction is quenched very quickly, other times less so. This could be due to the presence of impurities in solution, nanobubbles, or something else. One thing is clear, however, if the phenomena that give rise to the collision current signature are not understood, it is going to be very difficult to use this approach to correlate NP structure and function. Having said that, Stevenson's work has shown that the kinetics of aggregation can be measured using this method. In addition, interesting work by Zhang and coworkers has shown that fast scan cyclic voltammetry can be used to study how NPs evolve in time during and just after collisions.[14]

While there is still a long road ahead, ECA-type experiments do hold out promise for *in operando* analysis of NPs. For example, in their Discussion,

Tschulik and coworkers (DOI: 10.1039/c6fd00112b) described ambitious collision experiments, similar to those first reported by Compton,[15] aimed at analyzing bimetallic AgAu alloy NPs (most other collision studies have been performed using simple monometallic NPs). A recurring theme of the Discussion was that the information density of single-entity experiments can be increased by coupling electrochemistry to other analytical methods. Toward this goal, Alpuche-Aviles and coworkers showed that dye-sensitized semiconductor NPs also give rise to currents when they collide with electrodes (DOI: 10.1039/c6fd00100a). The idea of coupling spectroscopy to collision experiments was taken a step further by Kanoufi and coworkers (DOI: 10.1039/c6fd00098c), who showed that dark-field imaging could be used to follow the dissolution of individual AgNPs upon striking an electrode surface. Even more interestingly, through the use of some very clever chemistry, they were able to couple spectroscopy to the electrochemical experiments to distinguish between individual AgNPs and agglomerates. In a similar vein, Long and coworkers (DOI: 10.1039/c6fd00069j) coupled electrochemical measurements with dark-field microscopy, and they showed that individual Au nanorods exhibit differences in activity for peroxide oxidation. Continuing with the theme of coupling other analytical methods to electrochemistry to derive a better understanding of NPs, Kranz and coworkers (DOI: 10.1039/c6fd00128a) used SECM coupled with AFM to better understand the relationship between the structure and function of AuNPs having complex topographies.

Although not there yet, the types of experiments described in this section hold out the promise of correlating NP structure with electrochemical function. The drivers that have enabled this degree of progress are clearly clever experimental design, advances in instrumentation, and coupling of other analytical methods to the electrochemical measurements. These three approaches likely light the path forward as well.

Nanopores and nanofluidics

Electroanalytical chemists have historically confined their thinking about electrochemistry to something one does with an electrode made of Hg, Pt, Au, C, a few semiconductors like titania, and a handful of other solid materials. One of the most interesting aspects of the Discussion, however, was the innovative ways in which hollow pores of various sorts were used to probe nanoscale volumes, individual particles, and even single molecules. In this section, therefore, I want to take a little historical detour so that we can see how this field evolved and what led to the most seminal advances. Let me say at the outset, just as I did in the Introduction, that improvements in instrumentation, creative thinking, and a strong desire to probe smaller and smaller volumes and numbers of molecules are at the heart of the nanopore field.

Consider the Coulter counter. This device, depicted in Wallace H. Coulter's original 1953 patent in Fig. 2 and more schematically in Fig. 3, consists of two compartments separated by a thin tube or single-pore membrane. The compartments are filled with an electrolyte solution, and then a driving electrode is placed into each compartment. When a voltage is applied between the electrodes, an ionic current flows through the tube. Now, when entities having a size on the order of the tube diameter (blood cells in Coulter's original device) are

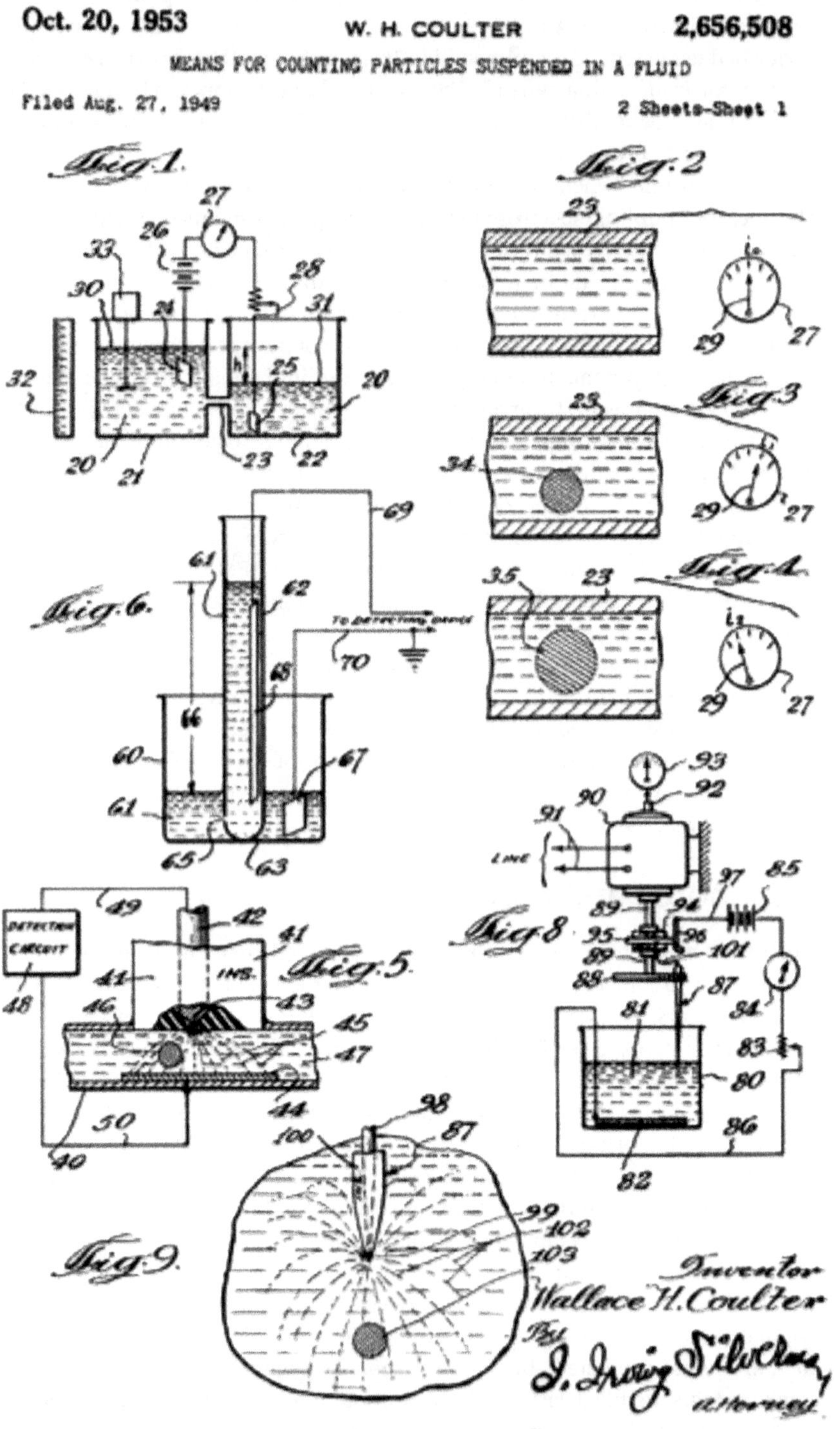

Fig. 2 A diagram from W. H. Coulter's original 1953 Coulter-counting patent.

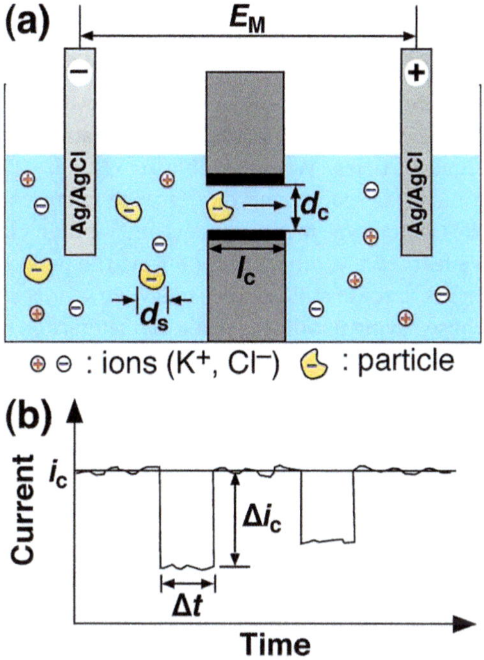

Fig. 3 (a) The general experimental approach for Coulter counting experiments. A membrane containing a single channel divides two chambers containing an electrolyte solution. When an appropriate potential is applied across the membrane, an ionic current is driven through the channel. If particles of an appropriate size and charge are present, they will enter the channel and reduce the ion current. (b) Coulter counter data consist of a series of current pulses associated with the presence of particles within the channel. The height of the pulse, Δi_c, is related to particle size and the width, Δt, corresponds to the particle transit time. In favorable cases, data such as these can provide information about the size, charge, and concentration of the particles. Reprinted from: R. R. Henriquez; T. Ito; L. Sun; R. M. Crooks "The Resurgence of Coulter Counting for Analyzing Nanoscale Objects" The *Analyst* 2004, **129**, 478–482.

placed into one of the compartments, they periodically find their way into tube. The entity then represents a volume constriction in the tube, and leads to a corresponding decrease in ionic current. By counting the resulting current pulses, one can back out the concentration of, for example, blood cells. Moreover, Coulter found that the magnitude of the pulses was closely related to the size of the cells. Pure genius: simple, effective, information dense, and still in daily use worldwide more than 60 years later.

Fast forward to 1970. A pair of scientists working at General Electric in Schenectady, New York, R. W. DeBlois and C. P. Bean, reduced the size of the pore separating the compartments of the electrochemical cell from the 10–100 μm used by Coulter, to just 450 nm by using track etched polycarbonate as a membrane material.[16] This made it possible for them to detect polystyrene particles as small as 90 nm. A few years later DeBlois used the same sort of device to detect individual viruses![17,18]

In 2000, my own group, building on the Coulter concept, used a multiwall carbon nanotube as the pore, and we were able to determine the concentration and size of particles having diameters around 50 nm.[19,20] Although this isn't much smaller than the particles used by DeBlois and Bean, there were some structural aspects of the carbon nanotube pore, primarily its uniformity and absence of permanent surface charge on the interior, that improved the resolution of the measurements and also made it possible to quantitatively assess the surface charge of individual particles. We were able to visualize DNA translocation through a carbon nanotube pore using fluorescence microscopy,[21] but the pore was too big to make Coulter-type measurements and we didn't know how to make a smaller pore. We learned two other things during the period we were carrying out these experiments that are still relevant today. First, it is surprising that the pores, especially those having nanometer-scale diameters, don't get clogged up by impurities in solution. The fact is, however, that clogging is rare. Second, Coulter-type experiments detect one entity at a time, but high entity concentrations are required or else the time between pulses is too long for practical applications. In other words, these devices are useful for single-entity analysis, but many entities are required to make measurements.

At about the same time our experiments were going on, Kasianowicz and coworkers were taking a very different approach for studying translocation of DNA.[22] Instead of using artificial pores, they were using a protein, called α-hemolysin, which has a hole in its center, as the pore. The protein assembles in a lipid bilayer, which is the high-resistance membrane in this experiment. The magic of this approach is threefold. First, the pore in α-hemolysin is very small: just 1.4 nm, so one can use it in a Coulter counting format to detect the passage of individual molecules. Second, because it is a protein, every α-hemolysin is structurally well defined and nearly identical. Third, the chemical structure and amino acid sequence of α-hemolysin is known, and therefore it is possible to change it using the tools of biotechnology. This means it is possible to introduce specificity into the pore.[23]

This trail of advances, one way or another, led to some remarkable discussions in York. While it is not 100% clear that it will be possible to routinely and accurately sequence DNA using α-hemolysin pores, they are being used to study the details of its molecular structure and dynamics (White *et al.*, DOI: 10.1039/c6fd00058d). Advances in the experimental design of nanopores is also leading to new understanding and applications. For example, faster electronics exhibiting lower noise have led to significant advances in contemporary translocation measurements (Albrecht *et al.*, DOI: 10.1039/c6fd00109b). Likewise, pores made of novel materials and fitted out with biorecognition elements (Platt *et al.*, DOI: 10.1039/c6fd00072j) to introduce specificity will likely lead to useful sensors for real-life applications.

Another important message from the Discussion is that theory can be of great value for understanding interactions between pores and the fluids they contain (Schmickler *et al.*, DOI: 10.1039/c6fd00076b; Eikerling *et al.*, DOI: 10.1039/c6fd00094k). Indeed, much work has been done using classical theories for this purpose, but for pores, like single-walled carbon nanotubes or α-hemolysin, it doesn't make sense to think of water as a continuum fluid. Rather, more advanced first-principles theory is required to obtain a molecular level interpretation of experimental results. Most electrochemists are familiar with classical theories of ion solvation, the electrical double layer, and so forth, but we would do well to collaborate with our theory colleagues to better understand the structure of water and ions in nanoscopic pores and other volume elements.

This brings up another important theme from the conference that I mentioned earlier: theory is no better than the experimental models used to test it. Therefore, it is incumbent on experimentalists to devise nanopore systems that as closely as possible mimic theoretical models. Carbon nanotubes probably represent such a model, but evaluating the structure of water or of small numbers of ions in such an environment is going to be tricky. We're getting there, however, as evidenced by the elegant experiments of Faez and coworkers (DOI: 10.1039/c6fd00097e), who discussed a well-defined capillary system for tracking individual (60 nm) gold nanoparticles. Their approach relied on light scattering for detection, and this brings up another important outcome of the meeting that I want to reinforce: it is profitable to couple non-electrochemical methods with electrochemistry to gain a fuller understanding of nanoscale phenomena. The key point is that if experimental models like this can be reduced in dimension by an order of magnitude, then we will be able to better understand and optimize many important types of energy conversion and storage devices where nanopores may well dominate function.

Complex surfaces and reactions at the nanoscale

One of the first publications dealing with single entity electrochemistry was published in 1969.[24] This work was carried out in Newcastle upon Tyne, just an hour up the road from York. In their article, Giles and coworkers showed that catalytic reduction of protons could be carried out on individual catalytic nuclei of different metals grown on Hg electrodes. One of the key points the authors made is that Hg is a great electrode material for studying the electrochemistry of electrocatalytic nuclei, because Hg itself is not catalytically active for hydrogen evolution and, more important, the surface of mercury is much more uniform than that of solid electrodes. This is a really interesting paper, but amazingly it has only been cited an average of once per year since its publication. This is a further indication that electrochemists weren't, until recently, paying attention to nanoscale reactions. Fortunately, we've made up for the delay with a range of remarkable experiments, discoveries, and emergent observations, some of which were discussed in York.

As Giles pointed out 50 years ago, a good approach for simplifying single-entity electrochemistry is to confine the entity to a surface: this makes it easier to find. The state of the art in this regard may be results reported at the Discussion by Magnussen and coworkers (DOI: 10.1039/c6fd00086j). They used STM to obtain high-resolution (both spatial and temporal) video during Bi electrodeposition onto single-crystal Au surfaces. The striking thing about their images were just how far the field has come since the pioneering work of Binnig and Rohrer. Surface confined structures were also discussed by Krischer and her colleagues (DOI: 10.1039/c6fd00115g). In their case, simulations and experiments were combined to better understand molecular domain formation.

Koper's article described the interesting phenomenon of cathodic corrosion (DOI: 10.1039/c6fd00078a). The process by which cathodic corrosion occurs is not well understood, but individual corroding defects certainly count as an important single entity. By combining their exacting electrochemical and microscopy results with density functional theory, they concluded that adsorbed ions play a central role in cathodic corrosion.

An emerging single entity of interests to electrochemists and others is nanobubbles.[25] White and his coworkers have been working on understanding these materials for a few years, and they presented their most recent findings at the Discussion (DOI: 10.1039/c6fd00099a). The question they addressed was the nature of electrogenerated nanobubbles of H_2, N_2, and O_2 at a Pt UME, and specifically the number of molecules required for nucleation of individual stable bubbles. The results showed that just a few thousand molecules are required. This study was a particularly good example of how the study of a single entity, a nanobubble in this case, can provide quantitative insights into bubble formation in important industrial processes ranging from the chloralkali production to energy conversion and storage.

Clausmeyer, Schuhmann, and their colleagues reported on the oxygen reduction reaction (ORR) at the three-phase boundary (DOI: 10.1039/c6fd00101g). The beauty of this experiment is its design, which consists of a well-defined model for studying the very complex electrochemical processes that occur in, for example, the gas diffusion electrodes used in fuel cells. As mentioned earlier in this article, these types of well-defined model systems, stripped to their essence, are required to disentangle the complexity of critically important industrial processes. This same general theme was found in the discussion of the work of Chen and coworkers (DOI: 10.1039/c6fd00087h), who reported on finite size effects of ions in the electric double layer.

Molecular electroanalysis: from single molecules to single cells

In 1996, Bard and coworkers reported the electrochemistry of single molecules in solution, and I think this publication really kicked off the modern era of single-entity electrochemistry.[26] Bard's experiment was carried out by trapping small numbers of molecules (1–10) between a sharpened metal tip and a planar electrode. The authors explained that the molecules were confined by a sheath of wax, and sufficient current for detection arose from redox recycling of the trapped molecules. A clear message from this work is that if one wants to study small numbers of molecules electrochemically, they have to be confined near the electrode. Another important message was that it is challenging to measure the very small currents associated with the electrochemistry of small number of molecules. This article is important background for another reason: the system was not very well-defined, and that meant that the experiment was difficult to reproduce. During the past 20 years the situation has gotten much better (in fact, as we will see shortly, there were some outstanding examples in this regard at the Discussion), but it is still difficult to construct well-defined volume elements for electroanalysis of small numbers of molecules.

As mentioned earlier, if one wants to study small numbers of molecules, then one has to know where to find them. There are two ways to do this: either trap the molecules within a small solution volume, as Bard did, or confine them to a surface. As he discussed at the meeting, Lemay and coworkers (DOI: 10.1039/c6fd00075d) implemented the basic Bard experiment but refined the confinement volume into a well-defined geometry using the tools of microfabrication. Specifically, they engineered an electrochemical cell having two parallel electrodes a little over 1 μm long but separated by only 40 nm (Fig. 4). Their interest ran more deeply than simply detecting molecules in this gap; rather, they were interested in measuring the dynamics of mass transport as the molecule moved from one end of the nanochannel to the other. As in Bard's experiment they relied on redox recycling to obtain measureable currents, but as with many of the experiments described at the Discussion advanced electronics (high sensitivity and fast rise times) were critical to success.

There are similarities between Lemay's approach and that of Bohn and his colleagues (DOI: 10.1039/c6fd00062b). Both studies rely on small volumes to confine small numbers of molecules, both involve difficult and exacting microfabrication to create well-defined volume elements, and both rely on redox cycling to generate measureable currents. However, there is one big difference: the Lemay approach involves a single electrochemical cell, whereas Bohn's system uses on a large array of electrochemical cells having effective volumes on the order of 1 aL. The average occupancy of a single cell in these arrays was less than one molecule: an amazing achievement.

As these examples show, with a good cleanroom and a few tens of millions of British Pounds worth of high-tech equipment, one can make a really small electrochemical cell. It turns out Nature is also pretty good at making small-volume containers: vesicles and liposomes are examples. Ewing described his group's efforts to better understand how these containers rupture in the presence of carbon electrodes. Unlike microfabricated devices, however, these systems are a little more akin to the original Bard experiment in that their structure and

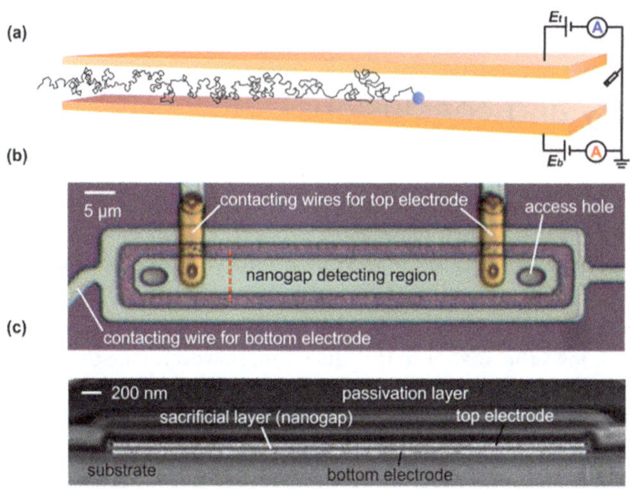

Fig. 4 (a) Schematic illustration of Lemay's nanogap device (DOI: 10.1039/c6fd00075d). (b) Optical microscope image (top view) of the device. The cyan rectangle in the center corresponds to the top electrode, below which lies the detection region. The remaining structures serve as electrical interconnects to the top and bottom electrodes (yellow and cyan lines, respectively). (c) Scanning electron microscopy image of the cross section of a device from a 52° viewing angle. To make this image the device was cut open using a focused ion beam along the line indicated by the dashed red line in (b).

dynamics are not well-defined. Actually, this is the whole point of these experiments: to better understand how molecules move in and out of natural systems. A lively discussion at the meeting emphasized the difficulty of drawing conclusions about such complicated systems using just electrochemical measurements.

Another way to confine a single molecule of electrochemical interest to a defined location is to attach it to a pair of electrodes. Higgins *et al.* (DOI: 10.1039/c6fd00080k) presented this approach, and showed that it could be used to distinguish between tunneling and redox hopping.

For electrochemical imaging,[6] SECM has been the go-to method for many years. Recently, however, the scanning ion conductance microscope (SICM)[27] has been adopted as a companion method for measuring and imaging ion fluxes. One of the leaders in this field is Lane Baker, and he presented remarkable findings in which one or more natural ion channels were placed over the tip of a glass capillary using the patch clamp method (DOI: 10.1039/c6fd00133e). This resulted in a kind of ionic transistor in which the current through the ion channel "gate" is controlled by a Ca^{2+} flux (akin to the gate voltage in a field-effect transistor). The key point, however, is that this method provides a means for understanding single ion channels and also for introducing specificity to SICM tips.

The original scanning probe method, STM, was also well represented at the Discussion and provided an example of studying single molecules by confining them to a two-dimensional surface. By using well-defined, single-crystal substrates and coupling atomically resolved STM with voltammetry, Ulstrup, Zhang, and coworkers showed that it is possible to correlate structure and function of a well-defined array of DNA quadruplexes (DOI: 10.1039/c6fd00091f). Although not strictly a single-entity method, the message here is that if one can be certain that every molecule in an array is identical, then one can learn as much as would be possible by studying a single molecule. This is an interesting idea, but I think there are limits in drawing this correlation, particularly for large biological molecules which often seem to have a life of their own.

Summary and conclusions

Up to this point I have described the findings discussed in York, and in the process of doing so have tried to include some general themes relevant to single-entity electrochemistry. I'm a list-maker, however, and so I want to call those out here explicitly. Before doing so, however, I promised in the Introduction to try to answer the question "why study single entities". I expect there are more answers to that question than there were discussants in York. Almost anyone who has ever been in love knows the answer to this question, however. One usually doesn't fall in love with the entire human race, one falls in love with a single entity (or in some cultures a small group of single entities). A lifetime probably isn't enough time to really understand that single entity (or small group of entities), but one sure can't complain about investing time in the study if they find an interesting entity (or entities)!

OK, here's my list.

• One has to locate single entities before one can study them electrochemically. The methods for doing this include: using large numbers of entities so that it is easy to find one of them; confining small numbers of entities to small volumes; tacking the entity down to a surface (preferably a nanoelectrode), thereby reducing the dimensionality of the search space; or using relatively large entities like microbeads.

• The heterogeneity of solid surfaces, which often changes as a function of time, is a real problem, because the entity under study is affected by, for example, the nature of an electrode surface. Impurities represent a kind of heterogeneity, and even atomic-level heterogeneity can affect measurements and result in incorrect conclusions. Smaller surfaces may reduce the impact of this effect. Biological pores and those based on perfect nanotubes may not suffer as much from this problem. Maybe we need all-bio electrodes that can be constructed by DNA programming?

• Many of the systems described in this Discussion volume were characterized only by electrochemical methods, and this inevitably results in some mechanistic guesswork. Coupling spectroscopy or microscopy to electrochemistry *in operando* can address this issue, but that is often difficult or impossible to do. New analytical methods are needed for coupling to electrochemistry.

• More effort should be put into coupling experiments with theory, and particularly for single entities first principles theory is an important companion to classical approaches. It might be productive to think of theory as yet another analytical tool. Developing experimental electrochemical models that very closely approximate constructs used in first-principles calculations is critical for improving theoretical approaches of any kind.

• Even if one can detect the presence of a single entity, it might not be possible to correlate its structure and function. This is a particularly serious shortcoming of most methods.

• Electrochemical measurements of single entities often require precision electrochemical cells, and fast, sensitive electronics. These refinements are evolving at a rapid rate, and it will be interesting to see how far they can be pushed.

Fig. 5 A cowpoke on his horse surveying a field of single entities. Imagine how difficult it would be to round up them dogies (that means "cattle" in Tex-speak) if they were in three dimensions!

At this Faraday Discussion we heard about advances in all of the areas mentioned above, but in many cases there is still too much complexity, too many unknowns, and too much guesswork. I reckon that if progress in addressing these issues proceeds as quickly during the next 20 years as it has during the past 20 the next Faraday Discussion on single entity electrochemistry is really going to be something to behold. I sure hope I'm around to see what the next generation of electrochemists cooks up!

Before closing, I'd like to add a personal reflection about my visit to England. I viewed York, and the parts of England I visited on my way there, through the eyes of a Texan (Fig. 5). For reading material en route, I was advised by an English friend to take along P. G. Wodehouse. He assured me that this would help to prepare me for the people I would meet on my journey. Imagine my surprise when I didn't hear a single "What ho!", did not run into any private clubs like the "Drones", did not have to "dress" for dinner, and was not greeted upon arrival at my lodgings by Jeeves! The other advice I received was to listen to Gary P. Nunn, a legendary Texas songwriter, and particularly his song "London Homesick Blues". This was also not helpful, because I made a lot of new friends (I didn't have to rely on my "only friends" being "a smoke and a cheap guitar"), no one stared at my boots ("And them Limey eyes, they were eyein' the prize that some people call manly footware"), and I will definitely not "substantiate the rumor that English sense of humor is drier than the Texas sand". It is true, however, that by the end of this excellent meeting old Gary P. had it right: my heart was "longin' to be home in a Texas bar".

Acknowledgements

I want to thank the organizing committee of this outstanding Faraday Discussion for inviting me to participate: Pat Unwin, Phil Bartlett, David Fermin, Justin Gooding, Marc Koper, and Kylie Vincent. I would also like to acknowledge the U.S. agencies who provided me and my research group with funding to work on nanoelectrochemistry projects over the past 30 or so years. These include the Department of Energy, the National Science Foundation, the Office of Naval Research, and the Robert A. Welch Foundation. I also thank the members of my research group, past and present, who pulled up their socks during the early days of nano and got us into the game. In this regard, I am especially grateful to Dr Li Sun and Dr Mingqi Zhao who started our translocation and NP studies, respectively, many years ago.

References

1 H. Y. Cheng, J. Schenk, R. Huff and R. N. Adams, *J. Electroanal. Chem. Interfacial Electrochem.*, 1979, **100**, 23–31.
2 G. Binnig and H. Rohrer, *IBM J. Res. Dev.*, 1986, **30**, 355–369.
3 G. Binnig, C. F. Quate and C. Gerber, *Phys. Rev. Lett.*, 1986, **56**, 930–933.
4 R. C. Engstrom, M. Weber, D. J. Wunder, R. Burgess and S. Winquist, *Anal. Chem.*, 1986, **58**, 844–848.
5 H. Y. Liu, F. R. F. Fan, C. W. Lin and A. J. Bard, *J. Am. Chem. Soc.*, 1986, **108**, 3838–3839.

6 M. Kang, D. Momotenko, A. Page, D. Perry and P. R. Unwin, *Langmuir*, 2016, **32**, 7993–8008.
7 J. T. Cox and B. Zhang, *Annu. Rev. Anal. Chem.*, 2012, **5**, 253–272.
8 L. Zhang, R. Iyyamperumal, D. F. Yancey, R. M. Crooks and G. Henkelman, *ACS Nano*, 2013, **7**, 9168–9172.
9 V. S. Myers, M. G. Weir, E. V. Carino, D. F. Yancey, S. Pande and R. M. Crooks, *Chem. Sci.*, 2011, **2**, 1632–1646.
10 Y. Zhang, N. Raman, J. K. Bailey, C. J. Brinker and R. M. Crooks, *J. Phys. Chem.*, 1992, **96**, 9098–9100.
11 B. M. Quinn, P. G. van't Hof and S. G. Lemay, *J. Am. Chem. Soc.*, 2004, **126**, 8360–8361.
12 X. Xiao and A. J. Bard, *J. Am. Chem. Soc.*, 2007, **129**, 9610–9612.
13 T. M. Alligrant, R. Dasari, K. J. Stevenson and R. M. Crooks, *Langmuir*, 2015, **31**, 11724–11733.
14 Z. Guo, S. J. Percival and B. Zhang, *J. Am. Chem. Soc.*, 2014, **136**, 8879–8882.
15 Y.-G. Zhou, N. V. Rees and R. G. Compton, *Angew. Chem., Int. Ed.*, 2011, **50**, 4219–4221.
16 R. W. DeBlois and C. P. Bean, *Rev. Sci. Instrum.*, 1970, **41**, 909–916.
17 R. W. DeBlois and R. K. A. Wesley, *J. Virol.*, 1977, **23**, 227–233.
18 R. W. DeBlois, E. E. Uzgiris, D. H. Cluxton and H. M. Mazzone, *Anal. Biochem.*, 1978, **90**, 273–288.
19 L. Sun and R. M. Crooks, *J. Am. Chem. Soc.*, 2000, **122**, 12340–12345.
20 T. Ito, L. Sun, R. R. Henriquez and R. M. Crooks, *Acc. Chem. Res.*, 2004, **37**, 937–945.
21 T. Ito, L. Sun and R. M. Crooks, *Chem. Commun.*, 2003, 1482–1483.
22 J. J. Kasianowicz, E. Brandin, D. Branton and D. W. Deamer, *Prog. Biophys. Mol. Biol.*, 1996, **65**, PB127.
23 H. Bayley and C. R. Martin, *Chem. Rev.*, 2000, **100**, 2575–2594.
24 R. D. Giles, J. A. Harrison and H. R. Thirsk, *J. Electroanal. Chem.*, 1969, **20**, 47–60.
25 Special issue on nanobubbles, *Langmuir*, 2016.
26 F.-R. F. Fan, J. Kwak and A. J. Bard, *J. Am. Chem. Soc.*, 1996, **118**, 9669–9675.
27 P. K. Hansma, B. Drake, O. Marti, S. A. Gould and C. B. Prater, *Science*, 1989, **243**, 641.

Poster titles

Dynamics of enzymatic degradation of oligonucleotides through an aerolysin nanopore, **M. Li, C. Cao and Y.-T. Long,** *East China University of Science and Technology, China*

Single molecule/nanoparticle analysis by biological nanopore, **Y. Wang, C. Cao and Y.-T. Long,** *East China University of Science and Technology, China*

Transmission surface diffraction for operando studies of heterogeneous interfaces, **F. Reikowski, T. Wiegmann, J. Stettner, J. Drnec, V. Honkimäki, F. Maroun and O. M. Magnussen,** *Kiel University, Germany*

Single-cell nanobiopsy, **P. Actis,** *University of Leeds, United Kingdom*

Corrosion processes of gold in aqueous electrolytes, **J. Rechmann, T. J. J. Müller, A. C. Götzinger, E. Dirksen and A. Erbe,** *Max-Planck-Institut für Eisenforschung GmbH, Germany*

Resistive and conductive pulses caused by particle translocation through a nanopore to detect surface charge, **L. Mayne, S. Christie and M. Platt,** *Loughborough University, United Kingdom*

Visualisation of cellular uptake using multifunctional scanning electrochemical probe microscopy, **A. Page, M. Kang, A. Armitstead and P. Unwin,** *University of Warwick, United Kingdom*

Localised Electrochemical Analysis for the Formation of Solid Electrolyte Interface in Graphite-Silicon-Polyacrylate Electrodes for Li-ion Batteries, **A. Kumatani, Y. Takahashi, T. Watanabe, K. Yamagiwa, K. Kubota, H. Inomata, H. Ida, H. Shiku, S. Komaba and T. Matsue,** *Tohoku University, Japan*

Real-time electrochemically monitoring formation of chemical bonds in solution, **N. Kong, D. Li, J. Liu, C. Barrow, J. He and W. Yang,** *Deakin University, Australia*

Stochastic charge fluctuations in bipolar electrodes, **Z. Kostiuchenko and S. Lemay,** *University of Twente, Netherlands*

Colloid particle dynamics in nano-capillary electrophoretic tracking: Numerical studies, **D. Baasanjav, S. Samin, A. Mosk and S. Faez,** *Utrecht University, Netherlands*

Multipass coulter counter characterisation of single nanoparticles, **M. Edwards, Y. Zhang, S. German and H. White,** *University of Utah, USA*

Hydrogen atom mediated electrochemistry with metal nanoparticles on thermally grown silicon oxide thin film, **J. Yun, J. G. Lee, J.-Y. Lee, S. J. Shin and T. K. Chung,** *Seoul National University, South Korea*

Electrochemical mapping of nanoscale features on sp^2 carbon surfaces: Steps, layers and defects, **M. O'Connell,** *National Physical Laboratory, United Kingdom*

Live, in situ STM observation of the oxidative roughening of Pt(111), **L. Jacobse, Y.-F. Huang, M. Rost and M. Koper,** *Leiden University, Netherlands*

Gateway state-mediated, low-attenuation long-range tunnelling in dialkylarene molecular wires, **N. Ferri, S. Sangtarash, A. Vezzoli, H. Sadeghi, H. O'Brien, I. Grace, L. Bouffier, R. Nichols, C. Lambert and S. Higgins,** *University of Liverpool, United Kingdom*

A metal mediated nanopore electrochemical-sensor based on an ultra-low noise current measurement system, **Y.-L. Ying, G. Zhen, Y.-X. Hu, C. Cao and Y.-T. Long,** *East China University of Science and Technology, China*

Modification of boron-doped diamond with gold nanoparticles for the enhancement of the mid-IR spectroelectrochemical response, **J. Izquierdo Pérez, A. I. López Lorente, B. Mizaikoff and C. Kranz,** *Ulm University, Germany*

Time-resolved detection of surface oxide formation at individual gold nanoparticles: role in electrocatalysis and new approach for sizing by electrochemical impacts, **C. Bentley, M. Kang and P. Unwin,** *University of Warwick, United Kingdom*

Stochastic phenomena in the electrodissolution of individual Ag nanoparticles, **M. Kang, J. Ustarroz, E. Bullions and P. R. Unwin,** *The University of Warwick, United Kingdom*

Diamond pores as a platform for single particle resistive-pulse sensing, **J. Webb, M. Joseph, M. Newton, I. Aharonovich, M. Toth, A. Martin and J. Macpherson,** *University of Warwick, United Kingdom*

Continuous evaluation of microvilli movements on cell membrane using fast scanning ion conductance microscopy, **H. Ida, Y. Takahashi, A. Kumatani, H. Shiku and T. Matsue,** *Tohoku University, Japan*

Intracellular electrochemical detection of reactive oxygen species with modified carbon nanoelectrodes, **M. Marquitan, J. Clausmeyer, A. Ernst, M. D. Mark, S. Herlitze and W. Schuhmann,** *Ruhr-Universität Bochum, Germany*

In operando optical monitoring of single particles during oxygen evolution reaction, **V. Brasiliense, J. Clausmeyer, C. Combellas, G. Tessier, W. Schuhmann and F. Kanoufi,** *Universite Paris Diderot, France*

Development of biomimetic nanopore platforms for investigating nucleocytoplasmic transport, **B. Akpinar, B. Nadappuram, J. Edel and B. Hoogenboom,** *Imperial College London, United Kingdom*

Investigation of nanopore arrays using scanning probe techniques, **A. Holzinger, P. Knittel, Y. Liu, D. W. M. Arrigan and C. Kranz,** *Ulm University, Germany*

Raman spectroscopy of molecular electronic junctions, **H. Arnolds, N. Wattanavichean and R. Nichols,** *University of Liverpool, United Kingdom*

Field-effect transistor nanosensors based on reduced graphene oxide, **A. Muhs, C. Andronescu, Y. Zhang, Y. Korchev, W. Schuhmann and J. Clausmeyer,** *Ruhr-Universität Bochum, Germany*

Nanoimpacts on bifunctional carbon nanoelectrodes, **D. Öhl, J. Clausmeyer, P. Wilde, A. Botz, E. Ventosa, K. Tschulik and W. Schuhmann,** *Ruhr-Universität-Bochum, Germany*

Oxygen reduction reaction at size-selected Pt_{923} clusters deposited on carbon-coated TEM grids, **I. Ornelas, J. Ustarroz, G. Zhang, P. Unwin and R. Palmer,** *Nanoscale Physics Research Laboratory, United Kingdom*

Electrocatalysis and energy storage at single $Ni(OH)_2$ Nanoparticles, **J. Clausmeyer, J. Masa, E. Ventosa, D. Öhl and W. Schuhmann,** *Ruhr-Universität Bochum, Germany*

Direct electrochemistry of bilirubin oxidase molecules, **M. Kurel, M. Filipiak, M. Michalak, M. Jönsson-Niedziolka, M. Opallo and W. Nogala,** *Institute of Physical Chemistry PAS, Poland*

In situ solid-state electrochemistry of mass selected ions at well-defined electrode–electrolyte interfaces using ion soft landing, **V. Prabhakaran, G. E. Johnson, B. Wang and J. Laskin,** *Pacific Northwest National Laboratory, USA*

Nondecaying hydrodynamic interactions in channels, **K. Misiunas,** *University of Cambridge, United Kingdom*

Towards biosensing at the single-cell level using colloidal AFM-SECM probes, **C. Kranz and P. Knittel,** *University of Ulm, Germany*

Single-crystal spectroelectrochemistry of redox proteins – infrared microspectroscopy using a synchrotron source, **P. Ash, S. F. Carr, H. Reeve, A. Skorupskaitė, J. S. Rowbotham, R. Shutt, M. D. Frogley, G. Cinque, F. A. Armstrong and K. A. Vincent,** *University of Oxford, United Kingdom*

Recognition of electro/chemical reactions at ultramicroelectrode surface by single-entity collision, **H. Ma, Y.-Y Peng, W. Ma, L.-J Zhao, Y.-L. Ying and Y.-T. Long,** *East China University of Science and Technology, China*

Pt nanoparticle mediated on chip catalytic combustion of methanol–air mixture, **S. Tian, T. Thundat and S. K. Mitra,** *York University, Canada*

The Faraday Discussions Poster Prize for the best poster was jointly awarded to Mr Finn Reikowski of Kiel University, Germany for his poster on 'Transmission surface diffraction for operando studies of heterogeneous interfaces' and Mr Vitor Brasiliense of Universite Paris Diderot, France for his poster on '*In operando* optical monitoring of single particles during oxygen evolution reaction'.

Second prize in the Faraday Discussions Poster Prize was awarded to Miss Na Kong of Deakin University, Australia for her poster on 'Real-time electrochemically monitoring formation of chemical bonds in solution'.

List of participants

Dr Paolo Actis, *University of Leeds, United Kingdom*
Miss Bernice Akpinar, *Imperial College London, United Kingdom*
Dr Tim Albrecht, *Imperial College London, United Kingdom*
Professor Mario A. Alpuche-Aviles, *University of Nevada, Reno, USA*
Dr Heike Arnolds, *University of Liverpool, United Kingdom*
Dr Philip Ash, *University of Oxford, United Kingdom*
Mr Dashdeleg Baasanjav, *Debye Institute for Nanomaterials Science, Utrecht University, Netherlands*
Professor Lane Baker, *Indiana University, USA*
Professor Philip Bartlett, *University of Southampton, United Kingdom*
Dr Cameron Bentley, *University of Warwick, United Kingdom*
Mrs Jenny Bergman, *Gothenburg University, Sweden*
Dr Paul Bohn, *University of Notre Dame, USA*
Ms Sage Bowser, *Royal Society of Chemistry, United Kingdom*
Mr Vitor Brasiliense, *Universite Paris Diderot, France*
Dr Samantha Catarelli, *Uniscan Instruments, United Kingdom*
Prof Dr Shengli Chen, *Wuhan University, China*
Dr Jan Clausmeyer, *Ruhr-Universität Bochum, Germany*
Mr Daniel Corken, *Queen Mary University of London, United Kingdom*
Professor Richard Crooks, *The University of Texas at Austin, USA*
Mr Gareth Davies, *Royal Society of Chemistry, United Kingdom*
Mr Mike Dawson, *EC-Lab Ltd, United Kingdom*
Mr Johan Dunevall, *Chalmers University of Technology, Sweden*
Dr Martin Edwards, *University of Utah, USA*
Prof Dr Michael Eikerling, *Simon Fraser University, Canada*
Professor Andrew Ewing, *Chalmers University and University of Gothenburg, Sweden*
Dr Sanli Faez, *Utrecht University, Netherlands*
Professor David Fermin, *University of Bristol, United Kingdom*
Mr Nicolò Ferri, *University of Liverpool, United Kingdom*
Professor Justin Gooding, *The University of New South Wales, Australia*
Dr Jingying Gu, *School of Chemistry, Newcastle Universtiy, United Kingdom*
Mr Thom Hersbach, *Leiden University: Leiden Institute of Chemistry, Netherlands*
Professor Simon Higgins, *University of Liverpool, United Kingdom*
Dr Joanne Holmes, *Metrohm, United Kingdom*
Mrs Angelika Holzinger, *Institute of Analytical and Bioanalytical Chemistry - Ulm University, Germany*
Mr Hiroki Ida, *Tohoku University, Japan*
Dr Javier Izquierdo Pérez, *Institute of Analytical and Bioanalytical Chemistry, Ulm University, Germany*
Mr Leon Jacobse, *Leiden University, Netherlands*
Dr Chao Jing, *East China University of Science and Technology, China*
Dr Robert Johnson, *University of Warwick, United Kingdom*
Miss Minkyung Kang, *The University of Warwick, United Kingdom*

Dr Frederic Kanoufi, *Université Paris Diderot*, *France*
Mr Peter Knittel, *Institute of Analytical and Bioanalytical Chemistry, Ulm University*, *Germany*
Miss Na Kong, *Deakin University*, *Australia*
Prof Dr Marc Koper, *Leiden University*, *Netherlands*
Ms Zinaida Kostiuchenko, *University of Twente*, *Netherlands*
Dr Christine Kranz, *Institute of Analytical and Bioanalytical Chemistry*, *Germany*
Prof Dr Katharina Krischer, *Technical University of Munich, Physics Department*, *Germany*
Dr Akichika Kumatani, *Tohoku University*, *Japan*
Miss Anna Larsson, *University of Gothenburg*, *Sweden*
Miss Sarah Latham, *Royal Society of Chemistry*, *United Kingdom*
Prof Dr Serge Lemay, *University of Twente*, *Netherlands*
Mr Mengyin Li, *East China University of Science and Technology*, *China*
Professor Yitao Long, *East China University of Science and Technology*, *China*
Miss Kirsty Lunnon, *Royal Society of Chemistry*, *United Kingdom*
Miss Hui Ma, *East China University Of Science and Technology*, *China*
Professor Julie MacPherson, *University of Warwick*, *United Kingdom*
Professor Olaf Magnussen, *Kiel University, Institute of Experimental and Applied Physics*, *Germany*
Miss Miriam Marquitan, *Analytical Chemistry, Center for Electrochemical Sciences, Ruhr-Universität Bochum*, *Germany*
Miss Laura Mayne, *Loughborough University*, *United Kingdom*
Mr Karolis Misiunas, *University of Cambridge*, *United Kingdom*
Professor Sushanta Mitra, *York University*, *Canada*
Professor Andrew Mount, *The University of Edinburgh, School of Chemistry*, *United Kingdom*
Miss Anna Muhs, *Analytical Chemistry, Center for Electrochemical Sciences, Ruhr-Universität Bochum*, *Germany*
Professor Richard Nichols, *Deparmtent of Chemistry, University of Liverpool*, *United Kingdom*
Dr Wojciech Nogala, *Institute of Physical Chemistry PAS*, *Poland*
Dr Mike O'Connell, *National Physical Laboratory*, *United Kingdom*
Mr Denis Öhl, *Ruhr-Universität Bochum*, *Germany*
Ms Isabel Ornelas, *Nanoscale Physics Research Laboratory*, *United Kingdom*
Mr Ashley Page, *University of Warwick*, *United Kingdom*
Mr Alex Pashley, *Royal Society of Chemistry*, *United Kingdom*
Dr Mark Platt, *Loughborough University*, *United Kingdom*
Dr Venkateshkumar Prabhakaran, *Pacific Northwest National Laboratory*, *USA*
Dr Jenny Pringle, *Deakin University*, *Australia*
Mr Julian Rechmann, *Max-Planck-Institut für Eisenforschung GmbH*, *Germany*
Mr Finn Reikowski, *Kiel University*, *Germany*
Prof Dr Wolfgang Schmickler, *Ulm University*, *Germany*
Prof Dr Wolfgang Schuhmann, *Ruhr-Universität Bochum*, *Germany*
Professor Keith Stevenson, *Skolkovo Institute of Science and Technology*, *Russia*
Professor Nongjian Tao, *Arizona State University*, *USA*
Mr Bradley Thomas, *School of Chemistry, University of Nottingham*, *United Kingdom*
Prof Dr Zhong-Qun Tian, *Xiamen University*, *China*
Mr Soichiro Tottori, *University of Cambridge*, *United Kingdom*
Prof Dr Kristina Tschulik, *Ruhr-Universität Bochum*, *Germany*
Prof Dr Jens Ulstrup, *DTU Chemistry*, *Denmark*
Professor Patrick Unwin, *University of Warwick, UK*, *United Kingdom*
Dr Jon Ustarroz, *Department of Chemistry, University of Warwick*, *United Kingdom*

Professor Kylie Vincent, *University of Oxford*, *United Kingdom*
Miss Yaqian Wang, *East China University of Science and Technology*, *China*
Ms Nungnit Wattanavichean, *University of Liverpool*, *United Kingdom*
Dr Jennifer Webb, *University of Warwick*, *United Kingdom*
Dr Henry White, *Univeristy of Utah*, *USA*
Dr Yilun Ying, *East China University of Science and Technology*, *China*
Mr Jeongse Yun, *Department of Chemistry, Seoul National University*, *South Korea*
Prof Dr Dongping Zhan, *Deparment of Chemistry, Xiamen University*, *China*